Richard F Carlson

AF895245

Ce volume a été publié avec l'aide financière de l'Association Canadienne-Française pour l'Avancement des Sciences (ACFAS)

This volume has been published with the financial support of the French Canadian Association for the Advancement of Sciences (ACFAS)

© 1975, LES PRESSES DE L'UNIVERSITÉ LAVAL
Tous droits réservés. Imprimé au Canada
Dépôt légal (Québec), 3ᵉ trimestre 1975
I.S.B.N. 0-7746-6752-4

PROBLÈMES
À PETIT NOMBRE DE CORPS
DANS LA PHYSIQUE DU NOYAU
ET DE PARTICULES ÉLÉMENTAIRES

CONFÉRENCE INTERNATIONALE TENUE À L'UNIVERSITÉ LAVAL,
QUÉBEC, CANADA,
DU 27 AU 31 AOÛT 1974

FEW BODY PROBLEMS
IN NUCLEAR AND PARTICLE PHYSICS

INTERNATIONAL CONFERENCE HELD AT LAVAL UNIVERSITY,
QUEBEC CITY, CANADA,
AUGUST 27 - 31, 1974

Éditeurs/Editors

R.J. SLOBODRIAN B. CUJEC K. RAMAVATARAM

LES PRESSES DE L'UNIVERSITÉ LAVAL

QUÉBEC, 1975

APPUI DES ORGANISMES

Union internationale de physique pure et appliquée (UIPPA)

Association canadienne des physiciens

Institut de physique des particules

Laboratoires nucléaires de Chalk River

Conseil national de recherches du Canada

Association canadienne-française pour l'avancement des sciences

Secrétariat d'Etat, Ottawa

Université Laval

GOUVERNEMENT DU QUEBEC:
Ministère de l'éducation
Ministère des communications
Ministère des richesses naturelles

SPONSORSHIP

International Union of Pure and Applied Physics (IUPAP)

Canadian Association of Physicists

Institute of Particle Physics

Chalk River Nuclear Laboratories

National Research Council of Canada

French Canadian Association for the Advancement of Sciences

State Department, Ottawa

Laval University

GOVERNMENT OF QUEBEC:
Ministry of Education
Ministry of Communications
Ministry of Natural Resources

APPUI DES ENTREPRISES PRIVEES

SPONSORSHIP BY PRIVATE ENTERPRISES

Dow Chemicals of Canada
Sarnia
Ontario

Electrostatics International Inc.
Box 117, Graber Road
Middleton, Wisconsin 53562
U.S.A.

High Voltage Engineering Corporation
Burlington, Massachussets 01803
U.S.A.

Northern Electric Company Limited
1600 boul. Dorchester
Montréal

Philips Electronics Industries Ltd.
116 Van Der Hoof Ave.
Toronto

Radionics Limitée
195 rue Graveline
Montréal

Steinberg Limitée
1500 rue Atwater
Montréal

La Banque Royale du Canada
Montréal

Xerox du Canada Limitée
2 Place Québec
Québec

Un grand nombre de personnes ont participé à la préparation de ce volume. Mesdemoiselles Jocelyne Blais, Carmelle Beaulieu et Linda Guillot méritent un remerciement spécial.

La publication de ce volume a été confiée aux soins diligents des 'Presses de l'Université Laval'; nous les remercions vivement de leur travail.

The assistance of many people has been invaluable in the preparation of this volume. Misses Jocelyne Blais, Carmelle Beaulieu and Linda Guillot deserve special gratitude.

The publication of this book was accomplished quickly by 'Les Presses de l'Université Laval' and we express our gratitude most heartily.

Les éditeurs
The Editors

PREFACE

Ce volume contient les textes des communications présentées à la Conférence sur les problèmes à petit nombre de corps dans la physique du noyau et des particules élémentaires ainsi que les rapports des séances de discussions, tels que soumis par leurs auteurs et reproduits par un procédé photographique. Les communications des conférenciers invités sont accompagnées de la discussion subséquente. Les rapports des périodes de discussions nucléon-nucléon ont été suivis d'une discussion générale qu'on trouvera aussi dans ce volume. Nous avons aussi inclus le programme détaillé de la Conférence. Un autre volume, déjà publié, contient les résumés de toutes les communications.

L'interaction nucléon-nucléon à toutes les énergies a été l'objet de séances de discussions organisées de façon remarquable par le professeur Michael J. Moravcsik. Dans ces dernières, on s'est écarté des procédures propres aux conférences internationales, ce qui a permis des échanges d'idées très francs et très stimulants.

Lors de l'inauguration de cette Conférence, des discours ont été prononcés par le recteur de l'Université Laval et secrétaire de l'IUPPA, le professeur Larkin Kerwin, par le doyen de la Faculté des sciences et de génie, le professeur Pierre Grenier, et par le directeur du Département de physique, le professeur Claude Saint-Pierre.

This volume contains the papers presented at the Conference on the Few Body Problems in Nuclear and Particle Physics, and the reports on the nucleon-nucleon discussion sessions, as they were submitted by their authors and reproduced by a photographic process. The papers of the invited speakers are followed by the subsequent discussion. The reports on the nucleon-nucleon discussion periods were followed by a general discussion which is also reproduced in this volume. We have also included the detailled programme of the Conference. Another volume, which has already been published, contains the summaries of all the contributed papers.

The nucleon-nucleon interaction at all energies was the object of a series of discussion sessions magnificently organized by Professor Michael J. Moravcsik. In these sessions, the traditional formality usual of international conferences was set aside and thus a stimulating and unhibited exchange of ideas took place.

The Conference was inaugurated with addresses by the Rector of Université Laval and Secretary of IUPAP, Professor Larkin Kerwin, by the Dean of the Faculty of Sciences and Engineering, Professor Pierre Grenier, and by the Director of the Department of Physics, Professor Claude Saint-Pierre.

COMITE D'ORGANISATION / ORGANISING COMMITTEE

Batty, C.J.	RHEL, ENGLAND
Bouchez, R.	Univ. de Grenoble, FRANCE
Breit, G.	SUNY, Buffalo, USA
Conzett, H.E.	Univ. of California, USA
Cüer, P.	Univ. de Strasbourg, FRANCE
Cujec, B.	Univ. Laval, Québec
Delves, L.M.	Univ. of Liverpool, ENGLAND
Depommier, P.	Univ. de Montréal, Montréal
Ferguson, A.J.	CRNL, Chalk River
Gross, E.E.	ORNL, USA
Hincks, E.P.	NRC, Ottawa, ONTARIO
Hird, B.	Univ. of Ottawa, ONTARIO
Ho-Kim, Q.	Univ. Laval, Québec
Jackson, D.	Univ. of Surrey, ENGLAND
Kerman, A.K.	MIT, USA
Komarov, V.V.	Univ. de Moscou, URSS
Levin, F.S.	Brown Univ., USA
Levinger, J.S.	Rensselaer, Poly. Inst., USA
Lovas, I.	Centr. Res. Inst. for Phys. HONGRIE
Magnac Vallette, D.	Univ. de Strasbourg, FRANCE
Margolis, B.	McGill Univ., Montréal
Marquez, L.	Centre d'Etudes Nucl. Bordeaux, FRANCE
Marty, N.	Univ. de Paris et Orsay, FRANCE
McKee, J.S.C.	Univ. of Birmingham, ENGLAND
Moravcsik, M.	Univ. of Oregon, USA
Neilson, C.G.	Univ. of Alberta, Edmonton
Pierre Noyes, H.	SLAC, USA
Pignotti, A.	Univ. of Buenos Aires, ARGENTINA
Pugh, H.	Univ. of Maryland, USA
Radvanyi, P.	Univ. de Paris et Orsay, FRANCE
Ramavataram, K.	Univ. Laval, Québec
Richardson, J.R.	TRIUMF, Vancouver
Resmini, F.	Univ. di Milano, ITALIE
Rosen, L.	LASL, USA
Sellschop, J.P.F.	Univ. of Witwatersrand, SOUTH AFRICA
Shin, Y.M.	Univ. of Saskatchewan, Saskatoon
Slaus, I.	Inst. Ruder Boskovic, YUGOSLAVIE
Slobodrian, R.J.	Univ. Laval, Québec
Teillac, J.	Inst. Nat. de Phys. de Particules, FRANCE
Tombrello, T.A.	Cal. Inst. of Tech., USA
van Oers, W.T.H.	Univ. of Manitoba, Winnipeg
Warren, J.B.	Univ. of British Columbia, Vancouver
Watson, K.	Univ. of California, USA
Zucker, A.	ORNL, USA

PRESIDENT DU COMITE
COMMITTEE CHAIRMAN : R.J. Slobodrian

COMITE DES ACTIVITES SOCIALES / COMMITTEE FOR SOCIAL ACTIVITIES

Mmes, G. Nadeau
C. St-Pierre
R.J. Slobodrian

Mlle P. Poitras

TABLE DES MATIERES — TABLE OF CONTENTS

PROGRAMME

MARDI LE 28 AOUT
TUESDAY, AUGUST 28TH

INAUGURATION — DISCOURS D'OUVERTURE
OPENING ADDRESSES

Larkin KERWIN, Recteur de l'Université Laval
Pierre GRENIER, Doyen de la Faculté des sciences et de génie
Claude SAINT-PIERRE, Directeur du Département de physique

Président: R.J. Slobodrian, Université Laval

SESSION I — LES UTILISATIONS DU PROBLEME A PETIT NOMBRE DE CORPS
THE USES OF THE FEW BODY PROBLEM

R.D. AMADO, University of Pennsylvania, U.S.A.

Président: J. Teillac, Institut national de physique nucléaire et de physique des particules, Paris
Secrétaires scientifiques: P. Amiot, Université Laval
T.C. Sharma, Université Laval

SESSION II — COMMUNICATIONS

Président: W.T.H. van Oers, University of Manitoba

PROGRAMME NUCLEON-NUCLEON, DIA, DIB, DIC

SESSION III — INFORMATION D'APRES L'ETUDE DES INTERACTIONS A L'ETAT FINAL
INFORMATION FROM THE STUDY OF FINAL STATE INTERACTIONS

J.S.C. McKEE, University of Birmingham, England

Président: A.J. Ferguson
Secrétaires scientifiques: P. Amiot, Université Laval
T.C. Sharma, Université Laval

PROGRAMME NUCLEON-NUCLEON, DIA, DIB, DIC

SESSION IV — COMMUNICATIONS

Président: H.E. Conzett, LBL, University of California, U.S.A.

SESSION V

LES FAISCEAUX, L'EQUIPEMENT ET LES RESSOURCES NECESSAIRES A L'AVENIR
BEAMS, EQUIPMENT AND RESOURCES DEEMED NECESSARY IN THE FUTURE

L. AGNEW, LASL, LAMPF, University of California, U.S.A.
J.M. CAMERON, University of Manitoba and TRIUMF
C.G. SCHUHL, Saclay, France
Y.M. SHIN, University of Saskatchewan

Président: P. Radvanyi, Université de Paris-Sud, France

SESSION VI

METHODES THEORIQUES POUR LES PROBLEMES A PETIT NOMBRE DE CORPS
THEORETICAL METHODS FOR FEW BODY PROBLEMS

Faddeev Solutions with non-seperable interactions
J.A. TJON, Université d'Utrecht, Hollande

A boundary condition approach to the three-body problem
D.D. BRAYSHAW, University of Maryland, U.S.A.

Interactions à trois corps à l'état lié, une nouvelle approche
B.T. DARLING et R.M. SHOUCRI, Université Laval

A new method in the theory of many body scattering
D.J. KOURI, Max-Planck Inst., Göttingen, W. Germany and
F.S. LEVIN, Brown University U.S.A.

Président: J.S. Levinger, Rensselaer Polytechnic Institute, U.S.A.
Secrétaires scientifiques: M. Boivin, Université Laval
R. Shoucri, Université Laval

MERCREDI LE 28 AOUT
WEDNESDAY, AUGUST 28TH

SESSION VII

RADIATION DE FREINAGE DANS LES SYSTEMES A PETIT NOMBRE DE CORPS
NUCLEAR BREMSSTRAHLUNG FROM FEW BODY SYSTEMS

A.M. GREEN, Université d'Helsinki, Finlande

Président: L. Heller, LASL, University of California, U.S.A.
Secrétaires scientifiques: B. Frois, CEN-Saclay, France
J. Birchall, LBL, Univ. of California

PROGRAMME NUCLEON-NUCLEON, D2A, D2B, D2C

SESSION VIII

COMMUNICATIONS

Président: D. Magnac-Valette, Université de Strasbourg, France

SESSION IX

PHENOMENES DE POLARISATION DANS LE SYSTEME A TROIS NUCLEONS
POLARIZATION PHENOMENA IN THE THREE-NUCLEON SYSTEM

H.E. CONZETT, LBL, University of California, U.S.A.

Président: J.E. Simmons, LASL, University of California
Secrétaires scientifiques: B. Frois, CEN-Saclay, France
J. Birchall, LBL, Univ. of California

SESSION X

CORRELATIONS A COURTE PORTEE DANS LA MATIERE NUCLEAIRE
SHORT RANGE CORRELATIONS IN NUCLEAR MATTER

O. KOFOED-HANSEN, CERN, Genève, Suisse

Président: H.G. Pugh, University of Maryland
Secrétaires scientifiques: B.K. Jain, Bhabha Atomic Res. C., India
C.L. Rao, Université Laval

PROGRAMME NUCLEON-NUCLEON, D2A, D2B, D2C

SESSION XI

COMMUNICATIONS

Président: T. Sasakawa, Université de Tohoku, Japon

SESSION XII

TRAITEMENT STATISTIQUE DES REACTIONS COHERENTES ET D'AUTRES REACTIONS
STATISTICAL TREATMENT OF COHERENT AND OTHER REACTIONS

B. MARGOLIS, McGill University

Président: P. Depommier, Université de Montréal
Secrétaires scientifiques: B.K. Jain, Bhabha Atomic Res. C., India
C.L. Rao, Université Laval

RECEPTION

JEUDI LE 29 AOUT
THURSDAY, AUGUST 29TH

SESSION XIII

LES SYSTEMES A TROIS PARTICULES: PROBLEMES ACTUELS ET PROGRES
THREE PARTICLE SYSTEMS: CURRENT PROBLEMS AND PROGRESS

L.M. DELVES, University of Liverpool, UK, and University of Victoria

Président: E. ELBAZ, Inst. de physique nucléaire de Lyon, France
Secrétaires scientifiques: R. Roy, LBL, Univ. of California, U.S.A.
M. Irshad, Université Laval

PROGRAMME NUCLEON NUCLEON, D3A, D3B, D3C

SESSION XIV COMMUNICATIONS

Président: Y.M. Shin, University of Saskatchewan

SESSION XV QUE SAVONS-NOUS SUR LE DEUTON ET L'ANTIDEUTON ?
WHAT DO WE KNOW ABOUT THE DEUTERON AND THE ANTI-DEUTERON ?

D.W.L. SPRUNG, McMaster University

Président: M.A. Preston, McMaster University
Secrétaires scientifiques: R.Roy, LBL, Univ. of California, U.S.A.
M. Irshad, Université Laval

VENDREDI LE 30 AOUT
FRIDAY, AUGUST 30TH

SESSION XVI LES PROBLEMES A PETIT NOMBRE DE CORPS DANS LA PHYSIQUE DES HADRONS A HAUTE ENERGIE
THE FEW BODY PROBLEMS IN HIGH ENERGY HADRON PHYSICS

L. VAN HOVE, CERN, Genève, Suisse

Président: A. Pignotti, Université de Buenos Aires, Argentine
Secrétaires scientifiques: M. High, Université Laval
A.M. Sourkes, University of Manitoba

PROGRAMME NUCLEON-NUCLEON, D3A, D3B, D3C

SESSION XVII COMMUNICATIONS

Président: A. Mitra, University of Delhi, India

SESSION XVIII MODELES COMPOSES DES HADRONS ET PROCESSUS A GRANDE IMPULSION TRANSVERSALE
COMPOSITE MODELS OF HADRONS AND LARGE TRANSVERSE MOMENTUM PROCESSES

S.J. BRODSKY, Stanford University

Président: H.P. Noyes, SLAC, Stanford University, U.S.A.
Secrétaires scientifiques: M. High, Université Laval
A.M. Sourkes, University of Manitoba

SESSION XIX DIFFUSION π-DEUTON
π-DEUTERON SCATTERING

D.V. BUGG, Queen Mary College, London, England

QUELQUES ASPECTS THEORIQUES DE LA DIFFUSION π-D A BASSE ENERGIE
SOME THEORETICAL ASPECTS OF LOW ENERGY π-D SCATTERING

A.W. THOMAS, University of British Columbia

SESSION XIX (suite)

Président: J.B. Warren, University of British Columbia
Secrétaires scientifiques: D.I. Bonbright, University of Manitoba
J.W. Watson, University of Manitoba

SESSION XX

PROGRAMME DE RECHERCHES DES USINES A MESONS
RESEARCH PROGRAMMES OF MESON FACILITIES

Table ronde / Panel discussion

P. CARRUTHERS, University of California, Los Alamos Scient. Lab., U.S.A
M. LOCHER, Schweizerisches Inst. f. Nuklearforschung, Suisse
R. MACEK, University of California, Los Alamos Scient. Lab., U.S.A.
K. MOLLER, JINR, Moscow, USSR
F. NICHITIU, JINR, Moscow, USSR
D. STORM, Columbia University, Nevis Lab., U.S.A.
W.T.H. van OERS, University of Manitoba and TRIUMF
E. VOGT, University of British Columbia
D. ZAIKIN, JINR, Moscow, USSR

Président: J.R. RICHARDSON, University of British Columbia, TRIUMF

BANQUET
Manoir Saint-Castin

L'UNION INTERNATIONALE DE PHYSIQUE
INTERNATIONAL UNION OF PHYSICS

Larkin KERWIN, Recteur de l'Université Laval

SAMEDI LE 31 AOUT
SATURDAY, AUGUST 31

SESSION XXI

INTERACTION NUCLEON-NUCLEON: RAPPORTS DES PERIODES DE DISCUSSION ET DISCUSSION GENERALE
NUCLEON-NUCLEON INTERACTION: REPORTS OF DISCUSSION PERIODS AND GENERAL DISCUSSION

Président: M.J. Moravcsik, University of Oregon, U.S.A.
Secrétaires scientifiques: J. Asai, University of Windsor, Ontario
M.S. Hussein, Univ. de Sao Paulo, Brésil
R. Pigeon, Université Laval
D. Shannon, U.C.L.A., U.S.A.

SESSION XXII

L'AMPLITUDE A DEUX PARTICULES.PROCESSUS NUCLEAIRES A PLUSIEURS PARTICULES
TWO-PARTICLE AMPLITUDE AND MANY-PARTICLE NUCLEAR PROCESSES

V.V. KOMAROV, Moscow State University, USSR

SOMMAIRE SUMMARY

A. PIGNOTTI, Université de Buenos Aires, Argentine
P. NOYES, SLAC, Stanford University, U.S.A.

Président: I. Slaus, Ruder Boskovic Institute, Yougoslavie
Secrétaires scientifiques: R. Cahill, Flinders Univ., Australia
A. Fonseca, Univ. Notre-Dame, U.S.A.

PROGRAMME

INTERACTION NUCLEON-NUCLEON, PERIODES DE DISCUSSIONS
NUCLEON-NUCLEON INTERACTION, DISCUSSION PERIODS

D1A PRODUCTION DE PIONS DANS LES COLLISIONS NUCLEON-NUCLEON
PION PRODUCTION IN NUCLEON-NUCLEON COLLISIONS
JOHN L. GAMMEL*, University of Calif., Los Alamos Sc. Lab., U.S.A.
IRAJ AFNAN**, Flinders University, Australia

D1B LA SITUATION DES POTENTIELS DANS L'ETUDE DE L'INTERACTION DE DEUX NUCLEONS
THE STATUS OF POTENTIALS IN THE STUDY OF THE TWO-NUCLEON INTERACTION
JOSEPH S. LEVINGER*, Rensselaer Polytechnic Inst., U.S.A.
MARK W. KERMODE**, University of Liverpool, England

D1C VERIFICATIONS DES SYMETRIES ET DES LOIS DE CONSERVATION
TESTS OF SYMMETRIES OF THE NUCLEAR INTERACTION
NELSON JARMIE*, University of Calif., Los Alamos Sc. Lab., U.S.A.
MARKUS SIMONIUS**, Lab. für Kernphysik, ETH, Suisse

D2A UN GUIDE EXPERIMENTAL POUR LES EXPERIENCES DE LA PROCHAINE DECENNIE EN PARTICULIER POUR LES EXPERIENCES DE POLARISATION
AN EXPERIMENTAL GUIDE TO EXPERIMENTS FOR THE NEXT DECADE WITH STRESS ON POLARIZATION.
JOVAN JOVANOVICH*, University of Manitoba
JOHN A. EDGINGTON**, Queen Mary College, London, England

D2B UNE SYNTHESE AUX ENERGIES "BASSES" ET "HAUTES"
A SYNTHESIS AT "LOW" AND "HIGH" ENERGIES
IVO SLAUS*, Inst. Ruder Boskovic, Zagreb, Yugoslavie
GERRIT SCHIERHOLZ**, Inst. Theor. Physik, Hamburg, W. Germany

D2C L'INTERACTION NUCLEON-NUCLEON A L'INTERIEUR DE 1 FM
THE NUCLEON-NUCLEON INTERACTION WITHIN 1 FM
KONRAD BLEULER*, Inst. Kernphysik, Bonn, W. Germany
FRANZ GROSS**, College of William and Mary, U.S.A.

D3A RADIATION DE FREINAGE
NUCLEON-NUCLEON BREMSSTRAHLUNG
LEON HELLER*, University of Calif., Los Alamos Sc. Lab., U.S.A.
MING-KUNG LIOU**, Brooklyn College, N.Y., U.S.A.

D3B DESCRIPTION AUX ENERGIES TRES BASSES
VERY LOW ENERGY PARAMATRIZATION
RAOUL D. VIOLLIER*, Stanford University, U.S.A.
JOHAN J. de SWART**, Inst. Theor. Fysica, Nijmegen, The Netherlands

D3C LES INTERACTIONS A TROIS ET PLUSIEURS CORPS COMME OUTILS POUR L'ETUDE DU PROBLEME A DEUX NUCLEONS
THE THREE AND MANY-BODY INTERACTIONS AS TOOLS IN LEARNING ABOUT THE TWO-NUCLEON PROBLEM.
DONALD SPRUNG*, McMaster University
HARRY FIEDELDEY**, University of South Africa, Pretoria, South Africa

---

* Modérateur ** Rapporteur

INDEX ANALYTIQUE — SUBJECT INDEX

I PROBLEMES GENERAUX A PETIT NOMBRE DE CORPS — I GENERAL FEW BODY PROBLEMS

RAPPORTS DES SEANCES DE DISCUSSIONS NUCLEON-NUCLEON

REPORTS ON NUCLEON-NUCLEON DISCUSSION PERIODS

VIII CORRELATIONS A COURTE PORTEE DIFFUSION QUASI-LIBRE — VIII SHORT RANGE CORRELATIONS QUASI-FREE SCATTERING 629

I

PROBLEMES GENERAUX A PETIT NOMBRE DE CORPS

GENERAL FEW BODY PROBLEMS

THE USES OF THE FEW BODY PROBLEM

R. D. Amado+
Department of Physics, University of Pennsylvania
Philadelphia, Pennsylvania 19174, U.S.A.

ABSTRACT

The few body system is a rich source of information for hadron dynamics. How some of this information manifests itself and how the general constraints of quantum mechanics help or restrict its extraction will be discussed. Examples will include the theory of final state interactions, of meson-baryon resonances and of meson degrees of freedom in the nucleus.

The hadron few body problem is useful in providing insight into a number of fundamental physics questions including the interaction of unstable particles, two and three-body forces and the importance of meson degrees of freedom, and reaction mechanisms. Few body systems are not unique in shedding light on these matters, rather they are useful because they are rich enough to give information on these questions, and still simple enough to be open to analysis. This analyzable complexity also gives the subject an intrinsic fascination apart from its "applications" to fundamental physics questions. In the past dozen years we have made remarkable progress in our ability to formulate and calculate in few body systems, particularly in the two and three-body case and much of this progress has been driven by that intrinsic intellectual interest. Here I would like to review some of the fundamental physics questions, to appraise our success, and to recall the basic physical situations that give rise to them. I shall perhaps spend more time stressing cautions than marking success in part to counteract earlier rounds of collective self-congratulations that we have enjoyed together.

We shall first take up nuclear dynamics - the nuclear force problem and meson degrees of freedom in the few-body system. Much heroic progress has been made lately (none by me) and we are beginning to see how off-shell effects, three-body forces, meson degrees of freedom, etc. play a role and how these questions interrelate. Next we shall turn to final state interactions, where there is less progress to report, but we have recently turned to an examination of this old problem using general quantum mechanical principles and I shall try to show how imposing these

---

+Supported in part by the National Science Foundation.

principles leads to the simplest form of the separable interaction equations and that that equation is therefore the minimal embodiment of the general constraints of quantum mechanics. Finally we shall take up reaction mechanisms where we are in the position of being able to use a tool developed in nuclear physics to shed light on the dynamics of particle physics.

A classic goal of few body physics is understanding nuclear forces and the classic place to look is the three nucleon bound state. Modern work has strengthened this idea by showing that three-body scattering is very insensitive to details of dynamical input.[1] (We shall have more to say about that later.) The traditional approach to the three-body bound state has been to take a good two-body potential and get the triton binding energy. For a modern nuclear interaction this is a very complicated problem, but it has been carried out now by a number of very strong groups.[2] They find the triton underbound by 1-2 MeV. Is this due to off-shell effects, three-body forces, relativistic effects or what? In fact phenomenological attempts to introduce off-shell changes yield little largely because of the relatively low density of the triton. Estimates of the longest range part of the three-body force do add 1-2 MeV of binding and again perhaps because of the low density of the system higher terms may not be important. There seems to be a discrepancy in the Coulomb energy of He^3 and H^3 of about 120 KeV that must be attributed to a charge asymmetry of nuclear forces.

The neatness of this picture is perhaps deceptive. There are still important discrepancies between calculations. In addition, Reid soft-core potential calculations seem to give some .7 MeV more binding than the hard-core Hamada Johnston case.[2] These two potentials are different only off-shell in the energy range of interest and this difference thus casts doubt on phenomenological claims about off-shell insensitivity. Furthermore, the nuclear force is a complex manifestation of the interaction of mesons and nucleons. In a non-relativistic potential formalism we freeze out the meson degrees of freedom. Off-shell effects, three-body forces, N^*'s in the nucleus, etc. are all attempts to correct for this initial simplification. Lacking a systematic way to take the initial step, the corrections are ad-hoc. Most done to date certainly are indicative of the order of magnitude of the effects, but much more work is needed to put them on a firm systematic basis. There is certainly some arbitrariness involved in doing this. One way of making the reduction may have three-body forces, another may describe the same thing with N^*'s in the nucleus, another may put part of the effect into off-shell effects. Just because the non-relativistic calculations correspond to a well-defined mathematical problem does not make them a full description of the physics and much work is needed to introduce a careful treatment of the meson degrees of freedom into the subject.

The meson degrees of freedom in the three-body system are also important for electromagnetic and weak probes, especially in the form factors at large momentum transfers. Unfortunately many attempts have been made to fit these form factors with the single particle density distributions determined from non-relativistic Schrodinger wave functions even at momentum transfers as large as 10-20 fm^{-2}.[2] The He^3 charge form factor has a dip and a pronounced second maximum in this region which wave function calculations have failed to fit. At the UCLA meeting many of us stressed the importance of meson degrees of freedom at these large momentum transfers in giving the leading form factor dependence.[3] Recently Kloet and Tjon[4] have put those ideas into a calculation and shown that both the dip and second maxima are largely controlled by the meson currents. Again this is an ad-hoc calculation that is more indicative than final, but it emphasizes the point that extracting detailed dynamical information about nuclear dynamics from three-body systems requires a new level of sophistication. A few years ago just doing a three-body calculation with a good potential seemed a nearly superhuman task. Now a few super-persons have emerged, but it seems we will have to go even further to shed more light on these classic questions of nuclear physics.

These same cautions apply to the deuteron. Estimates of meson effects are probably just that, but attempts to describe the deuteron at large momentum transfers or at small distances without them are probably even more suspect.[5] Even a quantity as familiar as the deuteron D-state probability has a degree of arbitrariness in its definition depending on what operator one is using to probe, and how one shares meson degrees of freedom between corrections to that operator and to the wave functions. In nucleon-nucleon Bremstrahlung such problems are even more complicated because of the difficulties in choosing electromagnetic currents.[6]

Another classic reason for studying few body systems is to obtain information on the interaction between unstable particles from final states involving more than two strongly interacting particles. There is considerable phenomenological lore for making this extraction based in some form of Watson's theorem.[7] It is easiest to visualize this approximation in a case where the pair resonate. If the resonance life-time is long, the resonance can escape from the multiparticle interaction volume before decay. This is the basic idea behind what is often called the sequential decay model in nuclear physics, or the iso-bar model in particle physics. But if the lifetime is not long, the various ways of arriving at the final state are coherent and one has difficulty deciding which pair interacted "last". In nuclear physics this is called the "order of emission" problem and in particle physics the problem of overlapping resonance bands in the Dalitz plot. If the final state interactions are strong but not resonant and/or if some of the final state particles are identical, the problem is compounded. Although this problem has been around for a long time, very little has been done to confront it in the context of phenomenology. On the other hand, there have been a number of very successful separable interaction calculations for three-body final states. These calculations after making simplifying assumptions about the interaction can solve the dynamics exactly. Hence they correctly embody all the coherence, etc., required by quantum mechanics, but they have two drawbacks. First they involve dynamical assumptions (potential or interaction form) which sometimes seem rather special. Second, they are very complicated to do and hence when they agree with experiment, which they often do with remarkable accuracy, it is not clear what precisely has been verified. What is needed is a general approach that while including the constraints of quantum mechanics, does not embody them in unnecessary detail. What then does quantum mechanics require?

Consider an amplitude $T_{2,3}$ for going from a two-body state to a three-body state. Make the standard iso-bar, sequential decay, Faddeev or multiple scattering disposition of $T_{2,3}$ into a sum of terms depending on which pair interacts last

$$T_{2,3} = \sum_i f_i \tau_i \tag{1}$$

where τ_i is the two-body scattering amplitude of the j-k pair $(i \neq j \neq k)$ and f_i is its coefficient in $T_{2,3}$. Usually τ_i is taken to be dominated by a particular resonant pair partial wave ℓ_i, and then f_i is the amplitude for making the ith particle and that resonant pair. This amplitude is usually taken to have no dependence on the ith pair's center of mass energy, σ_i. The hope behind this assumption is that the various quasi-two body amplitudes for forming a resonance and a particle are independent. But in the wide resonance case the various ways in (1) of arriving at the final state are all coherent and the f_i need not remain either constant or independent of one another. We are interested in any rapid or singular dependence of f_i on the pair sub-energy σ_i for fixed total energy E in the physical region. Unitarity is a useful tool for investigating such questions since it signals a singularity at each threshold. Substituting (1) in unitarity and picking out only those terms which have a sub-energy threshold, we obtain[8]

$$\text{Abs } f_j = \sqrt{\sigma_j} \sum_{k\neq j} f_k \tau_k \tag{2}$$

where we have written $f_j = \text{Disp } f_j + i\text{Abs } f_j$, the dispersive and absorptive parts of f_j being the analog of real and imaginary parts appropriate when both are themselves complex. Hence unitarity forces f_j to have a part with a rapid (singular like $\sqrt{\sigma_j}$) dependence on the sub-energy σ_j and the strength of that singularity is given by the "non-j" parts of the amplitude. f_j is neither constant nor independent of the other f's. When will this coherent, singular part of f_j be important? When the sum, which is understood to contain an integral over the phase space, is big. If the τ_k in (2) carry a resonance of width Γ_k, at energy E_{o_k}, (2) will give

$$\text{Abs } f_j \approx \sqrt{\sigma_j} \sum_{k\neq j} f_k \frac{\Gamma_k}{\sqrt{E_{ok}}\, Q} \tag{3}$$

as an order of magnitude estimate, where Q is the energy released. The singular term is big when the resonance band is wide compared with the phase space (measured by Q). When the final pair interactions are not resonant, some other measure of the relative importance of the other pair interactions must be found but in all cases f_j is rapidly varying and coherent in precisely those cases that more than one pair interaction is simultaneously large. In nuclear physics n-d break-up is a well known example of this.[9] In particle physics a costly analysis of the $\pi + N \rightarrow \pi + \pi + N$ reaction at moderate energies[10] is questionable because the f's have been taken as constant whereas in fact the unitarity contributions (2) are large.

Unitarity can shed light on the validity of the usual final state interaction phenomenology in few body states, but how can we incorporate this lesson into a better phenomenology? The key to this is another general principle, analyticity. The threshold singularity (2) represents a branch cut of f_j in σ_j and the discontinuity across that cut is just given by (2). If we know the discontinuity of a function across its cut, and if we assume the function has no further discontinuities, we can use Cauchy's theorem to write.

$$f_j(\sigma) = R_j(\sigma) + \frac{1}{\pi}\int \frac{d\sigma' \, \mathrm{Disc} f_j(\sigma')}{\sigma'-\sigma} \qquad (4)$$

where R_j is the part of f that does not have the unitarity cut. Since $\mathrm{Disc} f_j \equiv \mathrm{Abs} f_j$ is linear in the f_k, when we use (2) in (4) we get a set of coupled linear integral equations for the f_j. Some of the integrals implicit in (2) can be done since they involve delta functions and one finds, that (4) reduces to a simple form of the separable interaction scattering integral equation usually obtained from the Schrodinger equation.[11] The f_j are just the spectator functions of that approach in the case with unit vertex function. Hence this separable interaction theory is the minimal one compatible with unitarity and analyticity. To the extent that the answers of most separable interaction calculations are weakly dependent on the vertex form, this helps to explain why so little detailed dynamical information can be determined from these calculations- they are largely the implementation of on-shell nearly kinematic constraints. But also since even these "kinematic" constraints lead to equations that are technically complicated to solve, simple phenomenology is not easy.

The moral is that few-body final state theory is as difficult as had been feared. The separable potential calculations rather than being a complex special model are in fact the minimal realizations of the full constraints of quantum mechanics. It may be possible to find a phenomenology like effective range theory that embodies unitarity and some limited analyticity, but short of that, something like the full separable interaction approach is required if the simultaneous interaction of more than one pair in the final state is important.

Yet another place where the few-body problem can contribute to a broader understanding of physics in general is in the elucidation of reaction mechanisms. There have been a number of attempts in nuclear physics to use soluble three-body models to cast light on methods used for the study of direct reactions like DWBA or for the analysis of (p,2p) reactions, but I would like to take my example from the meson-baryon system in particle physics.

One of the striking features of the meson-baryon system is the many resonances. A number of imaginative "explanations" have been set out for these resonances. Some like the quark model are like the nuclear shell model and some like SU(3) are classification schemes. What can we say about these resonances by taking into account the fact that they are embedded in a complex continuum of many strongly coupled and rapidly opening channels? In the πN system this continuum is largely made up of elastic scattering and single π production through much of the resonance region. Hence we have a three-body problem ($\pi\pi$N). The π-π and π-N interactions are dominated by a very few states (ρ,σ ,N,Δ,etc.) and the usual separable interaction technology can be employed. There is a relativistic three-body formalism that is very similar to non-relativistic ones, and permits us to include two and three-body processes as well as "potentials" into a unitary amplitude.[12] A striking feature of these calculations is the important role coupled channels play. For example, the inelastic process π+N→ρ+N with a threshold at about 1700 MeV. total CM energy is well known to be very strong and to open rapidly at threshold.[13] This must mean ρ-N in s-waves and since the ρ is 1^-, J = $\frac{3}{2}^-$ or $\frac{1}{2}^-$. The $\frac{3}{2}^-$ state is the π-N $D_{3/2}$ state. Such a rapidly opening channel always acts like an energy dependent attraction below its threshold. In this case, strong enough to produce a π-N resonance, the well known D_{13}(1520).[14] The resonance so produced has all the credentials of a dynamical resonance. It is not a kinematic effect and it is certainly not a bound state in the upper channel (ρN), that becomes unstable when the lower channel (πN) is turned on. Its existence depends not on the ρ-N or π-N force but on the off-diagonal coupling between the channels. There are many other examples in the π-N system. Some involve much complex channel coupling and some the combined effects of channel couplings and "potentials" or left hand cuts. From this point of view these resonances have a complicated dynamical origin rather than being the manifestation of simple, kinematical classification schemes. Obviously a connection between the two viewpoints must be made. Eventually when the higher symmetry schemes are made to express the unitarity coupling of channels and the microscopic few body schemes made to reflect the higher symmetries we will have a very complete picture of the meson baryon resonances. Meanwhile the fact that few body methods allow us to do these calculations gives detailed and complementary insight to traditional particle physics approaches.

I have tried to indicate some of the ways we can use our three-body skills to answer some old questions. In the nuclear force problem we have made much progress, but we need to keep in mind that just because we get very good at solving a non-relativistic Schrodinger equation does not mean that that is what nature does. In the final state interaction problem we understand better why our separable interaction calculations work so well, but since they are at once the minimal embodiment of the general constraints of quantum mechanics and technically difficult to do, the prospects for simple phenomenology seem remote. Finally, we understand reaction mechanisms better and can use this to provide detailed insight into the meson-baryon resonance spectrum.

References

1. For a recent view of this see D. D. Brayshaw, Phys. Rev. Letters 32, 382 (1974). For an older view see R. D. Amado, Ann. Rev. Nucl. Science 1969, Vol. 19, p. 61.
2. For an up to date review of the modern bound state calculations and related matters discussed here see, Y. E. Kim and A. Tubis, Ann. Rev. Nucl. Science 1974, Vol. 24, to be published.
3. Few Particle Problems in the Nuclear Interaction, edited by Slaus, et al. (North Holland, Amsterdam, 1972) p. 133, 144.
4. W. M. Kloet and J. A. Tjon, Phys. Letters 49B, 419 (1974).
5. Cf. M. J. Moravcsik and P. Ghosh, Phys. Rev. Letters 32, 321 (1974).
6. Cf. L. Heller in The Two-Body Force in Nuclei, edited by M. Austin and G. M. Crawley (Plenum, New York, 1972) p. 79.
7. Cf. M. L. Goldberger and K. M. Watson, Collision Theory (John Wiley and Sons, New York, 1964) pp 540-553.
8. R. Aaron and R. D. Amado, Phys. Rev. Letters 31, 1157 (1973).
9. Cf. R. T. Cahill, Phys. Rev. C9, 473 (1974); Phys. Letters 49B, 150 (1974).
10. D. J. Herndon et al. Lawrence Berkeley Laboratory Report No. LBL-1065 and SLAC Report No. SLAC-PUB-1108, 1972 (unpublished).
11. R. D. Amado, Phys. Rev. Letters 33, 333 (1974).

12. R. Blankenbecler and R. Sugar, Phys. Rev. 142, 1051 (1966); D. Freedman, C. Lovelace and J. Namyslowski, Nuovo Cimento 43, 258 (1966); and R. Aaron, R. D. Amado and J. E. Young, Phys. Rev. 174, 2022 (1968).

13. Cf. L. F. Cook and B. W. Lee, Phys. Rev. 127, 297 (1962).

14. R. Aaron and R. D. Amado, Phys. Rev. Letters 27, 1316 (1971).

DISCUSSION

NOYES(SLAC): I am very pleased to see this derivation of the minimal content of analyticity and unitarity in terms of on-shell scattering. An alternative derivation which contains the same physics was presented at the Prague Symposium this spring (Czechoslovac Journal of Physics, in press) in which I showed that dropping the off-shell terms in the two-body input to the Faddeev equations leads to unitary and convergent integral equations. Presumably these are equivalent since they also consist of summing the on-shell scatterings.

I also indicated in a pre-print distributed this spring (SLAC-PUB 1401), that by utilizing the limitations imposed by four-body thresholds (anomalous rather than physical), it is possible to reduce the equations to matrix equations with coefficients of known analytic structure, thus avoiding the singularity problems which have plagued numerical work. Hopefully this will meet your second comment about the difficulty of solving these equations.

AMADO: Thank you very much. I have not yet seen that work. I would be interested in seeing it. Let me make one comment of basic philosophy: of course the Faddeev equations, which are just the Schrödinger equation, lead to a unitary answer. This is one of the great miracles that the equations obtained from Schrödinger's satisfy the non-linear constraints. What amuses me here is that I was able to derive these non-linear equations without ever saying anything about the Schrödinger equations nor assuming any linear equation; it just comes out of implementing analyticity and unitarity.

BLEULER(Bonn): Coming back to the three-body bound state, I was glad that you mentioned the mesonic degrees of freedom. In fact there are some calculations from our Institute in Bonn (D.Schütte et al.) which showed that these corrections are really large, (amazing accuracy, therefore of earlier calculations!). In addition I should also mention that the first excited state of the nucleus (the so-called Δ-resonance) should be taken into account (work of Professor G.Brown and Tony Green).

AMADO: Yes of course, as I mentioned, the Δ-degrees of freedom are important. However, in taking into account these degrees of freedom, one has to be very careful in understanding what he is doing. There may be great danger of double counting.

KOMAROV(Moscou): The separable-potential is of the form which permits one to use it easily for the many-body nuclear reactions. However, it is interesting to know about other potentials for the many-body problem. Do you have any idea about it in connection with your talk?

AMADO: I agree with what you said about separable potentials. The only place where I mentioned them is in the final state interaction theory, and more exactly, when this was dominated by a particular pair partial wave. This turns out, in the present formalism, to be essentially the same as saying that you have a separable interaction. Otherwise, you have to make a more complicated decomposition.

LEVINGER(Rensselaer): I was of course very happy to see another reason for the success of separable approximations. But we now know of failures in the separable interaction for tensor forces. Does this failure show up in your equations?

AMADO: At the present level of sophistication, certainly not. But eventually I am sure that it will show up. This approach is still very much in its infancy.

FEW-BODY PROBLEMS IN HIGH ENERGY HADRON PHYSICS *)

L. Van Hove
CERN - Geneva
and
Max-Planck Institute for Physics and Astrophysics
Munich

This review lecture deals briefly with two topics. The first concerns hadronic collisions with three and four particles in the final state. The second deals with the interpretation of inelastic pp collisions at centre-of-mass energies $\gtrsim 20$ GeV in terms of emission of a small number of clusters or fireballs, each composed of several secondaries.

---

*) Lecture at the International Conference on Few-Body Problems in Nuclear and Particle Physics, Université Laval (Québec, Canada), 27-31 August 1974.

PROBLEMES A PETIT NOMBRE DE CORPS EN PHYSIQUE HADRONIQUE A HAUTE ENERGIE *)

L. Van Hove
CERN - Geneva
et
Institut Max Planck pour la Physique et l'Astrophysique, Munich

L'exposé passe brièvement en revue deux sujets. Le premier concerne les collisions hadroniques à trois et quatre corps dans l'état final. Le second traite de l'interprétation de collisions inélastiques proton-proton aux énergies du centre de masse $\gtrsim 20$ GeV en termes d'émission d'un petit nombre d'essaims composés chacun de plusieurs particules sortantes.

---

*) Exposé à la Conférence Internationale Problèmes à Petit Nombre de Corps dans la Physique du Noyau et de Particules Elémentaires, Université Laval (Québec, Canada), 27-31 août 1974.

1. - INTRODUCTION

When originally planning the present lecture, I thought just to survey once more the main features of three- and four-body hadron collisions, as measured mainly in the centre-of-mass energy range $3 \text{ GeV} \lesssim E_{cm} \lesssim 7.5 \text{ GeV}$ corresponding to beam momentum $5 \text{ GeV} \lesssim p_{lab} \lesssim 30 \text{ GeV}$ [where at high energy $E_{cm} \simeq (2m\, p_{lab})^{\frac{1}{2}}$ with $c = 1$ and m the mass of the target particle, usually a proton]. Only very little is yet known about these collisions in the higher range of energies accessible at the FNAL accelerator ($p_{lab} \lesssim 400 \text{ GeV}$ corresponding to $E_{cm} \lesssim 27.5 \text{ GeV}$) and at the CERN Intersecting Storage Rings ($20 \text{ GeV} \lesssim E_{cm} \lesssim 60 \text{ GeV}$). The first generation of experiments with these new machines has revealed, however, that general inelastic collisions behave as if they were of few-body type in a new sense (the bodies now being mostly higher mass objects called clusters or fireballs). This important finding, which confirms and makes more precise many earlier indications from cosmic ray experiments, is of such interest that we shall devote the third section of the lecture to it. Section 2 deals with genuine few-body collisions. Both topics will be treated very concisely. More information can be obtained from the references and from the numerous papers quoted in the references.

2. - THREE- AND FOUR-BODY HADRON COLLISIONS

Our state of knowledge of hadron collisions of three- and four-body type ($pp \to pn\pi^+$, $pp \to \Delta^{++} p\pi^-$, $\pi^{\pm} p \to \pi^{\pm}\pi^+\pi^- p$, etc.) has not changed qualitatively in the last few years. The main phenomenological characteristics [for a concise review, see Section 5 of Ref. 1), esp. subsection 5.2] are, just as for two-body reactions, the smallness of all transverse momenta and the fact that the behaviour of the reaction is mainly controlled by the signs of the longitudinal centre-of-mass momenta of the outgoing particles ; these signs define which internal quantum numbers (baryon number, strangeness, charge, etc.) are exchanged in the reaction. The data show clearly that the energy dependence of the reaction cross-section is mainly controlled by these quantum number exchanges. Baryon exchange and double charge exchange decrease most rapidly with increasing energy, then comes strangeness exchange, then single exchange of isospin and G parity. The slowest energy variation, and hence the largest cross-section at high energy, is obtained for vacuum exchange, also called diffraction dissociation. It is very interesting to observe how a number of these exchange processes coexist in a single reaction (for example baryon exchange, charge exchange and vacuum exchange in $pp \to \Delta^{++} p\pi^-$) and how they separate in phase space as the incident energy increases. This is most clearly exhibited by analyzing the data in terms of longitudinal phase space.

For illustrations, we refer the reader to another 1972 review [2], especially its Section 4.

The question of the theoretical interpretation of the few-body reactions at medium and high energies is a complex one. The qualitative behaviour, especially the correlation between rate of energy decrease and exchange character of the collision, is well explained by the Regge model. The detailed quantitative properties, however, especially angular distributions and polarization features, require the model to become very complicated, with many ad hoc corrections and recipes. The present trend of work is determined by the realization that angular and polarization properties often look simpler and more universal when examined through an impact parameter analysis (the so-called s channel approach, Regge analysis being the t channel approach). As an example of this type of work we quote a recent study of Δ production reactions in meson-nucleon collisions by de Brion and Peschanski [3]. Recent developments in the s channel approach are found in the work of the Schrempps [4] and of Hara et al. [5].

Considerable progress can be reported in the study of a special reaction, single-pion exchange in meson-nucleon three-body collisions, especially $\pi^- p \to \pi^- \pi^+ n$. At very small angle (momentum transfer squared $t \lesssim 0.2$ GeV^2) pion exchange is found clearly to dominate. A good description of the data was obtained with a one-pion exchange model with absorptive corrections as originally proposed by Williams, and this led to very important progress in $\pi\pi$ phase shift analysis. Details and references can be found in recent papers of the CERN-MPI Munich Collaboration [6]. The situation at larger t values, where natural parity exchange is important, is less clear, and one does not know yet with certainty whether pion exchange is of elementary or Regge type.

3. - GENERAL COLLISIONS AT VERY HIGH ENERGY

As mentioned in the Introduction, the recent study of common inelastic pp collisions in the FNAL-ISR energy range $15 \text{ GeV} \lesssim E_{cm} \lesssim 60 \text{ GeV}$ has revealed very interesting properties which suggest that they are of few-body type in a new sense of the word. Their actual multiplicity is usually large (mean charged multiplicity $\simeq 7.5$ at $E_{cm} \simeq 20$ GeV and $\simeq 12$ at $E_{cm} \simeq 50$ GeV), but the measured distributions and correlations of secondaries can be simply understood if one makes the hypothesis that the observed secondaries occur in "clusters". Each cluster counts perhaps on the average three or four particles, of which two or three are charged. The mean number of clusters per collision is therefore small, of order 5 at ISR energies.

About 20-25% of inelastic pp collisions are of single diffraction dissociation type, the final state consisting of one proton and one "protonic cluster" (or excited proton) flying off in opposite directions in the centre-of-mass frame. In most of the remaining collisions, in addition to high energy protons or protonic clusters, so-called "central" secondaries come out at lower centre-of-mass energy and they have a positive short range correlation. The correlation data can be readily explained by assuming that the central secondaries occur in clusters. These central clusters are of mesonic type, are preferentially of zero charge and are assumed to be themselves emitted in bremsstrahlung-like fasion (i.e., in an uncorrelated way except for momentum energy conservation). Reference 7) gives a recent review of this so-called independent cluster production model. The evidence on which it rests is still fragmentary and the model is probably much too crude and oversimplified. Its great virtue, however, is that it gives a remarkable degree of physical insight in high multiplicity phenomena by reducing them to fewer effective degrees of freedom.

We illustrate by one example the success of the independent cluster production model. Under its assumptions one can relate the centre-of-mass energy spectrum $d\sigma/dE$ of outgoing protons (or protonic clusters) to the density ρ of central clusters in longitudinal rapidity, the relation being

$$d\sigma/dE \propto (\tfrac{1}{2} E_{cm} - E)^{\rho-1} \qquad (1)$$

The data give a flat $d\sigma/dE$ spectrum, hence $\rho \simeq 1$. But ρ is given by $\rho_c / < n_c >$, where ρ_c is the density of the charged secondaries in rapidity, and $< n_c >$ the mean multiplicity of charged secondaries per cluster. The short range correlation measurements lead to the estimate $< n_c > \simeq 2.5$, the density ρ_c is directly measured, and one finds again $\rho \simeq 1$. We refer to Section 4 of Ref. 7) for details.

A very recent development [8)] interprets the independent cluster production model in terms of the quark-gluon picture of the internal nucleon structure which has resulted from the interpretation of electron-nucleon and neutrino-nucleon deep inelastic scattering experiments. Excluding diffraction dissociation collisions, an inelastic proton-proton collision is interpreted as follows. The valence quarks of each proton fly through with the fraction of proton momentum they carried originally in the incident proton, while the glue of one incident proton interacts with the glue of the other. The latter process gives rise to the production of the central clusters. In the final state the quarks which flew through "dress up" again with glue and give rise to the outgoing protons or protonic clusters. A remarkable consequence of this interpretation is that the flat proton spectrum (1) with $\rho \simeq 1$ can be related to the shape of the quark distribution inside the nucleon (i.e., to the electromagnetic and weak structure functions νW of the nucleon as measured in the deep inelastic lepton-nucleon experiments). As shown in Ref. 8), excellent compatibility is found between the two distributions.

REFERENCES

1) L. Van Hove - in Trends in Physics, Volume of Plenary Lectures at the Second General Conference of the European Physical Society, 3-6 October 1972, Wiesbaden, published by European Physical Society, Geneva (1973), pp. 39-62.
2) L. Van Hove - in Developments in High Energy Physics, Volume of Lectures at the 1971 Enrico Fermi International School of Physics at Varenna, Academic Press Inc., New York (1972), pp. 130-152.
3) J.P. de Brion and R. Peschanski - "Amplitudes for ρ and A_2 Regge Exchanges in Δ Production Processes", CERN Preprint TH. 1848 (1974), to appear in Nuclear Physics B.
4) B. and F. Schrempp - Nuclear Phys. 54B, 525 (1973).
5) Y. Hara, T. Kuroda and Y. Takaiwa - "Peripheral Orbit Model", Preprint, Dept. of Physics, Tokyo University of Education (1974), submitted to Progress of Theoretical Physics.
6) G. Grayer et al. - Nuclear Phys. B50, 29 (1972) ;
B. Hyams et al. - Nuclear Phys. B64, 134 (1973).
7) S. Pokorski and L. Van Hove - Acta Physica Polonica B5, 229 (1974).
8) L. Van Hove and S. Pokorski - "High Energy Hadron-Hadron Collisions and Internal Hadron Structure", in preparation.

DISCUSSION

RINAT(Weizmann): i) Is it accidental that the function $M(b,s)$, appearing in the amplitude describing the universality, resembles the profile function of an object of size $\simeq 1$ fm ? ii) How should one view the binding of quarks in a nucleon when the individual quarks have the large momenta you discussed ? Or is the glue accounting for the binding ?

VAN HOVE: As to the first question, it is indeed true that the impact parameter analysis of hadron-hadron collisions leads to a distribution which looks like familiar distributions in light nuclei. One might hope that this is not accidental. There have been attempts: in particular the Yang-Wu model for proton-proton scattering can be quoted there. It will certainly be very interesting to think further about the analogy. Concerning the quarks, there is some misunderstanding if you believe (maybe you do not) that the quarks have a high momentum in the rest system of the proton --they don't. What I described is the sharing of the total momentum of a very energetic proton using its quarks. The quark momenta in the proton rest frame are supposed to be small. The most picturesque model is the 'bag' model developed at MIT, which says that the nucleon is a kind of bag inside which the quarks are moving around as massless fermions. They are not allowed to escape from the bag, the volume of which contributes an additional term in the hamiltonian and is fixed by minimization of the total energy.

BLEULER(Bonn) Summarizing, you put forward three different notions: (i) 'reggeism', (ii) quarks (rather hypothetical particles not seen so far) and (iii) the glue (which might be called the aether of the 20th Century). In view of the recent developments, I would, however, put more emphasis on the recent work of Feynman where the notion of partons is introduced. This leads to a successful model of hadrons which in turn has strong analogies to nuclear structure. At the same time I should also mention the spectacular increase of the N-N cross section with extremely high energies. This was anticipated by the ideas of Landau and Heisenberg whereas it was foreseen from the viewpoint of Regge trajectories.

VAN HOVE: The parton model of Feynman and Bjorken-Paschos had to be modified in view of the ep and νp deep inelastic data. The idea that the nucleon was exclusively composed of charged partons had to be corrected in the sense that one has to have a neutral matter inside the nucleon carrying on the average half of the proton momentum. Whether one likes to call it glue or not is another question. The name glue is rather natural because, since the partons do not escape, there must be a meson field to keep them together, hence the name glue. The quarks behave, in deep inelastic scattering, as very light spin 1/2 objects; still they cannot escape, which is very paradoxical. Both glue and quarks may be like the aether in the sense that the final theory will probably be very different. But, after all, the concept of aether may have helped Maxwell to make lasting contributions to physics.

The second point is on the total cross section increase. There are a number of possible explanations for this phenomenon, which was in fact first seen at SERPUKHOV in K^+p, several years ago, but is now known to be more general. The earliest explanation is the Heisenberg one, that at the edge of the nucleon there is more and more matter participating in the interaction as the energy increases, and this implies a growth of the total cross section like the square of the logarithm of the energy. The numbers in the Heisenberg model are compatible with the data because this model uses a scale given by the mass of the π, but the effect could still be qualitatively correct. Another possible explanation is that diffraction dissociation, which seems rather small at 20 GeV but creates bigger and bigger masses as you go up in energy, would be the cause of the increase in the total cross section. This explanation is favoured by Regge theory. To decide, one will have to turn to experiments and look at which inelastic processes are responsible for the increase in the total cross section.

PUGH(Maryland): I would like to ask about the extension of these ideas to nuclei. If one were to accelerate heavy ions at the Fermi National Accelerator Laboratory or the SPS, or better still to fill the rings of the ISR with heavy ions, one could produce collisions whose consequences are almost impossible to imagine without experiment. For example, if the quarks and glue from individual nucleons were to overflow into many other nucleons one might produce entirely new states of matter which could not be produced in any other way. Could you comment on this ?

VAN HOVE: This is an extremely interesting question. The question indeed is what would happen if you compressed a group of nucleons to such very high densities that the droplets of glue joined into a liquid of glue with the quarks swimming in it. A possible place to find that in the world might be at the centre of a neutron star, but we cannot easily go there and have a look. The question of whether you can achieve such states of nuclear matter by extremely energetic collisions of heavy ions is a highly disputed one. It is under active discussion because there are also other speculations about states of abnormal nuclear matter at high density (Lee-Wick). The difficulty that is present in the mind of many people, including myself, is that in very high energy collisions of nuclei there may be a considerable degree of trivial transparency. One fears that most of the time the nuclei largely ply through each other and that the collisions where very dense matter is produced may have a depressingly low cross section. I don't know whether this is so, but it is a problem of very great concern because of the cost of heavy ion acceleration at very high energy. However, as you say, it remains true that completely new phenomena might be revealed.

FADDEEV SOLUTIONS WITH NON-SEPARABLE INTERACTIONS

J.A. Tjon

Institute for Theoretical Physics, University of Utrecht,

The Netherlands

Although some variational calculations have been carried out a long time ago on the three-nucleon system[1)], the actual solutions with complicated realistic two-nucleon interactions have been achieved quite recently. The main reasons for this are that on the one hand there existed no correct mathematical framework of the problem until the work of Faddeev[2)] and on the other hand new computational techniques were needed to solve the resulting three-body equations. Here we would like to review one kind of computational methods which has proven to be successful. For the various results obtained so far using these methods we refer to reference 3 and the references quoted there.

Our starting point are the Faddeev equations which can be written formally as

$$U(s,\lambda) = U_o(s) + \lambda K(s)\, U(s,\lambda) \qquad (1)$$

where the kernel K and the inhomogeneous term U_o depend explicitly on the energy s of the three-body system. The physical solutions of eq. (1) are obtained by setting the parameter $\lambda=1$. The kernel K(s) can be expressed in terms of the two-body T-matrices T_i and the unperturbed resolvent $G_o(s) = [s-H_o]^{-1}$ through the relation

$$K = TB \qquad (2)$$

with

$$T = \begin{pmatrix} T_1 & & \emptyset \\ & T_2 & \\ \emptyset & & T_3 \end{pmatrix} \quad \text{and } B = G_o(s) \begin{pmatrix} 0 & 1 & 1 \\ 1 & 0 & 1 \\ 1 & 1 & 0 \end{pmatrix}$$

After partial wave decomposition the Faddeev equations become coupled linear integral equations in two continuous variables. In general they are difficult to solve by direct matrix inversion because of the occurrence of various singularities in the integrations and the dependence of the limits of integrations on the initial variables. Moreover, due to the two-dimensional character of the equations and the presence in general of coupling between various discrete channels the size of the matrix to be inverted becomes so large that it would take prohibitive long times even with present-day large computers. One exception to this is the situation where the two-nucleon interaction is assumed to be of a separable form. In this case the three-body equations reduce to one-dimensional integral equations.

The methods which we should like to discuss here donot depend on the specific form of the two-nucleon interaction. They are based on the fact that the kernel of the Faddeev equations is a compact operator. As a result the solution U (s,λ) is a meromorphic function of λ and furthermore it is analytic in $\lambda=0$. Let us assume that we can determine the Neumann solution of eq. (1)

$$U(s,\lambda) = \sum_{n=o}^{\infty} \lambda^n U_n(s) \tag{3}$$

where

$$U_n(s) = K^n U_o(s) \tag{4}$$

In practice, the coefficients (4) can be obtained by iterating numerically the integral equations with the singularities removed by subtraction procedures[4].

We may now introduce two methods to find the analytic continuation of eq. (3) to the physical value of $\lambda=1$. The first method is appropriate for finding the ground state solutions and consists of computing the the ratios $r_n = U_n/U_{n-1}$[5]. Since K(s) is compact, the spectrum of eigenvalues $\lambda_\alpha(s)$ form a denumerable set which have only infinity as a limit point[6]. As a consequence we may write

$$U(s,\lambda) = \sum_\alpha \frac{B_\alpha(s)}{\lambda_\alpha(s)-\lambda} + R(s,\lambda) \tag{5}$$

where $R(s,\lambda)$ is an entire function of λ . Expanding eq. (5) in a power series of λ and comparing the corresponding coefficients with those of eq. (3) one readily finds that

$$\lim_{n\to\infty} r_n = 1/\lambda_o(s) \tag{6}$$

where $\lambda_o(s)$ is the smallest eigenvalue in magnitude. In order to find the binding energy of the groundstate we merely have to form the ratios r_n as a function of s and determine the value of s for which r_n tend to unity for $n\to\infty$. In fig. 1 a typical example is shown of the dependence of $\lambda_o(s)$ on s . In practice, it turns out that the computation of about ten coefficients are sufficient to reach the limiting value of r_n . Since each iteration goes roughly as N^2 operations, where N is the number of meshpoints used in the integration, it is much less time consuming than a matrix inversion where N^3 operations are needed. Once the bound state energy has been found one gets as a direct byproduct the corresponding boundstate wavefunction, which is essentially related to $\lim_{n\to\infty} U_n(s)$. (For details see ref. 5).

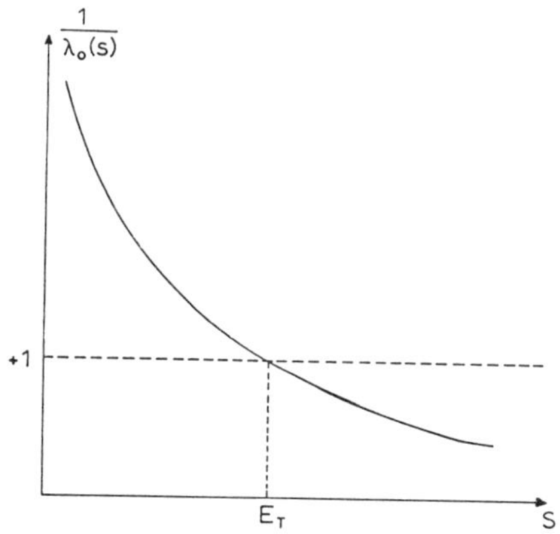

fig. 1

The second method which can be used not only in the scattering region but also for boundstates is the calculation of the Padé approximants (P.A.) of the series (3) i.e. $U(s,\lambda)$ is approximated by a ratio of polynominals of order M and N respectively.

$$U_{[N,M]} = P_M(\lambda)/Q_N(\lambda) \tag{7}$$

The coefficients of the polynominals can be found by comparing eq. (7) with eq. (3) up to order N+M+1. One finds[7)]

$$P_M(\lambda) = \begin{vmatrix} U_{M-N+1} & \cdots & U_{M+1} \\ \cdot & & \cdot \\ \cdot & & \cdot \\ \cdot & & \cdot \\ U_M & \cdots & U_{M+N} \\ \sum_{j=N}^{M} U_{j-N}\lambda^j & \cdots & \sum_{j=o}^{M} U_j\lambda^j \end{vmatrix} \tag{8a}$$

and

$$Q_N(\lambda) = \begin{vmatrix} U_{M-N+1} & \cdots & U_{M+1} \\ \cdot & & \cdot \\ \cdot & & \cdot \\ \cdot & & \cdot \\ U_M & \cdots & U_{M+N} \\ \lambda^N & \cdots & 1 \end{vmatrix} \tag{8b}$$

Although the convergence of PA's for solutions of linear integral equations have been conjectured and proven[8)] the rate of convergence should be fast enough in order to be of practical use. In the calculations where it has been applied up to now this indeed turns out to be the case.

The connection between the two different methods can easily be seen from eq. (8). Let us consider in particular the sequence of PA's $U_{[1,N]}$ with N = 1, 2,... From eq. (8b) we see that this sequence has only one pole given by $\lambda = U_N/U_{N+1}$ which is the same result as given by the ratiomethod. Moreover, if r_n has a limiting value the sequence of PA's $U_{[N,M]}$ with M = 1, 2,... tends to a function which has a pole at the position $\lambda = \lim_{n\to\infty} 1/r_n$, since the last two columns in (8b) become the same for this value of λ .

The above described methods have been applied extensively in the study of the three-nucleon problem. One of the puzzling differences which was found with variational methods[9] is that the dip in the resulting electromagnetic form factor of ^3He came out differently. It has sometimes been argued that the variational methods would be more sensitive to short range behaviour of the wavefunction. In this connection it seems worthwhile to point out the relationship between PA's and variational principles[10,11]. For this purpose, let us consider the boundstate solutions of the Faddeev equations. They satisfy the eigenvalue equation

$$U(s,\lambda) = \lambda\, K(s)\, U(s,\lambda) \tag{9}$$

It is easily seen that the expression

$$F = \lambda(U_t\, ,\, BKU_t) - (U_t\, ,\, BU_t)$$

is stationary under variations. If we now take for the trial function U_t a linear combination of terms of the Neumann series with a given U_0

$$U_t = \sum_{n=0}^{N-1} c_n U_n \tag{10}$$

with c_n the variational parameters, we readily find for $\delta F=0$

$$\sum_{n=0}^{N-1} [\lambda(U_m\, ,\, B\,K\,U_n) - (U_m\, ,\, B\,U_n)]\, c_n = 0 \tag{11}$$

In view of eq. (2) and the self-adjointness of the operators T and B we may rewrite eq. (11) as

$$\sum_{n=0}^{N-1} \left[\lambda(U_o \, , \, B\, U_{m+n+1}) - (U_o \, , \, B\, U_{m+n})\right] c_n = 0 \tag{12}$$

This admits a solution provided that det M=0 with

$$M_{mn} = \lambda(U_o \, , \, B\, U_{m+n+1}) - (U_o \, , \, B\, U_{m+n})$$

which is precisely the condition that $Q_N(\lambda)$ of eq. (8b) should vanish for M=N-1 . In a similar way it can be shown that in the scattering region the Schwinger variational principle with the use of the trial function (10) leads to the [N,N-1] approximant for the T-matrix.

References

1. See for example in N.F. Mott and H.S.W. Massey "The theory of Atomic Collisions", third edition, (Oxford at the Clarendon Press, 1965).
2. L.D. Faddeev, "Mathematical Aspects of the Three-Body Problem in Quantum Scattering Theory" (Jerusalem, 1965).
3. J.A. Tjon, Proceedings of the Symposium on "Theory of lightest nuclei" (Liblice, 1974).
4. W.M. Kloet and J.A. Tjon, Ann. Phys. 79, 407 (1973).
5. R.A. Malfliet and J.A. Tjon, Nucl. Phys. A127, 161 (1969); Phys. Lett. 29B, 391 (1969).
6. F. Smithies, "Integral Equations" Cambridge Tracts in Mathematics and Mathematical Physics, no. 49, Cambridge University Press, New York (1958).
7. G.A. Baker Jr., chapter 1 of "The Padé Approximant in Theoretical Physics", G.A. Baker Jr. and J.L. Gammel eds. (Academic Press,

New York, 1970).

8. J.S.R. Chisholm, J. Math. Phys. 4, 1506 (1964); J. Nuttall, J. Math. Anal. and Applications 31, 147 (1970); G.A. Baker Jr., preprint (1974).

9. A.D. Jackson, A. Landé and P.U. Sauer, Phys. Lett. 35B, 365 (1971); M.A. Hennell and L.M. Delves, Phys. Lett. 40B, 20 (1972).

10. C. Alabiso, P. Butera and G.M.Prosperi, Nucl. Phys. B31, 141 (1971); Lett. Nuovo Cimento 4, 561 (1972).

11. J. Nuttall, in "Padé Approximants and their Applications" P.R. Graves-Morris ed. (Academic Press, New York 1973).

DISCUSSION

AFNAN(Flinders): We have done some calculations on the tritium binding energy with OBEP; we find that the binding energy we get is about 8 MeV, which means there is only 1/2 MeV to explain, and I will be giving a contributed paper where I will discuss the details of the results. Briefly: it seems that non-locality that one gets from OBEP, allows one to get a small D-state probability for a deuteron and maintain a large quadrupole moment, which is in agreement with the experiment, and therefore the large binding energy.

TJON: Yes, but I think one should look at it in a very connected way, namely one should not look at only the binding energy, but one should question that if you would get the binding energy at the right place, how is the form factor going to be?

AFNAN: I cannot answer that until I finish the calculations of the form factor.

TJON: The reason I am asking this is that experience shows that if you increase the binding energy to 8 - 8.5 MeV, the dip shifts outwards.

AFNAN: That is true for local potentials but I don't know...

TJON: Also for separable ones ?

AFNAN: Well, as I said, I can't answer this question until I finish the calculations.

MORAVCSIK(Oregon): I understand from what you said that you are claiming that there is 2 MeV binding energy in the the triton which comes from essentially three-body forces of some sort, is that correct ?

TJON: Three-body forces, yes.

MORAVCSIK: Now that is a huge amount and if that is the case, presumably those three-body forces...

TJON: That might be 1 MeV, let me be a little more careful.

MORAVCSIK: O.K., but in any case, presumably that three-body forces also will show up in heavier nuclei and in nuclear matter calculations and in fact it would give, proportionately, I would guess, much bigger contribution here. Now, how come that this has not been seen ? I mean this huge effect you are talking about.

TJON: How do you want to see it ? I mean that all calculations that have been done up to now are shell-model type calculations, not really full scale calculations, where you have a hamiltonian, where you solve the Schrödinger equation. The question is just open I think.

MORAVCSIK: I am not really a specialist of nuclear matter calculation. But I would like to hear someone comment on it, because I am still a little dubious.

BLEULER(Bonn): Yes I would just like to comment about the comparison of these most interesting results with some calculations on nuclear matter; it seems to me that the situation there is very similar to the one in the three-body case: in our Institute, R. Machleidt calculated with the help of Brueckner's theory the properties of nuclear matter with both the Reid-potential and OBEP-expression as determined by Erkelenz et al.. It was seen, although the OBEP was slightly better that Reid, that a characteristic discrepancy remained. The main difficulty --also in the case of other potentials-- is a simultaneous fit of binding and density to the experimental values.

TJON: You mean that from this point of view there are some indications that something is missing too.

BLEULER: Yes. Absolutely. It seems that the two cases agree. I am therefore very interested in these results.

WIGNER(Princeton): I have one, two questions. The first is: could you repeat what you vary in order to obtain the Padé approximants ? It was not clear to me.

TJON: Yes, just because of the time, I have not gone through the calculations, but let me give you an idea. I have to find the right sheet.

WIGNER: Let me make the second remark in the meantime. We know that if we calculate the interaction between two Na atoms, and then try to calculate on that basis the energy of three Na atoms, we would get a terribly bad result. That in order to get a good result for a system of several particles when the interaction is not simple, one uses an entirely different method, namely, one puts a glue into the state --in the case of Na which I happen to know, the extra electron--; now it is possible that this method might have some advantages in nuclei. You see what I am trying to say, the glue in that case would be the π's, ρ's and partners . You know that would be a sea of π's and ρ's and the nuclei would sit in it. Does that make sense or does it not ? Dr. Noyes pointed out that Teller and Jahn made a similar remark a long time ago.

TJON: Well the happy feature about the 3-nucleon system, is, I think that if you would look at the pure 2-particle system,there is quite a bit of short-range correlation. Because of this the effect will be that you have to include, in the first approximation only, an additional pion to the 3-nucleon system, which means that, hopefully, the approach would be that you would only have to take the pion as a first step, and after that the presence of heavier mesons and additional pions, in contrast to the

Na case where you have essentially many electrons as a sea.

WIGNER: The question really is then: is it not that the triton is a too difficult problem and the heavier nuclei are the easiest problems ?

TJON: That might be the case.

WIGNER: But let us see about the Padé...

TJON: Yes, about the Padé approximants, it is so that the trial functions you take are linear combinations of the coefficients U_n which are just the n-th iterate of the integral equations, that is the U_n are just $K^n U_o$.

WIGNER: What is K ?

TJON: K is the kernel of the integral equation, and the c_m are the variational parameters. So if you vary with respect to these coefficients here,you end up with the Padé approximants.

WIGNER: I mean, if you vary something, I understand. But you vary it in an expression?

TJON: Yes.

WIGNER: And what is the expression ?

TJON: The expression is $F = \lambda(U_t, BKU_t) - (U_t, BU_t)$

WIGNER: And you vary the U_t ?

TJON: Yes.

WIGNER: Thank you very much. This answers my question.

A BOUNDARY-CONDITION APPROACH TO THE THREE-BODY PROBLEM

D. D. Brayshaw†

University of Maryland, Department of Physics and Astronomy
College Park, Maryland 20742 U.S.A.

It has generally been accepted that a study of three-particle systems can yield new information regarding details of the two-particle interactions, and in particular the short distance behavior of the nuclear force. This possibility has stimulated a great deal of work on the trinucleon system during the past decade. In view of the nature of the results which this program has produced, it is my belief that we must reassess both our expectations of what can be learned from such studies, and the techniques which we apply to the problem. A case in point is the triton ground state properties. Early calculations were cause for great optimism in several respects. On the one hand, a picture in which the dynamics are dominated by simple phenomenological pairwise forces was shown to produce quite reasonable results. On the other hand, considerable sensitivity was exhibited to such effects as the choice of form factor, presence of short-range repulsion, tensor vs. central forces, etc. However, these effects can also be distinguished to some extent by the two-nucleon data, and after a vast amount of labor involving increasingly more sophisticated models, it has begun to appear that the differences exhibited by competing "realistic" potentials are comparatively minor. Moreover, only the triton binding energy (E_T) and the minimum in the ^{3}He charge form factor are truly sensitive, and the interpretation of the latter is beclouded by unknown corrections due to meson currents. While it may be possible to pin down quantities such as the percent d-state in the deuteron via the weak and electromagnetic properties of the triton, there is little reason to believe that off-shell information can be extracted from such studies.

If we now turn to n-d scattering, it has been known for some time that the doublet scattering length a_2 is closely correlated with E_T, whereas a_4 is quite insensitive to the details of the forces. However, since there is an apparent wealth of experimental information above the breakup threshold, and theoretical studies in this regime are still in a rather preliminary phase, one might hope for the emergence of new, independently sensitive parameters which could yield new information. Unfortuately, the results I have obtained during the past year suggest that this hope may be largely futile. This brings me back to the two points I raised earlier: expectations and techniques. In the first case there is a high probability that a study of the trinucleon system will yield little, if any, new information concerning the basic N-N

†Alfred P. Sloan Foundation Fellow.

interaction. The other side of this coin is that ambiguities associated with off-shell effects can be virtually eliminated in quasi-three-body treatments of reactions such as (p,2p), deuteron stripping and the like, enhancing the possibility of probing nuclear structure.

In regard to the second point, it seems obvious that an attempt to pin down the content of the breakup reaction by standard techniques will involve an effort at least as great as that expended on the triton. By "standard" I mean simply the procedure of performing successive calculations with different potentials until there is a consensus that a definite picture has emerged. Inasmuch as the on-shell and off-shell properties are inextricably linked in the parameters characterizing the potential, this is a crude procedure at best. In practice it is slow, laborious and expensive. As a result, the pace is too slow to serve as a guide in designing experiments to yield maximum information; this is particularly critical in an era of limited resources. In contrast, the procedure I will describe is specifically designed to do this job with maximum efficiency and minimal expense.

The price one pays for these advantages is the unfamiliarity of a new formalism. I will thus concentrate in what follows on what I hope will be a clear exposition of the method and its relationship to the conventional approach. I want particularly to stress that the formalism is not based on any radical new assumptions. In fact, one assumes somewhat less than in potential theory, which is included as a special case. More specifically, given a potential there is a unique and well-defined procedure for generating the appropriate input, although it is much more efficient to isolate the characteristic properties of such input and parametrize it directly.

The basic idea of the boundary-condition (BC) approach is to characterize a three-body state in such a way as to completely isolate effects arising from the off-shell degrees of freedom. In spirit it closely parallels the ideas of H. P. Noyes[1)], who first stressed the advantages of such a procedure. The formalism is distinct from Noyes' work, however, in its treatment of the interior region. Also, neither half-on-shell t-matrices nor the Faddeev equations are required as a starting point.

The philosophy of this approach can best be illustrated by first examining a two-particle problem. Consider two spinless particles separated by a distance x, and described by a wave function $\psi_{>}(x)$ in the region $x > a$. This situation is depicted in Fig. 1.

Assuming that we are dealing with a particular partial-wave, it is convenient to write $\psi_{>}(x) = \psi^{ext}(x) + \psi^{pot}(x)$, where

$$\psi^{ext}(x) = j_\ell(\kappa x) + i\sin\delta_\ell e^{i\delta_\ell} h_\ell(\kappa x) \ , \tag{1}$$

and $\psi^{pot}(x)$ falls off like $\exp(-\mu x)$ for typical meson-theoretic potentials (we explicitly neglect the Coulomb potential in this discussion). An interior wave function $\psi_{<}(x)$ may exist in the region $x < a$; its sole effect on the exterior solution is via the BC

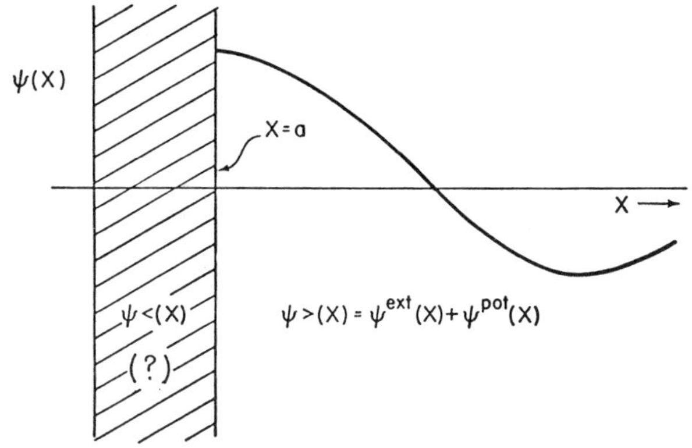

Fig. 1. Interior-exterior geometry for two-particle scattering.

$$(\psi'_{>}/\psi_{>})_{x=a} = (\psi'_{<}/\psi_{<})_{x=a} \equiv \lambda^{\rm int}(\kappa) \ . \tag{2}$$

One can thus account for the effect of the interior by a suitable $\lambda^{\rm int}(\kappa)$ without postulating a wave function description or a short-range interaction. This point appears to be important in applications to highly relativistic systems.

In a purely phenomenological description, one can go further and introduce an appropriate $\lambda^{\rm pot}(\kappa)$ such that

$$(\psi^{\rm ext\,\prime}/\psi^{\rm ext})_{x=a} = \lambda_{\ell}(\kappa) \equiv \lambda^{\rm int}(\kappa) + \lambda^{\rm pot}(\kappa) \ . \tag{3}$$

Given the trivial form of $\psi^{\rm ext}$ in Eq. (1), it is clear that a $\lambda_{\ell}(\kappa)$ can be chosen so as to reproduce the experimental phase-shift, providing an alternative representation of the data. An interesting empirical fact is that for many systems, there exists a particular value of the BC radius for which $\lambda_{\ell}(\kappa)$ appears to approach a constant for large κ[2]. In such systems this representation is particularly useful, and we shall assume it in what follows.

The essential advantage of a BC approach is that one deals with a particularly simple representation of the wave function, i.e., its asymptotic form. In a three-particle system, an analogous simplification occurs in the exterior region, defined by the requirement that each interparticle separation is greater than the (effective) range of forces. Thus, writing $\Psi^{\rm ext} = \Phi + \sum_{\alpha} \psi_{\alpha}^{\rm ext}$, where Φ is the initial plane-wave, and specializing to the case of s-wave forces and L=0 for simplicity, we have[3]

$$\psi_{\alpha}^{\rm ext}(x_{\alpha},y_{\alpha}) = -2\mu_{\alpha} \int_0^{\infty} dq q^2 j_0(q y_{\alpha}) \frac{e^{i\kappa_{\alpha} x_{\alpha}}}{x_{\alpha}} \tau_{\alpha}(q) \ . \tag{4}$$

Here $\vec{x}_\alpha$ is the relative displacement of particles β and γ, while $\vec{y}_\alpha$ locates particle α relative to the $\beta\gamma$ C.M. We also employ the reduced masses $\mu_\alpha^{-1} = m_\beta^{-1} + m_\gamma^{-1}$, and $M_\alpha^{-1} = m_\alpha^{-1} + (m_\beta + m_\gamma)^{-1}$. The on-shell relative momentum of β and γ, $\kappa_\alpha(q)$, is defined by the condition $\kappa_\alpha^2/2\mu_\alpha + q^2/2M_\alpha = E$ (the total C.M. energy).

In the case of finite range forces Ψ^{ext} is precisely the exterior wave function. Otherwise, it differs from the true wave function by a term Ψ^{pot} as in the two-particle problem; again, the effect of such a term can be taken into account by a corresponding BC on Ψ^{ext}. In any event, the observable properties of the scattering system are totally specified by $\tau_\alpha(q)$, the channel t-matrix. We now seek to determine the τ_α by appropriate BC's on Ψ^{ext}. Our initial concern will be to determine the constraints imposed on the system by our knowledge of two-particle scattering. In order to concentrate on the essential ingredients of the argument, let us suppose that we are dealing with the simplest possible case, $\lambda_\ell(\kappa) = \lambda_\ell$, a constant. The key observation is that as $y_\alpha \to \infty$, we must have $\Psi^{ext}(x_\alpha,y_\alpha) \to \psi^{ext}(x_\alpha)\,\Phi(y_\alpha)$; i.e., Ψ^{ext} must describe two-particle scattering as a special case when one particle is at infinite distance. We then deduce that

$$\lim_{y_\alpha \to \infty} \left[\partial_{x_\alpha} \Psi^{ext}(x_\alpha,y_\alpha)/\Psi^{ext}\right]_{x_\alpha = a_\alpha} = \lambda_0^{(\alpha)} , \qquad (5)$$

where $\lambda_0^{(\alpha)}$, a_α are the same quantities introduced to characterize the two-body scattering of β and γ. It is then clear that any BC of the form

$$\partial a_\alpha \Psi^{ext}(a_\alpha,y_\alpha) - \lambda_0^{(\alpha)}\Psi^{ext}(a_\alpha,y_\alpha) = F_\alpha(y_\alpha) ,$$
$$F_\alpha(y_\alpha) \to 0 \text{ as } y_\alpha \to \infty , \qquad (6)$$

will automatically "build in" the on-shell constraints of the system. A comparison with potential theory shows that $F_\alpha(y_\alpha) \propto \exp(-\mu y_\alpha)$, where μ is the lightest mass exchanged; it must furthermore be of a particular form in order to satisfy three-body unitarity. In fact, the most general statement of Eq. (6) consistent with unitarity is

$$\partial_{a_\alpha}\Psi^{ext}(a_\alpha,y_\alpha) - \lambda_0^{(\alpha)}\Psi^{ext}(a_\alpha,y_\alpha) = \sum_\beta \int_0^\infty dq q^2 B_{\alpha\beta}(y_\alpha,q)\tau_\beta(q) , \qquad (7)$$

where the function $B_{\alpha\beta}(y_\alpha,q)$ must be <u>real</u>, but is otherwise arbitrary.

It should be emphasized at this point that, in the context of a particular potential model, $B_{\alpha\beta}$ can be written down quite easily. For purposes of illustration, I have considered a class of off-shell t-matrices $t_\ell(p',p;\kappa) = f_\ell(p',\kappa)f_\ell(p,\kappa)t_\ell(\kappa)$, where $f_\ell(p,k)$ is the Kowalski-Noyes half-off-shell function[4]. The $B_{\alpha\beta}$ function is then trivial; for simplicity I shall represent it in the form

$$B_{\alpha\beta}(y_\alpha,q) = (1 - \delta_{\alpha\beta})\,\frac{B_{\alpha\beta}(y_\alpha)}{q^2 + \mu^2} , \qquad (8)$$

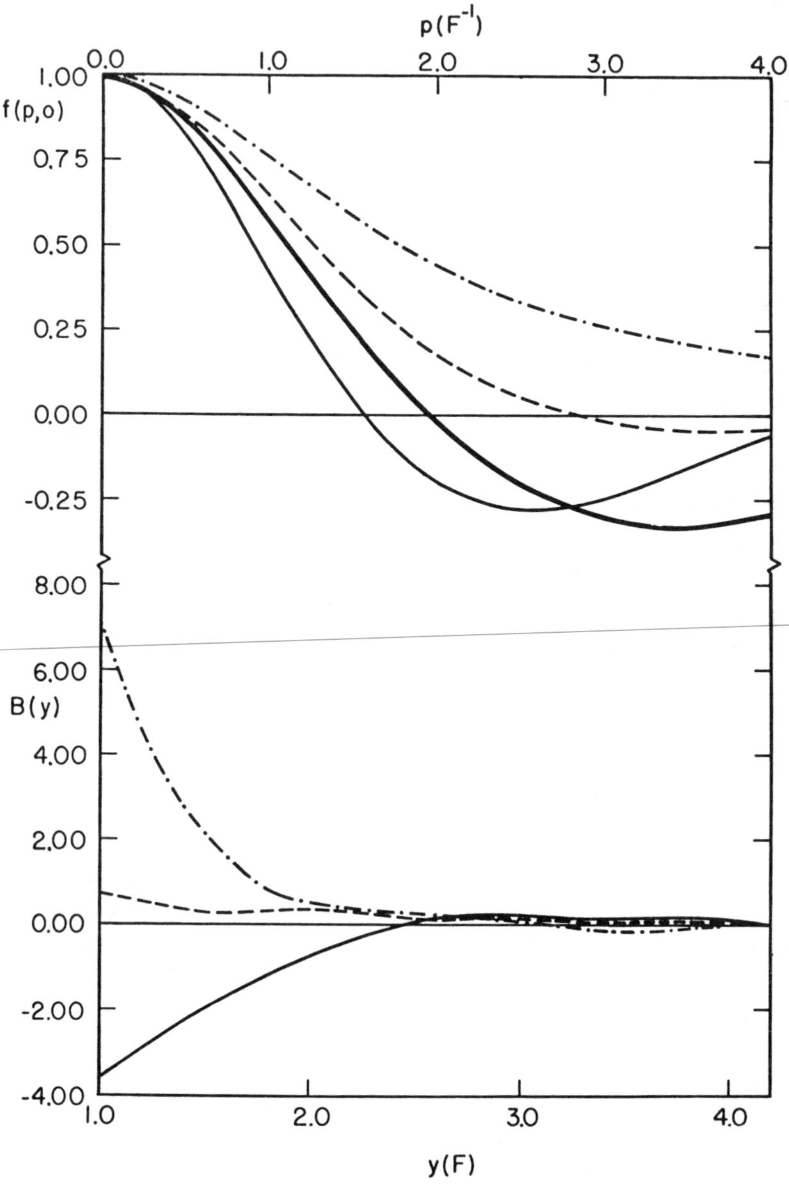

Fig. 2. Behavior of the function B(y) defined in Eq. (8) (lower curves) for various choices of $f_o(p,o)$ (upper curves). The heavy solid line represents the Reid SC.

although this is strictly true only as $q \to 0$. Typical behavior is illustrated in Fig. 2, which assumes identical particles and associates $B(y)$ with $f_0(p,0)$. For comparison, the form of the latter for the Reid SC is shown by the heavy solid line. Positivity of $B(y)$ is associated with attraction, and thus the trend is as one would expect given our experience with separable forms. However, it is clear that the BC type of representation strikingly emphasizes the differences between such models, making it particularly suitable for characterizing off-shell behavior.

We have thus exhibited a class of BC's consistent with the two-particle data, and correlated with the off-shell structure. Unfortunately, this BC is not in itself sufficient to uniquely determine $\tau_\alpha(q)$. The difficulty may be attributed to the fact that Eq. (7) is not applicable for arbitrarily small y_α. The reason is obvious if one considers Fig. 3, which portrays the geometry of the BC in the equal mass case.

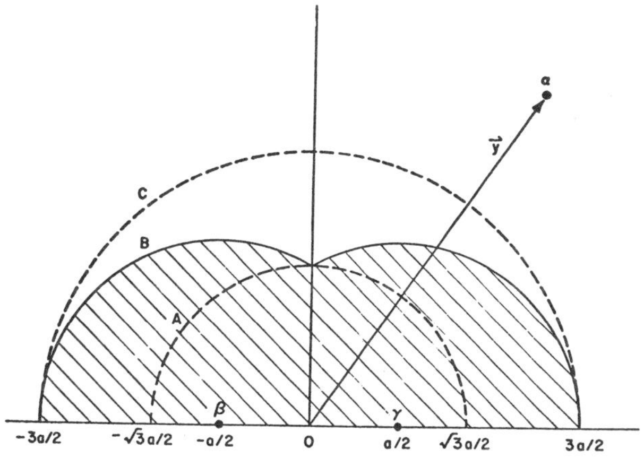

Fig. 3. Interior region (shaded area) and exterior region for three particles with $x_\alpha = a$.

The cross-hatched region in the figure corresponds to particle α being within the interaction radius of β or γ, and hence outside the domain in which Ψ^{ext} is a valid representation. Thus, Eq. (7) is meaningful (in equal mass geometry) only for $y_\alpha > \frac{3}{2}a$, although it may be extended to the region between curves B and C by taking Arg $(\vec{y}_\alpha)$ into account.

It is thus apparent that we have additional degrees of freedom in the problem, and one consequently seeks additional constraints in order to uniquely specify a solution. This is an interesting mathematical problem with classical parallels, but I will not consider it here. The key to a physically acceptable constraint is again unitarity; it can be stated in the form of an additional BC

$$\partial a_\alpha \psi_\alpha^{ext}(a_\alpha, y_\alpha) - \lambda_0^{(\alpha)} \psi_\alpha^{ext}(a_\alpha, y_\alpha) = \sum_\beta \int_0^\infty dq q^2 C_{\alpha\beta}(y_\alpha, q)\tau_\beta(q) \ , \tag{9}$$

which is valid for $y_\alpha < y_\alpha^{min}$ (defined as the smallest value of y_α contained in the exterior region). Note the use of ψ_α^{ext}, the channel wave function, in Eq. (9) rather than Ψ^{ext}. As in the above case, a unique function $C_{\alpha\beta}$ is associated with a given potential model, although the relationship is more complicated. In order to illustrate its behavior we again consider the representation of Eq. (8) and the t-matrix models employed above. As shown in Fig. (4), C(y) is rather structureless, although its magnitude is distinctive of the corresponding model.

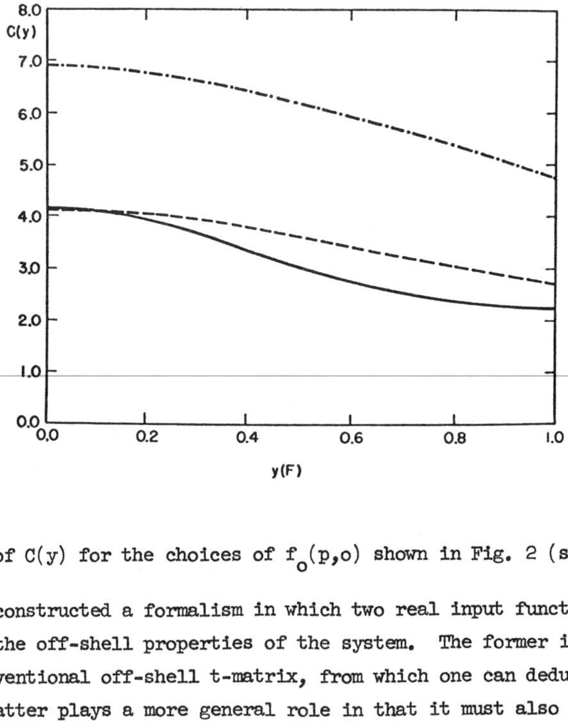

Fig. 4. Behavior of C(y) for the choices of $f_o(p,o)$ shown in Fig. 2 (same notation).

We have thus constructed a formalism in which two real input functions, $B_{\alpha\beta}$ and $C_{\alpha\beta}$, characterize the off-shell properties of the system. The former is directly related to the conventional off-shell t-matrix, from which one can deduce its general properties. The latter plays a more general role in that it must also specify the effect of three-body forces, if such are present. Together they can be represented as a single function $A_{\alpha\beta}(y_\alpha, q)$; this function has a specified discontinuity at $y_\alpha = y_\alpha^{min}$ if one assumes conventional interactions. By allowing $A_{\alpha\beta}$ to take on all possible forms, one can thus generate all possible three-body scattering amplitudes which are exactly unitary and correspond to a given set of two-particle data. In practice, application of the BC's to the explicit representation for Ψ^{ext} leads to a one-variable integral equation

$$\tau = Z + K\tau \ , \qquad K = K_1 + AK_2 \ , \tag{10}$$

where Z, K_1, K_2 are completely determined by on-shell information. Solution of this equation is equivalent to a typical Faddeev calculation with a rank one potential, even though separability has not been assumed and the phase-shift may be arbitrary. It is thus highly practical from a numerical standpoint. Moreover, by expanding A in some complete set, it is only necessary to solve an integral equation once for a given total energy; the off-shell variations then reduce to linear algebra. This makes it possible to compare a wide variety of possibilities with a minimal investment in computer time. Thus the formalism has been designed not only to clearly distinguish between the on-shell and off-shell characteristics, but to explore their consequences with maximal efficiency.

The utility of the BC formalism was demonstrated during the past year through an analysis of the n-d breakup reaction at 14.4 MeV.[5)] The method was applied for a fixed choice of on-shell input in order to test the sensitivities of various regions of phase space to variations in A. In doing so, it was clear that for A sufficiently large effects of arbitrary magnitude could be produced, just as would be the case for an arbitrarily large three-body force. The input was therefore constrained by requiring that each choice of A produce the identical value for a_2, the most sensitive scattering parameter. This was accomplished by picking an arbitrary "shape" f(y,q), setting $A = \lambda f$, and varying λ from zero to a value which produced $a_2 = .41F$ (chosen to best reproduce the low energy doublet scattering). In employing this input at 14.4 MeV, one is of course ignoring the implicit energy-dependence of A. However, it can be shown that the magnitude of this dependence should be considerably less than $M_n E/\mu^2$, with $\mu \geqslant 280$ MeV/c. This is confirmed by the models considered earlier; a representative example is shown in Fig. 5 for energies ranging to $E_L = 50$ MeV.

Fig. 5. Energy-dependence of $A(y) \equiv B + C$ for the model represented by the solid line in Fig. 2.

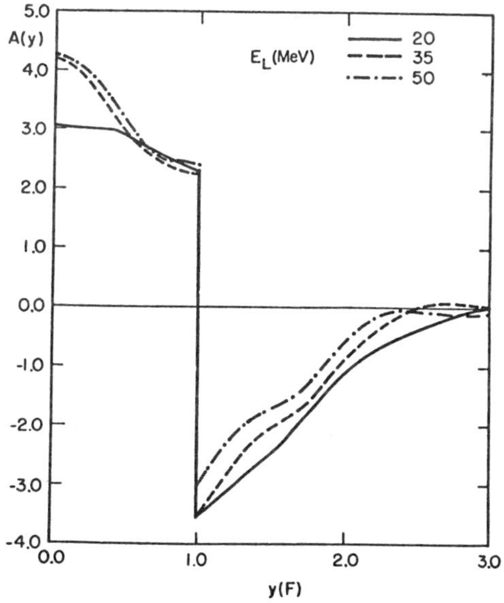

The variation is quite small except at $y=0$, where it is suppressed by a $y^2 dy$ integration factor. Compared to the differences exhibited by different choices of A, this effect is completely negligible.

Given this input, the formalism was applied to generate a class of predictions for n-d elastic scattering and breakup corresponding to the same low energy trinucleon parameters and two-nucleon observables. The nature of the results obtained is illustrated in Fig. 6, which corresponds to the most sensitive region reported by Kloet and Tjon[6] in their comparison of potential models. The mixed on-shell and off-shell differences implicit in such models result in the variations depicted in the lower half of the figure. However, the strong model-dependence at the minimum does not survive under the constraints we have imposed, as evidenced by the upper half of the figure (note that the two upper curves correspond to the <u>maximum</u> dispersion displayed over the entire A set). In order to obtain some feeling for just what is involved, two completely distinct choices of A for which the corresponding cross sections are indistinguishable are shown in Fig. 7. If one takes the examples quoted earlier as a guide, it is clear that the differences between these curves are considerably greater than one would expect from conventional interactions.

The results of this analysis, therefore, cast considerable doubt on the proposition that off-shell information can be extracted from this reaction. It appears that once a_2 is determined, the low energy scattering properties are effectively established. Since a_2 and E_T are strongly correlated, it would not seem possible to distinguish between an attractive three-body force and some exotic off-shell mechanism, for example. At higher energies, momenta capable of probing the appropriate distances ($y_\alpha \simeq 1F$) will be available, but the possibility of obtaining definitive information is then diminished by the need to include many more partial-waves in the interaction. Nevertheless, a recent calculation by Haftel[7] suggests that the sensitivity to the s-wave interaction does increase significantly between threshold and 50 MeV, and the possibility should certainly be explored. In view of the advantages I have stressed, it would seem that the BC technique offers the most practical hope of answering this question.

On the other hand, a lack of off-shell sensitivity does have its positive aspects. Thus, if quasi-three-body reactions are sensitive only to a small number of independent parameters which can be fixed by reference to the ground state or threshold properties, the prospects of obtaining detailed information regarding nuclear structure are greatly enhanced. For example, the nn FSI has been shown to be very sensitive to the value of the nn scattering length a_{nn}, as illustrated by the lower half of Fig. 8. However, there is also a background sensitivity on the order of 15%. As shown in the upper half of the figure, application of the BC technique can cut such off-shell ambiguities to the level of a few percent.

In conclusion, I'd like to mention several ways in which the present formalism can be generalized, and in which the BC approach appears to possess unique advantages. The first example concerns the Coulomb interaction, which is particularly easy to include in

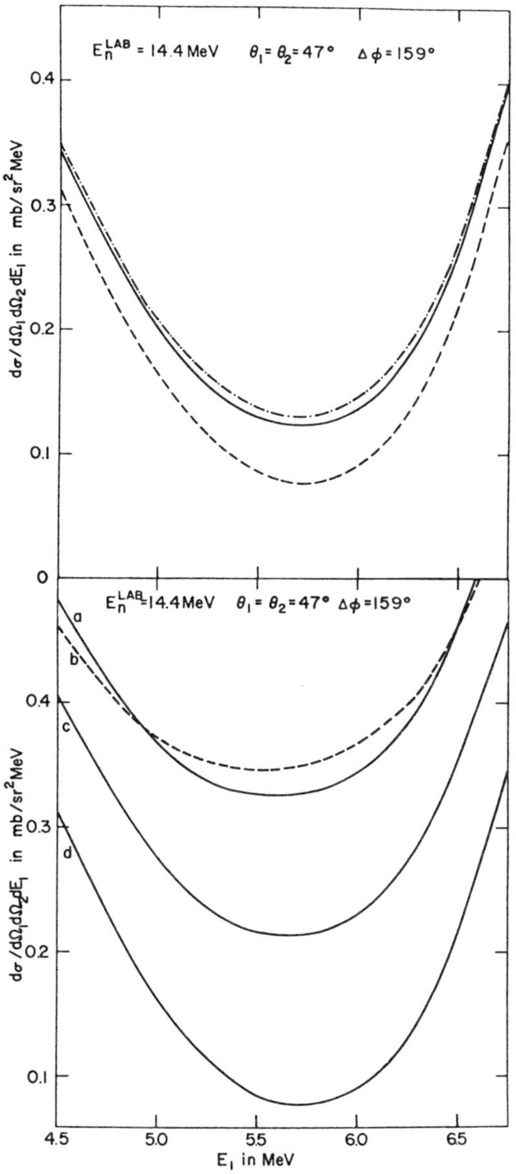

Fig. 6. Differential cross section for n-d breakup vs. the energy of an outcoming neutron. The lower set of curves arise from models which produce identical behavior in the low energy two-nucleon system. The solid and dash-dot curves exhibit the maximum difference over the entire Λ set.

Fig. 7. Example of contrasting forms of A(y) which produce indistinguishable trinucleon observables.

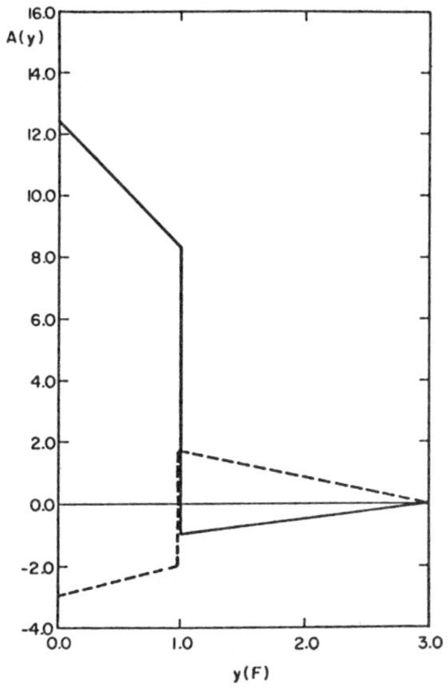

this formulation. This would permit one to take greater advantage of the more precise experiments. Secondly, the same technique can be applied to systems of more than three particles. In the case of four particles, for example, the resulting equations involve only two continous variables, bringing the problem within the scope of currently available techniques. Finally, a particularly exciting possibility lies in the application of this approach to elementary particle systems. The formalism permits a natural and unambiguous relativistic generalization, which I have recently applied to calculate the ω as a resonance in 3π scattering. I believe this to be the first successful three-body calculation in a highly relativistic system. The same formalism has been applied to π-d scattering in the 3,3 resonance region with highly encouraging results. I would therefore conclude that the approach not only provides a particularly efficient alternative to present techniques, but also opens the door to a wide variety of new and interesting possibilities.

REFERENCES

1) H. P. Noyes, Phys. Rev. D5 (1972) 1547, and earlier references quoted therein.
2) This was first noted by G. Breit and W. G. Bouricius, Phys. Rev. 75 (1949) 1029, and employed extensively by H. Feshbach and E. Lomon, Phys. Rev. 102 (1956) 891.
3) D. D. Brayshaw, Phys. Rev. D8 (1973) 952.
4) K. L. Kowalski, Phys. Rev. Lett. 15 (1965) 798, and H. P. Noyes, ibid (1965) 538.
5) D. D. Brayshaw, Phys. Rev. Lett. 32 (1974) 382.
6) W. M. Kloet and J. A. Tjon, Nucl. Phys. A210 (1973) 380.
7) M. Haftel, paper contributed to this conference.

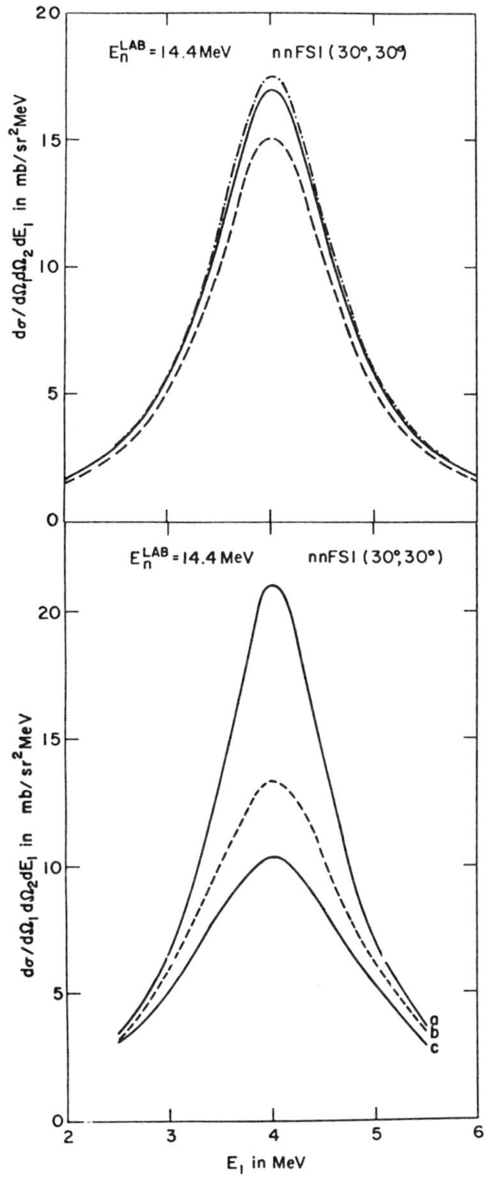

Fig. 8. The nnFSI at $\theta_1 = \theta_2 = 30^\circ$, coplanar geometry. The lower curves arise from potential models with different values of a_{nn}. All curves in the upper half have $a_{nn} = -23.7F$; the interpretation is as in Fig. 6.

DISCUSSION

TJON(UTRECHT): This is a question about Brayshaw's calculation in the deuteron break-up. I mentioned the experiment of the IKO-group from Amsterdam. If you take that, and you take your calculations, then the discrepancy would be even worse. How would you explain that ?

BRAYSHAW: I do not understand the question. Which discrepancy ?

TJON: Well I have shown a result of the IKO-experiment, where people have been looking at the break-up process of the p+d → p+p+n in precisely the most sensitive region which has been found by Kloet. For that region, the discrepancy with the simple local potential is already 50%, but if we would take your results, it would be even worse.

BRAYSHAW: Why do you say that ? I do not understand at all.

TJON: I assume the results from your paper as being right. If you take those...

BRAYSHAW: Are you talking about a theoretical comparison, or comparison with the experiment ?

TJON: Comparison with experiment.

BRAYSHAW: Well, this is a simple model calculation. I have chosen a set of phase shifts which are simple but have the right low energy property. There is no D-state, there are no other partial waves in the calculation. So I do not see any point in comparing with experiment at this point.

TJON: But it would mean that there is more information in that region. You are pointing out that the off-shell behaviour is insensitive, but there is certainly information lying there.

BRAYSHAW: No, I disagree entirely, I have entirely different on-shell information. If I fix my on-shell information to the same phase shift that you used, I can generate a curve centred about your results and I would not get any difference.

TJON: But there is still a discrepancy.

BRAYSHAW: There may be a discrepancy with experiment eventually, when people put in the appropriate interactions with higher partial waves.

TJON: Which means that there is more information there.

BRAYSHAW: Well, you are postulating there is a discrepancy. If and when a discrepancy is shown to exist, when a realistic calculation is done, then it would be clear that we need something different. But I am indicating that it is not going to come from off-shell effects of the normal type considered.

TJON: No, but there may be higher partial waves, for example.

BRAYSHAW: Yes, but I would not characterize that as off-shell information. I would simply say that you need a larger model to do the job properly. I am not saying there is no information; there is a lot of sensitivity to on-shell effects, including the need for other partial waves. But so-called off-shell effects are a different matter.

TJON: True.

GIBSON(LASL) In the 3-nucleon system, you use the doublet scattering length, though you could have used the three-body binding energy to determine your off-shell dependence. Right ?

BRAYSHAW: That's right.

GIBSON: In the 3-π calculation, where you are looking for the bound states, what do you use to fix this off-shell dependence ?

BRAYSHAW: You mean the ω or the actual bound-state ?

GIBSON: The ω.

BRAYSHAW: Well, actually, I have done two different things. I gave you an illustration that if you start with a potential model, you can generate the input which goes into the formalism. So in the 3-π calculation what I did was to look at the form of the A function for exchanges between the π's and tried to put that into the A that I used, rather that normalizing to another piece of data. What I had to do then was to change that parameter around and show that in fact the result I got was not dependent in a particular choice of it. The best way to put that kind of thing in, in the momentum-space language, is to introduce singularities far from the real axis, at distances which correspond to the masses of the exchanged particles, and thereby to build up from this minimal analytic function something which has the structure that you get from the typical exchange form of Faddeev or Blankenbecler-Sugar, whichever you want to consider. Making a calculation by varying this part of the input produces the first row of entries in this graph and the effect of them is to move the position of the 3-π resonance one sees around a bit, and to change its width. But I was not able to either move the resonance out of the region near the ω, nor was I able to get a width less than about 40 MeV, by considering the 3-π system in isolation. But in this kind of calculation, one has to look at the off-shell effects and see in a given system whether they are important or not. They do not seem to be important in n-d; in 3-π where there is as much kinetic energy as there is mass energy, one might expect to see more effect. There are larger effects, but basically the ω seems to come out as a consequence of the P-wave phase shift, that is of the ρ, and not of the particular choice of the correction effect. I have done something recently which is to improve on this calculation by considering the coupling to the K-$\bar{K}$ virtual channel. This is a nice advantage of the boundary condition approach. You do not have to say what happens to your particles when they all get into the interaction region, so you have a final state which sometimes consists of 3-π's, sometimes consists of K-$\bar{K}$. If you do the coupling there, you find that it gives you an additional attractive term in this formalism which brings the ω width down to something more like the experimental value and simultaneously gives you a prediction for the branching ratio $(\phi \to 3\pi)/(\phi \to K\bar{K})$. What I have done here is to change the parameters of the ρ in order to show how things move around. So A is a calculation with a so-called ρ of width 60MeV, and etc...

MITRA(Delhi): I would like to ask you the following: I am really surprised that you are getting such nice values of the ω-resonance energy with the input parameters of the ρ. About more than a dozen years ago, there were similar calculations by Schiff and other people, which indicated that if you take an input potential π-π that fits the ρ parameters, you get a very low energy for the 3-π system. It looks like a bound system as it was found by Schiff. It is surprising that you are getting rather a resonance value for the ω, almost one which is experimentally right. So I wonder what additional input information you have in you theory so as to give you such nice values.

BRAYSHAW: I am not familiar with the calculation of Schiff. From what you describe, I rather doubt it was a full three-body calculation.

MITRA: It was a three-body calculation.

BRAYSHAW: With separable potential, or what ?

MITRA: Well, Schiff had used the local potential and I happened to use the separable potential. Both of these have the same features namely you get a very tight binding for the 3-π system for a certain class of π-π potentials (see comment marked with *)

BRAYSHAW: You get binding in the P-waves.

MITRA: Yes, if you fit the potential of the π-π through the parameters ($\lambda>0$, β large) of the ρ, there is no escape from this that you get a very tighly bound 3-π system. Is there any additional feature in you assumption ?

BRAYSHAW: No, there is no additional feature, I have strong questions on the calculations you are describing. I am familiar with more recent 3-π calculations which also use the ρ parameters correctly and (they) got completely different results from those that you describe.

MITRA: Well, then it might be some special feature of the interaction you are using.

BRAYSHAW: No, I do not think so,

MITRA: Does your model give a positive or a negative value for Γ_ρ ?

BRAYSHAW: Well, I am very suspicious of using potentials to describe the π-π system. It is a nice virtue of this formalism that it does not try to mock up, by some interaction scheme, or potential scheme, interactions between elementary particles. All that goes into this is the phase shift, and as a check, the nearest singularity representing off-shell effects. The basic effect just comes from the two particle phase shift. I should point out that this is a completely different relativistic formalism than any three-body calculation which has been done before. The equations do not look anything like the equation of Omnes, or the equation of Lovelace, which were supposed to be the two choices for relativist three-body equations.

MITRA: The second question concerns the boundary condition model where you use the parameters $B_{\alpha\beta}$ and $C_{\alpha\beta}$. Have you used this also for the four-body calculation, say the binding energy of the four-body system ?

BRAYSHAW: Right. Well I should have mentioned, thank you for remainding me, that you can go on to treat the four-body system this way. You get a simplified set of equations in only two continuous variables. There is a nice approximation which suggests itself and I think it might be the more practical way to look at four-body scattering. The input which you put into the three-body problem can then be used in the four-body problem. You could fit the parameters from a 3-particle scattering experiment presuming that you could find one which is sensitive to them; then you could use the same input in the four-body problem.

MITRA: How is it different from the potential approach ? There also one uses the same kind of input information through the N-N potential.

BRAYSHAW: It is different from a potential approach in several aspects. The first aspect is that you get a one-variable equation in a three-body problem, and a two-variable equation in the four-body problem. So you always have one variable less in solving it. The second aspect is that the kernel is linear in the off-shell part which greatly facilitates studying it in doing calculations, because you only have to do a complicated calculation once, then you can do the off-shell analysis afterwards.

MITRA: I agree; for the three-body problem, it is true. But will it continue to be valid for 4, 5, 6-body problems, or will you have to re-enter a new input ?

BRAYSHAW: I believe that there is a natural assumption which one can make, which is equivalent to ignoring three-body forces in the three-body problem. This assumption in the four-body problem means that you can express the input for that by the same function that you used in the three-body problem. But that is a very technical question.

* Comment added by MITRA after the discussion

Several years ago, similar calculations were made with π-π interactions tuned to the ρ-parameters, using both variationals (L.Schiff-PR 126,1962) and Faddeev-type approaches with separable interactions (A.Mitra-PR 127, 1962; N.C.L. 1964; P.L. 1964 DeSwart-Kreps -PR 1965), all with relativistic kinematics. The essential results of the second approach which made use of Blankenbecler-Sugar type relativistic kernels may be summarized as follows. If one represents the π-π(p-wave) interaction as $-\lambda\vec{p}\cdot\vec{p}'\ g(p)g(p')\ ,\ g(p)=(\beta^2+p^2)^{-1}$, one can distinguish two cases: $\lambda>0$ ('attractive'), and $\lambda<0$ ('repulsive'). For the 'attractive' case ($\lambda>0$) one requires a relatively large β ($\sim$ few pion masses), while for the 'repulsive' case ($\lambda<0$) the value of β is small (a fraction of a pion mass), in order to account simultaneously for the (rather large) mass and width of ρ . (The words 'attractive' and 'repulsive' are put in quotes to emphasize the fact that a separable interaction is not wholly attractive or wholly repulsive in the sense of a local interaction). For the 'attractive' case ($\lambda>0$, β large) there is no escape from a tightly bound ω as a 3π state --not a resonance. Schiff's variational calculation (with relativistic kinematics) also confirms this conclusion. However, for the 'repulsive' case ($\lambda<0$, β small), it is possible to obtain a fairly reasonable position of the ω-mass (as a 3π resonance) (Mitra-NCL 1964, PL 1964). Both cases are consistent with a proper B-W structure for the ρ-resonance, as ensured through the energy variation of the π-π phase shift near 90°.

BRAYSHAW: This ambiguity is avoided by my formulation of the problem, and illustrates my prejudice against the use of separable forms in elementary particle calculations.

INTERACTION A TROIS CORPS A L'ETAT LIE. UNE NOUVELLE APPROCHE.

B.T. Darling et R.M. Shoucri
Département de Physique, Université Laval, Québec G1K 7P4, Canada

La dérivation de l'opérateur de l'énergie cinétique pour trois corps de masse égale de façon à montrer séparément le mouvement de translation, de rotation et de vibration a été faite par Linden[1] et Darling. Le résultat a été appliqué à l'étude de l'état fondamental des systèmes nucléaires ^3H, ^3He, en tenant compte des forces d'échange. (Voir aussi Derrick, Banville et Kunz[2]). Darling et Weiss[3]), Darling et Nadeau[4]), Darling et Lui[5]) ont appliqué cette méthode à l'étude de la molécule d'ozone O_3. Le cas de trois corps de masse quelconque a été étudié par De Celles et Darling[6]). L'opérateur de l'énergie cinétique prend la forme

$$\hat{T} = 1/2\left[p_R^2 + \frac{4}{R^2}\,p_\psi^2 + \frac{1}{R^2\cos^2\psi}\left(2p_\lambda + \sin\psi\, Pz\right)^2 - \frac{1}{4R^2} - \frac{1}{R^2\sin^2\psi\cos^2\psi}\right] + 1/2\left[\frac{Px^2}{R^2(1-\cos\psi)/2} + \frac{Py^2}{R\,(1+\cos\psi)/2} + \frac{Pz^2}{R^2}\right]$$

$$p_R = -\frac{\partial}{\partial R} \qquad p_\psi = -\frac{\partial}{\partial\psi} \qquad p_\lambda = -\frac{\partial}{\partial\lambda}$$

R est la racine carrée du moment d'inertie par rapport à l'axe z perpendiculaire au plan des trois corps. R, ψ, λ sont des variables qui décrivent le mouvement interne des trois corps. Px, Py, Pz sont des matrices de dimension 2L + 1, qui sont les représentations du groupe de rotation de l'espace euclidien. Les axes x, y sont pris dans le plan des trois corps le long des axes d'inertie principaux du système.

Pour les états de symétrie sphérique, Px = Py = Pz = 0. Dans ce cas l'opérateur de l'énergie cinétique est séparable. Les solutions caractéristiques des équations différentielles en ψ,λ conduisent à un système orthonormé de fonctions u_K^v,

$$\sqrt{\frac{K+2}{\pi^3}}\;\cos\lambda v\;\;\rho^v P^{v,0}_{\frac{1}{2}(K/2-v)}\,(1-2\rho^2)\;\begin{cases} 1 & v \neq 0 \\ \frac{1}{\sqrt{2}} & v = 0 \end{cases}$$

ou

$$\sqrt{\frac{K+2}{\pi^3}}\ \sin\lambda\ v\ \rho^{v}\ P^{v,0}_{\frac{1}{2}(K/2-v)}\ (1-2\rho^2)$$

$\rho = \cos\lambda$, $0<\rho<1$, $0\leq\lambda\leq 2\pi$, élément d'intégration $\pi^2\rho d\rho d\lambda$

En présence d'intéraction (non-tensorielle) la fonction d'onde Ψ de l'état de symétrie sphérique peut s'écrire sous la forme

$$\Psi(R,\ \psi,\ \lambda,) = \sum_{K,v} \chi^{v}_{K}\ (R)\ u^{v}_{K}\ (\psi,\ \lambda).$$

Substitution de cette fonction d'onde dans l'hamiltonien conduit à un système d'équations différentielles en une variable[7]

$$\left\{\frac{d^2}{dR^2} + 2E + - \frac{(K+3/2)(K+5/2)}{R^2}\right\} \chi^{v}_{K}\ (R) = - \sum_{K',v'} U^{vv'}_{KK'}\ \chi^{v'}_{K'}\ (R)$$

Dans le cas d'une intéraction coulombienne, ceci prend la forme

$$\left\{\frac{d^2}{dR^2} + 2E + \frac{C^{vv}_{KK}}{R} - \frac{(K+3/2)(K+5/2)}{R^2}\right\} \chi^{v}_{K}\ (R) = - \frac{1}{R} \sum_{K',v'} C^{vv'}_{KK'}\ \chi^{v'}_{K'}\ (R).$$

$C^{vv'}_{KK'}$ est indépendant de R. La partie homogène rappelle l'équation radiale de l'atome d'hydrogène. Application à l'étude des systèmes atomiques He et H^- est en cours. Une étude analogue pour le cas de trois corps de masse égale a été faire par Simonov[8].

BIBLIOGRAPHY

1) B.R. Linden, M. Sc. Thesis, The Ohio State University, 1949
Ph. D. Thesis, The Ohio State University, 1951

2) G.H. Derrick, Thesis, University of Sidney, 1959;
M. Banville and P.D. Kunz, Can. J. Phys. 44, 2095(1966)

3) B.T. Darling and J.A. Weiss, Scientific Report 1A-6, Contract No. AF19(122)65,
The Ohio State University Research Foundation (1953).

4) B.T. Darling and G. Nadeau, J. Mol. Spectry, 11, (1963)200

5) B.T. Darling and C.W. Lui, J. Mol. Spectry, 21, (1966)146
C.W. Lui and B.T. Darling, J. Mol. Spectry, 21, (1966)130

6) M. De Celles and B.T. Darling, J. Mol. Spectry, 29, (1969)66

7) B.T. Darling and R.M. Shoucri, to be published

8) Yu. A. Simonov, Soviet J. Nuc. Phys. 3, (1966)755

Ce travail a été appuyé par un octroi du Conseil National de Recherche du Canada.

DISCUSSION

SASAKAWA (Tohoku): All of these questions are written in Simonov's paper.

SHOUCRI: Simonov, in the Soviet Journal of Nuclear Physics, has not taken the same approach to the problem: he has not derived the kinetic energy hamiltonian in this way. He has used another approach for three-particles of equal masses.

SASAKAWA: Yes, but I would point out the difficulties of applying this approach to nuclear physics:

(1) To solve the differential equation, we have to impose the boundary conditions both of n-d channel and the break-up channel at the same time, even if one or both of these channels are closed. Since the asymptotic behaviour of these channels is quite different, it is very difficult to impose the correct boundary condition.

(2) The number of coupled equations is tremendous, especially for the Yukawa potential.

SHOUCRI: It could be. I know that Simonov has made some application in the case of the triton and he has used just three terms in the coupled equations to get a good result.

A NEW METHOD IN THE THEORY OF MANY-BODY SCATTERING

D. J. Kouri*, Max-Planck-Institut für Strömungsforschung, 3400 Göttingen, Federal Republic of Germany, and F. S. Levin†, Physics Department, Brown University, Providence, R. I. 02912

It is reasonable to expect that any method for describing many-body scattering processes should satisfy at least four conditions: (1), the dynamics must be properly taken into account, either via interactions or t matrices; (2), the kernel of the equation used in the method must have (finite) iterates that are connected; (3), the theory must have the requisite analytic structure, including proper treatment of threshold behavior and the correct unitarity (discontinuity) relations; and finally, (4), the effects of symmetries must be included. In addition, satisfaction of two other conditions is desirable, though not strictly necessary; (5), the equations should be free from spurious solutions [in Amado's terminology[1)], the equations should not suffer from Federbush disease[2,3)]; and (6), it should be possible to introduce unitary approximations in a relatively simple way.

The first four conditions may be summarized as follows. If the matrix equations for the operators of interest are of the form [z = a complex energy parameter]

$$U(z) = B(z) + C(z)\,U(z) \quad , \tag{1}$$

where B is the driving term and C is the kernel of the equation, then: B must properly represent the dynamics; C^M must be connected, for some $M \geq 1$;

$$\mathrm{Im}\,U(E + i0) = -\pi\,U(E + i0)\,\tilde{\Delta}(E)U(E - i0) \quad , \tag{2}$$

where E is the energy of interest and $\tilde{\Delta}$ is an on-shell projector; and effects of symmetries such as the Pauli principle must be either included in (1) or readily taken into account. Lack of Federbush disease means that (1 - C) is zero only at solutions of the relevant Schrödinger equation, while one way of satisfying condition (6) is that an auxiliary (real) operator Y(E) can be found such that (a) U and Y are related via a damping equation, e.g.,

$$U(E + i0) = Y(E) - i\pi\,Y(E)\tilde{\Delta}\,(E)\,U(E + i0) \quad , \tag{3}$$

and (b) Y(E) itself should obey a simple equation, e.g.,

$$Y(E) = B(E) + R(E)\,Y(E) \quad , \tag{4}$$

where R is a real operator, usually the real part of C(z). In the nonrelativistic case considered here, we may assume that U(z) = U*(z*), and that U is either equal to or is linear in the transition amplitude of interest.

The first many-body theory to satisfy conditions (1) - (4) was that of Faddeev[4] for the case where n = 3 [here n = number of particles, N = number of arrangement channels in the sense of Eckstein[5] or Newton[6]]. Other formulations, such as that of Alt, Grassberger and Sandhas[7] are well known; we do not need to consider any of them in detail. Extensions to the case of arbitrary n were first discussed by Yakubovski[8] and more recently for example by Bencze[9], by Karlsson and Zeiger[10] and by Redish[11]. In all of these works, the terms corresponding to B in Eq. (1) are t operators in n-body Hilbert space for 2, 3, ... and up to (n - 1)-body subsystems. Furthermore, the kernel C is normally a product of B and another operator: C(z) = B(z) D(z). One may identify three shortcomings to these methods. First, for many systems of interest, the relevant t operators may not be available. Second, and closely related to the first, even if on-shell values of some of the t-operators are known, their off-shell values generally are not known, and one must proceed using models. Thirdly, and of relevance to this work, is the fact that the t operators are complex, so that even if R(E) = Re C(E + i0), Y(E) of (4) will not be real because B(E) is not real. This means that unitary approximations cannot be introduced in a simple way [in particular, using (3)]. Cahill[12] and Kowalski[13], among others, have shown how to deal with this problem via hierarchies of coupled, integral equations for the case n = 3, but such solutions, while they do exist, are not simple. We know of no comparable studies for arbitrary n using as a basis the n-body formalisms noted above.

In contrast to these approaches using t operators, we have recently developed a new method for treating the n-body problem[14,15] which not only allows unitary approximations to be introduced in a simple fashion (at least from our point of view) but also satisfies the five other conditions listed above. In this treatment, the terms in our equations corresponding to the B(E) of equation (1) are the channel interactions rather than t operators, so that the auxiliary operators Y(E) of eq. (4) are real. An obvious shortcoming is that the formalism cannot be applied directly to hard-core potentials. However, most potentials in non-relativistic problems seem at worst to be soft-core, and in addition, it is generally true as implied above that it is the potential and not the t operators for which there is the most information. In the case of models, a potential is usually postulated first and its t-matrix elements determined second. Finally, by working directly with potentials, problems involving coulomb interactions can be handled relatively simply. An example is the electron-hydrogen scattering system, which despite its having been the object of study for more than 4 decades, is a problem that has not yet been solved. This is the first system to which our method has been applied: so far the results are extremely encouraging[16-18]; results of these calculations will be discussed below.

The key to our method is an N x N matrix W, denoted the channel coupling array, which links the arrangement channels together. Elements of W are fixed by requiring that the coupled equations in which they occur have a connected, iterated kernel (we assume that the potentials in H are sufficiently well-behaved that the iterated kernels of our equations are compact). In our earlier work[14,15], coupled equations were developed by beginning with the post-prior definition[6] of transition operators for rearrangement collisions, but more recently we have derived our results directly from the Schrödinger equation[19]. In addition we have shown that our equations do not suffer from Federbush disease[19] and have also established the relation between our results and those derived by Hahn[20] using the reduction technique. For simplicity, we use here the post-prior T operator approach to discuss our work.

To fix notation, consider a model three-body problem consisting of three distinguishable particles (A,B,C) interacting via pair potentials. For simplicity, the mass of particle B is taken as infinitely heavy. The Hamiltonian H is then the sum of the two kinetic energy operators and the three interactions:

$$H = \underbrace{K_A + K_C}_{H_0} + \underbrace{V_{AB} + V_{AC} + V_{BC}}_{V} . \tag{5}$$

We are interested in scattering processes initiated say, by A incident on a bound state of B and C: A + BC. The pair interactions are assumed to go to zero sufficiently rapidly as the pair separation becomes large. Thus there are four arrangement channels:

(1)	A + BC → A + BC	elastic
(2)	A + BC → AB + C	charge-exchange
(3)	A + BC → B + AC	pickup
(0)	A + BC → A + B + C	breakup

Corresponding to these four channels are four partitions of the Hamiltonian into channel Hamiltonians H_j and channel perturbations V_j $(H = H_j + V_j)$:

$$H_1 = H_0 + V_{BC} \quad , \quad V_1 = V_{AB} + V_{AC}$$

$$H_2 = H_0 + V_{AB} \quad , \quad V_2 = V_{AC} + V_{BC}$$

and similarly for H_3 and V_3, H_0 and $V_0 = V$.

In the general n-body, N-channel problem, we also have

$$H = H_j + V_j \quad , \quad 1 \leq j \leq N \quad , \tag{6}$$

and no restriction to pairwise interactions is assumed. The eigenstates of H_j, corresponding to observable, asymptotic states, will be denoted $|\Phi_E(j)\rangle$, where E is the total energy. These correspond to plane wave relative motion of the subsystems or fragments forming the channel j and also describe their internal bound states.

Amplitudes for transitions between states $|\Phi_E(k)\rangle$ and $\langle\Phi_E(j)|$ are determined[6] by taking matrix elements of the (post - form) rearrangement transition operator T'_{jk}:

$$T'_{jk}(E + i0) = V_j + V_j\, G(E + i0)V_k \quad , \tag{7}$$

where $G(z) = (z - H)^{-1}$ is the full resolvent operator. We also define $G_i(z)$ by $G_i(z) = (z - H_i)^{-1}$; G and G_i are related by

$$G(z) = G_i(z) + G_i(z)\, V_i\, G(z) \tag{8}$$

$$= G_i(z) + G(z)\, V_i\, G_i(z) \tag{9}$$

We shall develop a set of coupled, integral equations for the T'_{jk} [this can also be done for the (prior-form of the) operators[6] $\hat{T}'_{jk} = V_k + V_j G V_k$, but is not necessary for this development]. If we set i = k in (9) and substitute it in (7), we get (z = E + i0)

$$T'_{jk}(E + i0) = V_j + T'_{jk}(E + i0)G_k(E + i0)V_k \quad . \tag{10}$$

This is a standard[6] integral equation for T'_{jk} which suffers from the familiar drawback of not being connected; in addition, it does not have an analytic behavior that leads to the correct unitarity relation. In order to write out a more appropriate equation, we introduce the N x N channel coupling array, whose elements are initially restricted only by

$$\sum_{i=1}^{N} W_{i\ell} = 1 = \sum_{i=1}^{N} W_{\ell i} \quad . \tag{11}$$

Multiplying both sides of (8) by $W_{\ell i}$ and summing on i gives

$$G(z) = \sum_{i=1}^{N} W_{\ell i}[G_i(z) + G_i(z)\, V_i\, G(z)] \quad , \tag{12}$$

where the index ℓ may be chosen at our convenience.

Equation (12) partitions G(z) over all the arrangement channels. The elements $W_{\ell i}$ will determine which channels are explicitly coupled together.

Substitution of (12) into (10) gives

$$T'_{jk}(E + i0) = V_j + V_j \sum_i W_{\ell i}\, G_i(E + i0)\,(V_k - V_i)$$
$$+ V_j \sum_i W_{\ell i}\, G_i(E + i0)\, T'_{ik}(E + i0) \quad . \tag{13}$$

The key point in the derivation is now to notice that $T'_{jk}(E + i0)$ will always act on states $|\Phi_E(k)\rangle$, so that Lippmann's[21] equality can be used:

$$G_i(E + i0)(V_k - V_i) = \delta_{ik} - 1 \equiv -\bar{\delta}_{ik} \quad , \tag{14}$$

when acting on initial two-body states $|\Phi_E(k)\rangle$. Use of (14) in (13) leads to

$$T'_{jk}(E + i0) = V_j W_{\ell k} + V_j \sum_i W_{\ell i}\, G_i(E + i0)\, T'_{ik}\,(E + i0) \quad . \tag{15}$$

We now define new operators $T_{jk}(z)$ which obey (15) for <u>all</u> z, and equal the $T'_{jk}(E + i0)$ for $z = E + i0$:

$$T_{jk}(z) = V_j W_{\ell k} + V_j \sum_i W_{\ell i}\, G_i(z)\, T_{ik}(z) \quad . \tag{16}$$

Eq. (16) is the desired result. In matrix form it reads

$$T(z) = \mathcal{V} + \mathcal{V}\mathcal{G}(z)\, T(z) \tag{17}$$

where $\mathcal{V}_{ij} = V_i W_{\ell j}$, $\mathcal{G}_{ij}(z) = \delta_{ij}\, G_j(z)$. The preceding derivation, because it relies on eq. (14), has been questioned. However, not only is (14) a valid result when $|\Phi_E(k)\rangle$ is a two body state, but as noted above, we have recently derived eq. (15) directly from the Schrödinger equation, without making use of (14) at all, thus justifying (though unnecessarily) our use of (14). However, when $|\Phi_E(k)\rangle$ is a free-particle state, (14) no longer holds, and different equations result. We shall not consider such cases here.

The structure of (17) is such that the operators $T_{jk}(z)$ can satisfy the six conditions stated above. Let us consider connectedness first. The kernel of (17) is $\mathcal{V}\mathcal{G}(z)$, whose matrix element is $V_i W_{\ell k} G_k(z)$. By proper choice of the elements of W, a connected, iterated kernel can be obtained. For the 3-body problem, we have found a choice such that the first iterate of $\mathcal{V}\mathcal{G}(z)$ is connected[15], as in the Faddeev formalism. For arbitrary N, we have found a general class of (N - 1)! W's denoted channel permuting arrays (CPA), such that the (N - 1)st iterate of the kernel is connected[14]. Tobocman[22] first gave a prescription for choosing a W that would guarantee that the (N - 1)st iterate of $\mathcal{V}\mathcal{G}$ was connected; his prescription is an example of CPA. The general method for specifying W in this case is to set $\ell = j$ in (16) and then select N of the $W_{ij} = 1$ (all others = 0), one in each row and column with rows and columns linked successively, as follows:

$$W_{ii} = 0 \quad ,$$

$$W_{ij} = 1 \text{ or } W_{ji} = 1 \quad \text{but not both,}$$

$$W_{\ell m} = 1 \rightarrow W_{mn} = 1 \quad .$$

The first two conditions prevent only one or two T_{jk} being coupled together, while the last helps ensure connectedness of the (N - 1)st iterate. The channel permuting W's are such that $(W^N)_{ij} = \delta_{ij}$. It is easy to see that, although each of the (N - 1)! CPA's is a representation of the permutation group of order N, these W's (along with the N x N unit matrix) form neither a group nor a semi-group except for the cases N = 2 and N = 3.

We shall not give the proofs that the CPA's guarantee connectedness of the (N - 1)st iterate or that there are (N - 1)! of them. Instead we give a few examples and then write out the general set of equations for a specific choice. For N = 2 and 3, the following are the (unique) CPA's:

$$\begin{pmatrix} 0 & 1 \\ 1 & 0 \end{pmatrix}, \quad \begin{pmatrix} 0 & 1 & 0 \\ 0 & 0 & 1 \\ 1 & 0 & 0 \end{pmatrix}, \quad \begin{pmatrix} 0 & 0 & 1 \\ 1 & 0 & 0 \\ 0 & 1 & 0 \end{pmatrix}.$$

In the general case, we select as the initial channel k = 1; for W_{ij} we choose[14,22) $W_{ij} = \delta_{j,i+1}, 1 \le i \le N-1; W_{N1} = 1$; all others equal zero. Eq. (16) becomes

$$\begin{aligned} T_{11}(z) &= V_1\, G_2(z)\, T_{21}(z) \\ T_{21}(z) &= V_2\, G_3(z)\, T_{31}(z) \\ &\;\vdots \\ T_{N-1,1}(z) &= V_{N-1}\, G_N(z)\, T_{N1}(z) \\ T_{N1}(z) &= V_N + V_N\, G_1(z)\, T_{11}(z) \end{aligned} \tag{18}$$

Equation (18) is not in the usual form of coupled equations we are familiar with; nevertheless, it is a valid form. In particular, the symmetrized version of (18) has yielded unitary, S-wave $e^- + H$ scattering amplitudes while another set , similar to but not identical with (16), gave explicitly non-unitary amplitudes in the same problem[16)]. These results have been interpreted[17)] in light of the existence of the K operator and damping equations corresponding to (17) and their lack in other formulations of many-body scattering, even those utilizing the W array[22)].

Before commenting on the unitarity properties of our equations, we consider a further point connected with the specific set (18), viz, only the last equation appears to have a Born term. This is, of course, only an apparent lack, since it is evident that by back substitution, the N equations of (18) may be decoupled, and each of them will then display driving terms.

Before displaying the uncoupled equation for T_{11}, it is necessary to consider some analytic properties of our equations, since they are not yet in a form which properly takes account of unitarity restrictions. Eq. (16) [or (17)] is obtained from (15) simply by replacing $E + i0$ by z. We wish to determine if the matrix $\mathcal{G}(z)$ has the same analytic properties as the collection of its matrix elements $\{G_i(z)\}$, each taken individually. The answer is no: as shown in a paper contributed to this Conference[23], the resolvent $G_i(z)$ behaves differently when considered as an integral part of the equations for $T(z)$ [recall $\mathcal{G}_{ij}(z) = G_i(z)\delta_{ij}$] than when examined in isolation. The difference becomes noticeable when examining the discontinuity of $\mathcal{G}(z)$, $D\mathcal{G}(z)$, across the right hand cuts. Unless special precautions are taken, there will be multiple contributions from cuts occurring in more than one of the $G_i(z)$. Since $D\mathcal{G}(z) = 2i \operatorname{Im}\mathcal{G}(z)$, multiple counting is avoided by the simple device of multiplying each $G_i(z)$ by the projection operator $P(i)$ onto the set of bound internal states in channel i:

$$\operatorname{Im}\mathcal{G}_{ij}(z) \to \operatorname{Im}\tilde{\mathcal{G}}_{ij}(z) \equiv \delta_{ij}\,[\operatorname{Im} G_i(z)]P(i).$$

By Cauchy's theorem, we also find[23]

$$\operatorname{Re}\mathcal{G}_{ij}(z) \to \operatorname{Re}\tilde{\mathcal{G}}_{ij}(z) \equiv \delta_{ij}\,[\operatorname{Re} G_i(z)]\,P(i) \quad ,$$

so that we may in general replace $\mathcal{G}(z)$ in eq. (17) by $\tilde{\mathcal{G}}(z)$, where

$$\tilde{\mathcal{G}}_{ij}(z) = \delta_{ij}\,G_i(z)\,P(i) \equiv \delta_{ij}\,\tilde{G}_i(z) \quad . \tag{19}$$

This means that $T(z)$ is now defined by

$$T(z) = \mathcal{V} + \mathcal{V}\tilde{\mathcal{G}}(z)\,T(z) \quad , \tag{20}$$

thus ensuring that $T(z)$ obeys the proper unitarity relation[24], viz.,

$$\operatorname{Im} T(E + i0) = -\pi T\,(E + i0)\,\tilde{\Delta}(E)\,T(E - i0) \quad , \tag{21}$$

where $\tilde{\Delta}_{ij}(E) = \delta_{ij}\,\delta(E - H_j)P(j)$. The change $\mathcal{G}(z) \to \tilde{\mathcal{G}}(z)$ is, as we have shown[24], consistent with the effective orthogonality conditions found by Hahn to hold in multiparticle scattering[25].

In the form (17), $T(z)$ has the solution $T(z) = \mathcal{V} + \mathcal{V}\mathcal{Y}(z)\mathcal{V}$, where $\mathcal{Y}(z) = [\mathcal{G}^{-1}(z) - \mathcal{V}]^{-1}$. In addition, a matrix of K operators had been introduced[14,15] through $K(E) = \mathcal{V} + \mathcal{V}\mathcal{G}^P(E + i0)K(E)$, where $\mathcal{G}^P(E + i0) = \operatorname{Re}\mathcal{G}(E + i0)$, such that T and K were related by the damping equation[14,13] $T(E + i0) = K(E) - i\pi K(E)\Delta(E)T(E + i0)$. Here, $\Delta_{ij}(E) = \delta_{ij}\,\delta(E - H_j)$, so that $\tilde{\Delta}_{ij}(E) = \Delta_{ij}(E)P(j)$. This latter damping equation is incorrect, since there is multiple counting of the contributions from many of the unitarity cuts. With the new results, we find $T(z) = \mathcal{V} + \mathcal{V}\tilde{\mathcal{Y}}(z)\mathcal{V}$, where $\tilde{\mathcal{Y}}(z) = [\tilde{\mathcal{G}}^{-1}(z) - \mathcal{V}]^{-1}$, and also new K-operator equations and damping relations:

$$K(E) = \mathcal{V} + \mathcal{V}\tilde{\mathcal{G}}^{P}(E + i0)K(E) \tag{22}$$

and

$$T(E + i0) = K(E) - i\pi K(E)\tilde{\Delta}(E)\, T(E + i0) \quad . \tag{23}$$

Eq. (23) is clearly compatible with the unitarity relation (21), while eq. (22) allows for the introduction of unitary approximations in a simple way.

The implication of this modification is that in carrying out calculations, only bound internal states of the subsystems in each channel need be considered as intermediate states. The effects of a two-body breakup (continuum) state of one of the subsystems in an m-body channel are included in the contributions from bound states in an (m + 1)-body channel. As a result, the complexity of calculations can be greatly reduced. There is another point of interest in this connection. Our method counts contributions from unitarity cuts correctly by restricting the $G_i(z)$ as above. The same effect is obtained in other many-body scattering theories[7-11] by the device of using the n-body, free-particle Green's function $G_0(z) = (z - H_0)^{-1}$ only, but with the entire collection of 2-, 3-, ... (n - 1)-body t operators, which latter have the effects of all cuts other than the n-body free-particle one built in.

With the preceding modification, we may now consider the uncoupled equations for T_{11}, obtained by back-substitution of the equations from (18), modified by $G_i \to \tilde{G}_i$. The result is:

$$T_{11}(z) = U_1^{(N)}(z)\,[1 + \tilde{G}_1(z)\, T_{11}(z)] \quad , \tag{24}$$

where

$$U_1^{(N)} \equiv V_1\tilde{G}_2(z)\, V_2\tilde{G}_3(z)\, V_3 \;.\;.\;.\; \tilde{G}_N(z)V_N \tag{25}$$

is a quantity we have denoted[15] the direct-channel, optical-potential operator. It is a z-(energy-) dependent, complex, non-local, connected operator, expressing the physical result that flux may go into channels other than k = 1 . This has interesting implications for the theory of the optical model; in particular, even if the energy is too low to excite states in channel 1, it may be possible to have transitions to states in channels other than j = 1 , as for example in (d,p) reactions. In such cases, the optical potential will be complex even though in channel 1 only the ground state can be seen experimentally.

We now turn to the question of the effects of the Pauli principle. We have recently studied a model identical-particle problem in which one particle is scattered by a system of (n - 1) identical particles under the condition that the final state consists of (n - 1) particles bound together plus one free particle[24]. In this case, channel labels are identical to particle labels, and the amplitude is given, for example, as matrix elements of the operator $\mathcal{J} = T_{11} - (n - 1)P_{21}T_{21}$, where 1 (or 2) labels the unbound particle and P_{21} is the two-particle transposition operator. In this notation, 1 is the "direct" channel and 2 is the "exchange" channel. An integral equation for $\mathcal{J}$ based on eq. (16) and the symmetry of the problem is[24]

$$\mathcal{T} = [V_1W_{11} - (n-1)P_{21}V_2W_{21}](1 + \tilde{G}_1\mathcal{T}) \quad , \tag{26}$$

where the energy dependence is suppressed. For the case n = 2, (26) reduces to the general form of the equation solved in the $e^- + H$ case which gave unitary amplitudes[16-18]. The preferred choice for W_{ij} is $W_{11} = 0$, $W_{21} = 1$, although other values are allowed in this case because of the presence of $\tilde{G}_i$. For this choice, the "direct" driving term V_1W_{11} vanishes, leaving only the exchange term in (26). The results of approximate $e^- + H$ calculations based on (26) for the case n = 2, $V_i = -e^2/r_i + e^2/r_{12}$, are very encouraging, despite the lack of the usual Born term when $W_{11} = 0$. We emphasize that these equations, despite a possible superficial resemblance to the more familiar Hartree-Fock type of coupled equation used in electron-atom scattering[26), are in fact not only different in appearance, but represent a completely different approach to the physics of identical particle scattering problems. The results presented below for low energy $e^- + H$ scattering parameters calculated in the approximation of retaining only the 1s hydrogenic ground state as an expansion basis for evaluating $\tilde{G}_1$ are corroborative evidence for this conclusion.

The first calculations done for the $e^- + H$ system that gave unitary amplitudes using the W array were[16) for the values $W_{11} = 0$, $W_{21} = 1 - W_{11} = 1$. Singlet and triplet phase shifts from this and other calculations are compared in Tables 1 and 2. The results are seen to compare extremely well with the phase shifts computed from both the close-coupling approximation[27) and the (lower bound) variational method[28). For a detailed discussion see ref. 16 and 17.

Given these successes, additional low energy computational studies were made, in particular to examine the dependence of the singlet and triplet $e^- + H$ scattering lengths and E = 0.1 Rydberg phase shifts on W_{11}. A representative set of results is given in Table 3 for the scattering lengths $a^\pm$ (defined here as $\lim k \to 0\ [+k \cot \eta^\pm]^{-1}$, where $\eta^+(\eta^-)$ is the singlet (triplet) phase shift). The preferred choice of $W_{11} = 0$ is indicated by an asterisk, and results of several other calculations are included for comparison. Notice that a^+ and a^- must be considered together when examining these results, since we see that variation of W_{11} can yield a value for either one of $a^\pm$ close to, or even less than, Schwartz's variational result[28) while the other value ($a^\mp$) can then be arbitrary, both in magnitude and sign. (Schwartz's (upper bound) values are generally regarded as the biblical numbers against which all other calculations are to be measured.) The best pair occurs for the preferred values $W_{11} = 0$, $W_{21} = 1$.

More detailed results are given in Figure 1 for a^+ $(-0.3 \leq W_{11} \leq 1.3)$ and in Figure 2 for a^- $(-0.3 \leq W_{11} \leq 1.4)$. The use of W_{11} can be thought of as providing, in a certain sense, an effective-potential description of the scattering. The points A are where this effective potential turns from repulsive to attractive, while the singular points occur at the onset of a bound state at zero energy. The behavior for large, negative W_{11} seems smooth; analytic fits to these curves, extrapolated to $W_{11} = -\infty$ show no indication of further poles or zeros. It is tempting to associate the pole in a^+ with the well-known H^-, singlet bound state, but the occurrence of a pole in a^- where no bound, triplet H^- states exist, makes this association, at present, merely an interesting speculation. The E = 0.1 Rydberg phase shifts $\eta^{\pm}$ are shown in Figures 3 and 4. Their behavior corresponds quite closely to that of $a^{\pm}$; in particular, $\eta^{\pm}$ goes to zero at the points A where $a^{\pm}$ changes sign. A more detailed discussion of these calculations is given in ref. 18; other calculations, both for the $e^- + H$ and for the nuclear 3-body systems are in progress. As noted above, the results presented here are favorable numerical evidence that our method, even though it leads to unfamiliar equations, is a promising means for treating multi-channel scattering problems.

---

* Alfred P. Sloan Foundation Fellow, 1972-74; Alexander von Humboldt-Stiftung Senior Fellow, 1973-74. Permanent address: Departments of Chemistry and Physics, University of Houston, Houston, Texas 77004.

† Work supported in part by the U.S. Atomic Energy Commission. Address until June, 1975: Physics Department, University of Surrey, Guildford, Surrey GU2 5XH, England.

References

1. R. D. Amado, in Elementary Particle Physics and Scattering Theory, ed. by M. Chretien and S. S. Schweber (Gordon and Breach, New York, 1970).
2. P. Federbush, Phys. Rev. 148 (1966) 1551.
3. R. G. Newton, Phys. Rev. 153 (1967) 1562; J. V. Noble, ibid., 148 (1966) 1553; 161 (1967) 1495.
4. L. D. Faddeev, Sov. Phys.-JETP 12 (1961) 1014; Mathematical Aspects of the Three-Body Problem (D. Davey and Co., New York, 1965).
5. H. Eckstein, Phys. Rev. 101 (1956) 880.
6. R. G. Newton, Scattering Theory of Waves and Particles (McGraw-Hill, New York, 1966).
7. E. O. Alt, P. Grassberger, and W. Sandhas, Nucl. Phys. B2 (1967) 181.
8. O. A. Yakubovski, Sov. Journ. Nucl. Phys. 5 (1967) 937.
9. G. Bencze, Nucl. Phys. A210 (1973) 568.
10. B. R. Karlsson and E. M. Zeiger, Phys. Rev. D9 (1974) 1761.
11. E. Redish, Phys. Rev. C10 (1974) 67.
12. R. T. Cahill, Nucl. Phys. A194 (1972) 599.
13. K. L. Kowalski, Phys. Rev. D5 (1972) 395; D6 (1972) 3705; D7 (1973) 1806.
14. D. J. Kouri and F. S. Levin, Bull. Am. Phys. Soc. 19 (1974) 489; Phys. Letts. 50B (1974) 421.
15. D. J. Kouri and F. S. Levin, submitted for publication.
16. D. J. Kouri, M. Craigie and D. Secrest, J. Chem. Phys. 60 (1974) 1851.
17. D. J. Kouri, F. S. Levin, M. Craigie, and D. Secrest, J. Chem. Phys. 61 (1974) 17.
18. D. J. Kouri, F. S. Levin, M. Craigie, and D. Secrest, submitted for publication.
19. Y. Hahn, D. J. Kouri, and F. S. Levin, Phys. Rev. C, (in press) October, 1974.
20. Y. Hahn, Phys. Rev. 169 (1968) 794; Nucl. Phys. A132 (1969) 353; Y. Hahn and K. M. Watson, Phys. Rev. A5 (1972) 1718.
21. B. A. Lippmann, Phys. Rev. 102 (1956) 256.
22. W. Tobocman, Phys. Rev. C9 (1974) 2466.
23. D. J. Kouri and F. S. Levin, Unitarity Modifications of the Coupled Channel T and K Operators for Many-Body Scattering (Proceedings of this Conference), and Ref. 24.
24. D. J. Kouri and F. S. Levin, submitted for publication.
25. Y. Hahn, Physics Letters, 30B (1969) 595; Phys. Rev. C3 (1971) 1432.
26. Eg., N. F. Mott and H. S. Massey, Theory of Atomic Collisions, 3rd edition, (Oxford Univ. Press, Oxford, 1964).
27. P. G. Burke and K. Smith, Rev. Mod. Phys. 34 (1962) 458.
28. C. Schwartz, Phys. Rev. 124 (1961) 1468.

Table 1

Singlet e^- + H Phase Shifts

E(Ryd)	Hartree[27]	Static-Exchange[27]	Close-Coupled[27]	W Array[16]	Variational[27]
0.01	0.721	2.396	2.491	2.541	2.553
0.04	0.973	1.871	1.974	2.039	2.067
0.09	1.046	1.508		1.646	1.696
0.25	1.045	1.031	1.082	1.080	1.202
0.36	1.021	0.869	0.93	0.871	1.041

Table 2

Triplet e^- + H Phase Shifts

E(Ryd)	Hartree[27]	Static-Exchange[27]	Close-Coupled[27]	W Array[16]	Variational[27]
0.01	0.730	2.908	2.935	2.935	2.939
0.04	0.973	2.679	2.715	2.737	2.717
0.09	1.046	2.461	2.50	2.554	2.500
0.25	1.045	2.070	2.096	2.264	2.105
0.36	1.022	1.901	1.92	2.170	1.933

Table 3

Dependence of e^- + H Scattering Lengths on W_{11}

W_{11}	a^+	a^-
-0.4	5.792	2.267
0	6.230*	2.077*
0.4	7.376	1.745
0.8	17.253	0.598
0.9	193.568	-0.658
1.0	-9.442	-9.442
1.1	-2.140	7.655
Strong coupling[27]	7.5	2.33
Close coupling[27]	6.74	1.89
Variational[27]	5.96	1.77

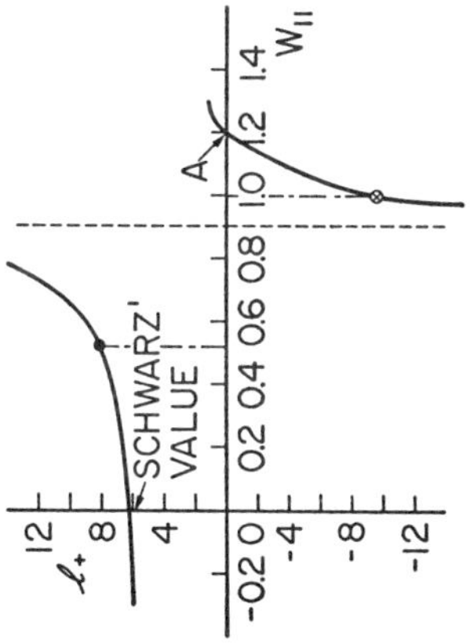

FIG. 1

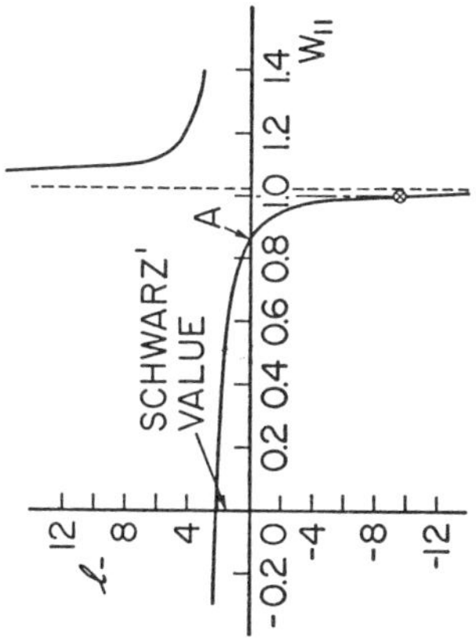

FIG. 2

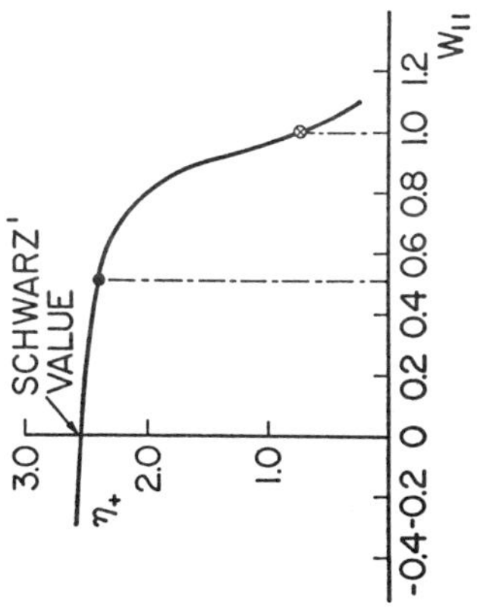

FIG. 3

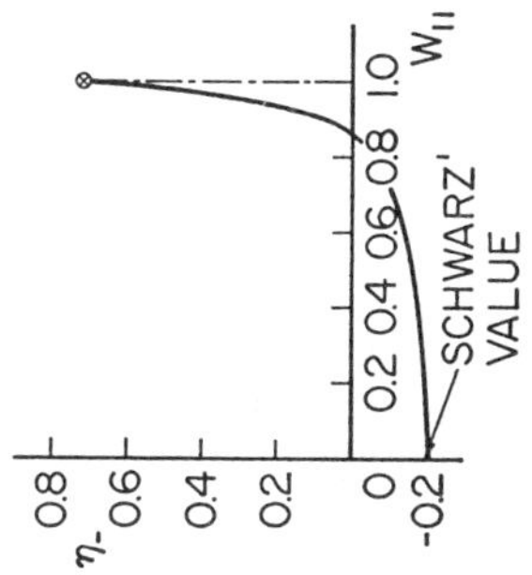

FIG. 4

R-MATRIX METHOD IN THE 3-BODY PROBLEM

W. Glöckle, Institute for Theoretical Physics of the
Ruhr-University Bochum, 463 Bochum, Germany

The solution of the Faddeev equations in coordinate space is a means to treat local interactions directly avoiding the search for separable approximations. To that aim the asymptotic form of the Faddeev amplitude $u(x_1,y_1)$ defined by

$$\left(-\frac{d^2}{dx_1^{\,2}} - \frac{3}{4}\frac{d^2}{dy_1^{\,2}} + V(x_1) - E\right) u(x_1,y_1) = - V(x_1) \int_{-1}^{1} d(\cos\vartheta_1) \frac{x_1,y_1}{x_2,y_2} u(x_2,y_2) \tag{1}$$

is derived [1] from eq. (1) up to terms of the order $O(\frac{1}{r^2})$, where r stands for the Jacobi variables x_1, y_1 (or x_2, y_2). Eq. (1) refers to the simplest case of spinless identical particles interacting by s-wave potentials [2]. The asymptotic form of u contains the on-shell 3-body T-matrix elements and can be assumed to be valid on boundary lines C_1: $x_1 = a$, $0 \leq y_1 \leq A$ and C_2: $0 \leq x_1 \leq a$, $y_1 = A$. The lines C_1 and C_2 confine a region D in the first quadrant of the x_1, y_1 plane.

One can introduce a complete orthonormal set of basis states $\varphi_\mu(x_1,y_1) = \varphi_{\mu 1}(x_1) \cdot \varphi_{\mu 2}(y_1)$ which are solutions of the left hand side of eq. (1) at the energy eigenvalues E_μ and which obey the boundary conditions

$$\varphi_\mu(0,y_1) = 0 = \varphi_\mu(x_1,0) \text{ and } \left.\frac{\partial\varphi_\mu}{\partial x_1}\right|_{x_1=a} = 0 = \left.\frac{\partial\varphi_\mu}{\partial y_1}\right|_{y_1=A} .$$

Furthermore due to the factor $V(x_1)$ on the right hand side of eq. (1) one can find numbers a and A such that x_2,y_2 remain in D if x_1,y_1 belong to D. Therefore one can use the expansion

$$u(x,y) = \sum_\mu C_\mu \varphi_\mu(x,y) \tag{2}$$

which is valid in D on both side of eq. (1). Eq. (1) is easily converted into

$$(E_\mu - E)C_\mu + \sum_{\mu'} C_{\mu'} V_{\mu\mu'} = \int_0^A dy\, \varphi_\mu(a,y) \left.\frac{du}{dx}\right|_{x=a} + \int_0^a dx\, \varphi_\mu(x,A) \left.\frac{du}{dy}\right|_{y=A} \tag{3}$$

which relates the C_μ's to the on-shell T-matrix elements T contained in $\frac{du}{dx}\Big|_{x=a}$ and $\frac{du}{dy}\Big|_{y=A}$ on the "asymptotic borderlines" of D. Furthermore

$$V_{\mu\mu'} = \iint_D dx_1\, dy_1\, \varphi_\mu(x_1,y_1)\, V(x_1) \int_{-1}^{1} d(\cos\vartheta_1)\, \frac{x_1\, y_1}{x_2\, y_2}\, \varphi_\mu(x_2,y_2) \qquad (4)$$

The additional necessary relations between C_μ and T are given by equating the asymptotic form to the interior form of u, see eq. (2), on the lines C_1, C_2. This set of equations allows the determination of C_μ and T, where the latter depends only on one energy variable, which measures how the total energy is split into the two relative motions.

The dimensionality of two of the problem is present in the calculation of the matrix elements $V_{\mu\mu'}$ of eq. (4), which are however energy independent and can be determined once for all.

REFERENCES

1) W. Glöckle, preprint March 1974, to be published

2) L. D. Faddeev, Soviet Phys. JETP 12 (1961) 1014

A MICROSCOPIC INTERPRETATION OF THE PROTON-INDUCED REACTIONS IN THE MASS-3 SYSTEM

C.L. Rao, S. Ramavataram and K. Ramavataram
Université Laval, Laboratoire de Physique Nucléaire,
Département de Physique, Québec G1K 7P4, Canada.

Extensive theoretical and experimental studies[1,7] have been carried out on the negative parity resonances in the A=4 system. Microscopic descriptions[6,7] based on a 1p-1h bound state model predict the observed spin sequences and energies of these resonances which arise from the configurations $(1s_{1/2})^{-1}$ $(1p_{3/2})$ and $(1s_{1/2})^{-1}$ $(1p_{1/2})$. In the case of the two T=1, 1^- states the calculated relative E1 transition strength showed certain descrepancies when compared with experiment. Theoretically most of the E1 strength was predicted to be in the higher energy state (30.5MeV) while experimentally it is seen to be in the 27.4 MeV state. Shell-model in the continuum calculations[8] showed that the introduction of the continuum can resolve this descrepancy to some extent. But the formalism of ref.8 used the so-called "separation approximation" which may not be a suitable approximation for a light nucleus[9]. A formalism which gives the total wave function in all space is preferable. We have therefore studied this system in the coupled-channel model of nuclear reactions[10]. The coupling of the (^{3}H+p) and (^{3}He+n) channels was calculated using a zero-range particle-hole interaction with the standard parameters having values similar to those found in literature[10,11]. The p-particle potential was fixed such that a $p_{3/2}$ resonance occurs at an incident proton energy of 1.5 MeV; this choice of potential was found to reproduce the experimentally observed threshold behaviour of the (γ,p) excitation. It was also found that the strength of the spin-orbit interaction determines whether the two T=1, 1^- resonances can be resolved in the (γ,p) reaction or not. The best fit to experiment is shown in FIG.1. The (γ,p)/(γ,n) ratio was found to be very nearly 1 in agreement with experiment[12,13].

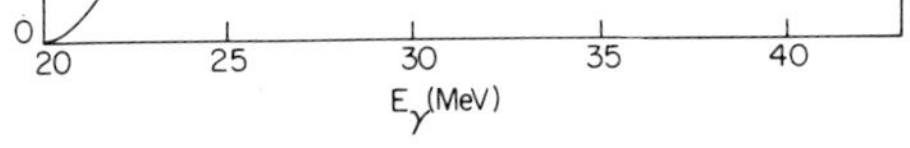

FIGURE 1. Comparison of the calculated ^{4}He total cross section with experimental data (solid circles from ref. 3; open circles from ref. 4).

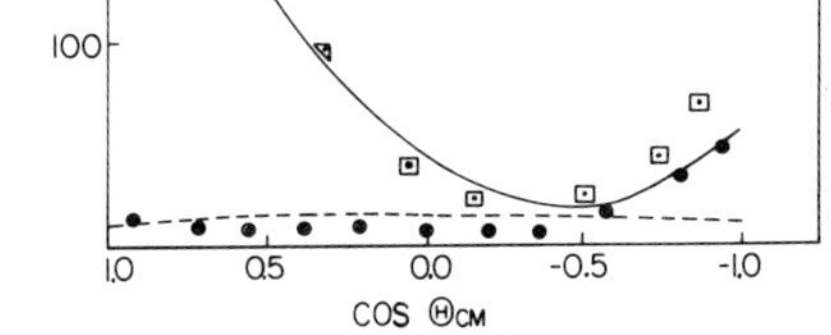

FIGURE 2. The experimental data for the differential cross section at E_p(lab)=8.0 MeV is taken from ref. 5.

Calculations were performed in the T=1 approximation and the phase-shifts in the $p+{}^3He$ scattering process for $J^\pi=0^-$, 1^- and 2^- were found to be in reasonable agreement with experiment[2]. For the choice of parameters mentioned earlier, the 2^- was found to have the lowest energy. Using the same reaction theory model[10] and parameters for the negative parity resonances we have also calculated the differential cross section for the ${}^3H(p,p){}^3H$ and ${}^3H(p,n){}^3He$ reactions at an incident proton energy of 8.0 MeV which corresponds to the primary peak in the (γ,p) excitation function. The scattering amplitudes for the s and d-waves were also included in the calculations. In the case of the ${}^3H(p,p){}^3H$ process, the agreement between theory and experiment is very satisfactory (FIG.2). The observed ${}^3H(p,n){}^3He$ cross-section is relatively much smaller than the ${}^3H(p,p){}^3H$; the descrepancy between theory and experiment in this case perhaps reflects the fact that in the calculation T=o admixtures are found to be negligible while experimentally this may not be the case.

REFERENCES

1) S. Fariman and W.E. Meyerhoff, Nucl. Phys. A206 (1973) 1.
2) T.A. Tombrello, Phys. Rev. 138 (1965) B40.
3) W.E. Meyerhoff, M. Suffert and W. Feldman, Nucl. Phys. A148 (1970) 211.
4) H.G. Clerc, R.J. Stewart and R.C. Morrison, Phys. Lett. 18 (1965) 316.
5) J.D. Seagrave: A compilation of the experimental A=4 scattering data in "Symposium on Few Body Problems, Light Nuclei and Nuclear Interactions" edited by G. Paic and I. Slaus (Gordon and Breach, New York, 1968) 787.
6) A. de Shalit and J.D. Walecka, Phys. Rev. 147 (1966) 763.
7) B.R. Barrett, Phys. Rev. 154 (1967) 955.
8) J. Hüfner and C.M. Shakin, Phys. Rev. 175 (1968) 1350.
9) C. Mahaux and H.A. Weidenmüller, Shell-Model Approach to Nuclear Reactions (North-Holland, Amsterdam 1969).
10) B. Buck and A.D. Hill, Nucl. Phys. A95 (1967) 271.
11) G.E. Brown, L. Castillejo and J.A. Evans, Nucl. Phys. 22 (1961) 1.
12) W.R. Dodge and J.J. Murphy II, Phys. Rev. Lett. 28 (1972) 839.
13) J.D. Irish et al. Phys. Rev. C8 (1973) 1211.
14) S. Ramavataram, C.L. Rao and K. Ramavataram (to be published in Nuclear Physics)

A POSSIBLE CHECK OF EFIMOV RESONANCES

M. S. HUSSEIN

Instituto de Física, Universidade de São Paulo, C.Postal 8219, S.Paulo, S.Paulo, Brasil.

Although Efimov states are thought to be threshold phenomena[1-2] there is the possibility of observing them, indirectly, by their potential influence of other more directly observed states in three-body reactions[3]. If seems that among such reactions is that when one has a pion produced in a two nucleon collision (fig. 1),

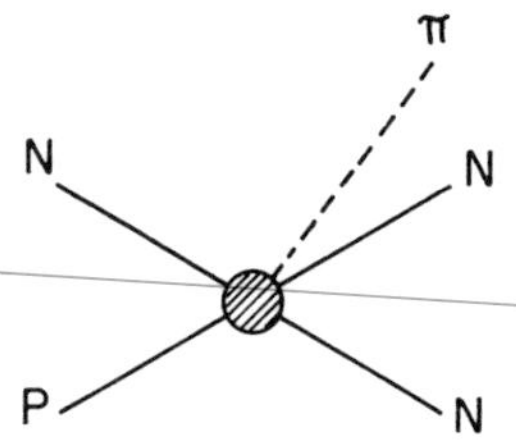

FIG. 1

where one looks at the "nucleon" as a resonance of one pion plus an elementary nucleon. The cause of collision may be looked at as due to this resonance reaction of the pions with the two nucleons i.e. in the blob the pion makes very many resonance collisions with both nucleons thus resulting in an over all nucleon-nucleon resonance. If one is looking for π^+, s from $p+n$ $n+n+\pi^+$ then one would expect that the three-body resonance (i.e. NN) to occur at somewhere around $E_r \approx E^n + E^n + 2/10\ (E^n)^{1/2}(E^n)^{1/2}$; E^n is the position of the N^* resonance in the πn system[4]. One usually associates a width to the resonance in an appropriate Breit-Wigner parametrization of the cross section in the vicinity of the resonance. Other states that resonate in the three-body system and are not observed can be thought of as being so many and overlapping that they are washed out completely[3]. However, if these states are somewhere in the energy range that encompasses the observed resonance at E_r then the total width of this resonance will contain two parts one $\Gamma\uparrow$ corresponding to the decay to open channels (including the three-body channel) the other part $\Gamma\downarrow$ is due to the coupling of the state that produces this resonance to these states resonating within the resonance it seft. What we suggest is that the measurement of this damping width $\Gamma\downarrow$ would constitute an indirect test of whether Efimov states resonate or not since apparently this is the only way that these resonances could manifest themselves in the cross section corresponding to the process envisioned in Fig. 1. If none of the Efimov states resonate then obviously $\Gamma\downarrow=0$. If some, say N_ε of them resonate within the energy $E \sim\Gamma\downarrow + \Gamma\uparrow$ around E_r then we argue that $\Gamma\downarrow$ is proportional to N_ε and thus N_ε can be found by careful analysic of the resonance region in question using the familiar intermediate structure theory of nuclear reaction that are used extensively at low energies[3]. In such a case the transition amplitude is written as follows $T_{fi} \sim \sqrt{\Gamma_f\uparrow}\ \sqrt{\Gamma_i\uparrow} \Big/ \left[(E-E_r') + i\frac{(\Gamma\uparrow+\Gamma\downarrow)}{2}\right]$ where $E_r'=E_r+\Delta$ and Δ is the continuum shift of the observed resonance, $\Gamma_i\uparrow$ $\Gamma_f\uparrow$ & $\Gamma\downarrow$ are complex quantities that are related to the partial widths of the three-body resonance. From reaction theory one finds that $\Gamma\downarrow = 2\pi \sum_\varepsilon |<\varepsilon\,|\mathcal{H}_{\varepsilon r}\,|\,r>|^2$ where r refers to the observed NN resonance at E_r and the sum is over all resonating efimov states within the width of the NN resonance. Thus $\Gamma\downarrow\sim N_\varepsilon$ as was suggested in the beginning.

REFERENCES

1) V.Efimov "Nuclear Physics A 210 (1973) 157.

2) R.Amado "Talk at the fifth Brazilian Symposium on theoretical physics Rio de Janeiro, January/1974 - Proceeding to be published.

3) M.S.Hussein "USP preprint IFUSP, p-10, Nov/1973, to be published.

4) D.D.Brayshaw and R.F.Peierls "Phys.Rev. 177, 2539 (1969).

THREE-PARTICLE CALCULATIONS ABOVE BREAKUP THRESHOLD FOR SEPARABLE, ALMOST SEPARABLE AND LOCAL POTENTIAL

H. Ziegelmann
Institut für Theoretische Physik der Universität Tübingen, GFR

Three-particle calculations for local potentials can be performed in a rather economic way using the quasi-particle method [1], which treats a separable part of the potential exactly and takes into account the non-separable part pertubatively. Above breakup threshold the half-off-shell solution T of the quasi-particle equation describes the production of a two-particle state with positive energy. This production amplitude is related in first quasi-Born approximation (1.QBA) to the breakup operator τ by the graphs shown in Fig. 1. In addition to the well-known graph of the separable model (type 1), there are graphs which are related to the spectator model (type 2), and further graphs which are related to type 1, but with modification of the formfactor (type 3) or rescattering (type 4). In Fig. 2 we show the results of a breakup calculation for three identical bosons interacting via Yukawa potentials. In contrast to three-nucleon calculations the finalstate interaction peaks are not so pronounced (sometimes they even disappear (Fig. 2b,c, E_p=0)) and instead of a minimum due to the vanishing of the quartet contribution in the symmetric point (Fig. 2c, E_p=0.5) we get a maximum. The differences between the dotted line (Yamaguchi separable potential with the same two-body low energy data) and dashed-dotted line (separable result plus spectator graph) indicate that the spectator correction can be quite important in some cases. It is a curious fact that in the spectator maximum (Fig. 2b, E_p=0.7) the spectator correction almost vanishes. The reason is that the non-separable part of the interaction has a zero near the spectator maximum. The differences between the full line (1.QBA) and the dashed-dotted line in Fig. 2 indicate that in most cases it is not sufficient to calculate the production process in separable approximation and then add the correction graphs due to the non-separability of the potential. The production process in our model is strongly dependent on the details of the potential and that is the main reason of the differences of the results for local and separable potential in Fig. 2. Preliminary calculations with graphs of type 4 indicate that rescattering effects are rather important in our model.

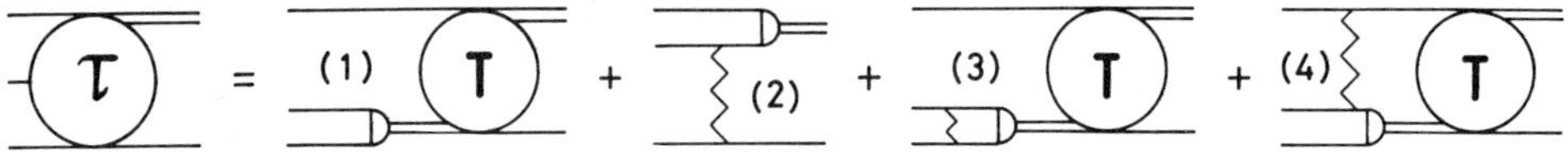

Fig. 1. Typical graphs contributing to the breakup operator τ in 1.QBA. The wavy lines represent the non-separable part of the interaction.

For local potentials with repulsion as, for instance, the Malfliet-Tjon potential, it seems to be doubtful whether a treatment with the quasi-particle method might be possible by subtracting only one separable term. The saving grace is, of course, that the UPA gives results very close to the exact ones [2]. Above threshold the UPA is also a good approximation. Fig. 3 shows the real part of the s-wave phase shift and the inelastic parameter in the doublet channel. The results for the other waves are almost identical with those of Kloet and Tjon [3]. It is currently studied how a 1.QBA can be calculated in this case although the quasi-Born series are divergent due to the repulsive eigenvalue [5].

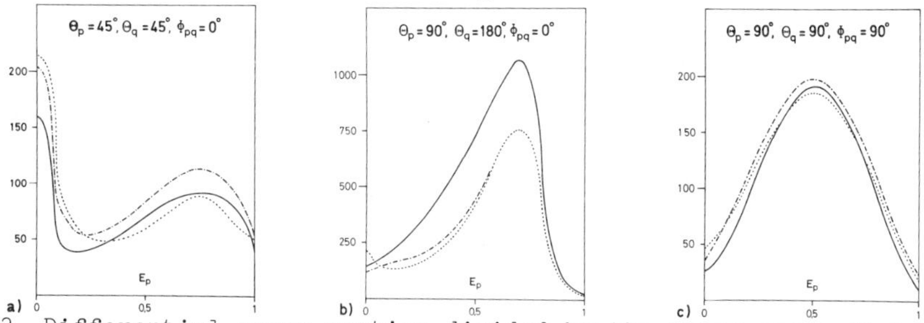

Fig. 2. Differential cross section divided by the phase-space factor in relation to the energy of a two-particle system $E_p=p^2/m$; c.m. energy $E=3q_0^2/4m+E^B=1$; two-particle binding energy $E^B=0.1$; units $\hbar=m=1$. The polar angles Θ_p, Θ_q and the azimuth angle Φ_{pq} define the orientation of the relative momentum in a subsystem $\vec{p}$ and the momentum of the corresponding third particle $\vec{q}$ with respect to the momentum of the incoming particle $\vec{q}_0$.

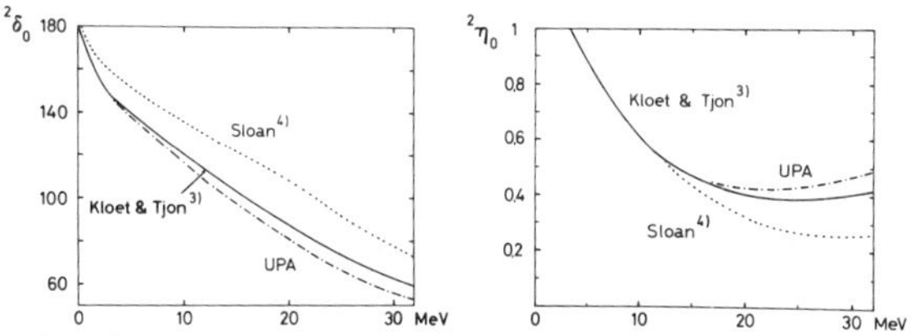

Fig. 3. Real part of the phase shift and inelastic parameter for the doublet s-wave for n-d scattering versus neutron laboratory energy.

REFERENCES

1) P. Grassberger and W. Sandhas, Z. f. Phys. 217 (1968) 9;
E.O. Alt, P. Grassberger, and W. Sandhas, Phys. Rev. D1 (1969) 2581;
H. Ziegelmann, Nucl. Phys. A192 (1972) 426.

2) E.O. Alt and B. Bakker, preprint, Univ. of Mainz and Amsterdam, 1973.

3) W.M. Kloet and J.A. Tjon, Ann. Phys. 79 (1973) 407.

4) I.H. Sloan, Nucl. Phys. A168 (1971) 211.

5) B. Bakker and W. Sandhas, private communication.

This work was supported by Deutsche Forschungsgemeinschaft.

ON THE COMPARISON BETWEEN QUANTUM MECHANICS AND LOCAL HIDDEN VARIABLE THEORIES: BELL'S TYPE INEQUALITIES FOR MULTI-VALUED OBSERVABLES

A. BARACCA - Istituto di Fisica dell'Università - Firenze
S. BERGIA - Istituto di Fisica dell'Università - Bologna
M. RESTIGNOLI - Istituto di Fisica dell'Università - Roma

It has been pointed out that, as far as dicotomic observables are concerned, an observable difference must exist between certain types of local hidden variable (H. V.) theories and quantum mechanics (Q. M.)(1) and experimental tests (2) suggest really a possible failure of Q. M. in this context. But the previous works dealed only with dicotomic (i. e. double-valued) observables. It seems conceivable, instead, that other observables might exibit a differennt behaviour.

In this connection we propose a simple (perhaps not unique) generalization of Bell's inequality to discrete multi-valued observables. Let A, B be two observables belonging respectively to the two subsystems and let their eigenvalues (in principle more than two) be such that

$$|A(\vec{a},\lambda)| \leq M_A \quad , \quad |B(\vec{b},\lambda)| \leq M_B$$

where $\vec{a}$, $\vec{b}$ refer to the settings of the experimental apparatuses and λ is the local h.v. (or set of h.v.); let moreover be

$$M^2 = M_A \cdot M_B$$

If we define the following quantity

$$P(\vec{a},\vec{b}) = \int \rho(\lambda)\, A(\vec{a},\lambda)\, B(\vec{b},\lambda)\, d\lambda$$

where $\rho(\lambda)$ is the probability density of the h.v., the following relation holds

$$\left|P(\vec{a},\vec{b}) - P(\vec{a},\vec{b}')\right| \pm \left\{P(\vec{a}',\vec{b}) + P(\vec{a}',\vec{b}')\right\} \leq 2M^2 \qquad \text{(a)}$$

which is a straightforward generalization of Bell's inequality(1).

We then have compared these results with Q.M., getting the following results:

Theorem 1 - The inequivalence (a) is always satisfied by the quantum mechanical mean values $\langle \vec{j}_A \cdot \vec{a}\ \vec{j}_B \cdot \vec{b} \rangle$ for "proper" mixtures(3).

In the case of "improper" mixtures(3), that is when the system $A \oplus B$ is in the state

$$|\Psi_{A\oplus B}\rangle = \sum_k \sqrt{w_k}\, |\varphi_{\mu_k}\rangle |\xi_{\nu_k}\rangle$$

we find the followin results:

Theorem 2 - In the (purely mathematical) case of a spin-0 particle decaying into two spin-j particles the quantum mechanical mean values of angular momentum correlations may violate the ineqivalence (a) only when $j = \frac{1}{2}$.

More physical cases are under investigation.

Theorem 3 - If in general we define the dicotomic observables

$$\hat{D}^{(i)}_{A,B} = 2\,\hat{P}^{(i)}_{A,B} - 1$$

with

$$\hat{P}^{(1)}_A = |\varphi_1\rangle\langle\varphi_1|$$
$$P^{(1)}_B = |\xi_1\rangle\langle\xi_1|$$
$$P^{(2)}_A = \left[\alpha_1|\varphi_1\rangle + \alpha_2|\varphi_2\rangle\right]\left[\alpha_1^*\langle\varphi_1| + \alpha_2^*\langle\varphi_2|\right]$$
$$P^{(2)}_B = \left[\beta_1|\xi_1\rangle + \beta_2|\xi_2\rangle\right]\left[\beta_1^*\langle\xi_1| + \beta_2^*\langle\xi_2|\right]$$

the Bell's type inequality

$$\left|\langle\hat{D}^{(1)}_A\otimes\hat{D}^{(1)}_B\rangle - \langle\hat{D}^{(1)}_A\otimes\hat{D}^{(2)}_B\rangle\right| + \left|\langle\hat{D}^{(2)}_A\otimes\hat{D}^{(1)}_B\rangle + \langle\hat{D}^{(2)}_A\otimes D^{(2)}_B\rangle\right| \leq 2$$

may always be violated for suitable choices of the observables $\hat{D}$.

We stress that the latter result is completely general.

REFERENCES

(1) - J. S. BELL, Physics 1, 195 (1966). See also L. DE LA PENA et al. Lettere al Nuovo Cimento 5, 177 (1972).

(2) - G. FARACI, D. GUTKOWSKI, S. NOTARRIGO, A. R. PENNISI "Angular correlations of scattered annihilation photons to test the possibility of h.v. in Q. M.", presented to the Third Intern. Conf. on Positron Annihilation, Helsinki, August 7-9/ 1973.

See however the previous negative result of S. J. FREEDMAN, J. F. CLAUSER Phys Rev. Letters 28, 938 (1972); the above mentioned authors propose a possible explanation of this disagreement.

(3) - Baracca, Bergia, Rivista del Nuovo Cimento, 4, 169 (1974).

THREE BODY BREAK-UP REACTIONS AND THE PADÉ APPROXIMANT

A.W. Thomas, Department of Physics,
University of British Columbia, Vancouver, B.C., Canada V6T 1W5.

The method of contour rotation for solving the three body scattering problem (in the Amado model[1]), has been known for many years. [2]For purely elastic scattering (e.g. n + d → n + d), this is undoubtedly the most straightforward approach. Unfortunately the break-up amplitude (e.g. n + d → n + n + p) is rather more difficult to obtain by this method. [3]Indeed these complications have led to recent attempts to solve the three body equations by either variational means, [4]or by direct integration along the real axis.[5]

We would like to suggest an alternative (and very simple) method for obtaining break-up amplitudes. This method involves making use of the information that is usually discarded after a contour rotation calculation - namely the values of the half-off-shell t-matrix ($\langle k'|T_{dd}|k\rangle$) along the rotated contour. (We use the notation and normalisation of Brady and Sloan.[6] As usual, k is the on-shell momentum for n - d elastic scattering, and k' is allowed to vary; at a laboratory energy of 14.4 MeV, k = .556 fm^{-1} and the break-up region is $0 \leq k' \leq (4E/3)^{\frac{1}{2}} =$ .487 fm^{-1}.)

For our purposes the analytic structure of this t-matrix is crucial. It is well known that it is an analytic function of k' in a region around the positive real axis (including the rotated contour), except for a square root branch point starting at $k' = (4E/3)^{\frac{1}{2}}$. [7]Even in the vicinity of this branch point T_{dd} is an analytic function - not of k' but of the new variable $q' = (E - 3k'^2/4)^{\frac{1}{2}}$.[8] (This new function will be written t(q')). We now know the values of an analytic function in one region, and would like to know it in a neighbouring region. This is clearly a situation where one can apply the techniques of analytic continuation developed by

TABLE I

Values of $\langle k'|T_{dd}|k\rangle$ (at 14.4 MeV) calculated from the method of analytic continuation described in the text. (The exact result is taken from Ref. 3)).

k' (fm^{-1})	Padé Approximant	"Exact"
0.05	-.910 + .371i	-.910 + .371i
0.15	-.909 + .546i	-.909 + .546i
0.25	-.880 + .878i	-.880 + .878i
0.35	-.697 + 1.260i	-.697 + 1.260i
0.40	-.463 + 1.390i	-.463 + 1.388i
0.4261	-.281 + 1.410i	-.280 + 1.411i
0.4444	-.118 + 1.386i	-.117 + 1.387i
0.4626	.087 + 1.304i	.089 + 1.304i
0.4748	.263 + 1.177i	.263 + 1.176i
0.4870	.546 + .647i	.546 + .675i

Schlessinger. [9] That is, one can construct a Padé approximant $t_N(q')$, which agrees with $t(q')$ at N points $\{q_i'\}$. [10] In practice the points $\{q_i'\}$ will correspond to the set of quadrature points $\{k_i'\}$ along the rotated contour, together with some points to the right of $(4E/3)^{\frac{1}{2}}$ - where $\langle k_i'|T_{dd}|k\rangle$ is evaluated just like the elastic amplitude.

In Table I we show the results of such a calculation for $\langle k'|T_{dd}|k\rangle$ at 14.4 MeV. As input we used values of this function at 14 (quadrature) points (with $|k'| < 1.0\ fm^{-1}$) along a contour rotated 0.4 radians into the fourth quadrant. (This data was supplied by R. Cahill from the calculations of Ref. 3)). In addition we used the elastic amplitude ($k' = .556\ fm^{-1}$), and two other values corresponding to $k' = 0.5\ fm^{-1}$ and $0.6\ fm^{-1}$ - that is, from the right of $(4E/3)^{\frac{1}{2}}$. (The latter were supplied by I.H. Sloan from the calculations of Ref. 4)). For comparison in Table I, we include the "exact" result taken from Ref. 3).

The result of our analytic continuation is clearly an excellent representation of the half-shell t-matrix through-out the break-up region. Thus, it would seem that the method of Padé approximants represents a very promising solution to this problem of three-body final states. While the example chosen above was n - d scattering, it should be a simple matter to extend the method to other systems, such as K - d and Λ - d scattering.

The author is indebted to Dr. R. Cahill, and Dr. I.H. Sloan for helpful discussions, and for providing tables of their results. It is also a pleasure to acknowledge the support of the Physics Department at Flinders University, where this work began.

References

1) R. Aaron, R.D. Amado, Y.Y. Yam, Phys. Rev. 140 (1965) B1291

2) J.H. Hetherington, L.H. Schick, Phys. Rev. 137 (1965) B935

3) R.T. Cahill, I.H. Sloan, Nucl. Phys. A165 (1971) 161

4) T.J. Brady, I.H. Sloan, Phys. Rev. C9 (1974) 4

5) F. Sohre, H. Ziegelmann, Phys. Lett. 34B (1971) 579
N.M. Larsen, J.H. Hetherington, preprint 1973

6) T.J. Brady, I.H. Sloan, P.364 "Proceedings of the International Conference on Few Particle Problems in the Nuclear Interaction" ed. I. Slaus et. al. (North - Holland 1973)

7) D.D. Brayshaw, Phys. Rev. 176 (1968) 1855

8) R.T. Cahill, Phys. Rev. C9 (1974) 473

9) R.W. Haymaker, L. Schlessinger, P.257 in "The Padé Approximant in Theoretical Physics" ed. G. Baker Jr., J. Gammel (Academic Press 1970)

10) A fast and effective method of calculation involves using continued fractions - see Pp. 263, 264 of reference 9). The formulae presented there are easily programmed, but the author will provide a Fortran program upon request.

GLAUBER REPRESENTATION FOR THE SCATTERING BY NONLOCAL POTENTIALS

by

B. S. BHAKAR

Department of Physics
University of Manitoba
Winnipeg, Canada

Considerable effort has been made in the study of eikonal approximation for the small angle scattering and is now known as Glauber [1] representation. In these studies, the local potential has played a crucial role in obtaining the main results. As we know, the nucleon-nucleon interaction is not purely local. Further, a standard phenomenological method of incorporating the effects of inelasticity, exchange and energy-dependence (e.g. near a resonance) is by means of a nonlocal potential. Thus, we see a strong motivation is present for the generalization of Glauber representation for the scattering by a nonlocal potential. Attempts in this direction have been made by Malenka[2] and Abarbanel[2] who generalized the representation for spin-orbit and $\vec{L}^2$ type potentials. But, these potentials depend on velocity to a finite power. A general nonlocal potential may be expressed [3]

$$\int d^3r' \, v(\vec{r},\vec{r}\,') \psi(\vec{r}\,') = V_0(\vec{r})\psi(\vec{r}) + V_1(\vec{r})(\vec{r}\cdot\vec{p})\psi(\vec{r}) + \cdots \tag{1}$$

and is equivalent to introducing terms of all powers in velocity. One cannot approximate the left hand side of eq.(1) by terminating it at second or third term as is usually done in effective mass approximation. The effective mass approximation is valid only at low energy. However, the Lippman-Schwinger equation in the momentum space has the same form for local as well as nonlocal potential. We therefore start from the Lippman-Schwinger equation and use Levi-Sucher[4] as discussed by Kamal Chavda.[5]

The Lippman-Schwinger eq. in the momentum space for the scattering of two identical particles (spinless and $\hbar = m = 1$) is

$$t(\vec{p},\vec{k}) = v(\vec{p},\vec{k}) + \int d^3q \, [k^2 - q^2 + i\epsilon]^{-1} v(\vec{p},\vec{q}) t(\vec{q},\vec{k}) \tag{2}$$

Introducing the Fourier Bessel transform (valid at all angles and energies) as

$$\begin{aligned} t(\vec{p},\vec{k}) &= \int_0^\infty b\,db \, J_0(2pb\sin\theta/2) \, H(p,b,k) \\ v(\vec{p},\vec{k}) &= \int_0^\infty b\,db \, J_0(2pb\sin\theta/2) \, B(p,b,k) \end{aligned} \tag{3}$$

One obtains in the high energy limit[5]

$$H(p,b,k) = B(p,b,k) + 2\pi \int_0^\infty dq \, [k^2 - q^2 + i\epsilon]^{-1} B(p,b,q) H(q, pb/q, k) \tag{4}$$

Solving the equation(4) following the procedure outlined by Levi-Sucher[5] one obtains

$$H(p,b,k) = B(p,b,k)\,(e^{i\chi} - 1)/i\chi \tag{5}$$

where

$$\chi(p,b,k) = B(p,b,k)/2p + B(k, pb/k, k)/2k$$

Finally one obtains on the energy shell (p = k)

$$t(k,\cos\theta) = -ik \int_0^\infty b\,db \, J_0(2kb\sin\theta/2)\,(e^{i\chi} - 1) \tag{6}$$

Note that equation (4) is valid at all angles at high energy. However to obtain expression (6) from (4), one has to make eikonal approximation.

Now if we take nonlocal separable potential

$$\mathcal{V}(\vec{p},\vec{p}\,') = \sum_{\ell} \lambda_\ell\, g_\ell(p) g_\ell(p')\, P_\ell(\hat{p}\cdot\hat{p}') \qquad (7)$$

we get

$$B(p,b,p') = 4p^2 \sum_{\ell} \lambda_\ell\, g_\ell(p) g_\ell(p')\, \frac{J_{2\ell+1}(2pb)}{2pb} \qquad (8)$$

The expression (8) is valid at all energies and angles, because in its derivation we have used eq. (5) and the property

$$P_\ell(\cos\theta) = \int_0^\infty d\alpha\, J_0(\alpha \sin\theta/2)\, J_{2\ell+1}(\alpha) \qquad (9)$$

However, in the case of nonlocal separable potentials, the equation (2) has an exact solution viz

$$t(\vec{p},\vec{k}) = \sum_{\ell} t_\ell(p,k)\, P_\ell(\hat{p}\cdot\hat{k}) \qquad (10)$$

where

$$t_\ell(p,k) = \lambda_\ell\, g_\ell(p) g_\ell(k) \left[1 + \lambda_\ell \int d^3q\, \frac{g_\ell^2(q)}{k^2-q^2+i\epsilon}\right]^{-1}$$

Following expressions (3), (9), and (10), one can express

$$H(p,b,k) = 4p^2 \sum_{\ell} \lambda_\ell\, g_\ell(p) g_\ell(k) \left[1+\lambda_\ell \int d^3q\, \frac{g_\ell^2(q)}{k^2-q^2+i\epsilon}\right]^{-1} \frac{J_{2\ell+1}(2pb)}{2pb} \qquad (11),$$

where the expression (11) is valid at all energies and angles and differs from expression (5). However, one can show that in the limit of high energy and small angles the expression (11) reduces to the form (5).

Finally, a similar extension was done[6] to obtain forward as well as backward amplitudes in the presence of exchange interaction.

REFERENCES

(1) R.J. Glauber, Lectures in theoretical Physics, Vol. 1, edited by W.F. Brittin and L.G. Dunham (Interscience Pub. Inc. N.Y. 1959), p. 315.

(2) B.J. Malenka, Phys. Rev. 95, 522 (1954); and H. Abarbanel, Lectures in Theoretical Physics, Vol. 14A, edited by A.O. Barut & W.E. Brittin (Colorado Ass. Univ. Press 1972).

(3) N.F. Mott & H.S.W. Massey, The Theory of Atomic Collisions (Clarendon Press, 1965), p. 181.

(4) M. Levy and J. Sucher, Phys. Rev. 186, 1656 (1969).

(5) A.N. Kamal and L.K. Chavda, Phys. Rev. D5, 994 (1972).

(6) J.M. Allerton & B.S. Bhakar, Can. J. Phys., Vol. 52, 830 (1974).

HOW TO ESTIMATE BREAKUP AMPLITUDES IN NUCLEAR REACTION CALCULATIONS

J. Schwager
Institut für Theoretische Physik der Universität Tübingen,
D-74 Tübingen, Germany

A microscopic treatment of nuclear reactions is possible within the framework of the nuclear shell or cluster models by variational methods in coordinate space. The resonating-group method which is based on the cluster model has been very successful in explaining scattering data for light nuclei [see, e.g. ref. [1]]. There are fundamental difficulties, however, in including three-body channels in the trial wave function. Fortunately, when they are only weakly coupled, their neglect does not significantly influence the quality of the results for the elastic and rearrangement amplitudes, except for the occurence of false resonances, which can be suppressed by the use of a least-squares principle [2]. This has been checked [3] in a three-particle problem where the results can be compared with solutions of the Faddeev equations.

Thus, the wave function obtained from a resonating-group calculation can be a rather good approximation to the true Schrödinger solution even if asymptotically the breakup components are missing. On the other hand, the breakup amplitude can be written as a functional of the wave function in such a way that the main contributions come from the interior region of coordinate space. To get an estimate for it, therefore, it appears reasonable to insert the approximate wave function in such an integral expression, the most convenient forms of which can be deduced from the generalized Kato identity [4].

As an example, let us consider a three-particle system with only one bound state in each subsystem. For a fixed total angular momentum we have three two-fragment channels (i=1,2,3) and the breakup channel (denoted by 0). Let the exact wave functions evolving from initial (or final) channel states Φ_i (or Φ_o) and satisfying outgoing (or incoming) wave boundary conditions be denoted by $\Psi_i^{(+)}$ or $\Psi_o^{(-)}$, respectively. Let $\overline{\Psi}_i^{(+)}$ be some approximation showing essentially the same asymptotic behaviour as $\Psi_i^{(+)}$, except that approximate amplitudes $\overline{T}_{oi}$ are read off from it. (In a resonating-group calculation they will be zero.) Then, the exact breakup amplitude can be expressed by the identity

$$T_{oi} = \overline{T}_{oi} + \langle \Psi_o^{(-)} | H - E | \overline{\Psi}_i^{(+)} \rangle .$$

Substituting for the function $\Psi_o^{(-)}$ the first term (or terms) of its Born series, we get an approximation $\hat{T}_{oi}$ for the breakup amplitude which can easily be evaluated numerically, since $(H-E)|\bar{\Psi}_i^{(+)}\rangle$ is different from zero only in a finite domain in coordinate space. The deviation $\Delta T_{oi} = T_{oi} - \hat{T}_{oi}$ will be of the order $\Delta\Psi_o^{(-)}$ times $\Delta\Psi_i^{(+)}$.

With the decomposition of the Hamiltonian, $H = H_o + V = H_o + \sum_{i=1}^{3} V_i$, it is easy to show that substitution of the undisturbed plane wave Φ_o for $\Psi_o^{(-)}$ yields an approximation $\hat{T}_{oi}^{(1)}$ which is identical to

$$\hat{T}_{oi}^{(1)} = \langle \Phi_o | V | \bar{\Psi}_i^{(+)} \rangle - \langle \Phi_o | V_i | \Phi_i \rangle .$$

This is the familiar expression for the breakup amplitude with $\bar{\Psi}_i^{(+)}$ inserted for $\Psi_i^{(+)}$. In this form, however, the integrand would contain parts of infinite range which otherwise are absorbed in the known quantity $\bar{T}_{oi}$.

Iterating the Faddeev equations for $\Psi_o^{(-)}$ and retaining only the single scattering terms we get

$$\hat{T}_{oi}^{(2)} = \hat{T}_{oi}^{(1)} + \sum_{k=1}^{3} \langle \tilde{\Phi}_{ok}^{(-)} | H-E | \bar{\Psi}_i^{(+)} \rangle .$$

Here, $\tilde{\Phi}_{ok}^{(-)} = \Phi_{ok}^{(-)} - \Phi_o$ is the scattered part of the wave function $\Phi_{ok}^{(-)}$ which would emerge from Φ_o if the potential V_k alone acted. Again, this can be rewritten in the alternative form

$$\hat{T}_{oi}^{(2)} = \sum_{k=1}^{3} \langle \tilde{\Phi}_{ok}^{(-)} | V - V_k | \bar{\Psi}_i^{(+)} \rangle ,$$

which is nothing other than the well known relation [compare, e.g. eq. (3.64) of ref. [5]] connecting the breakup amplitude with the half off-shell rearrangement amplitudes which are taken here from the variational calculation. There are good reasons for the assumption that the projection of $(H-E)|\bar{\Psi}_i^{(+)}\rangle$ on the omitted rest of $\Psi_o^{(-)}$ should be small (at least when $\bar{\Psi}_i^{(+)}$ is calculated by the resonating-group method). So we can stop improving $\Psi_o^{(-)}$ further at this stage.

Numerical calculations on a model three-body system are under way. They pose no other problems than those already present in the non-breakup case.

REFERENCES

1) F.S. Chwieroth, Y.C. Tang and D.R. Thompson, Phys. Rev. C9 (1974) 56; H.H. Hackenbroich, P. Heiss and Le-Chi-Niem, Nucl.Phys.A221 (1974) 461
2) E.W. Schmid, Nuovo Cim. 18A (1973) 771; J. Schwager, Nuovo Cim. 18A (1973) 787
3) J. Schwager, Nucl. Phys. A221 (1974) 371; F. Sohre and E.W. Schmid, Nuovo Cim., in press
4) M. Lieber, L. Rosenberg and L. Spruch, Phys. Rev. D5 (1972) 1347
5) E.W. Schmid and H. Ziegelmann, The quantum mechanical three-body problem (Vieweg, Braunschweig, 1974)

ON THE STRUCTURE OF LIGHT NUCLEI

K. B l e u l e r , University of Bonn, W-Germany

In view of the large number of detailed experiments on light nuclei it appears worthwhile to look into their quantum mechanical structure from a somewhat generalized view-point. As seen e.g. from some typical spectra in the realm of lighter nuclei - they are completely different from one another - it is clear that we have to deal with characteristic correlations between the nucleons which differ appreciably from nucleus to nucleus. In this respect one might even say that the experimental knowledge of the various level-schemes with the corresponding transition-probabilities gives a deep insight into the microscopical structure of these nuclei.

We will treat here a special class of lighter nuclei, namely those which may be visualized as being composed by a relatively stable core and one or several so-called valence-nucleons. Their spectra exhibit characteristic groups of spectral terms which correspond to different core-excitations. Typical examples are found for A = 5 and A = 6 where the excitation of the He^4 core corresponds to typical high-lying groups of levels whose characteristic displacement with respect to the excitation of the free core is easily understood (see below).

The case of a single valence nucleon has already been treated in a contribution to the conference on 'Correlations in Nuclear Physics'[1]. The main point was the generalized Fourier expansion of the complete (antisymmetrized) way function Ψ with the help of the orthonormal system corresponding to the states of the core φ_n:

$$(1) \qquad \Psi(x_1 \cdots x_N \; x) = \sum_n c_n(x)\, \varphi_n(x_1 \cdots x_N) + Excl.$$

($x_1 \cdots x_N$ stands for the combined particle variables for the core, x for the valence particle; the complete antisymmetrization of Ψ is reached by adding the exchange terms Exch.). The generalized Fourier coefficients c_n , which still depend on the variable x represent, so to speak, the single particle wavefunctions of the valence-particle. A folding procedure with the help of the core wave functions φ_n yields immediately (and still rigorously) the wave equation for the c_n which contains automatically charateristic matrix operators V^{op}_{nm} describing an 'average potential' through the diagonal elements and the core excitation through the non-diagonal ones. (T^{op} represents the kinetic energy, e_n the energy-eigenvalue of the core, compare[1]):

(2) $$T^{op} C_n + \sum_m \bar{V}^{op}_{nm} C_m = (E - e_n) C_n + \text{Excl.}$$

Already the 'unperturbed' solution of this equation in which only the diag. elements $\bar{V}_{nn}$ occur yields a better understanding of Nuclear and Atomic Alcalide spectra (C^{13}, O^{17}, Sodium Atom etc.) the Pauli principle being rigorously enshrined in the exchange-part of the potential $\bar{V}$; it represents so to speak the motion of the valence-particle in the field of the core. In this system it appears natural to treat the core excitation as a perturbation ($\bar{V}_{nm} \ll e_m - e_n$) of this unperturbed solution.

We discuss now the generalization to a larger number of valence-particles (in practice two or three). Very much the same developments as before can be made, but the new Fourier coefficients depend now on the variables of several valence-particles. After the folding one obtains a wave equation for the system of valence particles alone which has a most natural physical interpretation: These particles move - in a first approximation - in the field of the core, but they are, in addition, coupled by a direct interaction. The question arises whether a possible binding through this interaction, or whether the effects of core excitation must be considered first in the next step of the approximation. From inspection of some characteristic level schemes and from some very rough estimates it seems that the direct interaction should, in certain cases, be considered first, forming thus a 'bound state' which, in turn, moves in the field of the core. The core excitation then appears in a further step and becomes larger for a larger number of valence-particles. The lower levels of the isotopes A = 6, 14 and 18 are naturally interpreted through the formation of a bound Deuteron state of the 2 valence nucleons (apart from the typical 0^+0^- pair in F^{18}, which in turn may be interpreted through the direct Tensor-force-interaction between the valence particles). Finally the most characteristic low-lying levels of F^{19} and Ne^{19} (narrow parity doublets) could be visualized as single particle levels of the last valence-nucleon in the F^{18}-core which induces (through a strong mixing of the states of the $0^+ 0^-$ doublet) a so-called 'parity mixing' of the total structure[3].

---

1) Hungary 1973,

2) Compare e.g. Condon and Shortley, p. 142

3) K. Bleuler, Parity Mixing in Spherical Nuclei, Report of the Varenna School XXXVI Corso.

EXTENSION OF BRUECKNER'S THEORY FOR THE BOSON-NUCLEON-SYSTEM

K. B l e u l e r, University of Bonn, W-Germany

Boson exchange yields a physical understanding as well as a detailed numerical description of nuclear forces. In order to determine the force between 2 free nucleons one generally uses 6 different types of bosons, i.e. the $\pi, \eta, \sigma, \delta, \omega, \rho$-particles. An example of this kind of computation was worked out, among many others, in our institute by Erkelenz et al.[1]: In adjusting the values of the corresponding masses and coupling constants (within the limits determined from high energy data) one obtained a rather satisfactory agreement with the experimental scattering-phase-shifts of the 2-nucleon system in the relevant energy interval o-3oo Mev. It should, however, be remarked that the "unphysical" σ-meson used so far can be replaced by the introduction of the Δ-resonance of the nucleons in connection with the so-called block-diagram (compare ref.[2]). In addition, the relatively large cut-off used in our calculations in the case of the vector boson couplings can be partly replaced by the introduction of the physically well-known f-boson. It thus seems that the nuclear 2-body force may be understood in a relatively satisfactory way from facts known in high energy physics.

I would like to emphasize here that recent research in experimental nuclear physics - especially in connection with work in mesic atoms - yields a great number of additional checks of the bosonic origin of nuclear forces: (i) Accurate measurements in the μ-mesic atoms led to the determination of the compressibility of nuclear matter which, in turn, depends on the detailed properties of the so-called repulsive core, determined in our case from the exchange of the vector bosons ω and ρ. (This exchange is, in the same time, at the origin of the well-known spin-orbit coupling in finite nuclei). (ii) A recent analysis of the characteristic displacements of the energy levels of the anti-protonic atom points to the existence of a strongly attractive force between nucleon and anti-nucleon[3]. This is exactly what should be expected from the exchange of the vector bosons, because the hitherto considered repulsive core is now turned over into an attractive one. (iii) The possibility of the excitation of the Δ-resonance within a finite nuclear structure was recently revealed by a beautiful experiment by Argan et al.[4]. The detailed measurements of the $\gamma\pi$-process on He^4 showed in fact the existence of this excitation; there was, however, some characteristic splitting and a relatively large displacement which in turn have been theoretically interpreted by Huber et al.[5] (exchange of an "exiton"). (iv) The host of mesic corrections to all kinds of nuclear properties (moments, matrix elements,etc.) which are mainly due to the interaction of the π--meson with the nucleons within nuclear matter.

Although the situation appears thus to be quite satisfactory, the question arises, whether all these effects were determined in a consistent way: The bosons which originally gave rise to the nuclear 2-body forces were in fact eliminated through the subsequent introduction of these forces into nuclear structure. They have then to be reintroduced in order to explain the various bosonic effects risking thus certain double countings. In addition, the virtual excitation (through the block-diagram) of the Δ-resonance-state should again be explicitely taken into account as soon as nuclear matter is calculated because it introduces an essential change with respect to the Pauli principle (compare ref.[6]). In view of these facts, we propose to treat the various bosons together with the bare nucleons (containing some additional inner degree of freedom for the Δ -resonance) on equal standing. From such a view-point it becomes immediately clear, that it was not correct to insert - as usual - the 2--body interaction (as determined from the 2-nucleon system) into the many-body system without further corrections: The reaction properties of the nucleons as well as the mesonic self-energies have now been changed with respect to the case of the free particles. A systematic outline of this view-point has been given - as far as the mesonic part is concerned - in a recent paper by D. Schütte[7]: He shows that the Brueckner approach for nuclear matter can in fact be enlarged to the more general system of bosons and nucleons; in this framework it becomes quite evident that the nuclear forces as defined from boson exchange in the conventional way are changed in a characteristic way as soon as the nucleons are embedded into nuclear matter. In addition, the effect of the so-called meson condensation - as conjectured in the center of neutron stars - will appear as a direct consequence of this generalized treatment.

---

1) Nucl. Phys. A.198 (1972) p.598-6o8. 2) Green and Haapakoski, Reports from Research Institute Helsinki 18-73 and 28-73. 3) H. Koch, Report, Uppsala Conference, June 73. 4) Phys. Rev. Lett. 29 (1972) p.1191. 5) To be published in Conference Report of the Bonn Conference in remembrance of K. Erkelenz (Febr. 1974). 6) Green, Dahlblom and Kouki, Report of Research Institute Helsinki 7-73. 7) Nucl. Phys. A 2o9 (1974) p.17o--18o.

REACTIONS $^{3}H(^{3}He,d)^{4}He$ and $^{3}H(^{3}He,np)^{4}He$ AT LOW ENERGIES, BARSHAY-TEMMER THEOREM, AND ISOSPINSYMMETRY+

G. Schrieder, H. Genz and A. Richter
Institut für Experimentalphysik der Ruhr-Universität, Bochum, Germany

W. von Witsch
Institut für Strahlen- und Kernphysik der Universität, Bonn, Germany

H. Paffe and B. Ruth
Fachbereich Physik der Universität, Marburg, Germany

A strong violation of isospin has been found in the reaction $^{3}He+^{3}H \rightleftarrows ^{4}He+^{2}H$ in particular at low c.m. energiès in the $(^{3}He+^{3}H)$-system. The violation manifests itself in an asymmetric angular distribution with respect to 90° in the c.m. system. The Barshay-Temmer isospin theorem predicts for reactions involving isospin multiplet members in the case of unique and conserved isospin symmetric angular distributions[1)]. In order to investigate the origin of this isospin violation further, (i) measurements of angular distributions in the reaction $^{3}H(^{3}He,d)^{4}He$ have been extended to higher energies and (ii) the energy dependence of the n-p final state interaction in the reaction $^{3}H(^{3}He,np)^{4}He$ has been studied.

Angular distributions for the reaction $^{3}H(^{3}He,d)^{4}He$ have been measured in the energy range of $(4.25<E(^{3}He)<9.85)$ MeV in steps of 0.5 MeV at the 5 MV van de Graaff accelerator of the University of Marburg. Targets of AlTiT consisting of a layer of 200 $\mu g/cm^2$ Ti ($\sim$6 $\mu g/cm^2$ T) on a 200 $\mu g/cm^2$ Al backing were bombarded with $\sim$150 nA $^{3}He^{+}$ and $\sim$200 nA $^{3}He^{++}$ ions. All angular distributions are asymmetric about 90° in the c.m. system. The maximum values W_{max} of the asymmetry $W(\Theta) = [\sigma(\Theta)-\sigma(\pi-\Theta)]/[\sigma(\Theta)+\sigma(\pi-\Theta)]$ at each energy are shown in fig.1 together with the relevant data of other existing experiments[2-7)] as a function of the c.m. energy in the $(^{3}He+^{3}H)$-system. The sign convention of ref. 8 has been chosen.

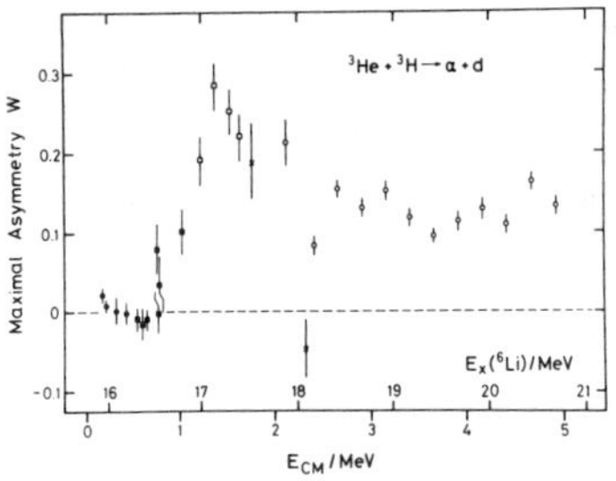

Fig.1 The largest asymmetries observed at a particular angle in the various existing experiments[2-7)] and in the present work (open circles) as a function of the c.m. energy of the $(^{3}He+^{3}H)$-system as well as of the excitation energy in ^{6}Li.

Of course, the behaviour of the asymmetry in fig.1, determined from complete angular distributions between 0° and 180° c.m., should only be used for qualitative arguments; however, three points are noted in connection with it: (i) W_{max} is strongly energy dependent with a maximum value at $E_{c.m.}$ ~1.4 MeV; (ii) a Legendre - polynominal expansion of all measured angular distributions yields a resonating behaviour of the L = 2 coefficient at almost this energy, and (iii) the level density in the compound nucleus ^{6}Li for excitation energies $\lesssim$21MeV is such that probably almost an equal number of broad isospin T=0

+Supported in part by Deutsche Forschungsgemeinschaft

and T=1 states are present and overlap. These arguments suggest that isospin mixing of compound nucleus levels might cause the sizable experimental asymmetries. This is also supported by simple model calculations[2].

The energy dependence of the T=1 n-p final state interaction in the reaction $^{3}H(^{3}He,np)^{4}He$ is presently being investigated with the ^{3}He-beam of the 400 kV Sames - and 4 MV Dynamitron tandem accelerators, respectively, at Bochum. The targets used were the same as described above. Protons and α-particles were detected in coincidence under angles for which the relative energy E_{pn} between n and p could be zero. Figure 2 shows preliminary results of spectra in which the coincidence yield is projected onto the energy axis of the proton detector. The experimental results (vertical bars) near $E_{pn}=0$ are fitted using the Watson-Migdal formalism[9] with a set of effective range- and scattering length parameters[10] for low energy proton-neutron scattering ($a_s=-23.75$ fm, $r_{os}=2.8$ fm, $a_t=5.426$ fm, $a_{ot}=1.763$ fm). The respective singlet (T=1) and triplet (T=0) final-state interaction contributions to the cross section are also given in the figure (dashed lines). It is assumed in the fit that possible background from other final state interactions (e.g. ^{5}He and ^{5}Li) or from simultaneous break up contributions is constant in the energy range investigated and has the same spectral shape as the T=0 n-p final state interaction. So far, no energy dependence of the T=1 final state interaction indicative of a T=1 compound nucleus state has been detected. It amounts to about 11% in all three spectra. More energies are presently investigated.

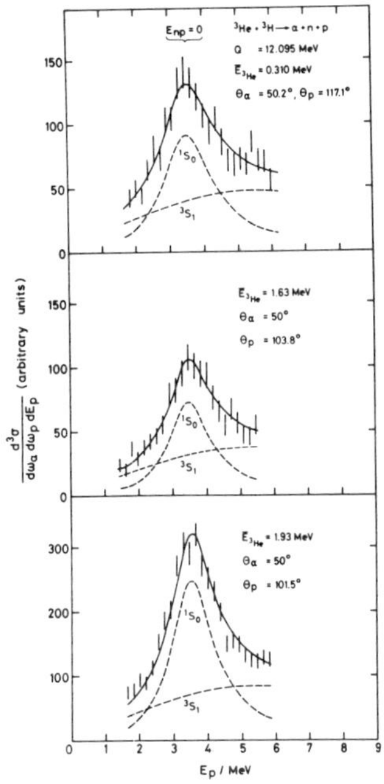

Fig.2

1) S. Barshay and G. M. Temmer, Phys. Rev. Lett. 12 (1964) 728

2) U. Nocken, U. Quast, A. Richter and G. Schrieder, Nucl. Phys. A213 (1973) 97

3) B. Kühn and B. Schlenk, Nucl. Phys. 48 (1963) 353

4) Kim Sin Nam, G. M. Osetinskii and V. A. Sergeev, Sov. J. Nucl. Phys. 10 (1970) 407

5) G. J. Wagner, C. C. Foster and B. Greenebaum, Nucl. Phys. A174 (1971) 123

6) E. E. Gross, E. Newman, M. B. Greenfield, R. W. Rutkowski, W. J. Roberts, and A. Zucker, Phys. Rev. C5 (1972) 602

7) G. J. Wagner, G. Maierle, P. Kleinagel and R. Bilwes, Int. Conf. on Few Part. Probl. in the Nucl. Interaction, ed. I. Slaus et al. (North Holland, Amsterdam,1972) p. 747

8) D. Robson and A. Richter, Ann. Phys. (N.Y.) 63 (1971) 261

9) K. M. Watson, Phys. Rev. 88 (1952) 1163; A. B. Migdal, JETP (Sov. Phys.) 1 (1955) 2

10) J. C. Davis and H. H. Barshall, Phys. Lett. 27B (1968) 636

COULOMB EFFECTS IN NUCLEON-DEUTERON SCATTERING

J.A. Edgington and C.J. Oram, Queen Mary College, London, England

Small differences, in both differential cross section and polarisation, between the n-d and p-d systems have been suspected for some time[1]. Measurements at UC, Davis at 36.0 and 46.3 MeV showed[2] the n-d cross section to have a deeper minimum than that found[3] at similar energies in p-d scattering. The same group later found[4] that the n-d polarisation at 35 MeV was more negative, in the region of this minimum, than p-d measurements had indicated[3]. Recently the Davis group have remeasured both the n-d and p-d cross sections, with improved normalisation, and find[5] similar results. In this paper we present an approximate treatment of Coulomb effects in an attempt to explain these differences.

The separation of Coulomb and nuclear effects in the scattering of charged particles is a problem which is in principle completely soluble. Conventionally the total amplitude is split into purely Coulomb, and "residual", terms. In an angular momentum decomposition, the phase shifts of the residual amplitude are not those of the nuclear force acting alone, i.e. the phase shifts are not additive. The magnitude of this effect depends upon the forms one takes for the Coulomb and the nuclear potentials. Thus, Bugg has treated[6] low energy πN scattering by the method of Auvil[7]; the calculations are feasible in this case because attention is concentrated on S and P waves only. In the N-d system up to 10 partial waves may contribute, rendering this approach rather laborious.

In an attempt to formulate the problem simply, we have been influenced by one particular aspect of the N-d system, namely, the rapid variation of cross section and polarisation with energy. One obvious difference between the two isospin states is that the relative interaction energy of proton and deuteron is reduced by the repulsive Coulomb field. Suppose one considers a nuclear potential with a sharp cut-off radius R (but otherwise arbitrary) and matches internal and external wave functions at R. Let the charge distribution be such that the Coulomb potential is constant for radii less than R. One can then show that, to a good approximation, the nuclear phase shifts in the presence of the Coulomb force are to be evaluated at an energy differing from the free particle energy by an amount equal to the Coulomb potential at R. This naive picture may not be an unreasonable representation of the N-d system, with its extended charge distribution.

We have used the partial wave analysis of Arvieux[8] and the formalism of van Oers and Brockman[9] to evaluate n-d and p-d cross sections below 50 MeV; the latter are found both conventionally (p-dI), and by using nuclear phase shifts appropriate to a lower energy, as above (p-dII). Figure 1 shows the results at 35.0 and 46.3 MeV, for two values of R, as well as an indication of the experimental ratios. The ratios are sensitive to the phase shifts chosen (those of van Oers and Brockman considerably enhance the maxima and minima) but in general p-dII yields greater cross sections than p-dI, by some 5% - 8% for reasonable values of R. At these energies, and others, this unsophisticated calculation seems at least partially to account for the observed cross section ratios.

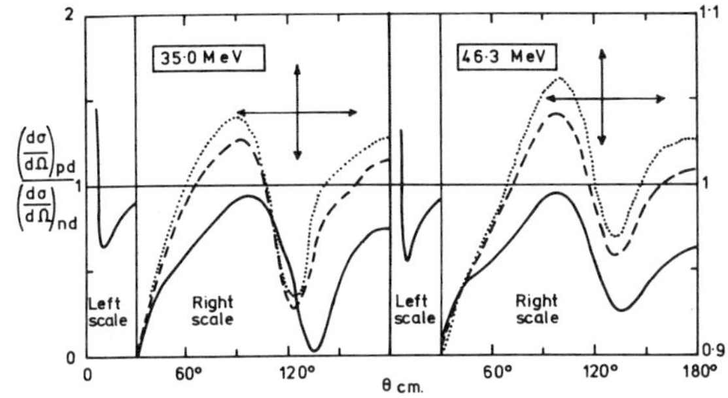

FIG. 1

Ratio of pd to nd cross sections, using phase shifts of Ref. 8.

The polarisation cannot be treated similarly, as present analyses do not allow of split phases. However, bearing in mind the extremely small spin dependence of the Coulomb interaction, it seems worthwhile to compare n-d and p-d polarisations, using a similar energy displacement. The p-d data are sufficiently plentiful for this to be done. Figure 2 shows the results at 35 MeV. The discrepancy is almost removed by an energy shift corresponding to R = 1.55 fm. At 23 MeV, where n-d and p-d polarisations also appear to differ[1], an effect of the right size to correct this is again found.

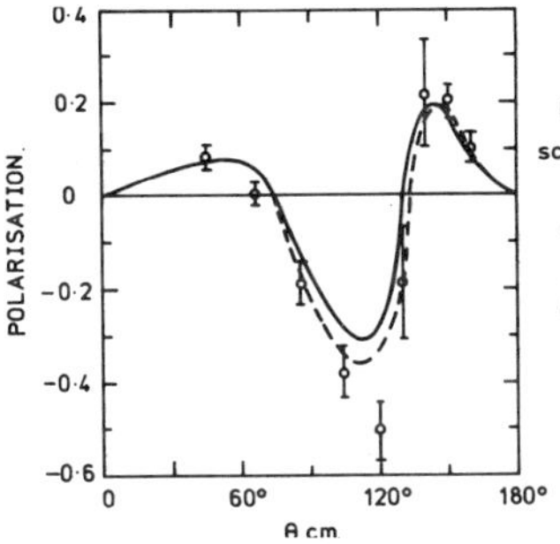

FIG. 2.

Polarisation in Nd scattering at 35 MeV

REFERENCES

1) W.Haeberli, Conf. on Three-Body Problems, Birmingham, 1970 (p.188)
2) J.L.Romero et al, Phys. Rev. C2,2134, (1970)
3) S.N.Bunker et al, Nucl. Phys. A113, 461 (1968)
4) J.Zamudio-Cristi et al, Phys. Rev. Lett. 31, 1009 (1973)
5) F.P.Brady et al, private communication
6) D.V.Bugg, Nucl. Phys. B58, 397 (1973)
7) P.P.Auvil, Phys. Rev. D4, 240 (1971)
8) J.Arvieux, Nucl. Phys. A221, 253 (1974)
9) W.T.H.van Oers and K.W.Brockman, Nucl. Phys. A92, 561 (1967)

THREE-BODY MODEL OF THE N-T SYSTEM

Shinsho Oryu
Department of Physics, Faculty of Science and Technology,
Science University of Tokyo, Noda-city, Chiba, 278 Japan

In this note, a deuteron in the n-T system was assumed to be a particle, for simplicity. Our problem becomes the three-body problem with two identical and one different particles. Two disconnected integral equations and one simple equation are obtained by some reduction of the Faddeev equations. One of the solutions corresponds to amplitude $M_{el}(n,n)=M(\text{diagonal})+M(\text{non-diagonal})$ and another to $M_x(n,n)=M(\text{non-diagonal})-M(\text{diagonal})$, where $|M_{el}|>>|M_x|$ may be assured in the cases of s-wave n-n and n-d interactions. Non-elastic amplitude M(n,d) is obtained automatically by simple calculation using the elastic amplitude M_{el}.[1)] We try a method to represent the elastic amplitude M_{el} without solving the Faddeev equations numerically. The shape of angular distribution for the n-T elastic scattering depends mainly upon some lower order terms of the Faddeev series ($Z^{(1)}, Z^{(2)}$, etc.) and the higher order terms may depend on the total energy and then they will contribute to the absolute value of cross section. The elastic scattering amplitude may be constructed by some selected and lower order terms which are multiplied by the complex factors, such as follows,

$$M_{el}(q',q;E)=a(E)Z^{(1)}(q',q;E)+b(E)Z^{(2)}(q',q;E)+c(E)Z^{(3)}(q',q;E)+ \dots .$$

Here, the complex parameters a(E), b(E), c(E), etc. are obtained by fitting the observed angular distributions for n-T elastic ($n+T\longrightarrow n+T$) and/or quasielastic ($n+T\longrightarrow n+T'$) scattering, and also they are related to the experimental total cross section by the optical theorem, and an ambiguity of the phase of M_{el} will be fitted by the other data (such as the polarization data). The analytical continuation of the derived on-shell amplitude M_{el} is made to off-the-energy-shell in the momentum q-plane with fixed E and q'. The first step of the calculation is the parameterization of the elastic amplitude in the following form,

$$M_{el}(\Theta_n)= aZ_{tt}(q',q;E) + bZ^{H}_{tt}(q',q;E) + cZ^{c}_{tt}, \qquad (1)$$

where, a, b and c are constants to be obtained by means of the χ^2-minimization to the experimental elastic data taking into account the constraint of the optical theorem. Z_{tt}, Z^{H}_{tt} and Z^{c}_{tt} correspond to the pick-up, the knock-on and the remained black-box amplitudes which are represented in Fig.1. Here, the remained amplitude Z^{c}_{tt} is tentatively assumed to be independent of angle $\hat{q}'\cdot\hat{q}=\cos\Theta_n$. The experimental data at the incident neutron energy E_n=14.1 MeV are refered. The discrepancy between the calculation and the experiment in Fig.2 would be due to our choice of only two angular dependent and one angular independent terms and/or be due to the n-d form factor of the Yamaguchi-type($\propto 1/(p^2+\gamma^2)$) which was derived from the three-body generalized separable potential approach.[2] The parameters a, b, c and γ ($\tilde{\gamma}=\gamma/\alpha_d$: $\alpha_d=0.2316\mathrm{fm}^{-1}$)may be owing to the structures of the higher order terms and to the four-body effects. It should be noted that our simple model taken in this work, however, reproduced the elastic data considerably well without solving very complicated four-body problem exactly. In the break-up differential cross sections: $d^2\sigma/dE_d d\Omega_d=(PSF)\int d\Omega_{12}|M|^2$ and $d^3\sigma/dE_1 d\Omega_1 d\Omega_2=(PSF)^*|M|^2$ for the T(n,d)2n reaction, the break-up amplitude M is expressed as follows, $M= Y^{(0)}\eta_3+ Y^{(1)}\eta_1+ Y^{(2)}\eta_2$ where each of the factors η_i is the spin function, labelling the two neutrons by 1 and 2 and the deuteron by 3, and with the phase space factors (PSF) and $(PSF)^*$, respectively. The amplitudes $Y^{(i)}$ are illustrated in Fig.1 by the Faddeev approach, using the elastic amplitude M_{el} in eq.(1).[1] The measured break-up deuteron spectrum is fairly well explained by our model. The result of the calculation is given elsewhere.[3]

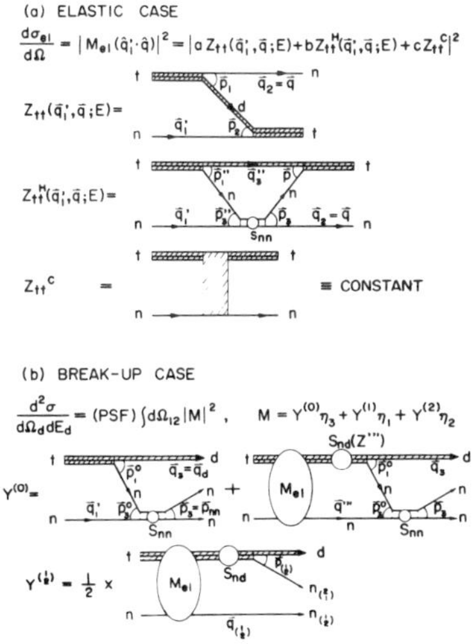

< Fig. 1 >

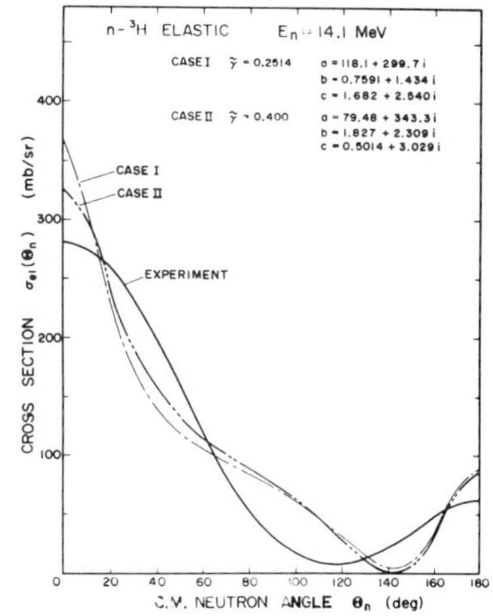

< Fig. 2 >

REFERENCES

1) S. Oryu, Prog. Theor. Phys. (Kyoto) 44 (1970) 1208; 45 (1971) 386.
2) E. O. Alt, P. Grassberger and W. Sandhas, Phys. Rev. C1 (1970) 85. S. Oryu, Prog. Theor. Phys. 52-2(1974); S. Oryu, T. Ishihara and S. Shioyama, ibid. 51-5 (1974).
3) S. Shirato, Y. Suda, S. Tsuruta and S. Oryu, contribution in this conference.

DOES THE FADDEEV THEORY REQUIRE OFF-SHELL INPUT?§

Bengt R. Karlsson+ and Enrique M. Zeiger
Stanford Linear Accelerator Center
Stanford University, Stanford, California 94305

If the three-body equations of Faddeev[1] are formulated in terms of transition amplitudes, one has the advantage of directly obtaining the three-body entities of physical interest. Such an approach has been pursued by Osborn and Kowalski[2], who advocate a set of equations for the transition amplitudes defined by Faddeev.

Here we present a new set of equations for a different pair of amplitudes which are also directly related to the transition amplitudes. While retaining the advantages of the previous approach[1,2], these equations are considerably simpler in structure.

The new pair of amplitudes to be used here are defined as follows:

$$\mathscr{H}_{\beta\alpha}(\vec{p}_\beta;\vec{p}_\alpha^{(0)};z) = \langle \vec{p}_\beta \phi_\kappa^\beta | V_\beta G_0(z)\, U_{\beta\alpha}(z)\, G_0(z)\, V_\alpha | \vec{p}_\alpha^{(0)} \phi_\kappa^\alpha \rangle$$

$$\mathscr{E}_{\beta\alpha}(\vec{p}_\beta \vec{q}_\beta;\vec{p}_\alpha^{(0)};z) = \langle \vec{p}_\beta \psi_{\vec{q}_\beta}^\beta | V_\beta G_0(z)\, U_{\beta\alpha}(z)\, G_0(z)\, V_\alpha | \vec{p}_\alpha^{(0)} \phi_\kappa^\alpha \rangle \quad . \qquad (1)$$

where $U_{\beta\alpha}$ is the AGS[3] transition operator, $\vec{q}_\beta$, $\vec{p}_\beta$ label the internal momentum of the β-pair and the relative momentum of the β-particle respectively, $|\phi_\kappa^\alpha\rangle$ is a two-body bound state wavefunction with energy $-\kappa_\alpha^2$, $|\psi_{\vec{q}_\beta}^\beta\rangle$ is an outgoing wave continuum two-body wavefunction with momentum $\vec{q}_\beta$, and $\vec{p}_\alpha^{(0)}$ is the relative momentum of the free particle in the initial state.

The first amplitude in (1) is the transition amplitude for elastic and rearrangement scattering, and the second is directly related to the breakup amplitude. The corresponding Faddeev equations are[4]:

$$\mathscr{H}_{\beta\alpha}(\vec{p}_\beta;\vec{p}_\alpha^{(0)};z) = \mathscr{H}_{\beta\alpha}^{(0)}(\vec{p}_\beta;\vec{p}_\alpha^{(0)};z) - \sum_{\gamma\neq\beta}\int d^3\vec{p}'_\gamma\, \mathscr{V}_{\beta\gamma}^{\mathscr{H}\mathscr{H}}(\vec{p}_\beta;\vec{p}'_\gamma)\,\frac{1}{\tilde{p}'^2_\gamma-\kappa_\gamma^2-z}\,\mathscr{H}_{\gamma\alpha}(\vec{p}'_\gamma;\vec{p}_\alpha^{(0)};z) - \sum_{\gamma\neq\beta}\iint d^3\vec{p}'_\gamma\, d^3\vec{q}'_\gamma\, \mathscr{V}_{\beta\gamma}^{\mathscr{H}\mathscr{E}}(\vec{p}_\beta;\vec{p}'_\gamma\vec{q}'_\gamma)\,\frac{1}{\tilde{p}'^2_\gamma+\tilde{q}'^2_\gamma-z}\,\mathscr{E}_{\gamma\alpha}(\vec{p}'_\gamma\vec{q}'_\gamma;\vec{p}_\alpha^{(0)};z)$$

$$\mathscr{E}_{\beta\alpha}(\vec{p}_\beta\vec{q}_\beta;\vec{p}_\alpha^{(0)};z) = \mathscr{E}_{\beta\alpha}^{(0)}(\vec{p}_\beta\vec{q}_\beta;\vec{p}_\alpha^{(0)};z) - \sum_{\gamma\neq\beta}\int d^3\vec{p}'_\gamma\, \mathscr{V}_{\beta\gamma}^{\mathscr{E}\mathscr{H}}(\vec{p}_\beta\vec{q}_\beta;\vec{p}'_\gamma)\,\frac{1}{\tilde{p}'^2_\gamma-\kappa_\gamma^2-z}\,\mathscr{H}_{\gamma\alpha}(\vec{p}'_\gamma;\vec{p}_\alpha^{(0)};z) - \sum_{\gamma\neq\beta}\iint d^3\vec{p}'_\gamma d^3\vec{q}'_\gamma\, \mathscr{V}_{\beta\gamma}^{\mathscr{E}\mathscr{E}}(\vec{p}_\beta\vec{q}_\beta;\vec{p}'_\gamma\vec{q}'_\gamma)\,\frac{1}{\tilde{p}'^2_\gamma+\tilde{q}'^2_\gamma-z}\,\mathscr{E}_{\gamma\alpha}(\vec{p}'_\gamma\vec{q}'_\gamma;\vec{p}_\alpha^{(0)};z), \qquad (2)$$

and the "effective potentials" $\mathscr{V}$ are given by

---

§ Work supported in part by the U.S. Atomic Energy Commission.

+Present address: Institute of Theoretical Physics, Fack, S-402 20 Göteborg 5, Sweden

$$\mathcal{V}^{\mathcal{H}\mathcal{H}}_{\beta\gamma}(\vec{p}_\beta;\vec{p}'_\gamma) = -\frac{\Phi^\beta_\kappa(\vec{q}^{(1)}_\beta)\,\Phi^\gamma_\kappa(\vec{q}^{(2)}_\gamma)}{(\tilde{q}^{(2)}_\gamma)^2+\kappa^2_\gamma} \qquad \mathcal{V}^{\mathcal{H}\mathcal{E}}_{\beta\gamma}(\vec{p}_\beta;\vec{p}'_\gamma\,\vec{q}'_\gamma) = \Phi^\beta_\kappa(\vec{q}^{(1)}_\beta)\;\psi^\gamma_{\vec{q}'_\gamma}(\vec{q}^{(2)}_\gamma)$$

$$\mathcal{V}^{\mathcal{E}\mathcal{H}}_{\beta\gamma}(\vec{p}_\beta\,\vec{q}_\beta;\vec{p}'_\gamma) = -\,t_\beta(\vec{q}_\beta;\vec{q}^{(1)}_\beta;\tilde{q}^2_\beta-i0)\,\frac{\Phi^\gamma_\kappa(\vec{q}^{(2)}_\gamma)}{(\tilde{q}^{(2)}_\gamma)^2+\kappa^2_\gamma} \tag{3}$$

$$\mathcal{V}^{\mathcal{E}\mathcal{E}}_{\beta\gamma}(\vec{p}_\beta\,\vec{q}_\beta;\vec{p}'_\gamma\vec{q}'_\gamma) = t_\beta(\vec{q}_\beta;\vec{q}^{(1)}_\beta;\tilde{q}^2_\beta-i0)\;\psi^\gamma_{\vec{q}'_\gamma}(\vec{q}^{(2)}_\gamma) \quad .$$

The driving term $\mathcal{H}^{(0)}_{\beta\alpha}$ $(\mathcal{E}^{(0)}_{\beta\alpha})$ vanishes if $\alpha=\beta$, and is otherwise obtained from the expression for $\mathcal{V}^{\mathcal{H}\mathcal{H}}_{\beta\gamma}(\mathcal{V}^{\mathcal{E}\mathcal{H}}_{\beta\gamma})$ in (3), taking $\gamma=\alpha$ and replacing κ^2_γ by $\tilde{p}^{(0)2}_\alpha-z$. In (3), $\vec{q}^{(1)}_\beta=(m_\gamma/m_\alpha+m_\gamma)\vec{p}_\beta+\vec{p}'_\gamma$, $\vec{q}^{(2)}_\gamma=-\vec{p}_\beta-(m_\beta/m_\alpha+m_\beta)\vec{p}'_\gamma$, and Φ^β_κ is the two-body bound state vertex function, defined as $\Phi^\beta_\kappa(\vec{q}_\beta)=-(\tilde{q}^2_\beta+\kappa^2_\beta)\,\phi^\beta_\kappa(\vec{q}_\beta)$.

Eqs. (2) have the following properties: (i) The input consists solely of two-body bound state wavefunctions and half-off-shell transition amplitudes $t_\beta(\vec{q}_\beta, \vec{q}'_\beta;\, \tilde{q}_\beta^{\,2}+io)$ (the latter of course directly yield the scattering wavefunctions). In this way, the fully-off-shell amplitudes -in particular for arbitrarily large negative energies- occurring in the usual treatments are completely eliminated; (ii) The effective potentials are all independent of the energy parameter z; (iii) The amplitudes (1) have no primary singularities[1].

Further, after an angular momentum decomposition of Eqs. (2), it can be seen that the amplitudes $\mathcal{E}_{\beta\alpha}$ can be redefined to explicitly exhibit a two-body Jost function. If this is done, the resulting effective potentials become <u>real</u> (all phase factors cancel out) and the redefined amplitudes differ from the physical amplitudes for breakup simply by a Watson fsi factor.

It is interesting to note that the amplitudes (1) are essentially the projections of the three-body Faddeev wavefunction components into the complete set of channel eigenfunctions. Such projections have been previously considered by H. P. Noyes[5]; we are indebted to him for many helpful discussions on his approach to the three-body problem.

REFERENCES

1) L. D. Faddeev, Mathematical Aspects of the Three-Body Problem in Quantum Scattering Theory (Davey, New York, 1965).

2) T. A. Osborn and K. L. Kowalski, Ann. Phys. (N.Y.) 68, 361 (1971).

3) E. O. Alt, P. Grassberger and W. Sandhas, Nucl. Phys. B2, 167 (1967).

4) B. R. Karlsson and E. M. Zeiger, Slac Report No. SLAC-PUB 1377 (1974).

5) H. P. Noyes, Phys. Rev. Letters 23, 1201 (1969) (Erratum 24, 493 (1970)).

CANONICAL TRANSFORMATIONS AND A SOLVABLE MODEL FOR THE LINEAR THREE-BODY PROBLEM

M. Moshinsky*, J. Patera and P. Winternitz

Centre de recherches mathématiques, Université de Montréal,
Montreal H3C 3J7, Quebec, Canada

The aim of this contribution is to report on some recently obtained results[1] concerning physical systems possessing nontrivial invariance groups of canonical transformations, which can serve as solvable models for few-body problems. A general approach to canonical transformations was developed[1-3] and is here applied to the problem of three particles, constrained to a line and interacting by means of pair potentials of the type

$$V_{ik}(x_i-x_k) = \alpha_{ik}(x_i-x_k)^2 + \frac{\beta_{ik}}{(x_i-x_k)^2} . \tag{1}$$

Physical systems with canonical transformation groups are of interest for the following reasons. They all demonstrate "accidental degeneracy" and correspond to non-ergodic motion in the classical limit. The corresponding Schrödinger (or Hamilton-Jacobi) equation allows the separation of variables in at least one system of coordinates and the obtained ordinary differential equations can be solved analytically. Lie algebras of integrals of motions can be constructed and the corresponding groups of canonical transformations mapping the problem onto a harmonic oscillator can be obtained.

We consider a general two-dimensional quantum system with the spectrum $E_{nN} = C(k_2n+k_1N) + D$ (C and D are real, k_1 and k_2 are mutually prime nonnegative integers, n and N are nonnegative integers). We construct raising operators a^+ and A^+ such that the states $|nN) = (n!N!)^{-\frac{1}{2}}(a^+)^n(A^+)^N|0)$ (where $|0)$ is the vacuum state) form a complete (but not orthogonal or normalized) set of states. We then construct an orthonormal set of states $|nN\rangle$ as eigenvectors of the Hamiltonian H and an integral of motion M^2. The states $a^+|nN\rangle$ and $A^+|nN\rangle$ will in general be linear combinations of orthonormal states corresponding to a raised energy. New raising operators $\hat{a}^+$ and $\hat{A}^+$ are constructed as functions of a^+, A^+, H, M^2 and their commutators, such that they act on the orthonormal states according to

$$\hat{a}^+|nN\rangle = (n+1)^{\frac{1}{2}}|n+1N\rangle , \qquad \hat{A}^+|nN\rangle = (N+1)^{\frac{1}{2}}|nN+1\rangle . \tag{2}$$

The hermitian conjugate operators $\hat{a}$ and $\hat{A}$ then act as lowering operators, the Hamiltonian is proportional to $k_1\hat{a}^+a + k_2\hat{A}^+A + D/C$ and the problem is reduced to that of a harmonic oscillator.

Consider now a linear three-body Hamiltonian with the sum of three potentials of type (1) introduced by Calogero[4]. Introducing Jacobi coordinates

$$x = r\sin\phi = \frac{1}{\sqrt{2}}(x_1-x_2), \qquad y = r\cos\phi = \frac{1}{\sqrt{6}}(x_1+x_2-2x_3), \qquad X = \frac{1}{\sqrt{3}}(x_1+x_2+x_3)$$

and separating out the center-of-mass, we find

---

* Instituto de Fisica, Apdo Postal 20364, Mexico 20, D.F. Mexico.

$$H = -\tfrac{1}{2}\left[\frac{1}{r}\frac{\partial}{\partial} r \frac{\partial}{\partial r} + \frac{1}{r^2}\frac{\partial^2}{2\phi^2} - r^2\right] - 1 + \frac{9g}{2r^2\sin^2 3\phi}\,. \tag{3}$$

It is easy to check

$$M^2 = -\frac{\partial^2}{\partial\phi^2} + \frac{9g}{\sin^2 3\phi} \tag{4}$$

is an integral of motion and that an orthonormal basis satisfies

$$H|nN\rangle = (2n+3N+3\tau)|nN\rangle, \qquad M^2|nN\rangle = 9(N+\tau)^2|nN\rangle \tag{5}$$

where $\tau = \tfrac{1}{2} + (\tfrac{1}{4}+g)^{\frac{1}{2}}$. The wave functions $|nN\rangle$ are given explicitly in the papers[1,4,5]. Putting

$$\eta_\pm = \frac{1}{\sqrt{2}}(x_\pm - ip_\pm), \qquad x_\pm = \frac{1}{\sqrt{2}}(x\pm iy), \qquad p_\pm = \frac{1}{\sqrt{2}}(p_x \pm ip_y)$$

we find the raising operators for nonorthogonal states

$$a^+ = \eta_+\eta_- - 36g\, r^2 \sin^2 3\phi$$

$$A^+ = \eta_+^3 + \eta_-^3 + \frac{27g}{4r^2\sin^2 3\phi}\left[\cos 3\phi\left(r - \frac{\partial}{\partial r}\right) + \frac{1}{r}\sin 3\phi\,\frac{\partial}{\partial\phi}\right]. \tag{6}$$

Introducing $|M| \equiv +(M^2)^{\frac{1}{2}}$ we obtain the raising operators (2) as

$$\hat{a}^+ = \tfrac{1}{2}[(H+|M|)]^{-\frac{1}{2}}a^+$$

$$\hat{A}^+ = (3\sqrt{6})^{-1}|M|^{\frac{1}{2}}\{(|M|-3)(|M|+3\tau-3)(H+|M|)(H+|M|-2)(H+|M|-4)\}^{-\frac{1}{2}}\ \{3A^+(2|M|-3)+[M^2,B^+]\}.$$

Expressing $\hat{a}^+$, $\hat{A}^+$ and their hermitian conjugates $\hat{a}$ and $\hat{A}$ in terms of the coordinates and momenta x, y, p_x and p_y we obtain the canonical transformations mapping the Calogero problem onto an anisotropic harmonic oscillator with 2/3 frequency ratio. For all details we refer to the original paper[1].

REFERENCES

1. M. Moshinsky, J. Patera and P. Winternitz. Canonical Transformations and Accidental Degeneracy. III. A Unified Approach to the Problem. J. Math. Phys. 15, 1974 (to be published.
2. J.D. Louck, M. Moshinsky and K.B. Wolf. J. Math. Phys. 14, 692 (1973).
3. P. Winternitz, Ya.A. Smorodinskiĭ, M. Uhlíř and I. Friš. Yad. Fiz. 4, 625 (1966) [Sov. J. Nucl. Phys. 4, 444 (1967)].
4. F. Calogero. J. Math. Phys. 12, 2191 (1969).
5. A.M. Perelomov. Teor. Mat. Fiz. 6, 364 (1971) [Theor. Math. Phys. 6, 263 (1971)].

FEW ASPECTS OF PION - NUCLEUS SCATTERING

F.Nichitiu+

Joint Institute for Nuclear Research - Dubna

It is well known that in the very near future meson factories will be able to give very good beams of $\pi^+$and π^-mesons of high intensity and energy resolution. The energy region available will be extended from 10 MeV up to 350 MeV.For this interval of energies a number of accurate experiments is now proposed in order to fill up the gap of experimental points for few "physical observables".

The general feature of these proposals is that they are uncorrelated, and, I will try, in the next few minutes to show you some few new problems of pion - light nuclei physics in order to convince you of the necessity of "priority" for experiments in this field.

New experimental data on the total cross section of $\pi^+$and π^-on nuclei are now available from CERN /1/ and Rutherford Laboratory /2/ (for He^4, Li^6, Li^7, Be^9, C^{12}, O^{16}, S^{32}). The total cross section has a maximum around the Δ_{33} resonance but shifted downward in energy (the shift seems to increase with the number of nucleons).

The new interesting aspect is in the large differences between $\sigma_{tot}^{\pi^-}$ and $\sigma_{tot}^{\pi^+}$ for zero isospin nuclei. The effect seems to be caused by the Coulomb barrier /3/ but the different shape or radii of the neutron and proton densities /4/ of these nuclei may produce the same effect.

On the other hand, these data on $\sigma_{tot}^{\pi^\pm}$ are obtained from transmission type experiments, where the knowledge of Ref(0) in the extrapolation procedure is very important. The values of Ref(0) were taken from optical models or forward dispersion relation calculations.

The problem is that there are still large discrepancies between the results of different model calculations /2/, and there are also, for example for πHe^4, discrepancies between experimental values of Ref(0) and dispersion relation or forward finite sum rules calculations /5/.

In order to have unambigous and accurate experimental data it is necessary to carry out a series of, I would say, "correlated experiments".

The first experiment of this series would include measurements of differential cross sections $d\sigma/d\Omega$ for both π^+ and π^-in an as large as possible angular interval, including special measurements in a very forward direction. In this way we may get some ideas about the Coulomb

---

+On leave of absence from the Institute for Atomic Physics,Bucharest.

interference, the general behaviour of elastic cross section, and, I suppose a good determination of Ref(0).

The second experiment would be the measurement of the total cross section by the transmission method (a very fast and "statistically accurate" method) using information from the first experiment.

To control the experimental data obtained by the transmission method (the possible systematical errors) the second experiment should be accompained by measurements (only at few energies) of the total cross sections by "classical" methods (with visualizing track chambers).

And the third experiment would be measurements at very low energy - up to the break-up threshold ($\sim$ 28 MeV for He^4) where it is possible to use some information continuing (for example by phase shift analysis) the results from medium energies.

The aim of this series of correlated experiments - combined into one "total" experiment - is to obtain as model independent as possible experimental results.

Accurate total cross sections and differential cross sections over a large angular interval will help in phase shift analysis, in understanding Coulomb corrections and will permit detailed nuclear structure analysis. Comparison of these results with the dispersion relation calculations will give usefull information about the unphysical region. For example it would be interesting to determine the effective pion - nucleus coupling constant. The theoretical estimations with different models gives for He^3 nucleus the values/6,7/

$$f^2_{\pi He^3 H^3} = (1. -- 1.94) f^2_{\pi NN}$$

whilst its determination for heavier nuclei is /8,9/

$$f^2_{\pi Be^9} \approx f^2_{\pi Li^7} \approx 0.05$$

Now new experimental data in this field are available for $\pi^{\pm} He^4$ elastic scattering differential cross sections up to 180 MeV /10/ (angular interval: 27° - 165°), $d\sigma/d\Omega$ for backward $\pi^- C^{12}$ elastic scattering /11/ (60 - 100 MeV) - very important in phase shift analysis, and $d\sigma/d\Omega$ for $\pi^{\pm} He^3$ at few energies/10,12/ (100,135,154 MeV).

In order to have useful and accurate experimental measurements one should continue the analysis of "old" type experimental data and use their results to plan the future series of experiments (maybe for different meson facilities) taking into account "physical priority".

REFERENCES

1) C.Wilkin,C.R.Cox,J.J.Domingo,K.Gabathuler,E.Pedroni,J.Rohlin, P.Schwaller,N.W.Tauner. Nucl. Phys. B62, 61 (1973).
2) B.W.Allardyce,C.J.Batty,D.J.Baugh,W.J.McDonald,R.A.J.Riddle, L.H.Watson,M.E.Cage,G.J.Pyle,G.T.A.Squier,A.S.Clough,G.K.Turner. Preprint RL-73-038,Rutherford Lab. (1973).
3) G.Fäldt,H.Pilkuhn. Phys. Lett. 40B, 613 (1972), Phys. Lett. 46B, 337 (1973).
4) F.Becker,Yu.A.Batusov. Revista del Nuovo Cimento 1,309 (1971).
5) L.Alexandrov,T.Angelescu,I.V.Falomkin,F.Nichitiu,Yu.A.Shcherbakov. contribution to this conference.
6) T.Ericson,M.Locher. Nucl. Phys. A178, 1 (1970).
7) B.Z.Kopelyovitsch. Preprint JINR P2-7086 Dubna (1973).
8) P.Osland. Ref Th-1613 CERN (1973).
9) G.T.A.Squier,M.E.Cage,G.I.Pyle,A.S.Clough,G.K.Turner,B.W.Allardyce, C.J.Batty,D.J.Baugh. Preprint RL-73-037,Rutherford Lab. (1973).
10) T.Angelescu private communication.
11) C.Guaraldo and G.Piragino private communication.
12) M.Albu,T.Besliu,R.Garfagnini,M.M.Kulyukin,V.I.Lyashenko,A.Mihul, F.Nichitiu,G.Piragino,G.Pontecorvo,I.V.Falomkin,Yu.A.Shcherbakov. Lett. Nuovo Cimento 5 1121 (1972), Preprint JINR P1-7742 (1974).

II

INTERACTION NUCLEON-NUCLEON A TOUTES LES ENERGIES

NUCLEON-NUCLEON INTERACTION AT ALL ENERGIES

INFLUENCE OF THE NEUTRON'S CHARGE DISTRIBUTION ON NUCLEAR COULOMB ENERGIES

C.R. Heimbach, American University and Harry Diamond Laboratories,
J.S. O'Connell, National Bureau of Standards, Washington, D.C.

Some models of the neutron[1] have a positively charged center and a negatively charged periphery due to a pi-minus cloud in a p-wave. The Fourier transform of this charge distribution can be interpreted as the neutron's electric form factor G_E^n, although this relation is not Lorentz invariant. A model charge distribution that is consistent with the form factor data is[2]

$$\rho_p(x)= N_a \exp(-ax),\ \rho_n(x)= N_b \exp(-bx) -N_c \exp(-cx)\ , \tag{1}$$

where each exponential is normalized to one. These density functions are plotted in Fig. 1. The parameters a,b,c were determined by the mean-square-charge radii, $\langle r^2\rangle_p = 0.64\ \text{fm}^2$, $\langle r^2\rangle_n = -0.12\ \text{fm}^2$ and fits to $G_E^n(q^2)$.

For the deuteron the Coulomb energy of a neutron in the electric field of the proton is

$$E_c = \int |\psi_o(\vec{r})|^2\ W(\vec{r})\ d^3r \tag{2}$$

where W, the average interaction potential, is

$$W(\vec{r}) = \frac{e^2}{2} \int \rho_n(|\vec{r}-\vec{y}|)\ V(|\vec{y}|)\ d^3y. \tag{3}$$

For a proton with an exponential charge distribution the electrostatic potential at the point $\vec{y}= \vec{r}+\vec{x}$ is

$$V(y)= y^{-1}(\ 1-(1+ay/2)\ \exp(-ay)\ ). \tag{4}$$

Two forms were used for the deuteron's wave function $\psi_o(r)$: the Hulthen and the Reid hard-core[3]. The calculated Coulomb energies are:

$$E_c(\text{Hulthen}) = 4\ \text{keV},$$

$$E_c(\text{Reid h.c.}) = 2\ \text{keV}.$$

The average potential W(r) shown in Fig. 2 is a short range interaction making the np Coulomb energy a sensitive function of the deuteron's wave function inside 1 fm.

The np contribution to the Coulomb energy of heavier nuclei should be characteristically NZ E_c(deuteron) since the short range np relative wave functions in nuclei are essentially deuteron-like. The np energy contribution to Coulomb energy shifts is (2T-1)E_c(deuteron).

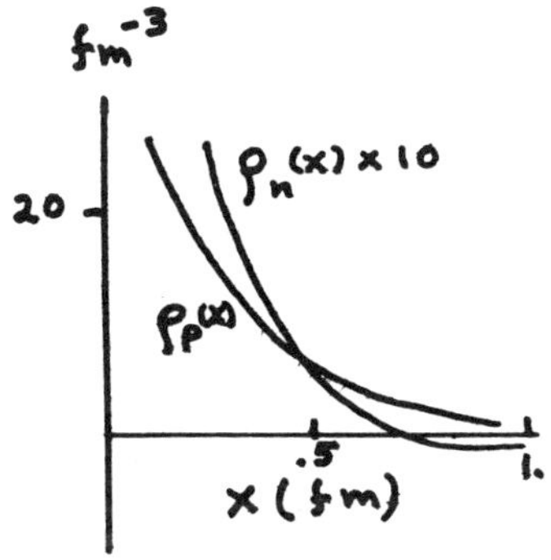

Fig. 1.

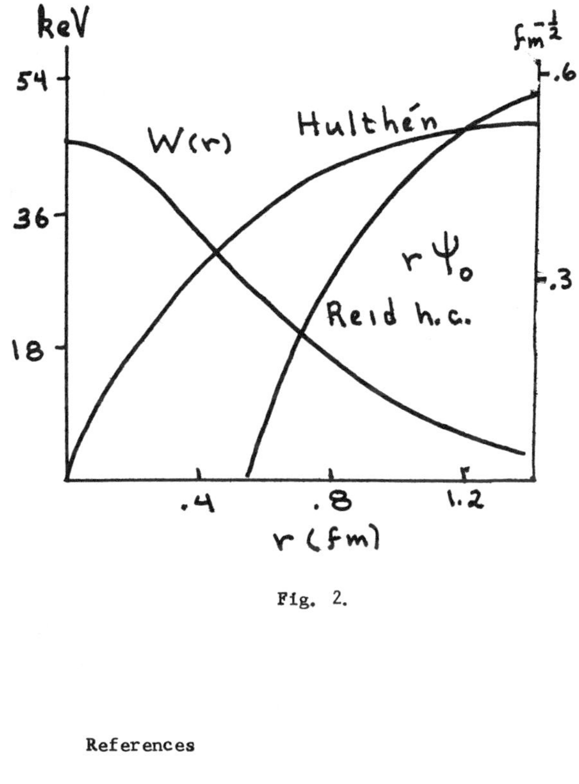

Fig. 2.

References

1. B.T. Feld, Models of Elementary Particles, Blaisdell Publishing Co., Waltham, Mass. 1969.

2. W. Bertozzi, J. Friar, J. Heisenberg, and J.W. Negele, Phys. Letters 41B, 408 (1972).

3. R.V. Reid,Jr., Annals of Phys. 50, 411 (1968).

PRECISION MEASUREMENT OF P-P ABSOLUTE CROSS SECTION NEAR THE INTERFERENCE MINIMUM

J.E.Benn, W.Maag, S.Münch, Ch.Thomann
University of Zürich, Physics Departement, Switzerland

With an apparatus as shown in the figure designed to achieve p-p elastic scattering angular distributions below 6 MeV with an absolute accuracy better than .2 % first cross section results are obtained in the energy region between 350 keV and 410 keV at the angles $\pm\ 45.020^{\circ}$ and $+\ 29.187^{\circ}$ in the laboratory system. From these data we deduced the energy of the cross section minimum

$$E_{min} = 382.75 \pm .11 \text{ keV} .$$

This result is about two standard deviations above the result of Brolley et al. (1), whose very precise and everywhere used value never was verified by an independent measurement. However an accurate knowledge of the minimum energy is important because of its influence on the scattering length a_{pp} of the effective range expansion.

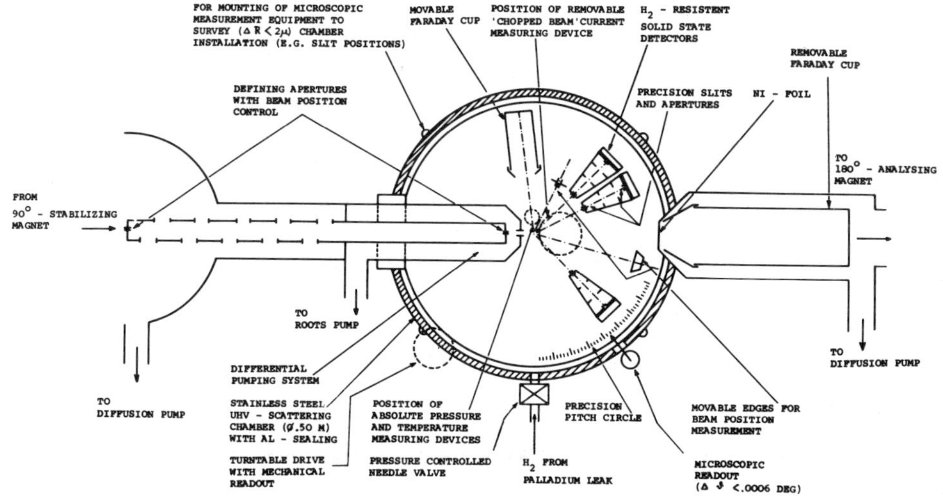

SCHEMATIC VIEW OF EXPERIMENTAL ARRANGEMENT

The energy region poses several experimental problems as small counting rates and consequently large background effects near the minimum, current measurement problems due to charge exchange effects, effects of multiple scattering of the scattered protons, large second derivative of the cross section with respect to the angle and corresponding errors in the geometry factor etc. To verify our estimated corrections it was tried in addition to determine the effects completely by experimental tests.
The following table shows the obtained data. The number in parenthesis gives the statistical error. The overall systematical error is .5 % . The error in energy is .04 keV, verified with a 180° analysing magnet (3).

Energy (keV)	$\sigma(45.020^o)$ (mbarn)	$\sigma(29.187^o)$ (mbarn)
350.33	3.825(28)	332.2(1.0)
360.35	2.229(17)	306.9(.9)
370.36	1.340(12)	283.0(.9)
380.37	.960(9)	264.3(.8)
390.39	1.055(9)	246.1(.7)
400.40	1.648(12)	230.2(.7)
410.41	2.522(18)	216.2(.6)

Present comparison with cross section values as obtained by the formulas of Gursky and Heller (2) indicates a general consistency with the theory except for a not yet understood .07 mbarn shift in the data at 45° . However this shift does not influence the position of the minimum.
Additional tests and measurements will be done to reduce the uncertainties in the data and the error of E_{min} to about .05 keV .

References

(1) J.E.Brolley,Jr., J.D.Seagrave, and J.G.Beery
Phys.Rev. 135, B1119 (1964)

(2) M.L.Gursky and L.Heller
Phys.Rev. 136, B1693 (1964)

(3) H.Gubler, S.Münch, and H.H.Staub
Helv.Phys.Acta 46, 772 (1974)

NUCLEON-NUCLEON INTERACTION WITH COULOMB MODIFICATIONS [x)]

L. Črepinšek, C. B. Lang, W. Plessas, and H. F. K. Zingl
Institut für Theoretische Physik, Universität Graz, Austria

Our potential [1)] in a channel α = J, S, T is given by

$$V_{LL'}^{\alpha}(p,\ p') = \lambda_{LL'}^{\alpha} g_L^{\alpha}(p)\, g_{L'}^{\alpha}(p') + \bar{\lambda}_{LL'}^{\alpha} h_L^{\alpha}(p)\, h_{L'}^{\alpha}(p') \tag{1}$$

where for the uncoupled channels

$$g_L(p) = \frac{p^L}{(p^2+\beta_L^2)^{L+1}} \qquad h_L(p) = \frac{p^{L+2}}{(p^2+\bar{\beta}_L^2)^{L+2}} \tag{2a}$$

and for coupled channels

$$g_\ell(p) = \frac{p^\ell}{(p^2+\beta_\ell^2)^{\ell+1}} \qquad h_\ell(p) = \frac{p^{\ell+2}}{(p^2+\bar{\beta}_\ell^2)^{\ell+2}}$$

$$g_{\ell'}(p) = \frac{p^{\ell'}}{(p^2+\beta_{\ell'}^2)^{\ell'}} \qquad h_{\ell'}(p) = \frac{p^{\ell'+2}}{(p^2+\bar{\beta}_{\ell'}^2)^{\ell'+1}} \tag{2b}$$

with ℓ = J - S and ℓ' = J + S.

In the proton-proton case [2)] we employ the Coulomb modified potential $\tilde{V}_{sc\ell}$ (p, p') analogous to (1) but with the modified form factors

$$g_{c\ell}(p) = g_\ell(p)\left[\frac{2\pi\xi}{e^{2\pi\xi}-1}\prod_{m=1}^{\ell}\left(1+\frac{\xi^2}{m^2}\right)\right]^{\frac{1}{2}} \exp\left(2\xi \arctan \frac{p}{\beta_\ell}\right)$$

$$h_{c\ell}(p) = h_\ell(p)\left[\frac{2\pi\xi}{e^{2\pi\xi}-1}\prod_{m=1}^{\ell}\left(1+\frac{\xi^2}{m^2}\right)\right]^{\frac{1}{2}} \left(1-\frac{\bar{\beta}_\ell\xi}{(\ell+1)p}\right)\cdot \exp\left(2\xi \arctan \frac{p}{\beta_\ell}\right) \tag{3}$$

Fitting the experimental phase shifts of the constraint solution of McGregor et al. [2)] and the deuteron data E_D and Q the open parameters of our potential were determined to:

---

x) Work supported by Fonds zur Förderung der wissenschaftlichen Forschung in Österreich, project 1410 and 1989.

	State	Rank	$\lambda\,(MeV/fm^{2L+1})$	$\bar{\lambda}\,(MeV/fm^{2L+1})$	$\beta\,(fm^{-1})$	$\bar{\beta}\,(fm^{-1})$
p-p interaction	1S0	2	-99.315	2407.4	1.2440	2.3601
	3P0	2	-102.01	39466	1.0710	2.3381
	3P1	1	3181.9	-	1.7954	-
	3P2	1	-10299	-	2.5407	-
	1D2	1	-18023	-	1.9275	-
n-p interaction	1S0	2	-106.63	∞	1.2663	4.6825
	1P1	1	2665.5	-	1.7404	-
	3D2	1	-11360	-	1.5623	-
	3D3	1	-8694.6	-	1.9365	-
	3S1-3D1	3	λ_{00} = -203.37 MeV/fm $\bar{\lambda}_{00}$ = 1909.4 MeV/fm λ_{02} = 221.54 MeV/fm λ_{22} = 276.14 MeV/fm		β_0 = 1.4530 $\bar{\beta}_0$ = 2.1542 β_2 = 1.2339	

Because of space limitation we shall demonstrate the quality of the fit by means of figures at another place [3]. The low-energy parameters were found to be [4]:

a_s = -7.67 fm $\quad$ r_s = 2.61 fm

a_s = -23.79 fm $\quad$ r_s = 2.53 fm

a_t = 5.445 fm $\quad$ r_t = 1.80 fm

The static properties of the 3S1 - 3D1 channel resulted as:

E_D = 2.2245 MeV $\quad$ p_D = 2.64 %

Q = 0.28749 fm^2 $\quad$ η = 0.034

Finally we state that the proposed potential gives saturation in infinite nuclear matter as shown by Meunier et al. [5].

References:

1) H. J. Faustmann, H. Oberhummer, and H. F. K. Zingl in Few Particle Problems in the Nuclear Interaction, ed. I. Slaus et al. (North-Holland Publishing Comp. Amsterdam, 1972) p. 34
and L. Črepinšek, H. Oberhummer, W. Plessas, H. F. K. Zingl, Acta Phys. Austr. 39 (1974) 345
2) M. H. Mac Gregor, R. A. Arndt, and R. M. Wright, Phys. Rev. 182 (1969) 1714
3) L. Črepinšek, C. B. Lang, H. Oberhummer, W. Plessas, H. F. K. Zingl, Acta Phys. Austr., to be published
4) L. Črepinšek, H. Oberhummer, H. F. K. Zingl, Acta Phys. Austr., to be published
5) J. P. Meunier, W. Plessas, J. Pigeon, R. Nahabetian, Phys. Rev. C, to be published.

DELTA FUNCTION POTENTIAL FOR THE $^3S_1$ - $^3D_1$ STATE

J. B. Aviles, Jr.

US Naval Research Laboratory, Washington, D. C. 20375, USA

A model of the N-N potential is taken in the form

$$\frac{M}{\hbar^2}\langle \alpha r L | V | \alpha r' L' \rangle = \delta(r-r') \sum_i \frac{v_i^{\alpha L L'}}{r_i} \delta(r - r_i)$$

The model parameters for uncoupled states, $J \leq 2$, have been previously determined.[1] Here we present one set of parameters for the $^3S_1$ - $^3D_1$ state.

i	r_i(fm)	$v_i^{(1,1)}$	$v_i^{(1,2)}$	$v_i^{(2,2)}$
1	0.600	3.488	2.400	0.000
2	1.278	-0.249	-3.007	0.970
3	2.511	-0.060	-0.844	0.336
4	3.600	-0.026	-0.194	0.111
5	4.800	-0.018	-0.109	0.060
6	6.000	-0.005	-0.026	0.013

The parameters at 4, 5, and 6 were chosen to give the same phase shift change as OPEP in the region $3.25 \leq r \leq \infty$. The deuteron properties are:

	Model		Experimental
Binding Energy	-2.225	MeV	-2.2246 ± 0.0005[2]
Scattering Length	5.425	fm	5.423 ± 0.004[3]
D-State Probability	6.8%		—
D/S Asymptotic Ratio	0.026		—
Quadrupole Moment	0.2875	fm^2	0.2875 ± 0.002[4]

The model phase shifts (Figs. 1 and 2) are compared with the Livermore X[5] analysis (circles). At 25MeV we also give the Harwell result[6] (squares). At 330 MeV we give the double solution found by Signell and Holdeman[7] (triangles).

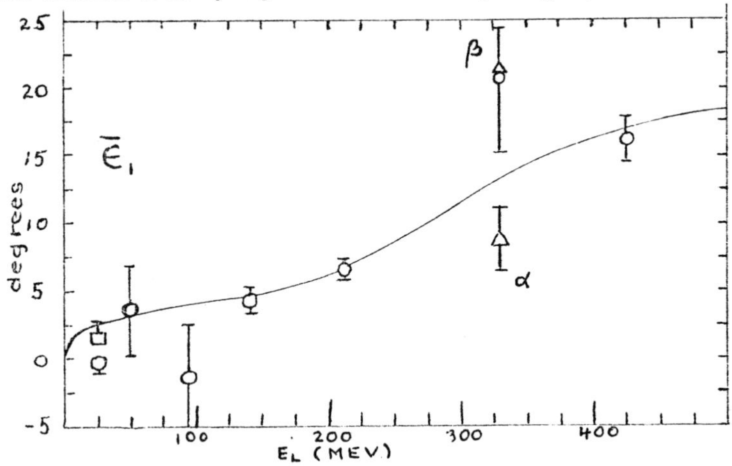

FIG. 1

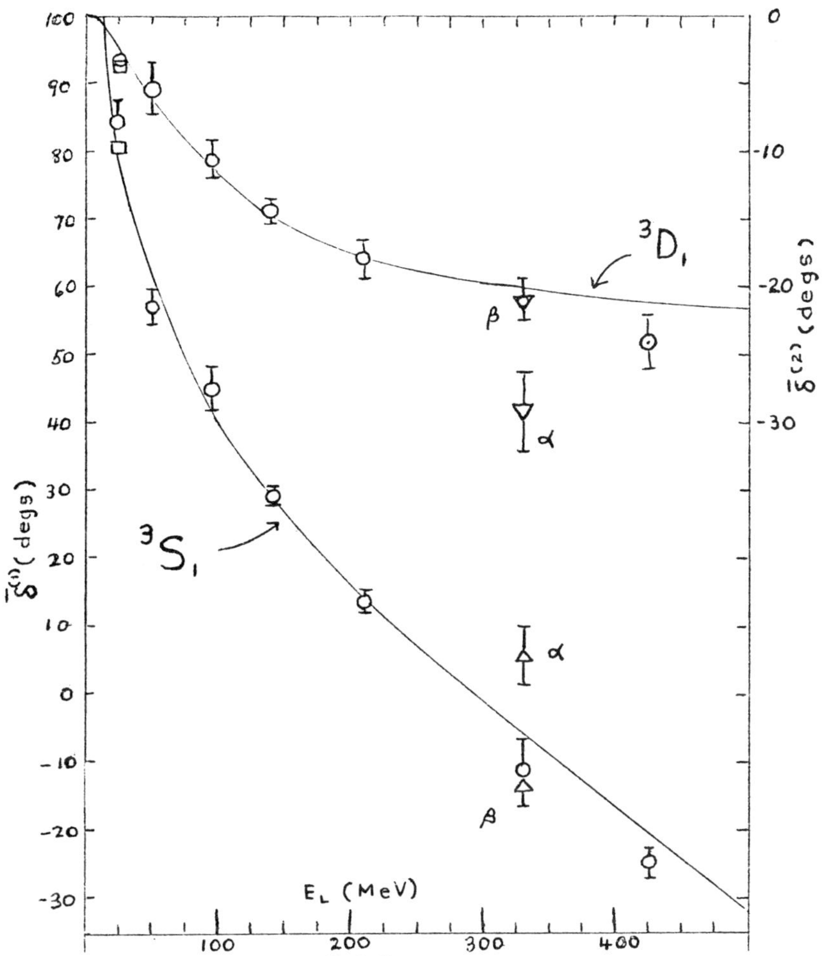

FIG. 2

REFERENCES

1. Joseph B. Aviles, Jr., Phys. Rev. C 6 (1972) 1467.

2. W. V. Prestwich, T. J. Kennett, L. B. Hughes and J. Fiedler, Can. J. Phys. 43 (1965) 2086.

3. H. Pierre Noyes, Ann. Rev. Nucl. Sci. (1972) 465.

4. R. V. Reed and M. L. Vaida, Phys. Rev. Lett. 29 (1972) 494.

5. M. H. MacGregor, R. A. Arndt, R. M. Wright, Phys. Rev. 182 (1969) 1714.

6. J. K. Perring, Rev. Mod. Phys. 39 (1965) 550.

7. P. Signell and J. Holdeman, Jr., Phys. Rev. Lett. 27 (1971) 1393.

A STUDY OF THE VARIATIONS IN THE ALLOWED VALUE OF a_{nn} OBTAINED FROM THE REACTION $\pi^- D \to \gamma nn$ DUE TO DIFFERENT UNITARILY-EQUIVALENT N-N WAVE FUNCTIONS *

W. R. Gibbs and B. F. Gibson
Los Alamos Scientific Laboratory, Los Alamos, New Mexico 87544

and

G. J. Stephenson, Jr. †
University of Maryland, College Park, Maryland 20742

In a recent letter [1] Sauer discusses the uncertainty in the Coulomb corrected p-p scattering length arising from the possibility of applying short-ranged unitary transforms to standard nucleon-nucleon potentials. In particular, he studies a rank two separable transform of the form

$$U = 1 - 2|g\rangle\langle g| \tag{1}$$

with

$$\langle r|g\rangle = Cr(1-\beta r)e^{-\alpha r} , \tag{2}$$

and C is a normalization constant.

Sauer presents results for the cases $\alpha=3\ \text{fm}^{-1}$ and β between 0 and 5 fm^{-1}, which correspond to transforms which heal in the range of 3 to 5 fm. When he includes the charge form factor of the proton in the calculation, he finds a_{pp} varying from about -17 fm to -34 fm.

We have recently completed an analysis of the uncertainties arising in the extraction of a_{nn} from the reaction $\pi^- D \to \gamma nn$ due to ambiguities in the theoretical assumptions which go into the construction of the transition amplitude.[2] In particular we have studied the dependence on the assumed short-range form of the n-n scattering wave function by calculating spectra with a standard value of a_{nn}(= -16 fm) and some model for the wave function and then analyzing those spectra with a standard model of the wave function, given by the work of Picker et al.[3] allowing a_{nn} to vary to give the best fit. For the phase-shifts arising from the Reid soft core (RSC) potential [4] our wave functions are similar to Reid's, however we are able to vary the phase-shift without refitting the potential while retaining the major physical features of the wave function. A kinematically complete geometry as described by Haddock et al. [5] is assumed and spectra are calculated for opening angles of 5°, 10°, 20° and 30° between the two neutrons.

For this work we have applied the unitary transform (1) to both our standard scattering state and the s-wave part of our standard (RSC) deuteron wave function. For consistency, only the s-wave part of the deuteron has been used in the analysis. Our results for $\alpha=3\ \text{fm}^{-1}$ and β = 1,2,3 and 4 fm^{-1} are presented in Table 1. The a_{pp} results are read from Sauer's Fig. 1. [1] For a more direct comparison with Sauer, we also include one case where the unitary transformation is applied only to the n-n scattering state. The difference in sign for a_{nn} and a_{pp} is only apparent as we calculate with the transformed wave function and analyze with the standard wave function while Sauer in effect does the opposite.

α	β		a_{nn} (-16 fm assumed)			a_{pp}
		θ_{nn} = 5°	10°	20°	30°	
3	1	-16.052	-16.064	-16.123	-16.234	-17
3	2	-16.034	-16.041	-16.083	-16.177	-33
3	3	-16.018	-16.022	-16.043	-16.095	-26
3	4	-16.018	-16.022	-16.046	-16.096	-25
		transform in scattering state only				
3	2	-15.693	-15.497	-14.951	-13.851	-33

Table 1. a_{nn} in fm extracted from spectra calculated with transformed wave functions. a_{pp} is read from Fig. 1 of Sauer [1].

It is apparent that the effect of such unitary transforms on the extraction of a_{nn} is very much less than the corresponding effect on a_{pp}. This is due to the fact that the absolute rate of the $\pi^- D \to \gamma nn$ reaction is not involved in the analysis, and the energy independence of the unitary transform guarantees a small effect on the relative spectra. In contrast, the calculation of the Coulomb effect depends very sensitively on the short ranged behavior of the scattering wave function.

REFERENCES

[*]Work supported in part by the U.S. Atomic Energy Commission.
[†]Present address: Los Alamos Scientific Laboratory, Los Alamos, New Mexico 87544.

1) P. U. Sauer, Phys. Rev. Lett. 32 (1974) 626.
2) W. R. Gibbs, B. F. Gibson and G. J. Stephenson Jr., to be published.
3) H. S. Picker, E. F. Redish and G. J. Stephenson Jr., Phys. Rev. C. 4 (1971) 2851.
4) R. V. Reid, Ann. Phys. (New York) 50 (1968) 411.
5) R. P. Haddock, et al., Phys. Rev. Lett. 14 (1965) 318.

PARITY-VIOLATING n-p PHOTOCAPTURE AND THE SHORT RANGE TWO-NUCLEON INTERACTION

G.N. Epstein
Department of Theoretical Physics,
12 Parks Road, Oxford, U.K.

It has been realized for some time now that due to the weak interaction, there is a weak parity violating nuclear force which may be recognized experimentally through violation of selection rules or through the existence of some non-zero pseudoscalar quantity (1,2). People have looked for and found irregular α-decay from ^{16}O to the ground state of ^{12}C, and non-zero circular polarization of γ's from the de-excitation of heavy nuclei or from neutron photo-capture reactions. So far, however, the emphasis has been on using these nuclear reactions to find out something about the weak interaction. We endeavour to indicate here that parity violations may also be fruitfully used to probe the short range strong interaction. In particular, we show that the γ-polarization (P) in the reaction $n+p \rightarrow d+\gamma$, is a sensitive measure of the low energy short-range NN wave function, and argue that to obtain agreement with experiment [3], (P(expt.) = -(1.30 ± 0.45) x 10^{-6}), a non-local strong NN interaction must be used.

We make extensive use of Danilov's work [4] and refer the reader there for details. His principal result is that the photocapture polarization, P, may be "read off" from the parity violating (p.v.) I = 0 2 NN scattering amplitude.

We have derived the following approximate formulae:

$$P = 0.6\lambda_1 m - 0.2\lambda_2 m \quad \text{where} \quad \begin{cases} \lambda_1 = \dfrac{2m G_\rho \mu_v}{m_\rho^2 a_t} \psi_{0t}^{(+)}(\underset{\sim}{0}) + \varepsilon_1 \\ \lambda_2 = \dfrac{2m G_\rho (2+\mu_v)}{m_\rho^2 a_s} \psi_{0s}^{(+)}(\underset{\sim}{0}) + \varepsilon_2 \end{cases} \tag{1}$$

$$\begin{Bmatrix} m = \text{nucleon mass} \\ \psi = \text{strong NN wave function} \\ a = \text{NN scattering length} \end{Bmatrix}$$

where m_ρ is the ρ-meson mass, μ_v the nucleon isovector anomalous magnetic moment, G_ρ the p.v. ρ-exchange strength defined in reference 1 (subscripts t , s = triplet, singlet respectively) and ε_1 , ε_2 are small error terms. For all the local strong potentials [5] such as Reid soft core (RSC), Yale, Hamada-Jonston (HJ) and the Sprung-de Tourreil supersoft core (SSC) model, the low energy short-range wave function is suppressed to zero and we find that

$$|P| \simeq 10^{-8} - 10^{-9} \tag{2}$$

These small polarizations will be obtained for any strong NN potential which at zero energy has strong short-range repulsion. To obtain agreement with experiment, we must insist that at low energies, the strong NN wave function is NOT suppressed at short distances. This immediately suggests a non-local NN strong force. Indeed, if we use the Tabakin [6] or Yamaguchi [7] separable potentials, we obtain

$$|P| = 0.2 \times 10^{-6} \tag{3}$$

which is much closer to the experimental number! When corrected, the more recent dispersion relation results of Danilov[8] support our work.

Our main point is that the usual local phenomenological potentials cannot produce a polarization large enough to agree with experiment. There is a 2-3 order of magnitude shortfall. Non-locality provides a means to bridge the gap between theory and experiment. Of course, it is true that local interactions have found widespread acceptance and require strong short range repulsion to reproduce the experimental data such as saturation and the two-nucleon phase shifts. However, it is also true that non-local models, such as that of Tabakin [6] both saturate and reproduce the experimental phases quite well too, even though much less effort has been expended on non-local fitting. In other words, a non-local interaction has sufficient flexibility to reproduce the usual nuclear physics experimental data <u>and</u> to allow a prediction of the photocapture polarization which resolves the current discrepancy with experiment.

REFERENCES

1) M. Gari, Phys. Reports <u>6C</u> (1973)
2) E. Fischbach and D. Tadic, Phys. Reports <u>6C</u> (1973)
3) V.M. Lobashov, D.M. Kaminker, G.I. Kharkevich, V.A. Kniazkov, N.A. Lozovoy, V.A. Nazarenko, L.F. Sayenko, L.M. Smotritsky and A.I. Yegorov, Nucl. Phys. <u>A197</u> (1972) 241
4) G.S. Danilov, Phys. Lett. <u>18</u> (1965) 40
5) R.V. Reid, Ann. Phys. <u>50</u> (1968); T. Hamada and I.D. Johnston, Nucl. Phys. <u>34</u> (1962) 382; K.E. Lassila, M.H. Hull, H.M. Ruppel, F.A. McDonald and G. Breit, Phys. Rev. <u>126</u> (1962) 881; R. de Tourreil and D.W.L. Sprung, Nucl. Phys. <u>A201</u> (1973) 193
6) F. Tabakin, Ann. Phys. (N.Y.) <u>30</u> (1964) 51
7) Y. Yamaguchi, Phys. Rev. <u>95</u> (1954) 1628
8) G. S Danilov Phys Lett. <u>35B</u> (1971) 579

On time reversal violating nucleon-nucleon potentials

M. Simonius, Lab. für Kernphysik
D. Wyler, Inst. f. theor. Physik
Eidgenössische Technische Hochschule
Zürich, Switzerland

In connection with proposed experiments on time reversal (T) invariance in polarized proton scattering[1], T violating potentials are reconsidered. Their general form, without specifying the radial dependence has been given by Herczeg[2]. The object of this work is to get more insight into these potentials by deriving them from various exchanges. Parity conservation is assumed.

Consider the angular momentum decomposition of the crossed channel, i.e. nucleon-antinucleon scattering. Exploiting the symmetry porperties of the transition amplitudes one shows: (J, π is the total angular momentum and parity of the exchanged state)

$J = 0$: No T violation possible (no T-violating one - π or η - exchange)

$\pi = (-1)^J$ (natural parity): T violation is possible only for charged states (e.g. $\rho^\pm$ - exchange). The isospin operator is $(\vec{\tau}_1 \times \vec{\tau}_2)^z$ which uniquely specifies the potential up to the radial depencence.

$\pi = -(-1)^J$ (unnatural parity): no further restrictions (e.g. A_1-exchange), all operators from ref.[2] possible except the one involving $(\vec{\tau}_1 \times \vec{\tau}_2)^z$.

If $\vec{r}$, $\vec{p}$, $\vec{\ell}$ are the relative distance, momentum and angular momentum operators, respectively, of the two nucleons, $\vec{\sigma}_{1,2}$, $\vec{\tau}_{1,2}$ their spin and isospin operators, m the mass of the exchanged meson, we get, allowing only for terms linear in p , the following T-violating (TV) nucleon-nucleon interactions.

$\rho^{\pm}$ - exchange: $V^{TV}_{\rho^{\pm}} = C\,\frac{e^{-mr}}{r}\left(\frac{1}{mr} + \frac{1}{(mr)^2}\right)(\vec{\sigma}_1 - \vec{\sigma}_2)\,\vec{\ell}\,(\vec{\tau}_1 \times \vec{\tau}_2)^z$

giving contributions to singlet-triplet transitions only.

A$_1$-exchange: $V^{TV}_{A_1} = C'\,\frac{e^{-mr}}{r}\left(\frac{1}{mr} + \frac{1}{(mr)^2}\right)\left((\vec{\sigma}_2\vec{p})(\vec{\sigma}_1\vec{r}) + (\vec{\sigma}_1\vec{p})(\vec{\sigma}_2\vec{r})\right) I$

$$+ C''\,\frac{e^{-mr}}{r}\left(\frac{1}{mr} + \frac{1}{(mr)^2}\right)(\vec{\sigma}_1 \times \vec{\sigma}_2)\,\vec{\ell}\,(\vec{\tau}_1 - \vec{\tau}_2)^z \quad {*)}$$

giving contributions to triplet-triplet and singlet-triplet transitions.

I can be an arbitrary (in the absence of a specific model) combination of the isospin operators $\vec{\tau}_1.\vec{\tau}_2$, $(\vec{\tau}_1+\tau_2)^z$, $\tau^z_1.\tau^z_2$.

(For an isosinglet particle I is a combination of the unity and $(\vec{\tau}_1+\vec{\tau}_2)^z$.)

Huffman's potentials derived from π - γ exchange[4] correspond to unnatural parity states in the π - γ system.

We note that for on shell nucleons the spin-part of the potential cannot be scalar (i.e. no $(\vec{p}\vec{r})$ or $(\vec{p}\vec{r})(\vec{\sigma}_1\vec{\sigma}_2)$ - term).

References:

1) M. Simonius, submitted to this conference
2) P. Herczeg, Nucl.Phys. 75 (1966) 655
3) Binstock, Bryan, Gersten Phys. Lett. 48B (1974) 77
4) A. Huffmann, Phys.Rev. D1 (1970) 882

---

*) $(\vec{\sigma}_1-\vec{\sigma}_2)\vec{\ell}(\vec{\tau}_1\times\vec{\tau}_2)^z$ has, up to a factor $(-1)^{\ell+1}$ the same matrix-elements as $(\vec{\sigma}_1\times\vec{\sigma}_2)\vec{\ell}(\vec{\tau}_1-\vec{\tau}_2)^z$

OFF-SHELL CONTINUATION OF THE PROTON-PROTON TRANSITION MATRIX

P. U. Sauer
Laboratory for Nuclear Science, M.I.T., Cambridge, Massachusetts, 02139, U.S.A.
H. Walliser
Fakultät für Physik der Universität, 78 Freiburg/Breisgau, Germany

The two-nucleon transition matrix T is defined in terms of the nuclear potential V,

$$T(\omega) = V + V(\omega - K)^{-1}T(\omega). \qquad (1)$$

K is the kinetic energy operator of relative motion in units of $\hbar^2/M$. T is closer to experiment and to the effective interaction of nuclear structure calculations than V. A parametrization of T in momentum space, which avoids the cumbersome potential fit, but is consistent with our complete knowledge of the nuclear force, has therefore great practical and theoretical advantages. The usual off-shell continuation[1] of T does, however, not directly apply to proton-proton (pp) partial waves. The on-shell pp T is not an observable, since pp scattering arises from the interference of the nuclear force with the Coulomb interaction V_C. How can the entire pp T of (1) be constructed? We restrict ourselves to the most important pp partial wave, ^1S_o. The extension to other partial waves is obvious.

In contrast to T, the modified transition matrix T^C,

$$T^C(\omega) = V + V(\omega - K - V_C)^{-1}T^C(\omega) , \qquad (2)$$

can, in the representation of the regular Coulomb functions $<r|f^C(k)> = \sqrt{(2/\pi)}F_o(\gamma(k),kr)$ with $\gamma(k) = e^2M/(2\hbar^2k)$, be continued off-shell according to the method of ref. 1. Once T^C is known, T follows from

$$T(\omega) = T^C(\omega) + T^C(\omega)[(\omega - K)^{-1} - (\omega - K - V_C)^{-1}]T(\omega). \qquad (3)$$

All matrix elements $<f^C(k_1)|T^C(\omega)|f^C(k_2)>$ are obtained from a symmetric continuation into two dimensions of $\sigma^C(kk) = -(2k/\pi)\sin\eta^C(k)$, where $\eta^C(k)$ is the experimental pp phase. The construction of T^C is based on the fact[1] that the scattering wave functions $|\psi^C(k)>$ are orthogonal and in the absence of a bound state complete, thus

$$<f^C(k_1)|\psi^C(k_2)> = \delta(k_2 - k_1)\cos\eta^C(k_2) + P(k_2^2 - k_1^2)^{-1}\phi^C(k_2k_1) \qquad (4)$$

is a real unitary operator. $|\psi^C(k)>$ is asymptotically normalized such that

$$|\chi^C(k)> = |\psi^C(k)> - |f^C(k)>\cos\eta^C(k) - |g^C(k)>\sin\eta^C(k) \qquad (5)$$

vanishes outside the range of the nuclear force. $<r|g^C(k)>$ stands for the irregular Coulomb function. ϕ^C is the real form of the half-shell T^C matrix, $\phi^C(k_2k_1) = <f^C(k_1)|T^C(k_2^2+io)|f^C(k_2)>\exp[-i\eta^C(k_2)]$. It is related to the scattering wave function by

$$\phi^C(k_2k_1) = -(2k_1/\pi)\sin\eta^C(k_2)C_o(\gamma(k_1))/C_o(\gamma(k_2)) + (k_2^2 - k_1^2)<f^C(k_1)|\chi^C(k_2)>, \qquad (6)$$

where $C_0^2(\gamma) = 2\pi\gamma/\left[\exp(2\pi\gamma)-1\right]$ is the Coulomb penetration factor.

σ^C is the symmetric part of ϕ^C, i.e. $\sigma^C(k_2k_1) = \left[\phi^C(k_2k_1) + \phi^C(k_1k_2)\right]/2$. σ^C provides a complete description of the nuclear interaction. The entire T^C and T can be derived from it. The diagonal elements of σ^C are given by the elastic pp scattering data. The low-energy (<100MeV) part of σ^C is uniquely determined, on-shell and off-shell, by the effective-range parameters of $\eta^C(k)$ and the requirement that the nuclear interaction be of short range. The remainder of σ^C has to be parametrized. The one-pion exchange tail does nowhere contribute significantly to the $^1S_0$ σ^C. The low-energy

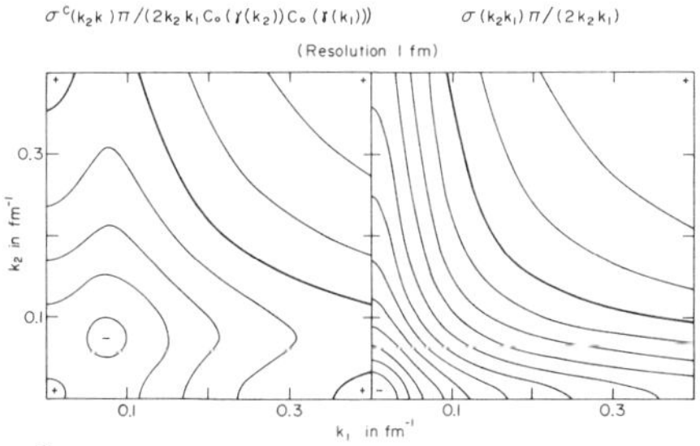

equal-value contours of σ^C are shown in the Figure. They are characteristic for the zero-range part ($\langle f^C(k_1)|\chi^C(k_2)\rangle = 0$) of (6). The heavy solid contours in the Figure correspond to the value - 5 fm. The σ of the nuclear $^1S_0$ T is also displayed for reference. σ^C and σ are different at low energies, however, they are quite similar elsewhere.

Though T^C could be employed directly, e.g. for the description of (p,2p) reactions, the purely nuclear T calculated from an assumed σ^C is of special interest. The reasons are: (i) Nuclear structure calculations with T could be performed in terms of a σ^C and the experimental pp phase. (ii) The dependence of the Coulomb subtraction in pp scattering data on the on- and off-shell parametrization of σ^C could be studied. (iii) The charge symmetry of the nuclear force may be used as a constraint on the off-shell parametrization of σ^C. The step from T^C to T is in numerical computations the most difficult one. We find it technically easiest to solve (3) indirectly, in an admittedly unelegant way. First, $\langle f^C(k_1)|V|f^C(k_2)\rangle$ is obtained from T^C and transformed to momentum space. Since the Fourier components $\langle k_1|f^C(k_2)\rangle$ have an unusual singularity structure, this transformation of the short-range operator V is carried out by going to cut-off Coulomb functions. Second, V in momentum space yields the nuclear T according to (1). It is emphasized that V is used in this formalism only as a technical means to solve (3). T^C and T are directly parametrized without explicit reference to a potential.

REFERENCE

1) M. Baranger, B. Giraud, S. K. Mukhopadhyay, and P. U. Sauer, Nucl. Phys. A138 (1969) 1.

POTENTIEL NUCLEON-NUCLEON A COEUR MOU TENANT COMPTE DE L'ECHANGE DE DEUX PIONS ET DE ω

R. de TOURREIL[+], D.W.L. SPRUNG[++*], B. ROUBEN[†*]

+ Division de Physique Théorique, Institut de Physique Nucléaire, 91406 - Orsay - France (Laboratoire Associé au C.N.R.S.)
++ Physics Department, Mc Master University, Hamilton, Ontario, Canada
† Laboratoire de Physique Nucléaire, Université de Montréal, Montréal, P.Q., Canada

Nous présentons un potentiel réaliste local à coeur très mou. Il diffère du potentiel de de Tourreil et Sprung (dTS)[1)] par l'inclusion des contributions des bosons ρ et ω au même titre que le π . De cette manière on tient compte d'une partie des effets dus à l'échange de deux et trois pions. Selon la théorie de Cottingham et al.[2)] les effets supplémentaires dus à l'échange de deux pions ne se prêtent pas à une représentation en forme d'échange de bosons. Nous avons donc traité cette partie du potentiel (de portée intermédiaire) d'une façon phénoménologique : elle se compose de fonctions du type OBEP avec une portée et des amplitudes libres dans chaque sous-espace de spin, isospin (S,T).

Le potentiel contient donc les termes habituels (central, tenseur, spin-orbite, spin-orbite quadratique et en L^2) dans chaque sous-espace (S,T). Pour les formes radiales on a choisi celles de Erkelenz et al[3)] :

$$Y_C(x) = e^{-x}/x$$

$$Y_{L12}(x) = \left(1 + \frac{2}{x}\right) \frac{e^{-x}}{x}$$

$$Y_{LS}(x) = -(1 + x) \frac{e^{-x}}{x^3}$$

$$Y_T(x) = \left(1 + \frac{3}{x} + \frac{3}{x^2}\right) \frac{e^{-x}}{x^3}$$

Toutes ces composantes sont régularisées à l'origine par une fonction $F(r) = (1.2r)^{20}/(1+(1.2r)^{20})$. Dans la région de très petits r ($r \lesssim 1$ fm) on a introduit des coeurs de la forme $C.[1-F(r)]$.

Les paramètres ne sont pas donnés ici faute d'espace. Les

masses et les constantes de couplage de ρ et ω ont été fixées aux valeurs de Cottingham et al[2]. La partie phénoménologique de la région intermédiaire en r a des portées inverses de $\sim$ 2.5 fm^{-1} pour les états pairs et de $\sim$ 2.0 fm^{-1} pour les impairs. Les coeurs centraux sont de l'ordre de 190 MeV pour tous les sous-espaces sauf le triplet-impair, où le coeur est de 92 MeV. La force tenseur dans les états triplet pairs est diminuée par rapport à dTS, tandis que la force centrale triplet impaire est plus répulsive.

L'accord avec les observables de diffusion N-N est excellent : le χ^2 par point étant de l'ordre de 2.0-2.4 comme pour dTS. Nous obtenons également de bons résultats pour les données de basse énergie. Le pourcentage d'état D dans le deuton est de 5.8 %. Le moment quadrupolaire du deuton est de .280 fm^2, soit plus petit que la dernière valeur expérimentale, .2875 fm^2 [4] qui semble très difficile à atteindre avec un potentiel local.

Le rapport des énergies potentielles des composantes centrales au deuxième ordre et au premier ordre des perturbations donne une idée de la rapidité de convergence de la série de perturbation. Ce rapport se situe dans les environs de 20 %. Les courbes de matière nucléaire, calculées en théorie de matrice G et en théorie des perturbations au deuxième ordre (avec une masse effective), sont très proches l'une de l'autre comme pour dTS. Nous avons réussi à améliorer quelque peu la saturation : au deuxième ordre de perturbation elle se trouve à $k_F = 1.57\ fm^{-1}$, E/A = - 15.3 MeV. En matrice G les valeurs sont 1.62 fm^{-1} et - 14.2 MeV. L'intégrale de défaut $\kappa = .077$ à $k_F = 1.4\ fm^{-1}$.

---

* Recherche appuyée par le Conseil National de Recherche du Canada

1) R. de Tourreil et D.W.L. Sprung, Nucl. Phys. A201 (1973) 193

2) W.N. Cottingham, M. Lacombe, B. Loiseau, J.M. Richard et R. Vinh Mau, Phys. Rev. D8 (1973) 800

3) K. Erkelenz, K. Holinde et K. Bleuler, Nucl. Phys. A139 (1969) 308; Phys. Lett. 49B (1974) 209

4) R.V. Reid, Jr. et M.L. Vaida, Phys. Rev. Lett. 29 (1972) 494.

PRECISION FITS TO THE $^1S_0$ PROTON-PROTON SHIFTS

Mark W. Kermode, Department of Applied Mathematics and Theoretical Physics, The University, P.O. Box 147, Liverpool L69 3BX, U.K.

In his review article on the interaction of two nucleons, Moravcsik [1]) stated that phenomenological potentials then in existence were not as successful as the phase parameters in fitting the scattering data. For example, the Reid soft core potential [2]) does not reproduce the phase shifts determined from the experimental data by MacGregor, Arndt and Wright [3]) within experimental error at all energies.

Recently, an equivalent local potential was introduced which gives a superb fit to the $^1S_0$ proton-proton phase shifts, the largest deviation being 0.04^o. This potential has the unusual feature of being positive between about 1.25 and 1.58 fm.[4])

More potentials have now been obtained which give equally good fits to the data. These potentials differ substantially in the region 0.0-0.75 fm but this one might expect since phase shifts are not known above a certain energy. Beyond 0.75 fm the potentials are practically identical.

There appears that there could be a simple explanation of the 'knoll' in the equivalent local potential, namely that the real interaction is nonlocal, a not unexpected result. Arnold and MacKellar [5]) have determined energy dependent equivalent local potentials from well known nonlocal potentials. The result for the Tabakin potential at a given energy, for example, shows a 'knoll'.

Also it is interesting to note that the solution of the inverse scattering problem obtained by Sprung and Srivastava [6]) contains a 'knoll'.

All the potentials give a neutron-neutron scattering length of -17.08 fm.

References

1) M.J. Moravcsik, Repts. Prog. Phys. 35 (1972) 587
2) R.V. Reid, Ann. of Phys. 50 (1968) 411
3) M.H. MacGregor, R.A. Arndt and R.N. Wright, Phys. Rev. 182 (1969) 1714
4) M.W. Kermode, J. Phys. A: Gen. Phys. 6 (1973) L154
5) L.G. Arnold and A.D. MacKellar, Phys. Rev. C3 (1971) 1095
6) D.W.L. Sprung and M.K. Srivastava, Nucl. Phys. A139 (1969) 605

POLARIZATION MEASUREMENTS IN p-p ELASTIC SCATTERING AT 462 MeV

D. Aebischer, B. Favier, G. Greeniaus*, R. Hess, A. Junod+,
C. Lechanoine, J.C. Niklès, D. Rapin, D. Werren.
University of Geneva, Switzerland

At intermediate energies, especially above 450 MeV, the phase shifts for p-p scattering are imprecise. The sparse experimental data[1)] must be supplemented with additional experiments, including polarization measurements. Almost no data exist for small angles ($\theta_{Lab} \simeq 10^0$) in the coulomb-nuclear interference region. We have measured $P(\theta)$ at 399, 462, 500 and 576 MeV at the CERN synchro-cyclotron. We present here only the results at 462 MeV.

The polarized proton beam was obtained by scattering the extracted proton beam at 7^0 on a carbon target. The scattered beam energy was degraded to 462 MeV by placing B_4C in the defining collimators. The momentum resolution was estimated to be ±1%. The experimental set-up is shown in fig. 1.

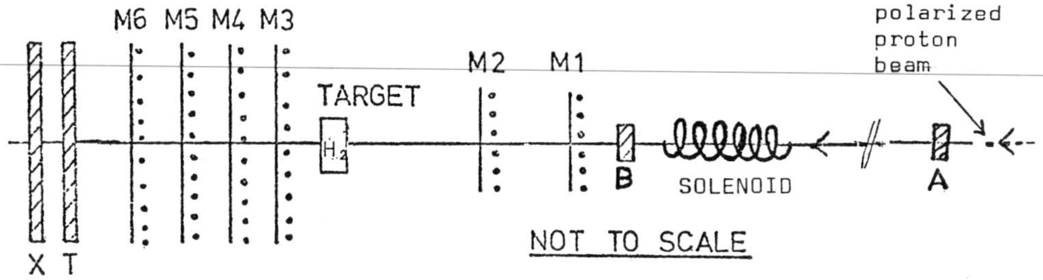

FIGURE 1 Experimental Set-up

Multiwire proportional chambers (M1 to M6) on either side of a 17 cm long liquid hydrogen target, placed directly in the low intensity ($<10^5$/sec) polarized beam, allowed the incident and outgoing particle trajectories to be determined. The time of flight between counters B and T, and energy losses in T and X were measured in order to eliminate background due to the inelastic reactions. A coincidence ABXT triggered coordinate read-out from the chambers to a coding and decision system[2)] which performed a rapid rejection of all events with scattering angle $\theta \leqslant 1.5^0$. Background produced in the target windows was estimated using a dummy target.

The azimuthal intensity distribution of the scattered particles is written

$$N(\theta,\phi) \propto N(\theta)(1+|\vec{P}o|.P(\theta).\cos\phi) \qquad (1)$$

where P_o is the transverse beam polarization, $N(\theta)$ the number of events scattered at angle θ, and ϕ is the angle between the polarization vector $\vec{P}_o$ and the normal to the scattering plane. The beam polarization, determined in a separate double scattering experiment was found to be $P_o = 0.374 \pm 0.013$. $P(\theta)$ was found both by fitting the distribution $N(\theta,\phi)$ with eq(1) and by performing a first order Fourier analysis. In the latter method, the statistics

$$S(\theta) = \Sigma \sin\phi_i = 0$$
$$C(\theta) = \Sigma \cos\phi_i = N(\theta) P_o A(\theta)/2$$

allow $P(\theta)$ to be determined and also provide a check on the shape of the angular distribution. The 2 methods give consistent results. The data are shown in fig. 2. The errors are purely statistical. Observation of unscattered events showed that systematic errors in the scattering angle determination were $\leqslant 0.01^o$. Systematic errors in the analysing power is estimated to be $\leqslant 0.01$. As a further protection against instrumental asymmetries, the direction of the beam polarization was frequently changed by reversing the sense of the polarizing scattering and by using the solenoid.

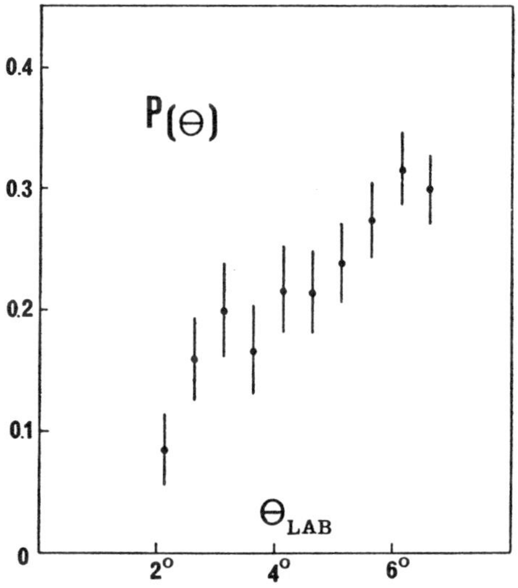

FIGURE 2

The data are consistent with a linear increase in analyzing power with the angle. We hope to extend the angular range from 1.5^o to 10° to be able to see the expected deviation from linear dependence.

References

1. U C R L 20000 NN August 1970
2. D. Aebischer et al. N.I.M. 99 (1972)405 and N.I.M. 117 (1974) 131.

STUDY OF THE REACTION $pp \rightarrow \pi^+ d$ AT 576 MeV

D. Aebischer, B. Favier, G. Greeniaus*, R. Hess, A. Junod+
C. Lechanoine, J.C. Niklès, D. Rapin, C. Serre, D. Werren
University of Geneva, Switzerland

The reaction $pp \rightarrow \pi^+ d$ has not been studied sufficiently at proton energies less than 600 MeV. C. Serre[1] has shown, for the inverse reaction $\pi d \rightarrow pp$, that the shape of the differential cross sections for incident proton energies $\gtrsim$ 570 MeV is described satisfactorily by an expression

$$\frac{d\sigma}{d\Omega}_{cm} = K\ (A + \cos^2\theta - B\cos^4\theta) \qquad (1)$$

Existing phenomenological models do not require the presence of the term $B\cos^4\theta$ which corresponds to the f-wave contribution of the π and is therefore expected to be less important at lower energies. The behaviour of this term is not well known. In fact, there are no measurements at very small angles, the region which is most sensitive to this term. Data obtained during a p-p elastic scattering experiment at small angles[2] at 576 MeV allowed the $pp \rightarrow \pi^+ d$ reaction to be studied for $5^0 < \theta_\pi < 30^0$ in the center of mass. The experimental set-up is described in ref.2.

The deuteron always has a laboratory scattering angle less than 12^0 at 576 MeV and was detected in a wire chamber-scintillation counter telescope with full efficiency up to $\theta_{LAB} \simeq 7^0$. Because of difficulties in estimating background from the p-p elastic and competing inelastic reactions, only deuterons emitted backwards in the center of mass are treated here. A simplified description of the analysis can be made by separating the events into three overlapping groups according to the pion laboratory production angle.

1) $3^0 < \theta < 10^0$. In this region the geometrical acceptance of the chambers for both the π and d was 100%. Background from the reactions $pp \rightarrow pp\pi^0$ and $pp \rightarrow pn\pi^+$ could be cleanly eliminated using the kinematic correlations between the π and d.

2) $13^0 < \theta < 16^0$. In this range, only the deuteron track was considered since the π was not detected in the scintillation counters. These events were identified by large time-of-flight and energy losses after the target. The background was estimated by observing the T-O-F and dE/dx distributions in regions kinematically forbidden for $pp \rightarrow \pi d$ and subtracting in an appropriate manner.

3) $8^{\circ} < \theta < 15^{\circ}$. This overlap region allowed the consistency between the procedures used for the other two regions to be checked.

Corrections for π-decay and nuclear interactions of the π in the liquid hydrogen target were also considered. Nuclear interactions of the deuteron in the target are common to all angular regions and do not affect the analysis.

The restricted angular range observed did not permit all the parameters in equation 1 to be determined simultaneously. However, A is relatively well known and has only a small effect on B. The value obtained in ref.1 was used.

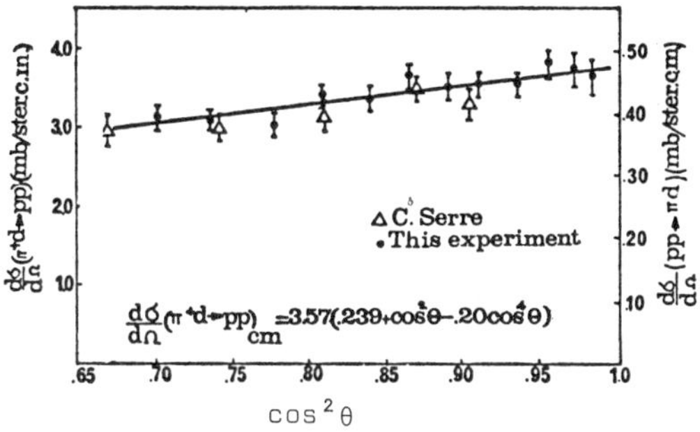

FIGURE 1

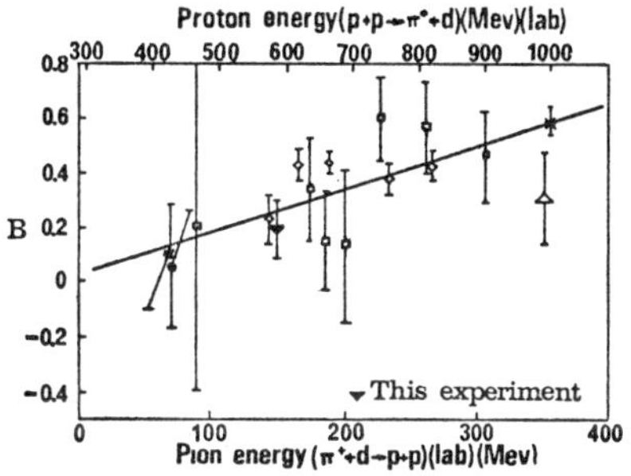

FIGURE 2

The differential cross sections in ref.1 for 572 MeV were used to normalize our data which are shown in figure 1. The value of A used and the best fit for B are also shown. Figure 2 shows the values of B which have been determined to this date (see ref.1).

This preliminary analysis seems to confirm the necessity of the term in $\cos^4\theta$ found by C. Serre at 572 MeV, even though it was obtained for the inverse reaction and a different angular range. A more refined analysis to follow will allow the angular range considered to be extended, will result in diminished error bars and include analysis of similar data at 500, 462 and 410 MeV incident proton energies.

References

1. C. Richard-Serre et al., CERN 68-40, Nucl.Phys.B 20 (1970)413.
2. Polarization measurements in p-p elastic scattering at 462 MeV, D. Aebischer et al., submitted to this conference.

MOMENTUM SPACE CALCULATION OF NEUTRON-PROTON BREMSSTRAHLUNG

Richard H. Thompson and Leon Heller
Los Alamos Scientific Laboratory, Los Alamos, New Mexico 87544

We describe a calculation of neutron-proton bremsstrahlung carried out in momentum space, which treats the double scattering term and the one-pion-exchange current exactly. The motivation for working in momentum space is that the exchange current has a much simpler form there than in coordinate space. (The normal current is also simpler since it does not involve any gradients.) A calculation which included the exchange current operator to zeroth order in the photon energy, k^0, has been carried out in coordinate space[1], and we have previously[2] written down the operator to order k^1. But the possibility of doing a calculation which does not require an expansion in powers of k/m_π has led us to momentum space.

By writing the initial and final wave functions as free waves plus scattered waves, the conventional decomposition into zero, single, and double scattering is obtained. While most calculations of the single scattering (with normal current) have been performed in momentum space, only one approximate calculation[3] of double scattering has been done that way. Other treatments of the double scattering have either been in coordinate space,[4,5] or to order k^0 in the matrix element using Low's theorem.

We work in the radiation gauge so that the polarization vector of the photon is given by $\varepsilon_\mu = (0, \hat{\varepsilon})$ with $\hat{\varepsilon}\cdot\vec{k} = 0$, and $\vec{k}$ is the momentum of the photon. Then the matrix element for the bremsstrahlung process is given by

$$(2\pi)^3 \delta^{(3)}(\vec{P}_f + \vec{k} - \vec{P}_i)\, \hat{\varepsilon}\cdot\vec{M} = \langle \psi_f^{(-)} | \int d\vec{x}\, e^{-i\vec{k}\cdot\vec{x}}\, \hat{\varepsilon}\cdot\vec{J}(\vec{x}) | \psi_i^{(+)} \rangle \tag{1}$$

where $\vec{j}(\vec{x})$ is the electromagnetic current operator, ψ_i and ψ_f are the initial and final n-p state vectors, and $\vec{P}_i$ and $\vec{P}_f$ are the initial and final total momenta for the n-p system. $\vec{M}$ has the following general form,

$$\vec{M} = \int \frac{d\vec{p}}{(2\pi)^3} \int \frac{d\vec{p}\,'}{(2\pi)^3}\, \psi_f^*(\vec{p})\, \vec{O}(\vec{p}\,', \vec{p}, \vec{k})\, \psi_i(\vec{p}\,') \tag{2}$$

where $\psi(\vec{p}) \equiv \langle \vec{p} | \psi \rangle$ is the wave function in the relative momentum, and $\vec{O}\cdot\hat{\varepsilon}$ is the momentum space matrix element of the electromagnetic Hamiltonian.

The operator $\vec{O}$ and the wave functions are expanded in terms of spherical harmonics (or in general irreducible tensors) and then the angular momenta are recoupled. The guide used in deriving these expansions is to require the result to contain the tensors involving $\hat{p}$ and $\hat{p}'$ to be coupled to each other. It is then possible to apply a form of the Wigner-Eckart theorm in taking matrix elements. As an example we show the formula for the convection current, where $\vec{O}$ is given by

$$\vec{O} = \frac{e\tau_1^P}{m} \vec{p}\, \delta^{(3)}(\vec{p}\,' - \vec{p} - \frac{\vec{k}}{2}) + \text{a corresponding term for particle 2 .} \tag{3}$$

where we have taken $\vec{P}_i$ (or, alternatively, $\vec{P}_f$) = 0. The angular momentum expansion for $\vec{O}$ is of the following form,[6]

$$\hat{e}\cdot\vec{O} = \sum_{\ell L L' L''} C^{\ell L L' L''}(p,p',k)\left[\left[Y_{L''}(\hat{p})\times Y_{L}(\hat{p}')\right]^{L'}\times\left[Y_{\ell}(\hat{k})\times Y_{1}(\hat{e})\right]^{L'}\right]^{0}_{0} \tag{4}$$

where due to the complexity of the C coefficient we do not define it here.

The expression for either the initial or final wave function in terms of spin angle functions is given by

$$\Psi_{\vec{k}Sm_S}(\vec{p}) = \sum_{\substack{J\ell\ell' \\ mM}} Y^{m*}_{\ell'}(\hat{k})\langle \ell' S m m_S | JM\rangle \mathcal{Y}^{\ell SJM}(\hat{p})\,\Psi^{JS}_{\ell\ell'}(p,k) \tag{5a}$$

where $$\Psi^{JS}_{\ell\ell'}(p,k) = \delta(p-k)\,\delta_{\ell\ell'}/k^2 + T^{JS}_{\ell\ell'}(p,k)/(E_{\vec{k}}-E_{\vec{p}}+i\eta) \tag{5b}$$

and $T^{JS}_{\ell'\ell}(p,k)$ is a solution of the partial wave Lippmann-Schwinger equation. Then using Eqs. (4), (5) and (2) the most complicated integral that we have to perform is a double integration when the exchange current is used in the double scattering term.

REFERENCES

† Work supported in part by the U. S. Atomic Energy Commission.

1) V. R. Brown and J. Franklin, International Conference on Few Particle Problems in Nuclear Interaction, edited by I. Slaus, S. A. Moszkowski, R. P. Haddock and W. T. H. van Oers (North-Holland, Amsterdam, 1972) p. 64.

2) R. H. Thompson and L. Heller, Phys. Rev. C 7, 2355 (1973).

3) M. I. Sobel, Phys. Rev. 152, 1385 (1966).

4) V. R. Brown, Phys. Rev. 177, 1498 (1969), and Phys. Rev. C 6, 1110 (1972).

5) L. Heller and M. Rich, "Proton-Proton Bremsstrahlung. Coulomb Effect" to be published in Phys. Rev. C, July 1974.

6) See reference 2, Appendix C for the angular momentum notation.

PROTON-PROTON POTENTIAL WITH A THEORETICAL INTERMEDIATE RANGE PART DEDUCED FROM DISPERSION RELATIONS

M. Lacombe, B. Loiseau, J.M. Richard, R. Vinh Mau

Division de Physique Théorique*, Institut de Physique Nucléaire, Paris**
and Laboratoire de Physique Théorique - Paris**

P. Pires, R. de Tourreil

Division de Physique Théorique*, Institut de Physique Nucléaire,
91406 Orsay

Recently, a theoretical nucleon-nucleon potential which describes the long and intermediate range part of the interaction, has been derived (1). This potential contains the one pion exchange contribution, the two pion exchange contribution and the ω meson exchange as part of the 3 pion exchange. The 2 pion exchange was calculated via dispersion relations from the physical pion-nucleon phase shifts and from the low partial wave pion-pion amplitudes and, therefore, does not contain any adjustable parameter.

In this work, we add to this theoretical potential a phenomenological soft core potential to describe the short range part, the parameters of which are adjusted to fit the proton-proton phases up to 350 MeV as well as the low energy parameters (scattering length and effective range parameter)

Our potential is of the form :

$$V = V_{phen}(1 - f(r)) + V_{theor} f(r) \qquad (1)$$

where V_{theor} is the solution G+NR of ref.(1)

$$f(r) = \frac{(pr)^\alpha}{1+(pr)^\alpha} \quad \text{is a cut-off function}$$

$$V_{phen} = \sum_{\alpha=c,ss,so,T,(so)^2} (V_\alpha + V'_\alpha E) \quad \text{(E: Lab. energy)}$$

Both V_{theor} and V_{phen} contain central, spin-orbit, spin-spin, tensor and quadratic spin-orbit terms. However, the quadratic spin-orbit effects turn out to be negligible.

V_{theor} is rather sharply cut off at $r \sim 0.9$ fm, i.e $\alpha=20$, $p=1.2$ fm^{-1} so that, in equation (1), the search of the phenomenological core does not perturb the theoretical part.

The parameters of V_{phen} were determined by fitting the Livermore energy independant phase shifts analysis (MAW) (2) and their values given in table I. The χ^2 values for the best fit to the MAW phases are given in table II and the phase shifts for $J \leq 4$ are drawn in the figure. First, the ω meson coupling constant is taken as $G^T_\omega / G^V_\omega = -0.12$ and $(G^V_\omega)^2/4\pi = 4.65$ (see ref.3). However, as expected, the χ^2 is significantly improved by increasing the value of G^V_ω and the best χ^2 value is obtained for $(G^V_\omega)^2/4\pi = 10.7$. The quality of this fit is somewhat better than the one obtained with the presently known nucleon-nucleon potentials (4). From the work of ref.(1), the theoretical potential has a small but significant energy dependence. We do take into account this energy dependence and our fit shows furthermore that the energy dependence is important in the core region.

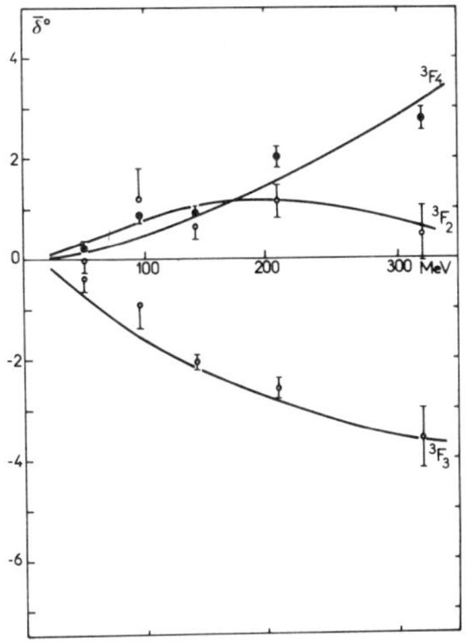

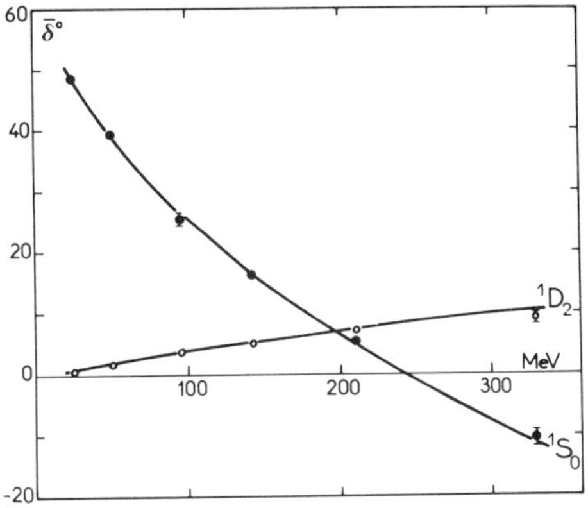

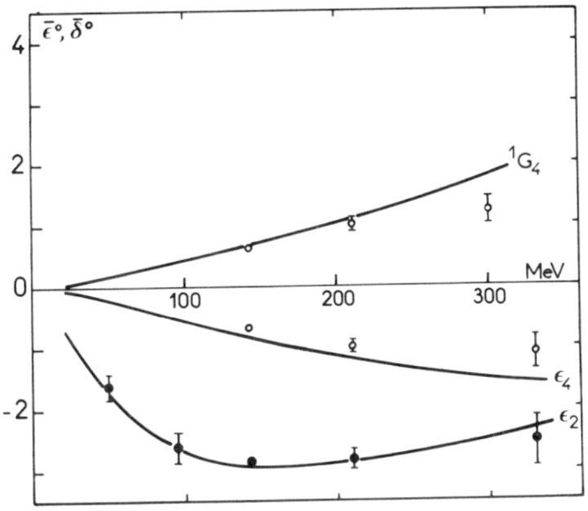

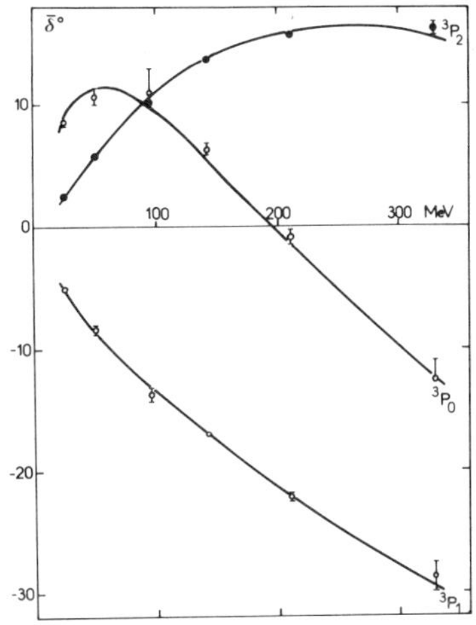

	$\alpha = C$		$\alpha = SO$		$\alpha = SS$		$\alpha = T$	
V_α	+335.5	+14.86	-550.0	-245.1	+82.41	-3.430	+82.89	+74.60
V'_α	+0.688	+0.790	-0.244	-0.378	+0.106	+0.128	0	0

Table I

- the values of the first column correspond to $(G^V_\omega)^2/4\pi = 4.65$
- " " " second " " " $= 10.7$

$\frac{(G^V_\omega)^2}{4\pi}$	4.65	10.7
$\frac{\chi^2}{\text{point}}$	4.77	1.88

Table II

References

(1) W.N. Cottingham, M. Lacombe, B. Loiseau, J.M. Richard, R. Vinh Mau Phys. Rev. D8 (1973) 800.

(2) M.H. Mc Gregor, R.A. Arndt, R.M. Wright Phys. Rev. 182 (1969) 1714

(3) R, Vinh Mau, J.M. Richard, B. Loiseau, M. Lacombe, W.N. Cottingham Physics Letters, 44B (1973) 1.

(4) See for example P. Signell, in "Few Particle Problems" ed. by J. Slaus et al. and the references cited therein.

AVERAGE VALUE OF THE NEUTRON-NEUTRON SCATERING LENGTH FROM A CRITICAL COMPILATION OF PUBLISHED MEASUREMENTS

B. Kühn

Zentralinstitut für Kernforschung der AdW der DDR Rossendorf, Bereich 2, DDR 8051 Dresden, Postfach 19

Up to now about 40 measurements of the neutron-neutron scattering length a_{nn} have been reported. As experiments on direct neutron-neutron scattering are still not feasible, in all these experiments a_{nn} was deduced from the final state interaction of two neutrons emitted simultanously in nuclear reactions. In table 1 are compiled the reactions, used in these measurements, and the numbers of kinematically incomplete N_{inc} and complete N_c experiments by each of the reactions. From the experimental spectra a_{nn} was extracted by means of the Born-approximation, the Migdal-Watson theory, the impuls approximation, diagramm techniques and theoretical complete descriptions, based on Faddeev-equations.

Table 1

Reaction	N_{inc}	N_c
$d + n = p + 2n$	13	6
$^3H + n = d + 2n$	3	-
$d + d = 2p + 2n$	-	1
$^3H + d = {}^3He + 2n$	5	2
$^3H + {}^3H = {}^4He + 2n$	1	1
$d + \pi^- = \gamma + 2n$	4	1

The values obtained are presented in fig. 1. The distribution of the results is peaked in the region of - 16 ÷ - 17 fm and in the region of - 23 fm. The last mentioned peak is populated by results from the (d + n) - reaction only. It could be shown, that the diagramm technique in second order approximation and the Migdal-Watson theory are inadequate to describe the kinematic incomplete spectra of this reaction. By these theories values at - 23 fm were obtained. Other experiments, from which $a_{nn} = -23$ fm was obtained, failed by experimental errors. Therefore the - 23 fm results were rejected before averaging. The remaining 30 results fulfill the conditions of a standard distribution.

An important point is, that complete measurements allow to choose the kinematic conditions more definite to favour certain reaction mechanisms. By this reason, in all cases the two dimensional spectra could be described more exactly than the spectra of incomplete measurements. Therefore in the averaging the results of complete experiments were taken into account with two fold weight. The weighted average and the standard deviation were found to be

$$\overline{a_{nn}} = (- 16.61 \pm 1.45) \text{ fm}.$$

The right value of a_{nn} is awaited to be found in the interval $(-16.61 - 0.54)\text{ fm} < a_{nn} < (-16.61 + 0.54)\text{ fm}$ with a confidence of 95 %.

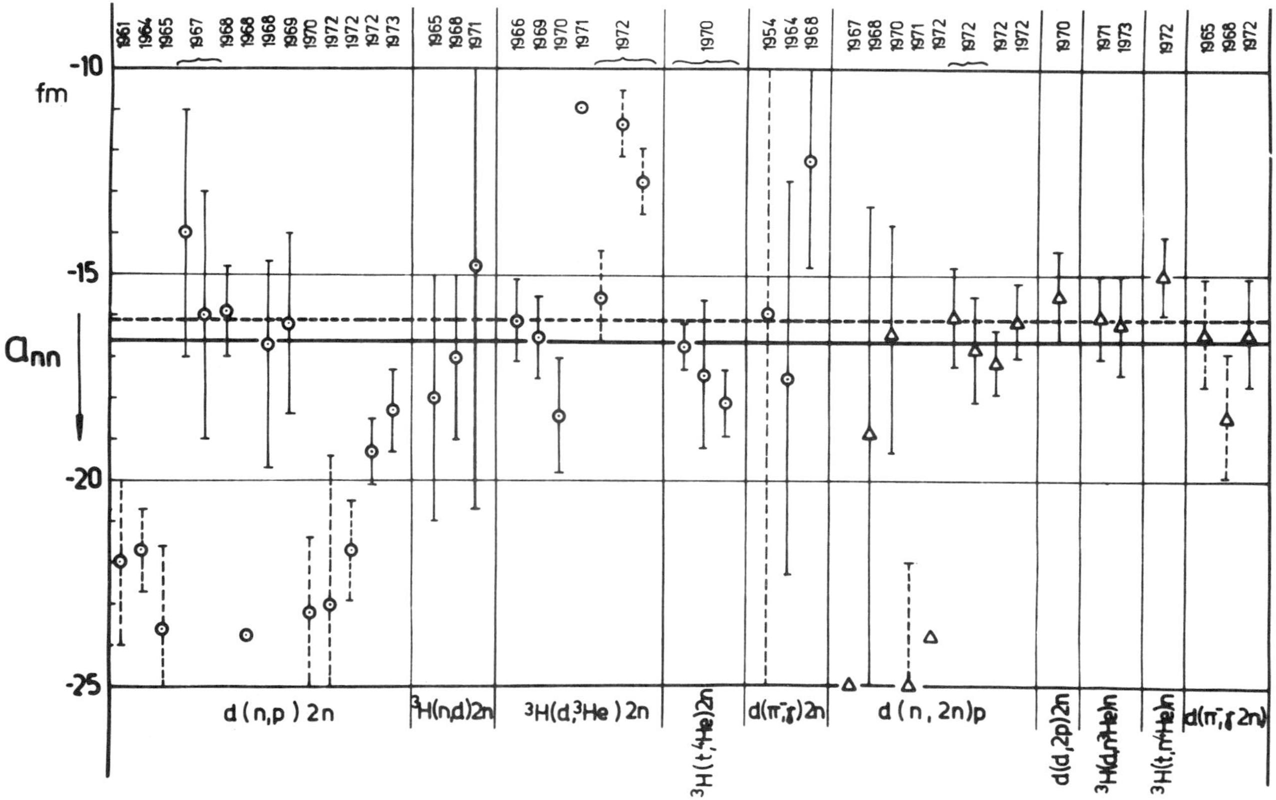

Fig. 1 The values of the n-n scattering length, obtained from the reactions indicated below. At the top of the figure the year of publication of the given result is noted. The full line gives the weighted average value. The broken line defines the weighted average value of the complete measurements alone.

ANALYSIS OF RESULTS OF FEYNMAN GRAPH CALCULATIONS FOR NUCLEON-NUCLEON SCATTERING*

J. L. Gammel and M. T. Menzel, Los Alamos Scientific Laboratory, University of California, Los Alamos, New Mexico 87544, USA

Fleischer,[1] Wortman,[2] and we have done calculations using the interaction

$$\mathcal{L}' \doteq i(4\pi)^{1/2} g\bar{\psi}\gamma_5\tau\psi\cdot\phi + 4\pi\lambda(\phi\cdot\phi)^2 \tag{1}$$

The results are contributions to coefficients A, B, C, D, E, where, for the $^3P_0$ state (say), the phase shift $\delta(^3P_0)$ is

$$\tan\delta(^3P_0) = A\left(\frac{g^2}{4\pi}\right) + B\left(\frac{g^2}{4\pi}\right)^2 + C\left(\frac{g^2}{4\pi}\right)\lambda + D\left(\frac{g^2}{4\pi}\right)^2\lambda^2 + E\left(\frac{g^2}{4\pi}\right)^3 + \ldots \tag{2}$$

One pion exchange dominates for high angular momentum except for graphs like

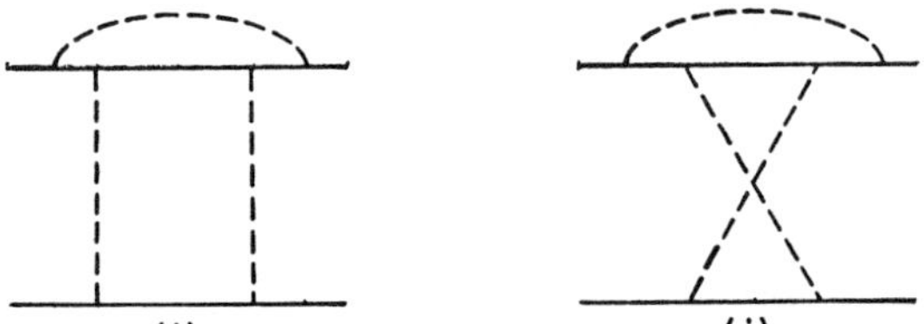

(i) (j)

Arguments have been advanced that the contributions of similar higher order graphs (two pion exchange with corrections to higher order scattering) damp the contributions of such graphs.[3]

Large classes of graphs which require renormalization contribute negligibly to the coefficients (see reference 4 for fourth order results). We report some sixth order contributions which confirm this observation. The graphs reported are

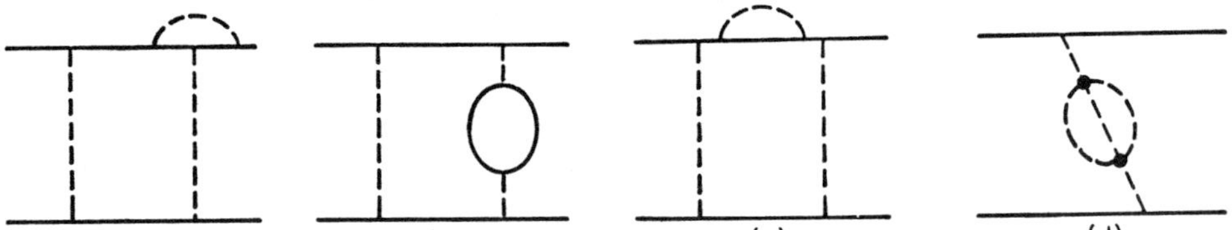

(a) (b) (c) (d)

The results are in Table I and should be compared to the results of reference 2.

The series in λ for a fixed power of g^2 appear to converge for $\lambda \approx 0.1$, the value commonly accepted. It would be of interest to have more calculations confirming this, especially for graphs related to the ω-meson.

If one sums up the contributions to tan $\delta(^3P_0)$ in Table II, he gets a result in spectacular agreement with experiment as shown in Table III. However, the $^3P_2$ phase shift is very bad. It is possible that the anomalous threshold dependence of the Born term of the $^3P_2$ phase shift, which is corrected by the use of matrix Padé approximants, is the source of the trouble.

REFERENCES

* Work done under the auspices of the U. S. Atomic Energy Commission.

1) J. Fleischer, J. L. Gammel and M. T. Menzel, Phys. Rev. D 8 (1973) 1545

2) J. L. Gammel and W. R. Wortman, J. Comp. Phys. 10 (1972) 568

3) E. M. Henley and M. A. Ruderman, Phys. Rev. 92 (1953) 1036

4) W. R. Wortman, Phys. Rev. 176 (1969) 1762

TABLE I

E_{LAB} (MeV)	graph (a) coeff E	(b) E	(c) E	E_{LAB} (MeV)	$(d)^*$ D
50	-3.62×10^{-5}	6.58×10^{-5}	2.28×10^{-5}	50	3.44×10^{-6}
100	-14.79×10^{-5}	18.28×10^{-5}	7.69×10^{-5}	100	8.99×10^{-6}
150	-31.47×10^{-5}	32.59×10^{-5}	15.28×10^{-5}	160	16.70×10^{-6}
200	-52.20×10^{-5}	48.66×10^{-5}	24.61×10^{-5}	200	22.16×10^{-6}
250	-76.53×10^{-5}	66.00×10^{-5}	35.55×10^{-5}	260	30.56×10^{-6}

$\lambda^2 = 36$ in the units used in preparing (d).

TABLE II. The factors involving $g^2/4\pi = 14.7$ and $\lambda = 0.1$ have been multiplied in.

	$A(g^2/4\pi)$	$B(g^3/4\pi)^2$	$C(g^2/4\pi)^2\lambda$	$D(g^2/4\pi)^2\lambda^2$
$^3P_1$	-0.350	0.243	-0.286	-0.101
$^3P_2$	0.041	0.284	-0.356	-0.169
$^1P_1$	-0.219	0.128	-0.321	-0.133
$^1D_2$	0.032	0.0382	-0.046	-0.012
$^3D_2$	0.267	0.088	-0.044	-0.011

TABLE III. The phase shifts are in degrees.

E_{LAB}	$^3P_0$		$^3P_1$		$^3P_2$		$^1P_1$		$^1D_2$		$^3D_2$	
(MeV)	th	exp	th	exp	th	exp	th	exp	th	exp	th	exp
50	15.2	11.2	-15.2	-8.5	-5.0	6.02	-0.3	-4°	0.8	1.77	8.8	9°
100	15.6	10.0	-26.3	-13.3	-11.3	11.0	-28.6	-15°	0.7	3.60	16.8	16°
200	-1.0	0.4	-41.1	-21.5	-22.1	15.9	-42.5	-20°	-0.7	6.66	26.7	25°

Two-Pion-Exchange Nucleon-Nucleon Potential*

A. D. Jackson, D. O. Riska and B. J. VerWest
Department of Physics, State University of New York, Stony Brook, N. Y. 11794, USA

We have constructed a nucleon-nucleon (NN) potential, which gives phase parameters in qualitative agreement with the results of phase shift analyses, from a model for the field theoretical elastic scattering amplitude. The model for the field theoretical elastic scattering amplitude is constructed using the framework of Chemtob, Durso and Riska[1]. In the elastic amplitude we have included the one-pion-exchange (OPE), two-pion-exchange (TPE), and one-omega-exchange contributions.

In the TPE amplitude we include the contributions from the two-nucleon intermediate state box and crossed box diagrams. To take into account the $\pi\pi$ interaction in the intermediate state, we have employed the crossed channel partial wave decomposition as suggested by Amati, Leader and Vitale[2]. The crucial quantities in the calculation are then the $N\bar{N}\to\pi\pi$ helicity amplitudes f_+^0 and $f_\pm^1$. For the amplitude f_+^0 we have constructed a model which gives values for $|f_+^0|$ which lie between the results of the analysis of Nielsen and Oades[3] and the model of Brown and Durso[4] as shown in figure 1. For the $f_\pm^1$ amplitudes we have made a model consisting of a

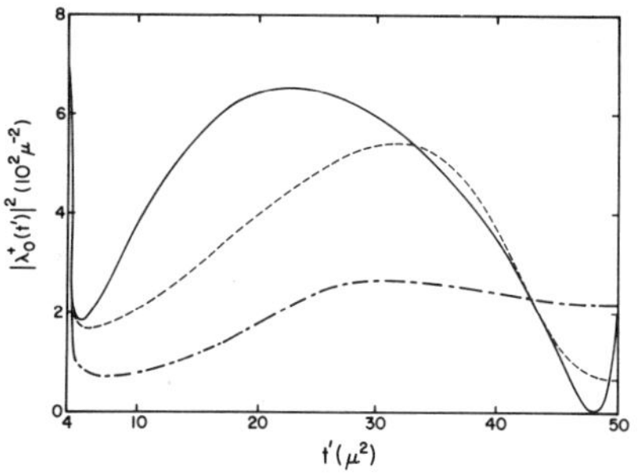

Figure 1. The $N\bar{N}\to\pi\pi$ s-wave amplitude $|\lambda_0^+|$, where $\lambda_0^+(t)=4\pi/(m^2-t/4)\ f_+^0(t)$, given by the Nielsen and Oades analysis (solid curve), the Brown-Durso model (dot-dash curve), and the model used in this calculation (dashed curve).

Breit-Wigner form describing the ρ-meson resonance and a smooth background term resulting from the nucleon pole. The ρN constants in the Breit-Wigner form are $g_\rho=2.73$ and $\kappa_V=3.25$. The iteration of the OPE potential was removed using the Blankenbecler-Sugar equation as described in ref. [1]. The ω-meson exchange was included with ωN coupling constants of $g_\omega=8.86$ and $\kappa_S=-0.12$.

The resulting momentum space potential is both non-local and energy-dependent. It has been regularized in the high momentum limit by the t- and u-channel cut off factors suggested by the relativistic eikonal approximation[5]. Using this regularized potential, we have solved the Lippmann-Schwinger equation numerically and

*Research supported by U.S.A.E.C. contract #AT(11-1)-3001.

have found that the phenomenological NN phase parameters are fairly well reproduced. The S- and low P-wave NN phase shifts are shown in figure 2.

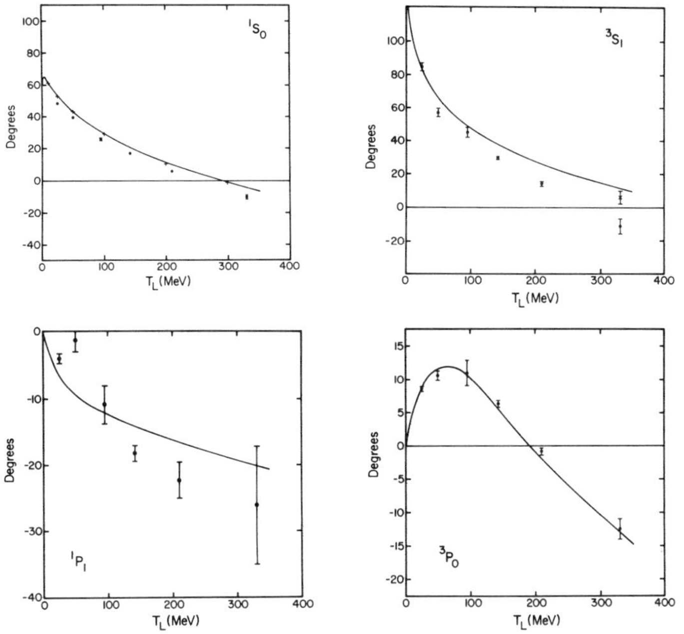

Figure 2. Low partial wave NN phase shift predictions from this potential model. The error bars represent the results of the MAW X energy independent analysis (dots) and the MAW X energy dependent np analysis (pluses)[7], and the Signell-Holdeman analysis (crosses)[8].

This potential gives a deuteron binding energy of 2.23 MeV and the deuteron wave functions in configuration space are shown in figure 3. The D state

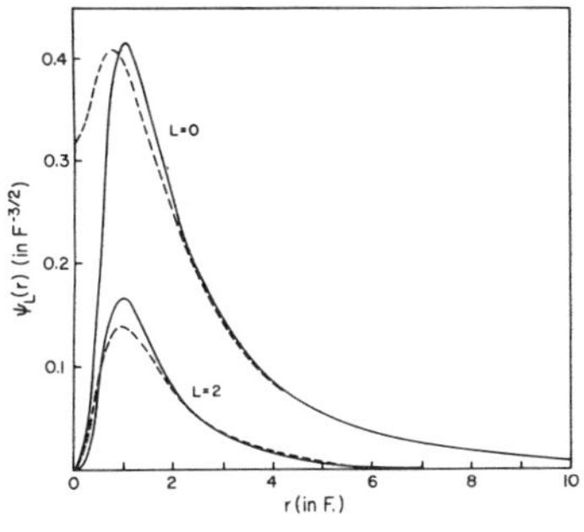

Figure 3. The configuration space deuteron wave functions given by this potential (dashed curve) and the Reid potential (solid curve).

probability is 6.1% and the quadrapole moment is 0.31 F^2. The potential contains much less short range repulsion than the conventional phenomenological local potentials and thus gives much weaker short range correlations. This can be seen in the comparison with the wave functions given by the Reid potential[6] which are also shown in figure 3.

REFERENCES

1. M. Chemtob, J. W. Durso and D. O. Riska, Nucl. Phys. B38,141(1972).
2. D. Amati, E. Leader and B. Vitale, Nuovo Cim. 17,68(1960); ibid. 18,409(1960); ibid. 18,458(1960); Phys. Rev. 130,750(1963).
3. H. Nielsen and G. C. Oades, Nucl. Phys. B49,586(1972).
4. G. E. Brown and J. W. Durso, Phys. Lett. 35B,120(1971).
5. R. Woloshyn and A. D. Jackson, Nucl. Phys. A185,131(1972).
6. R. V. Reid, Ann. Phys. (N.Y.) 50,411(1968).
7. M. H. MacGregor, R. A. Arndt and R. W. Wright, Phys. Rev. 182,1714(1969).
8. P. Signell and J. Holdeman, Phys. Rev. Lett. 27,1393(1971).

CROSS SECTION OF SLOW NEUTRONS ON PARAHYDROGEN

J. Callerame, D. J. Larson, S. J. Lipson, and R. Wilson
Department of Physics, Harvard University, Cambridge, Mass. 02138

The total cross section of neutrons on gaseous parahydrogen has been measured at several temperatures in the neighborhood of 20° Kelvin and at pressures ranging from one third of an atmosphere to one atmosphere. The runs were conducted at neutron energies of 4.2 and 4.6 milli-electron volts. At these temperatures and for the neutron energies used contamination of the coherent scattering of parahydrogen by the large, mostly incoherent scattering of orthohydrogen is kept to a minimum.

Our measurement was obtained by using slow neutrons from the M.I.T. Research Reactor which were doubly diffracted from two parallel graphite crystals and filtered through polycrystalline beryllium at 77°K to suppress higher order reflections. The neutrons were collimated by 5/8 inch holes in lithium loaded plastic and their transmission was observed through a 120-inch long scattering chamber filled with parahydrogen gas. The transmitted intensity was compared with that of an equivalent dummy approximately every four minutes. The temperature of the scattering gas was controlled by keeping the target in a constant termperature bath of liquid parahydrogen. The temperature was measured by measuring the vapor pressure of the bath. Both the vapor pressure in the bath and the pressure in the scattering sample were measured with an accurate quartz spiral manometer, calibrated to 1 part in 10^4.

From the transmission and the density of the scattering gas, a measured value of the cross section is obtained. The density was determined from the ideal gas law plus a second virial correction. In our analysis, we have chosen to use the virial coefficients given by Goodwin et al.[1] To extract a value of σ_s, the parahydrogen scattering cross section, from our measured value, σ_m, both neutron capture and the molecular motion of the gas must be properly accounted for. The former was subtracted from the measured cross section using the well-known 1/v law for capture at these energies.[2] The effect of the latter has been calculated by Hamermesh and Schwinger[3] as a function of neutron energy and gas temperature. With these corrections to our measured value and also taking into account the residual concentration of ortho-hydrogen, we get a value for σ_s, which is related to the coherent scattering length f by $\sigma_s = \frac{16}{9}\pi(2f)^2$ where $f=\frac{1}{2}(3a_t+a_s)$ and a_t and a_s are the triplet and singlet scattering lengths, respectively. Our results give a preliminary value of $f = -3.746 \pm .005 \times 10^{-13}$cm, in excellent agreement with the latest liquid mirror measurements of this parameter.[4] We observed no pressure dependence in the cross section at temperatures above 19.5°K, in contrast to the results of a previous experiment.[5]

---

[1]R.D. Goodwin, D.E. Diller, H.M. Roder, and L.A. Weber, Jour. of Res., Nat. Bur. of Standards, 68A, 121 (1964).

[2]A.E. Cox, S.A.R. Wynchank, and C.H. Collie, Nucl. Phys. 74, 497 (1965).

[3]M. Hamermesh and J. Schwinger, Phys. Rev. 71, 678 (1947).

[4]L. Koester and W. Nistler, Phys. Rev. Lett., 27, 956 (1971).

[5]T.L. Houk, D. Shambroom, and R. Wilson, Phys. Rev. Lett., 26, 1581 (1971).

NN POTENTIALS FROM ONE - BOSON EXCHANGES WITH HARD CORES.

M.M.Nagels[†], T.A.Rijken[†], and J.J. de Swart,
Institute for Theoretical Physics, University of Nijmegen, The Netherlands.

The present model E is almost the final stage in our series of local OBE potentials with hard cores. Models D and E are updated versions of models B[1)] and C[2)]. The potential forms, the treatment of broad mesons, and the hard core phenomenology are described in detail in a recent paper on model D[3)]. In all models we calculated NN and YN scattering due to the exchanges of nonets of pseudoscalar and vector mesons. In models B and D we have in addition a unitary singlet scalar meson ε, and in the models C and E a nonet of scalar mesons.

The main differences of D and E with the previous models are: i) η-X^0 mixing has been included; ii) the ε - pole parameters have been changed (see footnote 6 of ref. 2); iii) the potential forms have been changed slightly; iv) a drastic improvement of the NN fit to χ^2/data = 2.49 for model D, and χ^2/data = 2.33 for model E, compared to the χ^2 - surface data of the Livermore group[4)]; v) a larger value of α_{ps}= F/(F+D) ratio for the pseudoscalar mesons was needed in the fit to YN. Since the YN fit for model E has not yet been completed, we have taken α_{ps} for the time being from model D, bearing in mind that in the previous models B and C, without and with an octet of scalar mesons, the difference in the obtained values for α_{ps} from the fit to YN is only small.

The hard core radii are: x_s= 0.53674 fm for $^1S_0$; x_t= 0.50674 fm for $^3S_1$-$^3D_1$; $x_{L>0}$= 0.35323 fm for all higher L waves. The NN coupling constants are given in Table I, where the underlined numbers are constrained via SU(3). The main difference in the

	mass	$g^2/4\pi$		mass	$g^2/4\pi$	f/g		mass	$g^2/4\pi$
π	138.041	13.40	ρ	770(Γ=146)	0.694	6.07	δ	962	0.776
η	548.8	<u>7.43</u>	ϕ	1019.5	<u>0.466</u>	0.587	S^*	997	0.038
X^0	957.5	14.93	ω	783.9	12.00	0.689	ε	760(Γ=640)	26.58

Table I

coupling constants between E and D is the larger value of g_ρ in model E due to the nonzero value of g_δ. The additional central repulsion of the ρ is canceled largely by the central attraction of the δ, whereas the spin-orbit potentials reinforce each other. The low energy parameters for the s- and p- waves are given in Table II in powers of fm. Definitions are the same as in ref.5.

	$^1S_0$(pp)	$^3S_1$	$^1P_1$	$^3P_0$	$^3P_1$	$^3P_2$
a	-7.841	5.441	2.405	-3.002	1.824	-0.278
r	2.690	1.785	-7.095	3.437	-7.490	4.521
P	0.0341	-0.0097	-0.0019	-0.0064	-0.0006	-0.0106

For the deuteron we have: binding energy B = 2.224644 MeV (fixes x_t); d - state probability P_d = 6.09 % ; electric quadrupole moment Q = 0.2752 fm^2; $\rho(-B,-B)$ = 1.789 fm.

In table III the phaseshifts up to L = 3 are displayed.

T_{lab}	25	50	95	142	210	330
$^1S_0$	49.13	39.13	25.76	15.18	3.10	-13.21
$^3S_1$	79.79	61.49	42.89	30.25	17.16	1.25
ε_1	1.96	2.51	3.30	4.25	5.91	9.57
$^3P_0$	8.70	12.00	10.68	6.29	-1.01	-13.19
$^3P_1$	-4.78	-8.07	-12.28	-15.86	-20.42	-27.45
$^1P_1$	-5.69	-8.01	-10.27	-12.31	-15.46	-21.28
$^3P_2$	2.37	5.66	10.47	13.57	15.71	16.30
ε_2	-0.78	-1.71	-2.73	-3.14	-3.08	-2.21
$^3D_1$	-2.78	-6.42	-11.74	-15.84	-19.94	-23.54
$^3D_2$	3.75	9.39	18.75	26.24	33.02	37.52
$^1D_2$	0.64	1.60	3.54	5.71	8.67	12.29
$^3D_3$	0.06	0.38	1.63	3.39	5.90	8.84
ε_3	0.54	1.59	3.32	4.68	6.01	7.21
$^3F_2$	0.10	0.32	0.74	1.09	1.37	1.11
$^3F_3$	-0.22	-0.66	-1.42	-2.05	-2.70	-3.41
$^1F_3$	-0.41	-1.11	-2.06	-2.66	-3.15	-3.65
$^3F_4$	0.02	0.10	0.40	0.88	1.80	3.81
ε_4	-0.05	-0.18	-0.49	-0.79	-1.17	-1.65

Table III.

† F.O.M. research associate.

References:

1) M.M.Nagels, T.A.Rijken, and J.J.de Swart, in Proceedings of the International Conference on Few Particle Problems in Nuclear Physics, Los Angeles, California, 1972, edited by I.Slaus, S.A.Moszkowski, R.P.Haddock, and W.T.H.van Oers (North-Holland, Amsterdam, 1973), p.42.
2) M.M.Nagels, T.A.Rijken, and J.J.de Swart, Phys.Rev.Lett.31,569 (1973).
3) M.M.Nagels, T.A.Rijken, and J.J.de Swart, preprint University of Nijmegen, July 1974.
4) M.H.MacGregor, R.A.Arndt, and R.M.Wright, Phys.Rev. 182, 1714 (1969).
5) M.S.Sher, P.Signell, and L.Heller, Ann. Phys.(N.Y.) 58, 1 (1970).

Precision Measurement of n-p Charge-Exchange Cross Section at 640 MeV

L. C. Northcliffe, M. L. Evans, G. Glass, J. C. Hiebert, Mahavir Jain, Texas A&M U.*; H. C. Bryant, C. G. Cassapakis, S. Cohen, B. D. Dieterle, D. M. Wolfe, U. of New Mexico*; B. E. Bonner, J. E. Simmons, LASL*; D. W. Werren, SIN and U. of Geneva*; C. W. Bjork, P. J. Riley, U. of Texas*.

At LAMPF we have begun a program of n-p scattering experiments by studying n-p charge exchange scattering at 643 MeV. There is considerable disagreement between previous measurements in the neighborhood of 650 MeV, some of which are shown in Fig. 1. The Dubna values at 630 MeV[1] are much higher than the recent PPA values at 647 MeV,[2] while measurements at SACLAY at about 645 MeV[3] fall somewhere in between. The disagreement is well beyond the normalization uncertainty assigned to each experiment.

The monoenergetic neutron beam and the spectrometer used in our measurements are described elsewhere in these proceedings.[4] Briefly, neutrons from the $^2H(p,n)2p$ reaction at 645 MeV, collimated at 0° to a half-angle of 0.1°, struck a liquid hydrogen target and protons from the charge-exchange reaction were observed in a spectrometer oriented at 0°. The spectrometer included a large magnet and four multiwire proportional chambers which served to determine the particle trajectory ahead of and beyond the magnet with angular precision ~ 0.06°. Typical bend angles were ~ 22°. Separation of protons and deuterons was achieved by measuring their times-of-flight and momenta through the spectrometer. Where possible, helium-filled bags were used in gaps so as to reduce multiple scattering. Data were logged event by event on magnetic tape and analyzed off line by numerical integration of the path of the particle through the spectrometer to determine its momentum.

It is important to know the spectrometer acceptance as a function of polar scattering angle θ, azimuthal angle ϕ, momentum P, and magnet current I. Because of the finite diameter of the neutron beam the origin of the scattering is indefinite and consequently for certain regions of θ, ϕ, P, and I there is some chance of the particle striking a magnet pole or missing one of the multiwire proportional chambers. In order to exclude the first possibility we reject events for which $\sin\theta \sin\phi > .02$ and calculate the fraction of yield lost when $\sin\theta > .02$ using the efficiency formula $\varepsilon = (2/\pi) \arcsin (.02/\sin\theta)$. The possibility of a particle missing one of the wire planes was eliminated by determining empirically the safe regions of θ for left-scattered and right-scattered particles, from plots of θ vs hit-wire position in each of the planes, at many different momenta and all magnet currents. Only data in the safe θ-regions were used in the analysis.

The detection of deuterons from the reaction $^1H(n,d)\pi^0$ along with the charge-exchange protons provides a convenient basis for an absolute normalization of the n-p cross section, good to ~ 7%.[5] The same method of normalization has been used by the SACLAY group.

Data were obtained from independent runs at six different magnet currents. The results obtained from eight of the runs, independently analyzed, were reasonably consistent. On Fig. 1 are shown the cross section values obtained by pooling the data for all eight runs. The error bars are statistical and do not include a 7% normalization uncertainty. A background subtraction has been made for protons observed when the liquid hydrogen target is empty.

Our cross sections fall between the PPA values and the Dubna values. They are on the lower edge of the error band of the LRL phase shift analysis[6] which was based on the Dubna data. In the region 172°-176° they appear to be in agreement with a parameterization of the SACLAY results, but near 180° our values are 10-15% higher.

Some additional data at 643 MeV are available for analysis. In addition we have obtained extensive data at 800 MeV over a wide angular range (spectrometer angle 0-60° lab) which have not yet been analyzed. We plan to extend our 643 MeV measurements to a similar wide angular range in the near future.

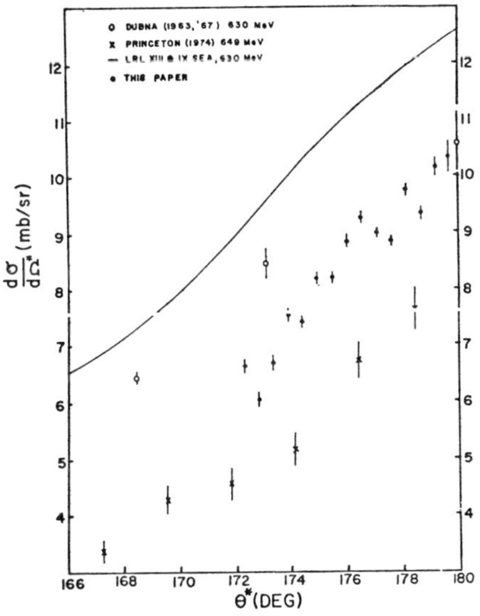

Fig. 1. Differential cross sections in the center of mass at backward angles for n-p scattering around 640 MeV. Also shown is an LRL phase shift fit at 630 MeV. Note displaced zero in both ordinate and abscissa.

*Work supported in part by the U. S. Atomic Energy Commission.

1. Z. Janout, et al., preprint, Joint Institute for Nuclear Research, Dubna, 1966.
2. P. F. Shepard, et al., Phys. Rev. (to be published).
3. G. Bizard, et al., contribution to IInd Aix-en-Provence International Conference on Elementary Particles (Sept. 1973).
4. C. Bjork, et al., contribution to the Few Body Conference at the Universite' Laval, Quebec (August, 1974).
5. C. Richard-Serre, et al., Nucl. Phys. B20, 413 (1970).
6. M. H. MacGregor, et al., Phys. Rev. 169, 1149 (1968), and Phys. Rev. 173, 1272 (1968).

OBE - Models and Regge Poles

T.A. Rijken†

Institute for Theoretical Physics, University of Nijmegen, The Netherlands.

The connection between OBE and Regge Poles can be derived in the framework of the Analytic S-matrix Theory using the "New Strip Approximation" (For details and references see [1,2]). For this purpose in particular the Khuri-Jones representation [3,4] for Regge Poles is useful. We restrict ourselves to the discussion of the "direct" nucleon-nucleon (NN) potential, and consider only that part of the NN-potential which is generated by the double spectral functions etc. in the regions where $s \geqslant s_R$ and $u \gtrsim s_R$, here s_R is such that both in the s- and u-channel Regge Pole behaviour has set in for $s = s_R$ and $u = u_R$ respectively. Making the usual Regge analysis (Sommerfeld-Watson transformation etc.) for the "direct"-potential $V(t,z_t)$ and assuming for the signatured $V^{\pm}(t,J)$ the bound

$$V^{\pm}(t,J) \sim \exp\left[-(J+\tfrac{1}{2})\right]\xi(z_R)\ , \quad |J| \to \infty \quad , \tag{1}$$

where $\xi(z_R) = \ln(z_R + \sqrt{z_R^2-1})$ with $z_R = -1-2s_R/(4M^2-t)$, one can include the background integral in the Regge Poles etc. (see [3]). Neglecting Regge cuts, one arrives this way at a representation of $V^{\pm}(t,J)$ in terms of Regge trajectories and residues. Performing the t-channel partial wave decomposition one gets for a single trajectory [3,4,5]

$$V_{\lambda_2\lambda_2';\lambda_1\lambda_1'} = \sum_{J=\lambda_{max}}^{\infty} (2J+1)\,\mathrm{sgn}(\eta;\lambda,\mu)\ \xi_J^{\sigma}\left[\frac{J(J+1)}{\alpha(\alpha+1)}\right]^{\frac{|\lambda|+|\mu|}{2}} \cdot$$

$$\cdot\ \beta_{\lambda\mu}^{\sigma\eta}(t)\ \frac{e^{-[J-\alpha]\xi(z_R)}}{\alpha(t) - J}\ d_{\lambda\mu}^{J}(z_t) \quad . \tag{2}$$

where $\lambda \equiv \lambda_1' - \lambda_1$, $\mu \equiv \lambda_2' - \lambda_2$, η = naturalness, $\mathrm{sgn}(\eta;\lambda\mu)$ is the appropriate sign-function, σ = signature, ξ_J^{σ} = signature factor.

Using the helicity crossing matrix one obtains from (2) the s-channel potential. The series (2) converges very fast in the low energy s-channel region [1], such that the $J = 0$ (for $\sigma = +1$) or $J = 1$ (for $\sigma = -1$) term in (2) is already a good approximation for $V_{\lambda_1\lambda_2';\lambda_1\lambda_1'}$. Taking linear trajectories this leads to the OBE terms in the potential, with form factors coming from $\beta(t).\exp\{-(J-\alpha)\xi(z_R)\}$. Note that for NN the unpleasant singularities due to $\xi(z_R)$ occur at large momentum transfers: $-\infty < t < -s_R + 4M^2$ ($s_R \simeq 8M^2$, M = nucleon mass), a region which anyhow has to be dealt with phenomenologically using form factors. So, unlike in the $\pi\pi$-system, this shortcoming of the Khuri-Jones representation is no obstacle for low energy NN-calculations.

Important in this derivation of OBE, is the appearance of the J = 0 components of the P, f, f' and A_2 trajectories, which not yet have been considered in the NN-work. This effect of the P and f has already been emphasized by Chew [6] and the corresponding potentials are commonly included in the $\pi\pi$-bootstrap work (see [2]). The resulting s-channel NN-potentials from the J = 0 component of the Pomeron P (for details see [5]) are

$$V^{(s)}_{++;++}(s,t) = 4M^2 \cos^2 \frac{\theta_s}{2} G_P^2 \Delta_P(t) \tag{3a}$$

$$V^{(s)}_{++;--}(s,t) = -4E^2 \sin^2 \frac{\theta_s}{2} G_P^2 \Delta_P(t) \tag{3b}$$

$$V^{(s)}_{++;+-}(s,t) = -4ME \sin \frac{\theta_s}{2} \cos \frac{\theta_s}{2} G_P^2 \Delta_P(t) \tag{3c}$$

$$V^{(s)}_{+-;+-} = V^{(s)}_{++;++} \; ; \; V^{(s)}_{+-;-+} = -V^{(s)}_{++;--} \; ; \; V^{(s)}_{++;-+} = -V^{(s)}_{++;+-} \tag{3d}$$

where $G_P^2 = \gamma_{oo}(o)\,(\bar{s}_R/M^2)^{\alpha_P(o)}$, $\Delta_P(t) = (1/M^2)\, e^{t/4m_P^2}$. Here $\gamma_{oo}(t)$ is the reduced residue in $\beta_{oo}(t) \equiv 4[1-t/4M^2]^{\alpha_P+1} \cdot \alpha_P(t)\gamma_{oo}(t)$ and the parameter m_P, having the dimension of a mass, is given by $m_P^2 = \frac{1}{4}[a_P + \alpha_P' \cdot \ln(\bar{s}_R/M^2)]^{-1}$ with $\bar{s}_R \equiv s_R - 2M^2$. a_P comes from the parametrization $\gamma_{oo}(t) = \gamma_{oo}(o) \exp(a_P t)$.

The potential (3) can be used in e.g. the Blanckenbecler and Sugar- or the Schrödinger equation. After Fourier transformation one finds the potential $\mathcal{V} = \frac{1}{4ME} V$ for the Schrödinger equation, after making a local approximation [7],

$$\mathcal{V}_P(r) = \frac{G_P^2}{\pi\sqrt{\pi}} \cdot \frac{m_P^3}{M^2} \{[1 - \frac{m_P^2}{4M^2} - \frac{m_P^4}{2M^2} r^2] + \frac{m_P^2}{M^2}(L.S) + \frac{m_P^4}{4M^4} Q_{12}\} \, e^{-m_P^2 r^2} \quad . \tag{4}$$

Because $\alpha_P' \ll 1$, m_P is determined by the width of the diffraction peak: $\frac{d\sigma}{dt} \sim e^{8t}$, giving $m_P \simeq 250$ MeV. From the fit of Rarita et al. [8] we estimate $G_P^2/4\pi \simeq 7.2$, giving $\mathcal{V}_P(o) \simeq 350$ MeV. Similarly one finds from the f-trajectory a potential like (4) with the parameters $m_f \simeq 250$ MeV and $G_f^2/4\pi \simeq 6.5$, showing that the P and f trajectory give a sizeable repulsive NN-potential.

Acknowledgements

I would like to thank Prof. J.J.de Swart for encouragements and a critical reading of this note. Also discussions with the other members of the Institute are gratefully acknowledged.

References: [1]. G.F. Chew "The Analytic S-Matrix", W.A. Benjamin, Inc., New York (1966). [2]. P.D.B. Collins and E.J. Squires,"Regge Poles in Particle Physics", Springer Tracts in Mod. Phys., Vol. 45 (1968). [3]. N.N. Khuri, Phys. Rev. 130, 429 (1963) [4].C.E.Jones report UCRL-10700 (1962), unpublished. [5]. T.A. Rijken, Thesis University of Nijmegen, The Netherlands (1974), unpublished. [6]. G.F. Chew, Suppl. of the Progr. of Theor.Phys. 1965, p.118, Phys. Rev. 140, B 1427 (1965). [7]. M.M. Nagels, T.A. Rijken, and J.J. de Swart, preprint University of Nijmegen, July 1974. [8]. W. Rarita, R.J. Riddell, C.B.Chiu and R.J.M. Phillips, Phys. Rev. 165, 1615 (1968).

A MEASUREMENT OF PARITY VIOLATION IN PROTON-PROTON SCATTERING AT 15 MeV†

J. M. Potter, J. D. Bowman, C. F. Hwang, J. L. McKibben,
R. E. Mischke, D. E. Nagle
University of California, Los Alamos Scientific Laboratory, Los Alamos, New Mexico 87544
P. B. Debrunner, H. Frauenfelder, L. B. Sorenson
University of Illinois, Urbana, Illinois 61801

We report here a recent measurement of the parity violation in proton-proton scattering at 15 MeV by measuring the helicity dependence of the total cross section. The importance of this measurement and the need of lower upper-limit in the experimental results are sufficiently well known that we shall only refer to the discussion by E. Henley, presented at the 1974 Summer Study on High Energy Physics with Polarized Beams (Argonne National Laboratory, July 22-26, 1974) for further details.

The results of measurements showed a possible effect of $(1 \pm 4) \times 10^{-7}$ which is consistent with the predictions of V. Brown, F. Krejs, and E. Henley.[1] In order to have confidence in measurements of such a small order of magnitude, it is imperative that the methods by which these measurements were taken be extremely carefully scrutinized for any possible systematic errors. Therefore the remainder of this presentation will be devoted to the discussion of method of experimentation and systematic errors.

At the initial planning stage of this experiment two years ago, we rejected the using of a polarized proton target in favor of the method suggested by Simonius[2] for three reasons. Firstly, the polarized beam yields a three-fold greater polarization than a typical 1.2 K polarized proton target. Secondly, the radiation damage effect limits the magnitude of the incident beam intensity, and hence the time of running to accumulate the necessary statistics would be impractical. Finally, the presence of bound protons in typical polarized proton target material renders a very large dilution over the polarizable free protons which would not be desirable for observing effects of the magnitude which we would expect. Thus we designed our experiment to be carried out in the polarized proton beam of the Los Alamos Scientific

Laboratory's Tandem Van de Graaff. We located our experiment at the straight-through port so that no further spin precession would occur once the H^- ions leave the ion source and enter the accelerator tank. This location also greatly reduce the change of incident beam position at our apparatus due to the energy change in the incident beam. This polarized proton beam was produced by a Lamb-shift source,[3)] and had a typical beam polarization of approximately 85%. Its intensity at the exit of the ion source was about 180 nanoamperes, of which nearly 80% survives the accelerator and its associates beam transport system to impinge on our target and detector assembly. A frequency spectrum of this beam's intensity fluctuation was made, and we found that at 1 kHz, the intensity fluctuation was at a minimum. Therefore we chose 1 kHz as the frequency for our synchronous detection.

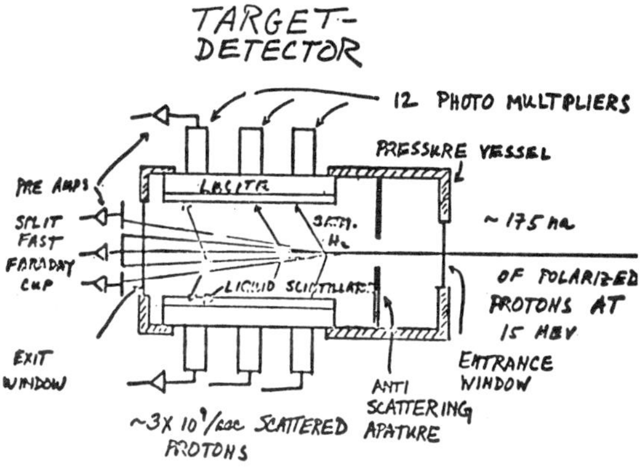

Target detector

FIGURE 1

Figure 1a depicts a schematic diagram of our target and detector assembly, and figure 1b is a photograph of the device. The target is hydrogen gas compressed to two to three atmospheres and contained in a pressure vessel whose ends were enclosed by thin Kapton windows. The sides of the pressure vessel were liquid scintillator cells separated from the gas target by 1/4 mil Mylar windows. Because the incident beam was about 150 nanoamperes at 15 MeV, it was inconvenient to utilize the ordinary method of measuring the transmission/incident beam intensity ratio to determine the total cross section. We therefore chose the method of measuring the scattered/incident beam intensity ratio as the alternative method to measure the total cross section. This method is entirely satisfactory so long as the solid angle of the scattered-particle detectors remained constant during the measurements. We shall go into more details in a moment to describe how we achieve this goal. The typical detected flux of the scattered particles as seen by our detectors were of the order of 0.2 nanoamperes which did not permit single particle counting. Moreover, if this flux is significantly reduced, the time required to accumulate sufficient statistics would be prohibitively long. Hence we decided upon the measure of an integrated anode current of the photomultipliers as a measure of the scattered particle flux. Twelve RCA6342 photomultipliers were placed in four rows of three each placed symmetrically about the incident beam. Preamplifiers were used one per each photomultiplier for current to voltage conversion prior to sending them to the counting room. These twelve signals were summed by the electronics in the counting room to form the (S) signal, and care was taken to insure each photomultiplier of contributing equally to (S).

After the proton beam passed through the gas cell and its kapton exit window, it was collected by a gold Faraday cup consisting of one center disk and four sector quadrants. The sectors permitted us to measure the deviation of incident beam position along two orthogonal axes which played an important role in the control of beam positions. Currents collected by these five gold pieces were also fed through individual preamplifiers and brought to the counting room. The sum of these five signals is called the (B) signal which was proportional to the transmitted beam intensity. A differential amplifier is used to form (S-B) which typically were set to zero

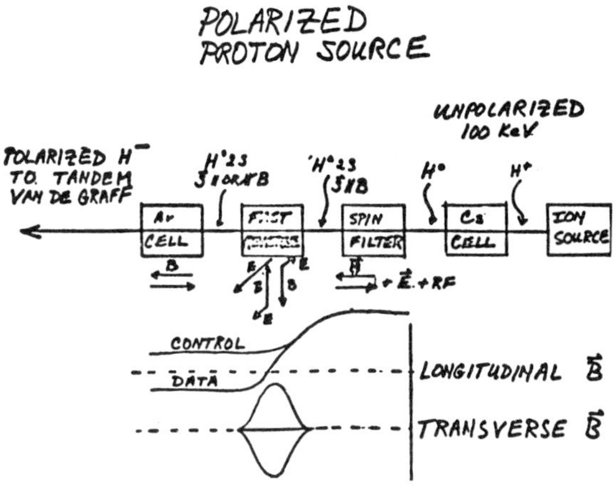

Polarized proton source

FIGURE 2

by adjusting the relative gains of (S) and (B) signals. An Analog Devices 425J divider was used to form Y=(S-B)/B, and the 1 kHz component of (Y) was measured by a PAR 186 lock-in amplifier. The output of this lock-in amplifier was read by a HP 2401C digital voltmeter a 1 second integration time intervals into the data acquisition computer. Figure 2 gives the schematic diagram of the data acquisition system.

Polarization reversal of the incident beam was accomplished by introducing a transverse magnetic field and a transverse electric field in the region of the Lamb-shift source between the spin filter and the argon cell. In addition, a d-c bias electric field deflects ions produced in this transition region due to the argon leakage from the argon cell. Figure 3 gives the longitudinal magnetic field pattern in the

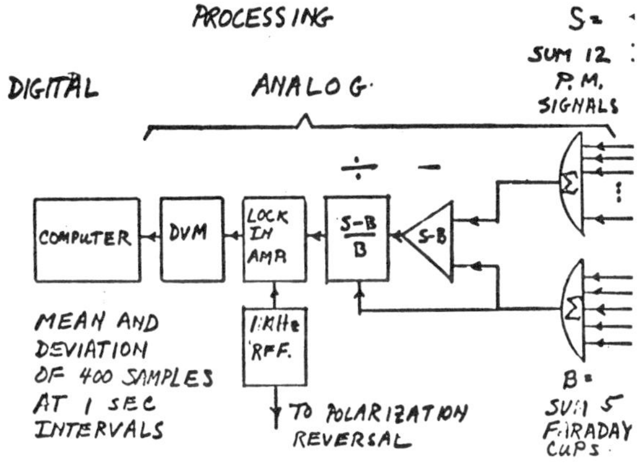

Processing

FIGURE 3

polarization reversal regions of the ion source. There were two combinations labelled as +,+ and -,- which did not change the beam polarization and two other combinations labelled +,- and -,+ which yielded beam polarizations of opposite directions. Thus the +,- and -,+ field configurations yielded polarization reversal as well as other systematic changes synchronous to the reversal frequency. However, the +,+ and -,- field configurations provided two control-conditions under which there were no helicity-dependent effect, yet almost all systematic-error effects remained. We shall see a little later how helpful this mode of polarization reversal proved to be in the final data reduction.

Of the many factors which contributed to the systematic errors, even taking into consideration the particular method of data acquisition used here, the most serious ones were the beam intensity changes, position changes, and incident angle changes at 1 kHz. Somewhat less severe was the transverse component of polarization present in the predominately longitudinally polarized beam. We shall now discuss each of these systematic errors in turn.

At the beginning of each run, the target and detector assembly was aligned to the best one can realize by optical methodes. The ion source was tuned for a known amount of intensity modulation at 1 kHz by the +,+ or -,- field pattern, and the asymmetry observed in the computer accumulated Y-signal was recorded and compared with the magnitude of the intensity modulation. From these observations, a beam intensity modulation rejection ratio of $2 \times 10^{+3}$ were obtained. Then the ion source was tuned for minimum beam intensity modulation which appeared in the (B) signal (typically less than 2×10^{-4}). In this way, we feel the systematic error due to beam intensity modulation in our final measurement was less than 10^{-7}.

To reduce the effect of phase space modulation in the incident beam which showed up as beam position and incident angle modulation at the target and detector assembly, we used a two stage feedback system to hold the beam position constant at two locations, one about 5 meters down stream from the vacuum tank of the Tandem Van de Graaff, and one at the Faraday cup of the detector assembly. Figure 4 shows the schematic diagram of the beam position control system. By measuring the differences of currents at the sectors in the Faraday cup, we estimated our typical beam movement to

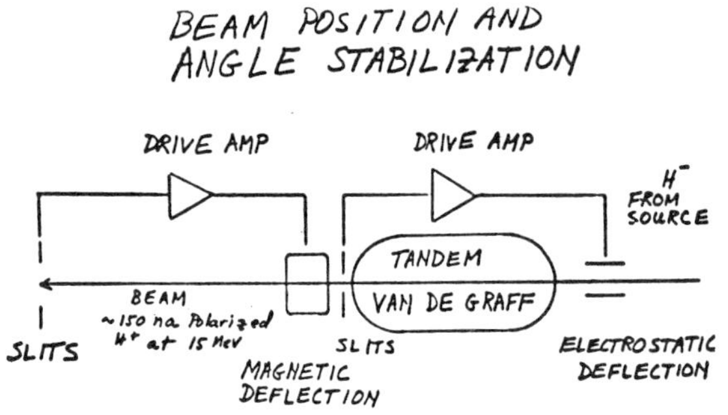

Beam position and angle stabilization

FIGURE 4

be nearly 1 micrometer during our data taking. By introducing deliberate position of 1 millimeter at the sectors, we observed an asymmetry of the order of 10^{-4} in the computer acquired Y-signal. Thus our overall rejection of the phase space modulation is again about 10^{-7}.

To determine the effect of transverse beam polarization, we precess the beam polarization until it was at its maximum value of transverse polarization. Then we measured the asymmetry in the computer acquired Y-signal and found a rejection ratio of 10^{-5}. Thus when we precessed the incident beam polarization to its longitudinal configuration, less than 1% transverse beam polarization was observed by using a carbon scatterer which could be flipped in and out of the incident beam at the center of the hydrogen target. Hence the contribution from the transverse beam polarization should also be of the order of 10^{-7}.

Finally, we measured electronically systematic errors due to ground loops and d-c offsets or non-linearities of the various electronic components and found them all to be less than 10^{-8}. We therefore believe that the predominate errors are statistical as exhibited by the control-experiments. By combining the results of the two spin-reversed runs, we arrive at our upper limit of a possible effect of $(1 \pm 4) \times 10^{-7}$. Figures 5 and 6 give the histograms of the control-runs and data-runs respectively plotted in data blocks of 400 seconds integration periods.

In order to explore the energy dependence of this parity violation effect, we have proposed to repeat these measurements at 6 GeV (ZGS, Argonne National Laboratory)

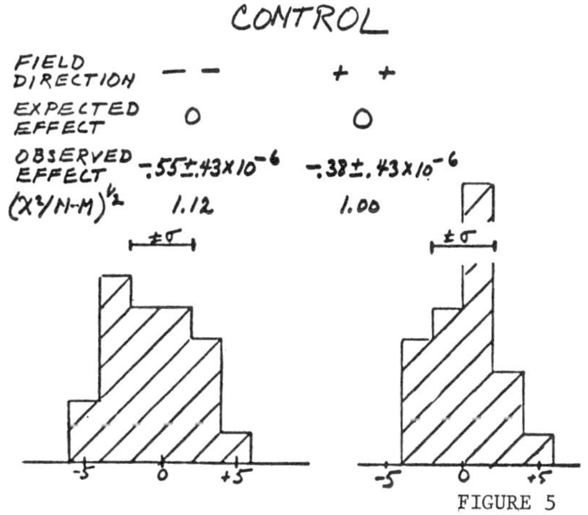

Control

FIGURE 5

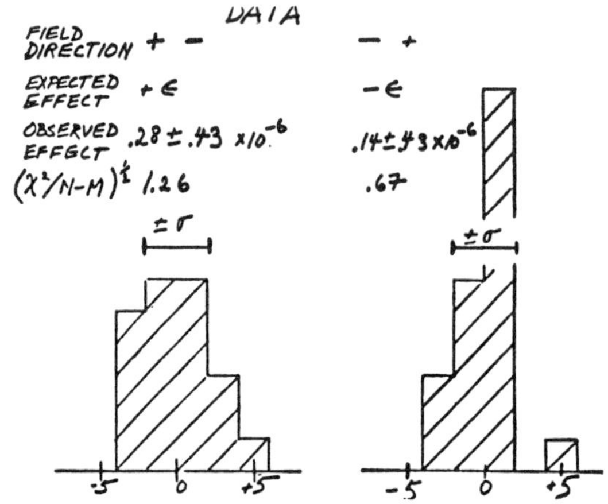

Data

FIGURE 6

in collaboration with the Anderson Group at University of Chicago. Preliminary runs are now scheduled in October of this year. We also have a proposal pending to repeat these measurements at 800 MeV (LAMPF, Los Alamos Scientific Laboratory) when its polarized ion source becomes operational.

REFERENCES

1) V. R. Brown, E. M. Henley, and F. R. Krejs, Phys. Rev. C 9, 935 (1974). Their prediction should be twice our results due to a difference in the definition of asymmetry.

2) M. Simonius, Phys. Letters 41B, 415 (1972).

3) G. P. Lawrence, G. G. Ohlsen, and J. L. McKibben, Phys. Letters 28B, 594 (1969).

---

† Work supported by the U. S. Atomic Energy Commission and NSF Contract No. GP 43671.

IMPROVED ONE-BOSON-EXCHANGE NUCLEON-NUCLEON INTERACTION

R. Machleidt, K. Holinde and K. Bleuler
Institut für Theoretische Kernphysik der Universität Bonn,
D - 53 Bonn, W. Germany

A one-boson-exchange-potential (OBEP) represented in the momentum space is constructed from the exchange of single $\pi, \eta, \sigma, \delta, \omega, \phi$ and ρ -mesons. The fit to the NN-scattering data is very satisfactory and the np low energy data and the deuteron data are reproduced accurately. The difference to the OBEPs, we presented before[1,2], is that this time we fitted the np data for $^1S_0$ instead of using the pp data.

The potential is regularized by strong interaction form factors of dipole type (see ref. 1). The meson parameters of the improved OBEP presented are given in Table 1.

Table 2 lists the results for the scattering, deuteron and nuclear matter data. The results are compared to those of the OBEP of Schierholz (SCH)[3], of Gersten, Thompson and Green (GTG)[4] and of Ueda, Nack and Green (UNG)[5] and to the Reid soft-core potential (REID)[6]. It is seen that the OBEPs of SCH, GTG and UNG show a great overbinding in nuclear matter. This is partly due to the bad fit to the scattering data (SCH), partly to the bad reproduction of the deuteron binding energy, and probably mainly due to too low lying ε_1-phase-shifts which is also reflected in the small $\varkappa$ of those OBEPs. The nuclear matter results are obtained in all cases by the same calculation in lowest order Brückner-Goldstone reaction matrix.

TABLE 1

Meson-parameters

	g^2	m (MeV)	Λ (MeV)		g^2	f/g	m (MeV)	Λ (MeV)	Λ_V (MeV)
π	14.	138.	2500.	ω	10.2	-	782.8	1300.	2500.
η	2.	548.5	2500.	ϕ	5.1	-	1020.	1300.	2500.
σ	5.215	500.	2000.	ρ	0.7	4.5	763.	1400.	2500.
δ	6.28	960.	2500.						

(for notation see ref. 2)

TABLE 2

Results for various momentum-space OBEPs and the Reid potential

		OBEP presented	SCH[3]	GTG[4]	UNG[5]	REID[6]	Experiment
χ^2/datum		3.4	25.0	5.3	4.3	3.8	-
effect. range	a_s(fm)	-23.69	-7.17	-24.41	-24.55	-17.17	-23.678 ±0.028
	r_s(fm)	2.68	3.40	2.68	2.71	2.79	2.51 ±0.15
	a_t(fm)	5.50	4.94	4.87	5.22	5.38	5.396 ±0.011
	r_t(fm)	1.86	1.79	1.76	1.81	1.72	1.726 ±0.014
Deuteron data	E_b(MeV)	2.224	2.910	2.985	2.510	2.235	2.22462±0.00006
	P_D (%)	5.74	4.85	4.88	4.40	6.43	2 - 8
	Q_D(fm^2)	0.282	0.249	0.249	0.264	0.281	0.278 ±0.008
minimum in nuc. matter	k_f(fm^{-1})	1.48	1.83	1.83	1.87	1.37	(≈1.36)
	E_b^f(MeV)	11.9	20.3	21.5	23.3	9.8	(≈16.0)
$\varkappa$ at k_f=1.5 fm^{-1}		0.118	0.090	0.067	0.059	0.163	-

REFERENCES:

1) K.Holinde, K. Erkelenz and R. Alzetta, Nuc. Phys. A194(1972)161

2) K. Erkelenz, K. Holinde and R. Machleidt, Phys. Lett. 49B(1974)209

3) G. Schierholz, Nuc. Phys. B40(1972)335

4) A. Gersten, R.H. Thompson and A.E.S. Green, Phys. Rev. D3(1971)2076

5) T. Ueda, M.L. Nack and A.E.S. Green, Phys. Rev. C8(1973)2061

6) R.V. Reid, Ann. of Phys. 50(1968)411

INTERACTION NUCLEON-NUCLEON
PERIODES DE DISCUSSIONS
RAPPORTS

NUCLEON-NUCLEON INTERACTION
DISCUSSION PERIODS
REPORTS

PION PRODUCTION IN NUCLEON-NUCLEON COLLISIONS

I.R. Afnan

School of Physical Sciences
Flinders University of South Australia
Bedford Park, S.A. 5042, Australia

Single pion production in nucleon-nucleon collisions is the most important inelastic channel below 1 GeV. In fact the pion production cross section is comparable in magnitude to the elastic cross section around 900 MeV.[1] Thus any attempt at understanding the nucleon-nucleon (N-N) interaction in the energy region from production to 1 GeV requires an understanding of pion production.

There are two ways pion production can help our understanding of the N-N interaction. First, as one is producing a particle of mass $\sim$ 140 MeV, one expects the production cross section to be sensitive to the short range behaviour of the N-N potential, and the off-shell behaviour of the scattering amplitude. In this way one may check the validity of different phenomenological potentials. Second, an understanding of the reaction mechanism may allow one to develop models which satisfy two- and three-body unitarity, and at the same time give a good representation of the production cross section. Such models may then be used to constrain the absorption parameters and thus reduce the number of solutions in any N-N phase shift analysis.

Before we can see how one may achieve the above aims, let us briefly consider the different pion production experiments that are in progress, or about to be started.

(i) $p + p \to p + p + \Pi^\circ$: This experiment to be performed at low energies, with a measurement of the γ-rays from the Π°. To distinguish between the Π° γ-ray and that produced by p-p bremsstrahlung, one needs to determine the spectrum of the γ-rays. The importance of this experiment is a direct measurement of the cross section σ_{11} (The indices refer to the initial and final iso-spin of the two nucleons.). The knowledge of σ_{11} allows one to determine σ_{01} from the reaction

$$n + p \rightarrow n + n + \Pi^{+} \qquad\qquad (1)$$
$$p + p + \Pi^{-},$$

since the cross sections for these two reactions is given by (1)

$$\tfrac{1}{2}(\sigma_{01} + \sigma_{11}). \qquad (2)$$

In the isobar model,[2] one assumes the production to go through an intermediate state of N Δ i.e.

$$N + N \rightarrow N + \Delta \rightarrow N + N + \Pi. \qquad (3)$$

An initial state of iso-spin zero is forbidden in this model since the Δ has iso-spin $^3/_2$. Thus a measurement of σ_{11} and $\tfrac{1}{2}(\sigma_{01} + \sigma_{11})$, will determine the importance of non isobar contributions to the production cross section. This is important if we are to use the isobar model in N-N phase shift analysis.

(ii) $p + p \rightleftarrows \Pi^{+} + d$: This reaction which dominates the production at low energies, has been shown to be sensitive to the deuteron wave function. In particular, the cross section is sensitive to the short range behaviour, and D-state probability of the deuteron[3]. Thus pion production and absorption can be used in conjunction with electron scattering to study the deuteron wave function. At the same time one can use this reaction to investigate the off-shell behaviour of both the N-N[4] and Π-N[5] scattering amplitude.

(iii) $n + p \rightarrow \Pi^{\circ} + d$: Assuming charge independence the cross section for this reaction should be ½ the cross section for $p + p \rightarrow \Pi^{+} + d$. The latest experiment,[6] which compares both total cross section and angular distribution, indicates that change independence holds to 1% in the energy region $E \leq 720$ MeV. Furthermore, this reaction is being considered for calibration of neutron beams in this energy region.

(iv) All reactions not included above, with the exception of elastic scattering, are of the form $N + N \rightarrow N + N + \Pi$. A proper analysis of such experiments, provided they are kinematically complete, can yield information that will be of value to N-N phase shift analysis. One may also be able to extract scattering amplitudes for $N + N \rightarrow N + \Delta$. Such amplitudes are useful in coupling the NN channel to the NΔ channel in a many nucleon system[7].

Having briefly surveyed the experiments discussed, we turn our attention to the theoretical models used for the analysis of pion production. The most commonly used model for pion production and determination of constraints on the absorption coefficients in N-N collisions is the isobar model[2]. Although a number of other methods such as soft pion theory, potential models, and dispersion relations have been used, the discussion centered about the isobar model.

Basically, the isobar model assumes that the production proceeds through an intermediate state of $N + \Delta$, where the Δ is the (3.3) Π-N resonance i.e.

$$N + N \rightarrow N + \Delta \rightarrow N + N + \Pi$$

The final N-N states are usually restricted to relative orbital angular momentum of 0 and 1. Finally, the amplitude for the process $N + N \rightarrow N + \Delta$ is assumed to be a slowly varying function of the relative Π-N momentum in the Δ. This last assumption allows one to take this scattering amplitude for a given angular momentum state to be constant.

It has recently been pointed out[8)] that such a model when applied to the reaction

$$\Pi + N \rightarrow \Pi + \Pi + N \tag{4}$$

can give the wrong sign for the $\Pi + N \rightarrow \Pi + \Delta$ scattering amplitude. The reason for this failure, is the fact that two- and three-body unitarity requires certain dependence for the $\Pi + N \rightarrow \Pi + \Delta$ scattering amplitude on the relative momentum of the Π-N in the Δ. If we write the production amplitude as

$$T_{2,3} = \sum_i f_i \tau_i \tag{5}$$

where τ_i is the two-body scattering amplitude for the j-k particles $(i \neq j \neq k)$. Then two- and three-body unitarity requires that f_i satisfy the equation

$$\text{Im } f_i = q_i \sum_{j \neq i} f_j \tau_j \tag{6}$$

where q_i is the relative momentum of the j-k pair. Since for reaction (4) f_i can be the amplitude for $\Pi + N \rightarrow \Pi + \Delta$ or $\Pi + N \rightarrow N + \rho$, then the imaginary part of f_i cannot be constant as the isobar model assumes. However, the importance of the momentum dependence of f_i is very much a function of how large is

$$\sum_{j \neq i} f_j \tau_j \ .$$

The above analysis can be extended to pion production in N - N collisions. In this case the sum over i in Eq.(5) will extend over the $N + \Delta$ and $\pi + d$ intermediate states. To determine quantitatively how good the isobar model is, may require a detail study of the effect of two- and three-body unitarity on the pion production cross section.

Another possible short coming of the isobar model, in its simplest form, is that it excludes initial state of the two-nucleons with isospin zero. In other words, σ_{01} is zero in the model. One can test this approximation experimentally as we have already discussed.

We now turn our attention to N - N scattering and phase shift analysis. Here some preliminary p - p phase shift analysis for energies between 100 - 700 MeV were presented[9)]. The interval 100 - 700 MeV was divided into six overlapping energy intervals. In each sub-interval an energy dependent phase shift analysis was performed. The resultant phases had some large discontinuities at the boundary between the different energy subinterval. The main cause for this, was the lack of sufficient data, particularly polarization cross sections and triple scattering parameters. One experiment that was suggested, was a total cross section measurement with polarized beams and polarized target.

One question raised as a result of the above phase shift analysis was the existence of $^1D_2$ resonance[10)]. The indication for the existence of such a resonance between 600 - 700 MeV was the sharp dip in the absorption coefficient with a possible sign change in the real part of the $^1D_2$ phase shift. If such a resonance does in fact exist, it should also appear in π-d elastic scattering, since the two channels are coupled. A relativistic three-body calculation of π-d scattering which includes both the Δ and the coupling to the N-N channel might reveal the existence of such a resonance.

The general conclusion was that one needs kinematically complete experiments to be able to extract interesting information from pion production in nucleon-nucleon collisions. Unfortunately, most of the present experiments on three-body final states are one arm experiments.

REFERENCES

1. W.O. Lock and D.F. Measday, "Intermediate Energy Nuclear Physics", Methuen & Co. Ltd. (1970).

2. S. Mandelstam, Proc. Roy. Soc. A244, 491 (1958); R.M. Sternheimer and S.J. Lindenbaum, Phys. Rev. 123, 333 (1961).

3. I.R. Afnan and A.W. Thomas, Phys. Rev. C10, 109 (1974).

4. H.C. Pradhan and Y. Singh, Can. J. Phys. 51, 343 (1973).

5. B. Goplen, W.R. Gibbs and E.L. Lomon, Phys. Rev. Letters 32, 1012 (1974).

6. S.S. Wilson et al., Phys. Letters 35B, 83 (1971).

7. A.M. Green and P. Haapakoski, "High Energy Phys. and Nuclear Structure", Ed. G. Tibell, 44 (1973).

8. R. Aaron and R.D. Amado, Phys. Rev. Letters, 31, 1157 (1973).

9. F. Lehar, presented at the Conference.

10. R.A. Arndt, Rev. Mod. Phys. 39, 710 (1967).

D1B. The status of potentials in the study of the two nucleon interaction.

Moderator: Joseph Levinger, Renselaer Polytechnic Inst., Troy, N.Y. 12181, U.S.A.

Rapporteur: Mark W. Kermode, University of Liverpool, Liverpool L69 3BX, England.

The main items discussed by the group were the concept of a potential, how is a potential obtained and how is it then handled, in particular in which equation is it to be used, and what are the alternatives ?

The concept of a potential in coordinate space has been extremely useful in gravitation, for example, and one is reluctant to abandon it for the two nucleon interaction unless there is strong pressure to do so. At present, it seems that any pressure is not sufficiently strong.

For the two nucleon case, it is customary to talk about the long, intermediate and short range parts of the interaction reflecting what we think we do know, might know and would like to know respectively. Of some of the potentials used in other calculations, the Yamaguchi potential (YA 54) neglects what we do know while the Reid potential (RE 68) takes account of one pion exchange but not of what we might know.

The description of the potential in the intermediate range, that is 1 - 3 fm, was the focus of attention of a number of the participants. Basically, their approach used knowledge obtained from high energy physics and relativistic field theory.

The Paris group (VI 73, LA 74) took into account the exchange of two pions, making use of known pion-pion and pion-nucleon data, together with the ω meson. Their potential, which was obtained from dispersion relation theory, is for use in the Schrodinger equation and contains no adjustable parameters for the intermediate range but to fit the experimental phase shifts a phenomenological soft core was added. The central and spin orbit components have small but significant energy dependence. A reasonable fit to the data with an omega coupling constant of 4.7 was much improved by increasing this constant to 10.7.

The problems associated with energy dependent, rather than velocity dependent potentials, for example the scattering states are not orthogonal, caused a little concern with the potential approach or at least with the potential plus Schrödinger equation approach.

A not too different report was given by the Stony Brook group (JA 74) who obtained a nucleon-nucleon potential also from one pion, two pion and one omega meson exchange using dispersion theory. The potential was constructed from a model for the field theoretical elastic scattering amplitude using the Blankenbecler-Sugar equation (CH 72). It was regularised using the relativistic eikonal approximation for the multiple vector meson exchange contributions (WO 72) and the appropriate parameter was determined from the nucleon electromagnetic form factors. In momentum space the potential is nonlocal and energy dependent. The potential in coordinate space was not calculated but the appropriate wave function indicates a weaker short range repulsion than most phenomenological potentials. The omega coupling constant was taken to be 6.24, which is not quite the same as its value in Paris. The phase shifts were calculated using the Lippman-Schwinger equation and are in good qualitative agreement with the experimental results. A quantitative fit was regarded to be unsatisfactory from their philosophical point of view since not all meson exchange contributions could be taken into account.

The question was raised of why it happened to be simpler to do these calculations im momentum space. Analyticity is simpler and one has integral equations rather than integro-differential equations to solve, but why should nature prefer momentum space ? No answer was given to this question but it was suggested that at present we do not have quite the right equations.

The problem of energy dependent potentials in coordinate space does not arise if we consider that the interaction is due solely to the exchange of single bosons of various kinds. This approach also satisfies ones desire to stay in coordinate space, although at the expense of a little phenomenology at short distances.

An offshoot of the Paris work took account of the exchanges of a pi, a rho and an omega meson together with a super soft core at short distances (TO 74a,b). The fit to the experimental data is good (χ^2/data = 2.3) and the potential has a form which is suitable for three- and more-body calculations. For comparison, the Reid potential gives χ^2/data = 2.7. It was stressed that quantitative fits should reveal χ^2/data rather than χ^2/phase shifts.

The Nijmegen OBE potential (NA 74, RI 74) on the other hand took cognisance of nine mesons and their widths for the intermediate range with the addition of an L-dependent hard core. It gives an equally good fit to the experimental data (χ^2/data = 2.3). Considerations of Regge trajectories and the pomeron lead to a Gaussian repulsive core in preference to a hard core and this work is still in progress.

One of the alternatives to a potential description of the nucleon-nucleon interaction suggested was the wave function, which is related to the potential if the wave equation is known, i.e. the wave equation, the wave function and the potential are inextricably connected. The wave function contains the information obtained by summing all effects. At low energies, we believe that the wave function satisfies the Schrödinger equation but the problem is that we do not know yet which equation to use at high energies. The recent relativistic equation of Gross (GR 74) which in the non-relativistic limit reduces to the Schrodinger equation with a potantial similar to the Reid soft core potential looks a promising candidate. One cannot obtain a complete knowledge of the wave function from phase shifts. Moreover, we should need to know the wave function at all distances and energies which seems far less economical than a potential and a wave equation. It was suggested that the potential V is really a kernel of some yet unknown integral equation.

In this context, the case of the deuteron was considered. From quantum mechanics, its wave function may be taken as a fundamental quantity and it is not really necessary to regard it as coming from the Schrödinger equation. Some knowledge of this wave function can be obtained from electron-deuteron scattering and this was discussed in some detail. In his lecture, Sprung (SP 74) gave an admirable account of these discussions.

In summary, the situation is that we are still attempting to obtain a two body description of a many body problem which has many internal degrees of freedom.

CH 72 Chemtob, M., Durso, J. W. and Riska, D. O., Nucl. Phys. B38 (1973) 141

GR 74 Gross, F., Phys. Rev. 10 (1974) 223

JA 74 Jackson, A. D., Riska, D. O., and Verwest, B., 1974, contributed paper at this conference.

LA 74 Lacombe, M., Loiseau, B., Richard, J. M., Vinh Mau, R., Pines, P. and de Tourreil, R., 1974, contributed paper at this conference.

NA 74 Nageld, M., M., Rijken, T. A. and de Swart, J. J., 1974, contributed paper at this conference.

RE 68 Reid, R. V., Ann. Phys. 50 (1968) 411.

RI 74 Rijken, T. A., 1974, contributed paper at this conference (also Nijmeken preprint).

SP 74 Sprung, D. W. L., 1974, invited paper at this conference.

TO 74a de Tourreil, R., Rouben, B. and Sprung, D. W. L., 1974, Orsay preprint IPNO/TH 74-14.

TO 74b de Tourreil, R., Sprung, D. W. L. and Rouben, B., contributed paper at this conference.

VI 73 Vinh Mau, R., RICHARD, J. M., Loiseau, B., Lacombe, M. and Cottingham, W. N., Phys. Lett. 44B (1973) 1.

WO 72 Woloshyn, R. M. and Jackson, A. D., Nucl. Phys. A185 (1972) 131.

YA 54 Yamaguchi, Y., Phys. Rev., 95 (1954) 1628.

TESTS OF SYMMETRIES OF THE NUCLEAR INTERACTION*

Markus Simonius
Laboratorium für Kernphysik, Eidg. Techn. Hochschule
Zürich, Switzerland

1. Parity violation in non-weak nuclear processes

First some history: The existence of a weak parity violating component in the nuclear interaction has been verified experimentally in many measurements, most of which are γ-decays of heavy nuclei[1]. Of particular interest for the subsequent discussion are Lobashov's measurement of the circular polarization P_γ of the emitted γ in thermal n-p capture[2]

$$n + p \rightarrow d + \gamma : P_\gamma = (1.3 \pm 0.45) \times 10^{-6} \tag{1}$$

and Wäfflers[3] latest analysis of the parity forbidden α-decay of $^{16}O^*(2^-, 8.87\ MeV)$

$$^{16}O^*(2^-) \rightarrow {}^{12}C + \alpha : \Gamma_\alpha = (1.03 \pm 0.28) \times 10^{-10}\ eV . \tag{2}$$

With the exception of the α-decay (2) the theoretical prediction using the "standard" π- and ρ-exchange parity violating nucleon-nucleon interaction are roughly two orders of magnitude smaller than the experimental values. The naturally arizing question who is at fault, the parity violating potential used or the nuclear theorists making the calculation (respectively the wavefunctions they use) is not settled yet. Obviously the most direct information would be obtained from observation of parity violating effects in nucleon-nucleon scattering as proposed and discussed two years ago[4]. The relevant observable is the analysing power A_L for longitudinally polarized nucleons

$$A_L = \frac{\sigma_+ - \sigma_-}{\sigma_+ + \sigma_-} \tag{3}$$

where σ_+ and σ_- are the cross sections for nucleons polarized completely parallel (positive helicity) and antiparallel (negative helicity) to the beam direction, respectively.

An experiment seeking A_L in p-p scattering is reported to this conference by a Los Alamos-Illinois collaboration[5]. The result they obtain is

$$A_L = (1 \pm 4) \times 10^{-7} \tag{4}$$

---

*Report on the discussion session on tests of symmetries and conservation laws, international conference on few body problems in the nuclear interaction, Québec, Aug.27-31, 1974.

at 15 MeV lab energy. However, as discussed in the appendix, there seems to be an error in the data reduction impliing that (4) has to be multiplied by two leading to

$$A_L = (2 \pm 8) \times 10^{-7} . \qquad (4')$$

The error shown is statistical. Based on Chester Hwang's presentation of these measurements a fairly large part of the discussion session was devoted to problems of systematic errors in this result. Their main sources are

- Influence of a nonzero component $P_{\perp}$ of the polarization perpendicular to the beam direction.
- Intensity and phase-space modulation produced together with the polarization reversal.

The sensibility to $P_{\perp}$, which is due to deviation from symmetry of the counting system with respect to the beam axis, can be tested with a beam polarized perpendicular to the beam and does not seem to make problems. Much more difficult are intensity and/or phase space modulations. They are an inherent feature of lamb-shift sources[4] as the one used to produce the polarized proton beam at the Los Alamos tandem accelerator. A carefully designed switching system allowed to reduce relative beam fluctuations to $\sim 2 \times 10^{-4}$ which is suppressed by additional three decades by the measuring system in which the scattered current is normalized electronically by the beam current. The problem is that this is not so easily tested - in particular for phase space modulations. The switching pattern used in the measurement is chosen such as to minimize and simultaneously continuously test the influence due to the beam modulations. The results are consistent with no systematic effect from this source with statistical uncertainty of the same magnitude as in A_L eq.(4) or (4'). They do, however, not completely exclude such an effect at the level of about twice the statistical error. In any case, this measurement, whose significance we shall discuss below, constitutes a great achievement. It is orders of magnitude more accurate than any scattering test of parity conservation done before and confirmes the feasibility of such experiments.

It is clear that these measurements have to be continued and improoved. One possible improovement is to use a ground state atomic beam source[4] for the polarized protons in which the two states of polarization are obtained by inducing two different rf transitions on the neutral beam. This is not expected to introduce beam modulations (at a level $\sim 10^{-7}$). Such a measurement is in an advanced state of preparation at the ETH tandem accelerator in Zürich (10-12 MeV). Appart from the different polarized ion source it uses also a different detector technique: A zylindrical ionization chamber of 40 cm diameter, filled with 2 - 4 atm. H_2 serves simultaneously as target and counter of the scattered beam. The ionization current from a 2 cm thick surface part of the zylinder as well as the Faraday cup signal are integrated and digitized about every .5 msec and transfered to an on-line computer for further evaluation.

What is the optimal energy? The energy behaviour of A_L is of the form[4,6,7]

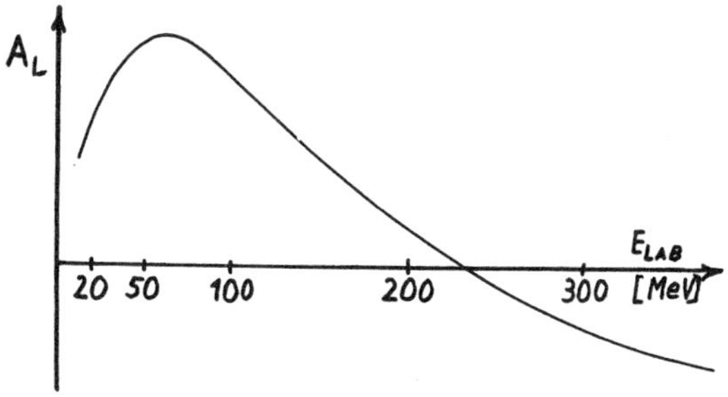

with maximum of the order 10^{-7} - 10^{-6}, depending on the choice of parity violating ρNN and ωNN couplings. This pronounced behaviour is essentially model independent and dictated by the behaviour of the well known $^1s_0$ and $^3p_0$ p-p phaseshifts which interfere with the dominant parity violating amplitude $f(^1s_0 \rightarrow {}^3p_0)$ in A_L and go through zero above 200 MeV. A qualitatively different picture is obtained only if $f(^1s_0 \rightarrow {}^3p_0) \approx 0$ due to accidental and improbable concellation of ρ^0- and ω-exchange in $f(^1s_0 \rightarrow {}^3p_0)$ in which case A_L is strongly suppressed below 200 MeV. At the maximum around 50 MeV one gaines, appart from a factor of two in effect, also in statistics since the target may be chosen much thicker than at 10 - 15 MeV and the polarized beam currents available are usually higher than for tandem accelerators (negative ions). In addition, available 50 MeV polarized proton beams are to our knowledge all from ground sate atomic beam sources with the advantages discussed above[4]. And last but not least the regular analysing power A which goes into the sensitivity to $P_\perp$ mentioned above is still small (~ 2%) at this energy, where we recall that the systematic error due to $P_\perp$ is

$$\sim A P_\perp \alpha \qquad (5)$$

with α the relative deviation of the counting system (efficiency and angular acceptance etc.) with respect to the (actual) beam axis[4]. There are plans also for such measurements at LAMPF and the Argonne ZGS (6 GeV)[5].

Tests in scattering of protons from other targets (planed in Los Alamos (C.Hwang) and at ETH Zürich) are of interest since they are sensitive also to the P V n-p interaction and in particular to one π exchange which has $\Delta I = 1$ isospin character and can not contribute to any of the measurements (1), (2) and (4). Of particular interest is low energy $\vec{p}$-d scattering since, in contrast to heavier nuclei, the regular analysing power A is reasonably small (systematic errors) and the n-p information can hopefully be extracted theoretically. n-p scattering itself seems, unfortunately, much more difficult.

In this connection the great interest should be reemphasized in measurements of PV γ-transitions or other reactions which are sensitive to the PV $\Delta I = 1$ interaction allone* which contains crucial information on neutral current contribution in the hadronic weak interaction[8] and could never be isolated yet. A new measurement was mentioned by A. Richter: $d + \alpha \rightarrow {}^6Li^*(0^+, T = 1{,}365\ MeV) \rightarrow {}^6Li\ (1^+, T = 0) + \gamma$ with a parity forbidden first transition. An upper limit of $|F|^2 \leq 2 \times 10^{-9}$ was found for the wrong parity admixture amplitude F to the intermediate state which, however, is still too large to yield the wanted information.

We now turn to the theoretical significance of the upper limit provided by (4) or 4') for A_L in p-p scattering in connection also with (1) and (2). The part of the parity violating nucleon-nucleon interaction relevant for the experiments (1), (2) and (4) is due to ρ- and ω-exchange with a presumably known strong coupling at one and an essentially unknown weak (parity violating) coupling at the other vertex. In the standard approach the latter is predicted from the weak hamiltonian using a factorization approximation. The usual Cabibbo current interaction yields then $P_\gamma \sim 10^{-8}$, i.e. a 2 order of magnitude to small result for thermal n-p capture (1). Different calculations do not agree very well. The latest one is ref.9 in which different models are discussed. Agreement is obtained[1], roughly, for the α-decay (2). For p-p scattering using {27} or $\Delta I = 2$ suppression in the factorization potential as first suggested in ref. 4, one obtaines $A_L \sim 10^{-7}$, depending somewhat on the ω-coupling and other details. This is an order of magnitude below the upper limit provided by (4'). However if one considers the possibility of an enhancement of ~ 2 orders of magnitude as suggested by the result on thermal n-p capture (and other γ-transitions) - which however would be at variance with (2)[1], though there are some questionmarks on the latter theory[6] - one sees that upper limits of the order of 10^{-6} allready give some useful restrictions.

For the evaluation of P_γ in thermal n-p capture or A_L in p-p scattering, matrix elements of the ρ-ω-exchange potential** discussed above have to be calculated with good

---

* Note that this could be obtained also from the difference of A_L in $\vec{p}$-d and $\vec{n}$-d scattering[4].

**One π-exchange is forbidden and 2 π exchanges not included in the ρ are negligible[6,7,9].

strong N-N wavefunctions. Assuming that these are reasonably well known one then obtaines from the experiments information on the three parity violating couplings $\rho NN^{\Delta I=0}$, $\rho NN^{\Delta I=2}$ and ωNN. However, due to the short range of ρ- and ω-exchange ($m_\rho^{-1} \approx m_\omega^{-1} \approx 0.26$ fm) it is clear that this information is sensitive to the less (or un-)known short range behaviour of the wavefunctions. Epstein[10] (supported by Noyes) therefor has proposed that one might use these effects to test these wavefunctions[†] and that the discrepancy between theory and experiment in thermal n-p capture indicates a strong short range enhancement of the n-p wavefunction rather than the supression obtained from the usual local repulsive core potentials. However, we want to stress here that this implies testing the N-N-wavefunctions with a certainly much less known interaction as probe. To begin with there's a problem of consistency: The general behaviour of the N-N system can be described by meson theory which yields the short range repulsion due to ω-exchange. Clearly where this description fails as drastically[*] as required by Epstein it is not possible to discribe the parity violating interaction by one-meson exchange either. However, even appart from these consistency questions we reemphasize that also for the one-boson exchange potential the parity violating couplings are not known. Already in the factorization approximation one obtains a wide variety of couplings from different models[9] and it is by no means clear wether the factorization approximation is applicable at all. The main object of these investigations is to determine these couplings. Only if the relevant ones are fixed from the data - and one number like the α-decay[(2)] is not enough for this[**] - is one in a position to test other assumptions. Deviations would then have to be analyzed taking into account the common meson theoretic features of the parity conserving and parity violating N-N interaction. However we emphasize that we do not expect large deviations since these would have considerable repercussions on other quantities, in particular the parity conserving n-p capture rate where the exchange current corrections are also sensitive to n-p short range correlations. It should be clear that the analysis has to take into account all available information also from other processes.

---

† For the energy dependence of A_L in N-N scattering this possibility was discussed in the first paper of ref.4.

* How drastic the proposed wavefunction enhancement in fact must be is seen from Epstein's result with nonlocal potentials which are 50 times larger even than the Born approximation[10] which contains no short range repulsion.

** Even appart from the uncertainty in the nuclear physics part of its interpretation.

2. Time reversal invariance in strong and electromagnetic interactions

Time reversal violation has been found up to now only in neutral K-decay[11]. All observations are consistent with Wolfensteins one-parameter superweak theory[12] which predicts that time reversal violating effects are unmeasurably small in all other known processes. However the possibility that time reversal violation originates in an electromagnetic or millistrong interaction is not excluded yet. This would lead to parity conserving time reversal violating effects of the order 10^{-3} in nonweak nuclear processes for which, however, no "safe" theory exists. The best upper limit is provided probably by the electric dipol moment of the neutron[13] $\leq 10^{-23}$ e·cm .

Thus nuclear physics tests aime to set upper bounds $\lesssim 10^{-3}$. The best upper limits obtained yet are ~ 3 times larger. The most important three classes of tests are (a) decay angular correlations, (b) detailed balance and (c) polarization tests in elastic scattering.

A new class (a) test has been published recently for the decay (501 keV) of $^{180}Hf^{m}$. From experience with parity violation a considerably enhanced effect might be expected. The observed interference phase angle $\sin\eta = 0.05 \pm 0.09$ thus implies perhaps a much stronger restriction for a T - violating interaction, though it seems hard to establish this "byond reasonable doubt".

Detailed balance tests are plagued by the fact that two absolute cross sections in inverse reactions have to be compared which implies that the normalization must be known to better than .1 % for a significant test. This difficulty may be circumvented by measuring relative differences in regions of strong fluctuations. A group in Bochum plans to redo this kind of measurements in the backward region of $^{27}Al + p \leftrightarrow {}^{24}Mg + \alpha$ aiming at a relative accuracy of ~ $.5 \times 10^{-3}$ (A. Richter). More accurate intermediate (or high) energy tests in simple systems - $n + p \leftrightarrow d + \gamma$, $p + p \leftrightarrow d + \pi^{+}$, $p + d \leftrightarrow T + \pi$ etc. - would, naturally, be of great interest. They are expected to be more sensitive to the short range part of the T-violating interaction (A_1-, and ρ-exchange[*16]). The γ-transitions are particularly sensitive to electro magnetic T-violation. Note that the latter yields also a long range N-N T-violating interaction due to $\pi\gamma$ - exchange to which nuclear physics tests are particularly sensitive[17]. (One π exchange does not contribute[16]).

---

* The extreme T-violating A_1-coupling postulated by Suderhan[14] gives large effects which are in conflict with n-p polarization data[15] (see below) and probably also with the upper limits obtained in other nuclear physics tests though no calculations have been done.

Polarization tests (c) of time reversal invariance in elastic scattering involve equality of the analysing power A for incoming polarized particles and the polarization P produced: P = A - tests. More complicated tests are possible also which have the advantage of not depending on the calibration of beam and analyser (Olsen)[18] which is paid for, however, by a loss in statistical accuracy. The polarization calibration may also be eliminated for P = A tests in $\vec{p}$ - ^3H and $\vec{p}$ - ^3He scattering proposed in refs. 18 and 19 by taking the ratio of P or A in the two reactions, which have the unique feature of having the same kinematics (up to the small ^3H - ^3He mass difference). An experimental accuracy of 10^{-3} or better should be possible here[18,19]. In general P and A should be measured for spin 1/2 particles in order to circumvent complications from higher order tensor components[19]. The simplest possible test systems have then the spin structure $1/2 + 1/2 \rightarrow 1/2 + 1/2$ since parity conservation forbids time reversal violation in elastic $1/2 + 0$ scattering. Intermediate energy (> 300 MeV) is required at least for n-p and even more so for p-p scattering (due to the exclusion principle, since the relevant partial wave transition amplitudes are strongly suppressed kinematically at lower energies[15,19]. The next simple systems are $\vec{p}$ - ^3H and $\vec{p}$ - ^3He scattering mentioned before for experimental convenience. The sensitivity of low energy tests in these reactions is expected to be fairly good[18] in particular for p - ^3He scattering.

In conclusion it is seen that upper limits of 10^{-3} or better might be reached by future accelerator tests of time reversal invariance. As was pointed out by A. Richter this brings us into a region where low energy bremsstrahlung starts playing a role yielding effects ($\sim 10^{-4}$) which look like T-violation (the γ is always outgoing). Hopefully the corrections can be calculated at least for the simple systems.

3. Charge and Isospin Symmetry

Isospin symmetry is broken by the electromagnetic but generally assumed to be conserved by the strong interactions. Apart from experimental difficulties (e.g. n-n scattering) the problem of its tests is the elimination of the coulomb interaction which is model dependent. It operates at several levels: coulomb potential between protons but also within the "strong" nuclear interaction through isospin breaking at the mesonic level. Thus the N-N interaction is well known to contain isospin breaking from the $\pi^{\pm}$ - π^0 mass difference, itself believed to be of electromagnetic origin. This does not, however affect the equality between "strong" n-n and p-p interaction, which leads to the stronger concept of charge symmetry.

One of the main testing grounds for charge symmetry are the low energy p-p and n-n scattering parameters, in particular the scattering lengths a_{nn} and a_{pp}. The main two problems are: extraction of a_{nn} from 3-body data and coulomb corrections for a_{pp}.

Within the uncertainty of ~ 1 fm in a_{nn} agreement is obtained with a_{pp} using standard local N-N potentials for the coulomb corrections. However, it was pointed out by Sauer[20], that these corrections, which depend on the not so well known short range behaviour of the N-N interaction, can be changed considerably by short range non localities which he introduces by unitary transformations. His conclusion is that one should not attempt to test charge symmetry in the low energy N-N scattering parameters but rather postulate it and use it as a constraint on the short range N-N interaction which has to be chosen such as to produce the right coulomb corrections. It should be remarked to this that it does not seem to be clear how the coulomb interaction has to be written in the presence of such "pathological" nonlocalities. Pure phenomenology, neglecting the underlying mesonic structure has too many unphysical possibilities. However, what Sauer's point shows is that charge symmetry or its breaking can only be tested or understood in conjunction with the meson-theoretic structure of the N-N interaction. Comparision of the scattering lengths a_{nn} and a_{pp} is of particular interest due to its sensitivity to details of the interaction.

Some of the problems discussed above can be bypassed by going to higher energies where the coulomb corrections are small. Zingl discussed a parametrization for the coulomb correction of the nuclear phase shifts of the form[21] $\delta_{pp} = \delta_{nn} + \alpha \frac{d}{dp} \delta_{nn}$ (p = CM - momentum) found empirically with a separable interaction [22]. $\alpha = \alpha_{\ell}$ depends on the angular momentum state and δ_{nn} denotes here the p-p sphaseshift without coulomb interaction.

4. Conclusions

In conclusion we emphasize that the aim of symmetry tests discussed are the understanding of the mechanisms of their braking rather than their general validity. As we have seen this mechanism consists in general of an interplay between the strong (symmetry conserving) N-N interaction and the specific symmetry braking interaction which itself depends in general on the strong interactions. However, these two can not be separated completely. As stressed also in other fields at this conference it is not possible to understand them on the basis of purly phenomenological N-N interactions without taking into account their common fundamental meson-theoretic structure (Noyes et al.).

Appendix: Discussion of the data reduction of ref.5

This discussion which leads to the result (4') instead of (4) is based on ref.5(II) which I obtained only at the end of the conference and was not available during the discussion session. The data reduction is not described in ref. 5(I). The relevant data are given in the following table copied from ref. 5(II):

TABLE 1. Experimental results for each ion source field configuration.

Source fields B_S	B_A	Beam helicity B_T OFF	B_T ON	Reversal	Ideal[a] form of data	Raw data $Y_f \times 10^7$	Corrected[b] Results $\times 10^7$
+	+	R	R	no	0	+ 6.0 ± 7.6	+ 4.7 ± 6.0
-	-	L	L	no	0	-10.6 ± 7.7	- 8.4 ± 6.1
+	-	R	L	yes	ε	+ 4.5 ± 7.7	+ 3.6 ± 6.1
-	+	L	R	yes	-ε	+ 2.9 ± 7.4	+ 2.3 ± 5.9

[a] ε is a hypothetical parity violation effect for purposes of illustration.

[b] Corrected for Coulomb scattering and net polarization change.

The lower two numbers of the last column are averaged, taking into account the sign of ε, to yield $F \equiv A_L$ given in eq.(4) where F is the notation used in ref.5.* Y_F is the effectively measured current asymetry. Up to the sign of ε in the fifth column, it is given by

$$Y_F = \frac{S_+ - S_-}{S_+ + S_-} = \bar{P}\,\frac{\sigma_+ - \sigma_-}{\sigma_+ - \sigma_-} = \bar{P}\,F \tag{6}$$

where $S_\pm$ is the measured scattered current and $\bar{P} = \frac{1}{2}(P_+ - P_-) = \frac{1}{2}(|P_+|+|P_-|)$ the average absolute beam polarization ± denoting polarization parallel and antiparallel to the beam (positive and negative helicity). The first equality in (6) is seen from an analysis of the measuring procedure of ref.5: The primary quantity is $Y = \frac{S - B}{B}$ where S = current from scattered protons and B = current from Faraday Cup with amplification adjusted to obtain S = B in the average. Y is measured by a lock-in amplifier in phase with the sigh of the polarization. It's output signal Y_F is thus given by (6). In the above table from ref.5(II) Y_F is multiplied by 1.35 to correct for the contribution from coulomb scattering and diveded by $2\,\bar{P} = 1.64$ instead of $\bar{P} = 0.82$ to obtain the last column which is thus a factor zwo to small as stated in the text to eq.(4').

* The data given in the figures of ref.5(I) differ from those in this table though according to Chester Hwang all the data have been taken in one measuring period.

References

1. M. Gari, Phys.Reports 6C(1973)317.
2. V.M. Lobashov, D.M. Káminker, G.I. Kharkevich, V.A. Kniazkov, N.A. Lozovoy, V.A. Nazarenko, L.F. Sayenko, L.M. Smotritsky and A.I. Yegorov, Nucl.Phys. A197(1972)241.
3. N. Neubeck, H. Schober and H. Wäffler, Phys.Rev. C10(1974)320.
4. M. Simonius, Phys.Lett. 41B(1972)415 and in Few Particle Problems in the Nuclear Interaction, ed. I. Slaus et al. (North Holland, Amsterdam 1972) p.249.
5. J.M. Potter, J.D. Bowman, C.F. Hwang, J.L. McKibben, R.E. Mischke, D.E. Nagle, P.B. Debrunner, H. Frauenfelder, L.B. Sorenson, proceedings of this conference (I) and submitted to Phys.Rev.Letters (II).
6. M. Simonius, Nucl.Phys. A220(1974)269.
7. V.R. Brown, E.M. Henly and F.R. Krejs, Phys.Rev.C9(1974)935. (The asymmetry A defined in this paper is 2 A_L).
8. R.J. Blin-Stoyle, Fundamental Interactions and the Nucleus (North Holland, Amsterdam, 1973).
9. H.J. Pirner and M.L. Rustgi, to be published.
10. G.N. Epstein, University of Oxford preprint, and paper submitted to this conference.
11. R.C. Casella, Phys.Rev.Lett.22(1969)554.
12. L. Wolfenstein, Phys.Rev.Letters 13(1964)180, T.D. Lee and L. Wolfenstein, Phys.Rev.B138(1965)1490.
13. L. Wolfenstein, Nucl.Phys. B77(1974)375.
14. E.C.G. Sudarshan, Proc.Roy.Soc. A305(1968)319.
15. J. Binstock, R. Bryan and A. Gersten, Phys.Lett 48B(1974)77.
16. M. Simonius and D. Wyler, contribution to this conference.
17. C.F. Clement, Ann.Phys.(N.Y.)75(1973)219.
18. M. Simonius, contribution to this conference.
19. M. Simonius, Lectures given at the Symposium on Nuclear Polarization Physics, Ebermannstadt, Oct. 1973 (Lecture Notes in Physics, in Press).
20. P.U. Sauer, contribution to this conference.
21. W. Plessas, L. Streit and H.Zingle, University of Graz, preprint.
22. L. Crepinsek, C.B.Lang, W. Plessas and H. Zingle, contribution to this conference.

Report on Nucleon-Nucleon Discussion Period D2A: "Experimental guide to experiments for the next decade, with stress on polarization"

Moderator: Jovan Jovanovich (University of Manitoba, Winnipeg, Canada)
Rapporteur: John Edgington (Queen Mary College, London, England)

Such a diffuse topic invited a wide-ranging discussion defeating even the valiant efforts of our moderator to guide the contributions into well-ordered and mutually exclusive channels. This attempt to synthesise the two sessions represents, therefore, a more-than-usually prejudiced selection, biassed heavily towards the polarization experiments of the title. The author's excuse is that without some such preference his report would be merely another compendium of possible experiments. The framework for this account is one of increasing beam energy.

N. Jarmie (LASL) drew our attention to serious discrepancies between recent proton-proton differential cross-section measurements between 10 and 20 MeV at LASL and Wisconsin, and earlier experiments. He made a case for accurate measurements of both cross-section and polarization in the p-p system, principally in the range 0-5 MeV. Uncertainties should, if possible, be reduced to less than 1% over the entire angular range. Besides their fundamental interest such data have practical applications in, for example, various energy resource programs involving fusion techniques. F. Seiler (LBL) agreed, adding that precise polarization measurements in other light element reactions could resolve present doubts about level schemes in certain A = 4 and 5 nuclei. As to whether such experiments were feasible, considering the small value of the polarization at low energies, G.R. Plattner (Basel) told us of measurements in the p-p system at 1, 1.5 and 2 MeV, which determined both $\frac{d\sigma}{d\Omega}$ and P to an accuracy of ±0.2%, at three angles. The motivation of these Basel experiments is the desire to extract the NNπ coupling constants in different charge states, using dispersion relations to extrapolate into the unphysical region. The polarization is used as a handle on the p-waves. M.J. Moravcsik (Oregon) pointed out that such a systematic program justified similar high-precision measurements on the n-p system, so as to extract both isosinglet and isotriplet phase-shifts in a manner independent of any p-p measurements.

At somewhat higher energies, but before the onset of copious meson production, conventional partial-wave analyses have been very successful in describing the data. Most experimental interest seems to lie in removing those uncertainties and ambiguities that remain, and in clarifying the energy-dependence of the phases, with the aim, presumably, of exposing the physics of the fundamental N-N interaction. It became clear that in achieving this aim a very close collaboration between the experimentalists, and the authors of phenomenological analyses, was essential to determine which variables were the most sensitive to uncertainties in the analysis, and thus most economical to measure. We were shown two ways in which this collaboration could work. In the first, one incorporates into an existing analysis certain extra "data", comprising arbitrary values of variables which are assumed to have been measured with a specified precision. Thus only a small amount of computation is needed to compare the usefulness of possible measurements. J.G. Rogers (U. of Alberta) illustrated this method at 50 MeV, where Bryan, Binstock and Arndt have shown that the n-p analysis would benefit most from a measurement of D_t. Also in the n-p system, we were shown how P. Signell had compared the usefulness of various proposed triple scattering and correlation parameter measurements near 330 MeV; the results of his work have considerably influenced the program of nucleon-nucleon scattering experiments to be carried out at LAMPF and TRIUMF.

The second method was illustrated by a written contribution from Ch. Weddigen (Karlsruhe). By working in terms of the scattering matrix rather than the phase-shifts, one obtains, in a rather perspicuous way, an idea of what experiments, to a given level of accuracy, are needed to resolve ambiguities arising from an "almost complete" set of measurements. This work, directed specifically to the p-p system between 400 and 600 MeV, is also relevant to future "meson factory" programs. A general comment by R.A. Arndt (Virginia Polytechnic Institute) was that more measurements were needed on the n-p system below 500 MeV, particularly to provide data points for his forthcoming energy-dependent analysis; and that above 500 MeV everything was obscure. He hoped to be better able shortly to incorporate inelastic effects at these energies, using a coupled channel approach.

L. Heller (LASL) made the point that, when only the elastic channel is open, the total cross-section is a most useful quantity, and relatively easy to measure precisely with present techniques. F. Lehar (Saclay) remarked that this conclusion remained valid at energies above the inelastic threshold, but that additional information was needed, namely, the total cross-sections with both beam and target polarized, in various orientations. Lehar showed, by reference to the phase shift analysis by Bystricky and himself (see the contributed paper in Section II) that appropriate combinations of these spin-dependent total cross sections could resolve ambiguities, particularly above about 500 MeV where conventional methods must introduce simplifying assumptions such as isobar dominance in order to be tractable. (Note that the beam and target spin combinations referred to here are quite distinct from those which, in the absence of P- and T- violating Hamiltonians, are necessarily zero. These other combinations were discussed at length in Discussion Period DIC.)

It thus became clear that in the next decade much effort will be devoted to measurements such as these, and of spin correlation parameters, which require that both beam and target be polarized. There was a lively discussion, led by C.F. Hwang (LASL) of the status of polarized target technology. The intense beams from future accelerators will probably require annealable targets, narrowing the practical choice to butanol. With this material, "frozen spin" targets and completely deuterated targets both appear possible. The use of "brute force" techniques to polarize HD at 0.02 K will probably have to await better refrigeration technology.

With this background, we were fortunate to hear a presentation by L.G. Ratner (U. of Michigan) of some experiments recently performed at the Argonne National

Laboratory. A high energy beam of polarized protons had been extracted from the ZGS, and the various spin-dependent total cross-sections for its interaction with a polarized target were measured. Data had been taken at 6 GeV/c and 3.5 GeV/c, and the measurements would soon be extended to lower momenta, possibly as low as 1 GeV/c. Both beam and target polarizations were at present restricted to being transverse to the scattering plane, but with a new magnet under construction, other relative orientations would become possible.

The data we were shown by Dr. Ratner were impressive, and in spite (or perhaps because of) the technological difficulties of this type of research, it seems that several laboratories intend to produce polarized targets and/or beams. The major obstacle would seem to be the matching together of suitable beams and targets, and the consensus of the participants seemed to be that this was a most promising area for inter-laboratory, and possibly international, collaboration.

Synthesis of the Nucleon - Nucleon Interaction at "Low" and "High" Energies[+]

G. Schierholz[++]

II. Institut für Theoretische Physik

der Universität Hamburg, Germany

[+] Rapporteur's talk given at the International Conference on Few Body Problems in Nuclear and Particle Physics, Quebec, August 27 - 31, 1974. Partly based on the discussion session held at this conference.

[++]Permanent address: CERN, Geneva, Switzerland

There is no question that any understanding of the two-nucleon interaction at low and medium energies will rest on our knowledge of the nucleon - nucleon (antinucleon) potential communicating with the crossed channel singularities. If we would only be interested describing the two-nucleon system such a potential is not necessary. However, it is most useful in various other aspects such as the deuteron problem, nucleon - nucleon bremsstrahlung, nuclear matter calculations, nucleon - antinucleon bound states, etc. Moreover, solely the potential approach is sufficiently comprehensive as to accumulate and put to account the several pieces of information we have on the two-nucleon interaction.

In the last couple of years the phenomenological potentials such as the Hamada-Johnston[1] and Reid[2] potentials have been successfully replaced by physical one-boson-exchange (OBE) potentials[3-9]. However, the status of the medium range forces and the actual shape of the core is still confusing, and it seems to me that any improvement of this situation will only come from higher energies (probing shorter distances) and adopting concepts of high energy physics.

So far there is no synthesis of the nucleon - nucleon interaction at low and high energies. But there is a great deal of interesting physics to be done.

Before I proceed presenting some concepts how this confusion might be overcome or resolved let me first briefly review where we stand theoretically. I understand that the phenomenological aspect of the present status of the nucleon - nucleon potentials (in particular its merits) will be reviewed in Kermode's talk. So I will ignore that topic.

In Table I I have compiled some candidates of OBE models[9] waiting for their final approval. It is striking (and I consider this a great success) that the various models now tend to agree on the magnitude of the vector meson (i. e., the ω and ρ) coupling constants, and even the coupling constants of everybody's favourite scalar mesons seem to agree (apart from Nagels et al.) when expressed in terms of some effective coupling constant G_{eff} (for the definition see Table I). What is surprising, however, is that the coupling constants do not take notice of the fine-structure of the various models, i.e., relativistic (off-shell) effects, short distance behavior (cut-off), parameterization of the (scalar) two-pion exchange, etc. In this respect hardly two

Ref.		$(g_\omega^V)^2/4\pi$	$(g_\rho^V)^2/4\pi$	g_ρ^T/g_ρ^V	$g_\sigma^2/4\pi$	m_σ	$g_\epsilon^2/4\pi$	m_ϵ	$g_\delta^2/4\pi$	G_{eff}^2 c)	cutt-off
3	Relativistic integral equation with generalized OBE potential	9.05	2.42	4.78	1.4	400	6.8	700	-	22.65	vector: $\exp\{(\alpha(t)-1)\ln s\}$ others: $(m^2-\Lambda)/(t-\Lambda)$
4	Schrödinger equation with regularized OBE potential	7.87	3.12	4.96	1.93	416	7.8	1070	1.83	18.05	$[\Lambda/(t-\Lambda)]^2$
5	Schrödinger equation with regularized OBE potential	10.00	2.33	5.18	-	-	13.9	783 b)	1.39	20.00	$[\Lambda/(t-\Lambda)]^2$
6	Blankenbecler - Sugar equation with regularized OBE potential	10.91	3.07	4.98	-	-	14.35	715 b)	2.41	23.08	$\Lambda/(t-\Lambda)$
7	Gross-equation	9.00	4.00	3.70 a)	2.14	364	-	-	-	16.20	no cut-off
8	Schrödinger equation with hard core	12.00	2.78	6.07	-	-	26.58	760 b)	0.78	35.00	individual hard cores for the different partial waves

Table I The most significant features of some recent OBE models

a) Input to the fit.

b) Finite width corrections.

c) $G_{eff}^2 = g_\sigma^2/4\pi\,(1/m_\sigma^2) + g_\epsilon^2/4\pi\,(1/m_\epsilon^2)$, or in case of finite width corrections $G_{eff}^2 = g_\epsilon^2/4\pi\,(1/(m_\epsilon^2 + 2\,m_\pi\Gamma_\epsilon))$; the dimension is $(\mathrm{GeV})^{-2}$.

of the present OBE models agree, and we may conclude that the low-energy nucleon - nucleon interaction ($\lesssim$ 400 MeV) is not sensitive to any of the details quoted above and, hence, cannot settle this confusion. In particular, this questions the use of any potential adjusted to the nucleon - nucleon phase shifts in other fields apart from improving its present status.

So we are led to ask how we can improve our understanding of the two-nucleon system and find a more convincing way of describing the nucleon - nucleon interaction at short distances and of handling the medium range (two-pion exchange) forces.

Let us first concentrate on the question how we can probe the core region and gain more insight in the nature of the short-range forces.

The short distance behavior of the nucleon - nucleon interaction is nicely reflected by the large-momentum-transfer behavior of the deuteron form factors[10] as, e. g., $G_c(q^2)$. If the deuteron is treated as a bound state of the Bethe-Salpeter equation or, what is equivalent here, of the Blankenbecler-Sugar equation[11] we obtain for large q^2 assuming the Bethe-Salpeter interaction being of the asymptotic form[12] $V((q-k)^2) \underset{(q-k)^2 \to \infty}{\simeq} [(q-k)^2]^{-1-\delta} \quad (\delta > 0)$:

$$G_C(q^2) \simeq G_E(q^2)\,[q^2]^{-\frac{3}{2}-\delta} \tag{1}$$

for the scalar and γ_5 coupling and

$$G_C(q^2) \simeq G_E(q^2)\,[q^2]^{-1-\delta} \tag{2}$$

for the vector coupling where G_E is the electric nucleon form factor, and similarly for the other form factors. The asymptotic behavior of the nucleon form factor is fairly well known[13] being $G_E(q^2) \simeq [q^2]^{-2}$ so that equs. (1) and (2) establish a direct relationship between the large-momentum-transfer behavior of the deuteron form factor and the nucleon - nucleon interaction. Unfortunately, the present status of the deuteron form factor data[14] does not allow any conclusions on the power δ . But we have learned[15] that a more informative experiment (at larger q^2) is considered at Stanford which certainly will be a sensitive probe of the nucleon - nucleon interaction. In equs. (1) and (2) double-scattering corrections are not taken into account as have been calculated by Blankenbecler and Gunion[16] and Chemtob, Moniz and Rho[17]. These contributions are expected to fall-off exponentially so that they do not survive asymptotically (provided that the potential is power behaved).

A further means of studying the nucleon - nucleon interaction at short distances is provided by the large-q^2 electro-disintegration of the deuteron (q is the photon four-momentum) near threshold. If we define the spin-averaged structure functions[18] W_1 and W_2 $(t = 0)$:

$$W_{\mu\nu} = \sum_{p,n} \langle d | j_\mu(0) | p,n \rangle \langle p,n | j_\nu(0) | d \rangle (2\pi)^4 \delta^4(q+d-p-n)$$
$$= \left(g_{\mu\nu} - \frac{q_\mu q_\nu}{q^2}\right) W_1 + \left(d_\mu - \frac{d \cdot q}{q^2} q_\mu\right)\left(d_\nu - \frac{d \cdot q}{q^2} q_\nu\right) \frac{1}{m_d^2} W_2 , \tag{3}$$

we find[19] for large q^2 and a Bethe-Salpeter interaction of the (asymptotic) form as given before

$$W_1 \underset{\omega' \approx 1}{\simeq} G_E^2(q^2)\, (\omega' - 1)^{2+2\delta} \tag{4}$$

for the scalar and γ_5 coupling and

$$W_1 \underset{\omega' \approx 1}{\simeq} G_E^2(q^2)\, (\omega' - 1)^{1+2\delta} \tag{5}$$

for the vector coupling where $\omega' = 1 - (d+q)^2/q^2$. Here we have assumed $G_E \simeq G_M$ for simplicity. A similar relation can also be derived for the structure function W_2. For small but fixed c.m. energy $s = (d+q)^2$ equs. (4) and (5) behave like $G_E^2(q^2)\,(q^2)^{-2-2\delta}$ and $G_E^2(q^2)\,(q^2)^{-1-2\delta}$ respectively as q^2 becomes large. In contrast to the situation of the deuteron form factor we have already good information here[20] up to $q^2 \approx 0.4$ GeV^2 which is worth analysing it with respect to δ even though q^2 might not cover an energy range being wide enough as to allow a precise conclusion. Experiments at higher q^2 would be welcome, and I understand there are various possibilities to do these kind of experiments (including the deuteron form factor, etc.), e.g., at DESY.

Another possibility of probing the core region and, perhaps, the most suggestive step towards a synthesis for the time being is to look at the nucleon - antinucleon system, a field which has been neglected theoretically but which is developing experimentally[21,22] in the last couple of years and occurs to me being very attractive vor various reasons.

It is well known that the negative G-parity exchange contributions such as, e. g., the π and ω exchange change sign when passing to the nucleon - antinucleon system. This means, e. g., that the ω exchange which is largely responsible for the repulsive core of the two-nucleon interaction will become attractive in case of the nucleon - antinucleon system. Hence, a combined analysis of the nucleon - nucleon and nucleon - antinucleon system would allow to single out the ω exchange contribution (i. e., the core), and I am sure this will have a drawback on various models.

We would conclude that the nucleon - antinucleon system is likely to have a couple of bound states lying considerably below the deuteron. Indeed, Gray et al.[21] found evidence for a $\bar{p}n$ bound state of 83.3 MeV binding energy corresponding to a meson of mass 1794.5 MeV. If one takes the one-boson-exchange model seriously there should even be many other bound states around the masses of the well established meson resonances.

It is tempting to pursue the alternative (to the quark model) that the mesons are nucleon - antinucleon (or, more general, baryon - antibaryon) bound states emerging in a bootstrap-like pattern. This, indeed, has proven to be a fruitful idea[23]. As a matter of fact the bootstrap principle puts some severe constraints on the dynamics of the two-nucleon system, in particular at short distances, as it turns out[23] that the baryon - antibaryon system has a self-consistent solution (where the baryon - antibaryon bound states are forced to coincide with the input mesons) incorporating the low-lying mesons (such as the π, ρ, ω, etc.) which correspond to binding energies of the order of 1 GeV. I believe this is a very good source of information on the two-nucleon system, especially on the nature of the short range forces.

With the existence of deeply bound nucleon - antinucleon states the hard-core models are ruled out. If one tries to fit the core parameters (Λ, α, etc., see Tab. I) to one of the nucleon - antinucleon bound states, one should be aware of the fact that there might be a whole series of bound states as suggested by the bootstrap model.

Everybody agrees that the one-boson-exchange diagrams have to be corrected at large-momentum transfer. There are many other methods how this region can be probed. But let us now ask for the dynamical origin of those short-range corrections.

It is time to face the truth: There are no elementary ρ, ω, etc. mesons, but all of them lie on Regge trajectories. The connection between one-boson and Regge-pole exchange is well established by means of the Khurie-Jones representation[24]. Using the "new" strip approximation[25], the Regge potentials result, in first approximation, in a "form factor" correction being of the form[26], e. g., for ω exchange

$$F(t) \approx \exp\left[(t - m_\omega^2)/m_\omega^2\right] . \tag{6}$$

In general, the Regge potential reads (neglecting the spin of the nucleons and assuming only one trajectory)

$$\begin{aligned} V(s,t) &= g^2 \frac{\pi}{\sin \pi\alpha(t)} \left(K^{\varphi}_{\alpha(t)}(-z_t) \pm K^{\varphi}_{\alpha(t)}(z_t) \right) \\ &= \sum_J (2J+1)\left(1 \pm \exp[-i\pi J]\right) \frac{g^2}{\alpha(t) - J} \exp\left[(\alpha(t)-J)\varphi\right] P_J(z_t) \end{aligned} \tag{7}$$

where K^{φ}_{α} are the so called Khurie functions[27], and the sum is over the (fixed J) resonance poles associated with the t-channel trajectory. The coupling constant g may, in general, be t-dependent. The parameter or function φ widely determines the analytic properties of the potential. Note that the Khurie functions $K^{\varphi}_{\alpha}(-z_t)$ have a cut at $-\infty < -z_t \leq \cosh\varphi$. There are various possibilities of determining φ. The only constraint is that the spurious singularities inherent in the Regge-exchange terms be removed in order to define a proper potential[27].

In the strip approximation, i. e., confining the dynamics requiring a potential to the strip $-s_R \leq s \leq s_R$, and let the potential solely describe the T-matrix beyond that region[25], we would have[28] $\varphi = \ln\left(z_R + \sqrt{z_R^2 - 1}\right)$ with $z_R = -1 - 2s_R/(4m^2 - t)$. In this case the potential becomes complex for $s > s_R$ (i.e., the Khurie function K^{φ}_{α} has a cut for $s > s_R$) which manifests itself in the partial-wave summation (7) becoming divergent for $s > s_R$. The disadvantage of this approach is that the potential has a (left-hand) cut in t for $t < -s_R + 4m^2$ arising out of $\varphi(t)$. This cut is supposed to be present in the t-channel partial-wave amplitude but not in the potential.

Another choice would be $z_R = -1 - 2s/(4m^2 - t_R)$ where $t_R = (2m_\pi)^2$ ($t_R = (3m_\pi)^2$) for, e.g., ρ (ω) exchange[3]. The Regge potential so obtained has the correct (t-channel) threshold behavior (i.e., a $(2\,m_\pi)^2$ and $(3\,m_\pi)^2$ branch point due to the 2π and 3π channels coupled to the ρ and ω mesons respectively) which should be of great importance at low and medium energies. Here $\varphi(s)$ gives rise to a left-hand cut in s which again should not be present in the potential. Such a cut is, however, to be expected in the partial wave projection of the potential (in the s-channel) so that this cut does not bother us as we are dealing with partial waves anyway.

In my opinion φ is most adequately chosen by imposing the proper double-spectral function boundaries on the potential[27]. If the double-spectral function boundary is chosen according to the diagram shown in Fig. 1, i.e.

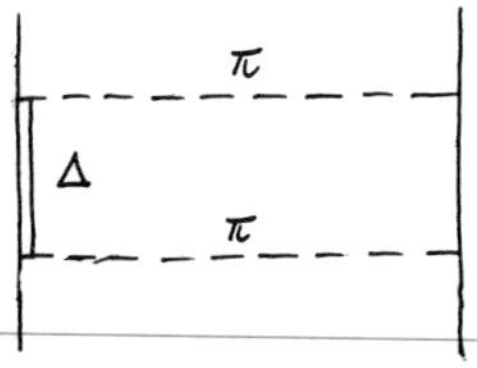

Fig. 1 Diagram contributing to the inelastic s-channel (elastic t-channel) double spectral function

$$s = b(t) = m^2 + m_\Delta^2 + \frac{2m^2 m_\Delta^2}{t - 4m_\pi^2} + \sqrt{\left(m^2 + m_\Delta^2 + \frac{2m^2 m_\Delta^2}{t - 4m_\pi^2}\right)^2 - m^4 - m_\Delta^4 + 2m^2 m_\Delta^2} \qquad (8)$$

which fairly well accounts for the elastic t-channel boundary, we would have $z_\mathrm{Z} = -1 - 2\,b(t)/(4m^2 - t)$.

Under normal circumstances the series (7) converges very rapidly and in most cases it is sufficient to consider only the first term corresponding to the original single meson exchange contributions. Treating the ρ, ω, etc. mesons as Regge poles thus provides a dynamical cut-off depending (more or less) on the Regge trajectory $\alpha(t)$ which is known over a fairly wide range from the high energy data.

The last topic I like to discuss concerns the medium range forces being widely described by σ, ϵ, δ, etc. exchange. This subject has caused all kinds of misunderstandings apart from its intrinsic problems how to incorporate finite width corrections.

Everybody agrees that a description of the medium range forces in terms of a few pole terms is a very poor approximation as the two-pion s-wave phase shift[29] indicates a very broad resonance (or even more). Several authors[5,6,8] have faced this problem replacing the propagator of a stable meson by a Breit-Wigner ansatz. This approximation is, however, very crude as it does not take into account any left-hand singularities.

The Regge-potentials discussed above are best suited as to incorporate finite width corrections. The problem then is to cook up any idea about the σ, etc. trajectories. In principle, the trajectories can be separated from the phase shifts by means of

$$-\mathrm{Im}\,\alpha(t)\,\xi_{\pi\pi} + \frac{1}{2}\ln\frac{\alpha(t)-J}{\alpha^*(t)-J} \underset{t\geq 4m^2}{=} \delta_J(t) \qquad (9)$$

being a consequence of ($\pi\pi$-)elastic unitarity and assuming that the coupling constant g (cf. equ. (7)) has no further cuts in t. Here $\xi_{\pi\pi}$ means the analogue of ξ for pion dynamics. This can be continued below threshold using

$$\alpha(t) = \alpha(\infty) + \frac{1}{\pi}\int dt' \frac{\mathrm{Im}\,\alpha(t')}{t'-t} . \qquad (10)$$

In practice, equ. (10) might, however, need some extra information on the large-t behavior of $\mathrm{Im}\,\alpha(t)$.

The main problem of the medium range forces lies, however, in the fact that the σ, etc. exchange is superimposed by the pomeron (having vacuum quantum numbers). To my knowledge the pomeron has not been taken into account explicitly though it is implicitly contained in the dispersion-theoretical two-pion exchange potentials to be discussed later on.

In the last couple of years the question of the dynamical origin of the pomeron has been largely settled. Drechsler[30] has shown that the pomeron exchange can be generated by multiple exchange of reggeons (or, e.g., pions) considering a sequence of excited intermediate states between the reggeons (pions)

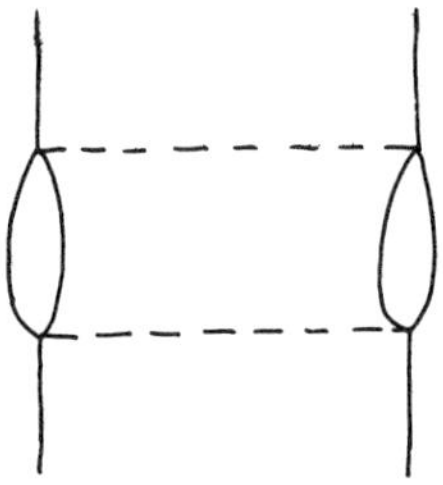

Fig. 2 Double-Regge (pion) exchange generating the pomeron. The dotted lines are reggeons (or, e. g., ordinary pions). The bubbles represent any excited intermediate state such as Δ(1236), etc.

as shown in Fig. 2. The lowest excited intermediate state one here can think of would be the Δ (1236).

Such inelastic contributions have been analysed with respect to the low-energy nucleon - nucleon interaction by several authors. It has been argued[31-33] that these contributions have the same effect as exchange of a scalar boson at least for not too short distances. This is, however, not to be confused with the "true" σ , etc. exchange contribution but has to be ascribed to the pomeron. Note that the pomeron has a $J = 0$ component (cf. equ. (7)) which has the same quantum numbers as the σ, etc.

We conclude that both the σ, etc. exchange and the pomeron contribution have to be included in the potential. The Regge potential (7) provides a canonical scheme of parameterizing the pomeron contribution as the pomeron coupling constant and the pomeron trajectory are well known from the high energy data (for details see Ref. 28).

Another (though much more elaborate) way of treating the medium range forces has been pursued by Cottingham and Vinh Mau and others[34-38] starting from a Mandelstam representation and the empirical πN- and $\pi\pi$-phase shifts as input. This approach avoids disentangling the various contributions to the two-pion exchange forces as it implicitly contains the σ , etc. and pomeron exchange. True, here are no free parameters involved. But if the once iterated pion-exchange is taken out the rest is not reliable at all as it depends very much on the actual πN- and $\pi\pi$-phase shifts. So, there will also be little hope that this method can be used to learn something about the shape of the pomeron. Nevertheless, it is a nice enterprise.

There are many other interesting features how the theory of low- and medium-energy nucleon - nucleon interaction could benefit from concepts of high energy physics and higher energy data than I could discuss in this talk. Even so not all the points I mentioned can be taken up immediately since we need better data at higher energies (say up to 1200 MeV) and higher q^2.

I like to thank the members of the session group for lively discussions.

References

1) T. Hamada and I.D. Johnston, Nucl. Phys. 34, 382 (1962).

2) R.V. Reid, Ann. Phys. 50, 411 (1968).

3) G. Schierholz, Nucl. Phys. B7, 483 (1968) and B40, 335 (1972).

4) T. Ueda and A.E.S. Green, Nucl. Phys. B10, 289 (1969).

5) R.W. Stagat, F. Riewe and A.E.S. Green, Phys. Rev. C3, 552 (1971).

6) R.A. Bryan and A. Gersten, Phys. Rev. D6, 341 (1972).

7) F. Gross, Stanford Linear Accelerator Center Report No. SLAC-PUB-1138 and Phys. Rev. D (in press).

8) M.M. Nagels, T.A. Rijken and J.J. de Swart, NN Potentials from One-Boson Exchanges with Hard Cores, contribution to this conference; see also Phys. Rev. Lett. 31, 569 (1973).

9) For a more complete survey see: G. Kramer, Springer Tracts in Modern Physics, Vol. 55, p. 152 (Springer, Berlin, Heidelberg, New York, 1970), M.J. Moravcsik, Reports on Progress in Physics, Vol. 35, p. 587 (London, 1972).

10) See, e. g., F. Gross, Phys. Rev. 142, 1025 (1966).

11) R. Blankenbecler and R. Sugar, Phys. Rev. 142, 1051 (1966).

12) C. Alabiso and G. Schierholz, Stanford Linear Accelerator Center Report (in press); for the spin-zero case see: C.Alabiso and G. Schierholz, Phys. Rev. D10, 960 (1974).

13) P.N. Kirk et al., Phys. Rev. D8, 63 (1973). For smaller q^2 the data is, however, not consistent with a dipole fit.

14) For a review see: U. Amaldi, Proceedings of the International Conference on Few Particle Problems in the Nucleon Interaction, Los Angeles, August 28 - September 1, 1972, p. 839 (North-Holland Publishing Co., Amsterdam, 1972).

15) S. Fiarman, private communication.

16) R. Blankenbecler and J.F. Gunion, Phys. Rev. D4, 718 (1971).

17) M. Chemtob, E.J. Moniz and M. Rho, Proceedings of the Aix-en-Provence Conference (1972).

18) See, e. g., J.D. Sullivan, Proceedings of the Summer Institut on Particle Physics, Stanford, July 9 - 28, 1973, Vol. 1, p. 289 (SLAC, 1973).

19) C. Alabiso and G. Schierholz, Stanford Linear Accelerator Center Report (in press).

20) See for example: R.E. Rand, R.F. Frosch, C.E. Littig and M.R. Yearian, Phys. Rev. Lett. 18, 469 (1967).

21) L. Gray, P. Hagerty and T. Kalogeropoulos, Phys. Rev. Lett. 26, 1491 (1971).

22) A.S. Carroll, I.-H. Chiang, T.F. Kycia, K.K. Li, P.O. Mazur, D.N. Michael, P. Mocket, D.C. Rahm and R. Rubinstein, Phys. Rev. Lett. 32, 247 (1974).

23) G. Schierholz, Nuovo Cim. 16A, 187 (1973).

24) N.N. Khurie, Phys. Rev. 130, 429 (1963), C.E. Jones, Lawrence Radiation Laboratory report UCRL-10700 (1962), unpublished.

25) See for example: P.V. Collins and E.J. Squires, Springer Tracts in Modern Physics, Vol. 45 (Springer, Berlin, Heidelberg, New York, 1968).

26) G.F. Chew, Progr. Theor. Phys., Suppl. 33-34, 118 (1965).

27) For details see: M. Kretzschmar, Zeitschr. Phys. 201, 33 (1967).

28) This case has been considered with respect to the nucleon - nucleon interaction by T.A. Rijken, OBE-Models and Regge Poles, contribution to this conference (communicated by J.J. de Swart).

29) S.D.Protopopescu, M. Alston Garnjost, A. Barbaro-Galtieri, S.M. Flatté, J.H. Friedman, T.A. Lasinski, G.R. Lynch, M.S. Rabin and F.T. Solmitz, Phys. Rev. D7, 1279 (1973).

30) W. Drechsler, Phys. Rev. D2, 364 (1970); see also the literature quoted there.

31) J.W. Durso and P. Signell, Phys. Rev. 135, B 1057 (1964).

32) S. Furuichi and W. Watari, Progr. Theor. Phys. 34, 594 (1965), 36, 348 (1966).

33) J.W. Durso, Phys. Rev. 149, 1234 (1966).

34) W.N. Cottingham and R. Vinh Mau, Phys. Rev. 130, 735 (1963).

35) M. Chemtob and D.O. Riska, Phys. Lett. 35B, 115 (1971).

36) G.E. Brown and J.W. Durso, Phys. Lett. 35B, 120 (1971).

37) M. Chemtob, J.W. Durso and D.O. Riska, Nucl. Phys. B38, 141 (1972).

38) W.N. Cottingham, M. Lacombe, B. Loiseau, J.M. Richard and R. Vinh Mau, Phys. Rev. D8, 800 (1973).

THE NUCLEON-NUCLEON INTERACTION WITHIN 1 FERMI

Franz Gross, Rapporteur, Department of Physics, College of William and Mary,
Williamsburg, Virginia 23185

In hindsight it appears that most of the activity in this section focused on discussion of two important questions:

1. How can the short distance behavior of the nucleon-nucleon wave function be measured directly?

2. In using the nucleon-nucleon interaction as input to the few or many body problems, is it sufficient to describe the interaction solely in terms of the NN channel (restricting the meson degrees of freedom to the potential), or must we explicitly introduce more than one channel (the NΔ and ΔΔ channels for example) in order to describe the problem satisfactory?

I will take up each of these questions in turn.

A. Direct Measurement of the Wave Functions at Short Distances.

Two possible means for measuring the nucleon-nucleon wave function at short distances were discussed.

1. Measurement of the parity violating circular polarization in $n + p \rightarrow \gamma + d$ and $n + d \rightarrow \gamma + {}^3H$.
2. Measurement of the tensor polarization in elastic electron-deuteron scattering.

1. Parity Violating Effects

Circular polarization P_γ of the emitting γ rays in neutron capture by protons has been observed by Lobashov et al.,[1] who obtain the result $P_\gamma = (-1.30 \pm 0.45) \times 10^{-6}$. Present calculations of this quantity give a result about 100 times smaller. It was suggested by Epstein[2] that this discrepancy was evidence that the nucleon-nucleon interaction at short distances was soft and non local, rather than strongly repulsive, and that in general measurement of P_γ could give us information about the two nucleon wave functions at short distances.

Before reporting on the discussion of this suggestion, I will sketch the theory briefly. The polarization arrises because the weak interaction generates a small parity violating contribution to the nuclear potential, V_{weak}. The exact nuclear states no longer have definite parity, so that, for example, the ${}^1S_0$ state picks up contributions from P waves:

$$|\psi_{scat}\rangle \cong |{}^1S_0\rangle + \eta|{}^3P_0\rangle \ . \tag{1}$$

From perturbation theory

$$\eta \cong Mp \left\langle {}^3P_0 \right| V_{weak} \left| {}^1S_0 \right\rangle \tag{2}$$

where M is the nucleon mass and p the momentum of the nucleon in the center of mass system. Similarly, the deuteron contains small admixtures of bound ${}^1P_1$ and ${}^3P_1$ states which do not exist when $V_{weak} = 0$ and must therefore be computed approximately solving the inhomogeneous Schrodingen equation:

$$(H_s + B) \mid d_- \rangle = - V_{weak} \mid d_+ \rangle \qquad (3)$$

where H_s is the strong Hamiltonian, B the deuteron binding energy and $|\psi_{deuteron}\rangle = |d_+\rangle + |d_-\rangle$ where $|d_+\rangle$ and $|d_-\rangle$ represent the positive and negative parity components of the deuteron ground state respectively.

The small, opposite parity components of the two nucleon ground states introduce a parity violating correlation into the neutron capture matrix element which results in a circular polarization. In skeleton form, the matrix element is

$$M = \left\langle \psi_{deuteron} \middle| [\vec{E}\cdot\vec{r} + (\mu_p\vec{\sigma}_p + \mu_n\vec{\sigma}_n)\cdot\vec{H}] \middle| \psi_{scat} \right\rangle \qquad (4)$$

where $\vec{E}$ and $\vec{H}$ are electric and magnetic field vectors and $\mu_p\vec{\sigma}_p$, $\mu_n\vec{\sigma}_n$ are the magnetic moment operators of the proton and neutron. Since $\vec{E}\cdot\vec{r}$ is odd under parity, both S waves contribute only to the M1 transition, giving the familar threshold result

$$M_o = \frac{\mu_p - \mu_n}{2} \left\langle d_+ \mid (\vec{\sigma}_p - \vec{\sigma}_n)\cdot\vec{H} \mid {}^1S_o \right\rangle \qquad (5)$$

However, the odd parity cross terms contribute to the E1 transition:

$$M = M_o + \left\langle d_- \left| \vec{E}\cdot\vec{r} \right| {}^1S_o \right\rangle + \left\langle d_+ \left| \vec{E}\cdot\vec{r} \right| \eta \, {}^3P_o \right\rangle \qquad (6)$$

It is the interference of the small terms in (6) with M_o which give the leading estimate of P_γ. But these interference terms are proportional to η (and a similar factor from $|d_-\rangle$), so that a measurement of P_γ can be considered a measurement of the matrix element (2).

Since the E1 transition operator cannot change the spin of the nuclear state, only the ${}^1P_1$ part of the negative parity deuteron state can contribute, and the odd parity states which contribute to (6) have the same isospin as their even parity counterparts, so that only the $\Delta I = 0$ part of the V_{weak} will contribute to circular polarization in neutron capture.

This last observation is crucial to Epstein's suggestion. If V_{weak} is approximated by a series of one boson exchange potentials, with one weak parity violating vertex and one strong vertex,[3)] calculations show that one pion exchange contributes only to $\Delta I = 1$ part. Hence the $\Delta I = 0$ part that we see in this experiment comes only from the ρ exchange, which is very short range, and the resulting matrix element is strongly dependent on the short range behavior of the nuclear wave functions.

However, the matrix element (2) is also strongly dependent on the details of V_{weak}. Unless we know V_{weak} better than we know the wave functions at short distances, it would be difficult to obtain much new information about the wave functions. This issue was discussed at some length. It was pointed out that theoretical calculations of V_{weak} are considerably uncertain, due not only to ignorance of the details of the weak interactions, but also to uncertainties in how to perform the strong "radiative corrections" to the weak interaction. The present discrepancy might be due simply to the fact that we have not calculated V_{weak} correctly. On the other hand, some limits on V_{weak} can be obtained from other parity violating decays, such as the decay of the 2^- state of ^{16}O by emission of an α particle to ^{12}C ground state. This decay also depends on the weak coupling of the ρ to the nucleon, and increasing this coupling to explain the circular polarization would greatly overestimate the ^{16}O decay rate.

No definite conclusion was reached, although there was considerable skepticism that these measurements could do more than place limits on the weak potential.

J. S. Levinger suggested a measurement of the ratio of circular polarization in $n + d \to \gamma + {}^3H$ to $n + p \to \gamma + d$.[4)] Although the capture rate for the former reaction is much smaller than the latter, the polarization is expected to be 100 times bigger and the ratio might prove to be less sensitive to uncertainities in the weak potential.

2. Tensor Polarization

Elastic electron-deuteron scattering has long been a source of information about the deuteron wave function. Early measurement of e - d scattering gave reasonably direct evidence for the existance of repulsive forces at short distances, but it has been difficult to obtain precise information about this repulsion. To see why this is so, recall that the usual measurements yield only $A(q^2)$,

$$A(q^2) = G_o^2(q^2) + G_2^2(q^2) \tag{7}$$

where q is the four momentum transferred by the electron and G_o and G_2 are the charge and quadrupole moment form factors respectively, given in the non-relativistic theory by:

$$G_i(q^2) = G_{ES}(q^2)\, F_i(q^2) \qquad i = 0,2$$

$$F_o(q^2) = \int_0^\infty (u^2 + w^2)\, j_o\left(\frac{qr}{2}\right) dr$$

$$F_2(q^2) = \int_0^\infty 2w\left(u - \frac{w}{\sqrt{8}}\right) j_2\left(\frac{qr}{2}\right) dr \tag{8}$$

$$G_{ES}(q^2) = G_{Ep}(q^2) + G_{En}(q^2)$$

where u and w are the deuteron S and D state wave functions, j_i is the ith spherical Bessel function and G_{Ep}, G_{En} are the electric form factors of the proton and neutron respectively.

There are three reasons why it is difficult to extract accurate knowledge of u and w from a measurement of $A(q^2)$.

(i) A depends on G_{En}, which is small but probably not zero and not easily measured in other experiments. Thus uncertainty in our knowledge of G_{En} translates into uncertainty in the values of F_o and F_2 which could be extracted from this experiment.

(ii) A gives only the sum of the squares of G_o and G_2, which washes out much of the detailed information which could be obtained from either one alone.

(iii) The equations (8) do not include meson exchange effects or relativistic corrections, which introduce further uncertainities in the extraction of u and w - particularly at small distances.

The measurement of high momentum transfer e - d scattering currently being done by Chertok et al. at SLAC may be dominated by meson exchange effects, and will be sensitive to relativistic corrections, but may also yield some indirect information about the short distance wave function.

However, discussion in this session focused on a recent observation by Moravcisk and Ghosh,[5)] and by Levinger et al.[6)] that measurement of the tensor polarization would separate G_o from G_2 and therefore yield more precise information.

The tensor polarization, P_t, is given by

$$P_t = \frac{3}{\sqrt{2}} \frac{(\frac{d\sigma}{d\Omega})_o - (\frac{d\sigma}{d\Omega})_1}{(\frac{d\sigma}{d\Omega})_u}$$

$$= 2y \frac{(1 + \frac{y}{2\sqrt{2}})}{1 + y^2} \; ; \; y = \frac{G_2}{G_o} \qquad (9)$$

where the subscript 0 and 1 refer to the polarization of the deuteron in the direction of the momentum transfer and the subscript u means "unpolarized".

Note that P_t depends only on the ratio y, so that it is independent of the unknown G_{En}. Also, the point at which $P_t = 0$ ($y = -2\sqrt{2}$) is very sensitive to the short range behavior of the deuteron wave function because it depends on the shape of the second "lobe" of G_o.

However, P_t should also depend in some way on relativistic corrections and meson exchange effects, which should be estimated before one can extract the desired information about u and w. These effects are probably not negligible in view of the fact that the zero of P_t is expected to be in the range of q^2 = 20 to 50 Fm^{-2}.

B. Evidence From Nuclear Matter Suggesting that Coupled Channels are Necessary at Short Distances.

Much of the discussion in the afternoon session was devoted to the question of whether there exists any two body potential correctly describing nucleon-nucleon scattering on shell which, when extrapolated off shell, gives the correct input to the few or many body problem. Evidence from the nuclear matter experts suggests that no such potential exists, and that it may be necessary to explicitly introduce other channels (such as the NΔ) into the nucleon-nucleon models in order to fit nuclear matter calculations.

Day pointed out that realistic nucleon-nucleon potentials do not give the right binding energy and density for nuclear matter. He displayed an interesting graph of the situation which is qualitatively sketched below.[7] The X's represent the results of the binding energy and nuclear density (proportional to the cube of the Fermi momentum k_F) obtained from various realistic nuclear potentials. The arrows show the changes which result from including 3 and 4 body corrections in the nuclear matter calculation. The "experimental" results are shown in the box. One concludes that none of the many realistic potentials available gives the correct binding energy and nuclear density - those which give stronger binding also give larger density.

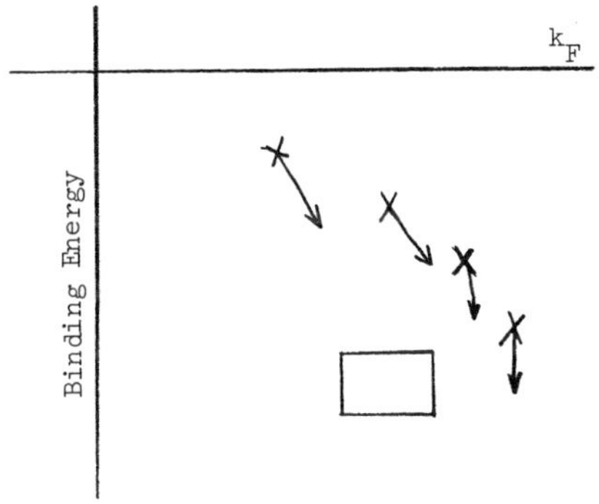

Figure 1

It is possible that the 3 or 4 body corrections are still not being done correctly, but Day believed it was more likely that the fault was with the two body potentials.

Green reported on work in progress which might remedy this situation.[8] He introduces an additional channel describing a free nucleon and Δ (1236 MEV) π - N resonance, and generalizes the potential to a matrix. His equations for the $^1S_0$ state are:

$$\left(-\frac{d^2}{dr^2} + V_{NN} + h^2\right)U = - V_{\Delta N}\, W \qquad (10)$$

$$\left(-\frac{d^2}{dr^2} + \overline{\Delta M}\right) W = - V_{\Delta N}\, U$$

where U and W are the reduced NN and NΔ wave functions, V_{NN} and $V_{\Delta N}$ are potentials, and $\overline{\Delta - N}$ is some average Δ - N energy difference; the diagonal potential in the NΔ channel is neglected. He next adjusts the potentials so that the effective one channel potential (obtained by eliminating W) is in approximate agreement with Reid's potential. Then, when he calculated the ^1S_o contribution to the potential energy (PE) of nuclear matter he obtains results sketched in Figure 2. The effect of the extra channel seems to be to introduce saturation at a lower density (there is no saturation from ^1S_o without this mechanism) which is in the right direction. However, if this mechanism decreases the overall binding energy at the same time, it may simply move us up and to the left along the curve in Figure 1 and not solve the problem at all. The contribution from other states must be calculated before it is clear which of these situations will prevail.

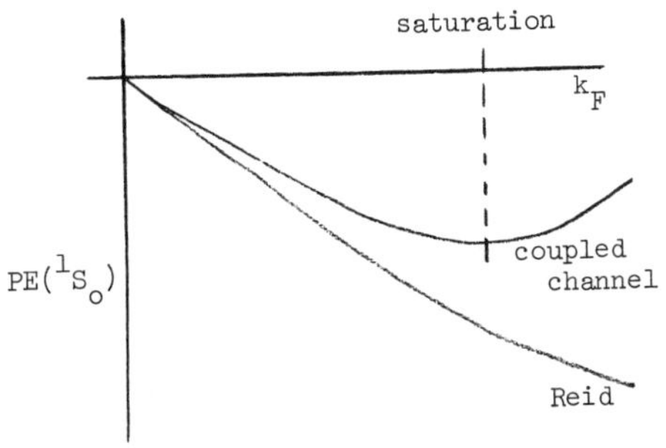

Figure 2

The session concluded with a discussion by Noyes[9] of his new method for calculating nucleon-nucleon interactions in which everything is on the mass shell. To do the nucleon-nucleon problem, he would calculate the πNN → πNN amplitude using 3 particle on-shell Faddeev equations with physical πN data, πN coupling, and the nucleon-nucleon singlet and triplet effective ranges as input. He believes that his approach would avoid the difficulties present in describing resonances (such as the Δ) off-shell.

REFERENCES

1. V. M. Lobashov et al. Nuc. Physics A197, 241 (1972).
2. Geoffrey N. Epstein, "Parity-Violating Neutron-Proton Photocapture as a Direct Probe of the Short-Range Two-Nucleon Interaction", Oxford preprint.
3. M. Gari, Physics Reports 6C, 317 (1973).
4. E. Hadjimichael, E. Harms, and V. Newton, Phys. Rev. Letters 27, 1322 (1971).
5. M. J. Moravcsik and P. Ghosh, Phys. Rev. Letters 32, 321 (1974).
6. J. S. Levinger, Springer Tracts in Modern Physics Vol. 71, p. 88 (1974) and T. Brady, E. Tomusiak, and J. S. Levinger, Canadian Journal of Physics 52, 1322 (1974).
7. F. Coester, S. Cohen, B. Day, C. M. Vincent, Phys. Rev. C1, 769 (1970).
8. G. E. Brown and A. M. Green, Nuclear Phys. A137, 1 (1969); A. M. Green, Physics Letters 34B, 451 (1971); A. M. Green, T. K. Dahlblom, T. Kouki, Nuclear Phys. A209, 52 (1973).
9. H. P. Noyes, Proceedings of the International Conference on Few Particle Problems in the Nuclear Interaction, p. 122 (1972).

Nucleon - Nucleon Bremsstrahlung - Rapporteur's Report

M. K. Liou

Department of Physics and Institute for Nuclear Theory
Brooklyn College of the City University of New York
Brooklyn, New York 11210

The topic of this discussion session is nucleon-nucleon bremsstrahlung which includes the following processes

$$p + p \to p + p + \gamma, \quad (pp\gamma)$$

$$n + p \to n + p + \gamma. \quad (np\gamma)$$

$$n + n \to n + n + \gamma. \quad (nn\gamma)$$

Because the last process has not been studied theoretically or experimentally, it is not included in this discussion. The reasons for studying these reactions are the following:

(i) To study the off-energy-shell behavior of the two-nucleon interactions.

(ii) To distinguish various phenomenological potentials, which are equivalent on the energy shell, so that the correct phenomenological potential can be selected by the bremsstrahlung experiments.

(iii) To learn about electromagnetic interaction.

About thirty five theorists and experimentalists participated in this session. The discussions were warm and pleasant. All participants were very anxious to know where we stood at present and what were the possible directions our research might lead us to follow in the future. They have taken advantage of this excellent opportunity to report their new results and developments, to discuss their problems and possible solutions, to exchange their ideas and opinions, and to formulate an outline for future research in this field.

The discussions covered both the experimental and theoretical aspects of the field, ranging from basic concepts and definitions to very specific technical problems. Unfortunately, time was too short to cover all the topics we intended to discuss. Therefore, in preparing this report, I have added much material which I believe is required for a complete description of the proceedings to emerge. Moreover, since it is essentially impossible and unnecessary to include all the technical details of the theory and experiment involved here, I have emphasized, instead, the underlying physics of the various topics. For the actual detail, the original papers and some fine review articles[1-4] are an excellent source.

Professor Levinger (RPI) has suggested that the discussion of bremsstrahlung should include real and virtual photons in general; e.g. the reactions: $\gamma + d \rightarrow n + p$ and $e + d \rightarrow e' + n + p$. Other process such as pion production from two-nucleon scattering were also suggested. However, due to the limited time available, the discussion on these processes was very brief and incomplete, and so these discussions have been omitted from this report.

The contents of this report are as follows:

I. Experimental background

Only a very brief summary will be given here. We refer to Halbert's excellent review article for more detailed discussions[2].

(1) Geometry (Coplanar or noncoplanar)

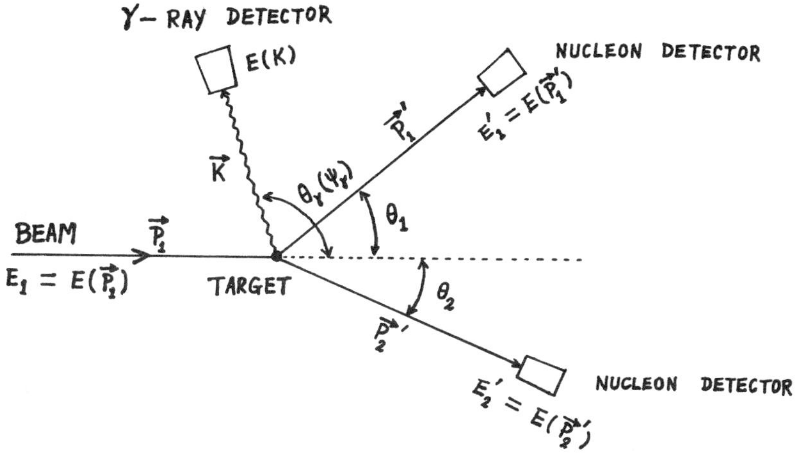

(A) Harvard geometry:

Harvard geometry refers to any experimental arrangement in which two final-state nucleons are detected and the photon momentum is inferred from their measured energy and direction.

Two most commonly used coordinate systems are the spherical coordinate system and a special coordinate system introduced by the Harvard group.[5-8] The main motivation for introducing this special coordinate system was to avoid unnecessary kinematical singularities which appear in the phase space factor if the ordinary spherical coordinate system is used for noncoplanar scattering. These singularities cause the theoretical differential cross sections to approach infinity at the limits of the photon angular distrubition, and introduce unnecessary difficulty when differential cross sections are integrated over the whole photon angular range.

In terms of the spherical coordinate system, the completely differential cross section can be expressed in the form:

$$\frac{d^3\sigma}{d\Omega_1\, d\Omega_2\, d\theta_\gamma} \quad (\mu b/sr^2 \cdot r)$$

which, as we have already mentioned, has kinematical singularities for the noncoplanar case. If the special Harvard coordinate system is used, then θ_γ is replaced by a special photon polar angle ψ_γ and the differential cross section can be written as $d^3\sigma / d\Omega_1 d\Omega_2 d\psi_\gamma$ ($\mu b / sr^2 \cdot r$) which is free of the kinematical singularities for both coplanar and noncoplanar cases. For the coplanar case, $d^3\sigma / d\Omega_1 d\Omega_2 d\psi_\gamma$ is equal to $d^3\sigma / d\Omega_1 d\Omega_2 d\theta_\gamma$. Another advantage of using the Harvard coordinate system is that the special photon polar angle ψ_γ defined in that system always runs through 2π, so that the cross section can be plotted as a function of ψ_γ over the entire range from 0° to 360°.

In some experiments the integrated cross section $\frac{d^2\sigma}{d\Omega_1 d\Omega_2}$ is measured.

(B) Rochester geometry:

Rochester geometry refers to an experimental arrangement in which all three final-state particles are detected. The differential cross section can be expressed in the form[9,6]

$$\frac{d^3\sigma}{d\tilde{\Omega}_1 d\tilde{\Omega}_\gamma d\tilde{E}_\gamma} \quad (\mu b / sr^2 \cdot MeV)$$

which is evaluated in the c.m. frame. The integrated cross section of the form

$$\frac{d^2\sigma}{d\tilde{\Omega}_\gamma d\tilde{E}_\gamma} \quad (\mu b / sr \cdot MeV)$$

is obtained by integrating $d^3\sigma / d\tilde{\Omega}_1 d\tilde{\Omega}_\gamma d\tilde{E}_\gamma$ over $d\tilde{\Omega}_1$ (the solid angle of $\vec{p}_1{}'$ in the c.m. system).

(2) Detector

(A) Counter telescope:

(i) $pp\gamma$ event rates are between 1 and 10 per hour.

(ii) The bremsstrahlung yield is about 10^6 times weaker than the elastic counting rate.

This shows that the conventional experiment is very time-consuming and tricky. The total number of events collected in most of the experiments performed (except

McGill's experiment) was always less than a few hundred. Statistics are low and in some cases only the integrated cross sections are measured.

Because of the finite height of the counters, experiments accept noncoplanar events, and the measured values are actually averages over the noncoplanarity angle. Thus the data must be corrected in order to be compared with theoretical calculations of the coplanar cross section. In order to correct these experimental data, one has to know the dependence of the integrated cross section on the noncoplanarity angle. Such information can be obtained experimentally or theoretically. Since only few such measurements are available, the correction relies mainly on the theoretical calculations.[6,8,10]

(B) Spark Wire chamber:

$pp\gamma$ event rates $\gtrsim 100\ ^{\text{events}}/\text{hour}$.

This new equipment can be used to obtain data simultaneously at many angles. See section IV (4) for more discussions.

(3) Target:

Hydrogen gas or liquid hydrogen are the most common targets used for $pp\gamma$ and $np\gamma$ experiments. D_2, $D_2O - H_2O$ and plastic have also been used as the target for the $np\gamma$ experiments.

(4) Beam energy:

$pp\gamma$: 3.5 MeV $\sim$ 204 MeV.

$np\gamma$: 14, 140, 197, 208 MeV.

(5) Angular range:

The angular range of detected protons is

$$8.5^\circ \leq \theta_i < 45^\circ \qquad (i=1,2)$$

Recently, forward angles ($\theta_i < 10^\circ$) have become of interest to both experimentalists and theorists.

II. Basic Theory

In the potential model bremsstrahlung calculation, the nuclear interaction is treated exactly by using a phenomenological potential and the electromagnetic interaction is treated only to first order as a perturbation. The bremsstrahlung amplitude consists of three main parts: the T-matrix, two-body Green's function and the electromagnetic Hamiltonian. We shall first discuss each of these basic parts and then put them together to give a complete bremsstrahlung amplitude.

(1) T-matrix

(A) On-shell T-matrix:

Elastic nucleon-nucleon scattering gives information only about on-shell T-matrix elements of the two-nucleon interaction. If we assume that two nucleons interact through a potential V_N, then the T-matrix elements can be obtained from the Lippmann-Schwinger equation:

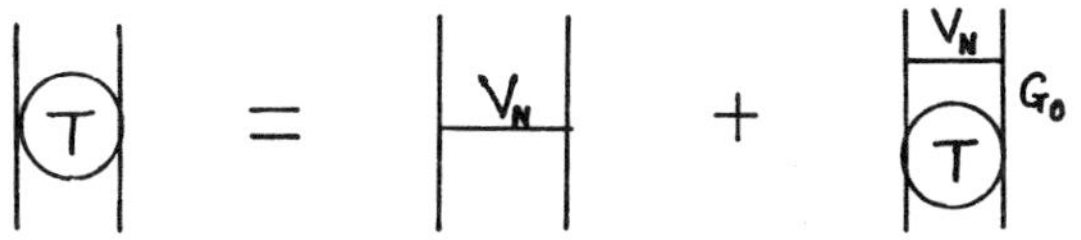

$$T = V_N + V_N G_0 T$$
$$= V_N + T G_0 V_N$$

Many different phenomenological potentials have been constructed by fitting to the elastic nucleon-nucleon scattering data. These potentials are, therefore, equivalent on the energy shell. However, existing potentials, such as Hamada-Johnston, Reid, Yale, Bryan-Scatt,...., etc., are not really equivalent on the energy shell as we expected.[11,12] Thus, if these potentials are used to predict the bremsstrahlung cross sections, one would expect to obtain different results partly due to this already-known differences in their on-shell behavior. Therefore, these potentials must be modified to give identical predictions on the energy shell before they are used to predict bremsstrahlung cross sections. Otherwise, the physically meaningful off-energy-shell effects can not be extracted from the bremsstrahlung experiments.

(B) Half off-shell T-matrix:

In bremsstrahlung calculation, one needs knowledge of the half off-shell T-matrix elements:

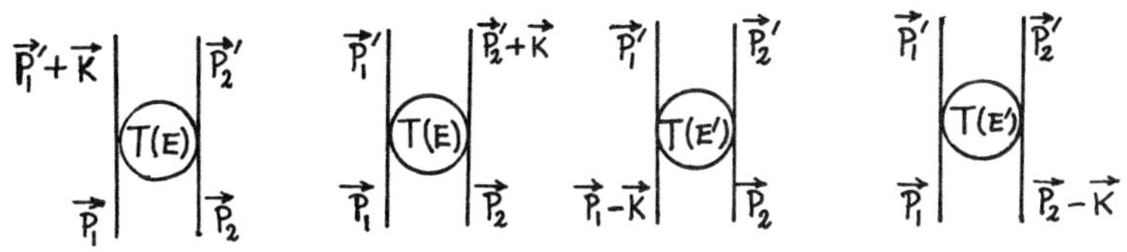

$\langle \vec{p}_2{}', \vec{p}_1{}'+\vec{K} | T(E) | \vec{p}_2, \vec{p}_1 \rangle$ $\langle \vec{p}_2{}'+\vec{K}, \vec{p}_1{}' | T(E) | \vec{p}_2, \vec{p}_1 \rangle$ $\langle \vec{p}_2{}', \vec{p}_1{}' | T(E') | \vec{p}_2, \vec{p}_1-\vec{K} \rangle$ $\langle \vec{p}_2{}', \vec{p}_1{}' | T(E') | \vec{p}_2-\vec{K}, \vec{p}_1 \rangle$

These half off-shell T-matrix elements can be calculated from a given phenomenological potential. Since different potentials will give different half off-shell T-matrix elements, various phenomenological potentials can be distinguished from the bremsstrahlung calculations. When these calculations are compared with the bremsstrahlung data, the best potential may be selected.

(2) Two-body Green's function:

Two-body Green's function $G_0(W)$ can be written as

$$G_0(W) = \frac{1}{W - H_0 + i\epsilon} .$$

Where H_o is the Hamiltonian for two free nucleons and W is the total energy of the two nucleon system. W can be either $E \equiv E(\vec{p}_1) + E(\vec{p}_2)$ or $E' \equiv E(\vec{p}_1{}') + E(\vec{p}_2{}')$, since the matrix elements used in the bremsstrahlung calculation are always half off the energy shell.

Nonrelativistically,

$$\langle \vec{p}_1{}'+\vec{K}, \vec{p}_2{}' | G_0(E) | \vec{p}_1, \vec{p}_2 \rangle = \left\{ \frac{\vec{p}_1^{\,2}}{2m} + \frac{\vec{p}_2^{\,2}}{2m} - \frac{(\vec{p}_1{}'+\vec{K})^2}{2m} - \frac{\vec{p}_2{}'^{2}}{2m} \right\}^{-1} .$$

The above expression may be modified and replaced by an invariant form

$$\langle \vec{p}_1{}'+\vec{K}, \vec{p}_2{}' | G_0(E) | \vec{p}_1, \vec{p}_2 \rangle = \frac{m}{P'_{1\mu} K^\mu} = \frac{m}{E_1' K - \vec{p}_1{}' \cdot \vec{K}} .$$

Arguments for using the relativistic expression were discussed recently by Nyman.[4,13]

(3) Electromagnetic Hamiltonian

The electromagnetic Hamiltonian V_{em} can be divided into two parts : V_{em}^{1} and V_{em}^{2} .

(A) V_{em}^{1} is the coupling of the electromagnetic field to the nucleon currents:

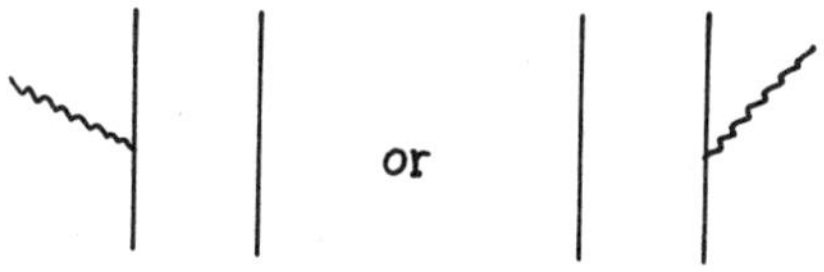

In the nonrelativistic case, V_{em}^{1} is obtained from the Pauli-Schrödinger equation (including an additional magnetic moment interaction representing the anomalous magnetic moments of the nucleon) and it has the form

$$\left(V_{em}^{1}\right)_{N.R.} = \varepsilon\left[\hat{e}\cdot\vec{p} + \frac{1}{2}i\mu\vec{\sigma}\cdot(\hat{e}\times\vec{K})\right].$$

Here, ε is the charge and μ is the magnetic moment of the proton, $\hat{e}$ is the polarization of the photon, $\vec{\sigma}$ is the Pauli spin operator and $\vec{p}$ and $\vec{k}$ are momenta of the proton and photon respectively. This Hamiltonian was used in most of the bremsstrahlung calculations and was found to be a good approximation except in the case of small angle scattering at high energy where the relativistic corrections to this nonrelativistic Hamiltonian become important, especially for $pp\gamma$ calculations.

The well-known Foldy-Wouthuysen (FW) reduction of the Dirac Hamiltonian describing the interaction of a nucleon with an external electromagnetic field can be applied to derive the relativistic correction terms provided a correction to the FW additivity is also included for the two-nucleon system.[14,8] The total electromagnetic Hamiltonian V_{em}^{1} is the sum of the FW electromagnetic Hamiltonian of the individual nucleons plus a correction term to FW additivity. Without discussing the detailed expression of V_{em}^{1}, (for $pp\gamma$, see Ref. 8) We shall just summarize some important features of the relativistic effects as follows:

(i) The relativistic corrections reduce the integrated cross sections for both $pp\gamma$ and $np\gamma$. The corrections are generally small except in the case of small angle scattering at high energy.

(ii) When differential cross sections are plotted as function of ψ_γ, the relativistic corrections are found to be larger around $\psi_\gamma = 0$, π and 2π, and smaller around $\psi_\gamma = \frac{\pi}{2}$ and $\frac{3}{2}\pi$. This feature is true for both

$pp\gamma$[8] and $np\gamma$[15].

(iii) Relativistic effects become larger as the nucleon scattering angles decrease.[8,16]

(iv) In the region above 100 MeV, relativistic effects are found to be more important for $pp\gamma$ than $np\gamma$ for the following reason: Magnetic terms dominate cross section for $pp\gamma$ due to considerable cancellation between the four electric terms. In the case of $np\gamma$, the magnetic term do not dominate the cross section at high energy since, for this case, none of the above mentioned cancellations occur. The main relativistic corrections to the magnetic terms are of order p/m (p = momentum of nucleon, m = mass of the nucleon) compared with the nonrelativistic magentic term. On the other hand, the main relativistic corrections to the electric term are of order $(p/m)^2$ compared to the nonrelativistic term.

(v) The contributions from the additional correction to the FW additivity are very small,[8,17] less than 1% at 157 MeV for the $pp\gamma$ calculation performed in the c.m. frame[8].

(vi) Relativistic corrections have very large effects on the $pp\gamma$ calculation performed in the lab system.[17,18]

Finally, further corrections to V_{em}^{1} have been discussed by Nyman.[4,19]

(B) V_{em}^{2} is the coupling of the electromagnetic field to the exchange (meson) currents:

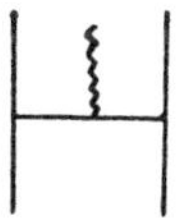

When photons are emitted from two interacting nucleons, a part of the probability amplitude for this process is due to exchange currents. This may be thought of as photon emission from charged mesons which have been exchanged between two nucleons. Such a contribution will also be found if one treats the N-N scattering in terms of a phenomenological potential

$$V_N = V_N^1 + \vec{\tau}_1 \cdot \vec{\tau}_2 \, V_N^2$$

with isospin dependence. V_N^1 and V_N^2 will be allowed to have a general dependence on the momentum operators $\vec{p}_1$ and $\vec{p}_2$ of the two nucleons. Thus the process of N-N bremsstrahlung can probe the inner regions of the N-N force by allowing one to observe radiation coming from exchange currents and nucleon currents.

The determination of the electromagnetic Hamiltonian V_{em}^2 is the most difficult part of the nucleon-nucleon bremsstrahlung calculation. This difficulty is partly due to some ambiguities involved in the derivation of V_{em}^2 from nonlocal and exchange potentials.[20] Although an exact solution to this complicated problem of exchange and nonlocal effects has not been found, some approximate solutions to this problem have been discussed by many groups.[20-23] The expression of V_{em}^2 within these approximations has been obtained and the result has been applied to the npγ calculation.

The contribution from V_{em}^2 has been ignored completely in most of the potential model ppγ calculations since it has generally been regarded as negligible. This assumption, however, must be tested especially for the case of forward-angle scattering which has attracted much attention recently. For those calculations using the low-energy theorem, the contribution from V_{em}^2 is partially included in the second term, i.e., K^o term, of the soft photon expansion. However, we can learn nothing about the off-energy-shell effects from these type of calculations.

For the npγ case, very large effects from V_{em}^2 term were reported in the recent calculations despite the fact that V_{em}^2 was ignored in some earlier calculations without justification. The inclusion of the V_{em}^2 term was found to increase the npγ cross section by about a factor of two.[23,15]

(4) Coulomb interaction:

If the two interacting nucleons are protons, they will interact via Coulomb forces as well as nuclear forces. In this case, a part of the measured ppγ cross section comes from the Coulomb interaction. Thus a Coulomb potential must be included in the ppγ calculation in order to take into account the Coulomb effects.

Coulomb effects have been considered in many ppγ calculations.[12,24-28] Heller and Rich have treated the effects exactly, and they have found that the Coulomb effects in ppγ calculations are not negligible for some of the existing experiments and the inclusion of the effects lowers the cross section.[24] They have also found that the Coulomb effects are quite important for those experiments performed at very small proton angles.[29]

(5) Bremsstrahlung amplitude:

The total bremsstrahlung amplitude M consists of the external scattering amplitude M_E and the internal scattering amplitude M_I.

(A) External scattering amplitude:

The external scattering amplitude M_E is the sum of those terms which describe photon emission from the external nucleon:

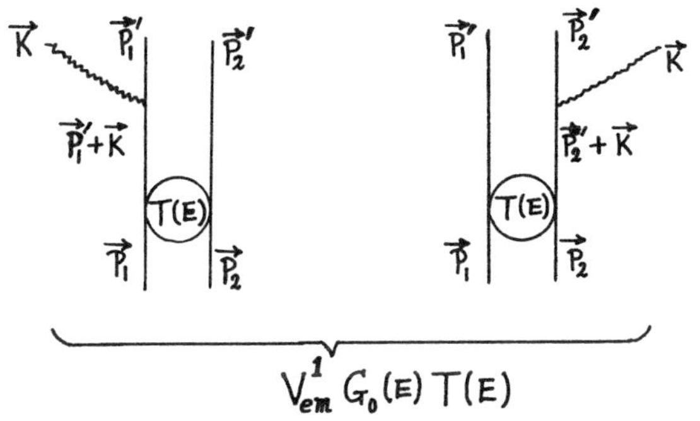

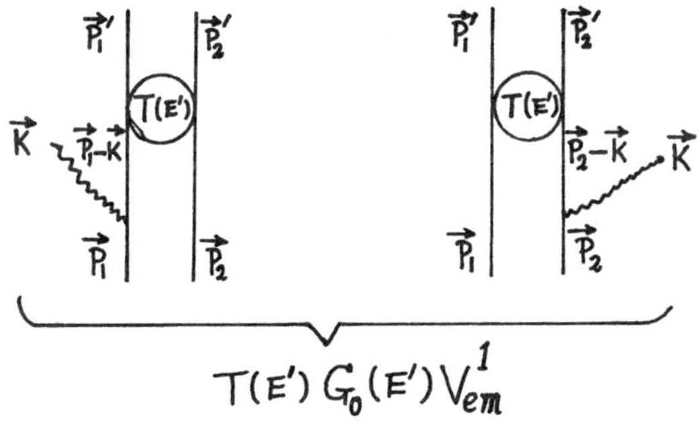

These terms have been treated exactly in most of the potential model bremsstrahlung calculations.

(B) Internal scattering amplitude:

The internal amplitude M_I is the sum of those terms which describe photon emission from the internal nucleons or mesons. We use the following diagram to represent M_I as a whole:

One part of M_I is due to photon emission from the internal nucleons; those terms can be written in terms of V_{em}^1, Go and T. The other part of M_I is due to the photon emission from mesons: those terms are expressed in terms of V_{em}^2, Go and T. The complete expression can be written as:

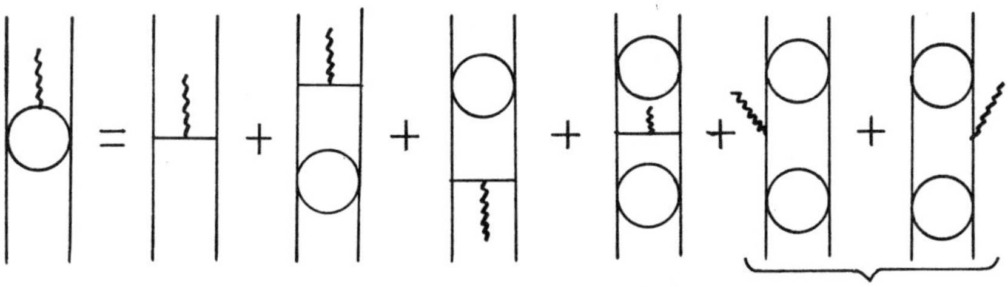

$$M_I = V_{em}^2 + V_{em}^2 G_0 T + T G_0 V_{em}^2 + T G_0 V_{em}^2 G_0 T + T G_0 V_{em}^1 G_0 T$$

The last two diagrams 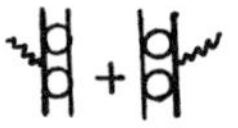 , $T G_0 V_{em}^1 G_0 T$,

are called rescattering amplitude (or double scattering amplitude) and the first four diagrams are called the exchange amplitudes.

For the ppγ calculation [24,30] at 157 MeV, inclusion of the rescattering amplitude is found to increase the cross section for proton angles of 30° and to decrease it for 10°, with the largest effect occurring when the photon is nearly forward or backward. The contribution to the cross section from this amplitude is less than 15% (maybe more for the forward angle scattering cases). As we have already mentioned in section II (3), the exchange amplitude has been ignored in the ppγ calculations.

For the npγ case, contributions from the internal scattering amplitude are found to be larger than those from the external scattering amplitude. It has also been found that many different calculations which use various approaches and approximations give quite different results. Different predictions given by these calculations do not imply that one may learn more about the off-energy-shell effects in the npγ case; it is simply that we do not yet have a sure way to calculate the internal scattering amplitude. The problem of combining strong and electromagnetic interactions is a very old one. Recently, progress has been made in connection with the exchange current in a one pion exchange model.

(6) Low energy theorem

The low energy theorem for photons is a consequence of electric current (charge) conservation. The theorem states that when total bremsstrahlung amplitude is expanded in powers of the photon momentum K,

$$M = A/K + B + CK + \cdots ,$$

the coefficients A and B can be determined from the elastic amplitude; i.e., they are independent of the off-energy-shell effects. This theorem was first derived by Low[31] and was generalized and extended later by many other authors[32-35,26] using the formalism of quantum field theory. Since bremsstrahlung can also be calculated from a potential theory, several authors have rederived the theorem using the Lippmann-Schwinger formalism[36-40]

It is well known that the external scattering amplitude M_E will contribute to all of the coefficients while the internal scattering amplitude will contribute to all coefficients except A. Today, we refer to all bremsstrahlung calculations which make use of only the observed elastic scattering amplitude as model independent approximations. Two types of model independent approximations are currently used in the N-N bremsstrahlung calculations. They are the soft photon approximation (the Low approximation) and the Feshbach-Yennie approximation (the on-shell approximation, or the two-energy approximation). These two approximations were studied and compared carefully by Signell and Marker.[1,11,12,41] The Low approximation is completely covariant; it is appropriate when the photon energy is small compared to the nucleon kinetic energy. On the other hand, the Feshbach-Yennie approximation is a non-relativistic potential model calculation; it is appropriate when the nucleon momentum is small compared to its mass. For example, Signell and Marker have found that the

Feshbach-Yennie approximation is much more accurate than the Low approximation at 10 MeV, $\theta_1 = \theta_2 = 30^o$.

(7) Further comments and discussions on bremsstrahlung theory

L. Heller (Los Alamos):

% CHANGE IN $\frac{d^2\sigma}{d\Omega_1 d\Omega_2}$

$V_{H.J.} - V_{S=3.5}$

E	$\theta_1 = \theta_2$ 10°	20°	30°
42	1.7	1.4	
99	9.4 (3.4)	4.3 (2.7)	1.0
158	17 (7.9)	5.2 (3.4)	0.4

ORDINARY V_{em}

$\vec{J}_1 + \vec{J}_2$ (MINIMAL)

In this table, we show the percentage change in the bremsstrahlung cross section between using the Hamada-Johnston potential and another which is phase equivalent. We obtain the new potential by making a Baker transformation which contained a parameter S. S = 7 is about the largest value permitted (with that particular form), and it produces changes which are approximately double those produced by S = 3.5 .

The new potential contains P^2 and L^2 dependence, which raises a question about the correct form of the electromagentic interaction. The numbers were obtained by using the ordinary current. (This corresponds to using the "Maximal" current for the P^2 and L^2 terms). The numbers in parenthesis were obtained by using the "Miminal" current for those terms. We do not know if the answer must be somewhere between those two extremes.

These calculations were performed together with Dr. M. Rich of this laboratory.

G. E. Bohannon (Michigan State University): The dominant relativistic corrections to the normal electromagnetic current used in the potential-model ppγ calculations are now well established. Part of these corrections, those given by the Foldy-Wouthysen reduction of the single-particle Dirac equation, were included in calculations by Liou, Cho and Sobel. They found that the contribution from the "correction to the F-W reduction" is small. The corrections were studied by Peter Signell and myself by making a c.m. frame (v/c) expansion of the covariant soft-photon ppγ amplitude derived by Nyman. We found corrections to the F-W reduction which agree with those found in studies of the electromagnetic interaction by Osborn, Primack, Brodsky, Krajcik and Foldy.

An interesting point we found is that for ppγ almost the total effect of these additional corrections is to just cancel in the c.m. frame a term of the F-W reduction proportional to $\vec{\sigma}_i \cdot (\vec{E}(\vec{x}_i) \times \vec{P}_i)$. The numerical effect of this term is small so that the corrections used by Liou et al. are sufficient. This cancellation does not occur for npγ.

R. Thompson (Los Alamos): Leon Heller and I are attempting to carry out an n-p bremsstrahlung calculation including the one-pion-exchange current, entirely in momentum space. This has the advantage that the current matrix elements have a much simpler form in momentum space than they do in coordinate space. After all angular integration are performed, there remains a numerical two-dimensional integration. For the double scattering part of the normal current matrix element, one of these integrations is limited to a finite range.

G. E. Bohannon (Michigan State University): I wish to emphasize that before npγ studies can yield definite information about the off-shell nucleon-nucleon interaction we must obtain more confidence in the validity of the theoretical exchange current contributions than I believe we now have.

G. L. Strobel (University of Georgia): I have calculated $n+p \rightarrow n+p+\gamma$ at 200 MeV at Paul Brady's suggestion. Calculation shows something can be learned from bremsstrahlung, I use a Glabuer theory to get a nuclear scattering amplitude from an assumed potential. A (many) one bosom exchange potential is parameterized (mass and coupling constants are the continuous variables) which in Born approximation reproduces the elastic scattering amplitude. We calculate relativistically (with no re-scattering term) using Bjorken and Drell appendix, for example, the unpolarized

$n+p \rightarrow n+p+\gamma$. Using only uncharged boson, the γ is predicted to come mostly on the proton side. Using only positively charged mesons, the γ is predicted to come mostly on the neutron side, in agreement with experiment. This suggests npγ differential cross sections offer a chance to learn something about exchange in n-p interactions. The calculation described is gauge invariant and satisfies the low energy theorem.

III. Comparison of theory and experiment

During the past decade, a great number of the NNγ experiments have been performed at various angles ($8.5^\circ \leq \theta_i < 45^\circ$) and energies (3.5 MeV $\leq$ E $\leq$ 208 MeV) using either the Harvard geometry or the Rochester geometry. Except for a few experiments using the spark wire chamber, most of these experiments have used the conventional detectors to measure the differential cross sections and/or the integrated cross sections. A tabulation of these cross sections can be found in Ref. 2, and some new results will be discussed in section IV of this report.

On the theoretical side, many authors have calculated the N-N bremsstrahlung cross section using various approaches and approximations. They are the potential model calculation, the model independent calculation, the one-boson-exchange model calculation, the S-matrix calculation, and so forth. An extensvie list of references covering the period from 1963 to 1971 can be obtained from Moravcsik's excellent review article.[3] Since the differentiation among various phenomenological potentials is one of the main motivations for the study of bremsstrahlung, most of these calculations are potential model calculations which use phenomenological potentials to predict the bremsstrahlung cross sections. It must be emphasized that different predictions from different types of calculations should not be interpreted entirely as the off-energy-shell effects, since part of the differences may be simply due to the different approximations used.

For ppγ, the agreement between theory and experiment is generally good except for some forward angle scattering cases. So far, the exact cause for the large discrepancy at the forward angle scattering is not clearly understood. However, several possibilities will be discussed in section IV (3). Since most of the experi-

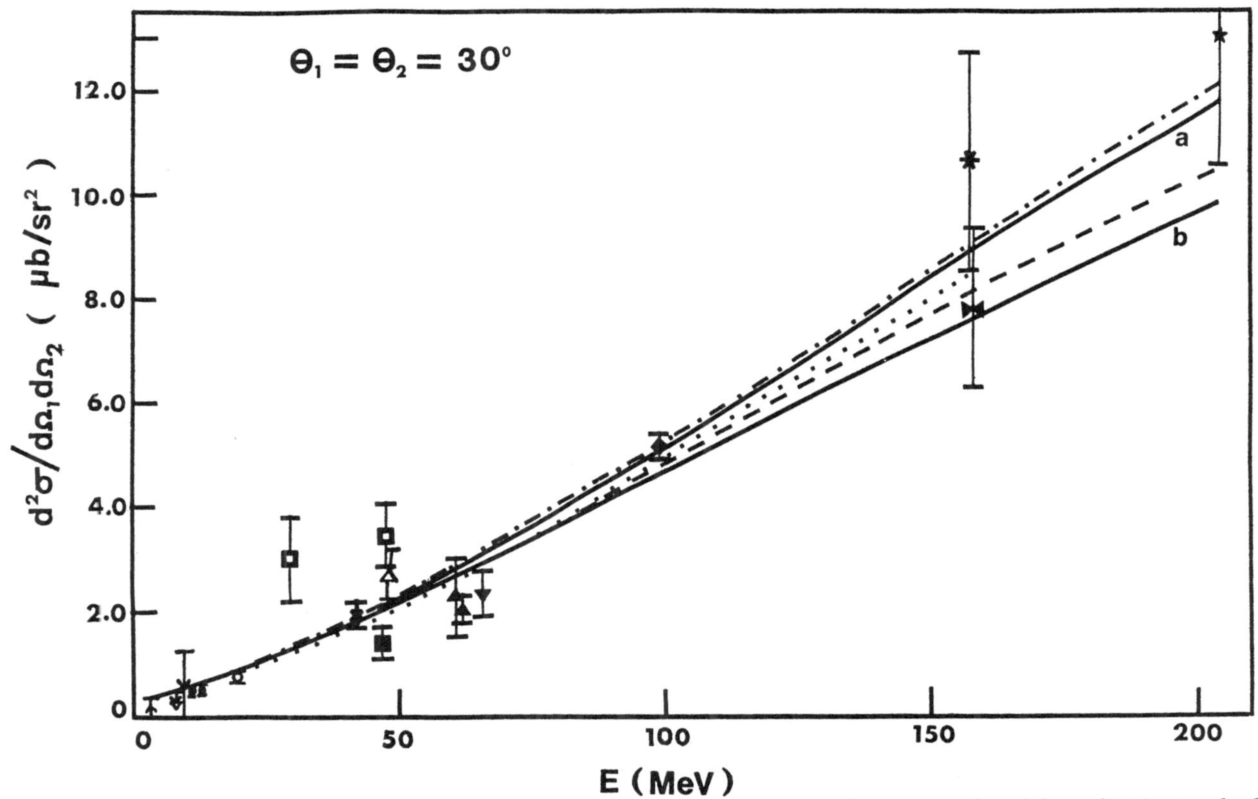

Fig. 1 Comparison of ppγ measurements at various energies with the potential model predictions calculated by many different groups using the Hamada-Johnston potential. The experimental points indicated by ∧, ⩔, ∨, ●, O, □, X, ■, △, ▲, ▼, ◆, ⋈, ✳, and ★ are those of Refs. 42, 43, 44, 45, 46, 47, 48, 49, 50, 51, 52, 53, 54, 5, 9, respectively. The references for the theoretical curves are as follows: ___a, Drechsel-Maximon, Ref. 10; ___b, Liou-Sobel, Ref. 8; -.-.-, Brown, Ref. 30; ---, Marker-Signell, Ref. 12; ■■■■■ , Heller-Rich, Ref. 24.

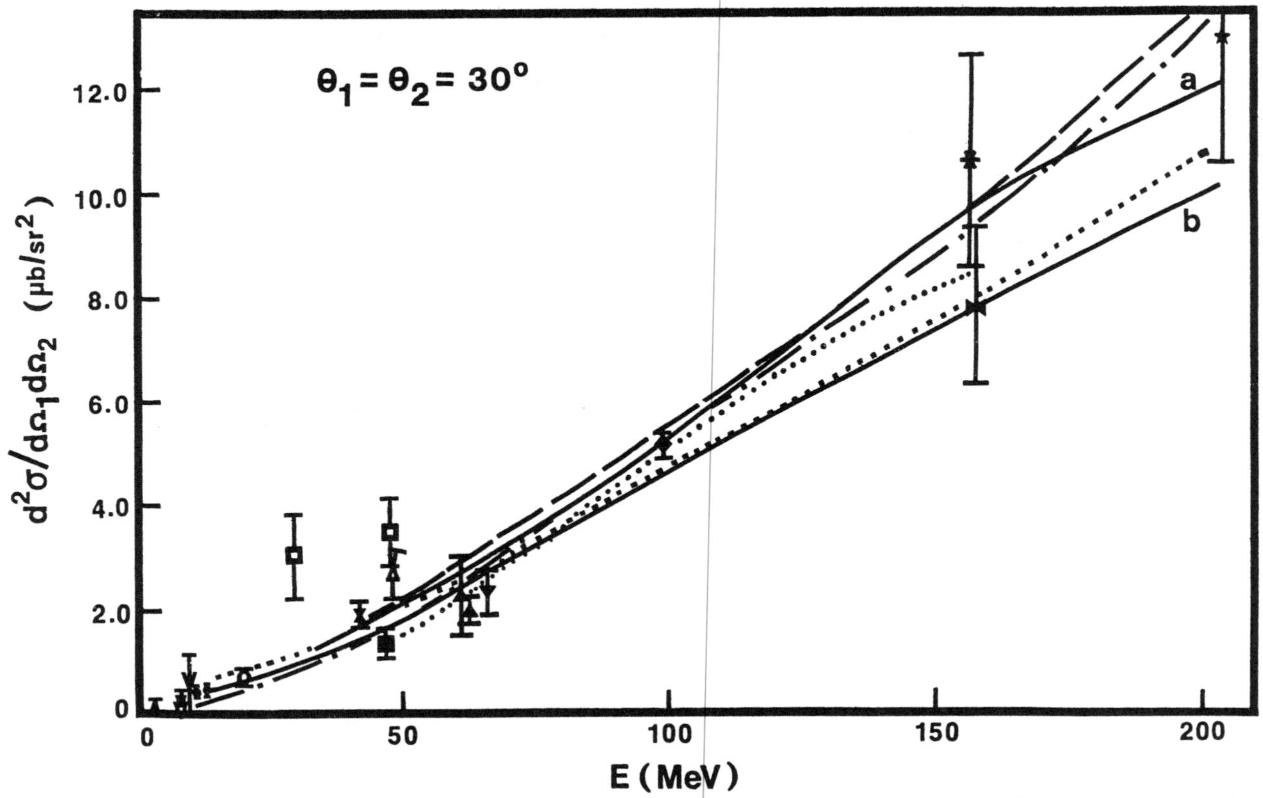

Fig. 2 Comparison of ppγ experiments at various energies with different types of calculations. The experimental points are the same as those of Fig. 1. The references for the theoretical curves are as follows: <u>a</u>, MCS, Ref. 55; <u>b</u>, Nyman, Ref. 26; -.-., OBE, Ref. 6; ---, Reid potential, Drechsel-Maximon, Ref. 10; ■■■ ■■■■, Bryan-Scott IV, Brown, Ref. 30; ...., Tabakin potential, PGD, Ref. 56.

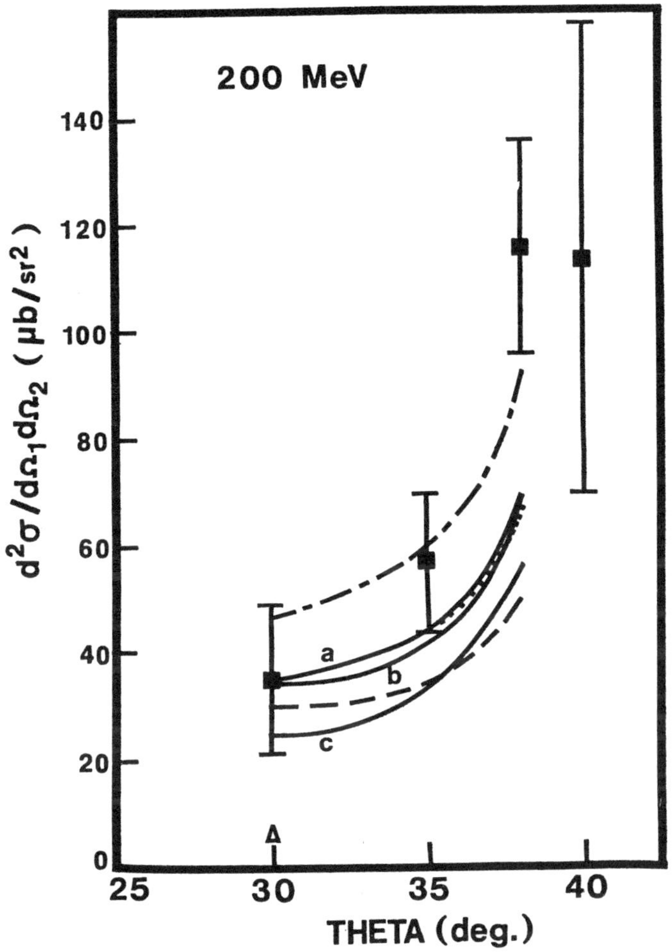

Fig. 3. Comparison of symmetric npγ measurements with various calculations. The experimental data were taken from Ref. 57. The references for the theoretical curves are as follows: a , Bryan-Scott III potential, Brown-Franklin Ref. 23; b McGuire, Ref. 58; c Nyman, Ref. 26; -.-.-., OBE, Ref. 6; ---, CLSG, Ref. 15; ....., Hamada-Johnston potential, Brown-Franklin, Ref. 23; Δ , PGD, Ref. 56.

ment have used the Harvard geometry and have measured the integrated cross section at $\theta_1 = \theta_2 = 30^o$. The results of these measurements[5,9,42-54] at different energy will be compared with the theoretical predictions.[6,8,10,12,24,26,30,55,56] Fig. 1 shows the comparison of the experimental results with the potential model predictions calculated by many different groups using the Hamada-Johnston potential.[8,10,12,24,30] In Fig. 2, the experimental results are compared with different types of calculations.[6,10,26,30,55,56] Comparison of the result at the forward angle will be shown in section IV.

For npγ, the agreement between theory and experiment is satisfactory for some cases. In Fig. 3, we show the comparison of Davis results[57] with various calculations.[6,15,23,26,56,58]

IV New experimental results

The bremsstrahlung experiment was the main subject of the whole discussion and the discussion of it was focussed on four recent experiments. These experiments are the Laval experiment, the ETH experiment, the Orsay experiment and the Manitoba experiment. A recent npγ experiment by Edgington et.al.[59] was not discussed.

(1) Laval experiment

This experiment was performed at Universite Laval, Canada by B. Frois, R. Roy, C. R. Lamontagne, U. von Moellendorff and R. J. Slobodrian. The experiment can be summarized as follows:

(i) Type of experiment: ppγ

(ii) Geometry: Harvard geometry

(iii) Beam energy: 6.92 MeV

(iv) Proton polar angles: $\theta_1 = \theta_2 = 38^o$

(v) Detector: counter telescopes

(vi) Cross section measured (corrected for noncoplanarity):

$$\frac{d^2\sigma}{d\Omega_1 d\Omega_2} = 1.7 \pm 0.7 \quad \mu b/sr^2$$

(vii) Theoretical prediction:

$$\frac{d^2\sigma}{d\Omega_1 d\Omega_2} = 1.1 \quad \mu b/sr^2$$

This was calculated by Dahblom and Green using the Feshbach-Yennie approximation. The measured cross section is also in good agreement with an estimation using Signell's low energy bremsstrahlung formula.

(2) ETH experiment

This experiment was performed at Eidgenossische Technische Hochschule, Switzerland by W. Wölfli, M. Suter, G. Bonani and Ch. Steller. The experiment was presented at the meeting by W. Wölfli and it can be summarized in the following

(i) Type of experiment: $pp\gamma$

(ii) Geometry: Harvard geometry

(iii) Beam energy: 11 MeV and 13 MeV

(iv) Proton polar angles: $\theta_1 = \theta_2 = 30^\circ$

(v) Detector: counter telescopes

(vi) Cross sections measured (corrected for noncoplanarity):

The new preliminary values are as follows:

$$\frac{d^2\sigma}{d\Omega_1 d\Omega_2} = \begin{cases} 0.44 \pm 0.04 \ \mu b/sr^2 & \text{for 11 MeV} \\ 0.46 \pm 0.04 \ \mu b/sr^2 & \text{for 13 MeV} \end{cases}$$

(vii) Comparison with theory:

E (MeV)	EXPERIMENT ($\mu b/sr^2$)	THEORY ($\mu b/sr^2$)	
		HELLER-RICH	DAHBLOM-GREEN
11.0	0.44±0.04	0.48	0.43
13.0	0.46±0.04	0.54	0.48

Discussions:

Justification for doing new ppγ and npγ experiments was discussed. Here, the problem is at which energies and angles these experiments should be performed in order to observe more off-energy-shell effects. In principle, this can be done by a theoretical search for a particular geometry and energy region where the off-energy-shell effects are largest. However, the calculation used should include all corrections since it is sensitive to the approximations and approaches used. The predictions from an imperfect calculation might mislead experimentalists to perform an experiment at a wrong place.

The discussion started when Wölfli showed a slide in which two different calculations (a potential model calculation using the Hamada-Johnston potential and a model independent calculation) had shown a larger difference at 30-60 MeV incident proton energy than at Ep $\geqslant$ 100 MeV for ppγ at 30°-30° angles. This difference was interpreted as entirely due to the off-energy-shell effects and therefore many people were surprised at the result. (As we have already mentioned the differences can result from the off-energy-shell effects and/or the approximations used in the calculations.) Then the following conversation followed:

J.V. Jovanovich (University of Manitoba): Can this slide be interpreted to mean that at 30°-30° it would be more profitable to do experiment at 42 MeV than at 100 MeV or 200 MeV?

W. Wölfli (ETH, Zürich): Yes, it is quite surprising but this is what these calculations show.

J. R. Richardson (TRIUMF): I would like to comment on it. Keeping in mind the past history of ppγ calculations, I would just be very cautious and would take results of these calculations with a grain of salt. I would be reluctant to justify an experiment on the basis of these calculations.

J. V. Jovanovich (University of Manitoba): This is perhaps so. But perhaps from the same reason the opposite should be also true. Namely, if some theoretical calculations do not predict the existance of some large and interesting effects in that particular part of phase space where these calculations were done we should not take this to mean that we should not do the experiment at all, not even in the part of phase space where calculation has not even been attempted. In fact, I believe that while it is interesting to look at the results of theoretical calculation before the experiment is started we should, at this stage, justify high accuracy nucleon-nucleon bremsstrahlung experiments on a differnt, quite general and "theoretical model" independent bases.

We physicists have studied p-p ant n-p elastic scattering for many years and have learned a lot about the nucleon-nucleon interaction on the energy-shell. Nucleon-nucleon bremsstrhalung is the simplest inelastic process and practically the only one that would enable us to study clearly the N-N interaction off-the-energy-shell. And, needless to say, off-the-energy-sehll part of interaction is just as fundamental as on-the-energy-shell part. Keeping in mind that off-the-energy-shell part has not yet been studied experimentally in more than just a qualitative manner, I ask a question: Is it not justified to start systematic study of this fundamental process irrespective of the present status of theory? High accurate $NN\gamma$ experiments are complex and difficult to do and it takes much longer to perform them than to perform a specific model dependent calculation. So, why to justify at all these experiments on basis of some model dependent calculations which will, the most likely, be obsolete, or at least incomplete, by the time the experiment is finished? Why not to start detailed and systematic study of this very fundamental and very promising process by doing accurate second generation bremsstrahlung experiments, even if original qualitative $pp\gamma$ experiments have not produced, from the theoretical point of view, any startling results?

Moravcsik made a statement at the session I moderated (D2a). His statement was effectively the following: If you want to find a shape of an object you can try standing in front of it in one place and taking pictures with better and better cameras. But soon you would come to a stage when it would be much more profitable to take a cheap box camera, walk around the object and take a poor picture from an entirely different angle. A new view of the object would likely reveal information that could never be obtained from the place we stood originally".

(3) Orsay Experiment:

This experiment was performed at Orsay, France by A. Willis, V. Comparat, R. Frascaria, N. Marty, M. Morlet and N. Willis.[60] The result of this experiment was presented by R.Frascaria, and we can summarize it in the following:

(i) Type of experiment: $pp\gamma$

(ii) Geometry: Rochester geometry

(iii) Beam energy: 156 MeV

(iv) Proton polar angles: $\theta_1 = 15^o$, $\theta_2 = 18^o, 21^o, 24^o$ and 27°.

(v) Detectors: protons : plastic scintillation counter telescopes
γ-ray : a lead-glass Cherenkov counter

(vi) Target : liquid-hydrogen

(iii) Results and comparison with theory

(θ_1, θ_2)	$\theta_{\gamma c}$	$\Delta\theta_\gamma$	BROWN'S CALCULATION Ref. 61	CLSG CALCULATION Ref. 16	EXPERIMENT Ref. 60
(15°, 18°)	126.5°	20.7°	1.22	1.03	1.17 ± 0.27
(15°, 21°)	133.0°	17.7°	1.48	1.21	0.95 ± 0.13
(15°, 24°)	135.2°	28.5°	1.41	1.16	0.83 ± 0.14
(15°, 27°)	133.7°	21.6°	1.45	1.20	0.68 ± 0.09

Discussions:

This experiment was discussed with great interest at the meeting mainly because the measured cross sections were very different from the theoretical predictions. Many people are surprised at the large discrepancy found in this particular case since the agreement between theory and experiment is generally good for most $pp\gamma$ cases. No one really knows what causes the discrepancy.

Experimentalists are quite aware that the bremsstrahlung experiments for forward angles are extremely difficult to perform especially when the emitted photons are also detected. They do not rule out the possibility that the discrepancy may result from experimental errors. For this reason, some technical details, such as the efficiency of the detectors, the treatment of the prompt background, the non-coplanarity effects and how the data were analysed, were questioned and discussed by some experienced experimentalists. But no conclusion has been reached from the discussion since no further information about this experiment other than in the published paper can be obtained.

On the theoretical side, it is obvious from the previous discussions that the relativistic effects and the Coulomb effects are quite large for the forward angle scattering. Contributions from other factors, such as the exchange effects and the gauge terms, are unknown. Since no calculation has included all of these corrections, it is quite possible that the discrepancy results from the imperfect theoretical calculations. Although it is also possible that the Hamada-Johnston potential used in these calculations may cause this lack of agreement, it is too early to conclude that the Hamada-Johnston potential is unable to fit the $pp\gamma$ process.

The discrepancy has not yet resolved. The problem is a real challenge to both theorists and experimentalists. To resolve this discrepancy more experiments at the forward angles should be done in order to check Orsay's results, and for theorists, it is time to do a calculation which includes all possible corrections. Until this work is accomplished, it seems unlikely that this discrepancy can be resolved.

Bernard Frois (CEN, Saclay): It is very hard to check the results of this experiment because the published results do not include the details of the calculation of the correlated efficiency of the three counters. In this experiment, noncoplanarity effects might be very large especially with three counters.

(4) Manitoba experiment

This experiment was performed at the University of Manitoba, Canada by J. V. Jovanovich, L. G. Greeniaus, R. Kerchner, J. McKeown, T. W. Millar, D. G. Peterson, R. Pogson, W. F. Prickett, C. A. Smith, K. F. Suen, and J. C. Thompson. The result of the experiment was presented at the meeting by Jovanovich. The experiment can be summarized as follows:

(i) Type of experiment: $pp\gamma$

(ii) Geometry: Harvard geometry

(iii) Beam energy: 42 MeV

(iv) Detector: Spark wire chambers and scintillation counters

(v) Target: H_2 gas

(vi) Angular range: $16^\circ \leq \theta_i \leq 36^\circ$

(vii) Equipment: The spectrometer used in this experiment is shown in Fig. 4.

(viii) Result: The results obtained are shown in Figs. 5,6,7 and 8. Data were analysed in two different ways: the conventional analysis and the Global analysis.

Discussions:

J. V. Jovanovich (The University of Manitoba)

(A) Conventional analysis:

The conventional analysis was described in our PRL paper.[48] The results shown in Figs. 5,6 and 7 are of the same kind as Figs. 1, 2 and 3 in that paper except that they are from a later experiment done in Nov. 1971. The experimental cross sections shown in Figs. 5, 6 and 7 are computed using equations (1), (2) and (3) of Ref. (48) except that the correction factors α and α' are computed using Liou's theoretical cross sections instead of phase space only. These cross sections (just as those presented in Ref. (48)) are not corrected for experimental resolutions. Histograms (block continuous lines) shown in Figs. 5 and 6 are due to Liou's theoretical calculations with all our experimental resolutions folded in. Therefore, these figures give fair compariosn of our experimental results with Liou's theoretical predictions. However, since the data are not correct for experimental resolutions, they are not precisely comparable with other theoretical calculations. Such comparisons could be made in future, but the effects of resolutions must be kept in mind.

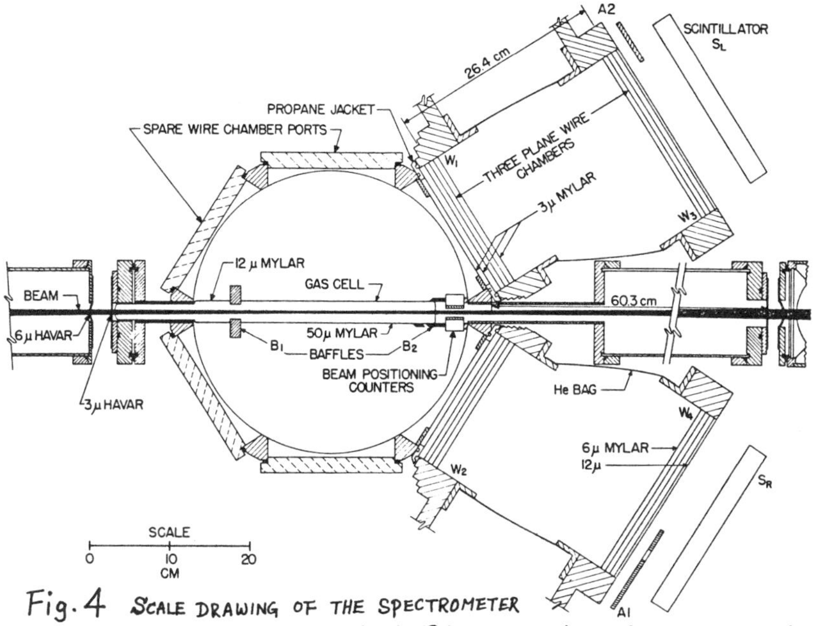

Fig. 4 SCALE DRAWING OF THE SPECTROMETER SHOWING THE SCATTERING CHAMBER, WIRE CHAMBERS AND SCINTILLATION COUNTERS AS USED IN MANITOBA EXPERIMENT. REF. 62.

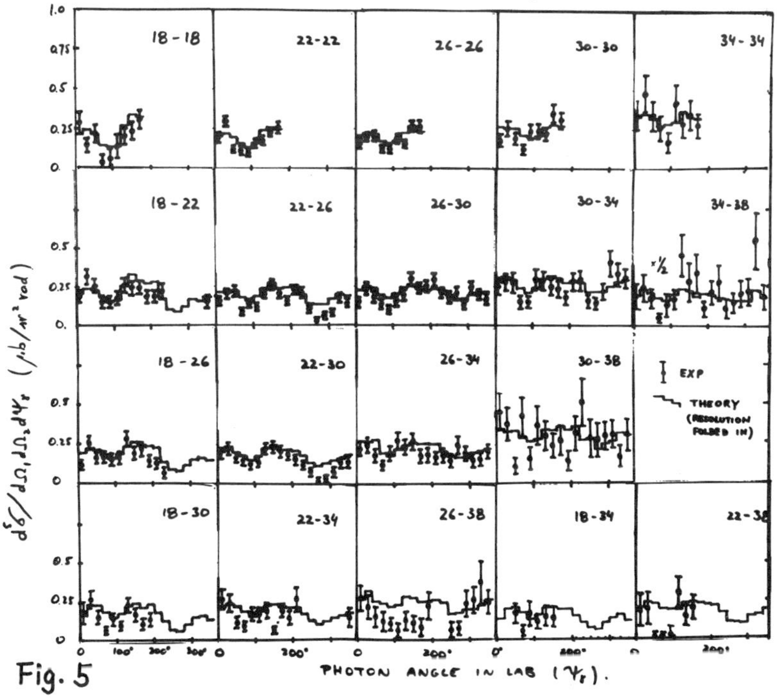

Fig. 5

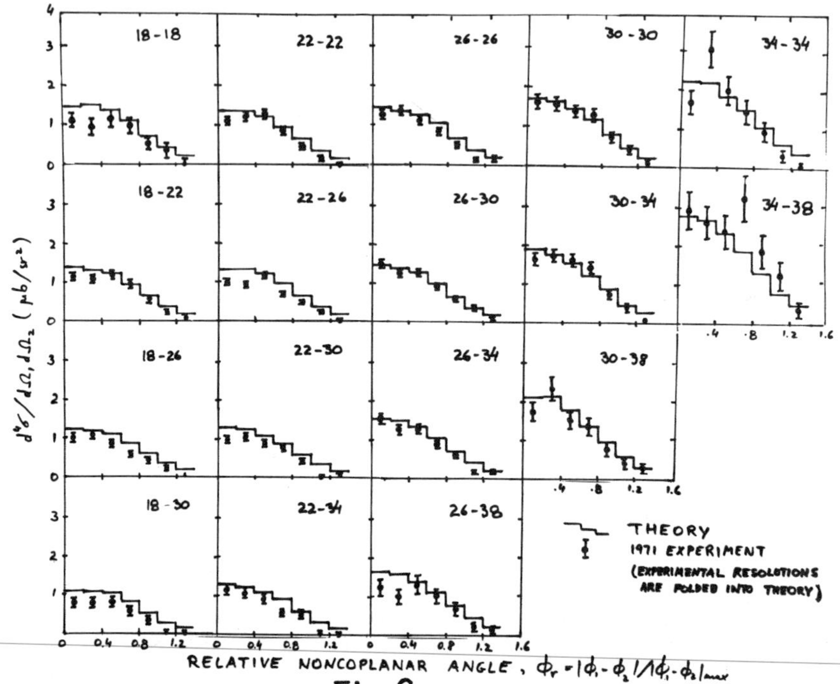
18-18
22-22
26-26
30-30
34-34
18-22
22-26
26-30
30-34
34-38
18-26
22-30
26-34
30-38
18-30
22-34
26-38
THEORY
1971 EXPERIMENT
(EXPERIMENTAL RESOLUTIONS ARE FOLDED INTO THEORY)
RELATIVE NONCOPLANAR ANGLE, $\phi_r = |\phi_1 - \phi_2| / |\phi_1 - \phi_2|_{max}$

Fig. 6

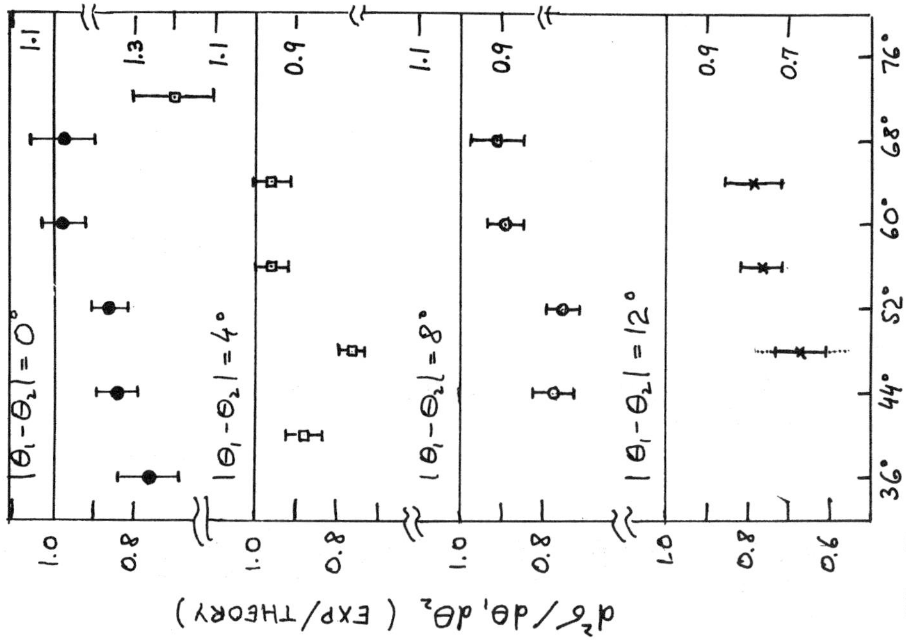
$|\theta_1 - \theta_2| = 0°$
$|\theta_1 - \theta_2| = 4°$
$|\theta_1 - \theta_2| = 8°$
$|\theta_1 - \theta_2| = 12°$
$\theta_s = \theta_1 + \theta_2$
$d^2\sigma / d\theta_1 d\theta_2$ (EXP/THEORY)

Fig. 7

It should be mentioned that corrections α and α' are, strictly speaking, theory dependent. But dependence on theory is rather weak as these correction factors are obtained as ratios of two numbers, each of which depends on the same theory. Only those points which had α and α' smaller than TWO were included.

We could in principle correct our data for experimental resolutions-and we shall probably do that - but this correction will again be theory dependent, and that would be an additional theory dependence (again fairly weak) of results. So the usefulness of such a correction may be questionable. We have not performed this correction yet, but I expect this correction to be less than 10% to 20%, so that theory dependent effects are still smaller than our statistical errors. Therefore, by introducing few percent of theory dependence into our experimental cross sections, we can correct our results for experimental resolutions and have them "CLEAN".

To summarize: Conventional results as shown in Figs. 5, 6 and 7 are (i) experimentally "DIRTY" up to ~20%, (ii) theoretically "DIRTY" up to ~ few %. We could correct these results so that they are (i) experimentally "CLEAN" up to ~2%, (ii) theoretically "DIRTY" up to ~(2-3)x few %. So, what do you think? Which is better?

(B) Global analysis:

Some of the characteristics of the performed ppγ experiment are: (i) large fraction of phase space available to the reaction was observed in the experiment, (ii) number of events detected was relatively small, on the order of 10^4, (iii) the equipment was unable to detect protons with energies smaller that certain limit.

As explained in last section, the data were first analysed in the conventional manner and the results are presented in Figs. 5, 6 and 7. While these results can be easily used by theorists, they suffer from the following disadvantages:

(i) Due to necessary splitting of data into many bins, each bin has rather low statistics (particularly in Fig. 5).

(ii) Due to existing energy cut-off, some bins have corrections applied and these corrections are (weakly) theory dependent.

(iii) Cross sections presented do not have experimental resolutions enfolded. This could be in principle done but the correction would be again theory dependent.

(iv) Cross sections shown in Figs. 6 and 7 are integrated cross sections. Therefore, it is not possible to express them as a function of some other variables.

We have tried to overcome these disadvantages by analyzing data in a different, unconventional manner, which we called "global analysis". The method used is in its principle a very old method of comparing data with theoretical calculations. It has been used before in many branches of physics, whenever the efficiency and response function of the experimental equipment was very complicated. Nevertheless, as far as we know, this method has not been applied before to the field of low energy nuclear physics on the scale we have done it, and certainly not to the analysis of ppγ experiments.

The essence of the method is described as follows: We first make few definitions.

(i) A set of five independent variables (P_i, with i = 1, ...5) used in theoretical calculations of cross sections will be called <u>theoretical variables</u>.

(ii) A set of five variables (q_i, i = 1,.....,5) measured in the experiment and in some way (i.e., via properties of the apparatus used) related to the theoretical variables will be called <u>experimental variables</u>. Corresponding q_i and P_i variables do not have to have same dimensions. (But, of course, they often do).

(iii) Using a certain theory (i.e., a supposedly known function of theoretical variables P_i), known (i.e., measured) properties of experimental equipment and randomizing (i.e., Monte Carlo) procedures, we can compute a set of variables (q_i', i = 1,...,5) resembling experimental variables in every respect except that they have been computer produced rather than experimentally measured. These variables are called <u>randomized variables</u>.

In an inelastic scattering experiment, we measure some experimental distributions which could be function of one experimental variable, $N_e(q_i)$, or two, $Ne\ (q_i,q_j)$, or more. For all practical purposes, in a coincident experiment these distributions are formed by subtracting "random" $N_{RB}(q_i)$ and "prompt" background $N_{PB}(q_i)$ from the "true" coincident events $N_T(q_i)$. The corresponding experimental distribution is

$$N_c(q_i) = N_T(q_i) - C_R N_{RB}(q_i) - C_P N_{PB}(q_i) \qquad (1)$$

Where C_R and C_P are properly chosen normalizing constants.

Randomized distribution $N_r(q_i')$ corresponding to the above experimental distribution $N_c(q_i)$, in case of an inelastic scattering experiment with three body final states, is obtained by integration over all theoretical variables, and all randomized variables which are not under investigation as follows:

$$N_r(q_i) = \iiiint \prod_{\substack{j=1 \\ j \neq i}}^{5} dq_i \iiiint K(q_1, \cdots, q_5; P_1, \cdots, P_5)\, d^5\sigma(P_1, \cdots, P_5) \qquad (2)$$

where $K(q_1,...,P_5)$ is the known responce and efficiency function of experimental apparatus connecting theoretical and randomized variables, and $d^5\sigma(p_1,...,p_5)$ is the theoretical cross section computed according to a certain theory. For the purposes of global analyses, it is convenient to present this cross section as a product of two factors.

$$d^5\sigma = W(P_1, \cdots, P_5)\, d^5R\,(P_1, \cdots, P_5)$$

where the first factor, $W\ (p_1,...p_5)$ is the invariant (square of) matrix element which we shall call "invariant weight", and $d^5R(p_1,...p_5)$ is the invariant element of phase space.

If the theory assumed in these calculations describes the measured process adequately (and if the experiment is correct) the experimental and randomized distribution should be identical within the errors of measurements and calculations. That would mean that

$$R(q_{ij}) = \frac{N_c(q_{ij})}{N_r(q_{ij})} = 1 \quad , \qquad j = 1, 2, \cdots, n$$

In this expression it was assumed that experimental and randomized variables correspond exactly to each other, so the prime was omitted from randomized variables. Subscript j was used to designate j-th element, i.e., bin, in the distribution considered. If $R(q_{ij})$ is not equal to one, then either theory, or experiment, or procedure of computing randomized distributions are wrong.

Form the discussion presented above, it can be concluded that if distributions of one single variable are produced, then one is comparing the theory and experimant at the level of single differential cross sections, $d\sigma(q_i)/dq_i$, but with all experimental resolutions and efficiencies folded in. If two-dimensional distributions $N_c(q_i,q_j)$ and $N_r(q_i', q_o')$ are made, then comparison would be at the level of $d^2\sigma(q_i,q_j)/dq_i dq_j$.

The advantages of this "global" procedure are the following:

(i) The comparison between the theory and experiment is essentially exact. This is so because the randomized distributions are obtained by computing integrals, which is a procedure that can be easily and almost exactly performed. In conventional analyses one attempts to solve an integral equation in the following way: One starts from equation (2) by replacing $N_r(q_i')$ with $N_c(q_i)$ as obtained from equation (1), assumes that he knows kernel function $K(q_1,\ldots,p_5)$ and solves the integral equation for $d\sigma(q_i)/dq_i$. As $K(q_1,\ldots,p_5)$ is not generally known in its analytical form, and we are dealing with multiple integrals, certain mathematical approximations must be made which often (as well as in our ppγ case) include some assumption about the solution search for. (In our case we have used Hamada-Johnston potential, i.e., Liou's calculation). Therefore, cross sections obtained in conventional procedure are not only mathematically approximate but are also "theory dependent" although they appear in a "theory independent" form.

(ii) In case of relatively low statistics experiments by integrating over variables which are not under investigation, one achieves high statistics per bin.

(iii) The global analysis used allows both in principle and in practice an easy change of variables. Therefore, one can use this procedure to easily make comparison between theory and experiment as a function of any variable. In fact, one could use this procedure to search for interesting variables. The main disadvantage of this procedure is that the comparison of experiment with another theory is impossible except by people who performed this original analysis as function K $(q_1 \cdots\cdots, p_5)$ can not be expressed in an analytic form.

Fig. 8 shows some preliminary results of the global analyses. These results are preliminary for two reasons:

(i) Experimental and randomized spectra (see Eq. (1) and Eq. (2)) were normalized so that the ratio $R(q_{ij})$ was on the average equal to one. Therefore, only shapes were compared.

(ii) All systematic errors have not yet been considered in detail. Although main points were considered and were found not to influence results appreciably.

Abscisa on five of the six graphs in Fig. 8 is self-explanatory. Angle γ_3^* is the photon angle in the reaction plane in c.m. system.

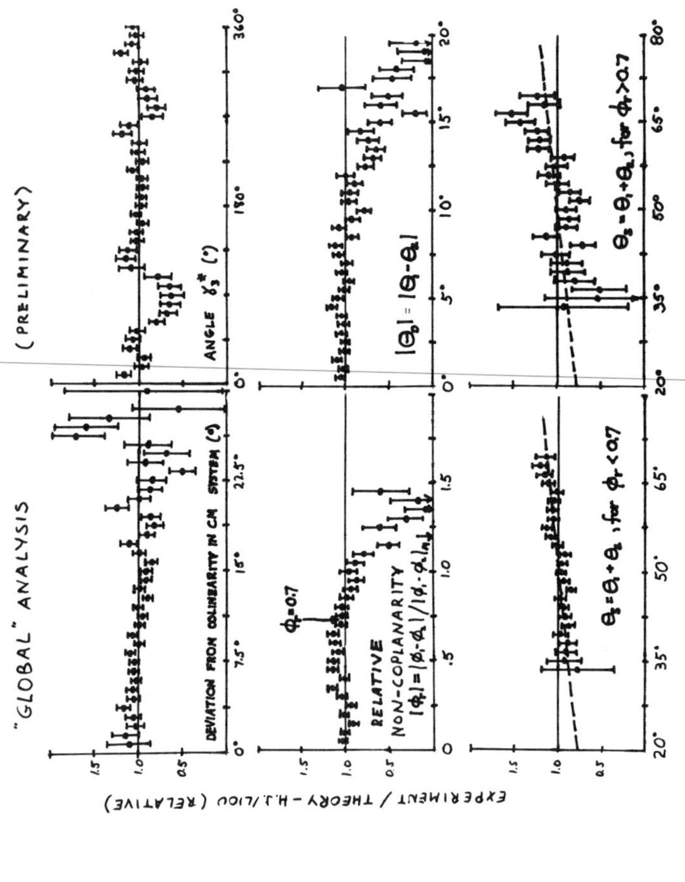
"GLOBAL" ANALYSIS
(PRELIMINARY)
EXPERIMENT / THEORY - H.J./LIOU (RELATIVE)
DEVIATION FROM COLINEARITY IN CM SYSTEM (°)
ANGLE δ_3^* (°)
ϕ_r=0.7
RELATIVE
NON-COPLANARITY
$|\phi_r| = |\phi_1 - \phi_2| / |\phi_1 - \phi_2|_{max}$
$|\Theta_0| = |\Theta_1 - \Theta_2|$
$\Theta_s = \Theta_1 + \Theta_2$, for $\phi_r < 0.7$
$\Theta_s = \Theta_1 + \Theta_2$, for $\phi_r > 0.7$

Fig. 8

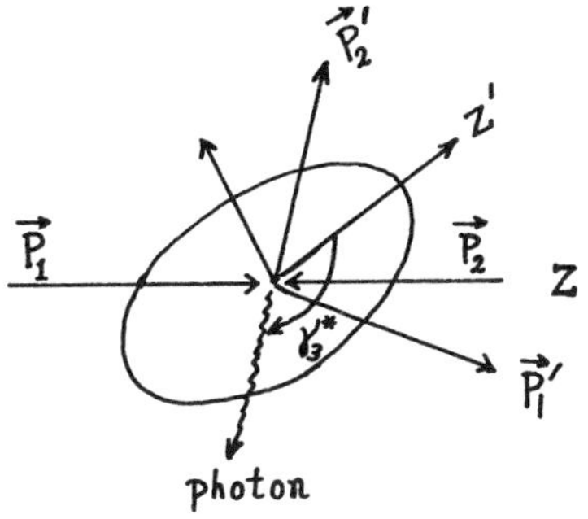

Conclusions are obvious from the drawings. In the lab system there is some discrepancy for large non-coplanarity and asymmetry. In c.m. system, there seem to be some discrepancy at $\gamma_3^* \approx 90^o$, and possibly at $\gamma_3^* \approx 270^o$.

We are continuing this analysis of data by investigating in detail the effect of some systematic errors, and by computing ratios as a function of some other variables, like photon transverse momentum, or others. We could make comparison for <u>any</u> variable. Therefore, <u>suggestions from the theorists concerning what variables to try will be the most appreciated.</u>

(C) <u>Proposed computerized data bank</u>

In brief, my belief is the following: If everything is organized properly it would be technically feasible for anybody to reanalyse our data along "global analysis" lines. For instance, you could work out another theory of your own, compute grid of cross sections, use our data, use some standardized computer programs (it would take you perhaps two days to learn how to use them) and perform global analysis, just as we have done it. As a matter of fact these procedures are so general that they could be applied to other three body reactions as well. Programs to do this, of course, exist. We have them but they have yet to be put into a well organized form so that they could be used by other people.

V <u>New experiment planned</u>

<u>TRIUMF experiment</u>

(i) Laboratory: TRIUMF, Canada

(ii) Type of experiment: ppγ

(iii) Beam energy: 175 MeV and up

(iv) Detector: Spark wire chamber

(v) Angular range: $\theta_1 \approx \theta_2 \approx 8-13^{\circ}$

Discussions:

J. G. Rogers (University of Alberta and TRIUMF):

A ppγ measurement will be performed beginning early next year at TRIUMF. The apparatus for this measurement is already constructed. We will measure over the angular range $\theta_1 \approx \theta_2 \approx 8\text{-}13^{\circ}$ for the proton lab angles and several different bombarding energies from 175 MeV upwards. The apparatus includes two MWPC's to define each proton's momentum. The work is motivated by Heller's calculations which show a large sensitivity to the off-shell characteristics of the NN interaction potential in these regions of phase space.

VI Conclusion

Without any doubt, nucleon-nucleon bremsstrahlung is the simplest and best process to study the off-energy-shell behavior of the two-nucleon interaction. The off-energy-shell effects have been observed from an approximately ten thousand bremsstrahlung data collected in the past decade. But these data do not permit us to select the best phenomenological potential because of the following two reasons: The first one is due to the experimental difficulty. The experiments are very time-comsuming and tricky, and consequently the data obtained are not accurate enough for the selection of a best potential. The second one is due to fundamental and practical difficulties associated with theory. As we know, in order to select a best potential, one has to know the electromagnetic interaction exactly. But the situation is not so simple since the bremsstrahlung process cannot be clearly divided into a part which represents the off-shell nuclear interaction and the other part which represents the electromagnetic interaction. It turns out that the determination of the electromagnetic interaction depends also upon the nature of the nuclear forces which we are searching for. Besides this fundamental difficulty, the present calculations are still far from perfect since no calculation in existence so far has included all possible corrections which we can treat.

Some remarkable progress in the theoretical study of the process have been made during the last few years. They are the studies of the Coulomb effects, relativistic effects and the exchange effects. The contribution from each of these effects to the ppγ and/or npγ cross sections has been investigated thoroughly by many different groups, but these effects have never been put together into a single calculation. The results of the calculation including all these corrections will be very important. We feel strongly that it is now the time for theorists to start doing this kind of more complete calculation.

A large discrepancy found between Orsay's experiment and theory has not yet resolved. A new ppγ experiment at the forward angle is already planned at TRIUMF, and we hope more experiments of this kind will be performed in near future so that the discrepancy could be resolved by new experimental results and new calculation which includes all corrections. Manitoba's experiment has almost completed. A great number of new data at 42 MeV will become available very soon. These data were analysed in two different ways: a conventional analysis and an unconventional "global" analysis.

npγ calculations have been improved recently by the inclusion of the exchange effects. We should now encourage experimentalists to perform more precise npγ experiments. Further study of the nucleon-nucleon bremsstrahlung process especially npγ will provide information not only the off-energy-shell behavior of the two-nucleon interaction but also the electromagnetic interactions of the interacting nucleons.

Finally, it was suggested that theorists and experimentalists should work togather to find out at which energies and angles the experiment should be performed. This can be done by a theoretical search for a particular geometry and energy region where the off-energy-shell effects are largest. Since the calculation used in this search must be the one which includes all possible corrections, this work has to wait until that kind of new calculation is accomplished. Jovanovich, on the other hand, argued that a detailed and systematic study of the bremsstrahlung process should be started irrespective of the present status of theory.

I would like to thank Dr. Leon Heller, Moderator for bremsstrahlung session, and Dr. J. V. Jovanovich for stimulating discussions which greatly increased my understanding of the bremsstrahlung process. I would also like to thank Mr. R. Larue, scientific secretary, for his assistance during the discussion period. Many useful discussions with Drs. G. E. Bohannon, B. F. Gibson, L. S. Celenza, B. Frois, Q. Hokim and W. Wölfli are gratefully acknowledged. I wish to thank Miss E. London and Mrs. M. Ferrantelli for their assistance in preparing and typing this report. Finally, I am grateful to Prof. M. I. Sobel for stimulating discussions, reading of the manuscript and making useful comments.

References

1. P. Signell, Advances in Nuclear Physics, Vol. 2 (Plenum Press, New York, 1969).
2. M. L. Halbert, Proceedings of the Gull Lake Symposium on the Two-Body Force in Nuclei, (Plenum Press, New York, 1972).
3. M. J. Moravcsik, Reports on Progress in Physics, 35, 587 (1972).
4. E. M. Nyman, Physics reports, 9, 179 (1974).
5. B. Gottschalk, W. J. Shlaer, and K. H. Wang, Nucl. Phys. A94, 491 (1967).
6. R. Baier, H. Kühnelt, and P. Urban, Nucl. Phys. B11, 675 (1969).
7. M. K. Liou, Ph.D. thesis, 1969, University of Manitoba (unpublished).
8. M. K. Liou and M. I. Sobel, Ann. of Phys. 72, 323 (1972).
9. K. W. Rothe, P. F. M. Koehler, and E. H. Thorndike, Phys. Rev. 157, 1247 (1967).
10. D. Drechsel and L. C. Maximon, Ann. Phys. (N.Y.) 49, 403 (1968).
11. P. Signell, Proceedings of the Midwest Theoretical Conference. (1970).
12. D. Marker and P. Signell, Phys. Rev. 185, 1286 (1969).
13. M. Fortes, A. D. Jackson and E. M. Nyman, preprint.
14. S. J. Brodsky and J. R. Primack, Ann. Phys. (N.Y.) 52, 315 (1969); Phys. Rev. 174, 2071 (1968);
    H. Osborn, Phys. Rev. 176, 1523 (1968);
    R. A. Krajcik and L. L. Foldy, Phys. Rev. Lett. 24, 545 (1970) and to be published;
    F. E. Close and L. A. Copley, Nucl. Phys. B19, 477 (1970);
    M. K. Liou and M. I. Sobel, Phys. Rev. C7, 2044 (1973) and M. K. Liou, preprint.
15. L. S. Celenza, B. F. Gibson, M. K. Liou and M. I. Sobel, Phys. Lett. 41B, 283 (1972); and proceedings of the International Conference on Few Particle Problems in the Nuclear Interaction, Los Angeles, (1972). Edited by I. Slaus et. al., American Elsevier Publishing Co., Inc., N.Y., p. 68.
16. L. S. Celenza, M. K. Liou, M. I. Sobel and B. F. Gibson, Phys. Rev. C8, 838 (1973).
17. M. K. Liou and K. S. Cho, Nucl. Phys. A160, 417 (1970).

18. J. H. McGuire, Phys. Lett. 32B, 73 (1970).

19. E. M. Nyman, Nucl. Phys. A160, 517 (1970).

20. L. Heller, Proceedings of the Gull Lake Symposium on the Two-Body Force in Nuclei. (Plenum, N.Y. 1972).

21. M. K. Liou and M. I. Sobel, Phys. Rev. C4, 1507 (1971).

22. R. H. Thompson and L. Heller, Phys. Rev. C7, 2355 (1973).

23. V. R. Brown and J. Franklin, Phys. Rev. C8, 1706 (1973).

24. L. Heller and M. Rich, Phys. Rev. C10, 479 (1974).

25. J. H. McGuire, Phys. Lett. 40B, 41 (1972).

26. E. M. Nyman, Phys. Rev. 170, 1628 (1968).

27. G. Brolley and L. K. Morrison, Proceedings of the International Conference on the Nuclear Structure, Tokyo, 1967, ed. J. Sanada, Suppl. J. Phys. Soc. Japan 24 (1968).

28. T. K. Dahlblom and A. M. Green, Phys. Lett. 41B, 23 (1972).

29. J. Sanada, K. Kondo, S. Seki, K. Kuriyama, N. Ryu, H. Hasai and K. Iwatani, Phys. Lett. 39B, 509 (1972).

30. V. R. Brown, Phys. Rev. 177, 1498 (1969).

31. F. E. Low, Phys. Rev. 110, 974 (1958).

32. S. L. Adler and Y. Dothan, Phys. Rev. 151, 1267 (1966).

33. T. H. Burnett and N. M. Knoll, Phys. Rev. Lett. 20, 86 (1968).

34. H. W. Fearing, Phys. Rev. C6, 1136 (1972); D7 243 (1973).

35. J. H. McGuire and B. J. Malenka, Nucl. Phys. B17, 641 (1970).

36. H. Feshbach and D. R. Yennie, Nucl. Phys. 31, 150 (1962).

37. L. Heller, Phys. Rev. 174. 1580 (1968).
Erratum: Phys. Rev. 180, 166 (1969).

38. M. K. Liou, Phys. Rev. C2, 131 (1970);
M. K. Liou and M. I. Sobel, Phys. Rev. C3, 1430 (1971); and Ref. 21.

39. E. M. Nyman, Phys. Lett. 40B, 323 (1972).

40. R. M. Woloshyn, Phys. Lett. 49B, 415 (1974).

41. P. Signell and D. Marker, Phys. Lett. 28B, 79 (1968); and 26B, 559 (1968).

42. E. A. Silverstein and K. G. Kibler, Phys. Rev. Lett. 21, 922 (1968).

43. A. Niiler, C. Joseph, V. Valkovic, R. Spiger, T. Canada, S. T. Emerson, J. Sandler, and G. D. Phillips, Phys. Rev. 178, 1621 (1969).

44. G. M. Crawley, D. L. Powell, and B. V. Narasimha Rao, Phys. Lett. 26B, 576 (1968).

45. W. Wölfli, M. Suter, G. Bonani and C. Stoller, Contribution paper to this conference.

46. D. W. Storm and R. Hellner, Phys. Rev. C4, 1541 (1971).

47. I. Slaus, J. W. Verba, J. R. Richardson, R. F. Carlson, W. T. H. van Oers, and L. S. August, Phys. Rev. Lett. 17, 536 (1966).

48. J. V. Jovanovich, L. G. Greeniaus, J. McKeown, T. W. Millar, D. G. Peterson, W. F. Prickett, K. F. Suen, and J. C. Thompson, Phys. Rev. Lett. 26 277 (1971).

49. D. L. Mason, M. L. Halbert and L. C. Northcliffe, Phys. Rev. 176, 1159 (1968).

50. R. E. Warner, Can. J. Phys. 44, 1225 (1966); and D. Drechsel, L. C. Maximon, and R. E. Warner, Phys. Rev. 181, 1720 (1969).

51. M. L. Halbert, D. L. Mason, and L. C. Northcliffe, Phys. Rev. 168, 1130 (1968).

52. D. L. Mason, M. L. Halbert, A. van der Woude, and L. C. Northcliffe, Phys. Rev. 179, 940 (1969); and D. O. Galde, M. L. Halbert, C. A. Ludemann, and A. van der Woude, Phys. Rev. Lett. 25, 1581 (1970).

53. F. Sannes, J. Trischuk, and D. G. Stairs Nucl. Phys. A146, 438 (1970); and Phys. Rev. Lett. 21, 1474 (1968).

54. B. Gottschalk, W. J. Shlaer, and K. H. Wang, Nucl. Phys. 75, 549 (1966).

55. J. H. McGuire, A. H. Cromer, and M. I. Sobel, Phys. Rev. 179, 948 (1969).

56. W. A. Pearce, W. A. Gale, and I. M. Duck, Nucl. Phys. B3, 241 (1967).

57. F. P. Brady and J. C. Young, Phys. Rev. C2, 1579 (1970); Phys. Rev. C7, 1707 (1973).

58. J. H. McGuire, Phys. Rev. C1, 371 (1970).

59. J. A. Edgington, V. J. Howard, I. M. Blair, B. E. Bonner, F. P. Brady and M. W. McNaughton, Nucl. Phys. A218, 151 (1974).

60. A. Willis, V. Comparat, R. Frascaria, N. Marty, M. Morlet, and N. Willis, Phys. Rev. Lett. 28, 1063 (1972).

61. V. R. Brown, Phys. Rev. C6, 1110 (1972).

62. J. McKeown, L. G. Greeniaus, J. V. Jovanovich, W. F. Prickett, K. F. Suen and J. C. Thompson, Nucl. Instr. and Meth. 104, 413 (1972).

Participants

J. Birchall	Lawrence Berkeley Laboratory
G. E. Bohannon	Michigan State University
B. E. Bonner	Los Alamos Scientific Laboratory
F. P. Brady	University of California, Davis
J. M. Cameron	University of Alberta/TRIUMF
J. Deregel	C.E.N. Saclay
J. A. Edgington	TRIUMF/Queen Mary College, London
A. M. Green	University of Helsinki
M. Hasinoff	University of British Columbia
S. Hayakawa	McGill University
L. Heller (Moderator)	Los Alamos Scientific Laboratory
Q. Hokim	Universite Laval
M. S. Hussein	Universidade De Sao Paulo
J. V. Jovanovich	University of Manitoba
M. Lamothe	University of Toronto
R. Larue (Scientific Secretary)	Universite Laval
J. Lee	McGill University
F. Lehar	CEN, Saclay
J. S. Levinger	Rensselaer Polytechnic Institute
M. K. Liou (Rapporteur)	Brooklyn College of the CUNY
B. Loiseau	IPN, Orsay
M. J. Moravcsik	University of Oregon
G. A. Moss	University of Alberta/TRIUMF
J. Owens	R. I. Junior College
J. R. Richardson	University of British Columbia/TRIUMF
D. O. Riska	State University of New York, Stony Brook
J. G. Rogers	University of Alberta/TRIUMF
H. S. Sherif	University of Alberta
D. W. Storm	Columbia University
G. L. Strobel	University of GEorgia
J. D. Svenne	University of Manitoba
R. H. Thompson	Los Alamos Scientific Laboratory
J. Vavra	University of British Columbia
D. W. Werren	University of Geneva
W. Wölfli	ETH, Zürich
B. Frois	CEN, Saclay

Very Low Energy Parametrization

J.J. de Swart

Institute for Theoretical Physics,

University of Nijmegen,

Nijmegen, The Netherlands.

1. Low energy pp-scattering.

It is generally assumed that the pp interaction at very low energies may be described by a potential V of the form

$$V = V_N + V_C + V_{VP} + V_{mod} \quad . \tag{1}$$

Here V_N describes the pure nuclear interaction, which has a range of $m_\pi^{-1} \simeq 1.4$ fm. $V_C = \alpha \frac{\lambda\!\!\!^-_e}{r} mc^2$ is the pure Coulomb interaction, due to point protons, which is of infinite range. V_{VP} is the modification of the pure Coulomb interaction due to the vacuum polarization [Ue 35, Du 57, He 60, Sh 70]. The range of this interaction is $(2m_e)^{-1} \simeq 200$ fm. The combination $V_E = V_C + V_{VP}$ is often referred to as the electric potential. V_{mod} is the modification of the pure Coulomb interaction due to the extended charge distribution of the protons. This potential is of finite range $\sim m_\pi^{-1}$. This V_{mod} also contains terms due to the extended magnetic moments of the protons. These contribute a spin-orbit and a tensor force of infinite range ($\frac{1}{r^3}$), and a central force of range m_π^{-1}.

In most analyses the long range contribution of the magnetic moment interaction has been neglected. The experiments seem to get, however, accurately enough that soon these terms need inclusion, especially for the polarization experiments. If we neglect the long range part of V_{mod}, then one has essentially three potentials of different range present. They are $V_N' = V_N + V_{mod}$ of range m_π^{-1}, V_{VP} of range $(2m_e)^{-1}$ and V_C of infinite range.

To describe the scattering one introduces the electric phase shifts δ_ℓ^E. These are the phase shifts due to V_N' calculated with respect to the wave functions of the electric potential V_E.

For practical reasons it is often convenient to introduce the Coulomb phaseshifts δ_ℓ^C. These are the phaseshifts due to V_N' calculated with respect to the wave functions of the Coulomb potential. These would be the phaseshifts for the hypothetical case that one can turn off the vacuum polarization. Using perturbation theory these Coulomb phaseshifts can be obtained from the electric phaseshifts. This is the Foldy correction [Fo 55]:

$$\delta_\ell^C = \delta_\ell^E + \tau_\ell - \Delta_\ell \qquad (2)$$

where τ_ℓ is phaseshift due to the vacuum polarization calculated with respect to the Coulomb wavefunctions [He 60] and Δ_ℓ is a somewhat model dependent correction [Sh 70, No 71, No 72].

For theoretical reasons it is also convenient to introduce the nuclear phaseshifts δ_ℓ^N.

These are the phaseshifts due to V_N or V'_N calculated with respect to the free field wavefunctions $j_\ell(kr)$ and $n_\ell(kr)$. We will take the phaseshifts δ^N_ℓ due to V'_N.

2. S-wave.

In the low energy region it is customary to introduce for the $^1S_0$ phaseshift the effective range functions $F(k^2)$, which are functions of the center of mass momentum k. They are for the nuclear phaseshifts [Sch 47]

$$F_N(k^2) = k \cot \delta^N_0 \quad , \tag{3}$$

for the Coulomb phaseshifts [Be 49, Ja 50]

$$F_C(k^2) = C_0^2\, k \cot \delta^C_0 + 2k\eta\, h(\eta) \quad , \tag{4}$$

and for the electric phaseshifts [He 60]

$$F_E(k^2) = \frac{C_0^2 k}{1-\phi_0} \left[(1+\chi_0)\cot\delta^E_0 - \tan\tau_0\right] + 2k\eta(h(\eta) + \ell_0(\eta)). \tag{5}$$

Here $C_0^2 = \frac{2\pi\eta}{e^{2\pi\eta}-1}$ and $\eta = \alpha \frac{c}{v}$. The function $h(\eta)$ is given by [Ja 50] and the functions ϕ_0, χ_0 and $\ell_0(\eta)$ by [He 60]. These effective range functions allow the expansions:

$$F(k^2) = -\frac{1}{a} + \frac{1}{2} rk^2 - Pr^3k^4 + Qr^5k^6 + \ldots .$$

where a is the scattering length, r the effective range, and P and Q the shape parameters.

Due to the one-pion-exchange branch cut this series expansion must diverge for laboratory energies beyond 9.71 MeV. Cini et al. [Ci 59] suggested to take this branch cut approximately into account. This lead to the CFS approximation

$$F_N(k^2) = -\frac{1}{a} + \frac{1}{2} rk^2 - \frac{pk^4}{1+qk^2} \qquad (6)$$

where the parameters p and q are determined by the pion-nucleon coupling constant,the pion mass and the effective range parameters a and r. Noyes [No 64] suggested to use the same formula also for $F_C(k^2)$ using the Wong-Noyes Coulomb correction [Wo 62, No 72]. Perhaps somebody should try to derive an analogous result for $F_E(k^2)$. This CFS approximation extends the radius of convergence to about 38.8 MeV. The effective range parameters are essentially determined by two measurements. The first one is the determination of the cross section near the interference minimum by Brolley et al. [Br 64]. This experiment was analyzed by Gursky and Heller [Gu 64] and the phaseshift δ_0^E at 0.38243 MeV was determined with a fantastic accuracy. At the higher energy end the experiment of Jarmie et al. [Ja 70] at 9.918 MeV fixes the phaseshift δ_0^E. Applying the Foldy correction and the Coulomb modified CFS approximation Noyes and Lipinski [No 71] obtained

$$a_{pp}^c = -\ 7.826 \pm 0.010 \text{ fm}$$

$$r_{pp}^c = 2.803 \pm 0.015 \text{ fm}$$

for $g^2_{pp\pi^0}/4\pi = 15.3$.

In a contribution to this conference a Zürich group [Be 74] reported a remeasurement of the position of the interference minimum. They found $E_{min} = 382.75 \pm .11$ keV. This should be compared with the value $E_{min} = 382.4 \pm .2$ keV of [Br 64]. Within the errors these experiments are compatible and this experiment will influence the quoted values of a and r only slightly.

In another contribution from a Kyoto group [Im 74] new angular distribution measurements at 5, 6, 7, and 8 MeV were reported. However, these still need analyzing in terms of the CFS effective range theory. Calculations by Arndt show that these new Kyoto data are in agreement with the latest VPI χ^2-surface [Ar 74]. From Basel [Mu 74] was heard about some new experiments at 0.5, 0.9, and 1.8 MeV. They found a slight disagreement with the existing s and p wave phaseshifts. This disagreement was the largest at 0.5 MeV and at the smallest angle (12°). No effect was seen at higher energies. In the discussion H.P. Noyes raised the question if perhaps the vacuum polarization contribution needs more theoretical work. This due to the fact that there seems to be some disagreement with the theory for as well the Basel data [Mu 74] as the older Wisconsin data [Kn 66].

3. P-waves

When the potential has the form

$$V = V_C + V_T\, S_{12} + V_{LS}\, \underline{L}\cdot\underline{S} \qquad (7)$$

then the ^3P_J phaseshifts δ_{1J} are given in Born approximation by

$$\delta_{10} = \delta_C - 4\delta_T - 2\delta_{LS}$$

$$\delta_{11} = \delta_C + 2\delta_T - \delta_{LS} \qquad (8)$$

$$\delta_{12} = \delta_C - \frac{2}{5}\delta_T + \delta_{LS} \quad .$$

In the analyses of the experiments the δ_{1J} phases appear in the combinations

$$\Delta_C = \frac{1}{9}(\delta_{10} + 3\delta_{11} + 5\delta_{12})$$

$$\Delta_T = \frac{5}{72}(-2\delta_{10} + 3\delta_{11} - \delta_{12}) \qquad (9)$$

$$\Delta_{LS} = \frac{1}{12}(-2\delta_{10} - 3\delta_{11} + 5\delta_{12}) \quad .$$

The angular distribution is given by [Sh 70]

$$\frac{d\sigma}{d\Omega} = \left(\frac{d\sigma}{d\Omega}\right)_E + \left(\frac{d\sigma}{d\Omega}\right)_N + \left(\frac{d\sigma}{d\Omega}\right)_{int} \qquad (10)$$

where $\left(\frac{d\sigma}{d\Omega}\right)_E$ is the electric cross section due to the Coulomb and vacuum polarization potentials, $\left(\frac{d\sigma}{d\Omega}\right)_N$ is the nuclear cross section which contains δ_0^E, Δ_T^E, and Δ_{LS}^E quadratically. The interference term $\left(\frac{d\sigma}{d\Omega}\right)_{int}$ contains the combination Δ_C^E linearly. This implies that Δ_C^E can be measured with the highest precision, while Δ_T and Δ_{LS} are less well known. Using (8) one can obtain so the ^3P_J phaseshifts.

For the $^3P_0$ and $^3P_1$ phases one uses the effective range expansion

$$[c_0^2k^3\cot\delta_{1J} + 2k^3\eta h(\eta)]\ (1+\eta^2) = -\frac{1}{a_{1J}} + \frac{1}{2}\, r_{1J}k^2 + \ldots \quad (11)$$

For the $^3P_2$ phase one uses the modified expansion [Sh 70]

$$[c_0^2k^3\cot\{\delta_{12}-c_0^2(1+\eta^2)\delta_{12}^{OPEC}\} + 2k^3\eta h(\eta)](1+\eta^2) =$$

$$= -\frac{1}{a_{12}} + \frac{1}{2}\, r_{12}k^2 + \ldots \quad .(12)$$

The reason for this modification is that at very low energies the $^3P_2$ phaseshift goes as k^5 instead of k^3 because

$$\delta_{12}^{OPEC} \simeq -\frac{g^2}{4\pi}\frac{8}{15}\left(\frac{k}{M}\right)^5 + \ldots. \quad (13)$$

Values for these p-wave effective range parameters are given by [Sh 70]. In a contribution to this conference [Na 74] they were also calculated in a potential model that has a good χ^2 with all the experimental data up to 330 MeV. The agreement with one of sets of [Sh 70] is given in table I and seen to be reasonable.

	Sh 70		Na 74	
	a_{1J}	r_{1J}	a_{1J}	r_{1J}
$^3P_0$	- 2.6 ± 2.0	4.3 ± 2.0	- 3.00	3.44
$^3P_1$	2.8 ± 1.3	- 9.0 ± 1.0	1.82	- 7.49
$^3P_2$	- 0.45 ± 0.28	15 ± 10	- 0.28	4.52

Table I: P-wave effective range parameters a_{1J} in fm^3 and r_{1J} in fm^{-1}.

The combination Δ_C has been determined by [Sh 70, Ar 74, Im 74, and Na 74]. In fig. 1 are sketched the error corridors as given by [Sh 70] and by [Ar 74] (cross hatched) for an analysis of all the data below 27.6 MeV, together with the energy dependent solution of [Ar 74] using all data between 1 and 500 MeV (upper corridor). Included is also theoretical curve of [Na 74] and the experimental determinations of the Kyoto group [Im 74]. Definite conclusions cannot be drawn, but it appears that the low energy polarization experiments as planned at Wisconsin and Kyoto will be helpful here.

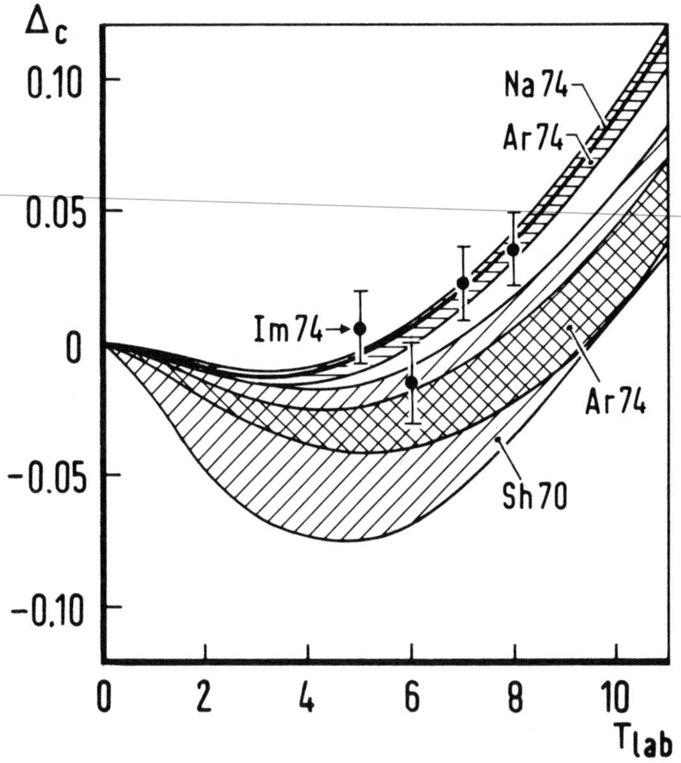

Fig. 1

4. Dispersion relations.

In a contribution to this conference Viollier [Vi 74] discussed Coulomb modified dispersion relations for the amplitudes on the spin basis: M_{SS}, M_{00}, M_{01}, M_{10}, M_{11} and M_{1-1}. Only the M_{SS} amplitude contains s-waves, the other waves (except M_{1-1}) start with p-waves. For each of these amplitudes M they can define and calculate a function R(E) which contains the singularities on the left hand side, as the pion pole, 2π cuts, etc. When they calculate R(E) for different E and then extrapolate to the pion pole $E \rightarrow -9.71$ MeV they obtain the residue of these modified amplitudes at the pole. This is essentially the pion nucleon coupling constant $g^2_{pp\pi^0}/4\pi$. They propose this as a method to define the charged coupling constant. From fig. 2 which contains their results for some of the phaseshift analyses one can deduce the following conclusions:

The Livermore X phaseshift analyses [Ma 69] extrapolates to $g^2_{pp\pi^0}/4\pi = 15.3 \pm .3$. At the same time the curves show the failure of the low energy treatment of the higher ℓ phaseshifts in this Livermore analysis.

The recent VPI analysis [Ar 74] treats the low energy p-waves properly as can be seen from fig. 2 , but their treatment of the s-wave is probably not correct. The VPI curves except R_{SS} extrapolate to $g^2_{pp\pi^0}/4\pi \approx 15$, but that was their input for this parameter.

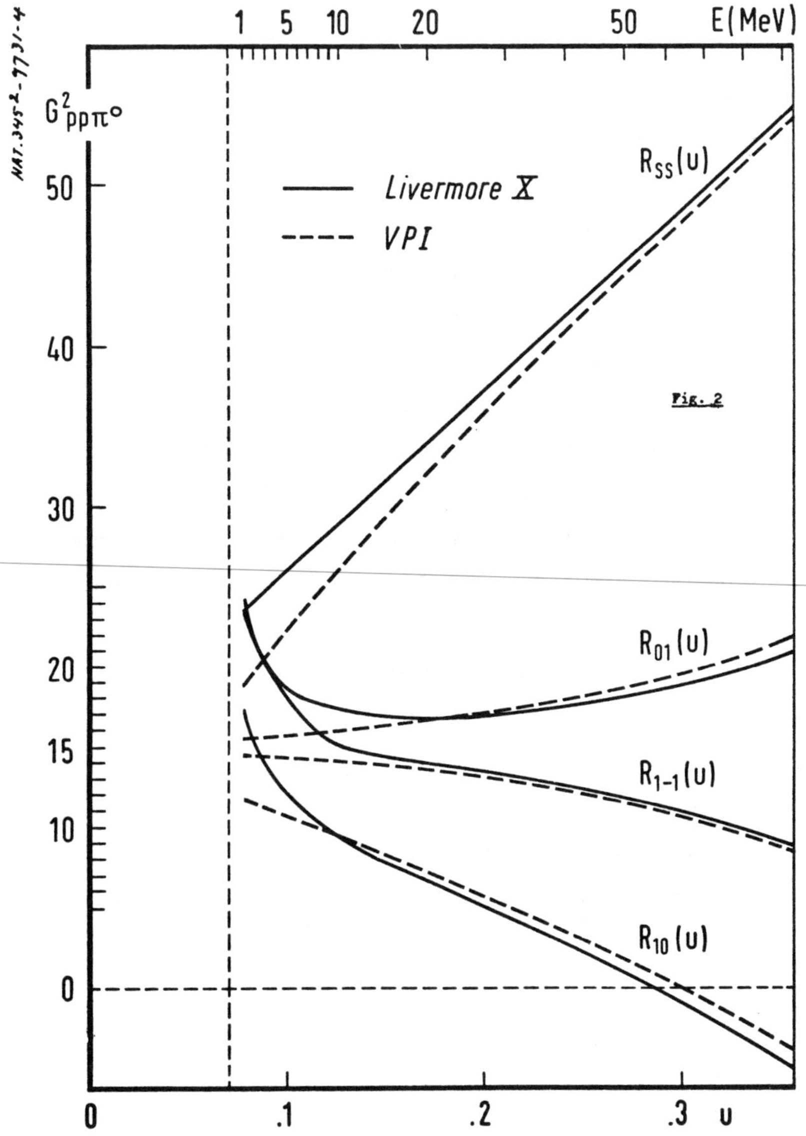

1 5 10 20 50 E(MeV)
$G^2_{pp\pi^0}$
50
40
30
20
15
10
0
Livermore X
VPI
$R_{SS}(u)$
$R_{01}(u)$
$R_{1-1}(u)$
$R_{10}(u)$
0 .1 .2 .3 u

Fig. 2

5. Low energy nn parameters.

Several determinations of the nn scattering length a_{nn} were discussed during this conference. Due to lack of time this topic was only shortly discussed in this particular session.

B. Kühn [Ku 74] presented a critical statistical evaluation of about 30 results published in the literature. He essentially assumes a combination of random theoretical and experimental errors in the different published values for a_{nn}. After throwing away some of the results that peak around $a_{nn} \simeq -23$ fm and giving the kinematically complete experiments some higher weight he finds a distribution with average value $a_{nn} = -16.61$ fm and a standard deviation 1.45 fm. This implies that the center of the distribution is determined with 95 % confidence level to lie at

$$a_{nn} = -16.61 \pm 0.54 \text{ fm} \quad .$$

In another contribution B. Zeitnitz [Ze 74] discussed the analysis of three kinematically complete experiments on the reaction $^{2}H(n,2n)H$. Their results from a complete charge dependent calculation using exponential form factors is

$$a_{nn} = -16.3 \pm 1.0 \text{ fm}$$

$$r_{nn} = 2.13 \pm .4 \text{ fm} \quad .$$

Their value of r_{nn} depends very much on the analysis, e.g. on the details of the form factors.

The better they fit the higher energies (up to 200 MeV) the better values for r_{nn} are obtained. It was pointed out by B.F. Gibson that the proposed measurements of the kinematically complete radiative π-capture and weak μ-capture reactions at Triumph and SIN should give rather accurate values for a_{nn} and r_{nn}. The theoretical analyses of these reactions should be less subject to uncertainties then the corresponding kinematically complete nd breakup reactions, because there will be only two strongly interacting particles in the final nn state. However, the results of these experiments will not be available for the first few years. In the meantime it is perhaps wise to stick by

$$a_{nn} = -16.5 \pm 0.8 \text{ fm}$$

and

$$r_{nn} = 2.86 \pm 0.03 \text{ fm}$$

6. Low energy np-scattering data.

A difficulty has been the value of the coherent scattering length a_{nH} for neutrons on parahydrogen. In a contribution [Ca 74] to this conference a Harvard group reported a new measurement which is in excellent agreement with the latest liquid mirror measurements. Averaging the latest values [Me 49, Ko 71, Ca 74] and forgetting the measurements [Ko 67, Ho 71] we get

$$a_{nH} = \frac{1}{2}(3a_t + a_s) = -3.742 \pm 0.002 \text{ fm}.$$

where a_t and a_s are the np-scattering lengths in the spin triplet and spin singlet s-states. Using the value of Houk [Ho 71b] for the incoherent total cross section

$$\sigma_o = \pi(3a_t^2 + a_s^2) = 2043.6 \pm 2.3 \text{ fm}^2$$

gives $|a_t - a_s| = 29.132 \pm .020$ fm

and $a_t = 5.412 \pm 0.006$ fm

$a_s = - 23.720 \pm 0.016$ fm

The radius R of the deuteron is determined by the binding energy of the deuteron [Ho 71b], one finds

$$R = 4.31767 \pm 0.00009 \text{ fm}$$

This implies for the mixed effective range

$$\rho(-\varepsilon,0) = 2R(1 - R/a_t) = 1.746 \pm 0.005 \text{ fm}$$

The determination of $r_t = \rho(0,0)$ from this value of $\rho(-\varepsilon,0)$ is model dependent [No 72]. We would suggest to use the values

$r_t = 1.733 \pm 0.006$ fm

$r_s = 2.63 \pm 0.06$ fm

7. Charge symmetry and charge independence.

Also this topic was only very superficially talked about due to lack of time.

The nuclear scattering length a_{pp}^N is related to the Coulomb scattering length a_{pp}^C by [Ja 50]

$$\frac{1}{a_{pp}^N} \simeq \frac{1}{a_{pp}^C} + 2k\eta \ln\{\frac{1}{2k\eta r_{pp}} - 0.33\} \quad .$$

This results in $a_{pp}^N \simeq -17$ fm. This subtraction of the Coulomb force is always thought to be rather model independent such that was assumed [He 72]

$$a_{pp}^N = -17.1 \pm 0.2 \text{ fm}$$

$$r_{pp}^N = 2.84 \pm 0.03 \text{ fm} \quad .$$

The most important development has been the disturbing paper by P.U. Sauer [Sa 74], who showed that the reduction from a_{pp}^C to a_{pp}^N is not so model independent as was assumed. He demonstrated this using the soft core Reid potential V_R. The Hamiltonian is then

$$H = T + V_R + V_C$$

where T is the kinetic energy.

Using unitary operators U which where of a separable form and of short range, such that they affect only the inner region of the potential, he constructed equivalent potentials

$$\tilde{V}_R = UHU^{-1} - T - V_C \quad ,$$

which gave exactly the same a^C_{pp} as the original soft core Reid potential. These equivalent potentials $\tilde{V}_R$, which supposed all still had the OPEP tail, gave a very wide range of values for a^N_{pp} from positive values down to - 17.4 fm. The question is: Are these $\tilde{V}_R$ still acceptable nn potentials? The corresponding wave functions $\tilde{\psi}$ seem to be peculiar. Are the singularities along the left hand cut still at the right place after the transformation? Perhaps highly strange things happen along this cut. Separable transformations are often very elegant and easy to handle, but are they really "physical" (What that may be)? It is very hard to draw any definite conclusion out of this work. Perhaps one could forget about these separable transformations and try to redo the work with more physical ones. The lesson of this work could be for other people (especially in nuclear structure calculations which do use such separable transformations)to be extremely cautious. Of course it does throw some light or doubt on what is called almost model independent. From the values of $a^N_{pp} = - 17.1 \pm .02$ fm, $a_{nn} = - 16.5 \pm .8$ fm, and $a^s_{np} = - 23.72 \pm .02$ fm it is clear that charge independence is broken, but nothing can be said about charge symmetry. A difficulty with the theory is the low value of r^s_{np}.

Acknowledgement:

I would like to thank, Dr. H.P. Noyes, Dr. R.D. Viollier, the people at Nijmegen and many other persons for discussions about these topics.

References:

Ar 74 ; R.A. Arndt, R.H. Hackman, and L.D. Roper, Phys. Rev. 9C(1974) 555.

Be 49 ; H.A. Bethe, Phys. Rev. 76 (1949) 38.

Be 74 ; J.E. Benn, W. Maag, S. Münch and Ch. Thomann, University of Zürich, Switserland. Contribution to this conference.

Br 64 ; J.E. Brolley, J.P. Seagrave and J.G. Beery Phys. Rev. B135 (1964) 1119.

Ca 74 ; J. Callerame, D.J. Larson, S.J. Lipson and R. Wilson, Harvard University, Cambridge, Mass. USA, Contribution to this conference.

Ci 59 ; M. Cini, S. Fubini, and A. Stanghellini, Phys. Rev. 114 (1959) 1633.

Du 57 ; L. Durand III, Phys. Rev. 108 (1957) 1597.

Gu 64 ; M. Gursky and L. Heller, Phys. Rev. B136 (1964) 1693.

He 60 ; L. Heller, Phys. Rev. 120 (1960) 627.

He 72 ; E.M. Henley, Proc. of the Int. Conf. on Few Particle Problems in the Nuclear Int. Ed. I. Slaus, S.A.Moszkowski, R.P. Haddock, W.T. van Oers, North Holland Publishing Company - Amsterdam (1972) 221 and 242.

Ho 71a ; T.L. Houk, D. Shanbroom, and R. Wilson
Phys. Rev. Letters 26 (1971) 1581.

Ho 71b ; T.L. Houk, Phys. Rev. 3C (1971) 1886.

Im 74 ; K. Imai, K. Nisimura, H. Sato and N. Tamura,
Kyoto University, Japan.
Contribution to this conference.

Ja 50 ; J.D. Jackson and J.M. Blatt,
Rev. Mod. Phys. 22 (1950) 77.

Ja 70 ; N. Jarmie, J.L. Jett, J.L. Detch Jr., and R.L. Hutson,
Phys. Rev. Letters 25 (1970) 34.

Kn 66 ; D.J. Knecht, P.F. Dahl, and S. Messelt,
Phys. Rev. 148 (1966) 1031.

Ko 67 ; L. Koester, Z. Physics. 198 (1967) 187, 203 (1967) 515.

Ko 71 ; L. Koester, W. Nistler. Phys.Rev.Letters 27 (1971) 956.

Ku 74 ; B. Kühn, Dresden, DDR. Contribution to this conference.

Ma 69 ; M.H. MacGregor, R.A. Arndt, and R.M. Wright,
Phys. Rev. 182 (1969) 1714.

Me 49 ; E. Melkonian, Phys. Rev. 76 (1949) 1744.

Mu 74 ; Mühry, Wassmer and Baumgartner, (Basel),
Helv. Phys. Acta (to be published).

Na 74 ; M.M. Nagels, T.A. Rijken, and J.J. de Swart,
University of Nijmegen, The Netherlands.
Contribution to this conference.

No 64 ; H.P. Noyes, C.R. du Congres International de Physique
Nucleaire, ed P. Gugenberger, Vol.II, 172 (1964), Paris.

No 71 ; H.P. Noyes and H.M. Lipinski, Phys. Rev. C4 (1971) 995.

No 72 ; H.P. Noyes, Ann. Rev. of Nuclear Science, 22 (1972) 465.

Sa 74 ; P.U. Sauer, Phys. Rev. Letters 32 (1974) 626.

Sch 47 ; J.S. Schwinger, Phys. Rev. 72 (1947) 742 A.

Sh 70 ; M.S. Sher, P. Signell and L. Heller,
Ann. of Physics 58 (1970) 1.

Ue 35 ; E.A. Uehling, Phys. Rev. 48 (1935) 55.

Vi 74 ; R.D. Viollier, G.R. Plattner and K. Alder,
Phys. Rev. Letters 48B (1974) 99.

Wo 62 ; D.Y. Wong and H.P. Noyes, Phys. Rev. 126 (1962) 1866.

Ze 74 ; B. Zeitnitz, University of Hamburg, BRD.
Contribution to this conference.
B. Zeitnitz, R. Mashuw, P. Suhr, W. Ebenhöh, J. Bruinsma,
and J.H. Stuivenberg, preprint.

THREE- AND MANY-BODY INTERACTIONS AS TOOLS IN LEARNING ABOUT THE TWO-NUCLEON PROBLEM

Moderator : Donald Sprung
McMaster University, Hamilton, Ontario, Canada.

Rapporteur : Harry Fiedeldey
University of South Africa, Pretoria, South Africa.

In this discussion period the principal topic of discussion was the information on the nucleon-nucleon interaction which could be extracted from three-nucleon scattering states. In a series of contributions to this conference Haftel, Petersen and Lambert[1-4] examined off-energy shell effects in deuteron breakup amplitudes and compared their results with those obtained previously by Brayshaw[5] using a three-body boundary condition model. They employed different S-wave separable potentials[1] ensuring phase shift equivalence by energy-dependent modifications of the two-nucleon T-matrix which also allow the adjustment of the triton binding energy. It is generally agreed that one must investigate those regions of phase space where the $S = \frac{1}{2}$, $L = o$ states dominate, to obtain off-shell information on the nucleon-nucleon interaction.

Brayshaw[5] investigated these regions of phase space at 14 MeV with his boundary condition model by keeping the two-body on-shell properties fixed and varying the short-range correlations, corresponding to unspecified combinations of off-shell effects and three-body forces. The conclusion was that n-d elastic scattering and breakup results show virtually no sensitivity to short-range correlations if the n-d doublet scattering length (^2a) is used as a constraint (or the strongly correlated triton binding energy E_T). He also found that in the most sensitive regions the cross section is small and difficult to measure. On the other hand this insensitivity to off-shell effects would make it possible to use three-body reactions as method to determine otherwise hard to measure two-body parameters. Brayshaw is confident that his results if anything overestimate the sensitivity to be expected from plausible physical mechanisms and that the variations included in the boundary condition model incorporate a broad variety of separable, local and semilocal interactions.

In the work by Haftel, Petersen et al[1-4] the conclusions of Brayshaw were tested for their model dependence by means of the above mentioned phase equivalent potentials and by examining deuteron breakup at various regions of phase space and at energies between 14 and 65 MeV. For the $^2H(n,nn)p$ cross section at 14 MeV, using their phase equivalent potentials, they obtain much larger off-shell effects than Brayshaw. They also find that the final-state region of phase space is the most practical region to search for off-shell effects and that the practicability of observing off-shell effects is a function of energy. At 14 MeV the effect of a change in the form factor (off-shell variation) is greatly reduced by a constraint on the triton binding energy E_T or ^2a as was found by Brayshaw. However, at higher energies the constraint on E_T still allows appreciable changes in magnitude and shape of the FSI (final state interaction) angular distributions for different off-shell variations of the phase equivalent potentials. The constraint of a fixed E_T eliminated only 30% of the off-shell effects at 45 MeV and 75% at 14 MeV in the deuteron breakup amplitudes. Since the sensitive states are those where all three particles are near to each other there appears to be an intrinsic difficulty in disentangling the effects of off-shell behaviour and three-body forces in breakup. Haftel and Petersen[2,3] plot their breakup amplitudes for potentials with a 3 MeV difference in E_T at 45 MeV, while those of Brayshaw[5] at 14 MeV allow a variation of only 1.5 MeV in E_T.

It was pointed out by Brayshaw during the discussion that if E_T is the experimental binding energy of the triton and E_M the value predicted by the theoretical model which is employed then off-shell effects in n-d scattering and breakup are nearly linear with $|E_M - E_T|$. The variation obtained by Haftel, Petersen and Lambert should therefore be correspondingly reduced. However this would not eliminate the discrepancy. A further point raised by Brayshaw was that the energy-dependent modifications employed by Haftel, Petersen et al for the two-nucleon T-matrix were not included in his variations of the short-range correlations of the three-body boundary condition model and could therefore not be directly compared to them. No agreement was obtained in the following discussion on the acceptability of such energy-dependent modifications of the two-nucleon T-matrix to which Brayshaw objected on the grounds that they did not fulfil criteria like analyticity and introduced additional singularities in the energy plane.

F. Gross remarked that the restriction of analyticity which Brayshaw would like to impose would also rule out a relativistic treatment of the three-nucleon system which he proposes. This proposal by Gross will be discussed later on. No such singularities are introduced by more conventional methods like unitary transformations (although in this case we have an unusual momentum dependency) and inverse scattering methods.

A few comments on this discussion could perhaps be made. In the first place it appears that there is general agreement that the constraint of fixing the triton binding energy (doublet scattering length) drastically reduces the off-shell sensitivity of n-d scattering and breakup although the size of the remaining off-shell effects is in some doubt. The influence of the constraint of a fixed E_T appears to diminish with increasing energy. In view of the controversy concerning the acceptability of the energy-dependent off-shell extensions of the two-nucleon T-matrix which were employed in reaching this conclusion, further calculations with energy-independent phase equivalent potentials would appear to be useful. This would also serve as an additional check on the model independence of Brayshaw's conclusions. In view of relative ease of calculations with separable potentials, phase equivalent separable potentials obtained by direct inversion of the two-nucleon scattering data[6], would appear to be possible candidates for such an investigation. Another question requiring some attention appears to be the finding of generally acceptable restrictions on the possible off-shell extensions of the two-nucleon T-matrix. However, there was a large measure of agreement concerning the dominance of on-shell effects in many regions of phase space for three-nucleon scattering, which would in principle allow them to be used as tools for learning about the on-shell properties of the nucleon-nucleon force.

As Conzett pointed out in his invited talk[7)] "Polarization Phenomena in the Three-Nucleon System" and during this discussion, it is clear that the n-d polarization data require the use of more realistic nucleon-nucleon interactions and are therefore better suited for obtaining information on them which has not been available from the nucleon-nucleon scattering data themselves. Recent work suggested that this possibility existed especially in the n-d polarization data. For instance Doleschall[8)] showed that the deuteron vector polarization data are very sensitive to the details of the tensor potential. The tensor interaction also has a substantial effect on the nucleon polarization in n-d scattering at 22.7 MeV. Recent measurements at Berkeley of the deuteron vector polarization in d-p scattering at E_d = 45.4 MeV, for direct comparison at the equivalent nucleon energy of 22.7 MeV, can also only be fitted by relatively realistic tensor interactions. The angular region forward of the center of mass angle $\theta_c = 120^0$ is the region with the highest sensitivity to the details of the $^3S_1$ - $^3D_1$ tensor potentials. It is also the region with the remaining discrepancies between theory and experiment. Conzett therefore proposed that it would be of great interest to carry out the calculation with a tensor force able to reproduce both the N-N $^3D_1$ phase shift and the mixing parameter $\varepsilon_1(k)$, and pointed out that the rank-4 potential recently constructed by Pieper[9)] fulfils this requirement.

The interest in the n-p mixing parameter $\varepsilon_1(k)$ originates from the fact that it is very poorly determined by n-p scattering data. For instance near 50 MeV data (σ_{Tot}, $\frac{d\sigma}{d\Omega}$, $P(\theta)$) leave $\varepsilon_1(k)$ undetermined between -10^0 and $+3^0$. The negative values of ε_1 which are presently allowed by the n-p scattering data are in disagreement with all potential models and with the one-pion-exchange potential (OPEP) in particular. Conzett discussed a possible method to improve the determination of ε_1 proposed by Arndt, Binstock and Bryan[10)] by measuring the neutron-to-proton polarization transfer coefficient D_t. However the absolute error on D_t must be small and the experiment accordingly difficult. Conzett proposed that in view of the demonstrated sensitivity of the vector polarizations to the N-N $^3S_1$ - $^3D_1$ tensor potential one could employ potentials simultaneously fitted to the $^3D_1$ phase shift and ε_1 and then vary ε_1 to obtain an improved fit to the vector polarization data. He suggests that in this way one could possibly obtain a better determination of the ε_1 at low energies than via the more difficult measurement of D_t.

One could perhaps comment that this method depends on the insensitivity of the polarization data to off-shell effects and that therefore a check of the residual off-shell sensitivity of the calculations should also be performed with strictly phase equivalent potentials. Since ε_1 has to be varied independently of the other on-shell nucleon-nucleon scattering data, separable potentials obtained by direct inversion of the scattering data appear to be well suited to such calculations from this point of view, and the calculations would probably still be manageable.

Conzett's proposal seems to the most promising possibility at present for using the 3-nucleon system as a tool for extracting information, (in this case on-shell information) on the two-nucleon interaction, that is very important for applications in other many-body systems and has until now been difficult to obtain directly from the two-nucleon scattering data.

Another interesting proposal to exploit the insensitivity of the 3N-scattering states to off-shell effects was presented by Šlaus[11]. The neutron-neutron effective-range parameters $a_{nn} = -16.4 \pm 0.9$ fm and $r_{nn} = 3.4 \pm 0.6$ fm[12] have been determined from final state interactions in multiparticle reactions. The possible violation of both charge independence and charge symmetry depends on the knowledge of a_{nn} and r_{nn}. That charge symmetry could be violated is for instance indicated by the well-known difference between the binding energies of ^3H and ^3He. Šlaus pointed out that in the region of phase space called the "quasifree region", for kinematic conditions where the proton has a very low momentum, the cross section is only weakly dependent on a_{nn} but much more strongly dependent on r_{nn}. It has been shown by Petersen, Haftel and Lambert[4] at this conference that QFS (quasifree scattering) is highly insensitive to off-shell effects. The reason is that QFS enhancement involves the coherent contribution of many partial waves which implies that the $(L = o, S = \frac{1}{2})$ M_{D2} amplitude for deuteron breakup, which is the one sensitive to off-shell effects, cannot be important. In the FSI region however the enhancement is due to the $^1S_0$ resonance in the N-N scattering amplitude practically at zero energy which makes it possible that both the cross section and the $L = o, S = \frac{1}{2}$ amplitude are large and shows that FSI could be sensitive to off-shell effects. Slaus therefore proposes[12] to use the predominance of the influence of r_{nn} in the QFS region to determine r_{nn} either by an absolute or a relative cross section measurement of breakup spectra. In the latter case an accuracy of about 2% could be obtained in an experiment lasting about a 1000 hours. As in the case of Conzett's proposal there appeared to be general agreement that the experiment proposed by Slaus would provide us with a very worthwhile piece of additional knowledge on the nucleon-nucleon interaction.

Sprung[13] discussed a proposal by Fuda[14] to employ the neutron exchange pole in n-d scattering to determine the higher partial waves which is the analogue of the one pion exchange pole in N-N scattering. It can then be shown that the pole term phase shift for n-d scattering depends on the asymptotic normalization of the deuteron wave function. During this session and in his invited talk Sprung[13] described a proposal to take Coulomb distortion effects into account. This method should in principle provide us with a method to determine the asymptotic deuteron normalization N_D, a very useful piece of information. However, there are difficulties in extracting N_D from the n-d and p-d scattering data. As was pointed out by Mitra during the general session the determination of N_D requires a contribution from nucleon isobars which is uncertain.

Another piece of important (off-shell) information on the deuteron wave function which was discussed during this session is the D-state probability P_D. Sprung[13] discussed Klarsfeld's[15] lower bound on the P_D under the assumption that beyond a certain range the force acting between two nucleons is OPEP (for $r \approx 3$ fm, $P_D \gtrsim 2\%$ and for $r \approx 2$ fm, $P_D \gtrsim 3.3\%$). However the uncertainty remaining in the accepted values of P_D(3-9%) remains embarrassingly large and it would be of great importance for a better determination of the off-shell behaviour of the nucleon-nucleon interaction if P_D could be fixed within much narrower error bounds, especially since nearly all local potentials give values of P_D of the order of 6-7%.

In this context the work of Remler and Miller[16] was discussed to some extent. They found a preference for a P_D between 7% and 9% from their analysis of elastic n-d scattering at intermediate energies (30-300 MeV). However no conclusion was reached on the reliability of this estimate. As was pointed out by Sprung[13] they employed certain deuteron wave functions in their analyses and this could influence the conclusions on the P_D.

Another possible method to determine P_D was discussed by Simonius and consists of an analysis of backward π-d scattering. The recent work of Thomas and Afnan[17] who performed a Faddeev calculation of the reaction $p+p \rightarrow d+\pi^+$ was also discussed. The cross section proved to be rather sensitive to the P_D in this case. However the possibility of a sensitivity to other off-shell effects (for instance to the off-shell behaviour of the pion-nucleon interaction) should also be analysed and its effect on the determination of P_D investigated. At present it seems that all that can be said is that the value of P_D is still uncertain between 3% to 9% with some preference for the higher values. However, there seems to be a good possibility that the attack made on this problem along several promising lines will lead to greater success within a few years.

A topic which also received some attention during this discussion period was the charge symmetry of nuclear forces. Sauer's recent work[18] on the charge symmetry of nuclear forces and his contribution to the conference[19] were discussed by (among others) Conzett, Šlaus, Sprung, Bleuler and Schick. Sauer questioned, by means of counter examples, the traditional belief that the electromagnetic effects can be removed from the p-p scattering data in an unambiguous way. Concentrating on the removal of the Coulomb potential he found that the resulting effective range parameters depend strongly on the nucleon-nucleon potential at short distances, which makes the comparison between the proton-proton and neutron-neutron scattering data ambiguous. In a contribution to the conference Sauer[19] proposed to employ charge symmetry as an off-shell constraint on the nucleon-nucleon interaction by requiring charge symmetry to hold. In a comment at the general session during the conference Sauer pointed out that charge symmetry would impose a constraint on the UPA formfactor in the $^1S_0$ state, and that many potentials obtained by unitary transformations from the Reid Soft Core potential would be ruled out by this constraint.

It was pointed out during the discussion that since the pion masses are slightly different one might also have slightly different coupling constants. Both these effects would lead to a certain amount of charge symmetry breaking. Another complication lies in the theoretical removal of the many other electromagnetic effects (finite size corrections etc.) for a proof of charge symmetry. However Sauer's point that the N-N wave function at short distances was the most unknown quantity was generally appreciated. Bleuler and Schick also discussed the importance of charge symmetry for elementary particle physics.

It seems quite clear that Sauer's constraint on the off-shell behaviour due to the requirement that charge symmetry holds, perhaps approximately only, is of great importance, perhaps as important as a knowledge of P_D, for the off-shell T-matrix, and that in the future nucleon-nucleon potentials employed in many-body calculations should include some form of this constraint. Further investigations along this line should be carried out.

There were also some comments on hypernuclear interactions by Schick[20], who pointed out that since the Λ-N interaction has no tail like the OPEP, it consists of a core interaction only. A study of Λ-N interactions could therefore be useful for understanding the properties of core interactions. Off-shell effects could be expected to be large in this case. However at present no definite information on the nucleon-nucleon force has become available from this direction.

Rouben[21] discussed a super-soft-core nucleon-nucleon potential including two-pion exchange and ω-exchange contributions with a percentage D-state of 5.8%. The core and the remainder of the intermediate range are treated phenomenologically. The potential fits the bound state and scattering data to a high degree of accuracy up to 350 MeV. Sasakawa[22] explained a low energy formula for n-d scattering based on an extension of the Jost function of potential scattering theory to the treatment of the three-body system.

Finally there was a discussion of the possibility that the relativistic treatment of the 3N-system might improve on the present non-relativistic calculation. Any relativistic theory which makes a direct connection with the fundamental interaction between nucleons and mesons would provide a theoretical framework in which to calculate 3-body forces, off-shell effects, and relativistic corrections. F. Gross[23] described briefly how his two-body theory could be extended to the 3N-system. The central idea would be to replace the Green's function describing 3 off-shell nucleons by a simpler Green's function, in which 2 of the nucleons are on-shell and only one (the interacting nucleon) is off-shell. The resulting Green's function is still relativistically invariant, but has only 6 variables as in the usual non-relativistic theory. Relativistic 3-body equations can now be constructed in a way analogous to the construction of non-relativistic 3-body equations, and when the non-relativistic limit of the relativistic equations is taken, one reproduces the non-relativistic equations together with new, relativistically generated, 3-body forces and off-shell effects. This work is still in its preliminary stages and it is not known how well it will work.

At first sight a surprising omission during this discussion period was the lack of any discussion on the information obtainable from the ground state of the triton, especially from the binding energy of the triton and the n-d doublet scattering length, which are rather sensitive to the off-shell behaviour of the nucleon-nucleon interaction. This omission was partly due to a lack of time. However the difficulty of using this information was probably also partly responsible for this omission. Unfortunately both the triton binding energy and the doublet scattering length are largely correlated leaving us with only one off-shell sensitive quantity. Although the charge form factor of ^3He and ^3H is also sensitive to the off-shell N-N T-matrix it appears according to Kloet and Tjon[24)] that the omission of mesonic exchange corrections could be responsible for the discrepancy with experiment in this case. Unfortunately it is very difficult to use the triton binding energy E_T (or the doublet scattering length ^2a) to restrict the off-shell behaviour of the N-N T-matrix because it appears to be very hard to disentangle the off-shell effects from the contribution of three-nucleon forces and to a lesser extent from relativistic effects. One should also be wary to draw firm conclusions from the agreement or disagreement between the theory and a single experimental number E_T. Although local potentials like the Reid-Soft-Core and Hamada-Johnson potentials clearly produce underbinding in the triton (of the order of -1.5 MeV for the RSC), it was pointed out by Delves at this conference that it seems to be possible to fit E_T by means of off-shell variations of the potential without any contribution from 3N-forces (see also Afnan and Read[25)], who found that several one-boson exchange potentials produced significantly closer values for E_T and ^2a than the realistic local potentials and the same is true for separable potentials obtained by direct inversion of the scattering data[6)]). However theoretically one does expect a contribution from 3N-forces and relativistic effects although the size of this contribution is unknown. This uncertainty very much diminishes the value of our knowledge of E_T and ^2a as constraints on the nucleon-nucleon potential. It has not even been possible yet to use this information to decide the question if purely local potentials (like the RSC potential) are compatible with the data or not, because of this uncertainty about the other contributions.

Several other topics and proposals were also briefly discussed during this discussion period and the author apologizes for not being able to include all of them (due to space restrictions, lack of references etc.) and for a perhaps too drastic abbreviation of several other discussions. The selection of subjects for this report was unavoidably somewhat arbitrary. However, it is hoped that the most important points related to the information obtainable from three- and many-body interactions about the nucleon-nucleon interaction have been included.

REFERENCES

1) M.I. Haftel and E.L. Petersen, Separable Potential Models for use in Deuteron Breakup Calculations.*

2) M.I. Haftel and E.L. Petersen, Sensitivity of Deuteron Breakup Amplitudes for Several Potentials.*

3) E.L. Petersen and M.I. Haftel, Examination of Off-Energy Shell Effects in Deuteron Breakup.*

4) E.L. Petersen, M.I. Haftel and J.M. Lambert, Potential-Dependence of Quasifree Scattering in Deuteron Breakup.*

5) D.D. Brayshaw, Physical Review Letters, 32 (1974) 382.

6) N.J. McGurk and H. Fiedeldey, The Phillips Plot for Phase Equivalent Separable Potentials.*

7) H.E. Conzett, Polarization Phenomena in the Three-Nuclear System.*

8) P. Doleschall, Nucl. Phys. A220 (1974) 491.

9) S.C. Pieper, Phys. Rev. C8 (1973) 1702.

10) R.A. Arndt, J. Binstock and R. Bryan, Phys. Rev. D8 (1973) 1397; J. Binstock and R. Bryan, Phys. Rev. D9 (1974) 2528.

11) D. Vranić, I. Šlaus, G. Paić, P. Tomaš, How accurately can r_{nn} be determined from Quasifree Scattering?*

12) In "Few Particle Problems in the Nuclear Interaction" (eds. I. Slaus et al), North Holland Co., 1972.

13) D.W.L. Sprung, What do we know about the Deuteron and the anti-Deuteron.*

14) M.G. Fuda, Phys. Lett. 47B (1973) 288.

15) S. Klarsfeld Orsay preprint IPNO/TH 74-5 (1974).

16) E.A. Remler and R.A. Miller, Ann. Phys. (NY) 82 (1974) 589.

17) T.R. Afnan and A.W. Thomas, Phys. Rev. C 10 (1974) 109.

18) P.U. Sauer, Phys. Rev. Lett. 32 (1974) 626.

19) P.U. Sauer, The Charge Symmetry of the Nuclear Force as an off-shell Constraint.*

20) L.H. Schick, Suppression of $V_{\Sigma\Sigma}$ in a separable $^3S_1$ - $^3D_1$ ΛN potential.*

21) R. de Tourreil, D.W.L. Sprung and B. Rouben, Super-Soft-Core Nucleon-Nucleon Interaction and ω-exchange Contributions.*

22) T. Sasakawa. Low Energy Formula of n-d Scattering.*

23) F. Gross, Phys. Rev. D10 (1974) 223.

24) W.M. Kloet and J.A. Tjon, The Mesonic Exchange Charge Formfactors of ^{3}He and ^{3}H.*

25) I.R. Afnan and J.M. Read, Trinucleon Results for several One-Boson Exchange Potentials.*

*See elsewhere in the proceedings of this conference.

DISCUSSION GENERALE PORTANT SUR LES COMMUNICATIONS QUI PRECEDENT

GENERAL DISCUSSION FOLLOWING THE PRECEEDING REPORTS

RINAT (Weizman Institute): I should like to make a remark about a possible resonance in π-D elastic scattering around the Δ. There is no doubt that the peak in the total cross section -and the rising phases from theoretical analyses- is a reflection of a Δ in the weakly bound deuteron. A genuine resonance should be coupled, amongst others, to the di-nucleon system. There is not the slightest indication of a resonance (or actually one corresponding to the several 'resonating' phases in π-D) in p-p scattering. Unless our information on p-p scattering around 500-600 MeV is erroneous, a 'width' of a pp resonance could not possibly satisfy the factorization theorem for p-p, π-D elastic **scattering** and $\pi D \to pp$.

I may add the historical note that Dyson once believed that a π-D resonance 'seen' in $\pi D \to pp$ could be fitted into a B=2 SU_6 multiplet. He withdrew his suggestion when it was pointed out to him that $\pi D \to pp$ was due to several partial waves.

SAUER(M.I.T.): During this conference we heard a lot about the off-shell t-matrix and the tool of phase-equivalent potentials to study off-shell effects. There is a problem in the isospin-triplet partial waves, because $^1S_0$ is over-looked. The phenomenological potentials are derived from proton-proton data, but off-shell variations are usually manufactured to leave the Coulomb-subtracted phase unaltered. These potentials are, however, not equivalent with respect to the experimental proton-proton data and, indeed, many do not fit them at all. It is very comforting to note that the largest $^1S_0$ off-shell effects in nuclear structure are derived from potentials which do not account for the proton-proton data. These off-shell effects are therefore of no physical relevance.

The second comment is related to the nucleon-nucleon interaction within one F. The usual folklore is that we cannot learn anything about the core unless we study nucleon-nucleon scattering at very high energies. However, I find it amusing to note that very low energy data can tell us something about the core in $^1S_0$. Assuming charge symmetry the combined experimental knowledge about the proton-proton and neutron effective-range parameters excludes a number of non local parametrizations of the core region in $^1S_0$, previously used in nuclear-structure calculations. Though the low-energy data surely do not teach us any detail about the core, they at least teach us how the core in $^1S_0$ should not be parametrized.

BLEULER(Bonn): I have some questions and remarks about the various bosons which constitute through these exchanges the very basis of the N-N interaction. In particular, the ω-boson (vector particle) is responsible -in a rather satisfactory way- for the so-called repulsive core. Now, it has been mentioned by one of the rapporteurs that this part of the N-N interaction is turned over into an attractive core as soon as the N-$\bar{N}$ system (nucleon-antinucleon) is considered (this fact might, amongst other things, be related to the Fermi-Yang theory of bosons). I would point out here that it appears by now possible to see the effects of this attractive interaction directly with the help of the study of antiprotonic atoms. In fact, there is already a strong indication from experimental results in this realm (Backenstoss, Knock et.al, at CERN). There exists a characteristic displacement of some critical levels of the antiproton in the Coulomb field of the nucleous. As a second point, I would like to ask a question about the necessity of introducing the f-boson (assignment 2^+) in the determinations of nuclear forces. Some calculations at our Institute (Machleidt et al.) show that the needed cut-off factor in the vector boson coupling has a range which corresponds exactly to the Compton wave-length of this f-boson, (other reasons are given by Prof. Salam in Trieste).

AARON (Northeastern): Goldberg and I performed a calculation of the (spin flip) $f_{\pi N\bar{N}}$ coupling constant several years ago. It was an extremely clean calculation which used a sum rule derived from backward π-N dispersion relations. The result was that the coupling constant was sufficiently large such that (without form factor effects) f-meson exchange could cancel the effects of ω-exchange in N-N s-waves.

BLEULER: I agree completely and I should mention in this repect some boot-strap calculations by S. Wagner (Bonn) which also yield a very strong coupling-constant.

LEHAR (CEN-Saclay): The situation of polarization measurements in the N-N scattering at energies higher than I GeV has not been treated in previous sessions. I want to show how the polarization in p-p scattering changes in the high energy region. (Fig. 2 and Fig. 3 from the following contribution, J. Deregel et. al)

EDGINGTON (TRIUMF): I think that we shall soon see measurements of scattering variables in the $\bar{N}$-N system, of quality similar to those of the N-N system. For instance, the Kycia group have recently published some total cross-section results obtained in the $\bar{p}$-p and p-D systems; these show structure near 450 MeV/c. We shall perhaps hear much more about the $\bar{N}$-N system at the next conference in this series.

HELLER (Los Alamos): A comment about the two-pion exchange contribution to the nucleon-nucleon interaction. I distinguish between the perturbation theory calculations and the on-shell calculations. The latter employ experimental pion-nucleon scattering information, analytically continued, as well as experimental pion-pion phase shifts. Precisely because they are on-shell, they do not directly provide a nucleon-nucleon potential. These amplitudes are functions of two variables only, whereas a potential is a function of three variables, $|\vec{k}|$, $|k^1|$ and θ. It is only by making some specific assumption about the dependence on $|\vec{k}| - |\vec{k}^1|$ that a potential is obtained. The standard assumption is that the potential does not depend on this variable. This may or may not be a good assumption, and this precaution should be kept in mind.

EBERHARD (Berkeley): What are the upper limits given by past experiments about violations of charge symmetry? For instance, $\pi^{\mp}$-D total cross sections have been shown to be equal within 1% after Coulomb effects have been subtracted. Is there a better experiment and is the interpretation of this experiment in doubt?

SIMONIUS (Zürich): I do not know exactly all the numbers, but the a_{nn} which is, from the point of view of measurement, the most contraversial one seems to have converged to 16.4± 1F . But the problem of charge symmetry is whether from the a_{pp} you can calculate the a_{nn} , and this is model-dependent. This is a point which has been raised several times, but probably we cannot do better. Even if you measure this constant to better accuracy you will not know the correction because it depends on the potential. Sauer's point is, and he made it once more today -in fact he made it several times- that one should rather take the data and see what restrictions are imposed on the potential if one believes in charge symmetry. That is probably the best one can do.

MITRA (Delhi): I have a comment on Dr. Fiedeldey's remarks on the possibility of using the nucleon exchange effect of N-d scattering as a means of probing the short distance behaviour of the deuteron wave function. In principle this is correct, but in practice the problem is mixed with N* exchange effects, even at moderately intermediate energies. Indeed, we believe, as we have shown (J. Shirma et al.) in some detail in a couple of recent papers, that N* exchanges are necessary for obtaining the correct magnitude as well as the energy and angular dependence of both backward p-D scattering and the process $pp \to \pi^+ D$. And this is not just the second Regge recurrence of the nucleon (Kerman-Kisslinger) which plays a comparatively small role, but a whole lot of N* 's in the 1.4 - 1.6 GeV region. Presumably their effects will have to be taken into account in probing the short distance behaviour of the deuteron wave-function.

GROSS (College of William and Mary): I agree completely. However, it is well impressed on most people's mind that there are nucleon resonances and they do contribute to N-D scattering. I think in fact that it has not been appreciated sufficiently that there will be contributions from the deuteron wave function at short distances which will also compete with these resonances. In fact the open question now is: which of these two effects, the most conventional deuteron wave function -off-shell or close to the short distance wave function- or the nucleon resonances, is more important? and how do they cooperate? how do they balance to give the right final result?

VINH MAU (Paris): I would like to comment on Dr. Heller's comment. For the nucleon-nucleon scattering itself, we need not go through a potential scheme to correlate the physical π-π, π-N and N-N phase shifts and, therefore, we need no off-shell extrapolation. Now of course, when one wishes to define an equivalent potential, one has to choose a definite prescription to go off-shell and there is, I think, no criteria for this. Anyway, I think that the 'arbitrariness' of this prescription is only a small part of the arbitrariness of the concept of potential which might be dubious, especially for small relative distances between the nucleons.

MASTERSON (British Columbia): I would like to ask a question which was raised but not discussed in the pion production session and that is: what can we expect to learn about the N-N interaction from the pion production experiments?

AFNAN (Flinders): For pion production above threshold in p-p scattering we have found that (i) the usual analysis of the experimental data has usually been done by writing the cross-section as $\sigma = \alpha\eta + \beta\eta^3$. Here η is the pion momentum in units of pion mass, α and β are the cross sections for S-wave and P-wave production which are assumed constant. The problem with this analysis is that our recent results show that α decreases with η for low pion energies. Thus one has to calculate both α and β and compare with the experimental cross-section. (ii) Our results also show that α is very sensitive to both short range behaviour and D-state probability of the deuteron. Thus $pp \rightarrow \pi D$ experiment at low energies may give us some information about the deuteron wave functions. Recent calculation by Lomon et al. have shown that pion production cross section near the (3,3) resonance is very sensitive to the off-shell behaviour of the $\pi\pi$-N interaction; this may introduce a problem in using pion production to study the N-N interaction.

SIMMONS (Los Alamos): We stand at a point of rebirth of interest in nucleon-nucleon experiments. We have heard of plans and of current experiments ranging from Van de Graaff energies to 400 GeV at the Fermi Lab. In order to reduce or eliminate step functions in systematic errors from different laboratories at different energies there is a pressing need for international cooperation to repeat the experiment at overlapping energies.

III

INTERACTIONS MESON-NUCLEON ET MESON-DEUTON	MESON-NUCLEON AND MESON-DEUTERON INTERACTIONS
PROGRAMMES DES USINES A MESONS	MESON FACTORIES PROGRAMMES

π - deuteron scattering

D. V. Bugg
Queen Mary College, London, El 4NS, England.

Abstract: What happens in πd scattering which cannot be understood directly from our knowledge of πN? Experimental data on total cross-sections and elastic scattering are reviewed from threshold to the GeV region, with particular attention to recent results. A new measurement of the scattering length, from the strong interaction shift in the ground state of the mesic atom, agrees with πN scattering lengths and multiple scattering theory. Elastic scattering at high energy and small angles now agrees well with Glauber theory. However, the magnitude of the contribution of inelastic states to Glauber shadowing in the total cross-section is confused both experimentally and theoretically. At large angles and low energy, the Glauber approximation fails. Recent calculations of Gabathuler and Wilkin using multiple scattering theory improve agreement with data in the N*(1230) region. Total cross-sections in the resonance region may now be explained quantitatively by collision broadening of the resonance on the spectator nucleon; a modification of Glauber theory to accommodate this is given. The choice of kinematics for πN collisions in the deuteron is settled by appeal to experiment. The way of handling the channel πd→NN is clarified. Other examples of collision broadening are discussed in π-nucleus total cross-sections and in high energy hadron collisions.

1. THE πd SCATTERING LENGTH

A new measurement[1] of the πd scattering length, a (πd), has been made from the strong interaction shift ΔE in the ground state of the πd atom, by measuring 2P→1S X-rays:

$$\frac{\Delta E}{E} = \frac{4a(\pi d)}{a_0}, \qquad (1)$$

where a_0 is the Bohr radius of the mesic atom.

The experimental problems are twofold. Firstly, one cannot use liquid deuterium as target, since the πd atom, being electrically neutral, suffers Stark mixing in the electric fields of neighbouring atoms; the Stark mixing induces S states in the πd atom from which nuclear capture is rapid, and the atom is lost before it de-excites to the 2P state[2]. Backenstoss' group at CERN has looked for π- and K-mesic X-rays from liquid hydrogen, but has so far failed to see any[3]. So a gas target must be used, and the X-ray rate is very low, about 150/hour in the experiment described below, and a crystal spectrometer cannot be used. Secondly $E \simeq 2.6$ KeV, and $\Delta E \simeq -5$eV, so one is looking at small effects in very soft X-rays.

Thus ΔE must be measured by a trick. An absorption edge is chosen which almost exactly coincides with the energy E, and which has a length of ~10eV, spanning the range of uncertainty ΔE. The attenuation of the X-rays is measured through a foil which has been calibrated using a crystal spectrometer (Fig.1), and E can be read off. The edge in question is the strong M_V edge of bismuth, i.e. the one corresponding to photoelectric absorption of a D shell J = 5/2 electron.

The experiment involves stopping a π^- beam in deuterium gas at 8 atmospheres. X-rays at 90^0 are detected in a proportional chamber separated from the deuterium by a thin window. Various foils can be inserted into the path of the X-rays, one being the bismuth foil used to measure ΔE. The X-rays are not purely $2P \rightarrow 1S$ (K_α), but there are some from $3P \rightarrow 1S$ (K_β) and $4P \rightarrow 1S$ (K_γ). The unwanted ones are eliminated by a foil containing chlorine, which has a strong K absorption edge at 2820 eV, between the K_α and K_β energies. The measured spectrum of X-rays is shown in Fig.2. It sits on a background, which seems to be due entirely to charged particles spraying from the region of the beam. This background is measured by inserting a foil of sulphur, which has a K absorption edge at 2470 eV, and kills the X-ray signal completely.

The measured absorption is shown cross-hatched on Fig.1. After allowing for vacuum polarisation and the charge radius of the deuteron, it gives

$$a(\pi d) = -\left(0.052 \begin{array}{l} +0.022 \\ -0.017 \end{array}\right) m_\pi^{-1} .$$

For free nucleons, the latest πN data[4] predict $a(\pi p) + a(\pi n) = -0.014 \pm 0.005 m_\pi^{-1}$. The difference is due to nuclear physics effects, and will be discussed in the next talk by Thomas. What I wish to point out is that, if there were sufficient interest, the experiment could probably be improved. Firstly, one should seek to reduce the background by using a cleaner beam and a proportional chamber with faster rise-time. Secondly, one sees in Fig.1 that the X-ray is on the tail of the absorption edge. One should examine chemical compounds of bismuth to see if any offers an appreciable chemical shift in the right direction. If the edge could be moved to coincide with the X-ray, one would gain a factor 3 directly from the slope of the edge.

2. $\pi^+ d$ AT 47.7 MeV

New results are available on $\pi^+ p$ and $\pi^+ d$ at 47.7 MeV from the University of British Columbia group working at Berkeley[5,6]. Their $\pi^+ p$ results[5] fit in well with the scattering length of Bugg et al.[4] and P wave phase shifts at higher energies[7]. Their $\pi^+ d$ results are shown in

Figs. 3 to 5. In Fig. 3 curves labelled M are <u>single</u> scattering predictions and indicate the range of uncertainties from pre-1973 phase shifts, although I suspect that newer phase shifts[7] fix them much better; the curve labelled Beder shows the effect of absorption via $\pi^+ d \to pp$. It will be interesting to see the effects of multiple scattering corrections. Fig.4 is in good agreement with Beder's analysis[8]. Fig.5 agrees well with the prediction of earlier data[9] (full line).

3. ELASTIC SCATTERING AT HIGH ENERGY.

To my disappointment, these are the only new πd data this year, so I turn now to theoretical developments. Everyone knows that at higher energies and small angles, Glauber theory[10] gives a reasonable account of the data. However, it fails at large angles, and here there is progress; it also fails to account for total cross-sections in the resonance region, and here again there is progress.

Sindhu and Quigg[11] have published what they intend as a "benchmark" calculation of total cross-sections and elastic scattering, using the latest deuteron wave functions and minor refinements over the work of Michael and Wilkin[12] and Alberi and Bertocchi[13]. However, they get themselves into trouble. Their prediction for the elastic scattering at 9 GeV/c is compared on Fig.6 with the data of CERN-Trieste[14]. There is good agreement in the single scattering region, but at large angles there is a discrepancy of up to a factor 4. Michael and Wilkin get much the same result. However, both authors rely on the Barger and Phillips πN amplitudes[15], which are only approximate. Zovko[16] points out very succintly that in the double scattering region, $d\sigma/dt$ depends on $|C^+(k,t/4)|^4$, and using the more recent and better πN amplitudes of Höhler and Jakob, he restores agreement with experiment. At 2.01, 3.77 and 5.33 GeV/c, Fellinger et al[17] analyse their own data, Fig.7, get this point right, and accordingly get agreement with experiment (except in the dip region at 3.77 GeV/c). So elastic scattering at high energies is in good shape.

4. GLAUBER SHADOWING BY INELASTIC STATES

Sindhu and Quigg also discuss Glauber shadowing in total cross-sections above 20 GeV. This is an open question both experimentally and theoretically. Experimentally, the problem is that one measures $\sigma_T - \int_0^\theta (d\sigma/d\Omega)\, d\Omega$ and extrapolates to $\theta = 0$. Any systematic errors in the extrapolation have a large effect on the small difference:

$$\delta\sigma = \sigma(\pi d) - \sigma(\pi p) - \sigma(\pi n) = \langle r^{-2} \rangle\, \sigma(\pi p)\sigma(\pi n)/4\pi . \qquad (2)$$

Let me remark that $\delta\sigma$ is typically 3% of $\sigma(\pi d)$, while the extrapolation from the smallest experimental angle to $\Theta = 0$ is typically 5 to 8%.

Serpukhov data[18] are shown in Fig.8, and those of Galbraith et al[19] at lower energy in Fig.9. Both show $\langle r^{-2} \rangle$ increasing with energy, but they are clearly inconsistent where they overlap. So we must await confirmation or refutation of the Serpukhov values at NAL. Many authors[13,20] have remarked that inelastic states will contribute at high energies (Fig.10). The reason they don't contribute at low energies is that, from the kinematics of their production, they have a lower momentum k' than that of elastic intermediate states, k. Thus the factor $e^{i(k-k')\cdot r}$ in their propagator makes them get rapidly out of phase, so that no coherent interference occurs with the incident wave, and $\delta\sigma$(inelastic)=0. Now $k'-k \simeq (m^{*2} - m^2)/2k$, where m and m* are masses of pion and the inelastic state; above 10 GeV they begin to be coherent. Sindhu and Quigg's calculation (Fig.11) of their contribution to $\delta\sigma$ agrees reasonably well with the Serpukhov result, Fig.12. However, there are two problems. Firstly, the "particle" in the interemediate state has a short lifetime; if it decays, say into three pions of momentum $k'' \simeq k'/3$, incoherence develops almost immediately. Secondly, Sindhu and Quigg take the angular distribution for the inelastic states to be $\frac{d\sigma}{dt} = \left(\frac{d\sigma}{dt}\right)_0 \exp\{A(t-t_{min})\}$ with $A = 2.5\ (\mathrm{GeV/c})^{-2}$. This is a much smaller value of A than is now believed[21], and they do not give enough detail to allow one to judge if their result is too big or too small. So one forsees more experiments[22] and more calculations on this topic.

5. ELASTIC SCATTERING NEAR THE BACKWARD DIRECTION.

I turn now to the breakdown of the Glauber approximation in backward scattering, and the recent work of Gabathuler and Wilkin[23]. They start from the multiple scattering formula of Brueckner[24]. To second order, the amplitude for scattering from proton and neutron separated by radius $\underline{r}$ is

$$F_{\pi d}(q, r) = f_p e^{iq\cdot\frac{1}{2}r} + f_n e^{-iq\cdot\frac{1}{2}r} + 2 f_p f_n \frac{e^{ikr}}{r} \cos(\underline{K}\cdot\underline{r}), \qquad (3)$$

where k and k' are initial and final momenta, q=k-k', and $K=\frac{1}{2}(k+k')$; I have ignored the angular dependence of the amplitudes for the moment, as I am interested in the phase factors. In the double scattering term, note the e^{ikr}/r factor for propagation of the radial wave from the first scatterer to the second, and the factor $\cos(\underline{K}\cdot\underline{r})$ which accounts for the coherence in the two scatters. At threshold, the double scattering term is $2 a_p a_n \langle r^{-1} \rangle$, where the brace indicates an average over the deuteron wave function. More generally the double scattering term is

$$F_{\pi d}^{(2)} = \frac{1}{4\pi}\int_0^{2\pi} d\phi \int_{-1}^{1} dz \int_0^{\infty} r^2 dr\, |\psi(r)|^2 f_p f_n e^{ikr}\left(e^{i\underline{K}.\underline{r}} + e^{-i\underline{K}.\underline{r}}\right)$$

Taking K as polar axis, and dropping the angular dependence of f_p and f_n for illustrative purposes,

$$F_{\pi d}^{(2)} = f_p f_n \int_0^{\infty} r^2 dr\, |\psi(r)|^2 \frac{e^{ikr}}{r}\int_{-1}^{1} dz\, e^{iKrz}$$

$$= f_p f_n \int_0^{\infty} r^2 dr\, |\psi(r)|^2 \frac{e^{ikr}}{iKr^2}\left(e^{iKr} - e^{-iKr}\right)$$

For large k and small angles, $\underline{K} \approx \underline{k}$, and the contribution from e^{iKr} oscillates rapidly with r and averages to zero. Finally, from the latter term in the bracket, $F_{\pi d}^{(2)} \simeq i f_p f_n \langle r^{-2} \rangle / k$.

This is the usual Glauber approximation. With a little more care in approximations, one gets $\langle r^{-2}\exp(-q^2r/8k)\rangle$ instead of $\langle r^{-2} \rangle$, the phase factor being the Fresnel correction of Gottfried[25].

However, at backward angles K=0, and the double scattering term is $\langle e^{ikr}/r \rangle$, which is totally different from the usual Glauber approximation. Gabathuler and Wilkin use this to calculate πd backward elastic scattering from 140 to 260 MeV. Their full curve on Fig.13 shows much better agreement with the data of Tanner's group[26], than the Glauber approximation. Their angular distribution at 256 MeV, Fig.14, is also in good agreement with experiment.

I suggest that simular multiple scattering calculations should be extended (a) to the total cross-section in the N*(1230) region, where k is not particularly big, and (b) to backward elastic scattering from 500 to 1000 MeV. There, Schroeder et al[27] observed a small backward peak showing a rise at T≃720 MeV (Fig.15). Its interpretation is obscure. It may have something to do with the suggestion of Kerman and Kisslinger[28] that the deuteron wave function contains NN* components, but until a proper multiple scattering calculation along the lines of Gabathuler and Wilkin is done, one cannot be sure.

6. CHARGE INDEPENDENCE

For completeness, I make some brief remarks on charge independence. Since the deuteron has isospin zero, π^+d and π^-d total cross-sections should be identical except for Coulomb effects. In the 1 to 2 GeV/c region, π^-d is larger than π^+d by 1.3 ± 0.3%[29]. From the Coulomb barrier penetration factor one accounts for 0.5%; the discrepancy could be due to experimental systematic error or could be because the calculation is too nievè[30]. Large Coulomb barrier effects have been observed in the width of the N*(1230) by Carter et al.[7]. A nicc treatment of this has been given by Pilkuhn[31]; an extension to the treatment of charge exchange and Coulomb mixing would be welcome. He also makes the point that the mass of π^+n resonances in the deuteron will be affected by the spectator nucleon by an amount $\pm \alpha \langle r^{-1} \rangle \simeq \pm 0.75$ MeV.

7. COLLISION BROADENING

Finally let us consider πd total cross-sections in the resonance region. In 1968, Carter et al.[29] presented precise measurements ($\pm$ 0.5%) of πd and πN cross-sections from 500 to 2500MeV/c. They tried to account for the difference $\{\sigma(\pi d) - \sigma(\pi n) - \sigma(\pi p)\}$ in terms of Glauber shielding. However, the values of $\langle r^{-2} \rangle$ they deduced fluctuated wildly with energy (Fig.16), and seemed to indicate failure of Glauber theory in the resonance region.

The discrepancy between $\sigma(\pi d)$ and Glauber theory using calculated values of $\langle r^{-2} \rangle$ is shown in the inset at the top of Fig.17. In this calculation, all the known refinements are included: (a) the D wave component of the deuteron wave function[32], (b) the flux factor[32], (c) real parts of forward amplitudes, (d) charge exchange in the intermediate state[33], (e) the variation of $\langle r^{-2} \rangle$ with energy from the variations of the πN elastic form factors, and (f) Fermi motion in both single and double scattering terms. All effects except Fermi motion are small, but a discrepancy well beyond experimental error persists. Fig.17 shows the Glauber correction itself. The discrepancy is similar in magnitude, but varies with momentum in a quite different fashion. The figure also displays $2\sigma^+ = \sigma(\pi p) + \sigma(\pi n)$. Here one sees a distinct correlation between peaks in σ^+ and dips in the discrepancy. It tells us clearly that resonances in the deuteron are lower and broader than anticipated. This cannot be due to an error in calculating the effects of Fermi motion, since all authors agree that uncertainties in the deuteron wave function produce effects $\lesssim$ 0.15mb.

I suggest that the explanation is collision broadening of the resonance on the spectator nucleon, Fig.18. In atomic physics one is familiar with three sources of the width of spectral lines: (a) natural width, (b) Doppler broadening (which in nuclear physics originates in the Fermi motion of the struck nucleon,) and (c) collision broadening. Remember the classical derivation of the Breit-Wigner formula. The amplitude of the pion radiated by a N* in free space is $f(t) = e^{-iE_0 t/\hbar} \times e^{-t/2\tau}$. Taking the Fourier transform one gets

$$k f(E) = \frac{\Gamma_e/2}{E_0 - E - i\Gamma_t/2}, \qquad (4)$$

where Γ_e is the elastic width, and Γ_t the total width $\hbar/\tau$. If τ is reduced by collisions, Γ_t increases.

A detailed discussion of this effect is given in a Rutherford Lab. preprint[34]. Here two simple treatments are outlined. First consider a plane wave of N*s of energy E, mass m, momentum p, propagating through a uniform nuclear medium of density ρ. The wave function of the N* is

$\psi_*(z,t) = A e^{-iEt/\hbar} e^{-\Gamma t/2\hbar\gamma} e^{ikz} e^{-\rho\sigma z/2}$

where the factor $\gamma = (1 - v^2/c^2)^{-\frac{1}{2}}$ accounts for the relativisitc time dilation, and σ is the NN* total cross-section. From a comparison of the second and last exponentials on the right-hand side,

$$\Delta\Gamma = \hbar\rho\sigma v\gamma = \hbar\rho\sigma p/m. \tag{5}$$

Note particularly that $\Delta\Gamma$ increases linearly with p; the origin of this is that the mean free path of the resonance for decay varies as $1/v\gamma$.

The deuteron is not a uniform medium, so ρ must be averaged over it:

$$\bar{\rho} = \frac{\frac{1}{4\pi}\int_0^\infty e^{-\Gamma r/2v\hbar\gamma}|\psi_D(r)|^2 dr}{\int_0^\infty e^{-\Gamma r/2v\hbar\gamma}dr} = \frac{\Gamma}{2v\hbar\gamma}\frac{\langle r^{-2}e^{-\Gamma r/2v\hbar\gamma}\rangle}{4\pi}.$$

$$\text{Hence}\quad \frac{\Delta\Gamma}{\Gamma} = \frac{\sigma(NN^*)}{8\pi}\langle r^{-2}e^{-\Gamma r/2v\hbar\gamma}\rangle. \tag{6}$$

The essential point is that the Breit-Wigner amplitude becomes

$$k f(E) = \frac{\Gamma_e/2}{E_0 - E - i(\Gamma_t + \Delta\Gamma)/2}. \tag{7}$$

An important result is that the product of the height of the resonance by its width is the same for free πN scattering and for N*s in the medium. The total cross-section at resonance is $\sigma_R = 4\pi(J+\frac{1}{2})\Gamma_e/(\Gamma + \Delta\Gamma)k^2$. But while the peak cross-section decreases, the width increases, and their product, which governs the area under the curve, remains constant:

$$\sigma_R(\Gamma + \Delta\Gamma) = 4\pi(J+\tfrac{1}{2})\Gamma_e/k^2, \tag{8}$$

independent of $\Delta\Gamma$.

A second more satisfactory treatment is to modify Glauber theory by treating the intermediate state, ψ, in the Wigner-Weisskopf approximation: $\psi = |N^*\rangle e^{-\Gamma r/2v\hbar\gamma} + |\pi N\rangle(1 - e^{-\Gamma r/v\hbar\gamma})^{1/2}$.

All the standard manipulations of Glauber theory survive this change, and at the peak of the resonance one finds

$$-\frac{\delta\sigma_R}{\sigma_R} = \langle r^{-2}e^{-\Gamma r/2v\hbar\gamma}\rangle\frac{\sigma(NN^*)}{8\pi} + \langle r^{-2}(1 - e^{-\Gamma r/v\hbar\gamma})^{1/2}\rangle\frac{\sigma(\pi N)}{8\pi}.$$

The latter term is the usual Glauber shadowing reduced in magnitude by a factor $(1 - e^{-\Gamma r/v\hbar\gamma})^{1/2}$. The former is due to collision broadening. Using (8), one recovers a value of Γ agreeing with (6).

So much for the formulae. Let us recapitulate the physics briefly by analogy with atomic physics. There the analogy to Glauber shielding is scattering of the radiated wave after decay of the atom; collision broadening occurs only within the lifetime of the excited state.

To apply (6) and (7) one has to guess $\sigma(NN^*)$. One expects it to be large, since de-excitation is exothermic; also the NN* interaction has the range of single pion exchange. Hence we guess

$$\sigma(NN^*) = \sigma(\bar{p}p) = A/p^*, \tag{9}$$

where p* is the momentum in the NN* centre of mass and A is chosen to give a $\bar{p}p$ cross-section of 120mb at a laboratory momentum of 1GeV/c. Fortunately results are not critically dependent on this guess: as σ increases, $\langle r^{-2} \rangle$ decreases through the effect of the strong interaction form factor. Fig.19(a) shows the calculated effect on $\sigma(\pi d)$; Fig.19(b), which shows much better agreement with experiment, is obtained if in evaluating $\sigma(\pi d)$ at beam energy E one uses $\sigma(\pi N)$ evaluated at energy E -8MeV. The agreement encourages belief in collision broadening as an explanation, and the 8MeV answers a long-standing question of how to do the kinematics of πN collisions in the deuteron.

In calculating Fig.19, the struck nucleon was treated on the mass shell. That is, if the spectator has momentum p_s, the struck nucleon has lab momentum $\vec{p} = -\vec{p}_s$ and energy $E = (M^2 + p_s^2)^{1/2}$, where M is the nucleon mass. The 8 MeV correction to the kinematics then allows for the 2.2 MeV binding energy of the deuteron, and 5.5 MeV mean kinetic energy of the spectator. This prescription agrees with that given for the non-relativistic case by Hetherington and Schick[35] from the Lippmann-Schwinger equation. An alternative prescription treating the struck nucleon off the mass shell has often been advocated: $\vec{p} = -\vec{p}_s$ and $E = M_d - (M^2 + p_s^2)^{1/2}$. For large p_s, this encounters the difficulty that $E < |p_s|$. Empirically one finds from it contributions to $\sigma(\pi d)$ equal to $\sigma(\pi N)$ evaluated at E - 19 MeV, which is in definite disagreement with experiment.

8. $\pi d \to NN$.

It has always been questionable how to treat the process, $\pi d \to NN$, which is kinematically forbidden for free nucleons. Does one just add its contribution to the free nucleon values? The answer is no. This reaction has a peak of 12mb at about the mass of the N*(1230), and I suggest the mechanism is that of Fig.20. One should therefore treat this channel as an inelastic channel of the resonance. In the deuteron, the N*(1230) has three couplings:

$N^* \to \pi N$, width Γ_e

$N^* \to N$, width Γ_N, (Fig.20)

$N^* \to N^*$, width Γ_* (from $N^* \to N \to N^*$).

If the NN* amplitude is that of diffraction from a black disc, $\Gamma_* = \Gamma_N$. The Breit-Wigner amplitude for any channel i is then

$$kf = \frac{\Gamma_i(k)/2}{E_0 - E - i\{\Gamma_e(k) + \Gamma_N(k) + \Gamma_*(k)\}/2} \qquad (10)$$

In particular, I predict a peak πd total cross-section in the N*(1230) region of 225 mb, made up of 262mb from the folded πN cross-sections, -15mb Glauber shielding, and -22mb collision broadening. This agrees well with the old data of Ashkin et al.[36], 221 ± 5mb. Note the large magnitude of the collision broadening effect.

So with these last three questions answered, namely how to do the kinematics, how to account for total cross-sections in the resonance region, and how to treat πd → NN, πd scattering is in much better shape than a year ago.

9. THE N* (1230) IN NUCLEI.

I wish to step marginally beyond the πd system, and consider π-nuclear cross-sections in the 1230 mass region. They are flatter and broader than πd, and become increasingly so as A increases. Attempts to calculate their shape using the optical model have failed[37]. Let me demonstrate a prescription which works well, namely

$$(\Gamma\sigma_R)_{Nucleus} = A\,(\Gamma\sigma_R)_{\pi N} \qquad (11)$$

Table 1 shows values of Γ obtained by fitting the nuclear cross-sections of Allardyce et al.[38] over the peak. The last column $(\Gamma\sigma_R)_{Nucleus}/(\Gamma\sigma_R)_{\pi N}$ agrees well with A.

Why should (11) be true? Suppose all nucleons inside the nucleus scatter the local pion amplitude like free nucleons. For collision broadening (11) follows from (8). For Doppler broadening, it follows to good approximation from the kinematics[34]. Finally one can show[34] that Glauber shielding of one resonance by an identical one has the effect of broadening the resonance just like collision broadening with $\Delta\Gamma/\Gamma = \sigma(\pi N)\langle r^{-2}\rangle/8\pi$. The broadening is obvious from Fig.21, which is a unitarity correction if the two πN vertices are identical. In atomic physics, the analogy is broadening of a spectral line by resonant self-absorption. Thus all 3 effects keep the product (height x width) of the resonant peak a constant per nucleon.

The following simple model of N*s in nuclei emerges. The pion wave is attenuated as it propagates into and out of the nucleus (nuclear skin effect), and the N* amplitude is strong only in the surface. The total cross-section therefore goes as $A^{2/3}$, and Γ goes as $A^{1/3}$; the very attenuation which confines N*s to the surface broadens them. The N*s radiate in phase over the surface, and produce a diffraction pattern with an angular width characterised by the nuclear radius, both in elastic scattering and excitation of low lying nuclear excitations.

10. DIFFRACTION DISSOCIATION.

Consider a pion with a momentum p of many GeV/c interacting with a nucleon. Suppose at a given radius it is attenuated with mean free path L. It suffers a broadening $m\Delta\Gamma = \hbar p/L$. It has a propagator $(m^2 - s - im\Delta\Gamma/2)^{-1}$. At 10 GeV/c, if L = 1F, $\Delta m^2 = 2\ \text{GeV}^2$. For such large values of Δm^2, the "pion" overlaps with other states of the same quantum numbers, and what propagates through the nucleon is a phase coherent mixture of these states. When the surviving wave emerges from the nucleon, Δm^2 shrinks to its free space value, and the equilibrium mixture of states will "freeze" adiabatically to the vacuum eigenstates, which are what one observes experimentally. This physical picture of diffraction dissociation is close to that given originally by Good and Walker[39]. However, I wish to pursue two conclusions beyond theirs. Firstly, one has a natural explanation of the observation[40] that 3π and 5π states formed by diffraction dissociation in nuclei have roughly the same total cross-section as pions, since within the nucleon all are described by wave packets with a large and overlapping energy spread.

The second point is serious for high energy physics, if correct. It has always been argued that to probe the nucleon to shorter and shorter distances, all one needs are particles of higher and higher momentum. However, one sees that as the wavelength of the hadron probe decreases, its resolution is blunted by collision broadening (or unitarity or radiation damping or the uncertainty principle, call it what you will). Unless eigenstates of hadrons exist which scatter weakly, it appears impossible in principle to use hadrons to study the deep interior of the nucleon. If such eigenstates do not exist, the uncertainty principle dictates that one cannot study the strong interaction structure of the nucleon at appreciably smaller distances than its Compton wavelength; this would be natural if its rest mass is indeed self-energy. It dictates correspondingly an energy window from MeV to a few GeV in which one can usefully study strong interaction dynamics.

The states radiated at high energy are properties of the surface rather than the interior of the nucleon. The situation is analogous to the more extreme case of diffusion of light within the sun, which radiates as a black body. If there were only one sun, and one wanted to learn about its interior, this could be done only by calculating what emerges from its surface by means of a model of the interior, or alternatively by studying neutrinos emerging directly from the interior. The same appears to be true of the deep interior of the nucleon.

Table 1. The broadening of the N* (1230) in light nuclei.

Nucleus	Peak cross section σ(mb)	(MeV)	$2(\sigma\Gamma)$Nucleus$/\sigma\Gamma$(H)
$H(\pi^+p+\pi^+n)$	280	135	2
D^2_1	221	172 ± 8	2.01 ± 0.10
Li^6_3	480	241 ± 23	6.1 ± 0.6
Li^7_3	530	260 ± 23	7.3 ± 0.6
Be^9_4	605	290 ± 25	9.3 ± 0.8
C^{12}_6	694	322 ± 25	11.8 ± 0.9
O^{16}_8	827	372 ± 45	16.3 ± 1.9

References

1) J. Bailey et al., Phys.Letters 50B (1974) 403.
2) J.B. Day, G.A. Snow, and J. Sucher, Phys.Rev.Letters 3 (1959) 61.
3) H. Koch, private communication.
4) D.V. Bugg, A.A. Carter and J.R. Carter, Phys.Letters, 44B (1973)278.
5) E.G. Auld et al., submitted to this conference.
6) D. Axen et al., submitted to this conference.
7) J.R. Carter, D.V. Bugg and A.A. Carter, Nucl.Phys. B58 (1973) 378.
8) D. Beder, Nucl.Phys. B34 (1971) 189.
9) C. Richard-Serre et al., Nucl.Phys. B20 (1970) 413.
10) R.J. Glauber, Phys.Rev. 100 (1955) 242.
11) D.P. Sidhu and C. Quigg, Phys.Rev. D7 (1973) 755.
12) C. Michael and C. Wilkin, Nucl.Phys. B11 (1969) 99.
13) G. Alberi and L. Bertocchi, Nu.Cim. 63A (1969) 285.
14) F. Bradamente et al., Phys.Letters 31B (1970) 87.
15) V. Barger and R.J.N. Phillips, Phys.Rev. 187 (1969) 2210.
16) N. Zovko, Karlsruhe preprint.
17) M. Fellinger et al., Phys.Rev.Letters 22 (1969) 1265.
18) J.V. Allaby et al., Phys.Letters 30B (1969) 500; S.P. Denisov et al. Phys.Letters, 36B (1971) 415.
19) W. Galbraith et al., Phys.Rev. 138 (1965) B913; J. Pumplin, Phys.Rev. 173 (1968) 1651.

20) J. Pumplin and M. Ross, Phys.Rev.Letters 21 (1968) 1778; V.N. Gribov, Sov.Phys. JETP 29 (1969) 483; D.R. Harrington, Phys.Rev. D1 (1970) 260; S.A. Guvits and M.S. Marinov, Phys. Letters 32B (1970) 55; O.V. Kancheli and S.G. Matinyan, JETP Lett. 12 (1970) 30; F. Henyey et al., Phys.Rev. 182 (1969) 1579; K.G. Boreskov et al., Sov.J.Nucl.Phys. 14 (1971) 457.
21) H.J. Miettinen, CERN preprint TH. 1864 (1974).
22) T.F. Kycia, private communication.
23) K. Gabathaler and C. Wilkin, Nucl.Phys. B70 (1974) 215.
24) K. Brueckner, Phys.Rev. 89 (1953) 834; ibid 90 (1953) 715.
25) K. Gottfried, Ann. of Phys. 66 (1971) 868.
26) K. Gabathuler et al., Nucl.Phys. B55 (1973) 397.
27) L.S. Schroeder et al., Phys.Rev.Letters 27 (1971) 1813.
28) A.K. Kerman and L.S. Kisslinger, Phys.Rev. 180 (1969) 1483.
29) A.A. Carter et al., Phys.Rev. 168 (1968) 1457.
30) G. Fäldt and H. Pilkuhn, Phys.Letters 40B (1972) 613; ibid 46B (1973) 337.
31) H. Pilkuhn, Karlsruhe preprint TKP 14/74 (1974).
32) G. Fäldt and T.E.O. Ericson, Nucl.Phys. B8 (1968) 1.
33) C. Wilkin, Phys.Rev. Letters 17 (1966) 571.
34) D.V. Bugg, "Collision broadening of resonances", Rutherford preprint, and to be published.
35) J.H. Hetherington and L.H. Schick, Phys.Rev. 137 (1965) B935.
36) J. Ashkin et al., Phys.Rev. 96 (1954) 1104.
37) A.S. Clough et al., Rutherford report RL-74-013 (1974).
38) B.W. Allardyce et al. Rutherford report RL-73-038 (1973).
39) M.L. Good and W.D. Walker, Phys.Rev. 120 (1960) 1855 and 1857.
40) C. Bemporad et al., Nucl.Phys. B33 (1971) 397.

FIGURE CAPTIONS

Fig. 1: The mass attenuation coefficient of bismuth through the M_V photoelectric absorption edge. The fitted curve is derived from a Lorenzian line shape. The cross-hatched region shows the measured value for πd K_α X-rays.

Fig. 2: Spectra of πd mesic X-rays. In the upper figure, the dashed curve shows all X-rays ($K_\alpha+K_\beta+K_\gamma$) with no foils; the full line shows K_α after $K_\beta+K_\gamma$ have been removed by a Saran foil containing chlorine; the dotted curve shows the background due to charged particles. In the lower figure, the background has been subtracted, and the two histograms are the K_α signal before and after attenuation by bismuth; the fitted curves are constructed from Gaussian line shapes, and include small correction terms, e.g. fluorescence.

Fig. 3: π^+d elastic scattering at 47.7 MeV; the curves are explained in the text.

Fig. 4: $\pi^+d \rightarrow \pi^+pn$ at 47.7 MeV. The curve is Beder's prediction[8].

Fig. 5: $\pi^+d \rightarrow pp$ at 47.7 MeV. The curve is from earlier analyses[9].

Fig. 6: The Sidhu-Quigg prediction of πd elastic scattering at 9GeV/c, compared with the data of the CERN-Trieste group[14].

Fig. 7: πd differential cross-sections of Fellinger et al.[17] compared with Glauber theory.

Fig. 8: Values of $\langle r^{-2} \rangle$ from the Serpukhov experiment[18]; the solid curve is the Glauber correction from elastic intermediate states.

Fig. 9: Values of $\langle r^{-2} \rangle$ from Galbraith et al[19].

Fig. 10: Inelastic intermediate states in Glauber shadowing.

Fig. 11: The Sidhu-Quigg screening correction from inelastic states.

Fig. 12: Serpukhov results[18] for the screening correction; the full line shows the calculated value for elastic intermediate states.

Fig. 13: Elastic π^+d scattering at 160°, compated with multiple scattering theory[23].

Fig. 14: Elastic π^+d scattering at 256 MeV, compared with multiple scattering theory[23].

Fig. 15: The backward elastic πd data of Schroeder et al[27].

Fig. 16: Values of $\langle r^{-2} \rangle$ deduced by Carter et al[29], with errors.

Fig. 17: The inset at the top shows the discrepancy between $\sigma(\pi d)$ and Glauber theory. In the lower figure, the full line shows $\sigma(\pi p)+\sigma(\pi n)$, the scale being at the left. The circles show the Glauber shielding term, the scale being at the right.

Fig. 18: Collision broadening of the N* in the deuteron.

Fig. 19: (a) Calculated corrections to $\sigma(\pi d)$ from the collision broadening using $\sigma(\pi N)$ at the same energy E as the beam. The full line is calculated from equation (9); the dotted lines indicate the effect of scaling $\sigma(NN^*)$ up or down a factor 1.5. (b) the calculated correction when $\sigma(\pi N)$ is evaluated at energy E - 8 MeV.

Fig. 20: The diagram describing $\pi d \rightarrow NN$ in the N* (1230) region.

Fig. 21: The Glauber double scattering term with elastic intermediate state.

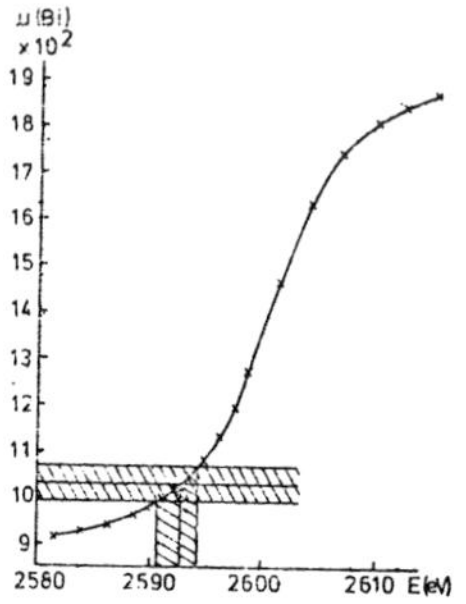

Fig. 1

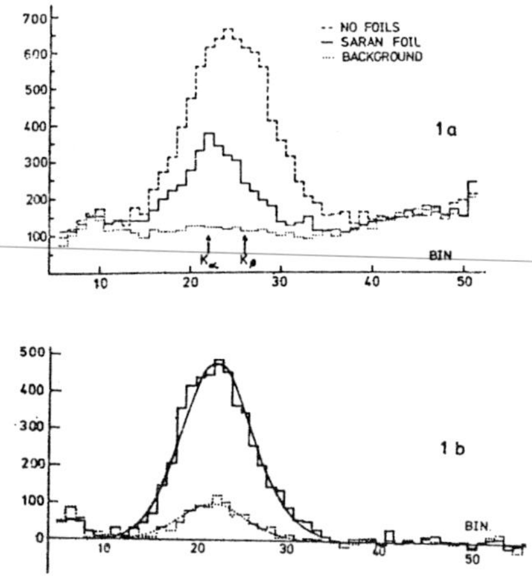

Fig. 2

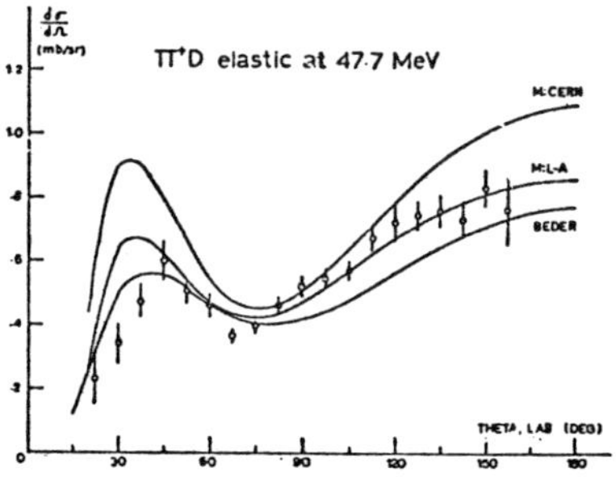

Fig. 3

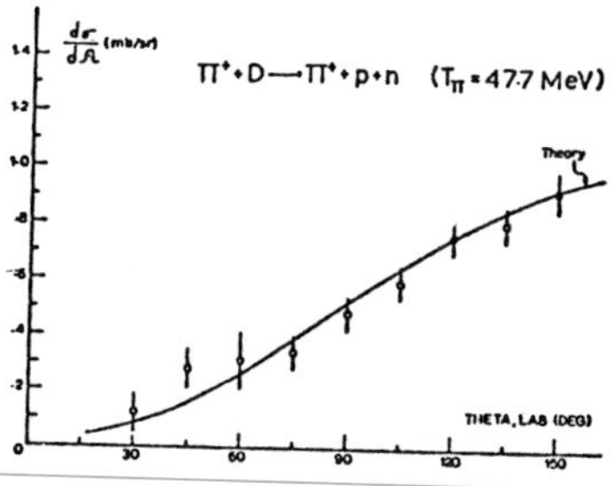

Fig. 4

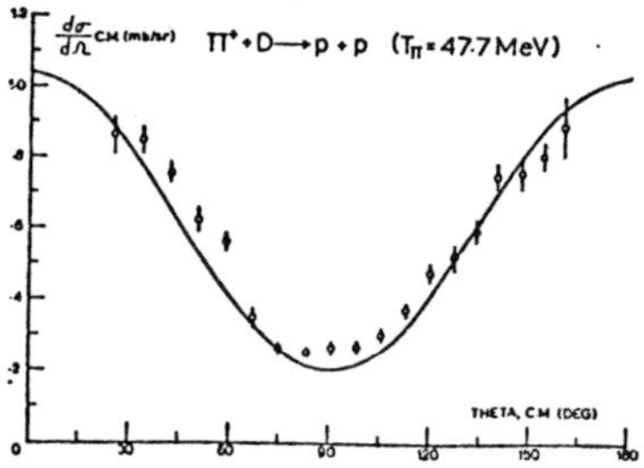

Fig. 5

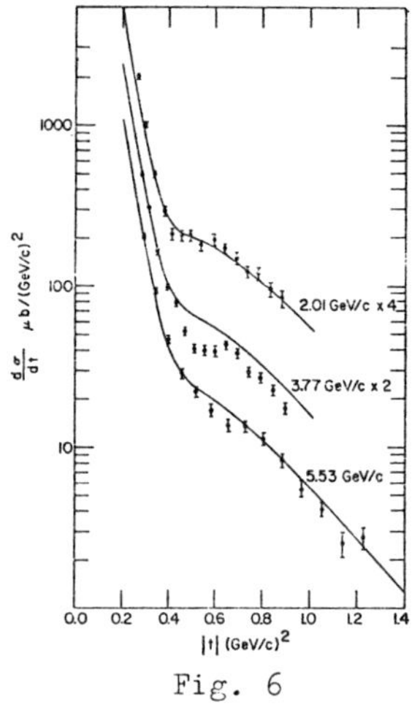

Fig. 6

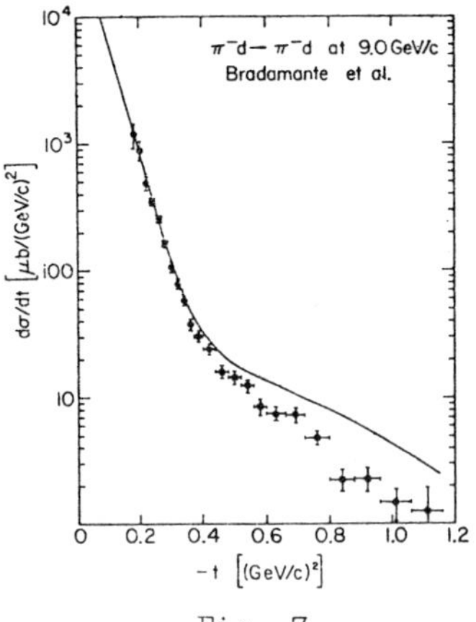

Fig. 7

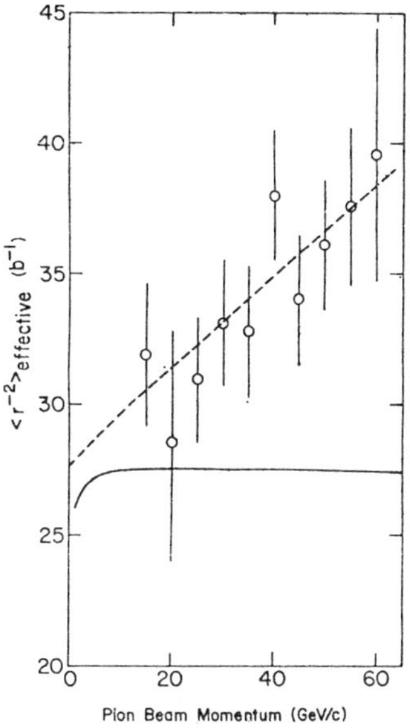

Fig. 8

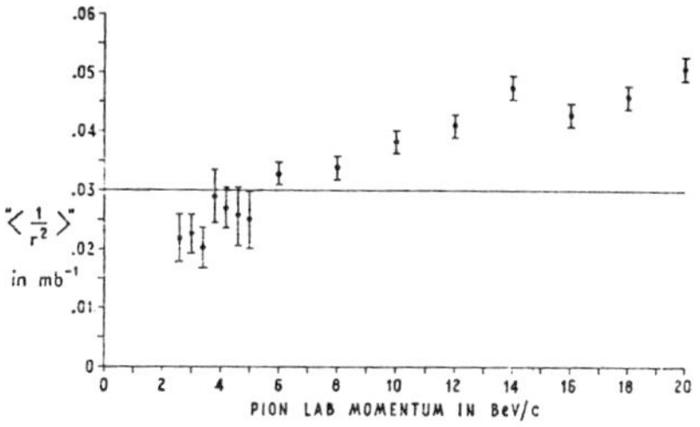

Fig. 9

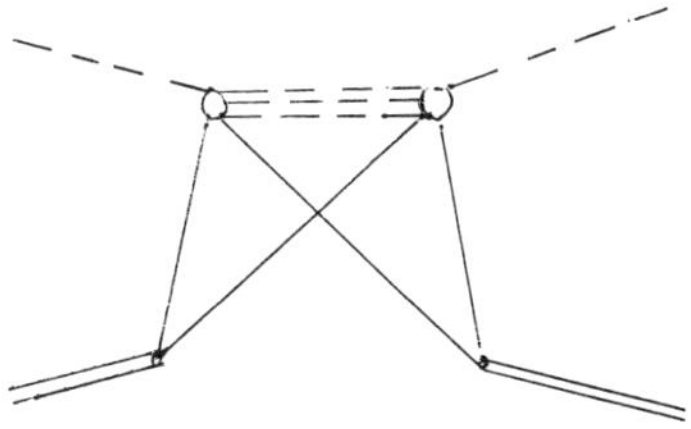

Fig. 10

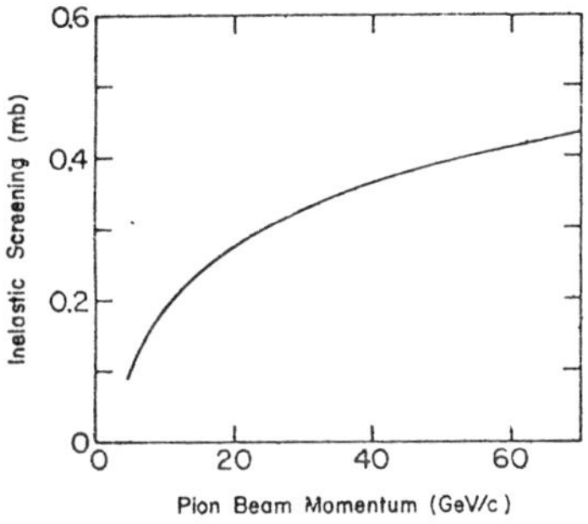

Fig. 11

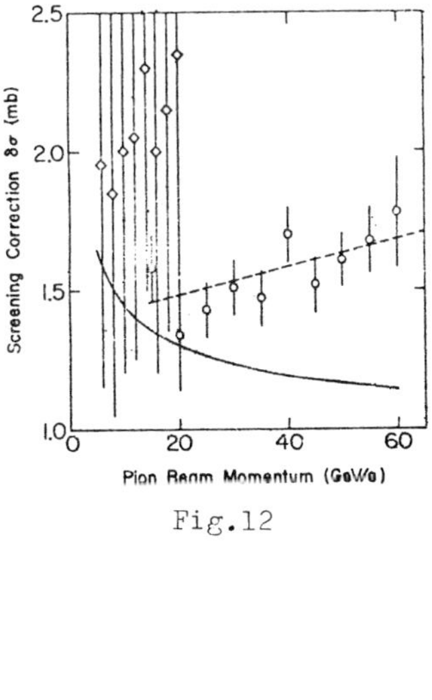

Fig. 12

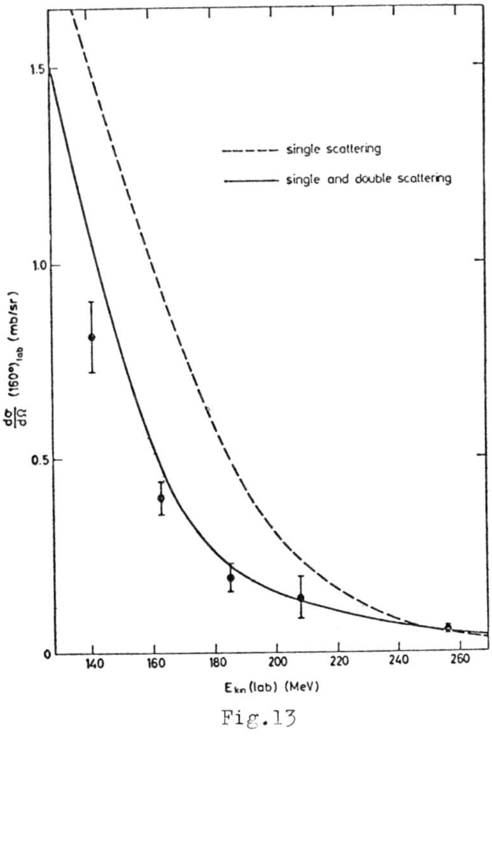

Fig. 13

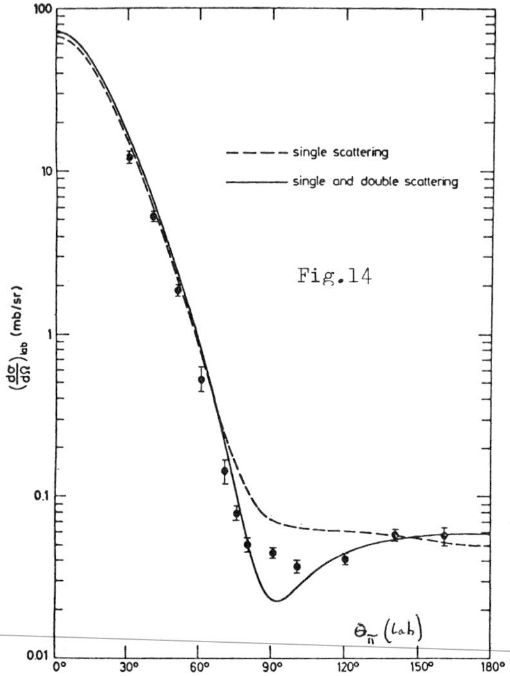

Fig.14

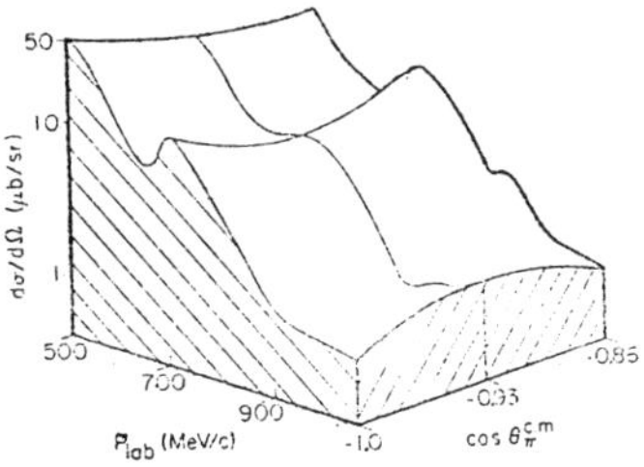

Fig.15

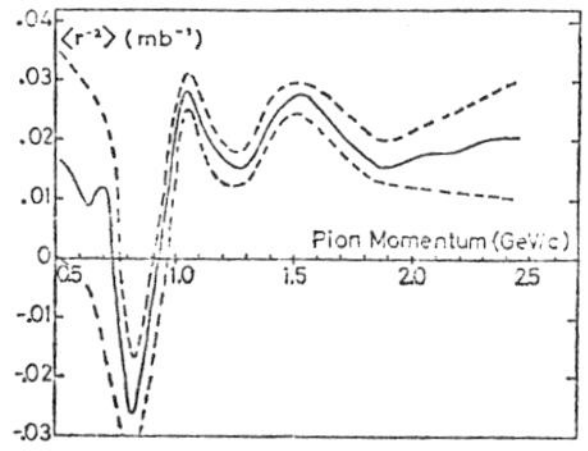

Fig.16

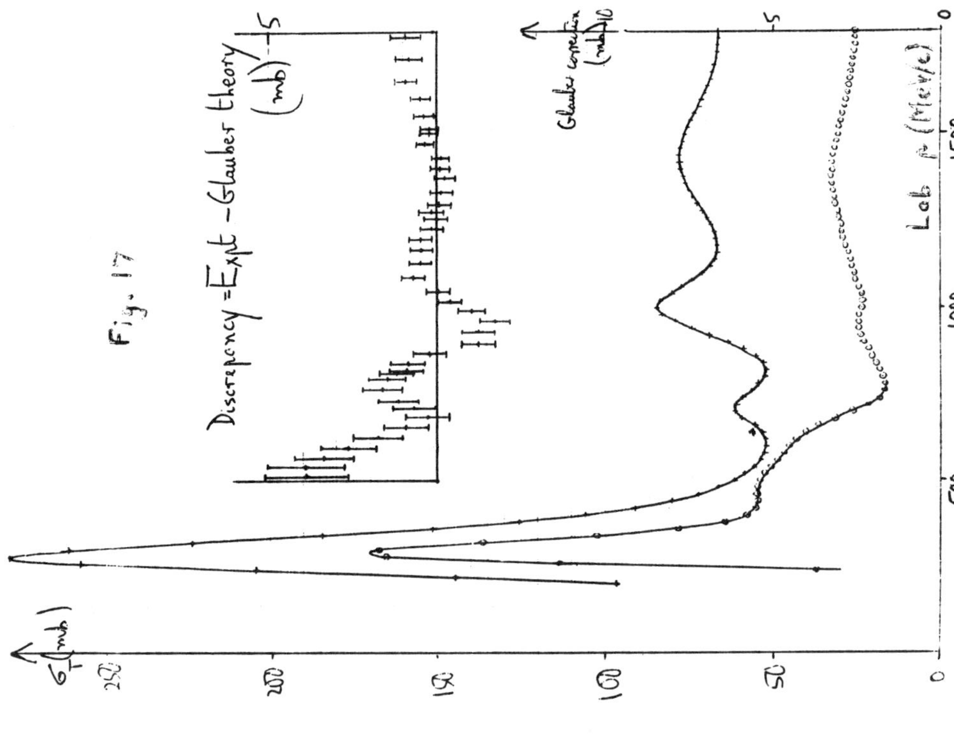

Fig. 17

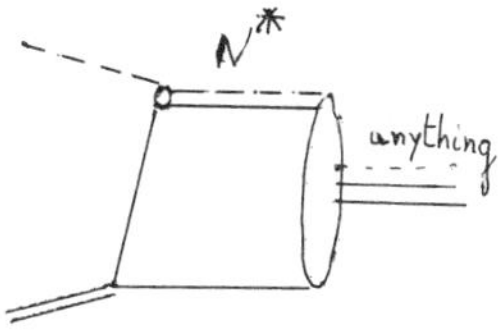

Fig.18

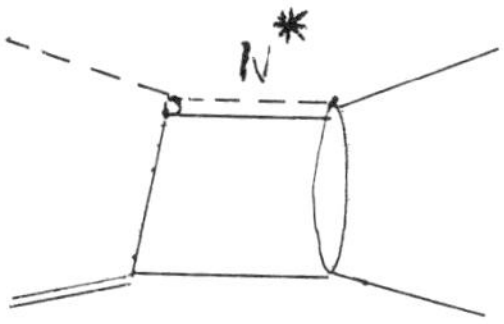

Fig.20

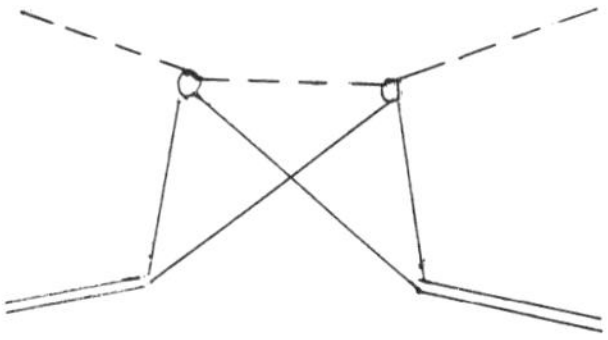

Fig.21

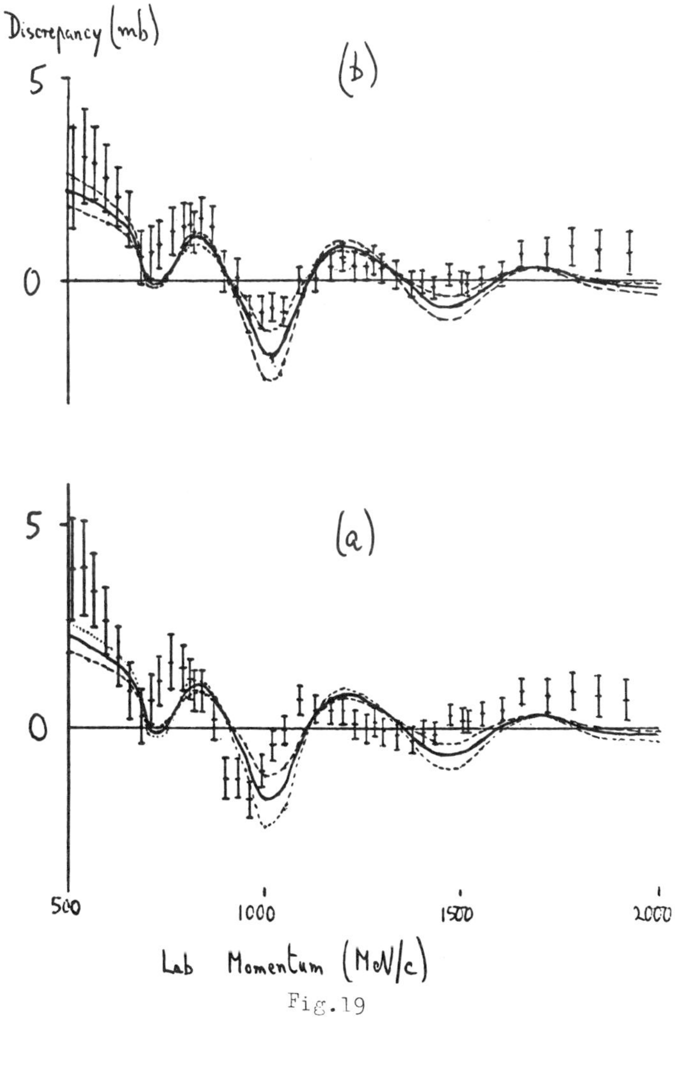

Fig.19

DISCUSSION

LEVIN (Brown): The collision due to N* production in π-nucleon scattering is very analogous to similar effects of channels strongly coupled to the elastic channels as seen in nuclear scattering. A beautiful example is given by scattering of α-particles by the samarium isotopes. ^{148}Sm is spherical and the spherical model fits the data well. The heavier isotopes become increasingly deformed (^{152}Sm, ^{154}Sm) and the spherical optical model cannot fit elastic scattering from these isotopes without drastically changing the parameters used for ^{148}Sm, changes inconsistent with the usual A dependence of the optical model parameters, However, the low-lying rotational states are easily excited (strongly coupled to the elastic channel). When these effects are explicitely included, the same optical model parameters as used for the spherical case (^{148}Sm) now work very well for the deformed isotopes. This is work of the LBL group reported in 1968; it demonstrates clearly (again) that strongly coupled channels must be explicitely taken into account when describing elastic processes.

BUGG: Yes, I think we are continually discovering that much of high energy physics has already been known in low energy nuclear physics.

LEVINGER:(Rennsselaer): Do you get an energy shift of the N* associated with the collision broadening you have discussed?

BUGG: Only this 8 MeV shift. Now something I haven't talked about: if there is a real part of the NN* forward scattering amplitude, one can get a mass-shift. The broadening is due to the imaginary part of the amplitude, and there might well be a real part. To put order of magnitude to this: one is accustomed in high energy physics to a ratio between the real and imaginary parts of the amplitude on the order of 0.3. And that would give you mass shifts of about 2 MeV, rather small.

BRAYSHAW (Maryland): The advantage of the π-d scattering is that one can treat it more exactly as a three-body problem. I have calculated tht p-wave π-d scattering in a relativistic three-body model, as shown in this slide. I have plotted η and δ for both this model and the quite different model of W. Gibbs at Los Alamos; both are quite similar and illustrate a basic difference in the π-d 'resonance' as compared to the underlying π-N resonance. That is, δ is always quite small (less than 20°); it is a sudden decrease in η which gives rise to the bump in the total cross section. Getting η right is all that is necessary to fit experiment and one can check that there is no nearby pole in the total energy. That is, π-d does not 'resonate', and there is probably little point in discussing Breit-Wigner resonance for such systems.

RINAT (Weizmann): 1) I am puzzled by your mentioning isobar admixtures in the deuteron wave functions. These may influence backward angle in p-D scattering but not in the πD case. 2) You mentioned Zovko's πD fit by changing the Barger-Phillips amplitudes. It has been Bradamente's observations that both p-D and π-D scattering for p≥5GeV/c and intermediate momentum transfers are ill-determined by Glauber theory. It is hard to believe that the Barger-Phillips parametrization would be wrong over a wide energy range and the argument anyhow does not hold for p-D scattering.

3) Finally, with regard to the Gabathuler-Wilkin fit of relatively low-energy πD scattering out to 180°. No fixed-scattered theory (neglecting nucleon recoil) can be trusted at those large angles. Irrespective of corrections to the double scattering amplitudes beyond small angles discussed this morning , that term contributes only 10 -15% and corrections to it hardly matter. On this account I should like to predict that Glauber theory may progressively fail to account for angular distribution for energies beyond $T > 250$ MeV, till one reaches the medium-energy range ($T > 700$-800 MeV ?) where Glauber theory or its modifications is expected to hold.

BUGG: Let me just comment very briefly on the second one. I'm only reporting what Zovko says: that Barger and Phillips amplitudes in fact do fail to fit quite a lot of data by 30%, so I could quite believe that what Zovko says is right. But of course, it is a matter of numerology.

CAHILL(Flinder's): I wish to emphasize Brayshaw's comments. With a resonant two-particle interaction, Faddeev theory leads to the result that the three-particle amplitude will have a related off-axis branch point singularity on the second energy sheet, instead of a pole as in the two particle amplitude. Hence we expect a broader bump in the cross-section.

BUGG: Yes, can I just come back to this. I agree completely. There is not a pole in any partial wave π-deuteron scattering. What I have done is to try to give a simple physical argument which will account for the width of the curve which you see, that's all. And the only 'resonance' is in the π-nucleon system. It's not a compound nuclear state, which is what you would have if you had a true resonance in one partial wave of π-d.

BHAKAR (Manitoba): You have given the impression that there is a discrepancy between Glauber theory and experiment. I want to know what sort of detailed calculations you have done, what sort of details you have put into the second order term, which is the shadow correction.

BUGG: I will tell you the ingredients that go into the calculation: Fermi motion, calculation of the parameter (r^{-2}), putting in the strong interaction form factors, flux factors, and picking a wave function for the deuteron. Many people have been through this calculation several times and I think we all agree.

BHAKAR: Since Glauber theory is not valid near resonance, is there any point in comparing the experiment and theory?

BUGG: That's just what I'm saying.

BHAKAR: As a matter of fact, your hypothesis of broadening the resonance is simply a many body effect, whether Glauber theory is valid or not.

BUGG: Well I have not seen any calculation in the literature which does explain the discrepancy up to now.

VAN HOVE (CERN): 1) In your treatment of collision broadening in π scattering you needed a very large $\sigma(N^*N)$. Isn't this in fact a free parameter that you could choose to fit the data? At your low energies, $\sigma(N^*N)$ is not really measurable in a direct way.

2) In your consideration of a pion diffraction dissociation, the width induced by the finite free path is not the only relevant parameter, also the coupling strengths of the pion to the 3π and the 5π states are relevant. These coupling strengths are rather weak as seen from the fact that $\pi p \to (3\pi)p$ and $\pi p \to (5\pi)p$ cross sections are small. To the degree that these couplings are weak the cross sections of (3π) or (5π) on nucleons can differ appreciably from the πp cross section.

BUGG: I accept your second point. One really wants to do detailed calculations with a particular model, and I have not done that.

VAN HOVE: But you agree that the strength of the diffraction dissociation amplitude is relevant.

BUGG: Yes, it is relevant, but not easily assessed without detailed calculation. Suppose at impact parameter $\underline{b}$ the transmitted amplitude is $a(\underline{b})$

The elastic diffraction is proportional to:

$$\int b \; db \; J_0 \; (kb \sin\theta) \; \left\{1 - a(\underline{b})\right\}$$

while the diffraction dissociation to inelastic states is:

$$\int b \; db \; J_0 \; (kb \sin\theta) \; a(\underline{b})\alpha(\underline{b}),$$

where α is the overlap between the pion and an inelastic state. The way I look at it is this: deep inside the nucleon, the interaction Hamiltonian is so strong compared to the free Hamiltonian, H_0, that the eigenstates are quite different to those of H_0. The latter are useful only near the surface or outside the nucleon. What one observes experimentally in the way of resonances and mass spectra must of course be eigenstates of H_0. Let me draw an analogy with the sun. Inside the sun the mean free path of electromagnetic radiation is very short, and multiple scattering robs the photon of any memory of its history. Likewise I suggest that states produced by diffraction dissociation forget in detail the mechanism by which they are created.

SOME THEORETICAL ASPECTS OF LOW ENERGY π D SCATTERING

A.W. Thomas
Department of Physics, University of British Columbia,
Vancouver, B.C., Canada. V6T 1W5

1. INTRODUCTION

The importance of understanding the π d system can not be overemphasised. Aside from the intrinsic interest in this fundamental system, many of the approximations that have been used to describe pion elastic scattering from complex nuclei should work best in the deuteron - where they can be tested by essentially exact calculation. From the point of view of pion absorption in nuclei, where the two nucleon mechanism seems to be important, an understanding of π d scattering is essential to understanding the reaction mechanism. Quite apart from these considerations, a study of π d scattering may provide some insight into the nature of the π N force.

Our aim is certainly not to provide a comprehensive review of all that has been done on π d scattering. There are a number of excellent reviews to which one should refer for such a discussion[1-4]. (Indeed we shall concentrate entirely on a Hamiltonian approach, to the exclusion of (covariant) dispersion relation methods[4].) Instead, we shall emphasise those areas of π d scattering where there has been some new development recently, or where some progress is to be expected soon.

Section two deals with the correct treatment of the off-energy-shell π N t-matrix in a single scattering approximation. Such considerations can have a significant effect in both pionic and kaonic atom calculations, as well as higher energy scattering. In the third section we show that the higher order corrections to the π d scattering length ($a_{\pi d}$) can be computed quite accurately. The point is that this is a good example of a case, where an accurate experimental measurement (of $a_{\pi d}$), would provide a direct verification of the need for correct off-energy-shell behaviour of the π N interaction. An important part of this discussion concerns the effect of pion absorption on the elastic scattering amplitude. This is intimately related to the problem of double counting. Finally, in section four we explain in detail why a series of π d elastic scattering experiments would be interesting. We emphasise, however, that a significant improvement in the low energy π N data is essential if one is to extract information from π d scattering.

2. SINGLE SCATTERING APPROXIMATION

It may seem surprising that there has been a recent development here! Nevertheless, there is an apparent problem with the single scattering approximation (SSA), which has been largely ignored until recently - namely the apparent freedom in choosing the π N interaction energy. We recall that within a consistent three body theory there is no ambiguity. Moreover, the correct choice may produce significant corrections not only in the calculation of the π d scattering length ($a_{\pi d}$), but throughout intermediate energy physics.

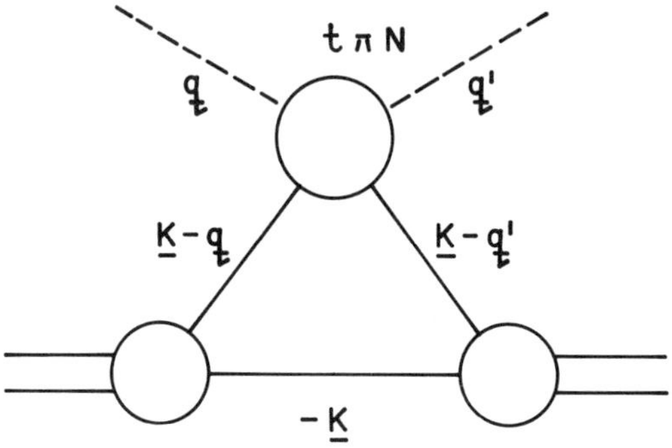

Figure 1: Single Scattering Approximation (in the π d c.m. system).

Consider the SSA to π d scattering - shown in Figure 1. The π N t-matrix ($t_{\pi N}$) which appears there in a correct three body treatment[5,6], is a fully off-shell function of three variables - the initial and final relative momenta, and the collision energy. The dependence on the relative momenta is quite familiar[7]

(Iso-scalar)

$$f_{\pi N} = b_0 + c_0 \underline{k}.\underline{k}' + i\, d_0\, \underline{\sigma}.(\underline{k} \times \underline{k}') \tag{1}$$

where $\underline{k}(\underline{k}')$ is the initial (final) relative momentum. There is also an isovector term with subscripts "1", and an overall isospin factor "$\underline{i}.\underline{\tau}$". While this (essentially on-shell approximation) has been used as a model for going off-shell to produce the very successful π-nucleus optical potentials[8,9], there have recently been some well founded objections to it based on the strange results produced in reactions sensitive to the high momentum components of the pion wave function[10]. For this reason, as well as consistency in a complete three body theory, we prefer the more reasonable off-shell behaviour one would obtain from a two body Lippmann-Schwinger equation[11].

Let us return to the main point of this section - the energy dependence of $t_{\pi N}$. If the π d system has a total c.m. energy s (excluding rest masses), then the energy at which $t_{\pi N}$ must be evaluated in Figure 1, is uniquely specified by Faddeev theory to be $(s - \kappa^2/2M)$ - where M is the reduced mass of the N - (πN) system ($M^{-1} = m_N^{-1} + (m_\pi + m_N)^{-1}$). Of course, one must integrate over $\underline{\kappa}$ to evaluate the SSA, which means that $t_{\pi N}$ is needed in the energy range (- ∞, s). Once again this presents no problem if one uses a potential in a Lippmann-Schwinger equation to generate the t-matrix[12].

$$t_{\pi N}(\underline{p}, \underline{p}'; \sigma) = v_{\pi N}(\underline{p}, \underline{p}') + \int d^3q \, \frac{v_{\pi N}(\underline{p}, \underline{q}) \; t_{\pi N}(\underline{q}, \underline{p}'; \sigma)}{(\sigma + m_\pi) - q^2/2m_N - (q^2 + m_\pi^{\,2})^{1/2}} \tag{2}$$

In summary, the SSA to π d elastic scattering is:

$$T_{\pi d}(\underline{q}, \underline{q}'; s) = \sum_{j=1,2} \int d^3\kappa \; \psi_d^*(\underline{\kappa} - \underline{q}/2) \; t_{\pi N}^{(j)}(\underline{k}_i, \underline{k}_f; s - \kappa^2/2M) \quad \psi_d(\underline{\kappa} - \underline{q}'/2) \qquad \ldots (3)$$

where $\underline{k}_i$ and $\underline{k}_f$ are the initial and final π N relative momenta, and the sum is over the two nucleons.

Now consider the scattering length calculation as a special case (i.e. $\underline{q} = \underline{q}' = 0$, and s = -2.225 MeV). The usual approach is to ignore p-wave effects, and take the s-wave t-matrix to be constant - as in equation (1). Then equation (3) reduces to

$$a_{\pi d}^{(0)} = \frac{2}{3} \frac{m_N + m_\pi}{m_N + m_\pi/2} (a_1 + 2a_3) \tag{4}$$

where the mass factor arises because we are in the π d not the π N c.m. As usual (a_1, a_3) are the isospin $(^1/2, ^3/2)$ π N scattering lengths. In fact, the combination $(a_1 + 2a_3)/3$ is the isoscalar scattering length (b_0 in (1)), which soft pion theory says is zero[13]. There is considerable interest in deriving the actual value of b_0 from dispersion theory[14]. We shall adopt Bugg and Carter's value of $-.005\ m_\pi^{-1}$, while bearing in mind that this small number is the result of a cancellation between two fairly large numbers ($a_1 = .170\ m_\pi^{-1}$ and $2a_3 = -.184\ m_\pi^{-1}$)[15]. It is as a result of this strong cancellation that corrections due to off-shell effects become very important, as has been observed by Afnan and Thomas[16], and independently by Myhrer and Silbar[17].

The important off-shell effect is the energy dependence of $t_{\pi N}$, since the relative momentum is just $(^{m_\pi}/m_N)$ times the fermi momentum in the deuteron - a small number. In Table I we show $f_{\pi N}(\underline{o}, \underline{o}; E)$ as a function of E along the negative energy axis. (These were calculated from separable t-matrices fitted to the π N phase shifts and scattering lengths, treating the pion relativistically - as in equation (2)[18].) For convenience we also show the isoscalar combination $(f_1 + 2f_3)$. Now, to a first approximation one can take

$$< f_1 + 2f_3 >_d \simeq f_1(-\varepsilon_d - <\kappa^2/2M>_d) + 2f_3(-\varepsilon_d - <\kappa^2/2M>_d) \tag{5}$$

where $< >_d$ denotes an average over the deuteron wave function. Using $<\kappa^2>_d \simeq (120\ \mathrm{MeV/c})^2$ we see that $(f_1 + 2f_3)$ is needed at about -14 MeV, which from Table I is $-.031\ m_\pi^{-1}$ - a correction of $-.019\ m_\pi^{-1}$ from the value at E = 0. If one actually

Table I

Illustration of the importance of off-energy-shell effects in the calculation of $a_{\pi d}$. (Units m_π^{-1}.)[18]

E (MeV)	$f_1(\underline{o}, \underline{o}; E)$	$f_3(\underline{o}, \underline{o}; E)$	$(f_1 + 2f_3)$
0	.1706	-.0913	-.012
-2	.1664	-.0928	-.019
-6	.1637	-.09412	-.024(5)
-10	.1620	-.09510	-.028
-14	.1607	-.09594	-.031
-20	.1591	-.09705	-.035

evaluates equation (3), the correction to $(f_1 + 2f_3)$ is $-.020\ m_\pi^{-1}$ [16,18]. Incidentally, Myhrer and Silbar offer an attractive explanation of the direction of this shift as a unitarity correction - we refer to their contribution to this conference for further discussion[17].

In closing this section, we shall briefly record some other places where this off-shell effect may be important. First, it has been long known that the isoscalar scattering length needed to fit pionic atom data, is much larger than experiment. Ericson and Krell find a value of $-.030\ m_\pi^{-1}$ [19], of which between $-.010$ and $-.020$ can be attributed to correlations. Clearly there should also be an off-shell correction here, at least as large as that found in the deuteron. Preliminary estimates using harmonic oscillator wave functions, suggest a correction of about $-.010\ m_\pi^{-1}$ for $A < 24$[20].

The situation for kaonic atoms is even more critically dependent on the correct off-shell dependence of the $\bar{K}N$ t-matrices. This is because of the $\Sigma\ \pi$ resonance just below the $\bar{K}N$ threshold. Although this off-shell dependence has been studied by Bardeen and Torigoe[21], and more recently by Myhrer[22], there is a clear need for more theoretical and experimental work - particularly in the K^-d system.

Finally we observe that this off-shell effect may also lead to corrections at higher energy. We shall return to this in § 4, but for the moment we note that this will tend to raise the resonance energy in π-nucleus scattering - as seems to be required in a realistic optical model calculation[23].

3. CORRECTIONS TO THE SCATTERING LENGTH

In the last section we saw there was a strong cancellation in the impulse approximation to $a_{\pi d}$. Consequently the usual prejudice, that low energy pion rescattering should be relatively small, does not hold here. Nevertheless, we aim to show in this section that it is possible to make an accurate estimate of all the corrections which arise. The first reason is that one would like to verify the (off-shell) correction to b_0 within a nucleus, which was discussed in § 2. This will only be possible through a measurement of $a_{\pi d}$ if the higher order corrections are known. Another important reason is that many approximations used in calculating $a_{\pi d}$ have been used elsewhere, and we intend to examine them where they can be checked.

The first systematic discussion of these corrections was made by Kolybasov and Kudryavtsev[24]. Our procedure will be to describe their method of calculation, and then indicate any subsequent modification. The notation for single, double and higher order s-wave scattering is $a^{(1S)}$, $a^{(2S)}$, $a^{>}$, while for single p-wave and double scattering through s- and p-waves we use $a^{(1P)}$, $a^{(SP)}$ [25].

3.1 S-wave Rescattering

Kolybasov and Kudryavtsev (KK) have generalised an earlier result of Brueckner[26], so that it is possible to sum the complete s-wave rescattering series (including spin and isospin). (A similar result was obtained by Bendix and Beder[25].) There are however, two major approximations in deriving this result - neglect of nucleon (recoil) kinetic energy, and the use of constant scattering amplitudes. By explicit calculation KK show that the inclusion of the deuteron binding energy and nucleon recoil in the double scattering diagram reduces its contribution by (40-70)%.

In Table II we summarise all calculations of $a_{\pi d}$ which explicitly separate $a^{(2S)}$ and $a^{>}$. The results of Bendix and Beder (BB) do not fit readily into this table, as

Table II

Results of several calculations of s-wave rescattering (Units m_π^{-1}).

	Constant t-matrices.		Non-Constant t-matrices.		
	KK[24]	Myhrer[27]	Myhrer[27]	A T[16]	Thomas[18]
$a^{(2S)}$	-.019	(-.018, -.021)	(-.013, -.015)	-.016	-.013
$a^{>}$	.002	(.002, .003)	(.001, .002)	.003	.003

they have not explicitly separated s- and p-wave contributions[25]. Further, their s-wave t-matrices are not strictly constant, since they include a relativistic transformation factor from the π d to the π N c.m. Nevertheless, their double scattering result (-.020 m_π^{-1} for a Hulthen wave function), is in excellent agreement with the result of KK. Returning to Table II, we notice there is also acceptable agreement between the constant t-matrix calculations of KK and Myhrer[27], the latter being a Faddeev calculation. (The range of results quoted by Myhrer corresponds to varying a_3 in the range (-.086, -.109)m_π^{-1}.) Another constant t-matrix Faddeev calculation (not shown in Table II), was made by Petrov and Peresypkin (PP)[28]. Although they do not separate higher order contributions, the final result agrees with Myhrer.

While the constant π N t-matrices fit the scattering lengths, one would also like to fit the phase shifts over a reasonable energy range. For the S_{11} (i.e. $L_{2T,2J}$) case, Myhrer found they could be fitted up to about 150 MeV using a Yamaguchi form factor ($g(p) = C\,(p^2 + \beta^2)^{-1}$)[29], with range ($\beta$) about 2.4 fm^{-1}. Although this simple form-factor could not fit the S_{31} data, Myhrer used the same value of β there too. The resultant value of $a_{\pi d}$ was considerably reduced from the constant t-matrix approximation.

The Faddeev calculation of Afnan and Thomas (AT)[16], differs from those of Myhrer and PP in several ways[27,28]. First, AT were able to fit both the S_{11} and S_{31} phase shifts. This meant using an S_{31} form factor with a dip at low momentum - see the discussion in ref. 16), and Fig. 3 of Landau and Tabakin (who solved the inverse scattering problem directly)[11]. A second difference is that AT included the deuteron d-state. (The results in Table II correspond to a Phillips potential with 4% d-state probability[30].) Incidentally, this results in an added complication, since one must include the possibility of an s-wave π N interaction with a d-wave (L=2) spectator nucleon. (Note that the calculation of AT was performed at -1 MeV rather than -2.225 MeV. This leads to a small correction ($-.016\ m_\pi^{-1}$ instead of $-.017\ m_\pi^{-1}$) to the value of $a^{(2S)}$ given in Ref. 16). This small numerical problem was removed in the later calculation by Thomas[18).])

The final column of Table II includes the effect of relativistic kinematics for the pion. Once again the separable π N t-matrices are fitted to the phase shifts past 200 MeV, as well as the latest scattering lengths[15)]. In view of the close agreement between the various calculations at different stages of approximation, and in view of the absence of other significant corrections, it is felt that $a^{(2S)} = -.013\ m_\pi^{-1}$ is probably accurate to better than 10%, unless there is a drastic change in the π N scattering lengths. (Double scattering is less sensitive to such changes because it depends essentially on b_1[24)], which is much better known than b_0.)

The higher order corrections are clearly small, and of order $.002\ m_\pi^{-1}$. We prefer the value $+.003\ m_\pi^{-1}$ of ref. 18), because it includes relativistic kinematics for the pion as well as two body interactions (π N *and* NN) which fit the available data.

3.2 P-wave Scattering.

There can be a contribution to $a_{\pi d}$ from p-wave π N scattering because of fermi motion. Using a scattering amplitude of form (1), KK have shown this contribution to be[24)]

$$a^{(1P)} = 2\,c_0\,(m_\pi/(m_\pi + m_N))^2 \quad < \kappa^2 >_d \tag{6}$$

$$c_0 = (4\alpha_{33} + \alpha_{11} + 2\alpha_{13} + 2\alpha_{31})/3 \tag{7}$$

where $\alpha_{2T,2J}$ is a p-wave scattering volume. Evaluating equations (6) and (7) leads to the result $a^{(1P)} = +.004\ m_\pi^{-1}$. (This value was also obtained by BB[25)].)

Our experience with s-wave rescattering suggests that correct off-shell behaviour may reduce this value a little. This is supported by a Faddeev calculation of $a^{(1P)}$ using only a P_{33} interaction - fitted to $\alpha_{33} = .210\ m_\pi^{-3}$ and the π N data through the resonance region (relativistic kinematics for the pion again)[18)]. That is, the Faddeev result is $+.004\ m_\pi^{-1}$, compared with $+.005\ m_\pi^{-1}$ which would be obtained by KK if $(\alpha_{11} + 2\alpha_{13} + 2\alpha_{31})$ was dropped from (7). We shall use the value $a^{(1P)} = +.003\ m_\pi^{-1}$.

While noting that double p-wave scattering is negligible, KK show that $a^{(SP)}$ is about the same size as $a^{(1P)}$ but opposite in sign. In fact, with their low energy parameters they find $a^{(SP)} = -.005\ m_\pi^{-1}$. Here, as in the case of s-wave double scattering, the important s-wave parameter is b_1, which is not so sensitive to the fluctuations of a_1 and a_3 with time (e.g. using Bugg and Carter's value of b_1, rather than KK's, would only reduce $a^{(SP)}$ by about 10%). Once again reasonable off-shell behaviour will reduce $a^{(SP)}$ a little, so we take as the total p-wave contribution $(a^{(1P)} + a^{(SP)}) = -.001\ m_\pi^{-1}$ - with an error of perhaps 50%.

In concluding this section we note that no account has been taken of Pauli

Effects. In particular, it has been shown recently that in a Chew-Low picture of π N scattering[31)], there can be significant changes from the free p-wave scattering volumes within a nuclear medium[32,33)]. Such corrections should be rather small in the case of the deuteron. Nevertheless, we shall show in § 4.2 that Pauli restrictions in the elementary π N interaction may significantly affect π d angular distributions.

3.3 Absorption.

An extremely interesting aspect of π d scattering, is that there is a coupled (open) inelastic channel (π d $\rightarrow$ NN), even at zero energy. Because of the large momentum transfers involved, it has long been hoped that the (π, 2N) reaction might be a useful probe of short range correlations within a nucleus[34)]. Attempts to unravel the (π, 2N) reaction mechanism by working in the much simpler π d system[35)], and to obtain information about the N-N force[36)], are fascinating, but there is no time for further discussion here. (For details we refer to the more recent articles just quoted, and to the proceedings of this conference.) Instead we shall examine the effect of pion absorption on the elastic channel.

There are two major problems in any discussion of this effect. First, one needs an unambiguous definition of the "effect of absorption". Brueckner first tackled this problem by evaluating Fig. 2(a),[37)] where the vertices include all diagrams which result in pion absorption (or production). Later it was suggested by Thouless that the crossed diagram (Fig. 2(b)) should also be included[38a)].

One promising method of distinguishing exactly what is meant by the "effect of absorption", is to start with a very simple field theoretic model - such as that described by Wick[39)]. (That is, we allow only non-relativistic nucleons, so that there is no $\bar{N}N$ production.) The techniques required are essentially diagrammatic summations of the type suggested by Zachariasen and Woodruff[35,40)]. In the spirit of this approach one assumes, for example, that meson exchange can generate the N-N force. The resultant (exact) expression for π d elastic scattering is[41)]

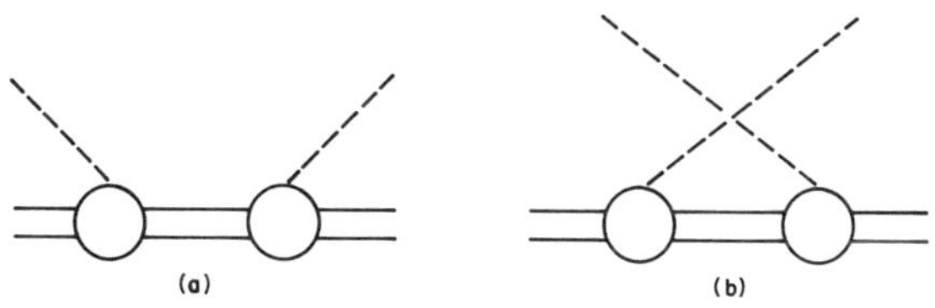

Figure 2: Absorptive contributions to π d scattering as suggested by (a) Brueckner[27)] and (b) Thouless[38a)].

$$<\pi d|T(E)|\pi d> = <\pi d|\tilde{T}|\pi d>_1 + P\sum_n \frac{|m_{\pi d,n}|^2}{E-\varepsilon_n} - i\pi\sum_n |m_{\pi d,n}|^2\,\delta(E-\varepsilon_n) \quad (8)$$

The last two terms of equation (8) include the sum of *all* π d scattering diagrams which have a two nucleon - zero meson intermediate state. (In fact, this provides a rigorous justification for Brueckner's use of Fig. 2(a).) It is our contention that these terms describe the major part of the contribution of absorption. The reasoning is simply that within this (non-relativistic) theory (which should not be bad at threshold), the simplest crossed graph one can think of in Fig. 2(b), corresponds to single scattering from one nucleon (through (say) the P_{33} interaction - see Fig. 3(a)). Certainly there will be some diagrams in Fig. 2(b) which are not strictly multiple scattering graphs. Nevertheless, these will correspond to many meson intermediate states, which one would expect to give only a small contribution. Therefore, any theory which calculates the crossed graph (Fig. 2(b)) as an "absorptive contribution", should (at least) subtract out the p-wave single scattering contribution. (Note that in the only realistic calculation of the contribution of Fig. 2(b) to $a_{\pi d}$, Nishimoto et. al. did not make this subtraction[38b)]. The overall contribution they found from Fig. 2(b) (after a very complex calculation), was very close to our estimate of $a^{(1P)}$).

Henceforth, we shall consider only the contribution of Fig. 2(a) (as expressed in equation (8)), to be an absorptive contribution. The sum over "n" in equation (8) includes all possible 2N intermediate states. Thus, in order to evaluate the real part of the absorptive contribution (Re $T_{\pi d}^{(abs)}$), one needs the pion absorption matrix element off the energy-shell. Of course, Im $T_{\pi d}^{(abs)}$ is simply related to the cross section for pion absorption (π d $\rightarrow$ NN), through unitarity.

We shall return to the absorptive term shortly. For the moment let us examine the first term on the right of equation (8) - viz: $<\pi d|\tilde{T}|\pi d>_1$. In this notation the subscript indicates that every intermediate state must contain at least one meson (to avoid double counting the absorptive effects); the "~" indicates that only

(a) (b)

Figure 3: Lowest order contributions to π N scattering in (a) the P_{33} state and (b) the P_{11} state.

diagrams with a π N interaction both first and last are included (since all diagrams with an initial (or final) N-N interaction have been summed to give the deuteron wave function). Clearly this is closely related to the usual π d multiple scattering series. Indeed, within a one meson theory, one could expect to evaluate this using (say) Faddeev theory, with phenomenological π N t-matrices. Once again this raises our second major problem, namely double counting.

The major advantage of using equation (8) to isolate this double counting problem is that equation (8) was derived from a theory containing only π N interactions - these then generated the N-N interaction. Consider now the single scattering diagram (Fig. 1) where there is a P_{11} interaction. If one evaluates this using a phenomenological t-matrix, it will contain (implicitly) a contribution of the type shown in Fig. 3(b)[31-33]. However, this means that the condition implied by the subscript "1" is being violated - that is, we have a 2N - zero meson intermediate state where there should be at least one meson. It is to be hoped that this very brief discussion will clarify how intimately related are these questions of "absorptive contributions", P_{11} scattering (in the direct case Fig. 2(a) - P_{33} etc. in the crossed case of Fig. 2(b)), and double counting. Fortunately, we have seen that the effect of P_{11} scattering in the calculation of $a_{\pi d}$ is very small (§ 3.2). Consequently, this question of double counting on the direct absorptive part of $a_{\pi d}$ is a theoretical, rather than a practical problem. However, as we shall see in § 4, this problem will become far more important in π d scattering at higher energy.

Within the 3-body model of π d scattering recently described by Afnan and Thomas, absorption occurs through a bound state in the P_{11} interaction[16]. There the effect of absorption can be very easily defined as the change in $a_{\pi d}$ caused by including the P_{11} interaction. By solving the Faddeev integral equations in that case one is, in effect, evaluating $T_{\pi d}^{(abs)}$ as given in equation (8). The important advantage is that the half-off-shell amplitudes ($m_{\pi d,n}$) are calculated in a consistent 3-body theory. Furthermore, this theory agrees with experiment on the energy shell.

To summarise, the best available estimate for the effect of absorption on the π d scattering length is $a_{\pi d}^{(abs)} = -.005 + .004i\ m_{\pi}^{-1}$ [42]. For comparison we recall that the total s-wave rescattering contribution ($a^{(2S)} + a^{>}$) is $-.010\ m_{\pi}^{-1}$. (Based on the results of Nishimoto et. al., it seems that inclusion of the crossed absorptive graph (Fig. 2(b)) might reduce this a little[38b]. However, if the subtraction of single p-wave scattering is made (as described earlier) to avoid double counting, this correction should be small.)

Finally, let us clear up one remaining correction to the usual multiple scattering series, which is in a sense an absorptive correction. In particular, we observe that the mass difference of π^{o} and $\pi^{\pm}$ can lead to positive energy intermediate states in the charge exchange process $\{\pi^{-}p \to \pi^{o}n \to \pi^{-}p\}$. Thus, there will be a small correction to Im $a_{\pi d}$, as well as a small shift in Re $a_{\pi d}$. The only estimate of this effect has been made by Kolybasov[43], and we use his result, namely $-.002 + .0003i\ m_\pi^{-1}$.

3.4 Summary

At last we are in a position to put together all the pieces, to form the total higher order correction to $a_{\pi d}$. The contribution from s-wave rescattering $(a^{(2S)} + a^{>})$ is $-.010\ m_\pi^{-1}$; p-wave scattering $(a^{(1P)} + a^{(SP)})$ gives $-.001\ m_\pi^{-1}$; the π^o mass defect gives $-.002 + .0003i\ m_\pi^{-1}$, and the contribution of true absorption is $-.005 + .004i\ m_\pi^{-1}$. Therefore the total correction to the s-wave single scattering approximation for $a_{\pi d}$ is $-.018 + .0045i\ m_\pi^{-1}$.

The uncertainty in each contribution has been discussed, but some overall comments seem appropriate. First, there is general agreement that $a_{\pi d}$ is not very sensitive to the deuteron wave function[16,22,28]. While all the estimates in this article (except $a_{\pi d}^{(abs)}$) are based on Hulthen wave functions (a Yamaguchi form in momentum space), AT have shown that using such realistic short range behaviour as given by the Reid Soft Core potential[44], only produces a change of order $-.001\ m_\pi^{-1}$. Second, there has been a rather crude study of the dependence of $a_{\pi d}$ on the off-shell behaviour of the π N t-matrices (basically $g(p) = C(p^2 + \beta^2)^{-n}$, with β arbitrary, n = (1,2,3), and C adjusted to fit the scattering length)[45]. The catch, of course, is that one really should fit the π N phase shifts not just the scattering lengths. This will place considerable restraint on the behaviour of g(p). (Indeed, if we know the phase shifts at all energies, with a one term separable potential g(p) is specified uniquely[11].) Consequently, even the rather small variation found by Peresypkin and Petrov must be regarded as an overestimate.

In conclusion, it is felt that the error in our estimate of higher order corrections is probably less than $.004\ m_\pi^{-1}$, and this can be reduced further with better π N data. Thus, if one takes a value of $-.010 \pm .003\ m_\pi^{-1}$ for $a_{\pi d}^{(o)}$ (equation (4))[15], and an off-shell correction to single scattering of $-.014\ m_\pi^{-1}$ (recall § 2), our result is $a_{\pi d} = -.042\ (\pm.005) + .0045\ (\pm.0005)i\ m_\pi^{-1}$. We emphasise that the major error in this value comes from the uncertainty in $a_{\pi d}^{(o)}$ (which is essentially $2b_o$)[27,42,45], which may lie outside the quoted errors.

A recent experimental determination of $a_{\pi d}$ by Bailey et. al. gave the result $\mathrm{Re}\ a_{\pi d} = -(.049\ {}^{+.023}_{-.016})\ m_\pi^{-1}$[46a]. While the errors are too large for definite conclusions, this does tend to support the need for the off-shell correction to b_o within a nucleus. (A rather attractive alternative method of obtaining $a_{\pi d}$ (rather than direct measurement) was recently suggested by Cheon and Egidy[46b]. The idea is to extrapolate to the A = 2,3 region using a semi-empirical formula which reproduces the (more easily) measured scattering lengths on nuclei with A > 4. Unfortunately, the extrapolation to $a_{\pi d}$ does not yet seem reliable. In particular the Im $a_{\pi d}$ derived this way ($+ .017\ m_\pi^{-1}$), is a factor of four larger than unitarity allows.)

4. π-d SCATTERING

There have been a number of recent reviews of π d elastic scattering below the (3,3) resonance[1-4]. No attempt will be made to re-review those earlier calculations here. Instead we ask what has been learned from them. Unfortunately the answer to this is very little! In part this is because the theoretical aims have been very limited. In addition, there has been a scarcity of good quality data.

The position taken here is that *now* the time is ripe for a drastic change in attitude. With the advent of good quality intense pion beams, it should be possible to extract data with errors of a few percent. There is no apparent reason why a similar improvement can not be made on the theoretical side. (We have already seen how one can systematically calculate higher order corrections to $a_{\pi d}$, for example.)

In this section we shall describe some more of the very interesting physics one can study in the π d system. Naturally this will be intimately related to the discussion of those difficulties which must be overcome. For illustration of the points made, we will use the recent 48 MeV π^+ d data of Axen et. al.[47]. (There has been a considerable theoretical effort at U.B.C. aimed at understanding this experiment[48-50].)

4.1 Some Theoretical Aims.

One of the most fascinating (and basic) questions to be raised in the last year or so, is whether the π N interaction changes in the presence of other nucleons[32,33]. This may sound familiar to nuclear physicists who use a G-matrix (including a Pauli operator "Q"), rather than the free N-N t-matrix. However, there is a more fundamental change possible in the π N system, where the basic π N "potential" can change drastically. For example, the nucleon in the intermediate state of Fig. 3(a) (the "potential" for (3,3) scattering in a sense) will be strongly blocked within nuclear matter. Note that one group has gone so far as to suggest Chew-Low theory may not be adequate for π N scattering because Pauli blocking would so change the effective value of the isoscalar scattering volume "c_0" (equations (1), (6)) within a nucleus, that it would lie outside the empirically determined range[33]. They suggest instead that one needs an elementary Δ. Whatever one may think of their evidence, the question raised is of fundamental significance in pion physics.

At first this topic may seem far removed from π d scattering - a system less like nuclear matter would be hard to imagine! In particular, the P_{33} interaction should not be significantly affected. Consider instead the case where (as in § 3.3) the pion scatters from one nucleon through the P_{11} interaction. Further suppose that the pion is in p-wave with respect to the deuteron, and that the total angular momentum (J) is 1^+. Then in order to conserve parity while the P_{11} interaction occurs, the spectator nucleon must have even angular momentum (at 50 MeV probably L = 0). To conserve J the channel spin is S = 1, and of course the total isospin is I = 1. (This is illustrated in Fig. 4.)

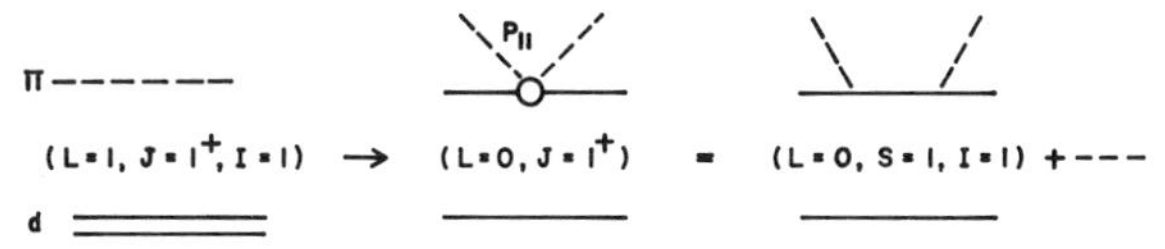

Figure 4: Pauli effects in π d scattering. We show how inclusion of the P_{11} interaction in single scattering can lead to forbidden states.

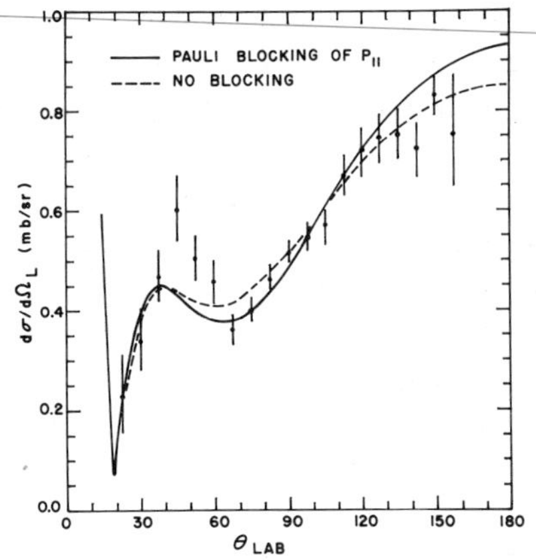

Figure 5: The effect of Pauli blocking of the P_{11} interaction in the (L = 1, J = 1^+) channel (48 MeV π^+d scattering - data from Axen et. al.[47]).

Now let us look inside the P_{11} interaction vertex (c.f. Fig. 3(b)). In Chew-Low theory a major contribution comes from the pole diagram. However this would correspond to having two nucleons in a forbidden state (L = 0, S = 1, I = 1). Figure 5 shows the results of a full Faddeev calculation of 48 MeV π^+ d scattering, with and without a P_{11} interaction in the (L = 1, J = 1^+) channel[50]. (While the P_{11} interaction used here does not have a bound state pole - it is in fact repulsive - it does reproduce the low energy phase shifts and scattering volume.) Although by turning off the whole P_{11} interaction we have overestimated this effect, there is clearly a significant difference between the curves. A good experiment could throw some light on this very basic question. (Note that in the Faddeev model there is no extra effort in dealing with partial wave amplitudes. For this 48 MeV calculation one typically solves sets of 8 or 9 coupled integral equations for each of 7 partial wave amplitudes viz: L = 0, J = 1^-; L = 1, J = 0^+, 1^+, 2^+; etc.)

A less fundamental, but nevertheless important reason for studying the π d system, is that one can test various approximations against the Faddeev results. Questions such as whether one can accurately approximate the correct off-energy shell behaviour by evaluating the π N t-matrix at some predetermined energy (recall § 2), or whether the form-factor approximation works, are not just relevant to the π d system. As an example, we show the 48 MeV π^+ d cross section for two different d-state probabilities (Phillips 4% and 7%[30] - see Fig. 6). Observe that the qualitative behaviour is rather different from what one might expect from a simple impulse approximation[49,51]. Clearly such questions must be answered before one can learn about (say) the small components of the ^{3}He wave function by pion scattering[49].

Finally we mention the old problem of multiple scattering corrections. A very common way of calculating a double scattering correction is to put the intermediate pion on the energy shell[2,3,51]. Now this approximation has been taken further by Sloan[52]. In the case of n - d scattering it was shown that unitarising the partial wave Born amplitudes ($T_\ell^{(B)}$), can improve the angular distribution (essentially one

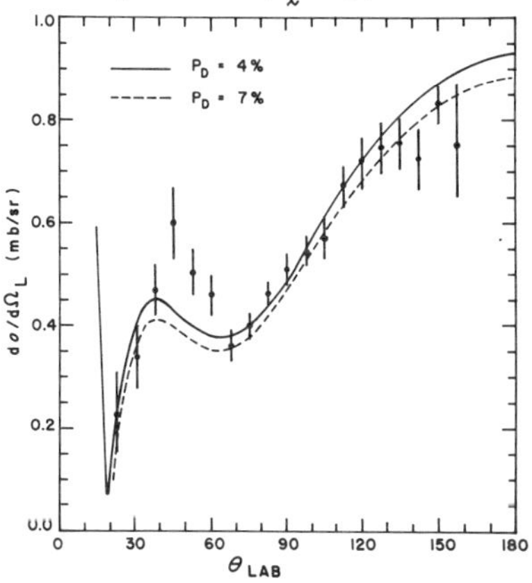

Figure 6: Dependence of the 48 MeV π^+ d differential cross section on deuteron d-state probability (P_D) - in a Faddeev calculation[50].

forms $T_\ell^{(U)} = T_\ell^{(B)}/[1 - i\,T_\ell^{(B)}]$). Our main point, however, is that this procedure is equivalent to summing the whole rescattering series, keeping the intermediate states on-shell[52,53]. Table III shows the Faddeev multiple scattering series for

Table III

Multiple scattering series for 48 MeV $\pi^+ d$ scattering in the ($L = 1$, $J = 2^+$) channel (Faddeev calculation, $P_D = 7\%$[45], where $f = e^{i\delta}\sin\delta/k$.)

Iteration	$f_{1,2^+}$		
1	.1699	+	.0099i
2	.1749	+	.0127i
3	.1792	+	.0190i
4	.1802	+	.0198i
∞	.1810	+	.0210i
Unitarised Born	.167	+	.026i

48 MeV $\pi^+ d$ scattering. (Only the ($L = 1$, $J = 2^+$) amplitude is shown because this dominates - most others (except ($L = 0$, $J = 1^-$)) converge in the first iteration.) Clearly this unitarisation procedure (and presumably the on-shell double scattering correction) does not work well here. This is probably because, unlike low energy n - d scattering, gross violation of unitarity in the multiple scattering series is not the problem here.

4.2 Problems.

We have seen that there are some very interesting and fundamental questions one can study in the π d system. However, one major stumbling block is shown in Fig. 7[49], where we show the 48 MeV $\pi^+ d$ differential cross section for two sets of currently acceptable π N phase shifts[54]. While the sensitivity to the input phase shifts may be overestimated by an impulse approximation, and while the variation was somewhat reduced at higher energy[49], it is clearly impossible to extract meaningful information from a π d scattering experiment until there is a drastic improvement in the basic π N data.

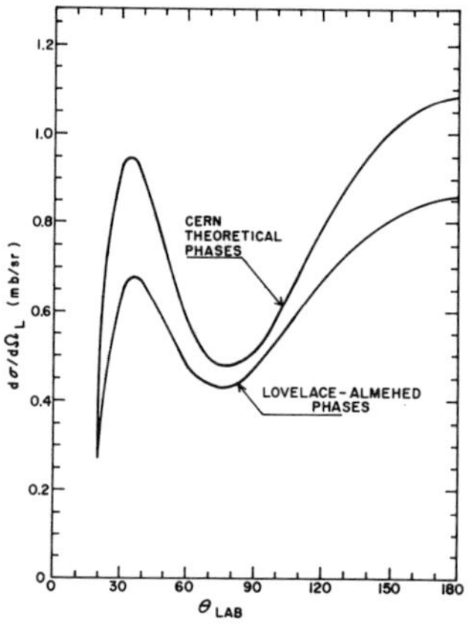

Figure 7: Dependence of the 48 MeV $\pi^+ d$ cross section on the π N phase shifts[49].

Therefore it is essential that there should be a program of careful π N experiments, either before, or in conjunction with, any program of π d experiments. The most interesting energy range for π d elastic scattering is from 50 MeV up to the (3,3) resonance region. Because of the smaller number of partial wave amplitudes which must be computed, the lower end of this range is preferable from the Faddeev viewpoint. As we saw in § 2, the most important energy range for the π N t-matrix in single scattering approximation, is from (10 - 20) MeV below the incident energy. Consequently, it would be useful if the energy range of these π N experiments extended (10 - 20) MeV below the lowest energy π d experiment. Furthermore, it is very important that one should be able to extract the small p-wave phase shifts (P_{11}, P_{13}, P_{31}) from this data - since these produce significant interference effects in the π d system. (The need to extract these phase shifts - should greatly influence the angles at which the π N data is taken.)

There are two other technical problems which must be tackled. First, in this relatively low energy region, coulomb effects must be handled carefully. Indeed, if the π N phase shifts (due to purely strong interactions) are to be accurately determined, some improved method of unravelling coulomb interference effects is needed[55). Furthermore, the accuracy of adding in coulomb interference as a small perturbation to the strong interaction in π-nucleus scattering, is of some concern. While we do not expect such corrections to be large in the case of the deuteron, this whole area deserves careful consideration in the near future[56).

The second problem is that we have so far ignored relativistic aspects of this problem - even at 50 MeV a pion is significantly relativistic. There have been several attempts to define a unique Lorentz transformation to the π N centre of mass, but this is only possible if the particles are taken off their mass shells. An excellent presentation of this problem has been given by C. Schmit et. al.[57). However the solution they suggest is at best a temporary one. Perhaps the direction to be taken is that suggested by L. Celenza et. al.[58), or by Aaron et. al.[59). Certainly this is an important area for future theoretical work.

5. CONCLUSION

In the two years since Eisenberg's talk at the last Few Body Conference, several of the ideas discussed there have born fruit. The application of Faddeev theory to the π d scattering length has revealed the importance of correct off-energy shell behaviour - resulting in a change of the π N isoscalar scattering length inside a nucleus. Further experimental work on an accurate determination of both the π d and $\bar{K}$ d scattering lengths would be very interesting.

A detailed study of positive energy π d scattering within Faddeev theory has just begun. In the next few years we can expect a considerable improvement in our understanding of this system. One expects that some of the benefits should also carry over to the theory of pion scattering from heavier nuclei. In addition, we have seen that good quality π d scattering experiments should shed some light on the nature of the basic π N interaction.

This optimism does not mean we are without problems. A correct treatment of coulomb effects is an important priority. More attention will also need to be given to relativistic effects. On the experimental side, it is essential that both good quality π N and π d data be available. These problems not withstanding, the detailed study of π d scattering in the next few years will be very rewarding.

Acknowledgements

It is a pleasure to acknowledge the support of the physics departments of both the University of British Columbia, and Flinders University, where this work has been carried out. I am particularly indepted to R.H. Landau and I.R. Afnan for many discussions on the subject matter discussed here.

References

1. D.S. Koltun, Adv. Nuc. Phys., 3 (1969) 71.
2. J.M. Eisenberg, P. 796 "Proc. Int. Conf. on Few Particle Problems in the Nuclear Interaction "ed. I. Slaus et. al. (North Holland, 1973).
3. C. Wilkin, P. 289 "Spring School on Pion Interactions.." CERN 71 - 14.
4. M.P. Locher, P. 155 "Spring School on Pion Interactions.." CERN 71 - 14.
5. L.D. Faddeev, Sov. Phys. J.E.T.P. 12 (1961) 1014.
6. C. Lovelace, Phys. Rev. 135 (1964) B 1225.
7. T.E.O. Ericson, CERN preprint TH-1093 (1969).
8. L.S. Kisslinger, Phys. Rev. 98 (1955) 761.
9. M. Ericson, T.E.O. Ericson, Ann. Phys. 36 (1966) 323.
10. M.P. Keating, J.G. Wills, Phys. Rev. C7 (1973) 1336.
11. R.H. Landau, F. Tabakin, Phys. Rev. D5 (1972) 2746.
12. M. Goldberger, K. Watson, "Collision Theory" Wiley, New York (1964).
13. S. Weinberg, Phys. Rev. Lett. 17 (1966) 616.
14. V.K. Samaranayake, W.S. Woolcock, Nucl. Phys. B48 (1972) 205.
15. D.V. Bugg, A.A. Carter, J.R. Carter, Phys. Lett. 44B (1973) 278.
16. I.R. Afnan, A.W. Thomas, Phys. Rev. C10 (1974) 109.
17. F. Myhrer, R.R. Silbar, "Unitarity Corrections to Pion-Nucleus Scattering Lengths". SIN Preprint (1974) - and this conference.
18. A.W. Thomas, to be published.
19. M. Krell, T.E.O. Ericson, Nucl. Phys. B11 (1969) 521.
20. R.H. Landau, A.W. Thomas, private communication.
21. W.A. Bardeen, E.W. Torigoe, Phys. Lett. 38B (1972) 135.
22. F. Myhrer, Phys. Lett. 45B (1973) 96.
23. R.H. Landau, S.C. Phatak, F. Tabakin, Ann. Phys. 78 (1973) 299.
24. V.M. Kolybasov, A.E. Kudryavtsev, Nucl. Phys. B41 (1972) 510.
25. P. Bendix, D. Beder, Phys. Lett. 49B (1974) 140, and private communication from D. Beder.
26. K.A. Brueckner, Phys. Rev. 89 (1953) 934.
27. F. Myhrer, "Pion-Deuteron Scattering Lengths via Faddeev", SIN Preprint (1974).
28. N.M. Petrov, V.V. Peresypkin, Phys. Lett. 44B (1973) 321.
29. Y. Yamaguchi, Y. Yamaguchi, Phys. Rev. 95 (1954) 1628.
30. A.C. Phillips, Nucl. Phys. A107 (1968) 209.
31. G.F. Chew, F.E. Low, Phys. Rev. 101 (1956) 1570.
32. C.B. Dover, D.J. Ernst, R.M. Thaler, Phys. Rev. Lett. 32 (1974) 557.
33. S. Barshay, G.E. Brown, M. Rho, Phys. Rev. Lett. 32 (1974) 787.
34. T.E.O. Ericson, Phys. Lett. 2 (1962) 278. A summary of the many developments

since 1962 may be found in Ref. 1) and in: J.W. Morris Jr., H.J. Weber, Ann. Phys. 79 (1973) 34.

35. D. Lichtenburg, Phys.Rev. 105 (1957) 1084. A.E. Woodruff, Phys. Rev. 117 (1960) 1113. D.S. Koltun, A. Reitan, Phys. Rev. 141 (1965) 1413. C. Lazard, J.L. Ballot, F. Becker, Il.N.C. 65 (1970) 117. I.R. Afnan, A.W. Thomas, Ref. 12).
36. A.W. Thomas, I.R. Afnan, Phys. Rev. Lett. 26 (1971) 906.
37. K. Brueckner, Phys. Rev. 98 (1955) 769.
38. (a) D.J. Thouless, Proc. Phys. Soc. 69 (1956) 280.
(b) K. Nishimoto, H. Ohtsubo, H. Narumi, Prog. Th. Phys. 46 (1971) 135.
39. G.C. Wick, Rev. Mod. Phys. 27 (1955) 339. see also G.F. Chew, F.E. Low, Phys. Rev. 101 (1956) 1570.
40. F. Zachariasen, Phys. Rev. 101 (1956) 371.
41. A.W. Thomas, Ch. 3 Ph.D. Thesis (Flinders University, 1973) and to be published.
42. A.W. Thomas, I.R. Afnan, Phys. Lett. 45B (1973) 437.
43. V.M. Kolybasov, private communication to D.V. Bugg. Apparently this effect was first suggested by T.E.O. Ericson and M. Ericson.
44. R.V. Reid, Ann. Phys. 50 (1968) 411.
45. V.V. Peresypkin, N.M. Petrov, Nucl. Phys. A220 (1974) 277.
46. (a) J. Bailey et. al., "The π d scattering length from Kα X-rays". Preprint, 1973.
(b) "Empirical Formula for Energy Level Shifts..." Il-Tong Cheon, T. von Egidy, Preprint (1974) and contribution to this conference.
47. D. Axen et. al., "The Interaction of 48 MeV π^+ with Deuterons", contributed to this conference.
48. D. Beder, Nucl. Phys. B34 (1971) 189, and private communication.
49. J.M. McMillan, R.H. Landau, TRIUMF report (1974).
50. A.W. Thomas, private communication. (Note that this calculation uses relativistic kinematics for the pion as described in § 2 and § 3.)
51. C. Carlson, Phys. Rev. C2 (1970) 1224.
52. I.H. Sloan, Phys. Rev. 185 (1969) 1361, 162 (1967) 855.
53. N.M. Queen, Nucl. Phys. 80 (1966) 593.
54. D.J. Herndon, A. Barbaro-Galtieri, A.H. Rosenfeld, UCRL-20030 π N (1970).
S. Almehed, C. Lovelace, Nucl. Phys. B40 (1972) 157.
55. D.V. Bugg, P. 379 "Spring School on Pion Interactions.." CERN 71 - 14 (1971).
56. C.M. Vincent, S.C. Phatak, Phys. Rev. C10 (1974) 391.
57. C. Schmit, J. Dedonder, P. Maillet, preprint IPNO/TH 74-17 (1974).
58. L.S. Celenza, L.C. Liu, C.M. Shakin, contribution to this conference, and references therein.
59. R. Aaron, R.D. Amado, J.E. Young, Phys. Rev. 174 (1968) 2022.

DISCUSSION

BUGG(Q.M.C.,London): One problem can be resolved immediatly. Many of the discrepancies between S-waves in different phase shift analyses depend purely on how the data are handled. Some analyses include $\pi^- p \to \gamma n$ which gives an anomalous threshold; some include the mass difference between $\pi^- p$ and $\pi^0 n$. Others don't. I believe that when these effects are included, the uncertainties in the long energy S-waves are much smaller that your last slide indicated.

AARON (Northwestern): This question is directed to Dr. Brayshaw. Could you please show your diagram again. Now, at first glance your phase shift (δ) and inelasticity parameter (η) seem to correspond to an inelastic Breit-Wigner term with the ratio elastic-width/total-width $\simeq$ 1/4. The fact that the phase shift peaks at about 20° is exactly what one would expect under the previously mentioned conditions. The only way to answer the question is to determine whether or not there is a nearby second sheet pole. Phase shift analyses usually try to answer this question by plotting the 'speed' and looking for a peak. Since you have an analytic theory, however, you can answer the question directly. Is there a second sheet pole ?

BRAYSHAW(Maryland): Yes, I did as a matter of fact. You can check the three-body calculations (which is one of its virtues) to see if there is any singularity there.

AARON: Well now, let's ask Dr. Thomas the same question. Have you looked at the second sheet ?

THOMAS: I can, but I have not done this yet. Let me emphasize that this is a very fundamental problem. In particular, I refer to Dr. Cahill's comment (after Dr. Bugg's talk) that one can expect different analytic structure in a three-body system. In a sense, the number of particles present really depends on how far you look inside a particle. For example, we have heard from Dr. Amado that one can treat πN scattering as a three-body reaction (where the nucleon is a πN), provided that we are below the 2π production threshold. How many particles one considers, really just depends on the energy range being considered (i.e. how hard you look). It would be a very interesting exercice to study the analytic structure of the resonant πN that have been found by Aaron and Amado.

STRENGTHS OF THE MESON FACTORIES IN FEW-PARTICLE PHYSICS

J. Reginald Richardson*, TRIUMF, U.B.C., Vancouver, Canada

For the purpose of this session of the Conference we will consider meson factories to be those facilities which produce intense beams of pi mesons, where these mesons originate from reactions produced by proton beams having an intensity of 50 μa or greater and energy above 400 MeV. Two meson factories have started operation and are now in the process of building up intensity: LAMPF(1) at Los Alamos, New Mexico, USA and SIN(2) at Zurich, Switzerland. Three others will soon be in operation: TRIUMF(3) at Vancouver, Canada, the synchrocyclotron conversion at Nevis(4), New York, USA and the synchrocyclotron conversion at JINR(5), Dubna, USSR. One or two additional meson factories are under design.

Since the accelerators associated with the present facilities are all different, they each have unique capabilities which turn out to be sources of strength in particular areas of research. Thus we can expect to find eventually that most of the meson factories will fulfill complementary roles in research.

Table 1 shows the important characteristics of the five meson factories, as defined here. (Sometimes the synchrocyclotron conversions are called mini-meson factories.) The energy of SIN, Nevis and Dubna are essentially fixed at the values shown, while the TRIUMF beam energy can be varied continuously within the limits given in the table. The LAMPF energy can be varied, but with greater difficulty and in discrete jumps. The maximum design output of LAMPF (1000 μa at 800 MeV) is far and away the largest for these facilities. On the other hand, LAMPF has by far the lowest macro duty factor (6%) for the accelerators discussed here.

Table 1

Facility	Range of Proton Energy	Maximum Planned Beam Intensity	Macro Duty Factor	Polarized Beam
	MeV	μa	%	na
SIN	590	100	100	300 (60%)
TRIUMF	180 - 520	100 - 200	100	60 (85%)
LAMPF	600 - 800	100 - 1000	6	15 (85%)
Nevis	560	10 - 40	60	
Dubna	700	25 - 50	50?	

It is beyond the scope of an introduction to describe all the secondary beam lines and experimental facilities under construction at the various meson factories. These will be brought out in succeeding talks. However, the characteristics of the accelerators themselves enable one to point out some of the areas of research strength of the different facilities, particularly as they relate to research in few-particle physics.

* on leave from UCLA

Areas of Strength

Neutrino experiments: LAMPF would have a clear-cut advantage here because of its high output and because of its natural time structure.

Neutron-scattering and neutron-induced reactions: Above 600 MeV, LAMPF would be the choice, but between 200 and 500 MeV, TRIUMF would be in the best position because of its easily variable energy. Between 500 and 570 MeV, SIN will have an excellent source of polarized neutrons.

Coincidence experiments in general: All the cyclotrons have a clear advantage here over LAMPF because of the macro duty factor.

Pion experiments: For experiments with pions above 300 MeV, LAMPF would obviously take the lead because of energy and intensity, while for experiments involving variable energy π^+ beams below 250 MeV, TRIUMF would be the choice, making use of the two-body reaction $p+p\rightarrow\pi^+ +d$.

π^- experiments: It should be pointed out that the synchrocyclotron conversions do have the advantage of multiple traversal with an internal target in the production of negative pions. However, the intensity is limited by the induced activity which can be tolerated in the cyclotron itself.

Stopping π and μ experiments: All of the meson factories will have good capabilities in this field and the relative strengths will depend in a major way on the quality of the secondary beam lines and associated experimental facilities. For example, it now appears that SIN, with its superconducting line, will have a major strength in experiments with stopping muons.

Proton induced reactions in few-particle physics: Because of their good duty factors, the cyclotrons will have an advantage over LAMPF in this field of research, except where the high LAMPF energy is essential. The expected good energy resolution and variable energy of TRIUMF will be particularly useful below 520 MeV.

Let us turn now to the consideration of the research programs of the individual meson factories, as they involve few-particle physics.

References

(1) D.C. Hagerman, IEEE Trans. Nuc. Sci. NS-20, 905 (1973).
(2) H.A. Willax, IEEE Trans. Nuc. Sci. NS-20, 202 (1973).
(3) J.R. Richardson, IEEE Trans. Nuc. Sci. NS-20, 207 (1973).
(4) J. Rainwater, AIP Conf. Proc. No. 9, 165 (1972).
(5) Y.N. Denisov, AIP Conf. Proc. No. 9, 179 (1972).

STUDIES OF FEW-PARTICLE PROBLEMS AT TRIUMF[†]

W.T.H. van Oers
Cyclotron Laboratory, Department of Physics
University of Manitoba
Winnipeg, Canada

In the short time available I would like to discuss some of the more salient aspects of those experiments planned at TRIUMF, that could be considered to lie within the scope of the present conference. Let me state that in the time available I cannot be complete and that the choice of topics is entirely a personal one. Consequently, there are many regrettable omissions.

The first experiment I would like to mention is a study of positive pion production in proton-proton and also proton-nucleus interactions, i.e., and investigation of the reactions $p + p \rightarrow d + \pi^+$, $p + p \rightarrow p + n + \pi^+$, and $p + {}^N_Z A \rightarrow {}^{N+1}{}_Z(A+1) + \pi^+$.[1] It is intended to study these reactions from threshold (∿290 MeV for hydrogen) up to 400 MeV. The interest in this study comes from recent theoretical calculations which have focussed on the use of the soft pion approximation.[2] Nambu and Lurié have pointed out that the partially conserved axial vector current (PCAC) hypothesis leads to a simple formula relating the matrix element for emission of a pion close to threshold (soft pion) to the matrix element for elastic scattering of the two particles in the initial state. Another theoretical approach valid at higher energies which is being used, is based on the isobar model.[4] At present there exists very little experimental information on the behavior of the total cross section for pion production in proton-proton interactions near threshold. There is also a need for information on the details of the angular distributions. Regarding pion production in proton-nucleus interactions, further experimental studies are required as has been recently emphasized at the Fifth International Conference on High-Energy Physics and Nuclear Structure.[5]

The original aim in studying pion production was to give information about the off-energy-shell behavior of the nucleon-nucleon interaction. As stated above most theoretical calculations so far have instead been concerned with explaining the pion production process. In this respect, a phase shift analysis of nucleon-nucleon scattering data together with pion production data, to give physical constraints to the inelastic parameters, should be considered a promising first step.

Information on the off-energy-shell nucleon-nucleon interaction can in principle be extracted more reliably from studies of nucleon-nucleon Bremsstrahlung. A proposal to measure proton-proton Bremsstrahlung ($p + p \rightarrow p + p + \gamma$) at two or three energies in the range 150-300 MeV has recently been submitted.[6] The experiment will look in particular at situations in which both final-state protons are emitted at extreme forward angles, i.e., with the sum of the two polar angles ($\theta_3 + \theta_4$) smaller than 30°. Present calculations of Heller [7] indicate strong dependence on the nu-

[†]Work supported in part by the Atomic Energy Control Board of Canada.

cleon-nucleon potential used in the calculations of the Bremsstrahlung cross section only for angles where one is far off-energy-shell, i.e., for incident proton energies above 100 MeV and for small polar angles of the two emitted protons. Although Heller's calculations show that the existing ppγ data cannot be used to differentiate between phase-shift-equivalent nucleon-nucleon potentials, they clearly indicate regions of phase space in which the calculated cross sections are sensitive to the off-energy-shell characteristics of the nucleon-nucleon potential. Figure 1 shows Heller's ppγ

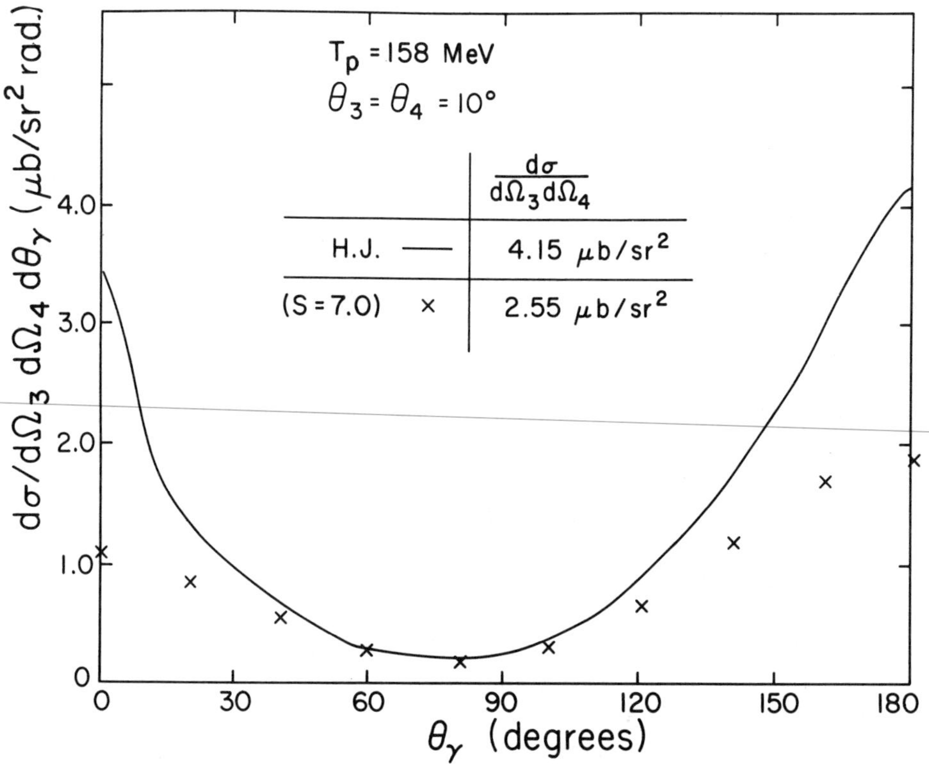

Fig. 1. Proton-proton Bremsstrahlung calculations at 156 MeV and $\theta_3 = \theta_4 = 10$ (ref. 7)

calculation at a bombarding energy of 156 MeV for symmetric coplanar scattering ($\theta_3 = \theta_4$, $\phi_3 - \phi_4 = 180^0$) $\theta = 10^0$. The solid curve is the predicted ppγ differential cross section for the Hamada-Johnston (HJ) potential, while the x's are the predicted cross sections for a potential which is phase-shift-equivalent to the Hamada-Johnston potential and which is constructed through a Baker's transformation [8] maximizing off-energy-shell differences. It should be noted that in the region of phase space corresponding to Fig. 1 off-energy-shell effects could vary the ppγ cross section by as much as a factor of two.

Next I would like to mention a series of experiments to measure neutron-proton observables in the energy region 200-500 MeV.[9] Measurements will be made of the differential cross sections, polarizations, and triple scattering parameters in n-p scattering at closely spaced energy intervals (50 MeV or so). Whereas the T = 1 phase shifts are fairly well determined up to 450 MeV (at which energy the inelastic parameters become important due to pion production), the T = 0 phase shifts are probably not unique and require considerable attention.[10] Assuming charge independence, simple formulae relate immeasurable T = 0 amplitudes to directly measured n-p and p-p quantities. A particularly interesting region in the n-p angular distribution is the one near 180^0 where there is a pronounced maximum in the slope parameter whose production mechanisms is not fully understood at present.[11] There is a possibility of considerable structure at intermediate energies and thus further experimental studies are needed.

Another study of interest is the scattering of medium-energy protons (or pions) off nuclei which should in principle be explainable by folding the nucleon (pion) - nucleon interaction into the matter distribution of the nucleus. A program of investigating the scattering of polarised and unpolarised protons from the very light nuclei ^2H, ^3H, ^3He, and ^4He has been proposed.[12] Here I would like to discuss only the present status of p-d and p-^4He scattering:

<u>p-d</u>

Without disregarding the importance of the forward part of the angular distributions I prefer to comment only on the backward scattering of protons by deuterons. In particular the anomalous energy behavior of the backward peak in the differential cross sections and the appearance of a second maximum in the 180^0 excitation function has created a great deal of theoretical interest resulting in many controversies. Figure 2 shows the 180^0 excitation function in a full logarithmic representation.

In the analysis of the 1 GeV data Kerman and Kisslinger [13] found that the backward part of the angular distribution could not be explained by nucleon exchange alone even though taking into account both S- and D-state components of the deuteron ground state wave function. They deduced the importance of the N*(1688 MeV, T = 1/2, J = $5/2^+$) nucleon resonance and estimated that the deuteron contains a N-N* component with a 1/2-1% probability accounting for about half of the backward scattering. Craigie and Wilkin [14], and Barry [15] used a triangle diagram stressing the importance of an intermediate Δ (1236) resonance to describe the process $p + d \rightarrow d + p$. This particular model reproduces the energy variation of the data as well as the backward part of the angular distributions fairly well. Later Sharma <u>et al</u> [16] suggested that all T = 1/2 nucleon isobars in principle should be considered. However, it was recently pointed out by Noble and Weber [17] that if a relativistically correct expression for the momentum transfer variable is used in the nucleon exchange diagram, appreciable larger cross sections are obtained. This makes it unnecessary to introduce either a N*-isobar exchange or a pion-nucleon exchange process, besides the at

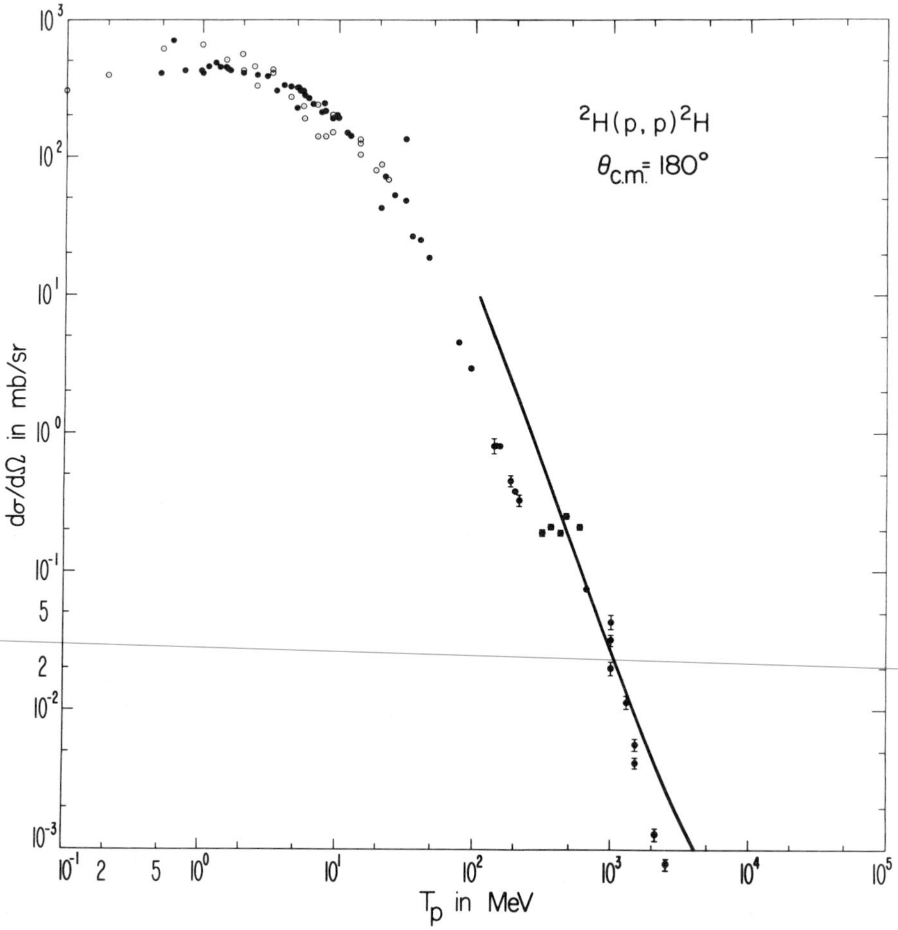

Fig. 2. Excitation function for p + d elastic scattering at 180^0 in the center of mass. The solid line corresponds to a prediction by Noble and Weber (ref. 17).

low-energy very prominent neutron transfer process. The solid line in Fig. 2 represents the excitation function calculated by Noble and Weber using Hulthen S- and D-state wavefunctions for the deuteron (the D-state admixture was taken to be 8%). It fails to reproduce the second maximum in the experimental excitation function.

In addition to the presence of a N-N* component in the deuteron and the importance of intermediate Δ resonances in order to explain backward scattering, a Δ-Δ component of the deuteron wave function has been hypothesized.[18] Isospin conservation requires an even number of Δ's. Thus the deuteron can consist of a Δ^{++} - Δ^- pair or a Δ^+ - Δ^0 pair with equal probability. Assuming Δ's are produced as spectator particles in collisions of high-energy projectiles with deuterons, the question arises whether one can distinguish between Δ's created in the collision and Δ's which were present in the deuteron at the time of the interaction. Two searches for a Δ-Δ component of the deuteron were reported on recently giving vastly different upper limits to the probability for the presence of such a component. (2.6% and 0.7%, respectively).[19] In summary, one can conclude that no quantitative explanation of the anomalous behavior of p-d backward scattering has been given. It should be noted that the 180^{0} excitation function is quite similar to the one for π-d elastic scattering at 180^{0}. Again the calculations of pion-deuteron elastic scattering in the forward direction could be understood in terms of two particle exchange mechanism involving conventional (dNN) vertices, but the backward scattering correction is found to depend on high momentum components of the deuteron wave functions which are scarcely available in the dNN vertex. Thus, to understand pion-deuteron elastic backward scattering, one needs to consider isobar Δ (1236) exchanges in the intermediate states, such that the initial or final deuteron is in a Δ-Δ configuration. Present calculations based on such a model produce qualitative agreement with the observed structure of the excitation function, and indicate a negligible presence of intermediate N*(1520) and N*(1535) states. The question of the possible existence of a N-N* component or a Δ-Δ component in the deuteron is still unresolved.

p-^{4}He

The current experimental situation regarding the differential cross sections is depicted in Fig. 3. Three observations can be made. First of all in the energy range of 100-1000 MeV the coulomb-nuclear interference changes from being destructive to constructive. Secondly there is, although less pronounced than in case of the deuteron, backward peaking of the angular distributions. Backward peaking should receive considerable experimental and theoretical attention in light of what discussed above about p-d backward scattering. Finally, the more recent 1.05 GeV p + ^{4}He elastic scattering data from CEN-Saclay [20] show a much less pronounced first minimum than observed earlier at Brookhaven National Laboratory [21] (See Fig. 4). It appears that the Saclay results would require the inclusion of nucleon-nucleon correlations in the nuclear wave function.

The p-^{4}He results have been analyzed using Glauber theory [22], using an optical potential [23,24] and using forward dispersion relations.[25] It should be noted that to explain p + ^{4}He scattering for the first time it appears necessary to include a pair correlation term. One can only speculate on the situation for ^{3}He because of the lack of experimental data. Spin effects have been taken into account only by Lambert and Feshbach [23] and by Lykasov and Tarasov [26],

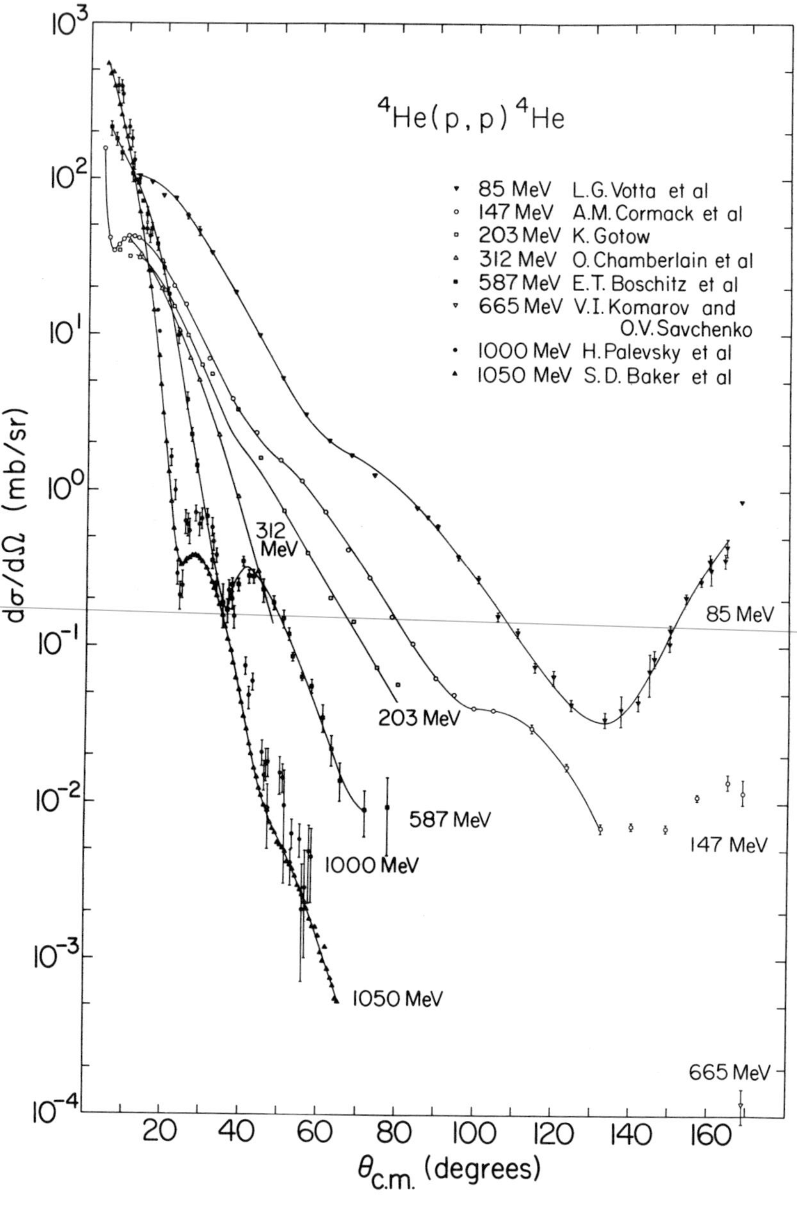

Fig. 3. Differential cross section angular distributions for p + ^4He elastic scattering. The solid lines are to guide the eye and have no significance.

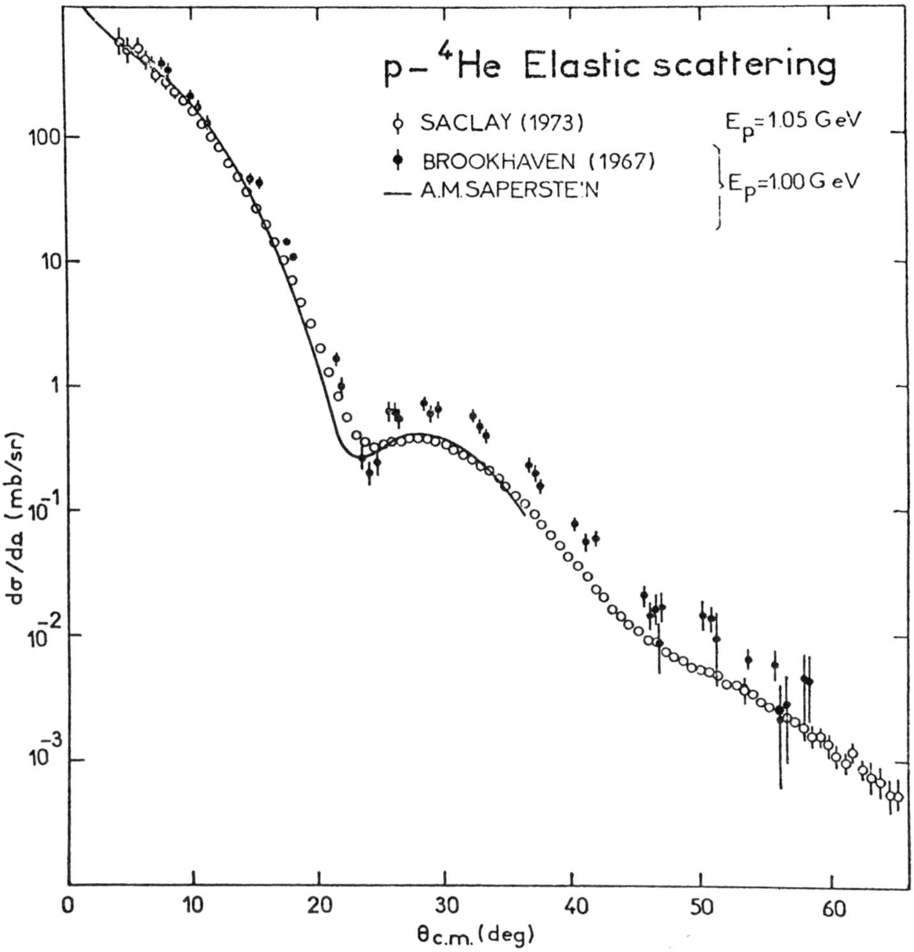

Fig. 4. Differential cross section angular distributions for p + ^{4}He elastic scattering at 1.00 and 1.05 GeV. The solid line corresponds to a prediction by Saperstein (ref. 24).

but no detailed polarization predictions have yet been made. It is apparent from the polarization contour diagram shown in Fig. 5 that appreciable polarizations result from p-^{4}He elastic scattering.

Finally, some comments about (p,2p) quasi-free scattering experiments on ^{2}H, ^{3}H, ^{3}He, and ^{4}He.[27] The aim in performing these type of experiments is to obtain information about the wave functions of the very light nuclei in the form of momentum distributions. The data might also give information

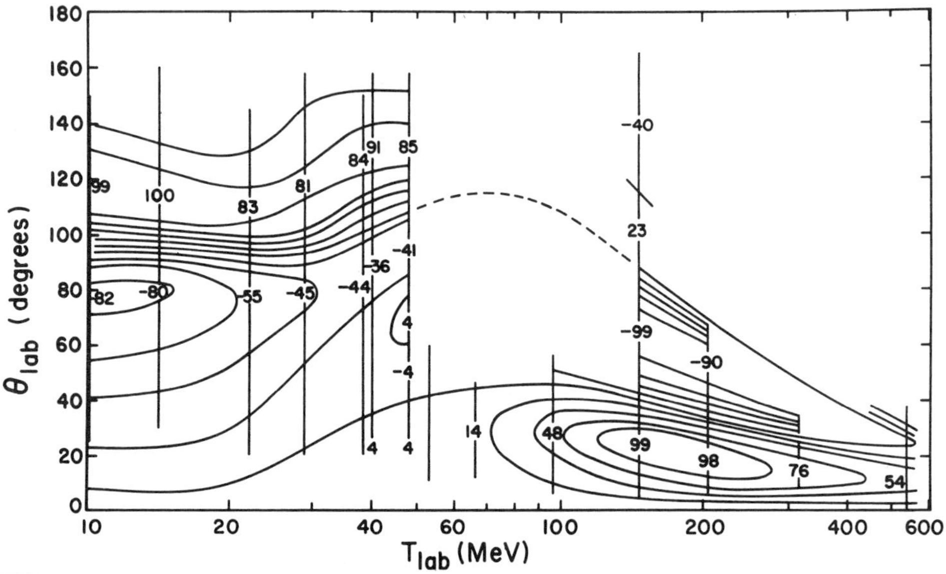

Fig. 5. Polarization contour diagram for p + He elastic scattering between 10 and 600 MeV

about the off-energy-shell scattering amplitude of the incident and struck particles. Quasi-free scattering data are usually analyzed in the framework of the plane-wave or distorted wave impulse approximations. The result is the square of the momentum wave function for the relative motion of the struck proton and the spectator particle which may also be a system of nucleons unbound in the final state. An example is given in Fig. 6. The data points are the extracted momentum distribution for the $^4He(p,2p)^3H$ reaction at an incident proton energy of 600 MeV from an experiment by Perdrisat et al.[28] The curves are theoretical momentum distributions calculated by Roos [29] using a 1S oscillator wave function with plane waves, a single particle wave function of Eckart form with plane waves, and the same wave function with distorted waves, respectively. There is an indication that the momentum distribution extracted from experiment is wider than the theoretical momentum distribution for large momenta.

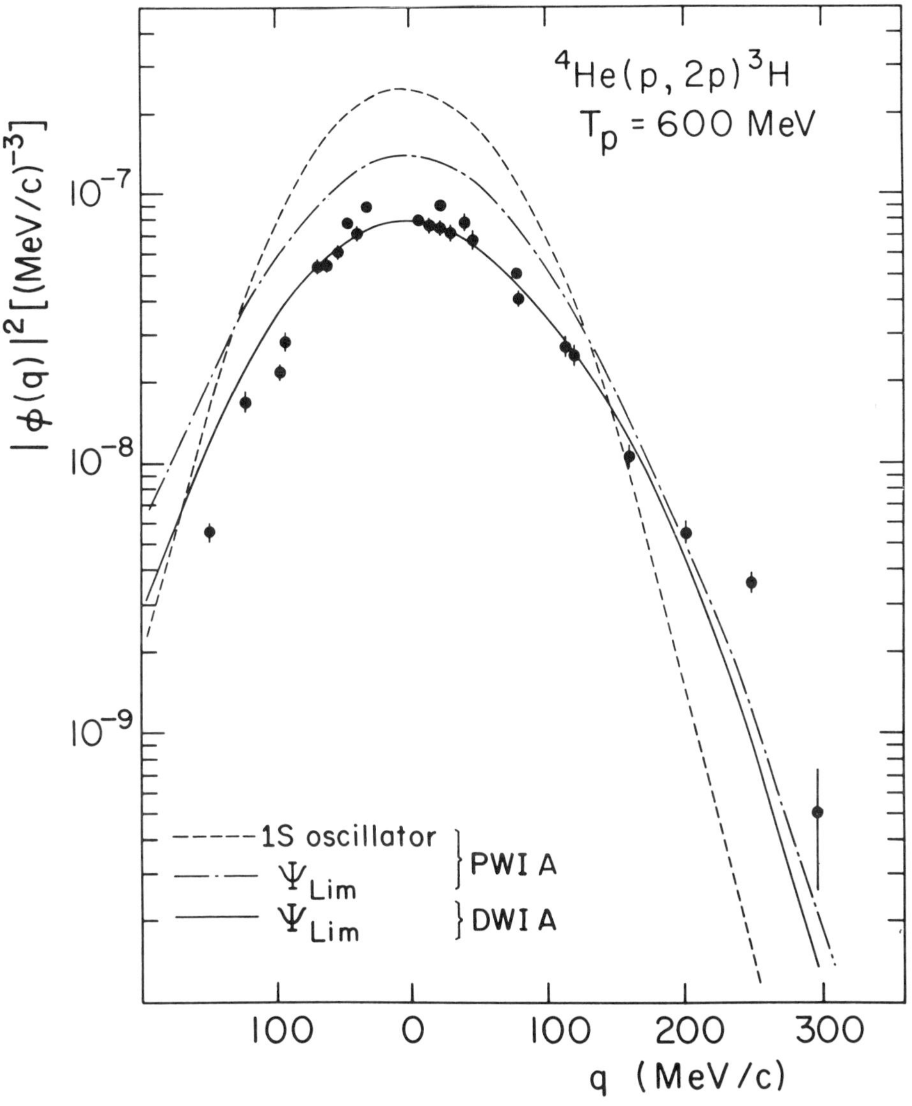

Fig. 6. Momentum distributions for the $^4He(p,2p)^3H$ reaction at 600 MeV

The author would like to thank Dr. B. S. Bhakar for stimulating discussions and valuable comments.

References

[1] G. Jones, D. Axen, R. Johnson, J. Warren, M. Salomon, L. Robertson, P. Kitching, and W. Olsen, Proposal #10 to the TRIUMF Experiments Evaluation Committee, unpublished.

[2] M. E. Schillaci, R. R. Silbar, and J. E. Young, Phys. Rev. Letters 21, 711 (1968); Phys. Rev. 179, 1539 (1969).

[3] Y. Nambu and D. Lurié, Phys. Rev. 125, 1429 (1962).

[4] D. Drechsel and H. J. Weber, Nucl. Phys. B25, 159 (1971).

[5] "High-Energy Physics and Nuclear Stucture", ed, by G. Tibell (North Holland Publishing Company, Amsterdam, 1974).

[6] J. G. Rogers, J. M. Cameron, J. V. Jovanovich, and A. W. Stetz, Proposal #66 to the TRIUMF Experiments Evaluation Committee, unpublished.

[7] L. Heller, Notes of the TRIUMF 1973 Summer School, unpublished.

[8] G. A. Baker, Jr., Phys. Rev. 128, 1485 (1962).

[9] E. G. Auld, D. F. Measday, L. P. Robertson, and G. Roy, Proposal #26 to the TRIUMF Experiments Evaluation Committee, unpublished; E. G. Auld, J. M. Cameron, D. F. Measday, L. P. Robertson, and G. Roy, Proposal #27 to the TRIUMF Experiments Evaluation Committee, unpublished; D. Axen, I. M. Blair, D. V. Bugg, A. C. Clough, J. A. Edgington, and N. M. Stewart, Proposal #40 to the TRIUMF Experiments Evaluation Committee, unpublished.

[10] P. Signell and J. Holdeman, Phys. Rev. Letters 27, 1393 (1971).

[11] R. E. Mischke, P. F. Shepard, and T. J. Devlin, Phys. Rev. Letters 23, 542 (1969).

[12] B. S. Bhakar, C. A. Goulding, W.T.H. van Oers, J. M. Cameron, R. McCamis, G. A. Moss, J. G. Rogers, G. Roy and A. W. Stetz, Proposal #14 to the TRIUMF Experiments Evaluation Committee, unpublished.

[13] A. K. Kerman and L. S. Kisslinger, Phys. Rev. 180, 1483 (1969).

[14] N. S. Craigie and C. Wilkin, Nucl. Phys. B14, 477 (1969).

[15] G. W. Barry, Ann. Phys. (N.Y.) 73, 482 (1972); Phys. Rev. D7, 1441 (1973).

[16] J. S. Sharma, V. S. Bhasin and A. N. Mitra, Nucl. Phys. 35B, 466 (1971); J. S. Sharma and A. N. Mitra, Phys. Rev. D9, 2547 (1974).

[17] J. V. Noble and H. J. Weber, Phys. Letters B50, 233 (1974).

[18] N. R. Nath, P. K. Kabir, and H.J. Weber, private communication.

[19] H. Braun, D. Brick, A. Fridman, J. P. Gerber, P. Juillot, G. Maurer, M. E. Michalon-Mentzer, R. Strub, and C. Voltolini, Phys. Rev. Letters 33, 312 (1974); C. P. Horne et al, Phys. Rev. Letters 33, 380 (1974).

[20] S. D. Baker, R. Beurtey, A. Chaumeaux, J. M. Durand, J. C. Faivre, J. M. Fontaine, D. Garreta, D. Legrand, J. Saudinos, J. Thirion, R. Bertini, F. Brochard, and F. Hibou, Phys. Rev. Letters 32, 839 (1974).

[21] H. Palevsky, J. L. Friedes, R. J. Sutter, G. W. Bennett, G. J. Igo, W. D. Simpson, G. C. Phillips, D. M. Curley, N. S. Wall, R. L. Stearns, and B. Gottschalk, Phys. Rev. Letters 18, 1200 (1967).

[22] R. H. Bassel and C. Wilkin, Phys. Rev. Letters 18, 871 (1967); Phys. Rev. 174, 1179 (1968).
W. Czyz and L. Lesniak, Phys. Letters 24B, 227 (1967); ibid, 25B, 319 (1967); W. Czyz and L. C. Maximon, Phys. Letters 27B, 354 (1968).

[23] E. Lambert and H. E. Feshbach, Phys. Letters 38B, 487 (1972).

[24] A. M. Saperstein, Phys. Rev. Letters 30, 1257 (1973).

[25] T.E.O. Ericson and M. P. Locher, Nucl. Phys. A148, 1 (1970).

[26] G. I. Lykasov and A. V. Tarasov, Sov. J. Nucl. Phys. 17, 153 (1973).

[27] B. S. Bhakar, J. M. Cameron, P. Kitching, G. A. Moss, J. G. Rogers and W.T.H. van Oers, Proposal #59 to the TRIUMF Experiments Evaluation Committee.

[28] C. F. Perdrisat, L. W. Swenson, P. G. Gugelot, E. T. Boschitz, W. K. Roberts, J. S. Vincent, and J. R. Priest, Phys. Rev. 187, 1201 (1969).

[29] P. G. Roos, Phys. Rev. C9, 2437 (1974).

The Experimental Program at SIN

M.P. Locher
SIN

Since the Few Body meeting in Los Angeles two years ago the high intensity proton accelerator of the Swiss Institute of Nuclear Research, SIN, has become operational. On January 18, 1974 protons in the second stage ring cyclotron have been accelerated for the first time to the full energy of 590 MeV. Further tuning of the ring proceeded smoothly and rapidly, for a more detailed description see [1]. On February 24 secondary pions of 300 MeV/c were produced from the thin target. The first stage, the Philips injector, produced an external proton beam of 100 μA (at 72 MeV) in mid March. In the beginning of June the pion beams πM_1 and πM_3 from the first target were ready for experiments. (See ref. [2] for the characteristics of the machine and ref. [3] for secondary beam lay-out.)

Due to the successful commissioning of the machine some of the experiments planned and briefly described [4] at the last Few Body meeting have had their first runs completed.

In Table 1 we give the list of proposals (in their order of submission) for the users of the ring and in Table 2 for the experiments at the multipurpose injector cyclotron. For completeness we have listed all the experiments in particle and nuclear physics which are presently planned. Proposals which have not been reported at the last Conference carry the numbers 73 and 74.

We mention the following experiments which are related to the few body problem. The deuteron wave function and the nucleon-nucleon interaction are relevant for the pion scattering experiments (R73-01 and R74-01) as well as for the muon capture on the deuteron (R73-04). The deuteron also provides the simplest nuclear target for testing scattering theory in the 33 resonance region. Intense pion beams offer the possibility of studying double charge exchange on nuclear targets. A kinematically complete experiment for the break-up of ^{4}He is proposed (R74-03) and should shed some light on the unsatisfactory state of the theory. An extension of the nucleon-nucleon scattering program proposes to measure up scattering with unpolarized neutrons in the available energy range (R73-06). Properties of elementary particles are at issue in a number of earlier experiments. The measurement of the magnetic moment of the positive muon has been added to the list of proposals (R73-02). Finally we mention the proposed search for the production of muonium in gases (R73-05).

In addition there are 11 proposals in solid state physics, chemistry, medicine and further applied fields, not given here. Except for some of the more recent entries in the tables these experiments have been accepted by the program committee.

Details on the first runs could not be given at the time these notes had to be prepared and are left to the oral presentation. However, an asterisk added at Conference time marks those experiments in the tables which are in actual progress.

References

1) CERN Courier 3 No. 14 (1974) 85.

2) J.P. Blaser, H.A. Willax and H.-J. Gerber in "High Energy Physics and Nuclear Structore", S. Devons ed. (N.Y. 1970) p. 556.

3) SIN Tätigkeitsbericht 1972, p. 91.

4) "Few Particle Problems in the Nuclear Structure Interaction", North Holland 1972, p. 766.

TABLE I

Number	Description	Experimental Group
R71-01 *	High resolution bent crystal spectrometer for experiments with mesic x-rays.	ETH Zürich-Fribourg B. Aas, W. Beer, J.C. Dousse, R. Eichler, J. Kern, H.J. Leisi W. Sapp, W. Schwitz
R71-02	High resolution, large solid angle gamma detection facility	Lausanne- München-Zürich J.C. Alder, C. Joseph, J.F. Loude, N. Morel, B. Vaucher, E. Windkelmann, W. Hering, H. Schmitt, C. Zupančič, F. Hoop, H. von Fellenberg, P. Truöl
R71-04	High resolution pion spectrometer for SIN. Investigation of the $(\pi^{\pm},\pi^{\pm}N)$ reaction in the 3,3-resonance	ETH Zürich-Neuchâtel-Grenoble-Karlsruhe-Heidelberg-SIN J. Arvieux, J. Beiner, E. Boschitz, R. Corfu, J. Domingo, J.P. Egger, C. Lunke, E. Pedroni, C. Perrin, M. Perrinjaquet, J. Piffaretti, E. Schwarz, C. Wiedner, J. Zichy
R71-06	Analysis of the decay $\mu \to e\gamma$ to test the conservation of muon number	ETH Zürich-SIN R. Engfer, H.J. Mahler, C. Petitjean, H.P. Povel, J.L. Vuilleumier, H.K. Walter

Table I (continued)

Number	Description	Experimental Group
R71-07	Elastic proton-proton scattering between 400 and 600 MeV (uses a polarized proton beam)	GENF-ETH Zürich G. Greeniaus, E. Heer, R. Hess, L. Lechanoine, D.W. Werren, B. Favier, D. Rapin, A. Junod
R71-08	Precise measurement of the μ^+ momentum in the decay of stopped pions	SIN M. Daum, L. Dubal, G.H. Eaton, R. Frosch, P. Funk, E. Steiner
R71-09	Experimental set-up for the analysis of nuclear structure by means of muonic and pionic atoms	ETH Zürich-SIN-Fribourg R. Engfer, R. Hartmann, C. Petitjean, J.L. Vuilleumier, H.K. Walter, L. Schellenberg, H. Schneuwly
R71-10 *	Measurement of the polarization in π^+p elastic scattering between 100 and 300 MeV	ETH Zürich-SIN C. Amsler, L. Dubal, G.H. Eaton, W. Fischer, R. Frosch, A. Janett, S. Mango, C. Perdrisat, F. Pozar, P. Weymuth
R71-12	Experiment for the simultaneous precise measurement of $m_{\nu_\mu}^2$ and $m_{\pi^\pm}^2$	ETH Zürich-BERN H. Anderhub, K. Borer, H. Hofer, P. Le Coultre, P.G. Seiler, M. Wälchli, P. Wolff

Table I (continued)

Number	Description	Experimental Group
R71-13	Rare muon reactions on nuclei ($\mu^-A \rightarrow Ae^-$ and $\mu^-A \rightarrow A'e^+$)	BERN-SIN K. Borer, G. Czapek, H. Hänni B. Hahn, J. Schacher, H. Scheidiger, G. Viertel, U. Schryber
R72-02	Experiments with unpolarized neutron beams on nuclear targets	Freiburg U. Fischer, W. Hagedorn, K. Kern, L. Lehmann, E. Rössle, K. Runge
R73-01	Measurement of the tensor polarization in the elastic π^+d scattering	ETH Zürich-SIN W. Grüebler, V. König, P.A. Schmelzbach, T. Stammbach
R73-02	Measurement of the magnetic moment of the positive muon to a precision of 0.5 ppm by means of a stroboscopic method	ETH Zürich-SIN M. Camani, A. Schenck, W. Rüegg, C. Petitjean
R73-03	Analysis of the formation of muonic helium atoms $He^{++}\mu^-e^-$ in gaseous Helium-Xenon mixtures	ETH Zürich M. Camani, W. Rüegg, A. Schenck

Table I (continued)

Number	Description	Experimental Group
R73-04	A study of neutron spectra and neutron-neutron correlations after μ-capture in μ^-D	Wien-Williamsburg-Fribourg-SIN-Amsterdam W.H. Breunlich, W.J. Kossler L.A. Schaller, C. Petitjean, R. van Dantzig
R73-05 *	Production of muonium in gases at low pressure (also μ^- capture in ^4He at low pressure)	ETH Zürich-SIN W. Hirt, H. Hofer, F. Kottmann, G. Pitzurra, B. Sapp, P. Shrager, P.G. Seiler
R73-06	Measurement of angular distributions in np-scattering in the range 100 MeV - 600 MeV using the SIN unpolarized neutron beam	Zürich-SIN-München-Lausanne F. Hoop, W. Reichart, P. Truöl, C. Tschalär, W. Hering, H. Schmitt, C. Zupančič, J.C. Alder, C. Joseph, B. Vaucher
R73-07	Experiments on nuclear pion absorption	Karlsruhe G. Backenstoss, D. Engelhardt, R. Hagelberg, H. Koch, Ch. Lewis, L. Tauscher, H. Ullrich
R74-01	Measurement of the $\pi^{\pm}d$ total cross sections and of the total production cross section of neutral particles in the π^-d interaction for π energies from 50 to 350 MeV	Karlsruhe-Neuchâtel-SIN-Oxford-Heidelberg-London-ETH Zürich E. Boschitz, J. Domingo, K. Gabathuler, C. Lunke, E. Pedroni, J. Piffaretti, P. Schwaller, E. Schwarz, N. Tanner, C. Wiedner, C. Wilkin

Table I (continued)

Number	Description	Experimental Group
R74-02	Search for Lyman α-radiation from excited pionium (π^+e^-) by beamfoil excitation technique (letter of intent)	EPF Lausanne-ETH Zürich-Zürich E. Geneux, M. Camani, W. Rüegg, A. Schenck, P. Truöl
R74-03	Analysis of the reaction $\pi^+ \, ^4He \rightarrow \pi^- \, 4p$ at 400 MeV/c by means of a streamer chambre	Neuchâtel-Lausanne D. Perrin, J.B. Jeanneret, E. Jeannet, J.C. Alder, C. Joseph, B. Vaucher
R74-04	Measurement of electron screening effects in some muonic atoms	ETH Zürich B. Aas, W. Dey, R. Eichler, H.J. Leisi, W. Sapp
R74-05	Polarization of electrons in μ-decay (letter of intent)	ETH Zürich-SIN H.-J. Gerber

Table II

Number	Description	Experimental Group
Z71-01	Nuclear spectroscopy by means of the reactions $(\alpha, xn\gamma)$	Fribourg J. Kern, V. Ionescu, S. Olbrich
Z71-02	Analysis of quartet states by means of heavy-ion transfer reactions	Basel I. Sick, G.R. Plattner, E. Baumgartner
Z71-03	Scattering of polarized particles on light nuclei	ETH Zürich-Basel W. Grüebler, V. König, P.A. Schmelzbach, G.R. Plattner, F.Seiler
Z71-04	Analysis of transfer reactions, in particular of (p,α) reactions on medium-heavy nuclei	Zürich-Basel H. Müller, V. Meyer, J. Benn, R. Wagner
Z72-01	Quasielastic three body reactions as a test of the impulse approximation	Neuchâtel-Zürich F. Foroughi, M. Ahmed, E. Bovet, H.H. Müller, C. Nussbaum, W. Reichhart, J. Rossel
Z72-02	Isotope production using the injector cyclotron at SIN	Federal Institute of Reactor Research (EIR) H. Hügli, I. Huszàr, F. Hegedüs

Table II (continued)

Number	Description	Experimental Group
Z72-03	Shell effects and "preequilibrium" emission from excitation functions in (p,xn) and (α,xn) reactions	Wien P. Hille, M. Uhl, G. Winkler
Z73-01	Measurement of delayed neutrons after fission of highly excited nuclei	Wien F. Bensch, G. Eder, H. Jasicek, H. Oberhummer, H. Rauch, P. Riehs
Z73-02	Determination of absolute production rates of stable and radioactive isotopes having important consequences for geophysics and cosmophysics	Bern M. Wahlen, H. Oeschger, J. Beer, P. Pulfer, F. Bühler, R.C. Finkel
Z74-01	Measurement of excitation functions for (p,xn) and (p,pxn) reactions to determine the optimal conditions for isotope production	EIR F. Widder, F. Hegedüs, S. Chakraborty

NON-ADDITIVITY OF PHASE SHIFTS AND THE PION-NUCLEUS SCATTERING AT MEDIUM ENERGIES

M.Błeszyński and T.Jaroszewicz
Institute of Nuclear Physics,Radzikowskiego 152, 31-342 Kraków,Poland.

There have been many attempts to apply the Glauber multiple scattering model in the case of the π-nucleus scattering at medium energies [1]. The crucial assumption of this model is the eikonal approximation which results in the additivity of phase shifts acquired by the projectile on the target constituents. At lower energies, however, the eikonal propagation of the projectile through the nucleus becomes questionable and the additivity of phase shifts is no longer valid.

In our recent paper [2] the following formula is proposed, which accounts for the deviations from the eikonal approximation for the small-angle potential scattering at medium energy:

$$F_c(\vec{\Delta}) = \frac{ik}{2\pi}\int d^2b\, e^{i\vec{\Delta}\cdot\vec{b}}\left\{1-\exp\left[-i\frac{m}{k}\int_{-\infty}^{+\infty}dz\, U(\vec{b},z)\right]\right\} \qquad (1)$$

where $U(\vec{b},z)= V(\vec{b},z)/(1+cV(\vec{b},z))$, k is the projectile initial momentum, V is the (local) potential acting between the projectile and the target and $c=m/2k^2$, m being the projectile mass.

In the case of the scattering of a single nucleon the above formula gives formally the same amplitude as Glauber's, the only modification being the replacement of the potential V by U, i.e. the reinterpretation of the potential.The essential difference arises in the case of scattering from several potential centres, i.e. when $V(\vec{r})=\sum V_i(\vec{r}-\vec{r}_i)$ where $\vec{r}_i$ are the target constituents coordinates. The resulting phase shift is no longer the sum of the individual phase shifts but also contains the two-body,three-body,..., etc. terms. This is simply due to the nonlinear dependence of

$$U = \sum_i U_i - c\sum_{i\neq j} U_i U_j \qquad (2)$$

on $U_i = V_i/(1+cV_i)$.

With the help of formula (1) we have calculated the corrections to the π-nucleus forward scattering amplitude as obtained in the Glauber approach. By substituting eq. (2) into eq. (1) and averaging over the nucleon positions with the independent particle model density we have obtained the following correction to the π-nucleus forward scattering amplitude:

$$\langle \delta F_c(0)\rangle \equiv \langle F_c(0)-F_{Gl}(0)\rangle = -\frac{A(A-1)}{4\pi}\left\{\int d^2s\,[1-\gamma(s)]\,\ln[1-\gamma(s)]\right\}^2 \cdot$$
$$\cdot\int d^2b\left[1+i\frac{2\pi}{k}f(0)\int dz\,\varrho(b,z)\right]^{A-2}\int dz\,\varrho^2(b,z) \qquad (3)$$

where only the terms of order k^{-1} with respect to F_{Gl} have been retained. Here A is the nucleus mass number, $\rho(\vec{r})$ is the nuclear ground state density, and f(q) and $\gamma(b)$ are the π-N amplitude and profile,

$$\gamma(b) = (2\pi i k)^{-1} \int d^2q \, e^{-i\vec{q}\cdot\vec{b}} f(q) .$$

The only approximation made when going from (1) to (3) was that the π-N profiles are much narrower than the density $\rho(\vec{r})$.

We have also derived the corresponding correction in the framework of the first-order eikonal expansion [3], which is as follows:

$$\langle \delta F(0) \rangle = -\langle \delta F_c(0) \rangle - A(A-1)\frac{\pi^2}{k^2} f^2(0) \int d^2b \left[1 + i\frac{2\pi}{k} f(0) \int dz\, \rho(b,z)\right]^{A-2} \int dz\, b \frac{d}{db} \rho^2(b,z). \qquad (4)$$

Similar corrections have been obtained in ref. [4].

In general, the correction (4) is larger than that of eq. (3). For light nuclei, when the phase shift $\approx \frac{2\pi}{k} A f(0) \int dz\, \rho(b,z)$ is relatively small, the two terms of eq. (4) partly cancel each other, reproducing eq. (3).

The correction (4) contributes mainly to the real part of the amplitude. In the energy range 500-1000 MeV it is roughly energy independent and e.g. for C^{12} equal about -0.05 fm (see also ref. [5]).

We stress that in the derivation of the microscopic formulas (3) and (4) no ad hoc assumptions about the off-shell behaviour of π-N amplitudes were needed; the off-shell continuation is uniquely determined by using the local potential in the description of the π-N interaction. This is to be contrasted with the derivation of the similar formula in ref [4], where it was effectively assumed that the π-N scattering amplitude does not depend on the longitudinal momenta, i.e. that the off-shell amplitude coincides with the on-shellone with the same transverse momentum transfer. This assumption is in fact equivalent to the so called ordered form of the Glauber formula, which, strictly speaking, is valid only when the individual potentials do not overlap.

---

[1] C.Wilkin, CERN preprint TH. 1711 (1973).
[2] M.Bľeszyński, T.Jaroszewicz, Lett. al N.Cim. 10, 19 (1974).
[3] S.J.Wallace, Ann. Phys. 78, 190 (1973).
[4] B.Shürmann, Nucl. Phys. B67, 425 (1973).
[5] G.Alberi, Z.Grossman, J.Leon, P.Osland, ICTP Trieste preprint IC/74/13.

π^+P Elastic Scattering at 48 MeV

E.G. Auld, D. Axen, J. Beveridge, G. Duesdieker, L. Felawka, C.H.Q. Ingram, R.R. Johnson, G. Jones, D. LePatourel, R. Orth, M. Salomon, W. Westlund, University of British Columbia and L.P. Robertson, University of Victoria.

The π^+p differential cross-section has been measured at 48 MeV using pions produced in the 747 MeV external proton beam of the 184 inch cyclotron at the Lawrence Berkeley Laboratory. A polythene product target and a production angle of 129° was used in order to reduce contamination in the pion beam. The pion beam line consisted of two 60° bending magnets two quadrupole doublets and a quadrupole singlet. Positional information from a twelve element hodoscope at the mid plane dispersed focus and a spark chamber in front of the scattering target was used to determine the incident pion momentum. The momentum bite was ± 3.5% and the resolution $\Delta p/p \approx 1\%$. Momentum analyzed π^+ were separated from μ^+ and e^+ by time of flight between counters near the mid plane and target. Muon contamination (≈ 4%) due to pion decay in the straight section in front of the target was estimated by a Monte Carlo calculation. The incident kinetic energy of 47.9 ± .4 MeV was measured by range.

The temperature and density of liquid hydrogen in a rotatable target 8 cms high 20 cms wide and 2.15 ± .05 cm thick was kept constant by maintaining the gas pressure to ± 1%.

Scattered pions were detected in a telescope consisting of two transmission counters, a spark chamber, and a stopping counter of NE 110 plastic scintillator 12.5 cms diameter and 30 cms long. The angular resolution of reconstructed pion trajectories was ± 4°. The pulse from the stopping scintillator was analyzed to determine the pion energy. The efficiency of this detector to measure the energy of elastically scattered particles was 85% the remainder of the pions being lost by decay or nuclear reactions in the scintillating plastic. The active diameter of the scintillator was reduced to 8 cms (using the reconstructed trajectories) to remove the possibility of pion scattering out of the detector before stopping. The energy loss resolution was less than 2 MeV. Similarly events due to scattering in the target frame, vacuum jacket walls and pion decay near the target were rejected in the subsequent analysis.

The average solid angle and target thickness was estimated by averaging the calculated solid angle for a large number of incident pions distributed in angle and position over the target.

The observed number of scattered pions was corrected for hodoscope efficiency, spark chamber efficiency (continuously monitored), dead time of the data acquisition system, stopping counter efficiency, and pion decay between the target and stopping counter.

The angular distribution is shown in Figure 1, the error bars representing a statistical accuracy ≈ ± 5%. The data has been binned in ± 7.5° angular bins in the range from 30° to 150°. The 160° datum point was taken with a single counter and the 172° datum point with an annular counter. In these two cases the forward going proton was observed in coincidence. The solid curve represents a best fit to the data with a χ^2 of 6 for 8 degrees of freedom. The three isospin triplet phase shifts obtained from a phase shift analysis using a relativistic coulomb amplitude,[1] $S_3 = -4.65° \pm .45$, $P_{31} = .53° \pm .19$, $P_{33} = 5.07° \pm .29$ are in agreement with an extrapolation of phase shifts obtained at higher energy.[2]

BIBLIOGRAPHY

1) L.D. Roper, R.M. Wright, B.T. Feld, Phys. Rev. 138, B190, (1965).

2) J.R. Carter, B.V. Bugg, A.A. Carter, RHEL report, RPP/H/108.

This research was supported by grants from both the National Research Council and the Institute of Particle Physics of Canada.

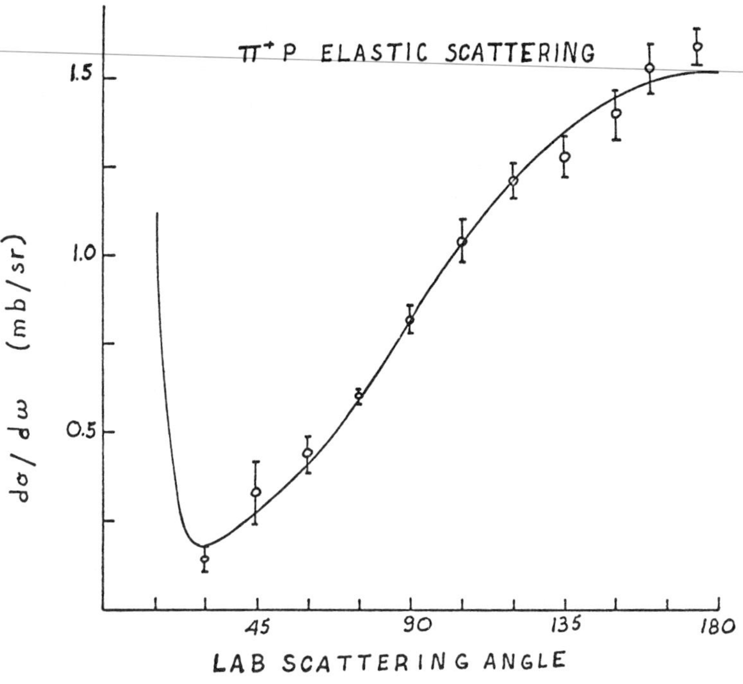

Figure 1. Angular Distribution for the Reaction $\pi^+ P \rightarrow \pi^+ P$ at 47.9 MeV.

NUCLEON-NUCLEON CORRELATIONS AND PION-NUCLEUS INTERACTION

L.S. Celenza, L.C. Liu and C.M. Shakin

Institute of Nuclear Theory and Department of Physics ,
Brooklyn College of the City University of New York, Brooklyn,N.Y. 11210

The optical potential for pion-nucleus scattering has been intensively studied in recent years. Having observed that most multiple-scattering theories on which these studies were based use completely non-relativistic dynamics and, consequently, are unsuited to treat the scattering of pions from a nucleus, we have proposed a covariant approach to the problem {1}. Since the four-momentum associated with each particle is unambiguously defined in our approach, it follows that multiple-scattering from correlated pairs leads to expressions containing highly off-shell amplitudes. Some of these off-shell effects in multiple-scattering theory have been discussed for nucleon-nucleus scattering using non-relativistic dynamics {2}. We would like to report here that our method reveals some important features which are totally ignored in a conventional multiple-scattering theory. Owing to the limited space, we shall present the detailed theoretical derivation as well as the numerical results elsewhere.

In an independent-pair approximation appropriate to the study of the role of pair-correlations in pion-nucleus interaction, we may represent the scattering of pions from a pair of correlated nucleons by the diagram shown in Fig.1. In this diagram, the dashed and the solid lines denote respectively the pion and the nucleons while the bold lines represent nuclei. The filled circle is the off-shell pion-nucleon scattering amplitude and the open circles are various nuclear vertices. Consequently, the two upper vertices represent the short-range nucleon-nucleon interaction and the two lower vertices describe the interaction between the center-of-mass of the di-nucleon cluster, D, and the residual nucleus, C (two-hole state).

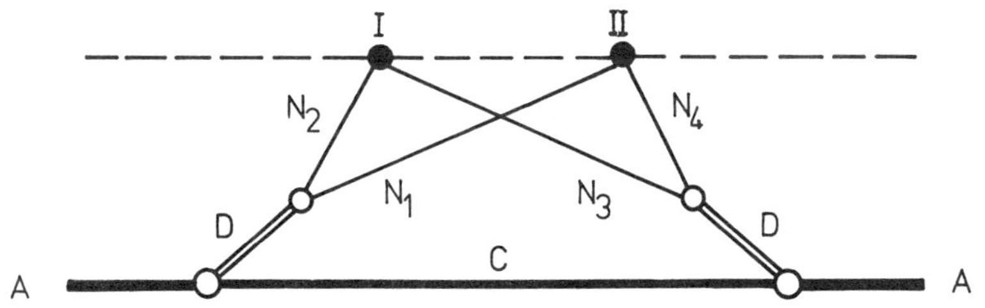

Fig. 1

For simplicity, we consider here the forward scattering of pions at very low energy, i.e. $\vec{p}_\pi \simeq 0$ in the lab. system (the pionic atom limit). If we use a specific covariant reduction procedure in which the " particles " marked by crosses in Fig.1 are put on the mass-shell {3}, then the various four-momenta in the lab. system can be parametrized as follows:

$$p_\pi = (M_\pi\ ,\ 0)\ ,\qquad p_A = (M_A\ ,\ 0)\ ,$$
$$p_C = (E_{C,\vec{K}}\ ,\ \vec{K})\ ,\qquad p_D = (M_A - E_{C,\vec{K}}\ ,\ -\vec{K})\ ,$$
$$p_{N_1} = (E_{N,(\frac{1}{2}\vec{K}+\vec{p})}\ ,\ -\tfrac{1}{2}\vec{K}-\vec{p}\ )\ \text{and}$$
$$p_{N_3} = (E_{N,(-\frac{1}{2}\vec{K}+\vec{p}+\vec{q}\,')}\ ,\ -\tfrac{1}{2}\vec{K}+\vec{p}+\vec{q}\,'\ )\qquad \text{etc.}$$

Here $E_{i,\vec{k}} = (M_i^2 + \vec{k}^2)^{\frac{1}{2}}$. The invariant energy for the pion-nucleon scattering process "I" is therefore equal to

$$s = (p_\pi + p_{N_2})^2 = (M_\pi + M_A - E_{C,\vec{K}} - E_{N,(\frac{1}{2}\vec{K}+\vec{p})})^2 - (-\tfrac{1}{2}\vec{K}+\vec{p})^2 \quad .$$

After introducing the binding energy defined by

$$\Delta = M_C + 2M_N - M_A \quad ,$$

we obtain when $\vec{K} = 0$ and $\vec{p}$ is large

$$s = (M_\pi - \Delta + 2M_N)(M_\pi - \Delta + 2M_N - 2E_{N,\vec{p}}) + M_N^2 < M_N^2 \quad .$$

This inequality is directly due to the high magnitude of the momentum $\vec{p}$ involved in the strong nucleon-nucleon correlation. (p is of the order of 3 fm^{-1} {4}). On the other hand, in the fixed-scatterer approximation, both $\vec{K}$ and $\vec{p}$ are set to be zero, it follows that

$$s^{FSA} = (M_\pi + M_N)^2 > M_N^2 \quad .$$

We conclude from the above consideration that the off-shell effects are particularly marked in the case of a scattering from correlated pairs so that the fixed-scatterer approximation, leading to a completely different kinematic region, is totally unreliable except at ultra-relativistic energies.

REFERENCES

1. L.S. Celenza, M.K. Liou, L.C. Liu and C.M. Shakin, in " Multiple Scattering Theory and the Relativistic Optical Model " and " Relativistic Wave Functions for Pion-Nucleon and Pion-Nucleus Scattering ", to be published in Phys. Rev. C .

2. R.R. Scheerbaum, C.M. Shakin and R.M. Thaler, Phys.Rev. C7, 2346 (1973). See also, D.J. Ernst, C.M. Shakin and R.M. Thaler, Phys.Rev. C9, 1370 (1974).

3. The external particles are on the mass-shell by definition.

4. P. Grange and M.A. Preston, Nucl.Phys. A204, 1 (1973).

THE INTERACTION OF 48 MeV π^+ WITH DEUTERONS

D. Axen, G. Duesdieker, L. Felawka, Q. Ingram, R. Johnson, G. Jones, D. LePatourel, M. Salomon, and W. Westlund, University of British Columbia, and L. Robertson, University of Victoria, British Columbia, Canada.

Study of the pion-nuclear interaction at low energies has several important advantages. Not only are very few partial waves involved, but also the basic pion-nucleon interaction is sufficiently weak that the impulse approximation should provide a reasonable description of the interaction.[1] In addition, investigation of the interaction of the pion with multi-nucleon systems provides information on certain features of the pion-nucleon interaction such as the off-mass shell amplitude. Because of our relatively good knowledge of the structure of the deuteron, the pion-deuteron system is the obvious initial candidate for such studies.

This paper presents the experimental results of a study carried out using the 184 inch cyclotron at LBL, Berkeley. In this experiment, positive pions of 47.7 MeV kinetic energy produced at back angles (about 130°) when the 750 MeV proton beam impinged on a polyethylene target were transmitted by a two-bend, achromatic channel to a 2.15 cm thick liquid deuterium target. The total useful pion flux was about $10^4 s^{-1}$. Reaction products were detected in a scintillation counter telescope consisting of two thin transmission counters, a spark chamber array for defining particle trajectories, and a 12.5 cm diameter full-energy stopping counter with an energy resolution of about 2% at 50 MeV.[2] In this way, data for each of the following three reactions were recorded simultaneously:

(1) $\pi^+ + d \rightarrow d + \pi^+$, (2) $\pi^+ + d \rightarrow p + n + \pi^+$, (3) $\pi^+ + d \rightarrow 2p$

Pions in the incident beam were selected by time-of-flight and their momenta determined to 1% dp/p. The outgoing particles were identified by E and dE/dx information, with the pions distinguished from muons by observation of the characteristic decay of the stopped pion in the full-energy counter. The energy-loss resolution of the whole system was typically about 2 MeV.

Although the telescope subtended a scattering angle of about 20° at the target, the use of spark chambers on both the incident and emergent beams permitted the definition of scattering angles with a resolution of about 4°. The differential cross-sections for all three reactions were measured over an angular range of 22.5 to 157.5 degrees to a statistical accuracy of about 6% when subdivided into 7.5° bins, and when the inelastic pions (reaction 2) were integrated over all energy.

The experimental results for the three reactions are displayed in Figs. 1 to 3. In Fig. 1, the curves labelled (M) are those obtained using single-scattering impulse approximation.[3] The two different curves illustrate the range of values accommodating the current uncertainties in the pion-nucleon phase shifts at these energies. The curve labelled (BEDER) indicates the effect of adding an absorptive amplitude.[1] It is interesting that a simple, single-scattering impulse approximation yields such a good fit to the data, although it is not yet clear how much effect multiple scattering and off-mass shell effects will have when incorporated into the simple model yielding curves (M). Faddeev calculations are currently being carried out to eliminate such dynamical uncertainties and so explore the relationship of this data to our understanding of the two-body interactions involved.[4]

The solid line in Fig. 2 is the theoretical prediction of Beder.[1] It is reassuring that such models provide a consistent treatment of both the elastic and inelastic pion scattering.

In Fig. 3, the solid line indicates the differential cross-section expected on the basis of a recent study of the energy dependence of this reaction.[5] The agreement with our results provides reassuring confirmation of the accuracy of our cross-section determinations.

BIBLIOGRAPHY

1) D. Beder, private communication, 1974; Nucl. Phys. B34 (1971) 189. R.M. Rockmore, Phys. Rev. 105 (1957) 256.
2) D. Axen, G. Duesdieker, L. Felawka, C.H.Q. Ingram, G. Jones, M. Salomon and W. Westlund, Nucl. Instr. and Methods, to be published.
3) J.M. McMillan, private communication, 1974.
4) A.W. Thomas, private communication, 1974.
5) C. Richard - Serre, W. Hirt, D.F. Measday, E.G. Michaelis, M.J.M. Sattmarsh, P. Skarek, Nucl. Phys. B20 (1970) 413.

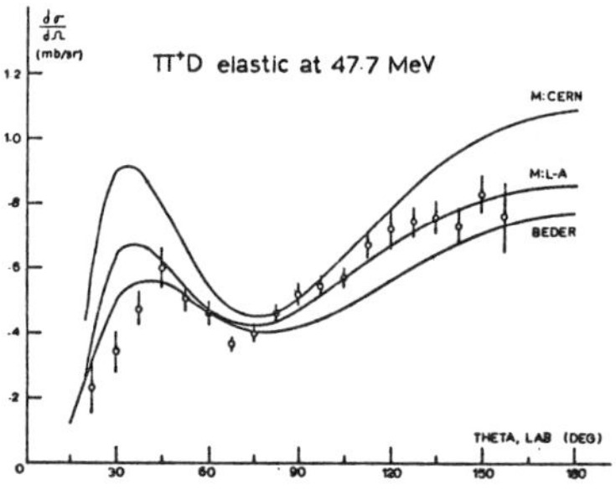

Fig. 1

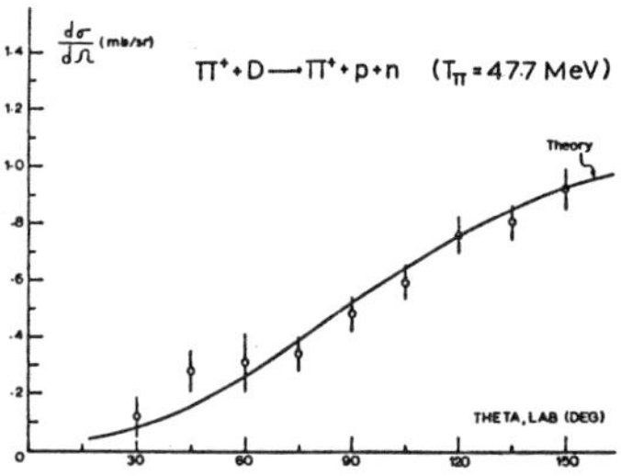

Fig. 2

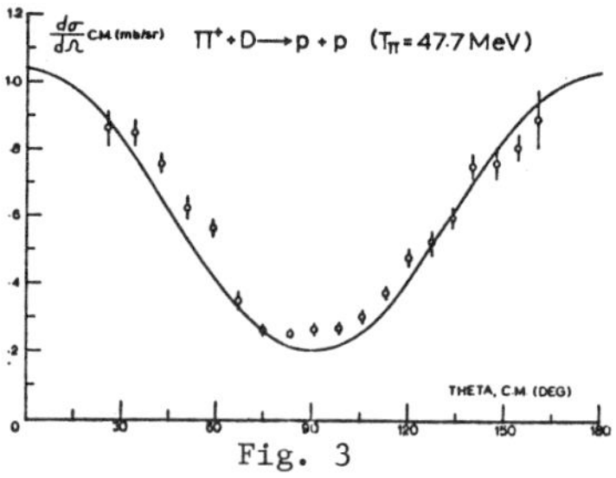

Fig. 3

This research was supported by grants from both the National Research Council and Institute of Particle Physics of Canada.

RELATIVISTIC THREE-BODY MODEL OF π-d SCATTERING

D. D. Brayshaw
University of Maryland
Dept. of Physics & Astronomy
College Park, Maryland 20742 U.S.A.

A relativistic generalization of the author's three-body boundary-condition formalism[1] has been applied to the π-d system. The eventual aim of such calculations is to test characteristic assumptions employed in current theories of pion-nucleus scattering. The present calculation is concerned with the $\ell = 1$, $J = 2$ scattering state, in which we ignore the P_{31} π-N interaction and the d-state of the deuteron. In the spin-independent approximation we may consider our result the π-d p-wave amplitude. The on-shell input is essentially the "exact" N-N and π-N phase-shifts, no off-shell corrections are considered.

The results plotted in Fig. 1. illustrate an important difference between the "resonance" observed in this system as compared to the underlying $\Delta(1236)$. That is, the phase-shift in π-d is quite small, the rapid energy-dependence being due to a sudden rise in inelasticity reflected in a dip in η. The $\Delta(1236)$ corresponds to the π-N phase going through 90° and is elastic. If one defines the position of the "resonance" as the point at which the elastic amplitude is purely imaginary, it corresponds in this model to $\delta = 0$ and is shifted downward as compared to the Δ.[2] The contribution to the total cross section assuming spin-independence is also plotted in Fig. 1.

The entries in Table I. illustrate convergence of the multiple-scattering series. Here SS represents single-scattering, DS represents single — plus double-scattering of the pion. Except at low energy the convergence is rather slow. As indicated, employing the SS term would overestimate σ_t and shift the peak to higher T_L.

TABLE I.

T_L (MeV)	Phase ($\delta^o_{2,1}$)			Elasticity (η)		
	SS	DS	T	SS	DS	T
24	1.59	1.60	1.68	.997	.996	.996
133	9.12	8.98	7.40	.864	.849	.768
174	5.90	6.56	-.45	.587	.571	.600
260	-4.02	-12.7	-7.97	.877	.768	.805

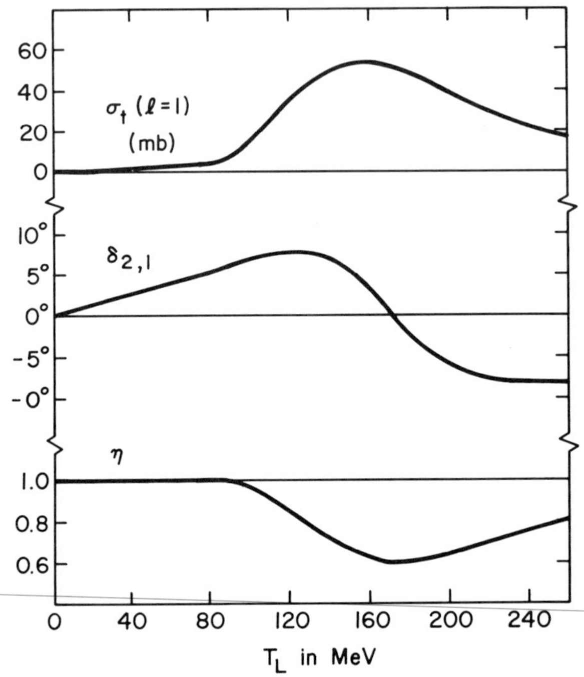

Fig. 1. Phase-shift and elasticity for J = 2, ℓ = 1 vs. laboratory energy. The (spin-independent) p-wave contribution to σ_t is also plotted.

The three-body nature of this model suggests that the rapid variation of the amplitude in the 3,3 region may be attributed to the rescattering singularities previously investigated by this author.[3] This suggests the possibility of summing the important terms via the analytic techniques introduced in that article.

REFERENCES

1) D. D. Brayshaw, Phys. Rev. D8(1973) 952.

2) An upward shift in a separable model Faddeev calculation (contrary to experiment) was reported by F. Myhrer and D. S. Koltun, Phys. Letters 46B(1973) 322.

3) D. D. Brayshaw and Ronald F. Peierls, Phys. Rev. 177(1969) 2539.

SCATTERING LENGTHS OF π-d, π-τ AND π-^{3}He DERIVED FROM ENERGY LEVEL SHIFTS AND WIDTHS OF PIONIC ATOMS

Il-Tong CHEON
Institut de Physique, Université de Liège au Sart-Tilman, 4000 Liège, Belgium.
T. von EGIDY
Physik-Department, Technische Universität München, Munich, Germany.

A considerable amount of experimental data on energy level shifts and widths of pionic atoms is available, and the optical potential model is known to be able to interpret successfully these data [1,2]. However, the optical potential model approach does not seem to be successful for very light nuclei. On the basis of the model of pion absorption by the nucleon-pair, an empirical formula for energy level widths of pionic atoms has recently been proposed [3]. Here, we present an empirical formula for energy level shifts of pionic atoms under consideration of the elastic [4] and inelastic processes [5]. Our formula is

$$\Delta E_{n\ell}(A,Z) = \frac{Z^{2\ell+5}}{A^{4/3}}\,\xi_{pn} + \frac{Z^{2\ell+3}}{A^{1/3}}\,(N-Z)\xi_n + 1.09\,\Gamma_{n\ell}(A,Z) \quad ,$$

where A and Z are mass- and atomic-numbers, ξ_{pn} and ξ_n ara parameters and the width $\Gamma_{n\ell}(A,Z)$ is given by

$$\Gamma_{n\ell}(A,Z) = \frac{1}{A^{4/3}}\left[\frac{1}{8}\{7+(-1)^{\ell}\}ZN + \frac{1}{4}\,Z(Z-1)\right]Z^{2\ell+3}\,\gamma_{n\ell}$$

with the effective width $\gamma_{n\ell}$. We obtain good agreements with experimental data [1] with $\gamma_{1S} = 0.008654$keV, $\xi_{pn} = 0.009$ keV and $\xi_n = 0.0125$ keV for ΔE_{1S} and Γ_{1S} of the nuclei for $A \geq 10$; $\gamma_{2p} = 0.1918 \times 10^{-6}$ keV, $\xi_{pn} = -0.42 \times 10^{-6}$ keV, and $\xi_n = 0.16 \times 10^{-6}$ keV for ΔE_{2p} and Γ_{2p} of the nuclei for A > 27 ; $\gamma_{3d} = 0.3503 \times 10^{-12}$ keV, $\xi_{pn} = -0.14 \times 10^{-11}$ keV and $\xi_n = 0.33 \times 10^{-13}$ keV for ΔE_{3d} and Γ_{3d} of the nuclei for A > 40 ; $\gamma_{4f} = 0.1233 \times 10^{-17}$ keV, $\xi_{pn} = -0.42 \times 10^{-17}$ keV and $\xi_n = 0.10 \times 10^{-18}$ keV for ΔE_{4f} and Γ_{4f} of the nuclei for A >181. In particular, for the nuclei of $A \leq 9$, we listed our results with experimental data and optical potential calculation in Table I. Table II shows the scattering lengths obtained by the formula

$$a_{\pi A} = -\frac{\Delta E_{1s}}{2\mu c^2\,(Z\alpha)^3\,(1-0.02A)}\left(\frac{\mu}{\mu_\pi}\right)^2\left(\frac{\hbar}{\mu c}\right),$$

where α is the fine structure constant, μ is the pion mass and μ_π is the reduced mass of the pion-nucleus system. We introduced an empirical factor (1 - 0.02A) [6, 7].

Table I. Energy level shifts and widths given in unit of eV. $\gamma_{1s} = 7.7$ eV, $\xi_{pn} = 4.8$ eV and $\xi_n = 20$ eV. ΔE^{exp}, Γ^{exp}, ΔE^{opt} and Γ^{opt} are taken from Refs. [8] and [9].

Element	ΔE_{1s}	Γ_{1s}	ΔE^{exp}_{1s}	Γ^{exp}_{1s}	ΔE^{opt}_{1s}	Γ^{opt}_{1s}
^2D	5.34	3.15				
^3H	18.97	3.67				
^3He	-35.46	36.68				
^4He	71.78	43.66	75.7±2.0	45. ± 3.		
^6Li	325.22	200.21	324. ±3.	195. ± 12.	451.	125.
^7Li	597.83	209.64	570. ±4.	195. ± 13.	672.	160.
^9Be	1537.97	605.51	1595. ±9.	591. ± 14.	1793.	495.

Table II. π-nucleus scattering lengths. $a^{no\ abs}_{\pi A}$ corresponds to the case without absorption effects. $a^{mult}_{\pi A}$ are the results of multiple scattering calculation.

Nucleus	$a_{\pi A}$(fm)	$a^{no\ abs}_{\pi A}$ (fm)	$a^{exp}_{\pi A}$ (fm)	$a^{mult}_{\pi A}$ (fm)
^2D	-0.083 + 0.024i	-0.030		-0.047 [10] -0.06 [11] -0.036 [12] -0.038 [13]
^3H	-0.287 + 0.028i	-0.226		
^3He	+0.067 + 0.035i	+0.143		
^4He	-0.135 + 0.041i	-0.046	(-0.143±0.004) +(0.042±0.003)i	-0.141±0.048$^+$
^6Li	-0.185 + 0.057i	-0.061	(-0.187±0.002) +(0.056±0.003)i	-0.166±0.070$^+$
^7Li	-0.346 + 0.061i	-0.214	(-0.330±0.002) +(0.056±0.004)i	-0.297±0.085$^+$
^9Be	-0.391 + 0.078i	-0.223	(-0.405±0.002) +(0.075±0.002)i	-0.342±0.103$^+$

$^+$ Ref. [7].

[1] G. Backenstoss, Ann. Rev. Nucl. Sci. 20(1970)467.
[2] M. Ericson and T.E.O. Ericson, Ann. Phys. 36(1966)323
M. Krell and T.E.O. Ericson, Nucl. Phys. B11(1969)521.
[3] Il-T. Cheon, preprint.
[4] S. Deser, K. Baumann and W. Thirring, Phys. Rev. 96(1954)774.
[5] K.A. Brueckner, Phys. Rev. 98(1955)769.
[6] R. Seki and A.H. Cromer, Phys. Rev. 152(1967)93.
[7] L. Moyer and D.S. Koltun, Phys. Rev. 182(1969)999.
[8] G. Backenstoss et al., preprint.
[9] G. Backenstoss et al., Nucl. Phys. B66(1974)125.
[10] V.M. Kolybasov and A.E. Kudryavtsev, Nucl. Phys. B41(1972)510.
[11] N.M. Petrov and V.V. Peresypkin, Phys. Letters 44B(1973)321.
[12] A.W. Thomas and Z.R. Afnan, Phys. Letters, 45B(1973)437.
[13] P. Bendix and D. Beder, Phys. Letters 49B(1974)140.

UNITARY CORRECTION TO PION-NUCLEUS SCATTERING LENGTHS

F. Myhrer and R.R. Silbar*
SIN, 5234 Villigen, Switzerland

In calculations of low-energy pion-nucleus elastic scattering one commonly uses as input the πN isospin scattering lengths a_I, $I = 1/2$ and 3/2. The "leading" contribution to the scattering length for any $T = 0$ nucleus is proportional to the isoscalar combination $a_0 = \frac{1}{3}a_{1/2} + \frac{2}{3}a_{3/2}$, which is quite small. It is well known that higher order multiple scattering will therefore be important for such nuclei[1]. Corrections to the single-scattering term may also enter, however, and such a correction arises from off-shell unitarity.

The single-scattering amplitude for a $T = 0$ nucleus is, supressing a kinematic factor,

$$a_{\pi A} = (\tfrac{1}{2})\ A < f_{\pi^- p} + f_{\pi^- n} > = (\tfrac{1}{3})\ A < f_{1/2} + 2f_{3/2} >, \tag{1}$$

where the f's are πN scattering amplitudes and brackets indicate an average over the nuclear wave function. In the impulse approximation, at threshold, $a_{\pi A}$ reduces to $A\ a_0$, which is small. The s-wave amplitudes f_I at low energies, however, satisfy

$$f_I = (\ q \cot \delta_I - iq\ )^{-1} = a_I + iqa_I^2 + O(q^2). \tag{2}$$

With this in eq. (1),

$$a_{\pi A} = A\ \ a_0 + (i/3) <q> (a_{1/2}^2 + 2a_{3/2}^2) + O(q^2)\ . \tag{3}$$

The importance of the second term depends on the value of $<q>$. As can be seen from the off-shell unitarity relation[2], q is related to the πN off-shell energy. We illustrate this here for the case of scattering with a separable potential,

$$V_I = \lambda_I\ v(k)\ v(k'), \qquad v(k) = (\beta^2 + k^2)^{-1}. \tag{4}$$

Solving for the fully-off-shell amplitude gives

$$f_I(k,k',\varepsilon) = (v(k)v(k')/v^2(p))\ (1/a_I - ip + O(p^2))^{-1}, \tag{5}$$

where k and k' are the initial and final c.m. momenta and $p = (2\mu\varepsilon)^{1/2}$. Clearly, q in eq. (3) is to be identified with p.

The off-shell energy ε is given by

$$\varepsilon = E - KE(\text{core}) - KE(\pi N \text{ c.m. system}), \tag{6}$$

where E is the total c.m. energy for the three bodies (π + N + core); E = 0 at breakup threshold. At scattering threshold, $E = E_B < 0$, the kinetic energy terms are related to the Fermi momentum of the struck nucleon, and $\varepsilon < 0$. Thus $<q> = <p>$ is purely imaginary (as it must be --

$a_{\pi A}$ is real in the absence of absorption). For the deuteron <p> ≃ +i60 MeV/c, and the second term in eq. (3) ≃ $-.01\ m_\pi^{-1}$, to be compared with[3] $-.028 < a_0 < +.007\ m_\pi^{-1}$. For a heavier nucleus the second term is even more important.

These somewhat qualitative remarks have been substantiated in a calculation of the single-scattering contribution to $a_{\pi d}$ using the Faddeev equations[4]. The difference between the impulse approximation and the full calculation (symbolized by eq.(3)) can be as big as a factor of two. For one choice of the a_I there is even a change of sign[5]. The multiple scattering contributions are, of course, also important, but do not "undo" the unitarity correction discussed here.

Finally, we make three comments:

1) The unitarity correction has its origins in the binding of the struck nucleon. Moyer and Koltun[1] also discuss a binding energy correction, which in spite of the very different language used, appears similar. However, it can be shown that they are distinct[6].

2) The Pauli principle gives a correlation correction to a_0 which has exactly the same form as the effect discussed here[7]. In magnitude it is smaller.

3) The unitarity correction persists even in the limit of zero πN range ($\beta \to \infty$). There is no contradiction, however, between this large off-shell contribution and the theorem on non-overlapping potentials[8], since that theorem assumes the nucleus to be frozen (i.e., the fixed nucleons have no Fermi motion or binding energy).

References

* Present address: Los Alamos Scientific Laboratory. Work supported in part by the U.S. Atomic Energy Commission.

1) L. Moyer and D.S. Koltun, Phys. Rev. 182 (1969) 999.
2) C. Lovelace, Phys. Rev. 135 (1964) B1275.
3) H. Pilkuhn et al., Nucl. Phys. B65 (1973) 460.
4) F. Myhrer and R.R. Silbar, Phys. Lett. (to be published; F. Myhrer, Nucl. Phys. B (to be published).
5) This point was also noted by A.W. Thomas (thesis, Flinders University, 1974) but not explained.
6) F. Myhrer, to be published.
7) See, e.g., L. Tauscher and W. Schneider, to be published, and references therein.
8) See, e.g., J. Hüfner, Nucl. Phys. B58 (1973) 55, and references therein.

OPTICAL MODEL ANALYSIS OF PION ELASTIC SCATTERING BY ^{3}He

R. Mach [+], F. Nichitiu [++], and Yu.A. Scherbakov
Joint Institute for Nuclear Research, Dubna, USSR

Recently, $\pi^{\pm}$ - ^{3}He elastic differential cross sections have been measured at 97 and 154 MeV [1,2]. It is interesting to know, whether it is possible to obtain for nonzero spin and isospin light nuclei a good qualitative description of elastic cross sections in terms of the first order optical potential. For A = 3 nuclei, the optical potential is constructed from the complete pion - nucleon amplitude and it has the following form [3,4]

$$\hat{V}\Phi(\vec{r}) = 2\Big\{2mQ^{\tau}_{\tau_z}V_c(r) + (B_o + \tfrac{2}{3}\vec{t}\vec{T}B_1)\varrho(r) + (C_o + \tfrac{2}{3}\vec{t}\vec{T}C_1)\cdot \cdot(-\vec{\nabla}\varrho(r)\vec{\nabla} + \tfrac{m}{3M}\nabla^2\varrho(r)) + \tfrac{2}{3}\vec{J}\vec{\ell}(1 + \tfrac{2m}{3M})(D_o - 2\vec{t}\vec{T}D_1)\tfrac{1}{r}\tfrac{\partial\varrho(r)}{\partial r}\Big\}\Phi(\vec{r}) \qquad (1)$$

where the channel isospin and its projection are denoted by τ and τ_z and $Q^{\tau}_{\tau_z}$ denotes the effective charge. The coefficients B_i, C_i and D_i, i = 0, 1 are related in the usual way [5] to the free pion - nucleon phase - shifts. Further, m is the reduced pion - nucleus mass and M is the mass of the nucleon. Finally, $\vec{t}$, $\vec{T}$, $\vec{J}$ and $\vec{\ell}$ are the pion isospin, nuclear isospin, nuclear spin and orbital momentum operators, respectively. The two terms proportional to $\frac{m}{M}$ in eq.(1) are obtained taking into account the Fermi motion [4,6]. The ^{3}He ground state wave function is considered to be the wave function of the pure symetric S - state only and the nuclear density is taken in the form

$$\varrho(r) = \frac{1}{(\pi a^2)^{1/2}}\cdot e^{-r^2/a^2}$$

The value a = 1.38 fm was obtained from a charge radius R = 1.88 fm [7].

The potential given by eq.(1) exhibits some interesting features :

- The mixed spin - isospin part in eq.(1) has the opposite sign compared to the corresponding term in the π - N amplitude.
- Since $2\vec{t}\vec{T} = -2$ for $\tau = 1/2$ and $2\vec{t}\vec{T} = 1$ for $\tau = 3/2$, the influence of the spin-orbital term in the optical potential will be much smaller in the $\tau = 3/2$ channel compared to the $\tau = 1/2$ one. Therefore, the minimum in the π^- - ^{3}He cross sections should be less deep than that for the π^+ - ^{3}He reaction.
- Due to the strong cancellations in the $\tau = 1/2$ channel optical potential, the π^+ - ^{3}He cross sections should be better described in our model than those for π^- - ^{3}He reactions.

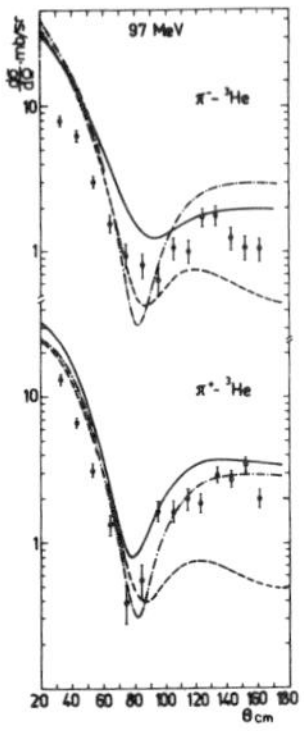

FIGURE 1

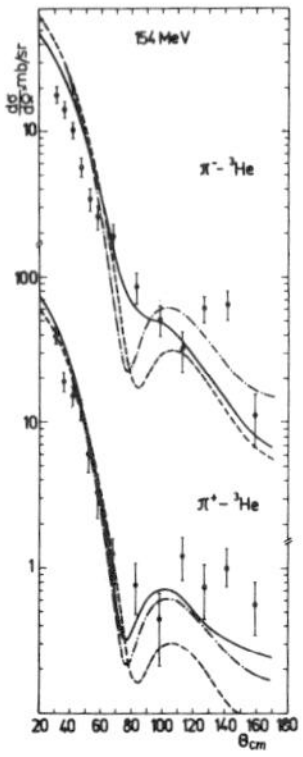

FIGURE 2

Results obtained using the potential given in eq.(1) and the experimental data of $\pi^{\pm}$-^{3}He differential cross section are displayed in Fig.1. (97 MeV) and Fig.2. (154 MeV). The dashed line shows the results obtained with only the Kisslinger model (Fig.1.), and with the Kisslinger model including the 1/A corrections (Fig.2.). The dot - dashed line is obtained if both the Fermi motion and 1/A corrections are taken into account in eq.(1). The full line was calculated using the complete potential eq.(1).

The following conclusions can be drawn: (i) The inclusion of the Fermi motion and 1/A correction terms improves the agreement of calculated cross sections with experimental data considerably at large scattering angles. (ii) The presence of spin and isospin dependent terms in eq.(1) affects strongly especially the π^--^{3}He cross sections. If we include them, the position of the minimum is shifted in the correct direction (at 97 MeV) and a better description of the cross section is obtained at large scattering angles (at 154 MeV). Both the experimental and theoretical minima are less deep for π^- - ^{3}He reactions. (iii) Nevertheless, the agreement between experimental data and theoretical curves is (namely for π^- - reactions) only qualitative. There remains large discrepancies at small scattering angles at 97 MeV, similar to those observed in π - ^{4}He cross sections [5,6] in the corresponding energy region.

A more elaborate description of the ^{3}He ground state wave function will be needed (e.g. the term proportional to the first power of D-state admixture parameter should be included in the spin - orbital part of eq.(1) [4] in order to make a conclusion to what degree of accuracy is the first order optical potential able to describe the pion elastic cross sections for scattering by light $J \neq 0$, $T \neq 0$ nuclei.

REFERENCES

1) I.V. Falomkin et al., Lett. Nuovo Cimento 55, 1121 (1972).
2) M. Albu et al., Preprint JINR Dubna P1-7742, 1974.
3) M. Ericson and T.E.O. Ericson, Ann. of Physics 36, 323 (1966).
4) R. Mach, will be published.
5) J.P. Dedonder, Nucl. Phys. A174, 251 (1971).
6) R. Mach, Nucl. Phys. A154, 56 (1973).
7) J.S. McCarthy et al., Phys. Rev. Lett. 25, 884 (1970).

Measurement of the 0° Neutron Spectrum from the Reaction $pp \rightarrow n p \pi^+$ at $T_p = 764$ MeV.

B. E. Bonner, J. E. Simmons, K. D. Williamson, Jr., Los Alamos Scientific Laboratory*; D. W. Werren, SIN and University of Geneva*; H. C. Bryant, C. G. Cassapakis, S. Cohen, B. D. Dieterle, D. M. Wolfe, University of New Mexico*; M. L. Evans, G. Glass, J. C. Hiebert, Mahavir Jain, L. C. Northcliffe, Texas A and M University*; C. W. Bjork, P. J. Riley, University of Texas*.

Previous experiments on the reaction $pp \rightarrow n p \pi^+$ in the energy range below 1 GeV have been of two types: 1) the measurement of the pion spectrum at several angles,[1] and 2) bubble chamber experiments[2,3] in which both the charges particles are detected, thus completely determining the events kinematically. While in principle the latter measurements contain complete information about the reaction, in practice they have suffered from large statistical uncertainty. Nonetheless calculations[4] have shown that the reaction is dominated by peripheral processes in which the neutron is a spectator to the recoiling Δ^{++}. Our measurement of the neutron spectrum was undertaken in order to provide a further test of this view of the reaction mechanism.

The apparatus used in the measurements was as described in a companion paper,[5] with the exception that the neutron production target was filled with liquid hydrogen instead of deuterium. The emerging neutron angle was 0° with an angular aperture of 0.2°. The measured quantity was the momentum of the recoiling protons resulting from n-p collisions in a LH_2 radiator. Neglecting small differences between the neutron and proton momentum, this is related to the neutron spectrum through the following equation

$$I_p(P)\Delta P = I_n(P)\Delta P \; n_H \int_{\theta_p = 0}^{0.04} \left(\frac{d\sigma}{d\Omega}\right)^{LAB}_{np \rightarrow pn} \varepsilon(\theta_p)\, d\Omega ,$$

where $I_p(P)\Delta P$ is the number of recoil protons in momentum interval ΔP, $I_n(P)\Delta P$ is the number of incident neutrons in momentum interval ΔP, n_H is the number of protons/cm^2 in the radiator, $(d\sigma/d\Omega)^{LAB}_{np \rightarrow pn}$ is the np laboratory elastic cross section at proton angle θ_p, and $\varepsilon(\theta_p)$ is the spectrometer acceptance function at polar angle θ_p.

In order to extract the desired quantity $I_n(P)\Delta P$, it is required to know the elastic cross section as a function of momentum over the angular acceptance of the spectrometer. Recent measurements[6,7] of the charge exchange cross section over the momentum range of interest differ by about 15%, but indicate that the total variation from P = 700 - 1100 MeV/c is about 20%. This correction has not been applied to the spectrum presented below.

A further source of uncertainty in the shape of the spectrum derives from the possibility of inelastic scattering in the radiator; namely the reactions $np \rightarrow pn\pi^0$ and $np \rightarrow pp\pi^-$. From the kinematics of these reactions the inelastic protons occur at $P < 1000$ MeV/c. An estimate of the magnitude of this effect was made from ratios of the published values of the total cross sections to the elastic cross sections and is $< 5\%$ over the region of $P > 400$ MeV/c. This effect will be measured empirically in the near future by determining the energy of the incoming neutron by a measurement of the time-of-flight.

Subject to the aforementioned uncertainties, in Fig. 1 is shown the measured recoil proton spectrum resulting from the 0° neutrons produced in the reaction $pp \rightarrow np\pi^+$ at an incident proton energy of 764 MeV. The representative errors shown on the figure result from counting statistics only. The strong peak observed around a momentum of 970 MeV/c can be attributed to the p-wave resonance in the π^+p final state corresponding to the Δ^{++}. A plot of the invariant mass in the π^+p system as a function of the measured momentum is also included in the figure. The observed peak momentum and FWHM correspond to an invariant mass in the π^+p system of approximately 1215 MeV with a width of about 95 MeV. Preliminary calculations of phase space and the momentum dependence of the production amplitude appear to account for the observed spectrum.[8]

By taking the slightly extrapolated value[6] of 68 mb/sr at $\theta p = 0.02$ for the elastic n-p differential cross sections, a peak cross section of $d\sigma/d\Omega_n dP_n = 0.06$ mb/sr-MeV/c at 970 MeV/c is obtained.

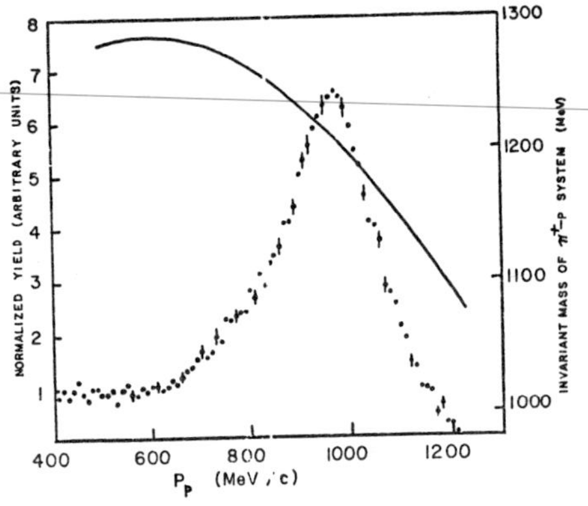

FIGURE 1. Momentum spectrum of recoil protons resulting from the 0° neutrons created in the reaction $pp \rightarrow np\pi^+$ at $T_p = 765$ MeV.

*Supported in part by the U.S. Atomic Energy Commission.

1. A. G. Meshkovskii, et al., Soviet Physics JETP 8, 46 (1959) and references cited therein.
2. B. Baldoni, et al., Nuovo Cimento 26, 1376 (1962).
3. V. E. Barnes, et al., Phys. Rev. Letters 7, 288 (1961).
4. E. Ferrari and F. Sellari, Nuovo Cimento 27, 1450 (1963).
5. C. W. Bjork, et al., paper submitted to this conference.
6. G. Bizard, et al., contribution to the IInd Aix-en-Provence Intl. Conf. on Elem. Particles (1973).
7. P. Shepard, et al., Phys. Rev. (to be published).
8. G. Stephenson and W. Gibbs, private communication.

SUPPRESSION OF NUCLEON ISOBAR EXCITATION IN LIGHT NUCLEI*

C. A. Heusch, R. V. Kline+, and K. T. McDonald++
University of California, Santa Cruz, California, U.S.A.

In a recent set of investigations designed to test time reversal invariance, we conducted a detailed study of differential cross-sections for the two-body photodisintegration of He^3. Since possible T invariance effects were most likely to occur through interference of a resonant with a leading non-resonant partial wave, we chose to design the experiment such that it swept the energy region of photoexcitation of one nucleon inside the He^3 to a Δ (1236) isobar.

The transition $\gamma N \to \Delta$ on free nucleons is well-known to proceed almost "elastically" through the M1 transition; a small E2 admixture of $\sim$ 5% has been consistently claimed, but is very hard to establish in any detail. Similarly, the photodisintegration of the deuteron is well known to equally proceed largely through an s-channel Δ resonance (graph 1). Inside the deuteron,

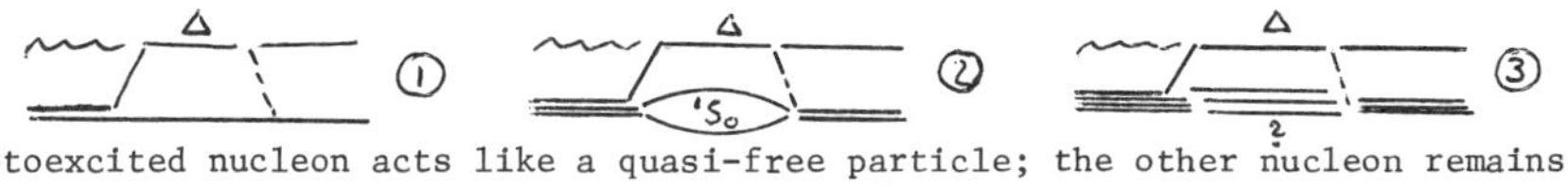

the photoexcited nucleon acts like a quasi-free particle; the other nucleon remains a spectator. To our astonishment, our study of $\gamma He^3 \to pd$ gave no indication of Δ excitation by an M1 transition at all. From 7 angular distributions (excluding the forward and backward directions) at energies between 250 and 550 MeV for the photon, we determined fifth-order polynomial fits in cos θ. These fits are motivated by considerations that the incoming He^3 is mostly in a $^2S_{1/2}$ state (neglecting the possible small $^4D_{1/2}$ admixture), and that one of its nucleons be subject to any possible E1, M1, and E2 transition from the photon field. This picture yields as the most likely transitions

$$E_1 \to {}^2P_{1/2} \qquad M_1 \to {}^4S_{3/2} \qquad E_2 \to {}^2D_{3/2}$$
$$\to {}^2P_{1/2} \qquad \to {}^2D_{3/2} \qquad \to {}^2D_{5/2}$$

where the pd final states are labeled according to total angular momentum, total spin, and relative orbital angular momentum. This leads to an angular distribution of the form

$$\sigma(\theta) = A + B\cos\theta + \sin^2\theta\,(C + D\cos\theta + E\cos^2\theta).$$

Photoexcitation of one nucleon to a Δ would show (graph 2) in the energy trend of the appropriate angular coefficients A...E. Our analysis shows that [1] nothing notable happens in the M1 transition at Δ excitation energy ($E_\gamma \sim$ 320 MeV), whereas the E2 transition shows changes at this energy. Notice that, in graph 2, the intermediate 2-nucleon system (lower line), for I-spin conservation, must be in a I=1 state, most probably a ^1S_o di-nucleon (non-resonant) state. The Δ deexcitation then demands that a π be absorbed by the di-nucleon system to form the loosely bound deuteron - which may be asking too much. (Conversely, $\pi d \to np$ may not prefer to lead to a final ^1S_o state).

We have a picture then where I-spin and J^P conservation effectively suppress the usually dominant $\gamma N \xrightarrow{M1} \Delta$ transition. The different angular behavior of E2 may allow the normally small $\gamma N \xrightarrow{E2} \Delta$ transition to persist in the constrained system. Fig. 1 shows the radically different behavior for $\gamma d \rightarrow np$ vs. $\gamma He^3 \rightarrow pd$, $\gamma He^4 \rightarrow pt$ at Δ energy.

We now look for other evidence for isobar excitation in light nuclei. These authors participated in a set of experiments destined to investigate $pd \rightarrow He^3\gamma$[(2)] and $\rightarrow He^3\pi^o$[(3)] in the Δ region. The first is linked to graph 2 by T invariance; the second would show, for a purely strong interaction, s channel resonance effects according to the graphs 4 and 5:

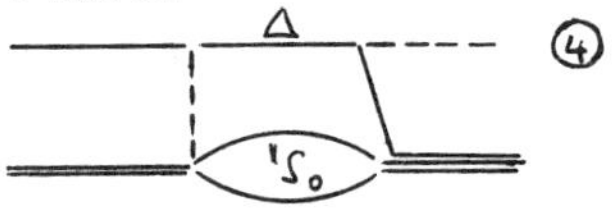

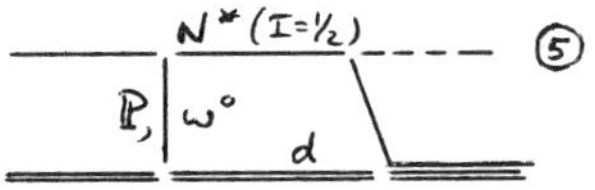

If the isobar is the Δ, the incoming deuteron has to go into an intermediate unbound di-nucleon state as above (2); for an intermediate I = ½ isobar like the $N^*(1460)$, Pomeron or ω^0 exchange leaves the deuteron undisturbed.

In fact, our collaboration[(2,3)] found no evidence for Δ excitation, in either of these reactions, with the strong interaction (graph 4) adding new information. However, a recent investigation at Saclay[(4)] of the process $dp \rightarrow He^3\pi^0$ suggests structure of the $180^0\pi^0$ production cross-section that can be interpreted as due to intermediate N^* formation (graph 5) according to the CM picture

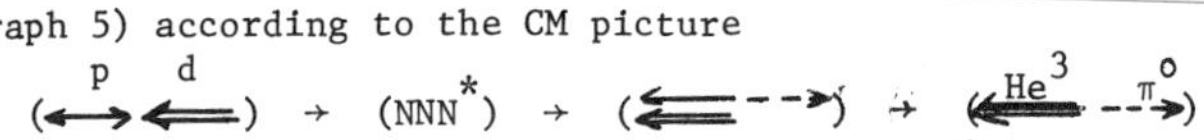

The energy at which the structure occurs indicates N^* (1460) as a likely candidate, but excludes the Δ (1238).

From all this evidence the following suggestive picture emerges:

1.) In the deuteron case, photoexcitation of nucleon isobars occurs roughly as on the free nucleon. The other nucleon acts as a spectator.

2.) 2-body He^3 photodisintegration and its inverse, radiative He^3 formation, are more heavily constrained systems where I conservation forbids Δ formation when the other nucleons are mere spectators. Δ excitation is strongly suppressed.

3.) The intriguing possibility exists that in 2-body He^3-photodisintegration, I and J^P conservation jointly conspire to allow only the E2 photoexcitation mode of the Δ, which, in the free nucleon, is only an insignificant admixture.

4.) In the strong interaction process $pd \rightarrow He^3\pi^o$, Δ formation is suppressed as in $pd \rightarrow He^3\gamma$; however, N^* (I=½) formation is allowed (by Pomeron or ω^0 exchange). In particular, diffraction-dissociation type processes like p+p (+ spectators) $\rightarrow p + N^*$ (+ spectators) may proceed much as in free nucleons.[(5)]

*Supported in part by the United States Atomic Energy Commission under Contract AT(04-3)-34, P.A. 197.

+Now at: Department of Physics, Harvard University, Cambridge, Mass.

++Now at: Nuclear Physics Division, CERN, Geneva, Switzerland.

References

1. C. A. Heusch, R. V. Kline, K. T. McDonald, C. Y. Prescott, UCSC 73-004 (1973); Phys. Rev. Lett.; to be published.

2. C. A. Heusch, R. V. Kline, K. T. McDonald, J. Carroll, D. H. Fredrickson, M. B. Goitein, B. MacDonald, V. Perez-Mendez, A. W. Stetz, UCSC 73-005 (1973); Phys. Rev. Lett.; to be published.

3. C. A. Heusch, R. V. Kline, K. T. McDonald, J. Carroll, D. H. Fredrickson, M. B. Goitein, B. MacDonald, V. Perez-Mendez, A. W. Stetz, UCSC 73-008 (1973); Phys. Rev. Lett.; to be published.

4. J. Banaigs, J. Berger, L. Goldzahl, T. Risser, L. Vu-Hai. Saclay preprint (1973).

5. In the non-relativistic quark model, this is particularly favored as the first radial quark excitation nucleon → $P_{11}(1460)$; its excitation shows a forward peak with a much steeper slope than other N^* excitation.

6. From C. Tzara, Proc. Int. Conference on Photonuclear Reactions and Applications, p. 573, Asilamar (1973).

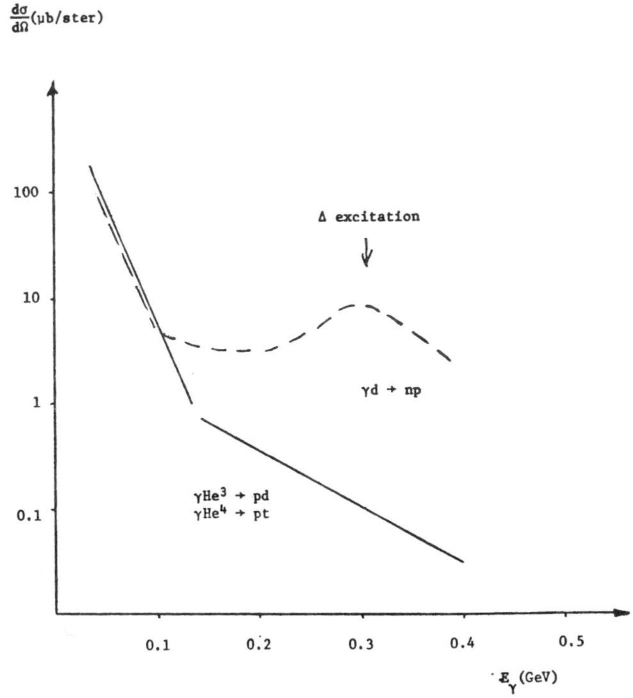

Fig. 1: Two-body photodisintegration of d, He^3, He^4: schematic behavior of 90° cross-sections. Dashed line (d) shows prominent Δ excitation peak. Solid line (He^3, He^4) shows no resonance effects up to E_γ = 0.4 GeV. Ref's. (1), (6).

ENERGY DEPENDENT PHASE SHIFT ANALYSIS OF $\pi^{\pm}He^4$ ELASTIC SCATTERING

L.Alexandrov, T.Angelescu, I.V.Falomkin, F.Nichitiu, Yu.A.Shcherbakov.

Joint Institute for Nuclear Research - Dubna
Institutul de Fizică Atomică - București

The energy interval of our energy-dependent phase-shift analysis is 24MeV-180MeV. The total amplitude for $\pi^{\pm}He^4$ elastic scattering is

(1) $f^{\pm}(\theta,K) = f_c^{\pm}(\theta,K) + f_s(\theta,K) + \frac{1}{K}\sum(2\ell+1)(e^{2i\zeta_\ell^{\pm}}-1)T_\ell(K)P_\ell(\cos\theta)$

where $T_\ell = (\eta_\ell e^{2i\delta_\ell}-1)/2i = 1/(\mathrm{ctg}\,\Delta_\ell - i)$ (2)

and $\zeta_\ell^{\pm} = \arg\Gamma(\ell+1+in_c^{\pm})$ (3)

are strong partial wave amplitudes and Coulomb phase shifts.
The pure nuclear and Coulomb amplitudes are

(4) $f_s(\theta,K) = \frac{1}{K}\sum(2\ell+1)T_\ell(K)P_\ell(\cos\theta)$

(5) $f_c^{\pm}(\theta,K) = \frac{-n_c^{\pm}}{2K\sin^2\frac{\theta}{2}}\exp(-in_c^{\pm}\ln\sin^2\frac{\theta}{2} - 2i\zeta_0^{\pm})\cdot F_\alpha(q^2)F_\pi(q^2)$

where $F_\alpha(q^2)$ and $F_\pi(q^2)$ are electromagnetic formfactors of He^4 and π, respectively. The non-relativistic Coulomb strength parameter is

(6) $n_c^{\pm} = Z_1Z_2\alpha m/K_{NR}$ $\quad m = m_1m_2/(m_1+m_2)$ $\quad \alpha = 1/137.$

In order to obtain similar equations for scattering amplitudes but with relativistic kinematics ($K_{NR}\to K_R$) we have used the idea of quasi-potential. The corresponding Schrodinger equation for the relativistic two-body problem /1/ is obtained with substitution (Todorov)

(7) $m \to m_W = m_1m_2/W$ $\qquad W = (m_1+m_2)^2 + 2Tm_2$
$E \to K_R^2/2m_W$

in the non-relativistic Schrodinger equation. In this work we have used the Todorov substitution (eq.7) in eqs.(1) - (6). The relativistic Coulomb strength parameter will be

(8) $n_c^{\pm} = Z_1Z_2\alpha m_1m_2/K_RW$

Prior to the energy-dependence phase-shift analysis we made a fit of all single-energy solutions /2/ to test different forms for energy behavior of partial wave amplitudes. This analysis was repeated including high partial wave amplitudes taken from ACE analysis /3/ and the new experimental data on total cross section /4/. The less χ^2 per data point and less number of free parameters is reached with the following parametrization of strong partial wave

(9) $K^{2\ell+1}\mathrm{ctg}\,\Delta_\ell = -1/a_\ell + b_\ell T + c_\ell T^2 + \cdots$ (T is kinetic energy in Lab.)

(a_ℓ, b_ℓ, c_ℓ are complex numbers). For S wave we have used 6 parameters, for P and D waves 8, and for F and G waves 6 parameters (34 physical parameters). The total number of experimental data is 378 for elastic differential cross section (154 π^+He^4 and 224 π^-He^4) and 15 data for total cross section. In our analysis we have used 16 more parameters for normalization factors (the same for π^+ and π^- at each energy) and one parameter for elm. pion radius (Gaussian formfactor). Total number of free parameters is 51 for 393 data points. For Chi-squared minimisation we have used a new algorithm for solving non-linear systems of equations, which is based on the regularized iteration processes of the Gauss-Newton type /5/. The χ^2/N_p was $\sim 4.$ without normalization factors, ~ 2.2 with 16 normalization parameters and 1.25 for fit without 28 points (with $\chi_i^2 \geqslant 8$).
The results are shown in Fig.1 (η_ℓ and δ_ℓ for S,P and D waves together with energy-independent phase-shift results) and in Fig.2 (Argand plot). The total and elastic cross sections are shown in Fig.3. The resonant behavior of the P,D,F and G waves is demonstrated in Fig.4 by velocity diagram of the Argand vector ($|df_\ell/dT|$). The smallest non-resonant

background is in P wave which has a clear Breit-Wigner form in partial total and elastic cross section: Fig.5. The $Ref_s(0)$ is shown in Fig.6 (dashed region is error corridor), our dispersion relations calculation (curve a) and finite sum-rules results /6/ (curve b).

The scattering length for S wave (in fm.) is

	From $\pi^{\pm}$He at 24 MeV /7/	Our En. Dep. Ph.Sh. Analysis	Mesic Atoms
Re a_0	-0.188	-0.160 +/- 0.010	-0.123
Im a_0	0.114	0.055 +/- 0.007	0.037

The electromagnetic pion radius obtained from our analysis using the method of ref./8/ is $\langle r_\pi^2 \rangle^{1/2} = 0.83 {+0.16 \atop -0.18}$ fm.

References.

1) I.T.Todorov. Preprint JINR E2-5813, Dubna 1971.
2) I.V.Falomkin, M.M.Kulyukin, V.I.Lyashenko, A.Mihul, F.Nichitiu, G.Piragino G.Pontecorvo, Yu.A.Shcherbakov. Preprint JINR E1-6534, Dubna 1972; Nuovo Cimento 21A, 168(1974); Lett.Nuovo Cimento 5, 1125(1972).
3) O.V.Dumbrais, F.Nichitiu, Yu.A.Shcherbakov. Rev.Roumaine de Physique 18, 1249 (1973).
4) C.Wilkin, C.R.Cox, J.J.Domingo, K.Gabathuler, E.Pedroni, J.Rohlin, P.Schwaller, M.W.Tanner. Nucl.Phys. 62B, 61 (1973).
5) L.Alexandrov. Preprints JINR P5-7258, P2-7259, Dubna 1973.
6) G.Alberi, Z.Grossman, J.Leon, P.Osland. Preprint IC/74/13, Trieste 1974.
7) M.E.Nordberg, K.F.Kinsey. Phys.Lett. 20, 692 (1966).
8) F.Nichitiu, Yu.A.Shcherbakov. Nucl.Phys. 61B, 429 (1973).

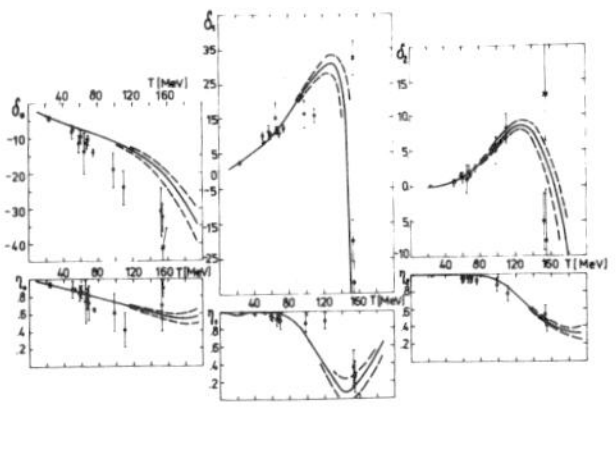

Fig.1.

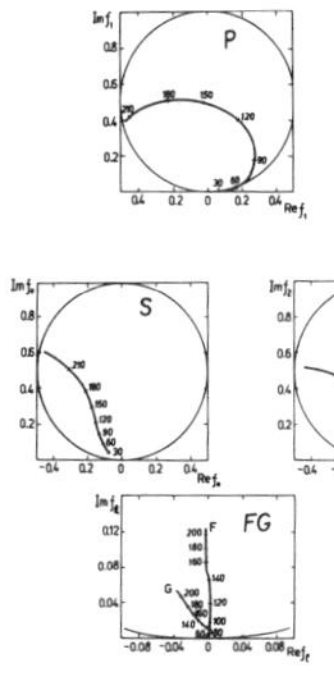

Fig.2.

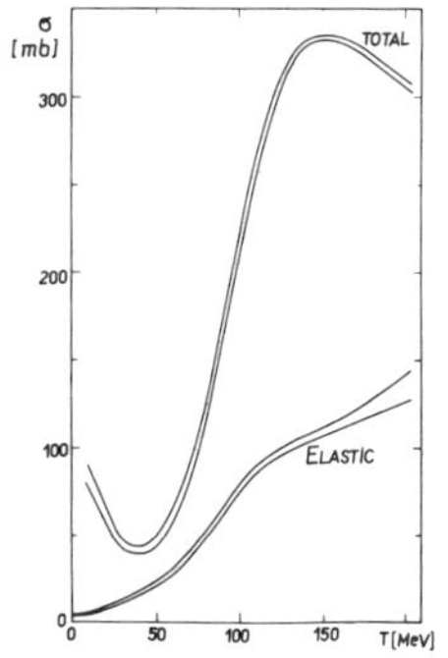

Fig.3.

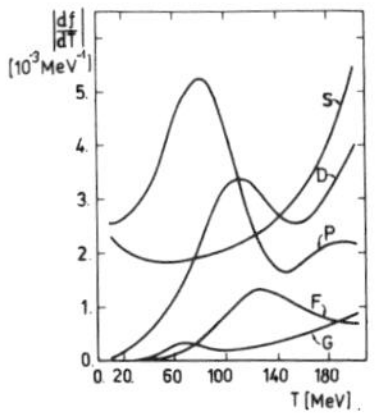

Fig.4.

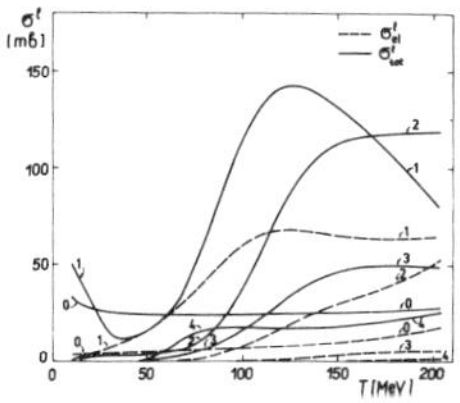

Fig.5.

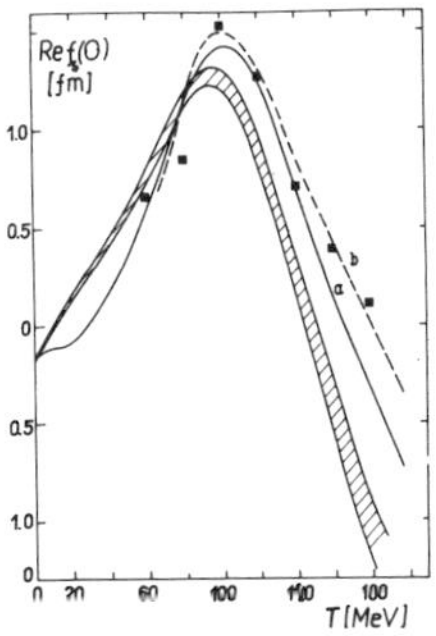

Fig.6.

IV

RADIATION DE FREINAGE NUCLEAIRE
INTERACTIONS SUR COUCHE ET
HORS COUCHE D'ENERGIE

NUCLEAR BREMSSTRAHLUNG
ON-SHELL AND OFF-SHELL
INTERACTIONS

NUCLEAR BREMSSTRAHLUNG FROM FEW BODY SYSTEMS

A.M. Green, Research Institute for Theoretical Physics, University of Helsinki, Finland

This talk will be restricted mainly to those bremsstrahlung processes

$$A + B \rightarrow A + B + \gamma \tag{1}$$

in which at least one of the particles (A,B) is not a nucleon e.g. d+p, p+α, He^3+α and α+α.

As a first attempt to understand the recent experiments the theory introduced by Feshbach and Yennie[1)] has been used and proves to be quite successful. This theory expresses the bremsstrahlung cross section in terms of phase shifts describing the corresponding elastic process

$$A + B \rightarrow A + B \tag{2}$$

When A and B scatter before emitting the γ-ray, the elastic scattering amplitude (T) is needed at an energy E_i (see fig. 1a). When A and B scatter after the γ-emission, T is needed at an energy E_f (see fig. 1b), where $E_f = E_i - E_\gamma$. The bremsstrahlung cross section can then be expressed as

$$\frac{d^2\sigma}{d\Omega_A \, d\Omega_B} \propto \left| M(E_i)\,\beta_f + \beta_i\, M(E_f) \right|^2 \tag{3}$$

where $\beta_{f,i}$ are known functions representing the γ emission from A and B; and

$$M(E) = T(E) + K \frac{\partial}{\partial\theta} T(E) \tag{4}$$

represents the scattering of A and B. In this last expression K is a

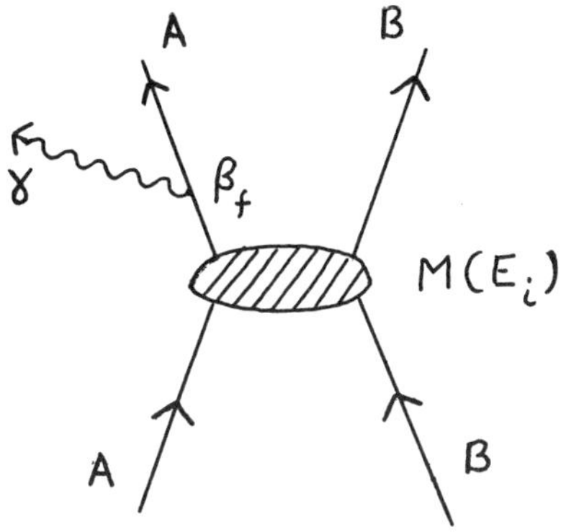

Fig. 1a

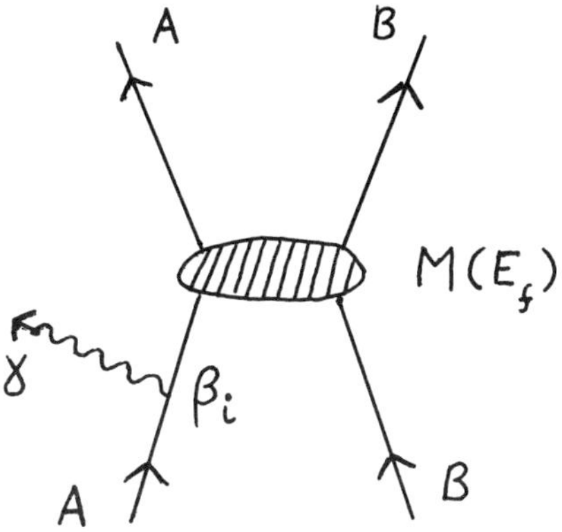

Fig. 1b

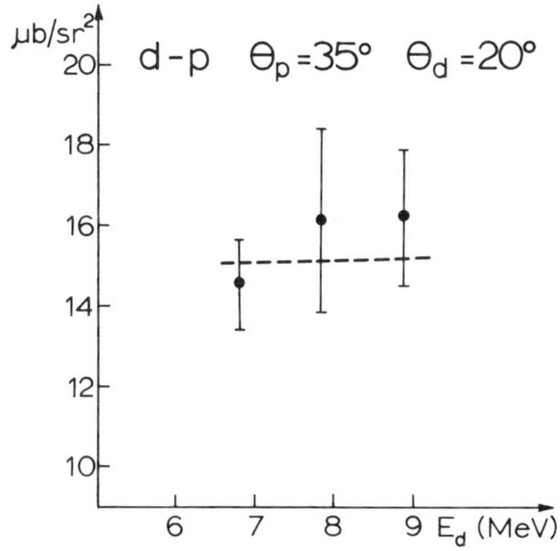

Fig. 2

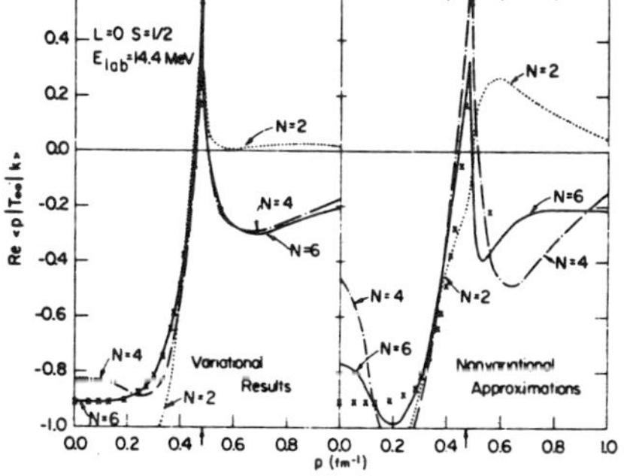

Fig. 2a

known kinematical factor, and θ the centre of mass scattering angle. The two contributions to the cross section can be visualized as figures 1a and 1b.

The two terms in eq. 4 are the first terms in an expansion of M(E) in powers of E_γ, and depend only on the on-energy-shell scattering amplitude T. The higher order terms depend on the off-energy-shell amplitude, and so cannot be obtained directly from T. This approach of Feshbach and Yennie is an extension of the work of Low[6], which relates the bremsstrahlung cross section to the elastic scattering amplitude at an energy $\frac{E_i+E_f}{2}$. It is thought that the theory of Feshbach and Yennie is more appropriate when there are resonances in the two body system.

$d+p \rightarrow d+p+\gamma$

The only experiments measuring the bremsstrahlung from d-p scattering[2] have been done with the coplanar geometry for $\theta_p=35^\circ$ and $\theta_d=20^\circ$. At the energies considered the elastic scattering is mainly s-wave. Therefore, the second term in eq. 4, which is zero for pure s-wave scattering, gives only a small contribution to M(E). In fig. 2 it is seen that this simple theory[3] is in good agreement with the present experimental data.

An interesting feature of the d-p elastic scattering amplitude, unlike the nucleon-nucleon amplitude, is that it varies rapidly when you go away from the energy shell (see fig. 2a). This is associated with the deuteron having a small binding energy and the variation is more or less model independent[8]. However, since this variation is due to a square root singularity, eq. 4 probably cannot be extended by simply including a model independent term of order E^2_γ. It would be of interest to see whether this enhanced off-energy-shell effect could be observed, since in the nucleon-nucleon case this has so far proven to be rather difficult.

$p+\alpha \rightarrow p+\alpha+\gamma$

The experiments are few in number[4,5] the only geometry being

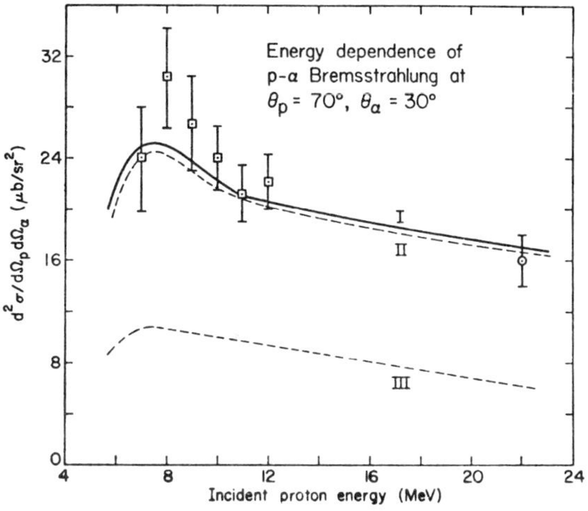

Fig. 3

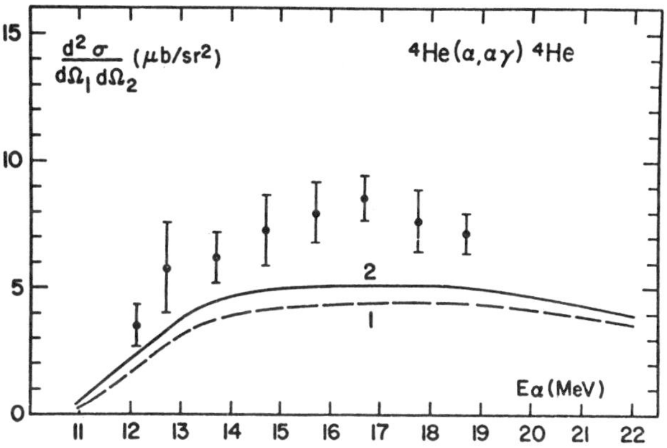

Fig. 4

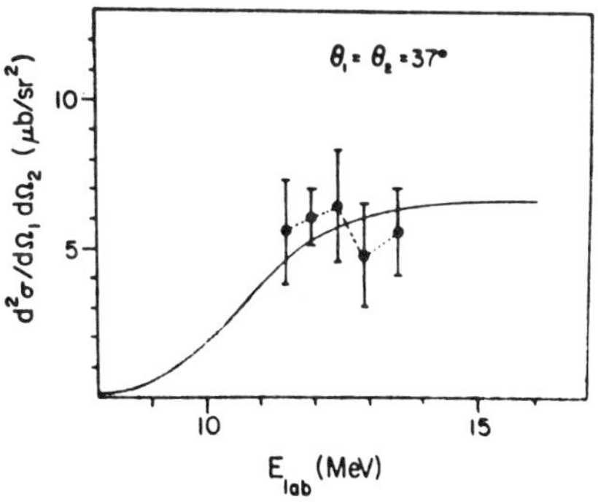

Fig. 5

$\theta_p=70^o$ and $\theta_\alpha=30^o$. This corresponds to a centre of mass scattering angle of about 90°. Unlike the d+p case p-waves play an important role in p+α scattering. This means that both terms in eq. 4 are important as can be seen in fig. 3. Curve III shows the cross section when only the s-wave dominated first term of eq. 4 is used. Curve II includes the p-wave dominated second term in eq. 4 and Curve I contains a Coulomb correction compared with II. As can be seen the agreement between theory and experiment is good. This is still true at $E_p=45$ MeV where inelastic processes can take place[5]. The p+α bremsstrahlung now seems sufficiently well understood to employ it as a tool in the study of nuclear reactions as will be described later.

$He^3+\alpha\rightarrow He^3+\alpha+\gamma$

This reaction has been performed for only one geometry[9] ($\theta_\alpha=\theta_{He^3}=$ $=37^o$) and for a narrow range of energies $E(He^3)$ from 7.4 to 8 MeV. At 7.4 MeV the measured cross section is 12.6±3.4 $\mu b/sr^2$ in good agreement with the theoretical estimate of 14.3 $\mu b/sr^2$ obtained using only the first term of eq. 4.

$\alpha+\alpha\rightarrow\alpha+\alpha+\gamma$

The previous reactions have involved dipole radiation and in all cases there seems to be good agreement between theory and experiment. However, the α-α reaction, which generates quadrupole radiation, shows a slight disagreement for $\theta_\alpha=35^o$ as shown[10] in fig. 4. The dashed curve is obtained using eq. 4 and the solid curve includes a potential dependent correction[11], which is seen to be small. For $\theta_\alpha=37^o$ this disagreement[9] seems to disappear as seen in fig. 5.

A distinguishing feature of the α-α system at these energies is the presence of resonances - the 4^+ and 2^+ excited states of Be^8 at 11.4 MeV and 2.9 MeV respectively. So far only bremsstrahlung from external emission has been employed in the comparison with experiment. This is the process depicted in fig. 1. However, internal bremsstrahlung shown in fig. 6 may also contribute. For p-p scattering this contribution is

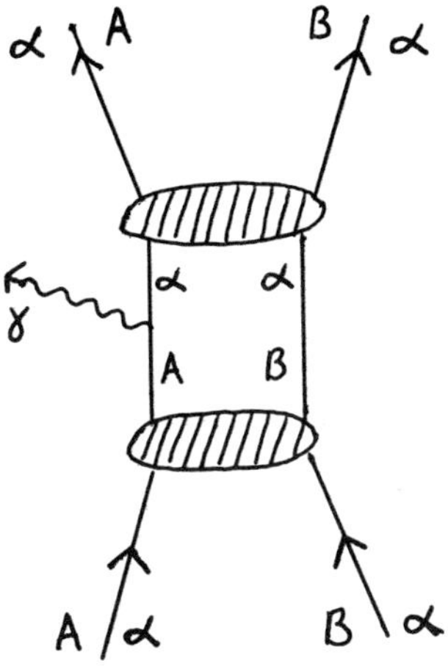

Fig. 6

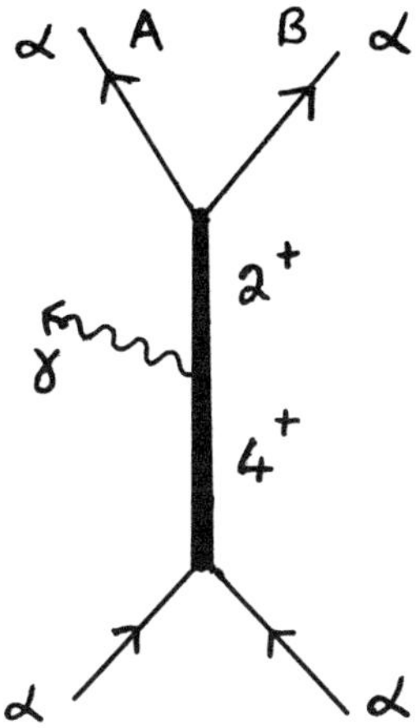

Fig. 7

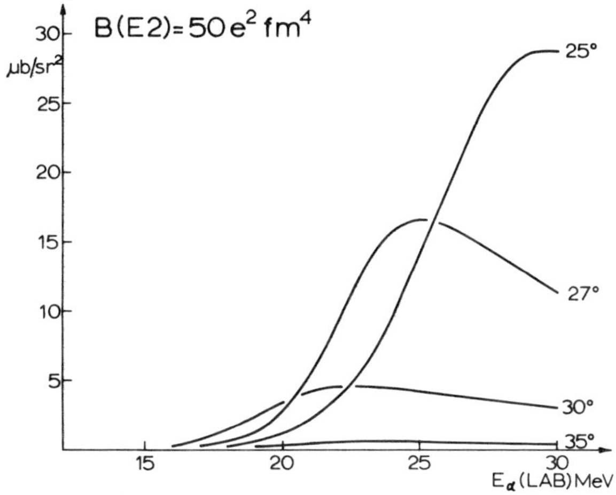

Fig. 8

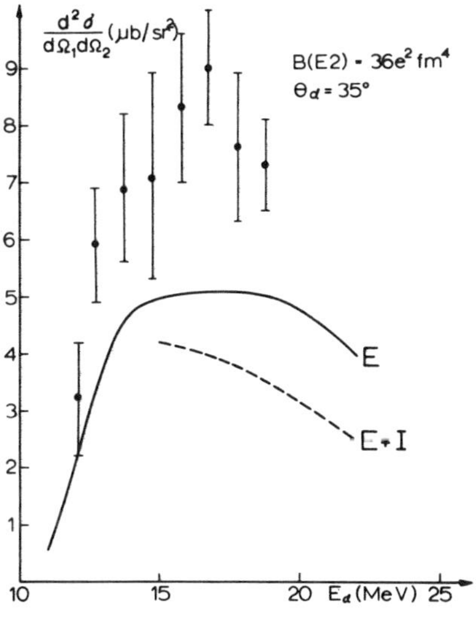

Fig. 9

small; but there is no reason to believe this is so for α-α scattering since there are now resonances present. The effect of the 4^+ and 2^+ resonances has been estimated[12] in terms of a quantity that can be interpreted as the B(E2) between the two resonances. It is then more appropriate to illustrate the effect by fig. 7 and not fig. 6. The amplitude for the initial α to form the 4^+ in Be^8 is given by the width Γ_4 of the 4^+. The width (Γ_2) of the 2^+ gives the amplitude for this state to break up into two α's again. This process, in fact, goes beyond what is usually meant by internal bremsstrahlung (fig. 6). The transition between the 4^+ and 2^+ levels involves the eight nucleon wavefunctions of Be^8 and these wavefunctions are more than just two α particles. Of course, the spectroscopic factor is near unity and so the non-α components are small. Even so these additional components may add to the collectivity of the 4^+ to 2^+ transition. To estimate the bremsstrahlung cross section the widths Γ_2 and Γ_4 are taken from elastic α-α scattering which is well described in terms of these resonances in this energy region. The $B(E2, 4^+ \to 2^+)$ is taken to be 50 $e^2 fm^4$ or 36 $e^2 fm^4$ in agreement with microscopic calculations[13]. The contribution to the bremsstrahlung cross section from the resonance to resonance transition can then be written as

$$\frac{d^2\sigma}{d\Omega_1\, d\Omega_2} \propto \frac{\Gamma_4}{D_4^2}\, B(E2, 4^+ \to 2^+)\, \frac{\Gamma_2}{D_2^2} \qquad (5)$$

where $D_4 = E(4^+) - E_i - i\frac{\Gamma_4}{2}$

and $D_2 = E(2^+) - E_f - i\frac{\Gamma_2}{2}$

are the resonance propagators.

In fig. 8, are shown the results from eq. 5 for various values of $\theta_{\alpha 1} = \theta_{\alpha 2}$. When this resonance to resonance contribution is added to the external bremsstrahlung the cross sections shown in figs. 9 and 10 are obtained. The curve E is the for the external bremsstrahlung calculated from eq. 4 and includes a small potential dependent correction[11]. The curve E+I is the coherent sum of the external (E) and internal (I) processes. Unfortunately for $\theta_\alpha=35^\circ$ the effect of I is to increase the discrepancy with experiment. For $\theta_\alpha=30^\circ$, I reduces E by over a factor of three. For $\theta_\alpha=25^\circ$, I dominates and the total cross section is more or less that shown in fig. 8.

If the above theoretical estimates for the B(E2) are reasonable, the results show that the B(E2) could be measured by performing bremsstrahlung experiments with $\theta_\alpha < 30^\circ$ and $20 < E_\alpha < 30$ MeV. If this method proves to be practicable, perhaps the same ideas could be extended to, for example, $\alpha+C^{12}$ to extract electromagnetic transition rates between states in O^{16} that have a large $\alpha+C^{12}$ parentage.

As a side comment, it should be mentioned that a similar idea has been proposed[7)] as a means of measuring the magnetic moment of the Δ(1236). The experiment is

$$\pi^+ + p \rightarrow \pi^+ + p + \gamma$$

in the region of the Δ(1236). Since the Δ(1236) has a large width (≈120 MeV), the π+p system can emit γ-radiation and still remain within the resonance after the emission. However, there seem to be problems since the only experiment performed[23)] does not agree with the predictions of ref. 7.

Non-coplanarity

With α-α bremsstrahlung there is an additional complication not present for dipole radiation. This is the problem of non-coplanarity. For d-p, p-α and He^3-α the radiation is dipole and the non-coplanarity factor F defined by the equation

$$\frac{d^2\sigma}{d\Omega_1\, d\Omega_2}(\theta,\phi) = F\,\frac{d^2\sigma}{d\Omega_1\, d\Omega_2}(\theta,0)$$

has the approximate simple form[3)]

$$F = \begin{cases} 1+y^2 & y \leq 1 \\ 0 & y > 1 \end{cases}$$

where $y=\phi/\phi(\max)$. $\phi(\max)$ the maximum angle of non-coplanarity is only

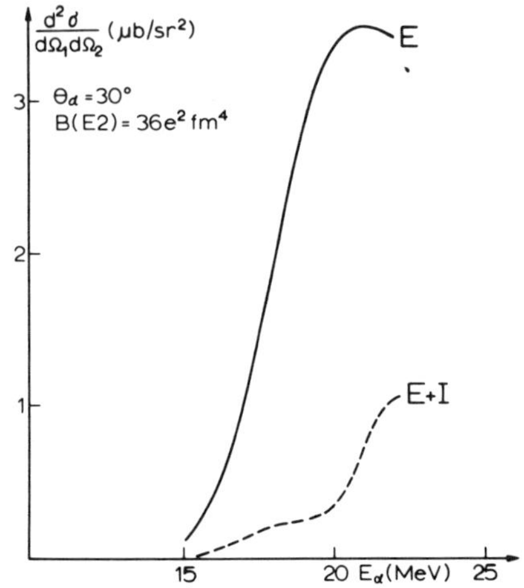

Fig. 10

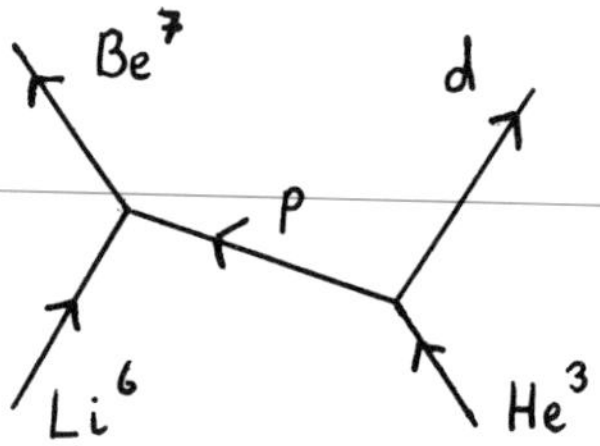

Fig. 11a

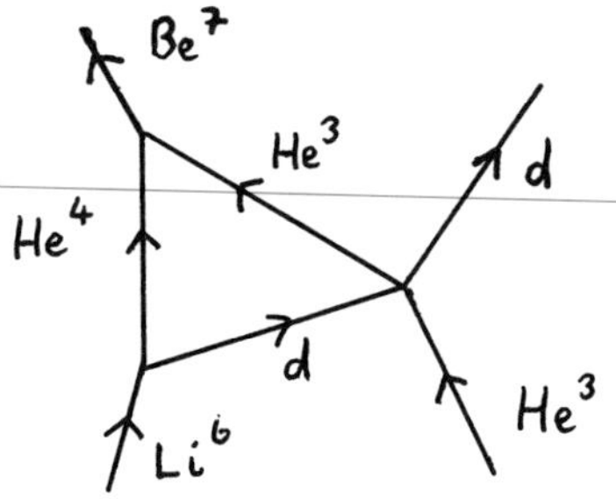

Fig. 11b

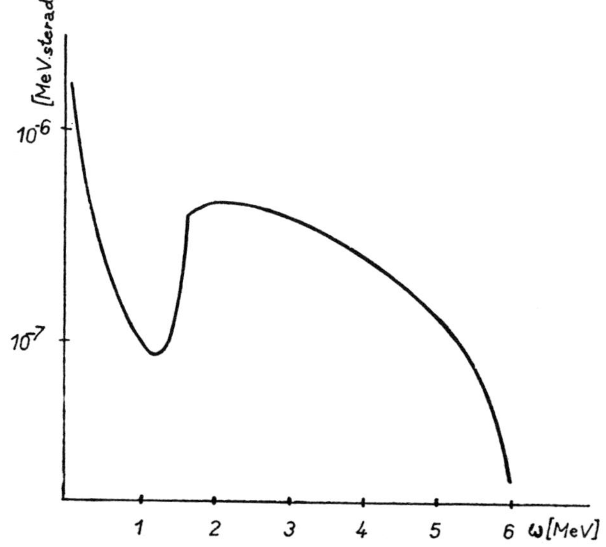

Fig. 12

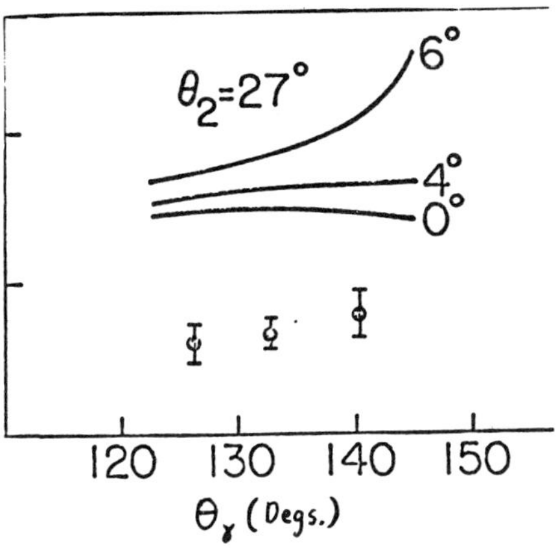

Fig. 13

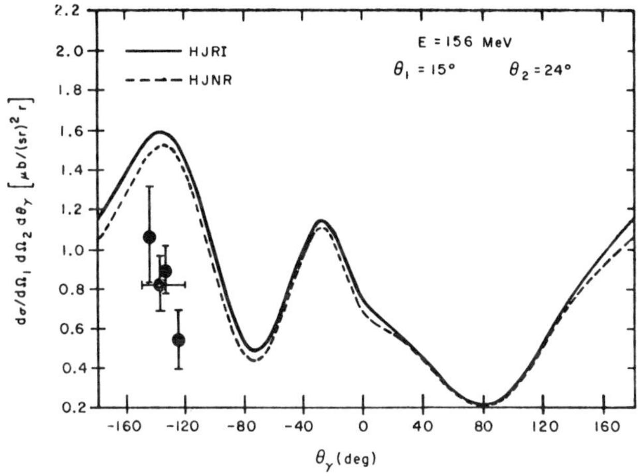

Fig. 14

The coplanar asymmetric $pp\gamma$ cross section $d\sigma/d\Omega_1 d\Omega_2 d\theta_\gamma$ at 156 MeV (15°, 24°), where negative θ_γ is taken on the 15° side of the beam. The results are presented for the Hamada-Johnston potential both including rescattering (HJRI) and with no rescattering (HJNR).

a few degrees for the situations considered here. However, for quadrupole radiation the form of F is much more complicated[3)]. This creates problems when experimentalists want to extract the coplanar cross section from their data, since the angular size of their counters is often comparable to ϕ(max). The form of F is then crucial, with different "reasonable" forms giving coplanar cross sections differing by a factor of two. Perhaps experimentalists should also quote their bremsstrahlung cross sections integrated over ϕ, since ϕ(max) is a perfectly well known angle and not subject to the uncertainties plagueing F.

The study of nuclear reactions

The original reason for considering nucleon-nucleus and nucleus-nucleus bremsstrahlung was to see whether it could be used as a tool for studying nuclear reactions[1,14)]. The idea was that bremsstrahlung could measure the time delay between the beginning and the end of a nuclear reaction and so allow an unambiguous separation between compound nucleus reactions and direct interactions. General arguments show that the bremsstrahlung intensity as a function of emitted photon energy makes a rapid change at a photon energy of about $\hbar/\tau$, where τ is the time delay. However, estimates of τ put the rapid change at a photon energy of a few KeV. Early attempts to measure this effect were defeated by the background.

Another theoretical proposal in some way similar to the one just discussed, is the attempt to distinguish different microscopic processes for direct nuclear reactions[15)]. This is best illustrated by the example $Li^6(He^3,d)Be^7$, which in principle involves the two processes shown in fig. 11. It is thought that the mechanism in fig. 11b dominates leading to characteristic cusps in the bremsstrahlung cross section as shown in fig. 12. These cusps arise from radiation being emitted from the particles in the triangle of fig. 11b. No experiments have been performed to demonstrate this effect.

<u>$p+p \to p+p+\gamma$</u>

Proton-proton bremsstrahlung has a long sad history[16]. The original hope of being able to measure off-the-energy-shell matrix elements of the two nucleon potential, and so eliminate some of the many phenomenological potentials, has yet to be fulfilled. Most experiments have not been sufficiently exotic to go far enough off-shell to differentiate between different potentials. A possible exception to this is the Orsay experiment[17] at 156 MeV, where one counter is fixed at 15° and the other varied from 18° to 27° on the other side of the beam. Calculations[18,19] with the Hamada-Johnston and Bryan-Scott potentials give results differing by several standard deviations from the experimental one (See fig. 13 and 14). However, this is not sufficient to eliminate these potentials as being unrealistic. There still remain some uncertainties, in particular how to deal with non-local potentials in the presence of the electromagnetic field. Also relativistic corrections[24] should be added. Even so, it would be surprising if these effects could account for the observed discrepancy.

The only experimental papers submitted to this conference concerning bremsstrahlung are two dealing with low energy p-p bremsstrahlung. The first[25] is for 6.92 MeV protons with 38° equal angle Harvard geometry and the result is in good agreement with theory. The second contribution[22] is more "exotic" having a 30° geometry and 11 and 13 MeV protons. [The ratio $\frac{E_f}{E_i} = \mathrm{Tan}^2\,\theta$ (see eqn. 3) is 0.61 for ref. 25 and 0.33 for ref. 22 and is a measure of how far the reaction is off-shell.] In this case, the experimental cross sections are well below the theoretical estimated as was found in the Orsay experiment at 156 MeV. Possibly the discrepancy is also due to the same reason as in the Orsay experiment. However, at these lower energies Coulomb effects and the proximity of E_f to the $^1S_0$ resonance lying at a few hundred KeV could also be responsible.

Interest in p-p bremsstrahlung was also raised when a calculation by McGuire[20] seemed to show a large Coulomb effect for proton angles of 9° and a proton energy of 52 MeV. However, a subsequent experiment[21] and later calculations[18] apparently did not substantiate McGuire's prediction. At present it is not clear what the true story is concerning this.

The author wishes to acknowledge valuable discussions with Dr. E. Nyman and people at Los Alamos (where much of this manuscript was prepared) in particular Drs. R. Amado, H. Feshbach and L. Heller.

References

1) H. Feshbach and D.R. Yennie, Nucl. Phys. 37 (1962) 150
2) J. Hall, W. Wölfli and R. Müller, Phys. Lett. 37B (1971) 53
3) A.M. Green and A. Prodon, Nucl. Phys. A183 (1972) 225
4) W. Wölfli, J. Hall and R. Müller, Phys. Rev. Lett. 27 (1971) 271
5) G. Anzelon, I. Slaus, S.Y. Tin, W.T.H. van Oers, M.B. Epstein, R.M. Eisberg, M. Makino and C.N. Waddell, Los Angeles Conference on Few Particle Problems in the Nuclear Interaction (North Holland 1972) p 54
6) F.E. Low, Phys. Rev. 110 (1958) 974
7) W.F. Fisher and P. Minkowski, Nucl. Phys. B36 (1972) 519
8) T.J. Brady and I.H. Sloan, Phys. Rev. C9 (1974) 4
9) B. Frois, J. Birchall, C.R. Lamontagne, U. von Muellendorff, R. Roy and R.J. Slobodrian, Phys. Rev. C8 (1973) 2132
10) U. Peyer, J. Hall, R. Müller, M. Suter and W. Wölfli, Phys. Lett. 41B (1972) 151
11) A.M. Green, R. Müller and U. Peyer, Nucl. Phys. A203 (1973) 42
12) A.M. Green and J. Niskanen to be published
13) M. Bouten, M.C. Bouten and P. van Leuven, Nucl. Phys. 168 (1971) 438
14) R.M. Eisberg, D.R. Yennie and D.H. Wilkinson, Nucl. Phys. 18 (1960) 338
15) F. Janouch and R. Mach, Nucl. Phys. A158 (1970) 193

Addendum

In $\alpha+\alpha$ Bremsstrahlung the contribution from the decay between the 4^+ and 2^+ resonances in ^{8}Be had a sign error. The agreement with the experimental data at $\Theta_\alpha = 35^\circ$ (fig. 9) is now much better. At $\Theta_\alpha = 30^\circ$ the two contributions E and I add constructively to give a cross-section of about 12 $\mu b/sr^2$ at $E \approx 21$ MeV .

$\Theta_\alpha = 35^\circ$

E_α (MeV)	$\sigma(E)$ ($\mu b/sr^2$)	$\sigma(I)$	$\sigma(E+I)$	$\left\lvert\sqrt{\sigma(E)} + \sqrt{\sigma(I)}\right\rvert^2$
15	4•98	.03	5•8	5•8
16	5•05	.05	6•1	6•1
17	5•08	.08	6•4	6•4
18	5•08	.11	6•7	6•7
19	4•99	.16	6•8	6•9
20	4•78	.21	6•9	7•0
21	4•43	.27	6•6	6•9
22	3•99	.32	6•1	6•6

$\Theta_\alpha = 30^\circ$

E_α	$\sigma(E)$	$\sigma(I)$	$\sigma(E+I)$	$\left\lvert\sqrt{\sigma(E)} + \sqrt{\sigma(I)}\right\rvert^2$
15	0•12	.06	0•3	•3
16	0•42	.17	1•1	1•1
17	1•03	.45	2•8	2•9
18	1•93	1.00	5•7	5•8
19	2•79	1.75	8•8	9•0
20	3•31	2.48	11•3	11•5
21	3•48	3.01	12•4	13•0
22	3•42	3.28	12•4	13•4

The column $\sigma(E)$ shows the results of ref. 11, $\sigma(I)$ from eq. 5, and $\sigma(E+I)$ the coherent sum of the two amplitudes E and I . The last column $\left\lvert\sqrt{\sigma(E)} + \sqrt{\sigma(I)}\right\rvert^2$ shows that there is almost maximum constructive interference. The column $\sigma(E+I)$ replaces the results shown in figs. 9 and 10.

We identify the internal wavefunction obtained in the neighborhood of a resonance as

$$|\psi(R_\ell)\rangle \propto \frac{\sqrt{\Gamma_\ell}}{D_\ell}\,|R_\ell\rangle\, i^\ell\, Y_\ell$$

where the constant of proportionality is real, and the radial part of the resonance wavefunction $|R_\ell\rangle$ can also be chosen as real - see Mahaux and Weidenmüller, Shell Model approach to Nuclear Reactions, page 82. Eq. 5 is then obtained from

$$\langle \psi(R_2)\,|\,\theta(E2)\,|\,\psi(R_4)\rangle$$

where θ(E2) is the electromagnetic E2 operator. The corresponding reduced matrix element is expected to be positive for the angular momentum considered here, and so can be related to $+\sqrt{B(E2,\ 4^+ \to 2^+)}$. For example, in a single particle model the reduced matrix element becomes

$$(-1)^{\ell_i}\sqrt{\frac{\hat{\ell}_i\,\hat{\ell}_f\,5}{4\pi}}\begin{pmatrix}\ell_i & 2 & \ell_f\\ 0 & 0 & 0\end{pmatrix}\langle R_2|r^2|R_4\rangle$$

Since R_2 and R_4 correspond to the lowest eigenvalues of angular momentum l_i and l_f , the radial wavefunctions will not contain nodes and so can be chosen as positive. On the other hand for the collective model the reduced matrix element contains the intrinsic quadrupole moment $Q_0 = \langle X|\theta(E\,2)\,|X\rangle$, where X is the intrinsic wavefunction from which $\psi(R_{2,4})$ are obtained by projection. For ^{8}Be we expect Q_0 to be positive and so again the reduced matrix element needed is positive.

16) M.L. Halbert - Contribution to Gull Lake symposium, published in "The Two Body Force in Nuclei" p. 53 (S.M. Austin and Crawley-Plenum Press 1972)

D. Drechsel - Contribution to Rome symposium 1972, published in "The Nuclear Many-Body Problem" (International Physics Series -4)

E. Nyman - Nordita Lectures and Preprint (1973)

17) A. Willis, V. Comparat, R. Frascaria, N. Marty, M. Morlet and N.Willis Phys. Rev. Lett. 28 (1972) 1063 and Los Angeles Conf. (1972) page 79

18) L. Heller and M. Rich to be publisehd in Phys. Rev. C and preprint LA-UR-73-1298

19) V.R. Brown, Phys. Rev. C6 (1972) 1110

20) J.H. McGuire, Phys. Lett. 40B (1972) 41

21) J. Sanada, K. Kondo, S. Seki, K. Kuriyama, N. Ryu, H. Hasai and K. Iwatani, Phys. Lett. 39B (1972) 509

22) W. Wölfli, M. Suter, G. Bonani and Ch. Stoller - submitted to this conference

23) M. Arman, D.J. Blasberg, R.P. Haddock, K.C. Leung, B.M.K. Nefkens, B.L. Schrock, D.I. Sober and J.M. Sperinde, Phys. Rev. Lett. 29 (1972) 962 and Uppsala Conf. 1973 p. 51

24) L.S. Celenza, M.K. Liou, M.I. Sobel and B.F. Gibson, Phys. Rev. C8 (1973) 838

25) B. Frois, R. Roy, C.R. Lamontagne, U. von Moellendorff and R.J. Slobodrian - submitted to this conference

DISCUSSION

CAHILL(Flinders): In discussing d-p bremsstrahlung you mentioned the strong singularity in the off-shell d-p amplitude. This singularity is also discussed by Amado, Thomas and myself at this conference. The position of the singularity is kinetical and is a square root branch point in the variable shown. This singularity may be removed by a correct choice of momentum variable. Even after removal of the singularity, the off-shell amplitude is rapidly changing (but of course in a non-singular manner). This rapid change is related to the small deuteron binding energy (ref:R.T.Cahill, Phys.Rev.

C9(1974) 473).

GREEN: You would say that this strong square root singularity should not show up directly in the bremsstrahlung ?

CAHILL: No, I am making no inference in that direction. I am just clarifying the nature of the singularity and you might well see it; but it depends on what variable is appropriate to bremsstrahlung.

GREEN: When I was in Los Alamos this summer, I discussed this with Heller and Amado. We all agreed that this problem was interesting, but for different reasons: Amado, the three-body expert, thought this would greatly interest Heller; and Heller, the bremsstrahlung expert, thought it would greatly interest Amado.

NOYES(SLAC): I very much like the direct test of quantum mechanics suggested by measuring the time delay in bremsstrahlung. In this connection it is worth noting that a rigorous theory of the time delay has been developed by Osborn, using the full Faddeev theory. It is, roughly speaking, simply the energy derivative of the reduced S-matrix; so it can be directly computed from the experimental quantities. The result presumably generalizes to N particles. This work has just been completed this summer, and is available as a SLAC preprint by Osborn and Bollé.

WIGNER(Princeton): I am glad you pointed out that the internal bremsstrahlung is the same thing as capture radiation from a resonance. However, the angular distribution of this should be very different from that of the usual bremsstrahlung: it results from a $4 \rightarrow 2$ angular momentum transition, whereas the usual bremsstrahlung involves all angular momenta. Why can the two not be distinguished by this angular distribution ?

GREEN: Because for the moment, only the integrated cross-section over the γ-ray angular distribution has been measured.

WIGNER: So the difficulty is they did not measure the angular distribution of the γ-ray.

O'CONNELL(NBS): How do you choose the relative phase between internal ($4^+ \rightarrow 2^+$) and external bremsstrahlung in the α-α reaction ?

GREEN: We were very careful about that. Clearly we expected it to go the other way. We went through that many times. It is just a technical detail. I would hope that we have done it correctly.

O'CONNELL: What is the prescription for choosing the phase ? There is no absolute phase. How do you choose the relative phase?

GREEN: You mean when adding the two processes?

O'CONNELL: Yes, you wrote down a B(E2), and that is already a matrix element squared. When you took the square root of that, you put a minus relative phase. How did you know to do that ?

GREEN: We were always guided by what would happen if we were considering a straight internal bremmstrahlung situation, and by saying that this internal bremsstrahlung arising from a 4^+ to 2^+ transition is simply a modification of that. As I said, the transition we are discussing is almost the emission of a γ-ray from an α-particle in the intermediate state. But it is slightly more than that. We get the phase from the usual internal bremsstrahlung matrix element.

HELLER(LASL): I would like to ask a somewhat related question. The Feshbach-Yennie formula does not include some internal radiation of the order of k^0. In the α-α case, if you work in the right frame and in the right gauge, there would be no such contribution. Is it clear that the way you did the calculation, you were not adding an extra internal radiation ?

GREEN: In the case of θ=30^0, there is perhaps a problem of double counting. Except that the contribution we add is of much higher order in k than in k^0 .

HELLER: That's what I thought.

GREEN: Remember that at the smaller angles the internal term is completely dominating, so that if we are double counting something, it is something very small.

HELLER: But if you are going to add terms of higher order in k coming from the internal emission, then you would also have to add terms of higher order in k coming from the external emission .

GREEN: I think that you should consider these as two different processes. One is bremsstrahlung; the other is a transition between states. The bremsstrahlung part, we treat as Feshbach and Yennie --it is some internal plus the external contribution--, and then there is another process which gets confused with bremsstrahlung. And there is probably no other way of looking at it.

MORAVCSIK(Oregon): What about the other internal contributions; for example, processes in which the proton is emitted by the intermediate state in the t-channel. I mean, the discrepancy does not frighten me greatly because there are things that have been left out.

GREEN: In ^{8}Be there are these two strong resonances and you might suspect that something odd could happen if you come in on one of these and go out on the other. We have simply calculated this effect and found it to be large.

LIOU(Brooklyn): The internal scattering term that you have included in your calculation is not complete. As we know there are many other terms that should be included. It has been shown in my earlier work that the rest of the terms, if they had been included, would cause a cancellation with the so-called rescattering term that you have included in your calculation. If we ignore these terms, then the low energy theorem would be violated. Are these terms important? We don't know. But certainly they are important in n-p bremsstrahlung. Now those terms are mainly due to the non-locality of the potential, and to the exchange effects. I think that exchange effects might be important in this case.

BRAYSHAW(Maryland): You showed a curve for the p-p bremsstrahlung calculated for the Hamada-Johnston potential, and also mentioned that it had been done for Bryan-Scott. How does the Bryan-Scott curve compare with Hamada-Johnston and with experiment ?

HELLER: It was comparable.

BRAYSHAW: But in the other direction.

LEVIN(BROWN): Given that inclusion of the internal bremsstrahlung in the α-α case worsens the fit to the data, can one really expect to use this as a method for determining B(E2)'s ? In particular, does changing the sign in the internal portion give better agreement with the α-α data ?

GREEN: Clearly it would be much better. But I think the important point is that the non-coplanarity form used to correct the data is simply the one of Feshbach and Yennie . Now one should use the best theory to correct the data and so include corrections from internal bremsstrahlung. When you analyse the data, you have to insert some theoretical input to extract a coplanar value. I think that it is better to give the average value over the finite geometry of the detectors and let the theoricians make their own mistakes.

LEVIN: It still leaves you a little bit up in the air about the phase of the extra matrix element which is coming in. Maybe you end up with imaginary amplitudes that you extract from the data.

GREEN: I am sure that there would be no question if we had put the phase the other way. But we tried to be as honest as possible and it goes the wrong way. I wish someone could find a physical reason for it.

SLOBODRIAN(Laval): With respect to the discrepancy in α-α bremsstrahlung between the Zurich data and theory, taking into account both external and internal emissions you mentioned in passing the off-shell effects, do you believe that the factor of two will be explained by such effects ?

GREEN: No. I frankly expect to explain that by a different form factor for the non-coplanarity.

AMADO(Pennsylvania) In two situations, the ^{8}Be one and the Δ one, the simple ideas seem not to work. Is this a technical accident, or is there a systematic difficulty that can be explained in general ?

GREEN: In our case we go from one resonance to another while in the other you go from one side of a resonance to the other, which is more difficult.

HELLER: I would make one comment on Thomson's calculation for the π-N bremsstrahlung case. Using a reasonable π-N off-shell T-matrix he comes out with no serious discrepancy with experiment. Have you considered doing an actual dynamical calculation in the α-α case, assuming some potential between the two α's which is consistent with what is known of ^{8}Be ,and do both external and internal bremsstrahlung behave that way?

GREEN: No. But I am sure that people like Virginia Brown and yourself have programs waiting for that.

HELLER: I asked the wrong question.

RADIATION DE FREINAGE DANS L'INTERACTION p-p à 6.92 MeV

B. Frois †, R. Roy ††, C.R. Lamontagne, U. von Moellendorff ††† et R.J. Slobodrian, Université Laval, Département de Physique, Laboratoire de Physique Nucléaire, Québec G1K 7P4, Canada.

Le problème de la radiation de freinage dans le système p-p a été l'objet d'un grand nombre de travaux expérimentaux et théoriques, aussi bien que de travaux de revue [1,2,3]. Récemment Dahlblom et Green [4] ont calculé la section efficace pour un nombre d'angles en fonction de l'énergie entre 1 et 13 MeV dans le système du laboratoire en utilisant le formalisme de Feshbach-Yennie [5] et ils ont remarqué le manque d'accord entre les mesures faites par Wölfli et al. [6] à 11 MeV et leurs calculs, aussi bien que le désaccord avec Felsner [7] dans l'estimation des termes correctifs.

L'étude de la radiation de freinage à 6.92 MeV a été faite pour déterminer si l'anomalie apparente à 11 MeV a lieu aux énergies plus basses. Les expériences ont été faites à l'aide du van de Graaff CN de 7 MV possédant un tube à haut gradient. Les techniques expérimentales ont été décrites précédemment [8]. Les deux compteurs étaient plaçés à $\theta = 38^o$. Le pic temps relatif aux événements de la région cinématique s'étendait sur 1.5 ns. Le fond moyen des coincidences fortuites a été évalué sur 70 ns. Les données expérimentales ont été corrigées pour les effets de non-coplanarité et pour la géométrie finie des détecteurs de façon analogue à la description contenue dans la référence 8. La section efficace coplanaire est égale à $1.7 \pm 0.7 \mu b\ sr^{-2}$. La figure ci-jointe contient

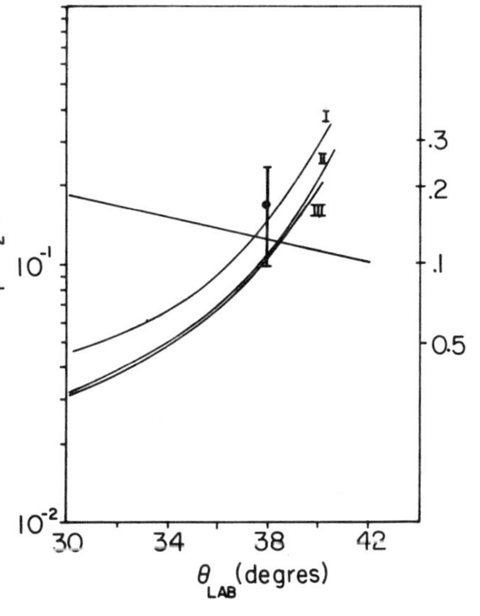

FIGURE I

une comparaison entre le résultat expérimental et les différentes prédictions théoriques. La courbe I est obtenue selon la méthode de Signell: l'amplitude élastique est calculée dans l'approximation de la portée effective

$$t = \frac{1}{-ip + \frac{1}{a_s} + r_s \frac{p^2}{2}}$$

p est l'impulsion relative des deux protons, a_s est la longueur de diffusion et r_s la portée effective. Signell avait utilisé dans ses calculs les paramètres de l'interaction p-p neutralisée $a_s = -17$ fm et $r_s = 2.8$ fm. Il nous avait semblé peu réaliste d'annuler l'effet coulombien et nous avons effectué le même calcul en utilisant directement les valeurs issues de l'expérience: $a_s = 7.69$ fm et $r_s = 2.65$ fm (voir courbe II). La courbe III a été tirée des résultats récents de Dahlbloom et Green [4], développés comme extension du formalisme de Feshbach et Yennie. Les courbes II et III sont très voisines et montrent l'importance de l'effet coulombien. Néanmoins notre précision expérimentale est insuffisante pour distinguer de façon concluante entres les différents modèles. Néanmoins nous pouvons affirmer qu'il n'est pas possible d'expliquer le désaccord observé à 11 MeV. Nous remercions avec plaisir l'aide précieuse du personnel du van de Graaff à l'Université Laval et l'appui de la Commission de Contrôle de l'Energie Atomique du Canada.

---

† A présent au Service de Physique Nucléaire à Haute Energie, CEN Saclay, B.p. N°2 91190 Gif-sur-Yvette, FRANCE.

† A présent au Lawrence Berkeley Laboratory, Berkeley, Ca 94720, U.S.A.

† A présent au IASR, Kernforschungszentrum, 75 Karlsruhe, Postfach 3640, R.F. Allemande.

BIBLIOGRAPHIE

1) M. Moravcsik, Rep. Prog. Phys. 35, 587 (1972)

2) P. Signell, Advances in Nucl. Phys. Vol.2 (1969)

3) M.L. Halbert, Gull Lake Symposium on the Two Body Force in Nuclei (1971)

4) T.K. Dahlbloom and A.M. Green, Phys. Letters 41B, 23 (1972)

5) H. Feshbach and D.R. Yennie, Nucl. Phys. 37, 150 (1962)

6) W. Wölfli, R. Müller and P. Marmier, Helv. Phys. Acta (separatum) 43 (1970)

7) G. Felsner, Phys. Letters 25 B, 290 (1967)

8) B. Frois, J. Birchall, C.R. Lamontagne, U. von Moellendorff, R. Roy and R.J. Slobodrian, Phys. Rev. C8, 2132 (1973).

BREMSSTRAHLUNG FROM LOW ENERGY PROTON-PROTON SCATTERING

W.Wölfli, M.Suter, G.Bonani and Ch.Stoller
Laboratorium für Kernphysik, Eidgenössische Technische Hochschule
8049 Zürich, Switzerland

Earlier measurements on the p-p bremsstrahlung (PPB) process at low energies delivered only upper limits for the cross sections, which in some cases[1,2], are not compatible with the most reliable theoretical predictions[3,4,5].

In an attempt to clarify this problem, the PPB cross sections were measured at two beam energies of 11 and 13 MeV using the equal angle Harvard geometry with $\theta_{lab} = 30^o$. The energies of the scattered and recoil protons as well as the time difference between the fast detector signals were measured simultaneously by means of a triple-parameter system implemented for use on line with a PDP 15 computer and associated software[6].

The time resolution of the fast coincidence circuit was better than 0.5 ns over the whole energy range of interest. The time calibration procedure and the evaluation of the true and random events are described elsewhere[7].

The PPB cross sections were obtained by comparison with the elastic scattering process, the differential cross section of which is well known[8]. In order to get the coplanar values given in table 1, the measured data were corrected for the finite beam size (4 mm dia) and the angular spread of the detector collimator system ($\pm 4^o$). The azimuthal dependence of the PPB cross section was approximated for Ø less than $Ø_m$ by an expression of the form $F(Ø) = \{1 - (Ø/Ø_m)^2\}\{1 + k(Ø/Ø_m)^2\}$, where $Ø_m$ is the maximum azimuthal angle allowed by the kinematics of this process and k an energy dependent parameter derived by Green et al.[9]. The necessary multidimensional integration were performed by means of the Monte Carlo method. The errors quoted reflect only the statistics and the uncertainty of the numerical integration procedure.

Table 1. Experimental and theoretical coplanar PPB cross sections for $\theta_1 = \theta_2 = 30^o$ at low energies.

E_p (MeV)	Experiment ($\mu b/sr^2$)	Theory 1[5] ($\mu b/sr^2$)	Theory 2[4] ($\mu b/sr^2$)
11.0	0.44 ± 0.04	0.47	0.43
13.0	0.46 ± 0.04	0.54	0.48

The data are compared with two theoretical approaches: Theory 1 is the nonrelativistic Hamada-Johnston potential model by Heller and Rich[5] in which the coulomb and double scattering effects are treated exactly; theory 2 is an extension of the model independent Feshbach-Yennie approach[10] discussed by Dahlblom and Green[4]. It is interesting to note that the two theories differ by about 10% even at these low energies. Our data favour the model independent values.

This work was supported by the Swiss Foundation for Scientific Research.

References

1) C. Joseph, A. Niiler, V. Valkovic, R. Spiger, S.T. Emerson, T. Canada, J. Sandler and G.C. Phillips, Phys. Rev. 178(1969)1621.

2) W. Wölfli, R. Müller and P. Marmier, Helv. Phys. Acta 43(1970)423.

3) P. Signell and D. Marker, Phys.Lett.28B(1968)79.

4) T.K. Dahlblom and A.M. Green, Nucl.Phys.A183(1972)225.

5) L. Heller and M. Rich, LA-Report UR-73-1298.

6) R. Müller, U. Peyer, W. Wölfli, J.Hall and M. Suter, Nucl. Instr. Meth. 104 (1972)125.

7) W. Wölfli, J. Hall and R. Müller, Phys. Rev. Lett. 27(1971)271.

8) N. Jarmie, J.H. Jett, J.I. Detch,Jr. and R.L. Hutson, Phys. Rev. C3(1971)10.

9) A.M. Green and A. Prodon, Nucl. Phys. A183(1972)225.

10) H. Feshbach and D.R. Yennie, Nucl. Phys. 37(1962)150.

OFF-SHELL OPTICAL POTENTIAL AMBIGUITIES: A THREE-BODY MODEL*

E. F. Redish, University of Surrey and U. of Md., and S. K. Young, U.C.L.A.

Grassberger and Sandhas (Z.f.Ph. 220 29 (69)) have shown how the AGS three-body equations may be separated into an L-S equation and an AGS - type equation for the effective interaction (optical potential) by separation of the pole part from the 2-particle T matrices. The AGS equations and the optical potential separation may be generalized to the N-body problem (Gy. Bencze, Nucl. Phys. A210 568 (1973) and E. F. Redish, Saclay preprints D.Ph. T. 74/3 and 74/34). This separation is ambiguous corresponding to the fact that one may perform phase equivalent transformations of the optical potential without affecting the elastic scattering amplitudes. Explicitly, in either the 3- or the N-body case, the equations for the scattering operators may be written symbolically as matrices (on the 2-cluster channel indices) of operators on the Hilbert space as

$$U = B + KU. \tag{1}$$

The kernel of these equations, has a cut corresponding to the 2-cluster states of the available subsystems. (This cut is usually referred to as the "pole" part.) The kernel may be written in general

$$K = B\Theta\Gamma\Omega + K' \tag{2}$$

where Θ and Ω are energy-dependent matrix-diagonal operators equal to the identity on the energy shell and Γ is the propagator for 2 bound non-interacting clusters. This leads to the equations

$$u = u^{(1)} + u^{(1)}\Gamma u \tag{3}$$

$$u^{(1)} = B + \Omega K'\Omega^{-1}u^{(1)} \tag{4}$$

Eq. (3) is an L-S equation with optical potential $u^{(1)}$ and eq. (4) is a many-body equation for the optical potential. The italic operators indicate transforms by Θ and Ω, thus $u = \Omega U\Theta$, etc.

Iteration of (4) leads to a multiple scattering series for the optical potential. The convergence rate of this series depends on the kernel of (4) and hence on the operators Θ and Ω. Otherwise said, *the convergence rate of the series for the optical potential depends on the off-shell continuation chosen.*

A particularly striking example of this can be obtained by a consideration of a 3-body model with separable forces. The model considered has 2 identical protons and an inert triton with s-wave forces chosen to fit 2-nucleon and nucleon-^{3}He data (see Young and Redish, Md. preprint 74-014 for details). We find that when the pole separation is made following Grassberger and Sandhas, an iteration of eq. (4) diverges whenever the iterated AGS eqn. does and does so more rapidly. However the full kernel of (1) has the form $K = B\Theta\Gamma\Omega$ for a particular choice of Θ and Ω so an off-shell transformation can be made which makes the kernel of (4) identically zero. The exact optical potential is therefore obtained after one iteration (elimination of the pickup channel).

A study of the trajectories of the eigenvalues of the kernel of (4) for the diverging case suggests a physical explanation of the divergence. There is a single eigenvalue outside the unit circle which appears to be the analytic continuation of the bound state pole to the complex coupling constant plane (Padé variable). We know that (1) must diverge at the 3-body bound state pole. When the equation is separated into (3) and (4) the pole may either arise from iteration of (3) (the bound state is a bound state in the optical potential) or from an iteration of (4) (the bound state is not a bound state of the optical potential but the optical potential itself has a pole at the bound state energy). In order to get convergence of the series for the optical potential this warns us that it is necessary to calculate that optical potential whose analytic continuation to negative energy binds the bound state. This will be particularly true at low energies when the bound state trajectory will not have had time to return to inside the unit circle.

*Work supported in part by the U.S. A.E.C..

PROJECTILE NUCLEUS BREMSSTRAHLUNG AT RELATIVISTIC ENERGIES

George L. Strobel, University of Georgia, Department of Physics and Astronomy, Athens, Georgia, 30602, U.S.A.

Gauge invariance and Low's low energy theorem have been shown to be important (Nyman[1]) in calculating nucleon-nucleon bremsstrahlung. The problem of calculating proton-nucleus bremsstrahlung is more difficult because of many body complications. These include, among others, exchange of identical particles, numerous exited states possible in the nucleus, other nuclear reactions, what the proper nucleon-nucleon force is within a nucleus, and how to calculate relativistically. We utilize a model which ignores most of these complications. We treat the nucleus as an elementary particle and ignore exchange effects. The nuclear interaction is approximated by use of a coulomb plus a complex nuclear optical potential which therefore allows the approximate treatment of some many body effects. We use Glauber theory to calculate the elastic scattering amplitude from this potential. Then a many boson exchange amplitude is constructed which, in first order, reproduces this amplitude, somewhat analogous to Baier's potential[2] for nucleon-nucleon scattering. Glauber theory provides the approximate three momentum transfer dependence $(\vec{q})^2$ for the scattering amplitude. We choose the boson exchange amplitude such that the four momentum transfer $q^2 = (q_o^2 - \vec{q}^2)$ dependence of the boson exchange amplitude matches the three momentum transfer dependence of the Glauber theory amplitude. The bosons are assumed electrically neutral, consistent with no exchange effects, and all to be either scalar or pseudo-scalar mesons. We treat the electromagnetic interaction to first order only, and ignore nuclear rescattering effects, as Brown showed[3] these were small. We calculate in the overall center of mass frame, with relativistic propagators for photon, projectile, nucleus, and mesons. The spin, charge, and anomalous magnetic moment of the projectile and target enter the calculation but no form factors are required. This is so because $q^2 = 0$ for all the emitted photons. The calculation is gauge invariant and satisfies Low's low energy theorem.

Bremsstrahlung cross sections have been calculated for 200 to 800 MeV protons on a variety of carbon isotopes. Assuming exchange of scalar or pseudo-scalar mesons in the nuclear amplitude results in very small differences in the calculated bremsstrahlung, showing that off-shell effects are present, but small. The optical potential used[4] reproduced 140 MeV elastic scattering data on ^{12}C as well as some bremsstrahlung data at 200 MeV. The differential cross section calculated is similar to a relativistically shifted classical dipole or quadrupole radiation pattern. The quadrupole pattern results when the charge/mass ratio of the projectile matches the target. For gamma ray energies less than about 1/5 the projectile laboratory energy, the cross section shows a 1/E dependence. For gamma energies in the above region, the non-coplanar cross section is nearly independent of the angle the gamma makes with the nuclear reaction plane. The maximum coplanar cross section occurs when the gamma scattering angle equals one half the projectile scattering angle. Such cross sections are shown in the figure. The maximum cross section is of the order of 10^{-6} f^2/MeVsr^2.

REFERENCES

1) E. Nyman, Phys. Rev. 170 (1968) 1628

2) R. Baier, H. Kuhnelt, and P. Urban, Nucl. Phys. B11 (1969) 675

3) V. Brown, Phys. Lett. 52B (1970) 259

4) C. Dale, Nuclear Physics A156 (1970) 600

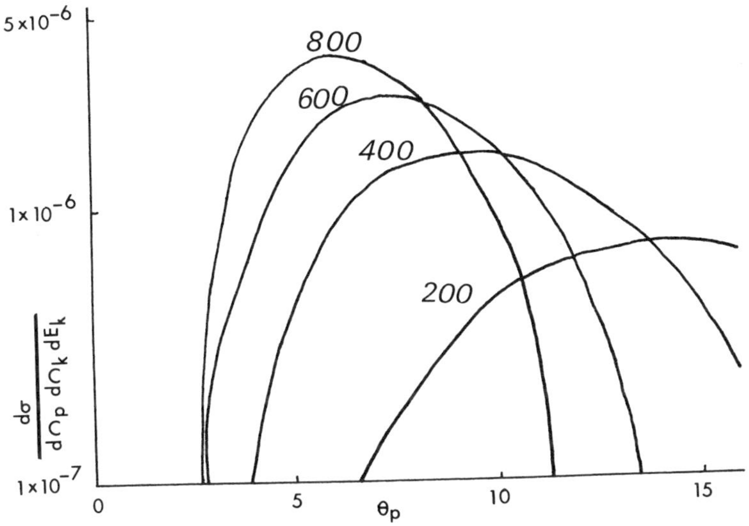

COPLANAR $^{12}C(p,p\gamma)^{12}C$ WITH $\theta_k = \theta p/2$ FOR VARIOUS PROTON LABORATORY ENERGIES IN f^2/MeVsr^2, FOR E_k = 29.6 MeV

THE CHARGE SYMMETRY OF THE NUCLEAR FORCE AS OFF-SHELL CONSTRAINT

P. U. Sauer

Laboratory for Nuclear Science, M.I.T., Cambridge, Massachusetts 02139, U.S.A.

Equivalent proton-proton (pp) potentials

$$\tilde{V}_R = U(K + V_C + V_R)U^+ - K - V_C \tag{1}$$

are generated from the Reid soft-core potential V_R by off-shell changes. K is the kinetic energy of relative motion (k denotes the corresponding momentum), V_C is the Coulomb interaction and U an arbitrary unitary transformation of short range. The potentials $\tilde{V}_R$ are purely nuclear. They are nonlocal at small pp distances r, but reduce to the local long-range tail of V_R where U reduces to unity. In the ^1S_o partial wave the simple form $U = 1 - 2|g\rangle\langle g|$ is employed with $\langle r|g\rangle = Cr(1-\beta r)\exp(-\alpha r)$ normalized and the range parameter α^{-1} chosen to be sufficiently small. E.g. $\alpha = 3\ \text{fm}^{-1}$ for the results of Fig. 1. The potentials $\tilde{V}_R$ are, together with the Coulomb force V_C, equivalent to each other and to V_R. They have exactly the same pp phase shifts $\eta^C_{ppR}(k)$ at all energies, thus preserve the fit of V_R to the pp data. Do the potentials $\tilde{V}_R$ also account for the known neutron-neutron (nn) data?

The Schrodinger equation of the Coulomb subtracted hamiltonian $K + \tilde{V}_R$ is solved for the ^1S_o scattering states $|\tilde{\psi}(k)\rangle$. The ^1S_o nuclear pp scattering length a_{pp} (solid line in Fig. 1) of the resulting non-experimental phase $\eta_{pp}(k)$ varies dramatically with $\tilde{V}_R$. It is in general not close to the value a_{ppR} of V_R and to the experimental value[1] of the ^1S_o nn scattering length, $a_{nn} = -16.4 + 0.9$ fm (dotted line in Fig. 1). Thus, most off-shell variations $\tilde{V}_R$ of V_R are not acceptable, if charge symmetry is required to hold for the nuclear part of the two-nucleon potential.

Can those unitary transformations U of (1) which conserve charge symmetry be recognized without lengthy calculations? The relation

$$k\cot\eta_{pp}(k) \simeq k\cot\eta_{ppR}(k) - \frac{M}{\hbar^2}\langle\psi(k)|U^+V_CU-V_C|\psi(k)\rangle \tag{2}$$

between the nuclear phases $\eta_{pp}(k)$ of $\tilde{V}_R$ and $\eta_{ppR}(k)$ of V_R and the wave function $|\psi(k)\rangle$ of V_R in the effective-range normalization[2] is a reliable approximation, as Fig. 1 demonstrates for a_{pp}. (2) is derived from the exact two-potential formula for the scattering amplitude by equating $|\tilde{\psi}(k)\rangle$ to $U|\psi(k)\rangle$. According to Fig. 2 these two wave functions are indeed very similar over the range of U. If U is chosen to leave the zero-energy wave function unchanged, i.e. $U|\psi(o)\rangle = |\psi(o)\rangle$, then $a_{pp} \simeq a_{ppR}$ according to (2), furthermore $\eta_{pp}(k) \simeq \eta_{ppR}(k)$ even at all energies. Thus, the acceptable off-shell variations of any ^1S_o two-nucleon potential V_R, which accounts for the pp and nn data with the same accuracy as the Reid soft-core potential, are obtained from (1) with unitary transformations satisfying $U|\psi(o)\rangle = |\psi(o)\rangle$. The charge symmetry of the nuclear force yields an effective constraint for the off-shell parametrization in ^1S_o. The

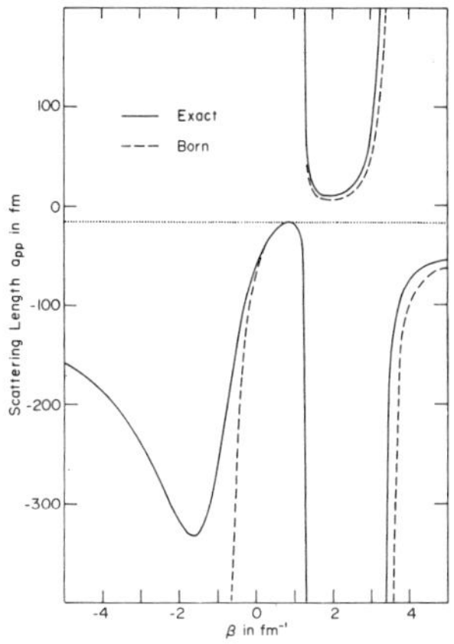

Fig. 1

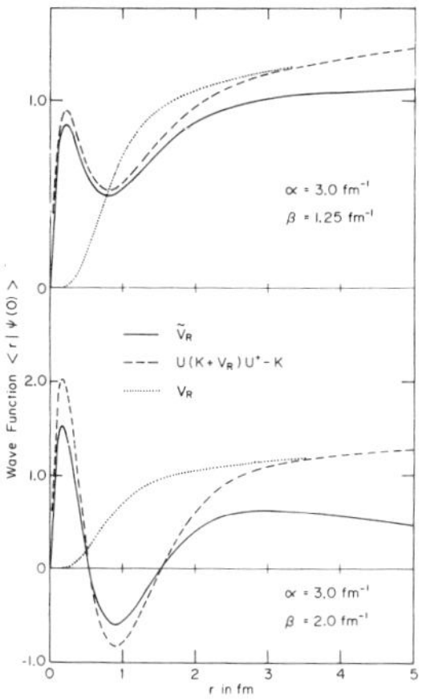

Fig. 2

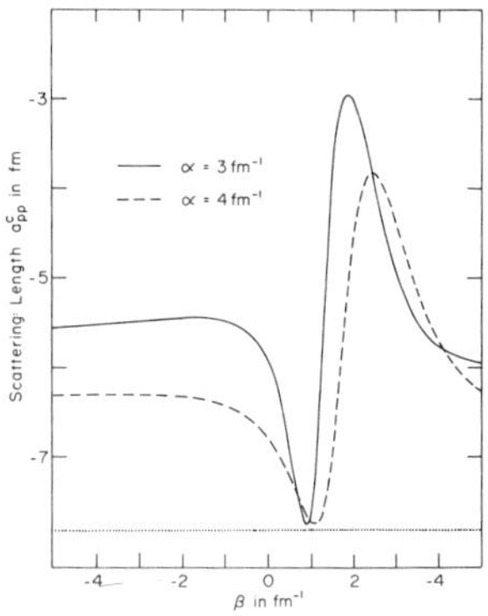

Fig. 3

constraint can easily be implemented in any U.

In contrast to (1), the potentials[3] $\bar{V}_R = U(K + V_R)U^+ - K$ have ordinarily been used in nuclear structure calculations to study off-shell effects. By construction they account for the nn data as well as V_R. Do the potentials $\bar{V}_R$ together with V_C also fit the experimental pp phase shifts? Fig. 3 proves for the ^1S_o pp scattering length with Coulomb, a^C_{pp} (the experimental value is indicated by the dotted line), that many potentials $\bar{V}_R$ fail to do so. They therefore have to be discarded. The results of Fig. 3 are obtained from

$$C_o^2(\gamma)k\cot\eta^C_{pp}(k) \simeq C_o^2(\gamma)k\cot\eta^C_{ppR}(k) + \frac{M}{\hbar^2}\langle\phi(k)|U^+V_CU-V_C|\phi(k)\rangle, \qquad (3)$$

which approximately relates the phase shifts $\eta^C_{pp}(k)$ of $\bar{V}_R + V_C$ to quantities of $V_R + V_C$, to its phase $\eta^C_{ppR}(k)$ and to its ^1S_o wave function $|\phi(k)\rangle$ in the effective-range normalization[2]. $C_o(\gamma)$ is the Coulomb penetration factor of orbital momentum zero. (3) is based on the same approximation as (2) and is expected to be as reliable as (2).

The results of Figs. 1-3 have been derived with the point Coulomb interaction. The variation of a_{pp} and a^C_{pp} with the parameters of U is greatly reduced, if the proton finite size is taken into account. But even in this case, most ^1S_o potentials $\tilde{V}_R$ violate charge symmetry and most ^1S_o potentials $\bar{V}_R$ do not fit the pp data. Most of them are not acceptable off-shell variations of V_R.

REFERENCES

1) E. M. Henley and D. H. Wilkinson, "Few Particle Problems in the Nuclear Interaction" I. Slaus et al, Eds., (North-Holland, Amsterdam, 1972), p. 242.

2) P. U. Sauer, Phys. Rev. Letters 32 (1974) 626.

3) F. Tabakin, "The Two Body Force in Nuclei", S. M. Austin and G. M. Crawley, Eds., (Plenum Press, New York - London, 1972), p. 101.

V

INTERACTIONS A L'ETAT FINAL

FINAL STATE INTERACTIONS

INFORMATION FROM THE STUDY OF FINAL STATE INTERACTIONS

J.S.C. McKee

Department of Physics, University of Birmingham

Processes leading to a final state of three nucleons are of great relevance to the dynamics of the three-body system, and involve particle-particle interactions both on- and off- the energy shell. The breakup of the deuteron by nucleon bombardment is an obvious case for study, and a great deal of research, both theoretical and experimental, has already been carried out on this subject. A major advance in this field was described briefly by Ebenhöh at the Los Angeles Conference on Few Particle Problems (1972), and later published in Nuclear Physics [1)]. This advance was the development of a new technique for calculating the breakup cross -sections with separable potentials. Calculations for kinetically complete n + d and p + d breakup experiments were made which agreed very well with high quality experimental data of Zeitnitz et al. (1972)[2)], and Bruckmann et al. (1970) [3)]. It was shown possible, using the Ebenhöh code, to extract a precise value for the neutron-neutron scattering length from the complete experimental data of Zeitnitz (1972) [2)]. The exact calculations with separable potentials included all coherent and incoherent breakup amplitudes, and predicted absolute values of the cross-sections to be obtained. This was of course a great advance over the limitations of the pole dominant Watson-Migdal and spectator models in general use up to the time.

The success of the new technique lay in the fact that once the two-body separable potential was adjusted to reproduce accurately the S-wave nucleon-nucleon phase shift, the exact three-body calculations gave results which were in good agreement with observed experimental data. Off-shell effects only entered as a small perturbation to the problem, and the breakup processes studied were not sensitive to these aspects.

The question which immediately arises from this assessment of the situation is whether elastic nucleon deuteron scattering and deuteron breakup are after all in any way sensitive to the details of the nucleon-nucleon force.

Brayshaw (1974) [4)] has suggested that these processes can yield no more off-shell information on the force than is already implicit in the value of n - d doublet scattering length, which is of course correlated with the triton binding energy as noted by Phillips (1968). Indeed he has commented that the observed absence of off-shell effects is in itself a valuable piece of information demanding theoretical attention. In summary, Brayshaw believes that model dependent variations are restricted exclusively to three-body ^{2}S-waves when the potentials to be compared exhibit identical on-shell two-nucleon scattering behaviour.

Haftel and Peterson (1974) [5], while agreeing substantially with the conclusions of Brayshaw find that the ^{2}S partial wave should nonetheless exhibit significant model sensitivity, even when the doublet scattering length is fixed. The region of maximum sensitivity seems to reside in the vicinity of the singlet F.S.I. peak, and a study of the angular distributions and spectra of such peaks may help to distinguish between potential models to represent the basic nucleon-nucleon interaction.

Several contributions to the present Conference develop and test these ideas. Rad, Saylor and Jain [6] suggest that a measurement of the vector to vector spin transfer parameters $K_{y}^{y_1}$ in ^{1}H $(\vec{d}, \vec{p})$ and ^{2}H $(\vec{p}, \vec{p})$ for slightly inelastic p - d scattering would be particularly valuable in isolating the neutron-proton singlet F.S.I., and in providing checks on the theory of final state interactions. Peterson and collaborators [7] in a study of the angular distribution of the neutron-proton F.S.I. in deuteron breakup at 23 MeV present data which exhibits considerable sensitivity to the off-shell behaviour of the potentials studied in the region between 100° and 140° for the centre of mass angle of the proton. Two potentials with identical on-shell characteristics are clearly distinguishable on the basis of the analysis, which is in the spirit of the Haftel-Peterson work. (see fig.1)

Two further contributions to this Conference from Kuhn et al. [8] (Dresden), and Bonbright et al. [9] (Manitoba) are relevant to the analysis of the breakup process. Both concern breakup by protons, the former of 8.5 MeV incident energy, and the latter of energies between 20 and 40 MeV.

The Dresden group show a good fit to a neutron-proton F.S.I. peak using an Ebenhöh code, while the Manitoba group demonstrate that the variation in the peak cross-section with energy, in the case of the proton-proton F.S.I., is also well represented by the Ebenhöh calculation. It would be useful to examine an angular distribution of F.S.I. at these energies.

A typical spectrum showing the proton-proton F.S.I. is given in fig. 2, and the the variation of cross-section with energy shown in fig. 3. No errors in the cross-sections are quoted. Spectra for the same reaction at higher energies (156 MeV) are presented in a paper from Orsay by Yuasa [10], and are compared with the results of several theoretical calculations. A single particle spectrum of protons from n - D breakup at 14.4 MeV is shown in a paper from Montrouge, Cocu et al. [(11)], but no analysis of the data has been attempted.

Finally, as a conclusion to this part of the talk, a paper by Cahill [12] is of some interest. He studies the breakup state for which one pair of nucleons is in the FSI region, while another is in the quasi-free scattering (Q.F.S.) region, kinematically speaking. Using a general theory of FSI and Q.F.S. based on the Faddeev equations, Cahill investigates the consequences of such overlap for single particle spectra. He finds himself able to identify those features of the calculation that result in a sharpening of the neutron-neutron FSI peak, and the forward peak in neutron-deuteron elastic scattering.

Little work on the extraction of low energy nucleon-nucleon scattering parameters has been presented at this Conference. Madame Ligatto de Slobodrian has shown that the decay of the ground state of ^{5}Li can have an effect even in the region of the proton-proton FSI peak for the reaction ^{3}He(τ,α)2p. The resulting error in the calculation of the proton-proton scattering length is found to be at least 0.2 fm. This contribution emphasises how difficult it is to assess accurately in retrospect the precision with which a published quantity has in fact been measured, All methods and techniques so far adopted in the extraction of a unique value for a_{nn} have been model dependent in some respect. Although the n^-d reaction has no mirror reactions for comparison, this desadvantage is outweighted by the fact that the final state contains only two strongly interacting particles. The μ^-d reaction has still to produce a result, and the theory for the reaction may not be above reproach. Other nuclear reactions such as ^{3}H(d,^{3}He)2n, ^{2}H(n,p)2n among others, are difficult to analyse, even today, and the methods adopted are not equivalent for each case. There does seem therefore to be much value in producing a weighted average ot the results contained in a time ordered set of complete and incomplete experiments as Kuhn has done [14].

In some of these experiments the errors quoted are 'optimistic' to say the least, while in others, the method of analysis itself may not be above question. Perhaps, as Wilkinson [15] has suggested we should await the remeasurement of a_{nn} from the $\pi^- + d \rightarrow 2n + \gamma$ reaction with high precision at one of the meson factories now coming into full production. Once this has been accomplished, and the theoretical uncertainty in this analysis reduced, other measurements of this quantity should become historical milestones rather than equivalent data to be averaged in an 'ad hoc' way.

Three-body final states form of the d - ^{4}He reaction have been the subject of two contributions to this Conference. Segawa et al. [16] find that the modified off energy shell impulse approximation (M.I.A.) of Nakamura which has been successful in reproducing experimental data for alpha energies from 2 to 165 MeV is less satisfactory at the energy of the present experiment (deuteron bombardment energy 9 MeV). In this work the proton and alpha particle are detected in coincidence, but no allowance is made in M.I.A. for the strong neutron proton interaction which at this energy must contribute to all regions of phase space. It is perhaps surprising that their spectator model fits the F.S.I. peaks so well, for the same reason. The multi-channel cluster model of Hackenbroich [17] on the other hand which has recently had considerable success in interpreting elastic α-deuteron scattering is now applied to the three-body αnp final state from the the ^{4}He - d reaction (Heiss et al.) [18]. It deals exclusively with the sequential decay process through an intermediate ^{5}He - p structure. While his data are in reasonable agreement with experiment (no energy quoted), it is clear that at large proton angles other competing processes are playing an important role. The paper by Ferguson and collaborators [19] at Chalk River which looks at the τpp final state from the d - ^{3}He reaction, shows a discrepancy between theory and experiment if it is assumed that the Watson-Migdal approximation is valid for the analysis of this system. The suggestion that a charge-exchange mechanism is contributing to the scattering in the region of the discrepancy might be examined through a Treiman-Yang test of pole-dominance. The discrepancy is clearly shown in fig. 4.

The decay of states of ^{12}C into three alpha particles has been a subject of study for groups at Zagreb and Québec. The Zagreb paper (Antolkovic) [20] succeeds in assigning a spin-parity of 1^- to the 11.83 MeV state of C in the ^{12}C(n,n)3α. Excitation energies in ^{12}C above 15.96 MeV can of course be reached by less esoteric means, and Roy et al. [21] have succeeded in assigning a spin parity of 2^+ to the 33.5 MeV level studied previously by Jacquot [22], from examination of the ^{9}Be + ^{3}He → 3α reaction. They do not however exclude the presence of α1^-level at 33.4 MeV and suggest the presence of a 1^+ level at 33.7 MeV.

The alpha particles in these 3α final states do not necessarily have similar histories, as two usually come from the sequential decay of an intermediate ^{8}Be system. Roy et al.[23] in a further paper comment on the mechanisms by which states of ^{8}Be are produced in the reactions ^{9}Be(τ,α)αα and ^{10}B(d,α)αα, and investigate those factors relevant to the width of the resonant states. It appears that a direct reaction process is incapable of reproducing the experimental shapes of the first two states of ^{8}Be. In other works, Brown, Chien and Yang [24] take the two alpha scattering problem and investigate the influence of the Pauli exclusion principle on relative cluster-cluster motion at short distances. They produce an α - α potential which is shown to be 'realistic' in its ability to fit existing data on alpha alpha scattering, and this may now be applicable to cluster models of other nuclei.

Oryu [25], in a theoretical paper concerning a model for the neutron-triton system reduces his four-body problem to a three-body one by assuming that the deuteron constituent of the triton is an elementary particle. The calculation includes several simplifications which may not in the end be justified. The fits to n-^{3}H elastic scattering data are not impressive when compared with that of the optical potential of Sherif (1972)[26], see fig. 5.

Warner et al.[27] have bravely tackled the study of double final state-interactions in the ^{2}H(^{2}H,pp)nn reaction for conditions where both proton-proton and neutron-neutron pairs emerge with low relative energy. Calculations using an antisymmetrical DWBA theory are in progress with the object of fitting the experimental data. No results are available to date. The need for information on the electric-quadrupole contribution to the ^{4}He(γ,n)^{3}He reaction mentioned at the Los Angeles Conference (1972) was recognised by Erdas and Pompei[28] who have presented preliminary calculations suggesting that the E_2 part of the (γ,n) cross-section is much smaller than that of the (γ,p) cross-section. The Born cross-section shows a resonance with a maximum at E≃4 MeV when the isoscalar contribution to the E_2 amplitude is included in the calculation. This paper is the only one concerned with the photodisintegration of simple systems.

The first excited state of ^{6}He might not be considered in any sense a simple system but Kolalis et al.[29] have studied the sequential decay process $^6He \rightarrow {}^5He + n \rightarrow \alpha + 2n$, and deduced that at least 30% of all decays occur through this means. No consideration is given in their paper to interference between competing processes in the exit channel and it seems difficult to extend the calculations to include other modes of decay. Two theoretical papers concerning rate for muon capture in deuterium (Quang Ho-Kim et al.)[30] and helium (Eramzhyam et al.)[31] have been received.

For the reaction $\mu^- + d \rightarrow n + n + \gamma$ it is shown that the final state interaction between the two neutrons enhances the capture rate by around 12% and that this change is independent of the model used to interpret the F.S.I. In fact, because the neutrons interact at low relative momentum, their interaction is independent of the details of the short range potential. In contrast, the total rate for capture of muons in helium is shown to be quite sensitive to the details of the potential used to describe the neutron-triton interaction.

A calculated rate of 220 S^{-1} obtained using the optical potential of Kim becomes 450 S^{-1} when the Sherif potential is used. It seems clear from comparison of these papers that a knowledge of the relative angular momenta of particles in the final state is important in assessing the sensitivity to the nature of the potential used.

The detailed study of quasi free and final state processes in the ^{4}He+^{3}H reaction at an incident energy of 70 MeV by Allas et.al.[32], will be discussed later in this Conference. Three and four-body final states from this reaction were examined for kinematic conditions favouring the quasi free scattering and quasi free reaction channels. In each case momentum transfers up to ~ 0.6 $fm.^{-1}$ were studied. In addition to the excited states which might be expected to show up in this analysis, a broad level in ^{6}He at 15 MeV excitation was indicated, while no evidence was found for the suggested 28.4 MeV level in ^{4}He.

Finally, in this varied assortment of nuclear phenomena we find an experiment of relevance to neutrino astrophysics.

It is well known that the fundamental nuclear reaction in astrophysics is taken to be: $P + P \rightarrow {}^{2}H + e + \nu$, and that this reaction has a very low cross-section. For low energy collisions the two protons are taken to be in an S state and their spin wave function is therefore odd.. In the final state the deuteron has odd isospin, the space wave function contains S and D waves, and the spin wave function is even. Both initial and final states have even parity, so that this transition is of the allowed Gamon-Teller type. The cross-section for this process can readily be calculated in the absence of resonances, and is expected to be of the order of picobarns.

The experiment of Pigeon et al. [33)] reported to this Conference however concerns an examination of the β-decay of diproton pairs, and the conclusion reached by the authors is that identifiable events for this decay indicate a cross-section not in excess of 1 nanobarn.

The pseudo-resonant diproton state as observed in experiments specifically designed for its study, has an average life time for sequential decay which is large compared with the time between proton collisions in the scattering state. The effect of this dilation on the β-decay rate can perhaps be significant. Analogously, the three body reaction rate $^{4}He + {}^{4}He + {}^{4}He \rightarrow {}^{12}C$ which can play a small part in the stellar creation of carbon has a slower rate than the two-step reaction

$$\left.\begin{array}{l} ^{4}He + {}^{4}He \rightarrow {}^{8}Be \\ ^{8}Be + {}^{4}He \rightarrow {}^{12}C \end{array}\right]$$

because ^{8}Be, despite being unstable has a longer life time than the time between ^{4}He collisions.

Whether the cross-section for β-decay of the diproton is a nanobarn or a picobarn could have important implications for neutrino astrophysics.

If we assume that stellar sources preferentially produce diprotons in the low energy proton gas, and that this is reflected in an enhanced β-decay rate for the $p + p \rightarrow d + e^{+} + \nu$ system, then the sun itself could be colder than is generally believed and still give the same total level of solar radiation.

If the sun is colder than normally expected, then light element syntheses of the kind:

$$\left.\begin{array}{l} ^{3}He + {}^{4}He \rightarrow {}^{7}Be \\ ^{7}Be + e^{-} \rightarrow {}^{7}Li + \gamma \end{array}\right\}$$

would be inhibited. The result of this would be a greatly reduced cross-section for production of high energy solar neutrinos thus explaining the 'observed 'absence' of such neutrinos in recent experiments. For these reasons it seems of interest to establish whether or not the diproton decay cross-section is as large as 1 nanobarn, and if it is, to decide whether the solar proton gas is likely to favour such diproton states whithin it.

Finally, it is a pity that no multiparticle states from high energy physics are to be discussed at this session, and related to their low energy neighbours. The absence of data from those 'inclusive' and 'exclusive' reactions that we read about so frequently in the literature is unfortunate. It is certainly evident that final state interactions play a major role in elementary particle physics at the present time, and their interpretation in many cases is far from clear.

Perhaps the most disapointing feature of those papers that have been contributed to this session is the generally poor quality of the experimental data presented, and the lack of fundamental progress in the extraction of model dependent features from such data as do exist. The future task for experimentalists is to ensure that they put pressure on their theoretical counterparts through the medium of high precision multi-analyses of few body final states in complete experiments.

REFERENCES

1. W. Ebenhoh. Nucl. Phys. A191, (1972), 97.

2. B. Zeitnits, R. Maschav, P.Suhr and E. Ebenhoh. Proc.Int. Conf. on Few Particle Problems, Los Angeles (North-Holland) (1972), 117.

3. H. Bruckmann, W.Kluge. H.Mattay, L. Schanzler and K. Wick. Nucl. Phys. 157 (1970), 209.

4. D.D. Brayshaw. Phys. Rev. Letts. 32 (1974), 382.

5. M. I. Haftel and E.L. Peterson. Bull. Am. Phys. Soc. 19 (1974), 505.

6. F.N. Rad, D.P. Saylor and M. Jain. Proceedings of this Conference (1974).

7. E.L. Peterson, M.I. Haftel, R.G. Allas, L.A. Beach, R.D. Bondelid, E.M. Diener, P.A. Treado and J.M. Lambert. Proceedings of Conference (1974).

8. B. Kuhn, H. Kumpf, J. Mosner, W. Neubert and G. Schmidt. Proceedings of Conference (1974).

9. D. Bonbright, J.W. Watson and D.J. Roberts. Proceedings of Conference (1974).

10. T. Yuasa. Proceedings of Conference (1974)

11. F. Cocu, G. Ambrosino and D. Guerreau. Proceedings of Conference (1974).

12. R.T. Cahill. Proceedings of Conference (1974).

13. M.L.V. de Slobodrian. Proceedings of Conference (1974).

14. B. Kuhn. Proceedings of Conference (1974).

15. D.H. Wilkinson. Proc. Int. Conf. on Few Particle Problems, Los Angeles (North-Holland) (1972), 191.

16. K. Sagara, Mr. Hara, T. Motobayashi, N. Takahashi, F. Soga, F. Takeutchi and Y. Nogami. Proceedings of Conference (1974).

17. H.H. Hackenbovich, P. Heiss and Le Chi Nien. Nucl. Phys. A22 (1974) 461.

18. P. Heiss and H.H. Hackenbroich. Proceedings of Conference (1974).

19. A.J. Ferguson, G.C. Ball, W.G. Davies and J.S. Forster. Proceedings of Conference (1974).

20. B. Antolkovic. Proceedings of Conference (1974).

21. R. Roy, G. Turcotte, R.J. Slobodrian and G. Goulard. Proceedings of Conference (1974).

22. C. Jacquot. Nucl. Phys. A201, (1973), 247.

23. R. Roy, J. Birchall, B. Frois, U.von Moellendorff, C.R. Lamontagne and R.J. Slobodrian. Proceedings of Conference (1974).

24. R.E. Brown, W.S. Chien and YC. Tang. Proceedings of Conference (1974).

25. S. Oryu. Proceedings of Conference (1974).

26. H.S. Sherif and B.S. Padmore. Proc. Int. Conf. on Few Particle Problems. Angeles (North-Holland) (1972) 691.

27. R.E. Warner, S.B. Di Cenzo, G.C. Ball, W.G. Davies, A.J. Ferguson and J. S. Forster. Proceedings of Conference (1974).

28. F. Erdas and A. Pompei. Proceedings of Conference (1974).

29. R. Kolalis, R. Seltz and D. Magnac. Proceedings of Conference (1974).

30. Quang Ho-Kim, J.P. Lavine and H.S. Picker. Proceedings of Conference (1974).

31. R.A. Eramshyan, G.E. Dogotar and Yu.A. Salganic. Proceedings of Conference (1974).

32. R.G. Allas, L.A. Beach, R.O Bondelid, E.M. Diener, E.L. Peterson, P.A. Treado, J.M. Lambert, R.A. Moyle, L.T. Myers, Ivo Slaus and P. Thomas. Proceedings of Conference (1974).

33. R. Pigeon, U. von Moellendorff, C.R. Lamontagne, M. Irshad and R.J. Slobodrian. Proceedings of Conference (1974).

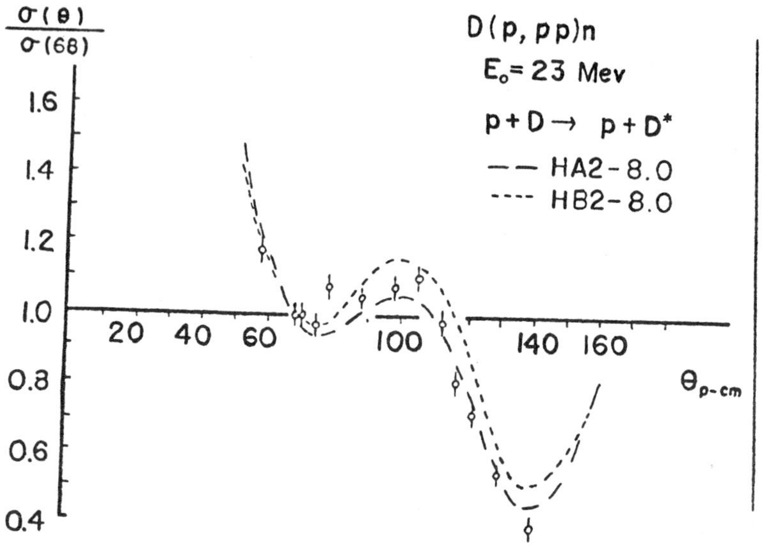

Fig. 1 The ratio $\frac{d^3\sigma(\theta)}{d^3\sigma(68^o)}$ is shown for the experiment $p + d \rightarrow p + d^*$. The two sets of predictions differ only in off-shell behaviour of the corresponding potentials. HA2 - 8.0 has off-shell behaviour similar to the Reid soft-core potential, and HB2 - 8.0 has off-shell behaviour similar to the Yamagachi potential.

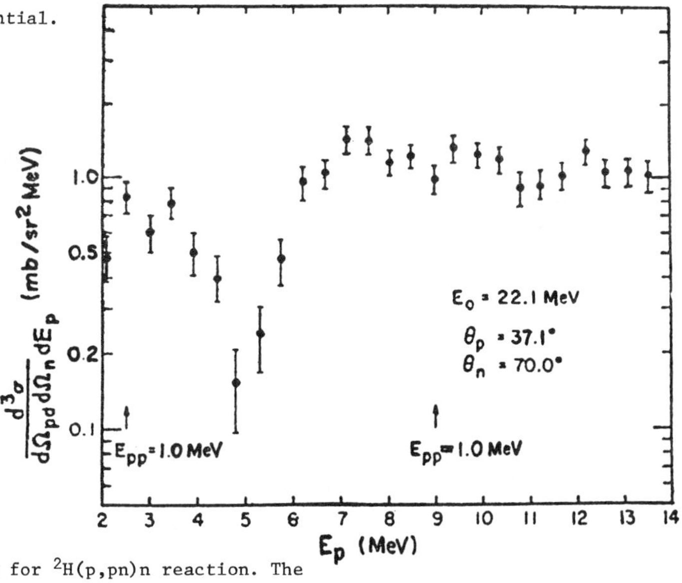

Fig. 2 Typical spectrum for $^2H(p,pn)n$ reaction. The deep minimum is due to Coulomb repulsion of two protons with low relative energy.

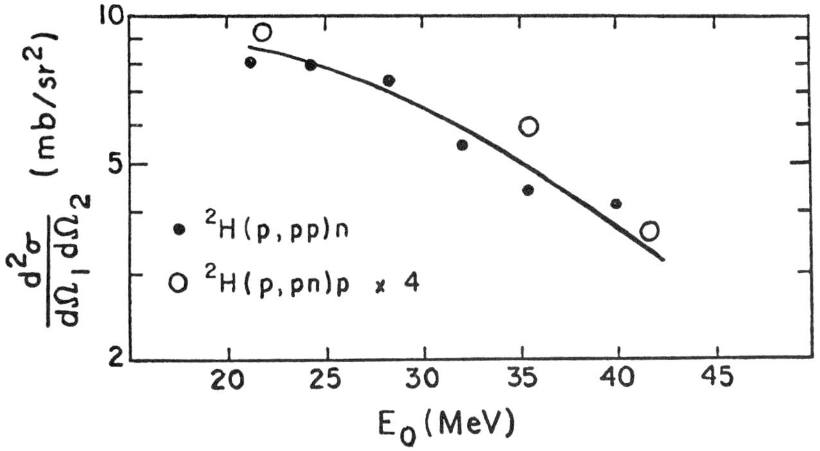

Fig. 3 Energy behaviour of the cross-section for the reactions shown when integrated over the F.S.I. peak,

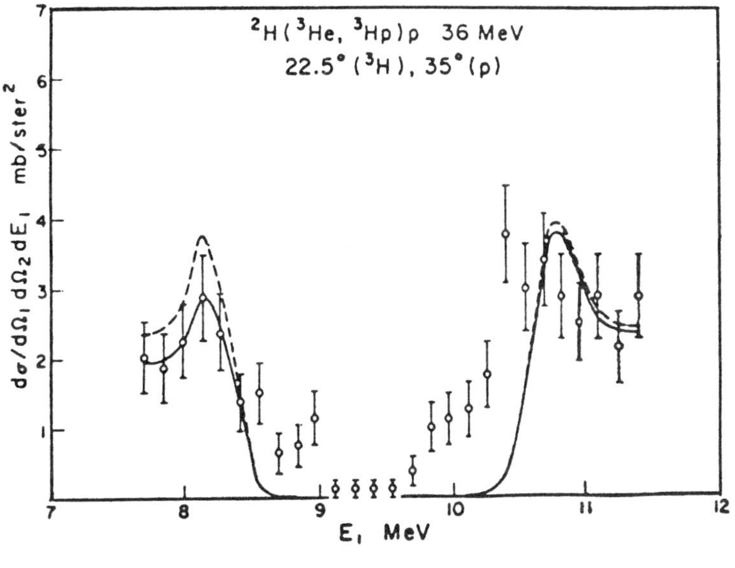

Fig. 4 Least squares fit to data on the left of the cross-section minimum for the reaction $^2H(^3He,^3Hp)p$ at 36 MeV, for the angles shown in the figure. The dashed curve represents the yield assuming no losses due to the unobserved proton entering the triton counter. Curves on the right of the spectrum are obtained using the parameters obtained for the left.

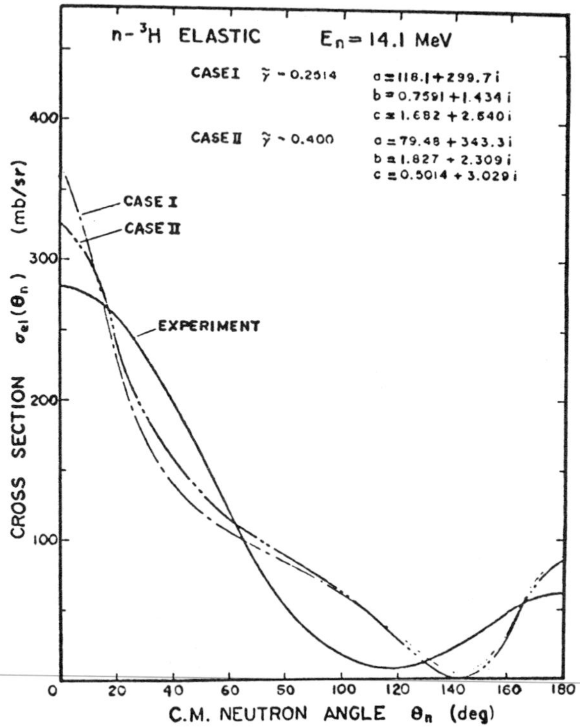

Fig. 5 Fits obtained by the method of Oryu to experimental data for n^3H elastic scattering. a,b, and c are constants appearing in his elastic scattering amplitude, while γ is related to the n - d Yamagachi form factor $\alpha \; \frac{1}{p^2+\gamma^2}$ $\quad \gamma \; \frac{\gamma}{.2316}$

DISCUSSION

KÜHN: (Rossendorf): Dr. McKee has cited our contribution to this Conference (B. Kuhn, J. Mosner, W. Nembert and G. Schmidt: Kinematically complete measurements of the deuteron breakup by protons at 8.5 MeV) on kinematically complete measurements of the deuteron breakup by protons at 8.5 MeV. In this contribution is presented only one spectrum for the quasifree scattering case as an example, showing good agreement between experiment and an Ebenhöh calculation with Yamaguchi potentials. I would like to show you in two examples that in the case of optimal conditions for FSI the differential cross sections in the FSI peaks appear to a factor of 1.6 to 1.8 greater than the calculated ones, (see fig.1 and 2) while the cross sections of the remaining parts of the spectra are reproduced very well. Such great differences were not observed at higher energies. So calculations with different form factors of the used potential give rather strong variations in the amplitude of the FSI peaks, and it seems to us, that FSI spectra at small energies are especially sensitive to the form factors.

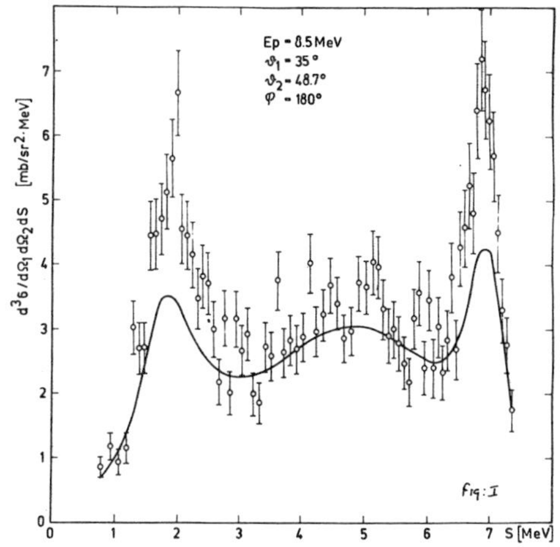

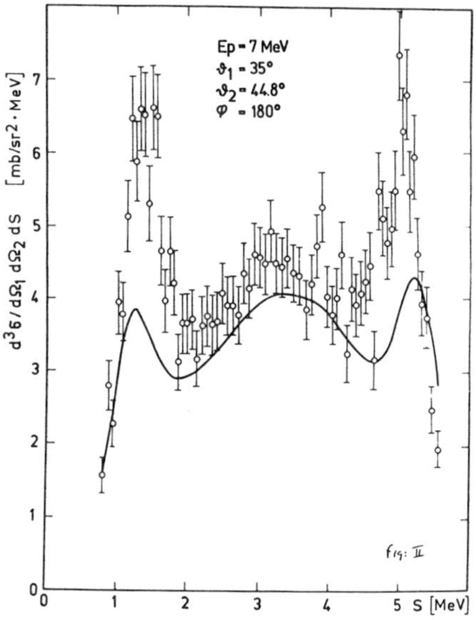

ZEITNITZ (Hamburg): Recently there were two important improvements of the Ebenhöh code. The first is that one can use form factors which now fit the nucleon-nucleon phase shifts much better that the Yamaguchi form factor. A second improvement was performed by Stuivenberg and Bruisma in Amsterdam. They took into account the complete charge dependance also in the kernels of the integral equations. Changes of the cross sections are between about 0 and twenty percent.

The analysis of our earlier experiment on the n + d breakup reaction using the new code with exponential form factors changed the extracted value of the neutron-neutron effective range from 3.6 to 2.1 fm.

MCNAUGHTON (UC, Davis): Considering Dr. McKee's final remark, I would like to advertise the fact that I have extensive, kinematically complete, neutron-deuteron breakup data at about 130 MeV, hoping to interest a theoretician with a graduate student in extending more sophisticated models to higher energies.

SELTZ (Strasbourg): La réaction de décomposition de ^{6}He* (1.8 MeV) a ceci de particulier qu'elle ne peut donner ^{5}He + n que grâce à la largeur de l'état fondamental de ^{5}He. D'autre part les neutrons émis ont une énergie très faible (1 MeV). On peut donc s'attendre à être dans des conditions qui favorisent d'interaction n-n au détriment de la formation de ^{5}He. Or nos résultats expérimentaux montrent qu'ils ne peuvent être expliqués seulement par l'interaction n-n. Nous espérons que l'analyse en cours permettra d'introduire quantitativement l'influence de la résonance ^{5}He.

The n-p FSI Angular Distribution in Deuteron Breakup at 23 MeV

E. L. Petersen, M. I. Haftel, R. G. Allas, L. A. Beach,
R. O. Bondelid, and E. M. Diener*
Naval Research Laboratory, Washington, D. C. 20375

and

P. A. Treado† and J. M. Lambert†
Georgetown University, Washington, D. C. 20007
and Naval Research Laboratory

The Final State Interaction (FSI) region of phase space is interesting because of the sensitivity of the cross section to both on-shell and off-shell effects. Previous measurements, although informative, lack the precision needed for detailed comparison with potentials.[1,2,3] This experiment was performed with the NRL Sector Focusing Cyclotron and standard coincidence electronics for kinematically complete experiments. Particle identification was not used, as it was necessary to detect low energy protons. Any strong effects of accidental deuteron coincidences were always well-separated from the final state peak of interest and were nearly eliminated by background subtraction.

Figure 1 shows the measured cross section at the final state peak as a function of angle. Also shown are the predictions of the 4 potentials discussed in the preceeding series of papers.

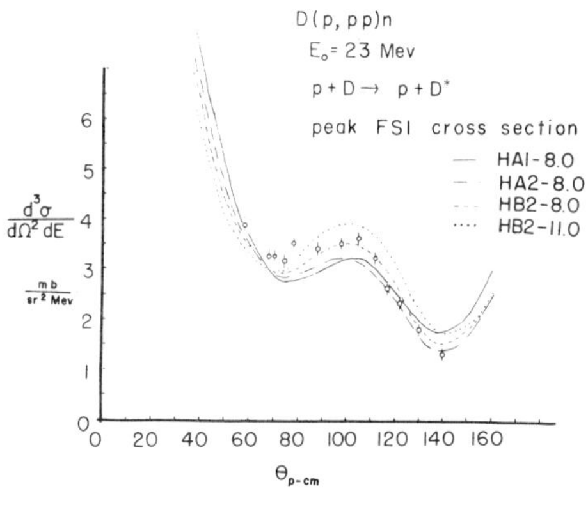

Figure 1. Angular dependence of the FSI peak. A comparison of predictions of the various potentials shows changes due to:

a) on-shell changes HA1-8 HA2-8

b) off-shell changes HA2-8 HB2-8

c) triton binding energy HB2-8 HB2-11

This data is consistent with the Karlsruhe data obtained using 52 MeV deuterons on protons.[2,3] Both sets of data are slightly higher than predicted in the region 60°-90° and slightly lower in the region 130°-140°.

An obvious point of comparison of theory and experiment is the ratio of the cross section at 100° to that at 140°. However, this ratio is not sensitive to the off-shell change in potentials but only to the on-shell changes. The experiment appears to be closest to HA2 and HB2 potentials, which underestimate both the free proton-proton cross section and the S-wave contribution to the cross section. The reason why the data here favors S-wave potentials that do not fit the N-N data should be investigated.

Perhaps the influence of higher (two-nucleon) partial waves is qualitatively different here than in two-nucleon scattering.

All of the potentials predict nearly the same cross section in the vicinity of 68°, and the off-shell changes in potential appear to show up primarily in the ratio of the cross section in the region 100°-140° to that at 68°. The source of the 8% difference between theory and experiment at 68° is not clear. In an effort to isolate off-shell changes, we plot in Fig. 2 the ratio $d^3\sigma(\theta)/d^3\sigma(68^\circ)$ for experiment and for the two potentials that differ only off-shell.

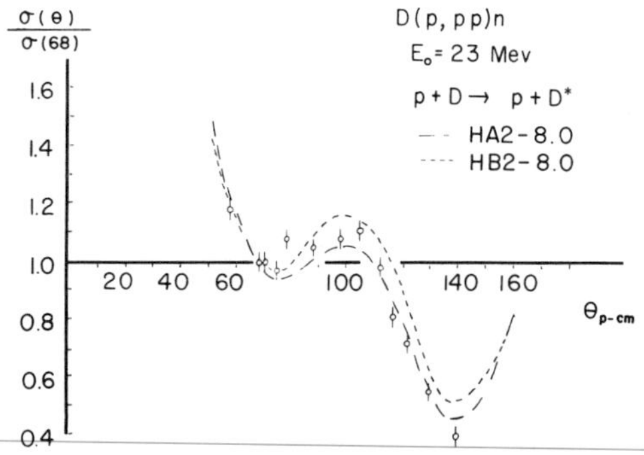

Figure 2. Ratio $d^3\sigma(\theta)/d^3\sigma(68^\circ)$ for the experiment $p + d \rightarrow p + d^*$. The two sets of predictions differ only in the off-shell behavior of the corresponding potentials.

The ratio appears to follow potential HA2-8.0, which has off-shell behavior similar to that of the Reid Soft Core potential, rather than the potential HB2-8.0, which is similar off-shell to the Yamaguchi potential. Contrary to Brayshaw,[4] this work suggests that further experimental and theoretical investigations of this type should lead to better knowledge of nuclear off-shell behavior.

<u>References</u>

*National Research Council - Naval Research Laboratory Research Associate

†Work Partially Supported by the National Science Foundation

1. D. J. Margaziotis et. al., Phys. Rev. <u>C8</u>, 870 (1973).

2. H. Bruckmann et. al., Nuc. Phys. <u>A157</u>, 209 (1970).

3. W. Kluge, R. Schlufter, and W. Ebenhöh, preprint.

4. D. D. Brayshaw, Phys. Rev. Letters <u>32</u>, 382 (1974).

LA REACTION $^{9}Be+^{3}He \rightarrow \alpha+\alpha+\alpha$ ET LE NIVEAU D'ENERGIE DU ^{12}C VERS 33.5 MeV D'EXCITATION.

R. Roy, G. Turcotte, R.J. Slobodrian, Université Laval, Département de Physique, Laboratoire de Physique Nucléaire, Québec G1K 7P4, Canada.

G. Goulard, C.M.R., Département de Physique, St-Jean J0J 1R0, Québec, Canada.

L'énergie d'excitation du noyau composé de ^{12}C formé dans la réaction $^{9}Be(^{3}He,\alpha)^{8}Be^{*}$, évaluée aux énergies incidentes de 9.44 MeV 9.94 MeV est de 33.4 MeV et 33.7 MeV. Un niveau d'énergie prévu par Jacquot et al.[1] est situé à 33.5 MeV avec un spin J = 4 ,de plus des niveaux $J^{\pi} = 1^{+}$ existent de chaque côté. Dans un autre article[2] relatif en partie aux états du $^{12}C(J^{\pi}=1^{-}, T=1)$ est annonçé la possibilité d'un niveau à 33.47 MeV. En supposant un mélange (T=0, T=1) la désintégration en trois alphas n'est pas impossible. Des expériences ont été faites à l'aide de l'accélérateur Van de Graaff CN de 7 MV de l'Université Laval, possédant un tube à haut gradient. Les particules α en corrélation ont été détectées en utilisant un télescope avec identification de masse et un détecteur simple. Un spectre du temps de corrélation a été obtenu et a permi de choisir la région d'événements réels sans ambiguité. Les détails de l'électronique modulaire employée peuvent être trouvés dans plusieurs publications précédentes[3,4]. Les spectres en corrélation ont été accumulés en ligne sur un ordinateur PDP-9, et ils sont remarquablement propres à cause de l'identification de masse et les restrictions dans le spectre temps. Pour calculer la section efficace σ_L qui correspond à la détection de deux alphas de l'état final, le mécanisme suivant est utilisé:

$$^{9}Be+^{3}He \rightarrow {}^{12}C^{*} \rightarrow \alpha_1+{}^{8}Be^{*} \rightarrow \alpha_1+\alpha_2+\alpha_3$$

L'expression de σ_L dans le laboratoire a déjà été dérivée[5] ,ses principales caractéristiques étant: les noyaux de ^{12}C et ^{8}Be sont considérés comme des noyaux composés et le formalisme[6] de la matrice R sert à développer l'amplitude de diffusion f . Les variables habituelles du problème à trois corps et trois systèmes de coordonnées sont nécessaires pour obtenir f . D'autre part, l'identité des trois particules dans l'état final conduit à symétriser f . Les sommations dans σ_L sur les spins totaux des voies d'entrée et de sortie sont chacune des sommes incohérentes, le spin de la voie $(\alpha_1+{}^{8}Be^{*})$ étant celui du $^{8}Be^{*}$. Les calculs numériques ont été effectués en supposant plusieurs valeurs J^{π} du spin et de la parité pour le ^{12}C $(0^{+}, 1^{\pm}, 2^{\pm}, 3^{\pm}, 4^{\pm})$, les largeurs réduites $\gamma_{1\,s}^{J}$ du ^{12}C sont prises égales à un en première approximation. Les niveaux du ^{8}Be intervenant ici sont les deux premiers états excités 2^{+} 4^{+} et le doublet 2^{+}, d'autres états peuvent être aussi considérés mais ne semblent pas contribuer énormément

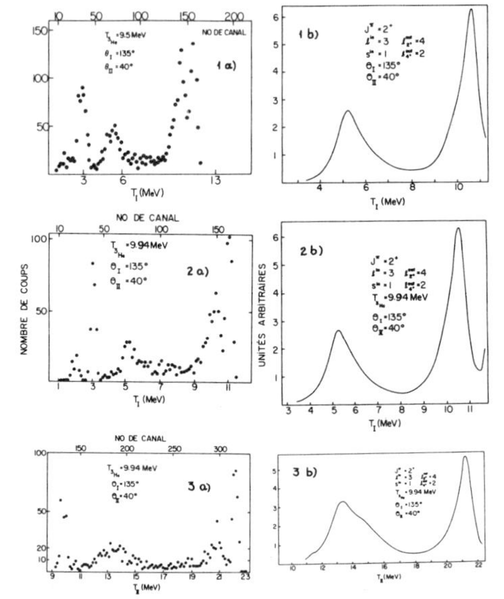

Spectres expérimentaux | Spectres théoriques

Figs. 1,2 et 3

ici. Les paramètres du ^{8}Be pour les 2^{+} et 4^{+} sont les largeurs réduites 0.96 MeV, 0.97 MeV, le rayon d'interaction 4.6 fermis (tirés de la diffusion α-α) [7] et pour le doublet 2^{+} ,les largeurs [8] sont 90 keV et 70 keV. Les facteurs de pénétrabilité du ^{12}C et ^{8}Be ont été pris variables avec les énergies relatives de chaque voie.

Les résultats des calculs sont les suivants: les valeurs $J^{\pi}(^{12}C) = 4^{\pm}$ ne donnent pas de bons accords avec l'expérience. La meilleure comparaison (figs. 1 et 2) a lieu pour $J^{\pi} = 2^{+}$ aux deux énergies incidentes, à plusieurs angles de détection, sur plusieurs ensembles de nombres quantiques dans les voies d'entrée et de sortie pour $J^{\pi} = 2^{+}$. Il faut cependant ajouter que la contribution d'un niveau 1^{-} à 33.4 MeV d'excitation n'est pas complètement exclue et qu'à 33.7 MeV, un niveau 1^{+} semble intervenir. Nous remercions l'aide précieuse du personnel du van de Graaff et l'appui de la Commission de Contrôle de l'Energie Atomique du Canada.

BIBLIOGRAPHIE

1) C. Jacquot et al., Nucl. Phys. A201, 247 (1973).

2) H.D. Shay et al., Phys. Rev. C9, 76 (1974).

3) R. Roy and R.J. Slobodrian, Nucl. Inst. Meth. 87, 1 (1970).

4) B. Frois, J. Birchall, C.R. Lamontagne, U. von Moellendorff, R. Roy and R.J. Slobodrian, Phys. Rev. C.8, 2132 (1973) and refs. therein.

5) G. Goulard, Nucl. Phys. A140, 225 (1970).

6) A. M. Lane, R.G. Thomas, Rev. Mod. Phys. 30, 334 (1958).

7) M.W. Kermode, Nucl. Phys. A134, 336 (1969).

8) W.D. Callender, C.P. Browne, Phys. Rev. C2 , 1 (1970).

LA REACTION $^{3}He(\tau, \alpha)2p$ ET LA DETERMINATION DE LA LONGUEUR DE DIFFUSSION p-p

Maria L.V. Ligatto de Slobodrian, Université Laval
Laboratoire de Physique Nucléaire, Département de Physique
Québec, G1K 7P4, Canada

L'utilisation des spectres de particule alpha dans la réaction $^{3}He(\tau, \alpha)2p$ et son spéculaire $^{3}H(t, \alpha)2n$ pour la détermination[1] de la longueur de diffussion p-p, implique qu'il soit important de déterminer à quel point on est capable d'isoler l'information de l'intéraction de deux protons (ou neutrons) dans l'état $^{1}S_{o}$ à partir des résultats expérimentaux obtenus. Dans des travaux antérieurs, on a considéré que le pic qui apparaît dans les spectres des particules alpha est dû seulement à l'intéraction p-p (n-n). Même dans les cas où la présence d'autres processus significatifs est admise[2], on suppose qu'ils sont important seulement quand la section efficace pour cette intéraction est un minimum.

Afin de déterminer l'erreur provenant de telles suppositions, on a calculé les spectres des particules α dans la réaction $^{3}He(\tau, \alpha)2p$, à 43.7 et 53 MeV (ref. 3). On a confirmé que les résultats expérimentaux sont bien représentés en utilisant[4] la somme incohérente des intéractions p-p et p-α (^{5}Li). A ces énergies, la contribution de l'intéraction p-p est plus importante qu'aux énergies plus basses et pourtant on va être capable d'établir une limite inférieure de l'erreur en choisissant pour le calcul de la longueur de diffussion un spectre correspondant à un maximum de la distribution angulaire (E_{LAB} = 53 MeV, θ_{LAB} = 18°). Le spectre a E_{LAB} = 43.7 MeV et θ_{LAB} = 12°, correspondant à un minimum de la distribution angulaire a été aussi analysé, dans le but de comparer les valeurs de la longueur de diffussion obtenu dans les deux cas.

En supposant que le pic est dû seulement à l'intéraction p-p, les valeurs de la longueur de diffussion obtenues sont

... 2

$6.77 \pm .42$fm pour $\theta_{LAB} = 12^o$

et $7.76 \pm .25$fm pour $\theta_{LAB} = 18^o$

Par contre, si on soustrait premièrement la contribution du ^5Li, on obtient les valeurs concordantes

$7.96 \pm .24$fm

$7.96 \pm .43$fm

On peut conclure que l'erreur systématique introduit dans le calcul de la longueur de diffussion en négligeant des contributions significatives est d'au moins .2 fm.

1) Slobodrian, R.J., Rep. Prog. Phys. 34 (1971) 175.

2) Roberts, W.J., Gross E.E., Newman, E., Phys. Rev. C9 (1974) 149.

3) Slobodrian, R.J., McKee, J.S.C., Tivol W.F., Clark, D.J., Tombrello, T.A., Phys. Letters 25B (1967) 19.

4) M.L.V.L. de Slobodrian, B. Cujec, R. Roy et R.J. Slobodrian, Nucl. Phys. A194 (1972) 577.

CASSURE DU DEUTERON DANS LA REACTION n + D ⟶ n + p + n

F. Coçu, G. Ambrosino, D. Guerreau[+], Service de Physique Nucléaire,
Centre d'Etudes de Bruyères-le-Châtel, B.P n° 61, 92120 Montrouge, France

L'étude de la réaction n + D ⟶ n + n + p a été réalisée par la mesure du spectre en énergie des protons à $\langle\theta\rangle$ = 3° . Les neutrons rapides incidents d'énergie 14,1 MeV sont produits par la réaction d + T ⟶ n + He^4 . Les deutérons incidents étaient fournis par un accélérateur Samès de Bruyères-le-Châtel avec une énergie de 200 keV et un courant de 3 μA . La mesure relative du flux de neutrons incidents a été réalisée en comptant les particules alpha associées aux neutrons .

La cible de deutérium (CD_2) avait une épaisseur de 7,4 mg/cm2 et une surface de 1,54 cm2 . Les protons issus de la réaction étudiée étaient détectés par un télescope composé de deux détecteurs à barrière de surface d'épaisseur respectives 45 μm et 1500 μm . Une analyse bidimentionnelle (E - ΔE) a été faite afin d'extraire du bruit de fond les évènements correspondant aux protons issus de la réaction étudiée .

Les résultats expérimentaux sont présentés ci-contre, les barres d'erreur tiennent compte uniquement des erreurs statistiques .

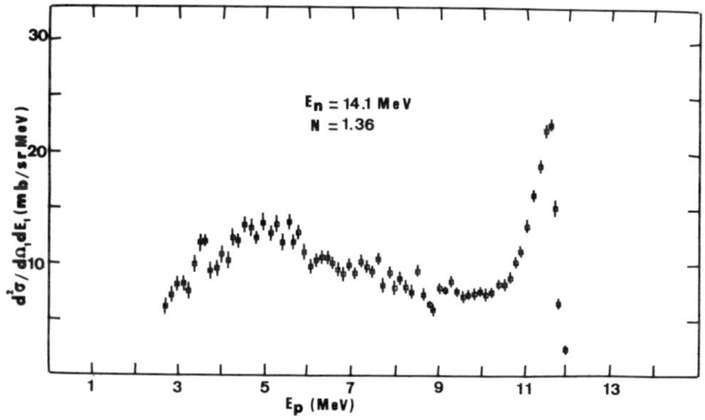

Spectre des énergies proton dans la réaction n + D ⟶ n + n + p pour $\langle\theta\rangle$= 3° et pour l'énergie des neutrons incidents de 14,1 MeV

RESONANCE ANALYSIS OF THE ^{3}Hpp 3-BODY FINAL STATE

A.J. Ferguson, G.C. Ball, W.G. Davies & J.S. Forster
AECL, Chalk River Nuclear Laboratories
Chalk River, Ontario, Canada KOJ 1J0
and R.E. Warner, Oberlin College, Oberlin, Ohio 44074

A number of recent experiments[1,2] have been performed on the Chalk River MP tandem accelerator to study final state interactions of the various 3-body products resulting from the bombardment of ^{2}H by ^{3}He. In the case of the reaction ^{2}H(^{3}He,^{3}Hp)p, it is known[3] that there are two resonances in the ^{3}H + p system at low relative energies of these two particles. This kinematic condition is appropriate for the application of the Watson-Migdal approximation[4] and an extension of this formalism is presented to include the resonances.

The Watson-Migdal theory assumes that the 3-particle scattering amplitude can be expressed as a product of several factors one of which is the amplitude representing the motion of the two particles with small relative velocity. It is assumed further that this factor is the principal one in describing the variation of the cross section as a function of the relative motion of the products and that it is given simply by the two particle free scattering amplitude.

Instead of the usual effective range approximation, we use a scattering amplitude obtained from Wigner-Eisenbud resonance theory[5]. Only s-wave orbital states are considered and a singlet and a triplet channel spin state are present which combine incoherently. We thus assume that the triple differential cross section is given by:

$$d^3\sigma/d\Omega_1 d\Omega_2 dE_t = N\left[\,|{}^1T_{tp}|^2 + 3|{}^3T_{tp}|^2\right],$$

where ${}^1T_{tp}$ and ${}^3T_{tp}$ are the scattering amplitudes for the singlet and triplet spin states, N is an arbitrary normalization factor having dimensions $[L]^2/[E]$, E_t is the energy of the triton and Ω_1 and Ω_2 represent the directions of the triton and observed proton.

Both scattering amplitudes are given by $T_{tp} = (1-U_{tp})/2\pi i$, where U_{tp} is the t-p scattering matrix. The Wigner-Eisenbud theory[5] gives:

$$U_{tp} = \frac{I-R(dI/dr+bI)}{O-R(dO/dr+bO)},$$

with I = G-iF, O = G+iF, I and O represent incoming and outgoing waves and F and G the regular and irregular Coulomb wave functions. In a many level case:

$$R = \sum_\lambda \frac{\gamma_\lambda^2}{E_\lambda - E},$$

where γ_λ^2 and E_λ are the reduced width and resonant energy of the λth resonance. For the singlet triplet channels we assumed that

$$^1R = {^1R_o} + \frac{\gamma_1^2}{E_1 - E} \quad \text{and} \quad {^3R} = {^3R_o}.$$

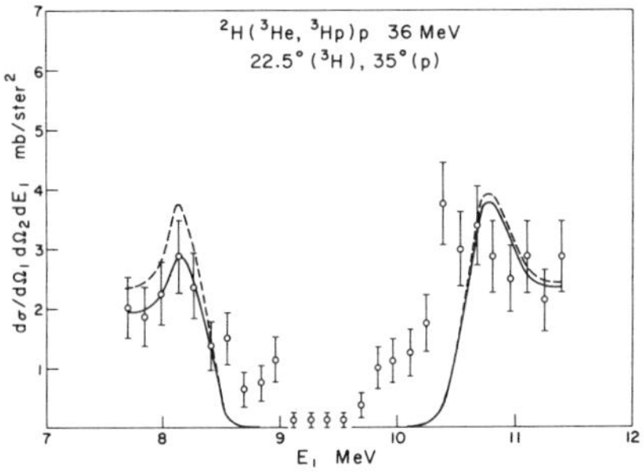

Fig. 1 shows a least squares fit to one of the measured geometries. The parameters obtained are E_1 = 114 ± 30 keV, γ_1^2 = 250 ± 80 keV, ^1R_o = -0.87 ± 0.70, ^3R_o = 0.8 ± 1.0, N = 4.63 ± 1.17. The fit was made only to the data on the left of the cross section minimum. The fit is represented by the solid curve and contains a correction for losses due to the unobserved proton entering the triton counter. The dashed curve represents the yield that would be obtained if this loss did not occur. The curves on the right side of the spectrum were made with the same parameters as for the left. The above parameters are in excellent agreement with those of Parker et al.[3)] for the same reaction.

Systematic deviations between the measurements and theory evident in Fig. 1 suggest either an unidentified source of error in the data or else a process, such as charge exchange, that is omitted from this simple theory. The possibility of applying the Faddeev equations to this data is being studied.

References

1) R.E. Warner, G.C. Ball, W.G. Davies, A.J. Ferguson, J.S. Forster, Phys. Rev. Lett. 27 (1971) 961.

2) R.E. Warner, G.R. Flierl, W.G. Davies, G.C. Ball, A.J. Ferguson, J.S. Forster, S.A. Gottlieb, Nucl. Phys. A192 (1972) 341.

3) P.D. Parker, P.F. Donovan, J.V. Kane, J.F. Mollenauer, Phys. Rev. Lett. 14(1965)15.

4) K.M. Watson, Phys. Rev. 88 (1952) 1163, and A.B. Migdal, Soviet Physics - JETP 1 (1955) 2.

5) E.P. Wigner, L. Eisenbud, Phys. Rev. 72 (1947) 29 and A.M. Lane, R.G. Thomas, Rev. Mod. Phys. 30 (1958) 257.

DOUBLE FINAL-STATE INTERACTIONS IN THE $^2H(^2H,pp)nn$ REACTION

R.E. Warner† and S.B. Di Cenzo††, Oberlin College, Ohio 44074
and G.C. Ball, W.G. Davies, A.J. Ferguson and J.S. Forster
AECL, Chalk River Nuclear Laboratories
Chalk River, Ontario, Canada KOJ 1JO

Recently Fritts and Parker[1] have studied the 2p + 2n four-body final state of the $^3H(p,pp)nn$ reaction under conditions such that the final n-n pair had low relative energy ε_{nn}. They could not fit their p-p coincidence spectra with any reasonable value of the n-n scattering length, using Watson-Migdal theory. Their work has motivated us to make a similar study of the $^2H(^2H,pp)nn$ reaction, under conditions where the p-p and n-n pairs both emerge with low relative energies ε_{pp} and ε_{nn}.

A CD_2 target was bombarded with 23.2 MeV deuterons from the upgraded Chalk River MP tandem accelerator and fast p-p coincidences were recorded between two 2-counter telescopes, coplanar with and on the same side of the beam. Two geometries, ($\theta_1 = 18^o$, $\theta_2 = 36^o$) and (25°, 43°) were investigated. These geometries excluded the contaminant reaction $^1H(^2H,pp)n$ which would have been observed at $\theta_1 < 18^o$ and it was verified that a CH_2 target gave no yield. Other techniques used for charged-particle coincidence studies at this laboratory have been summarized elsewhere[2].

Our data at (18°, 36°), plotted vs. both ε_{pp} and ε_{nn}, appear in Fig. 1. Predic-

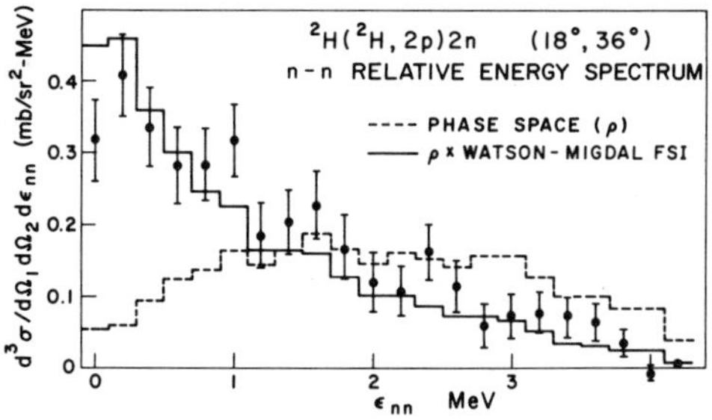

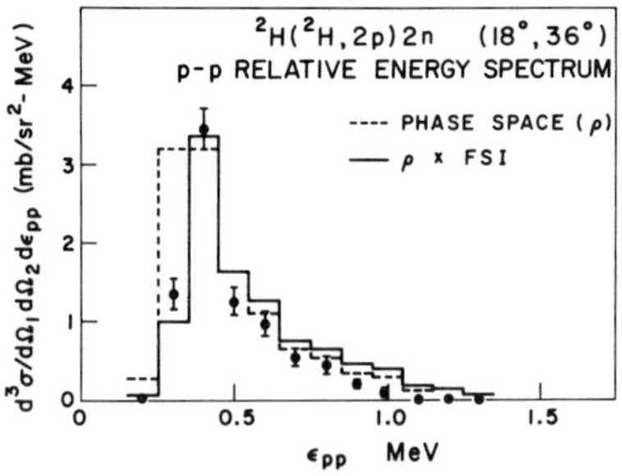

† Research supported by the National Science Foundation
†† NSF Undergraduate Research Participant

tions are given for phase space only[3], and for phase space times Watson-Migdal factors[2] for both the p-p and n-n interactions, using accepted values[4] of the effective range parameters for the singlet p-p and n-n interactions. The latter predictions give reasonably good fits. The enhancement relative to phase space at low ε_{nn} and the depression at low ε_{pp} show clearly the presence of both kinds of final-state interactions. The (25°, 43°) data support similar conclusions.

The integrated cross sections for detecting two protons, both above 3 MeV in the laboratory system, are (0.84 ± 0.04) and (0.63 ± 0.03) mb/sr^2 at (18°, 36°) and (25°, 43°), respectively.

Calculations using an antisymmetrized DWBA theory[2,5] are in progress and will be reported. A more comprehensive account of this work will be published elsewhere.

References

1) M. Fritts and P.D. Parker, Nucl. Phys. A198 (1972) 109.

2) R.E. Warner, S.A. Gottlieb, G.C. Ball, W.G. Davies, A.J. Ferguson, and J.S. Forster Nucl. Phys. A221 (1974) 593.

3) M. Furic and H.H. Forster, Nucl. Inst. and Meth. 98 (1972) 301.

4) M.A. Preston, Physics of the Nucleus (Addison-Wesley, Reading, Mass.) 1962,pp.27-33.

5) R.E. Warner and E.W. Vogt, Nucl. Phys. A204 (1973) 433.

INTERACTION FINALE DANS LA DECOMPOSITION $^{6}He^{*}$ (1.80).

R.Kolalis* , R.Seltz et D.Magnac - Valette
Laboratoire des Basses Energies - C.R.N. - Cronenbourg et U.L.P.
67037 - Strasbourg - Cedex.

Le premier état excité de ^{6}He à 1.80 MeV se désexcite soit par tripartition directe soit par une réaction séquentielle $^{5}He + n \rightarrow \alpha + 2n$[1,2]. Pour préciser le mode de désintégration nous avons répété avec une statistique améliorée l'étude de la réaction $^{7}Li(t,\alpha_1)^{6}He^{*}$ (1.80); $^{6}He \rightarrow \alpha_2 + 2n$.

Les expériences ont été faites à l'accélérateur vertical de 3 MeV du C.R.N. de Strasbourg. L'énergie du faisceau de tritons était de 2.4 MeV et son intensité de 300 nA. La cible a été obtenue par évaporation de ^{7}Li isotopique sur support de carbone mince.

Les particules α_1 ont été détectées à 85° en coincidence avec α_2 , l'ouverture totale du second détecteur étant de 6° . La résolution en temps des coincidences rapides était de l'ordre de 20 nS. A cause de la largeur du niveau excité et de la cinématique, la résolution en énergie était de 200 KeV. Nous avons enregistré ∿ 2.000.000 de particules et ∿ 74.000 coincidences.

Le dispositif expérimental a été vérifié à l'aide de la réaction $^{11}B(p,\alpha_0)^{8}Be$; $^{8}Be \rightarrow 2\alpha$ par l'étude des coincidences entre les particules α_0 et les particules α de la désintégration du niveau fondamental du ^{8}Be dans la direction de vol du noyau de recul.

Nous avons trouvé que dans le domaine d'énergie incidente de protons de 2.0 à 2.6 MeV , la désintégration du niveau fondamental du ^{8}Be est isotrope , ou plus précisément que la distribution angulaire des particules α ne change pas dans la limite des erreurs statistiques (1%) .

La figure 1 donne le spectre des particules α_2 de la décomposition de ^{6}He en coincidences avec α_1 . Les flèches sur les flancs du spectre indiquent la position des reculs ^{6}He et ^{15}N provenant respectivement des réactions $^{7}Li(t,\alpha)^{6}He$ et $^{16}O(t,\alpha)^{15}N$ (fond.). En effet, les pics correspondants dans le spectre simple présentent des queues dont une faible partie se superpose au pic des particules α_1 .

La courbe 1 est obtenue dans le formalisme de Migdal - Watson en supposant une longueur de diffusion a_{nn} =-16 fm. , un rayon d'interaction r_{nn} = 2.8 fm. et une résolution globale ΔE = 200 KeV .

La courbe 2 représente un lissage avec un polynome de la forme :

$$y = a_0 + a_1x + a_2x^2 \text{ r } a_3x^3 + a_4x^4$$

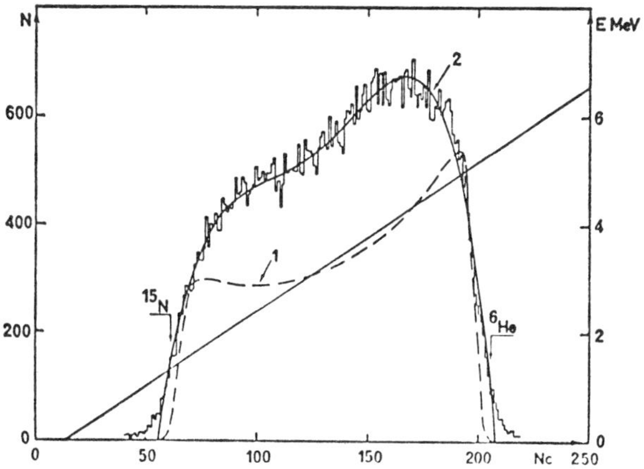

avec $x = 3.63E$ (MeV) $a_0 = (-0.306 \pm 0.008)10^4$

$a_1 = (0.128 \pm 0.003)10^4$ $a_2 = (-0.175 \pm 0.005)10^3$

$a_3 = (0.106 \pm 0.003)10^2$ $a_4 = (-0.234 \pm 0.006)$

Le lissage de la courbe expérimentale sans tenir compte de l'interaction α-n dû au passage par ^5He nécessite une valeur a_{nn} de l'ordre de -10 fm. . Une telle valeur serait en désaccord important avec les meilleures estimations récentes , avancées par exemple par Henley et Wilkinson , soit a_{nn} = -16.4 fm. .

Si nous acceptons cette dernière valeur , au moins 30% des évènements devraient être attribués au passage par ^5He .

Des calculs plus détaillés permettant de déterminer la forme du spectre due à l'interaction α-n , ainsi que la proportion exacte d'évènements due à cet effet sont en cours.

BIBLIOGRAPHIE

* Détaché de l'Université de Leningrad (U.R.S.S.)

1) I.Linck, J.Linck, R.Seltz et D.Magnac-Valette Jour. Phys., Colloque n°1, Lyon (1970)

2) S.N.Abramovich, L.A.Morkin, V.A.Pereshivkin et V.I.Serov, Bull. Acad. U.R.S.S., Vol. 34 (1970) 1532

3) E.M.Henley and D.H.Wilkinson, Few particle Problems in the Nuclear Interaction (Int. Conf. Los Angles 1972) (Noth Holland) p.242

THE $^4He(d,p\alpha)n$ REACTION AT LOW ENERGY

K. Sagara, M. Hara, T. Motobayashi, N. Takahashi,
F. Soga*, F. Takéutchi** and Y. Nogami

Department of Physics, University of Tokyo,
Bunkyo, Tokyo, Japan.

The experimental studies of the $^2H(\alpha,p\alpha)n$ reaction have been done in the E_α = 22 - 165 MeV region[1-5]. Fits with simple models such as the spectator model (SM) or the plane wave impulse approximation (PWIA), have succeeded, to some extent, in reproducing the shape of the p-α quasi-free scattering (QFS) peak at higher energies[4,5] and that of the N-α final state interaction (FSI) peak at lower energies[1,2]. Nakamura has been able to reproduce the data ranging E_α = 29 - 165 MeV both in shape and in absolute value of the whole spectra with the modified off-energy-shell impulse approximation (MIA)[5-7].

We report here the experimental study of the $^4He(d,p\alpha)n$ reaction at E_d = 9.0 MeV, the lowest energy up to now. At this low energy, i) it is expected that the FSI peak is better fitted in shape with the simple two-body model i.e. SM, and ii) it was already suggested by Nakamura that some correction to MIA might be neccessary.

The present $^4He(d,p\alpha)n$ experiment was performed with a beam of the University of Tokyo tandem accelerator. The target gas of natural helium was circulated through a cell of 18 mm in diameter, which had windows of 1.2 μm formvar foils reinforced with 2.5 μm Havar foils as beam ports. The gas pressure of 50 to 80 Torr was monitored during the experiment by counting elastically scattered α-particles. One countor (700 μm Si) detected the proton energy, and the other (7.6 μm Si, totally depleted) was used for the detection of α-particle. The difference of flight times of proton and α-particle (T_p-T_α) and E_p were used to identify the reaction. The two dimensional spectra (only the loci of larger E_α) were projected onto the E_p axis (fig.1). The uncertainty in the absolute cross section was ±15%.

Results of calculations with the proton spectator model ($SM_{n-\alpha}$), the neutron spectator model ($SM_{p-\alpha}$) and their coherent sum, i.e. PWIA are shown in fig.1 together with those with MIA by Nakamura. The shapes of the n-α FSI peaks are fairly well fitted with $SM_{n-\alpha}$, rather than with PWIA. This success with SM, compared with the similar results at E = 21.9 - 23.7 MeV[1]) and some discrepancy at E_α = 29.2 MeV[2]), indicates that the FSI plays so important a role at lower energies that the simple two-body model such as SM is very useful in reproducing the shapes of the spectra.

Fits with MIA[8], though the parameters used there are varied to find the best fit, are less satisfactory than they are at higher energies[6]. This suggests that effects neglected so far in MIA such as the final n-p interaction or the spin dependent force, become important at this low energy.

The authors wish to thank Professor H. Nakamura for his informative discussions and calculations.

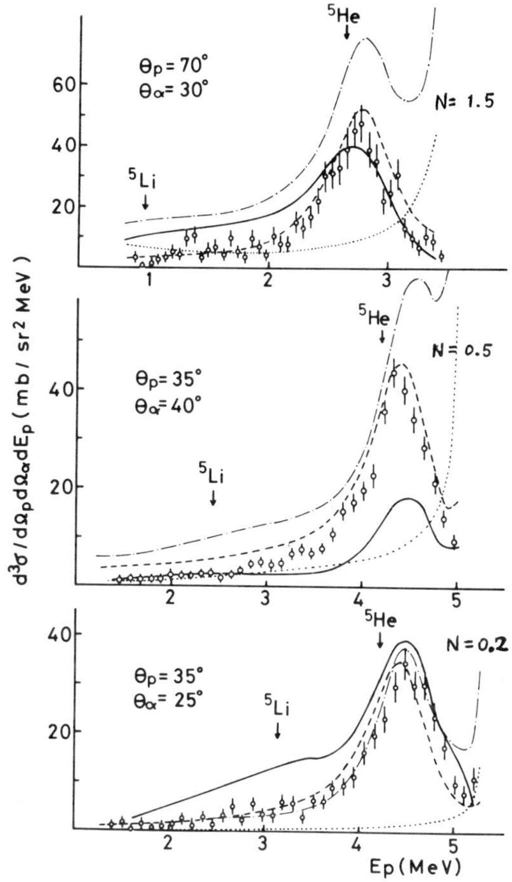

Fig.1; Cross sections for the ^{4}He(d,pα)n reaction at E = 9.0 MeV. Curves are
———— MIA,
—·—·— PWIA x N,
— — — — $SM_{n-\alpha}$ x N,
·········· $SM_{p-\alpha}$ x N.

REFERENCES

1) T. Rausch, H. Zell, D. Wallenwein and W. von Witsch, Nucl. Phys. A222 (1974) 429.
2) T. Tanabe, J. Phys. Soc. Japan 25 (1968) 21.
3) R. E. Warner and R. W. Bercaw, Nucl. Phys. A109 (1968) 205.
4) D. I. Bonbright, Ph. D. Thesis, Univ. of Maryland (1970).
5) E. Hourany, H. Nakamura, F. Takéutchi and T. Yuasa, Nucl. Phys. A222 (1974) 537.
6) H. Nakamura, Nucl. Phys. A208 (1973) 207.
7) H. Nakamura, private communication and to be published.
8) H. Nakamura, private communication.

THE ELECTRIC-QUADRUPOLE PART OF THE $^4He(\gamma,n)^3He$ REACTION+

F. Erdas, A. Pompei
Istituto di Fisica dell'Università, Cagliari, Italy

P. Quarati
Istituto di Fisica dell'Università, Cagliari, Italy
Istituto di Fisica del Politecnico, Torino, Italy

Recent measurements of the $^4He(\gamma,n)^3He$ differential cross section(1), together with a report on the photodisintegration of ^4He(2) and an analysis of the energy levels of the A=4 nuclei(3), have emphasized the importance of the proper treatment of the E2 differential cross section of the $^4He(\gamma,n)^3He$ reaction. The E2 part of the (γ,p) cross section can be experimentally determined(4), whereas similar data are not available(2) for (γ,n).

The asymmetry coefficient of the total (γ,n) angular distribution has an energy dependence which supports the existence of a 2^+ state at $\simeq$ 35 MeV with a width of 5 MeV in ^4He(1). The existence of this state, with isospin T=0, has been already confirmed by us in a study of the E2 reaction $^4He(\gamma,d)d$ by means of a dispersive approach(5).

In this work we have used the same method to evaluate the E2 part of the (γ,n) reaction. We have separated in the matrix element the isoscalar and the isovector amplitude; we have calculated the corrected isoscalar matrix element as the general solution of an Omnès equation.

These preliminary calculations have been computed by using for the ^3He and ^4He Gaussian wave functions with equal parameters. The n-^3He phase shift $\delta(E)$ is given by a hard-sphere phase shift plus a nuclear resonant phase shift, calculated by assuming a 2^+, T=0 state in ^4He at $\simeq$ 30 MeV of excitation and with a width of 3.2 MeV. A complete survey of more extended calculations will be published elsewhere.

The corrected cross section can be written in the form:

$$\sigma(E) = .34\ 10^{-34}\ E_\gamma\{|3M(E)\ (1+d)|^2 + |2M(E)|^2 - 12|M^2(E)\ (1+d)|\}\ \text{cm}^2 \qquad (1)$$

where the first term represents the isoscalar contribution, the second the isovector, the third the interference term contribution. The quantity E is the relative energy of the n-^3He system; $E_\gamma = E + 20.578$ is the photon energy and the correction to the Born matrix element is d and is expressed as:

$$d = e^{i\delta(E)}\{\cos\delta(E) + \frac{e^{\rho(E)}}{\pi M(E)}\ P\int_{\epsilon_0}^{\infty} \frac{M(E')\ \sin\delta(E')\ e^{-\rho(E')}}{E' - E}\ dE' + \frac{1}{M(E)}\ Q(E)\ e^{\rho(E)}\} \qquad (2)$$

where $e^{\rho(E)}$ is the Omnès function and Q(E) a polynomial(5). The quantity M(E) is given by:

$$M(E) = E_\gamma\left(\frac{E}{\varepsilon}\right)^{5/4} e^{-\frac{E}{2\varepsilon}} \qquad (\varepsilon = 6.\ \text{MeV}) \qquad (3)$$

In the figure we report, up to E = 30 MeV, the Born cross section (curve 1) and the cross section (1) (curve 2) corrected in the isoscalar amplitude only. Also the corrected cross section computed without separating the isoscalar and the isovector part is reported (curve 3).

+ Supported in part by I.N.F.N.

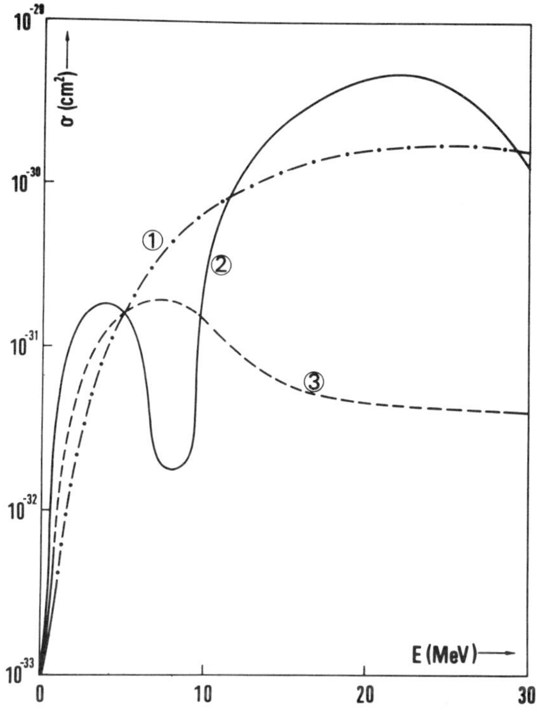

Curve 2 clearly shows a resonating behaviour with a maximum at E ≃ 4 MeV.

A comparison with the E2 part of the (γ,p) cross section, which has a maximum of ≃ 10^{-28} cm^2, shows that the (γ,n) contribution is much smaller than the (γ,p) one.

REFERENCES

([1]) C.K. Malcom, D.V. Webb, Y.M. Shin and D.M. Skopik, Phys.Lett. 47B (1974) 433.

([2]) W.E. Meyerhof and S. Fiarman, Proceedings of the Intern.Conf. on Photonuclear Reactions and Applications, Asilomar, 1973, Ed. B.L. Berman (Lawrence Livermore Laboratory, 1973) Vol. I, p. 385.

([3]) S. Fiarman and W.E. Meyerhof, Nucl.Phys. A206 (1973) 1.

([4]) W.E. Meyerhof, M. Suffert, W. Feldman, Nucl.Phys. A148 (1970) 211.

([5]) F. Erdas, A. Pompei, P. Quarati and B. Mosconi, Nucl.Phys. A174 (1971) 657.

MUON CAPTURE IN ^{4}He

R.A.Eramzhyan, G.E.Dogotar, Yu.A.Salganic

Joint Institute for Nuclear Research, Dubna, USSR

We report some results on muon capture in ^{4}He, concerning only the dominating channel

$$\mu^- + {}^4He \rightarrow {}^3H + n + \nu \qquad (1)$$

The impulse approximation is used for the muon-nuclear Hamiltonian. The relativistic terms of Hamiltonian (they are proportional to nucleon velocity inside the nucleus) are taken into account.

The main problem in the description of process (1) is the construction of the final state nuclear wave function. Recently the experimental data on neutron scattering on tritium have been analyzed in the framework of the distorted waves approximation and the optical potential was deduced [1,2]. With the potential obtained in [1,2] one succeeds to describe neutron scattering on ^{3}H in the energy region from 4.5 MeV up to 17.25 MeV.

Basing on optical potential we have calculated the wave function of the relative motion of the nuclear system and have employed it for calculating the capture rate, energy spectrum and asymmetry of the angular distribution, for the reaction (1) when polarized muons are captured. The remaining parameters of the wave functions are the same as in the paper [3]. The parameters of the optical potential obtained by Sherif [1] were extrapolated to the high and low energy region. As to the former region the parameters used were the parameters at E=17.25 MeV. When the energy tends to E=0, the strength of the imaginary part of the potential became equal to zero, and the strength of the real one 51 MeV.

The results of the calculation are given in Fig.1 (the energy spectrum) and Fig.2 (asymmetry). At the same time the results obtained are compared with the plane wave approximation and with the calculation based on the optical potential obtained by Kim [2], deduced from fitting the scattering data at E= 23.5 MeV.

1. The total capture rate in the reaction (1) appears to be equal to 230 sec^{-1} in the plane wave approximation, 220 sec^{-1} when the optical potential of Kim [2] is used, and 450 sec^{-1} when we use the potential of Sherif [1]. The capture rate depends significantly on the parameters of potential.

II. Energy spectrum in the plane wave approximation is spreaded up to 40 MeV. Final state interaction between n and ^{3}H results in a

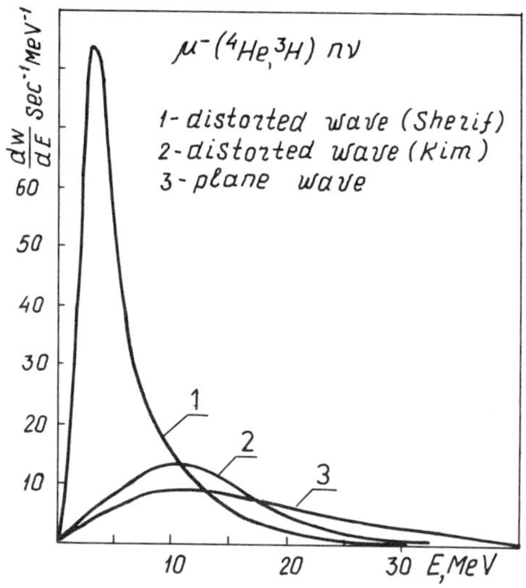

Fig.1

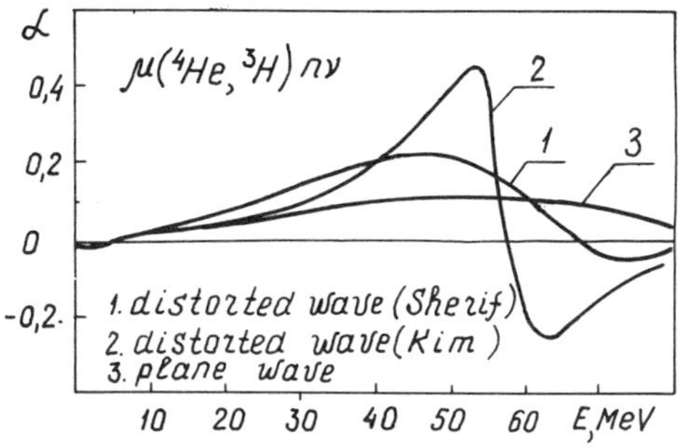

Fig.2

formation of a sharp maximum. The position of the maximum depends on the position of the resonance in p-wave. (The Sherif 's potential [1]) predicts the resonance at E ~ 3 MeV).

III. Asymmetry of the angular distribution undergoes the quantitative variation, dependent on the choice of optical potential.

Thus the data obtained, show that the investigation of different characteristics of the reaction (1) enables one to obtain information on the structure of the (n^3H) system.

BIBLIOGRAPHY

1. H.S.Sherif, B.S.Padmore, Proc.Int.Conf. on Few Particle Problems in Nucl. Interaction, p. 691, North Holland Publishing Company-Amsterdam-London, 1972.
2. C.C.Kim, S.M.Bunch, D.W.Devins, H.H.Forster, Nucl. Phys. 58 (1964) 32
3. R.A.Eramzhyan, V.N.Fetisov, Yu.A.Salganic, Nucl.Phys. B39 (1972) 216

FINAL STATE INTERACTIONS IN MUON CAPTURE BY DEUTERONS

Quang Ho-Kim[†*], J.P. Lavine[††], and H.S. Picker[†††]

† Physics Department, Indiana University, Bloomington, Indiana 47401.

†† Département de Physique, Université Laval, Québec, G1K 7P4, Canada.

††† Physics Department, Trinity College, Hartford, Conn. 06106.

We examine the reaction $\mu^- + d \rightarrow n + n + \nu$ in order to investigate the effect of final-state interactions (FSI) in the $^1S_0$ n-n partial wave. For this state, we use n-n wave-functions that correspond to Bargmann potentials[1], whose s-wave Jost function $F(k) = (2k + i\alpha_1)(2k + i\alpha_2) / \{(2k + i\beta_1)(2k + i\beta_2)\}$. We have chosen purely real α_i and β_i so that the potential will not contain long-range oscillations. Three of these parameters are determined by the scattering length, the effective range, and the energy at which the phase shift changes sign. The fourth parameter is varied: It is related to the rate at which the phase shift goes to zero at high energy, or equivalently, the value of the potential at the origin. The associated potentials, V(r), are positive and finite at r=0, change sign at about 0.7 fm. and then remain attractive as $r \rightarrow \infty$. Since the $^1S_0$ n-n wave-functions are obtained in analytic form[2], they are easy to evaluate.

The formalism we use for μ-capture is standard and the details may be found in Ref. 3. Only the $^1S_0$ state is corrected for the FSI, since we assume all higher partial waves are non-interacting. We use a deuteron wave-function developed by Gourdin et al.[4] which has 6.5% D-state. Our weak interaction coupling constants are $G \cos\theta = 1.1484 \times 10^{-11}(\mathrm{MeV})^{-2}$, $g_V = 0.968$, $g_M = 3.6$, $g_A = -1.226$, and g_p equal either to -8.54 or to -10.27. It turns out that the numerical integrations are sensitive to the number and distribution of the sample points. The energy integrations require a large number of sample points at low energy before a stable result is obtained. We used 120 sample points for the energy integration with the larger range. The r-integration was done accurately with a 15-point Laguerre formula.

Table I shows our results for the capture rates (Γ) without FSI. The first row of each set has g_p=-8.54, while the second row has g_p=-10.27. Table II presents our results for the capture rates with various FSI models. These calculations include the D-state of the deuteron and relativistic corrections. For each FSI model the two rows of capture rates correspond to g_p=-8.54 and g_p=-10.27 respectively. Each FSI model is characterized in the first row by the n-n scattering length (a), the effective range (r_{eff}), the energy at which the phase shift changes sign ($E(\delta=0)$), and the value of the potential at r=0 (V(0)). The second row gives the values of $\alpha_1, \alpha_2, \beta_1$ and β_2 in fm^{-1} for the model.

We note that the FSI enhances the rate by roughly 12%. The capture rate is unchanged as we go from one FSI model to another. We attribute this to the low relative momenta at which the neutrons interact, which prevents them from sampling the variations in the short-range potential which are the most important changes from one model to the next. The results do show a dependence on the value of g_p. We are in agreement with the work of Truhlík[5] both with regard to the approximate size and the sign of the FSI effect and with respect to the lack of sensitivity of the capture rates to small changes in the scattering length. The FSI models all give very similar neutron spectra which differ from the spectrum without FSI. Finally, we are investigating the way unitary transforms change the ^1S_o contribution to $\Gamma_{1/2}$ from 232 sec^{-1} to, for example, 156 sec^{-1}. This result is obtained with the transform used in reference 6 and with parameters of (3.0, 1.0, 0.0).

TABLE I

$\Gamma_{1/2}$ (sec^{-1})	$\Gamma_{3/2}$(sec^{-1})	Γ (sec^{-1})
	S-state deuteron	
331.07	5.77	114.21
319.75	7.29	111.44
	S + D-state deuteron	
329.46	7.93	115.10
317.96	9.26	112.16
	S + D-state deuteron and relativistic corrections	
334.49	7.71	116.64
322.91	9.06	113.68

TABLE II

a/α_1	r_{eff}/α_2	$E(\delta=0)/\beta_1$	$V(0)f_m^{-2}/\beta_2$	$\Gamma_{1/2}$(sec^{-1}) g_p=-8.54	$\Gamma_{3/2}$(sec^{-1}) /	Γ(sec^{-1}) -10.27
-16.5	2.8	280.0	63.56	374.59	8.44	130.49
-12.0	-0.112381	-2.27691	-3.41990	361.45	10.02	127.16
-16.5	2.8	280.0	780.54	374.90	8.44	130.60
-40.0	-0.112380	-1.98288	-5.91558	361.75	10.02	127.27
-16.5	2.6	240.0	637.29	375.55	8.46	130.82
-36.0	-0.112932	-2.74447	-3.72712	362.38	10.04	127.49
-16.5	2.8	240.0	190.11	373.86	8.43	130.24
-20.0	-0.112382	-2.29479	-3.81015	360.75	10.00	126.92
-17.1	2.8	240.0	190.10	374.07	8.43	130.31
-20.0	-0.108698	-2.28265	-3.82071	360.96	10.00	126.99
-17.1	2.8	240.0	786.07	374.14	8.43	130.33
-40.0	-0.108697	-2.13835	-4.82606	361.02	10.00	127.01
-17.1	2.8	280.0	63.56	374.80	8.44	130.56
-12.0	-0.108697	-2.26407	-3.43049	361.66	10.02	127.23
-17.1	2.8	280.0	780.52	375.12	8.45	130.67
-40.0	-0.108696	-1.97599	-5.92154	361.96	10.03	127.34

REFERENCES

1) V. Bargmann, Rev. Mod. Phys. 21, 488 (1949).
2) R.G. Newton, Phys. Rev. 107, 1025 (1957). M. Blažek, Czech, J.Phys. B12,258 (1962).
3) P. Pascual, R. Tarrach and F. Vidal, Nuovo Cimento 12A,241(1972).
A. Fujii and H. Primakoff, Nuovo Cimento 12, 327 (1959).
4) M. Gourdin, M. LeBellac, F.M. Renard, and J. Tran Thanh Van,Nuovo Cimento 37,524(1965).
5) E. Truhlík, Nucl. Phys. B45, 303 (1972).
6) M. Sotona and E. Truhlík, Phys. Letters 43B, 362 (1973).

Work supported in part by the National Research Council and the Atomic Energy Control Board of Canada. One of us (Q. H.-K.) wishes to thank Professors R. Newton and G. Walker and the other members of the Physics Department of Indiana University for their kind hospitality.

A TREIMAN-YANG TEST FOR THE D+P→P+P+N REACTION AT 12.2MeV

N. T. Okumusoglu*, N. Berovic, C. O. Blyth, P. B. Dunscombe and J. S. C. McKee. Department of Physics, The University of Birmingham, B15 2TT, England.

The Treiman-Yang test[1], which is used to indicate the dominance of a spectator pole in a reaction, has so far had limited application to low energy few nucleon studies [2,3,4]. It was originally believed that the test was applicable only when the transferred particle or cluster had spin zero. However, Shapiro[5,6] has shown that, provided the reaction can be treated non-relativistically, the test is also valid for a transferred system of spin half. This discovery clearly broadens the scope of the test and confirms its importance in the determination of reaction mechanisms in few nucleon studies at low energy.

The reaction $^1H(d,pn)p$ has been examined previously by Corfu et al.[3] who conclude that the Treiman-Yang test is not satisfied under their conditions. If their datum point, $\varepsilon = 45^0$ (ε denoting Treiman-Yang angle) (see Fig.3 of ref.3) is ignored, however, the conclusion drawn can be quite different. Also, it is known[2] that the validity of the test depends sensitively on the extent of the momentum transfer.

Bearing these points in mind a Treiman-Yang test for the reaction $^1H(d,pn)p$ at 12.2MeV incident deuteron energy was carried out at Birmingham for conditions of low momentum transfer. The dominant pole was expected to be that shown in Fig.1. The definition of the Treiman-Yang angle is included.

A 3.45mg cm^{-2} polythene target was positioned in the middle of a specially constructed scattering chamber capable of rotation about the beam direction. The proton detector, supported on a movable arm inside the chamber could reach any required angle in or out of the reaction plane by simple adjustment of both arm and chamber positions. Coincident neutrons were detected in an organic scintillator at 5^0 to the incident deuteron beam. Pulse shape discrimination eliminated most of the γ-rays detected. The in-coming data was displayed as a 2 dimensional plot of proton energy versus the time difference between the detection of a proton and the detection of a neutron. This time difference covered several beam bursts of cyclotron acceleration, and the technique permitted the accurate measurement of the random background. Each spectrum was normalised to a monitor counter kept at a fixed scattering angle throughout the experiment.

The results with statistical errors only, are shown in Fig.2. where the square of the reaction amplitude normalized to the phase space factor at $\varepsilon = 40^0$ is plotted against Treiman-Yang angle. It should be remembered that it is the quantity $|M|^2$ that determine whether or not the test is satisfied. Two independent measurements were made at $\varepsilon = 0^0$ and 40^0 to ensure that there were no systematic errors in the experiment.

A straight line fit to the measured values of $|M|^2$ is consistent with a confidence level of 62.5%. The indications from this experiment are therefore that the criterion of Treiman-Yang invariance is satisfied for the nucleon exchange in the $^1H(d,pn)p$ reaction in the vicinity of the quasi-free scattering peak.

REFERENCES

1. S. B. TREIMAN and C. N. YANG. Phys. Rev. Lett. 8, 140 (1962)
2. A. D. IJPENBERG et al., in the 'Few-particle problem in the nuclear interaction'. Edited by I. Slaus et al., (North-Holland, Amsterdam, 1972), p.651.
3. R. CORFU et al., Phys. Rev. Lett. 27, 1661 (1971)
4. R. B. LIEBERT et al., Nucl. Phys. A216, 335 (1973)
5. I. S. SHAPIRO et al., Nucl. Phys. 61, 353 (1965)
6. I. S. SHAPIRO in 'interactions of high-energy particles with nuclei' proceeding of the International School of Physics. 'Enrico Fermi', Course 38'. Edited by T. E. Ericson, (Academic Press, New York, 1967), p.232.

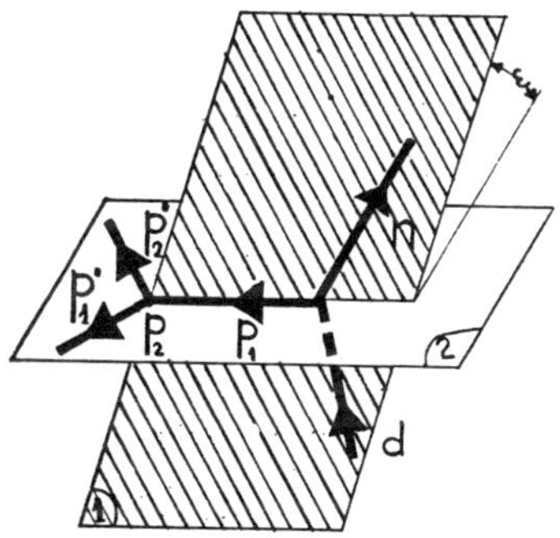

$$\cos \varepsilon = \hat{n}_1 \cdot \hat{n}_2$$

FIGURE 1

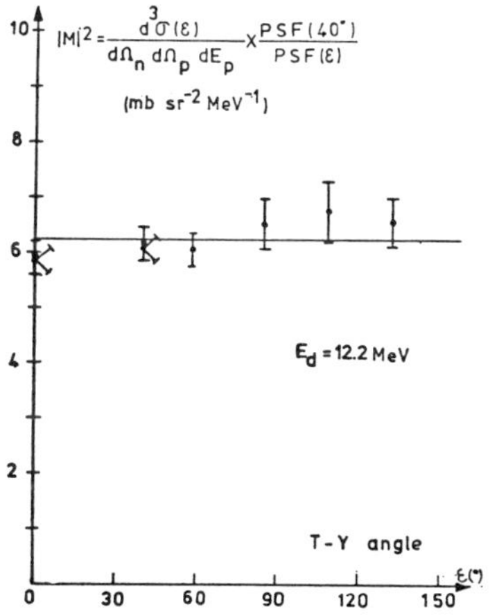

FIGURE 2

PAULI REPULSIVE CORE IN $\alpha+\alpha$ SCATTERING

Ronald E. Brown, W. S. Chien†, and Y. C. Tang
School of Physics, University of Minnesota, Minneapolis, Minnesota 55455, U.S.A.

Interactions between clusters of nucleons are often treated by representing the clusters as point objects which interact by means of an effective potential V. In such a formalism for a two-cluster interaction, the state function has the form $\hat{\psi}=(\mathcal{A}_1\Phi_1)(\mathcal{A}_2\Phi_2)\chi$, where Φ_1 and Φ_2 are individual cluster functions, $\mathcal{A}_1$ and $\mathcal{A}_2$ antisymmetrize Φ_1 and Φ_2, and the relative motion function χ is determined from the Schrödinger equation with the effective potential V. In $\hat{\psi}$ no antisymmetrization is carried out between the clusters, and in such a treatment it is desirable that V contain features which can represent the many-fermion nature of the system, such as the operation of the Pauli exclusion principle. Thus V should be chosen such that physical quantities calculated with $\hat{\psi}$ differ as little as possible from those calculated with a fully antisymmetrized state function $\psi=\mathcal{A}(\Phi_1\Phi_2F)$. Features which have been included in V to represent Pauli-principle effects are an explicit angular-momentum dependence and repulsive cores in some angular momentum states[1-3]. Here we discuss, in the context of $\alpha+\alpha$ scattering, how a repulsive core for V can be constructed from a consideration of the cluster-cluster relative motion functions F and χ at small distances. We term this a "Pauli core" because it represents effects of the exclusion principle.

The nature of F and χ at small intercluster separation r can be adequately discussed in the harmonic oscillator approximation[2,3] in which the ground and first two excited states of ^{8}Be have relative motion functions of the types 3s, 2d, and 1g, respectively. For example, the 3s relative motion function is proportional to $g_{3s}=(1-8\alpha r^2/3+16\alpha^2r^4/15)\exp(-\alpha r^2)$, where α is the usual[2] Gaussian size parameter for the α particle. However, in a totally antisymmetrized formulation, a relative motion function F taken to be $F=g_4=r^4\exp(-\alpha r^2)$ yields the same result for ψ, except for a normalization constant, as does $F=g_{3s}$. This equivalence in using either g_4 or g_{3s} does not hold when the state function $\hat{\psi}$ is used, and, therefore, it is appropriate to inquire whether the use of one of the functions g_{3s} or g_4 is significantly better than the other in reproducing antisymmetrization effects when used as the relative function χ in $\hat{\psi}$. In this respect we note that the proton density has been computed[4] for the $\alpha+\alpha$ system using both $\hat{\psi}$ and ψ. The density obtained using $\hat{\psi}$ is much closer to that obtained using ψ when $\chi=g_4$ is used in $\hat{\psi}$ than when $\chi=g_{3s}$ is used in $\hat{\psi}$. This happens because for $\chi=g_4$ the α clusters do not approach each other as closely as for $\chi=g_{3s}$; therefore the choice $\chi=g_4$ contains some features associated with antisymmetrization between the clusters. Similar considerations are valid for the 2d relative motion function. Consequently, it is suggested that a reasonable requirement for V is that it produce, in the low angular-momentum states of the $\alpha+\alpha$ system, a function χ

---

†Present address: Michigan State University, East Lansing, Michigan 48823, U.S.A.

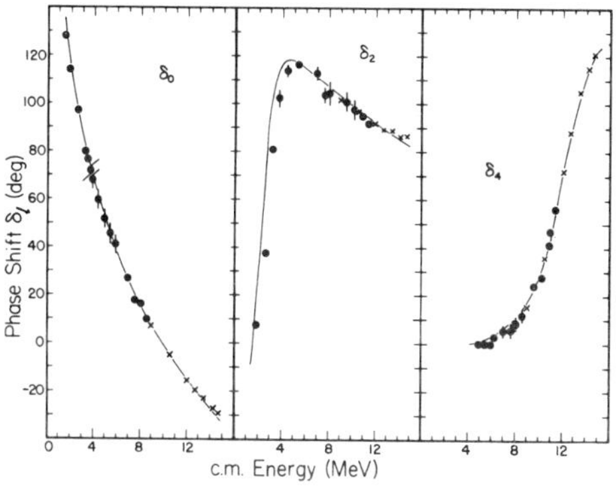

Fig. 1. $\alpha+\alpha$ phase shifts. Data are from ref. 5) (crosses) and ref. 6). Curves show results of the present calculation.

which is proportional to r^4 for small r.

If we write $V=V_D+V_R$, where V_D is an ℓ-independent attractive potential and V_R is a repulsive potential, then the following V_R has the required property:

$$V_R(r)=V_{R\ell}(r)= \hbar^2(2\mu r^2)^{-1}[N(N+1)-\ell(\ell+1)] \text{ for } r<r_{c\ell},$$

$$\text{and} \quad V_R=0 \text{ for } r>r_{c\ell},$$

where $\mu=m_\alpha/2$ is the reduced mass, N=4, and ℓ=0, 2, 4. Note that the strength of $V_{R\ell}$ is given by the r^4 requirement, and only the cutoff radii $r_{c\ell}$ are parameters to be adjusted.

Phase-shift data[5,6] for $\alpha+\alpha$ scattering in the c.m. energy range from 2 to 15 MeV have been fitted as follows. The attractive potential V_D was obtained by folding Gaussian α-particle densities together with the direct part of a nucleon-nucleon potential of Yukawa form. With this $V_D(V_{R4}=0)$ the Yukawa parameters were adjusted to fit δ_4. The excellent fit obtained is shown in fig. 1. The V_D obtained in this way was then added to V_R, and r_{c0} and r_{c2} were varied to obtain least-squared fits to δ_0 and δ_2. The core cutoff radii obtained are r_{c0}=2.82 fm and r_{c2}=2.78 fm, and the fits are shown in fig. 1. It is interesting that r_{c0} and r_{c2} are so nearly equal; this indicates that the r^4 requirement contained in V_R automatically yields the strength of the repulsion in the ℓ=2 state as equal to about .7 of that in the ℓ=0 state, a feature found empirically by Ali and Bodmer[1].

The $\alpha+\alpha$ potential we have described here should be quite useful in considering α cluster models of nuclei, and the general ideas associated with the construction of V_R should be applicable to other systems.

REFERENCES

1) S. Ali and A. R. Bodmer, Nucl. Phys. 80, (1966) 99

2) D. R. Thompson, I. Reichstein, W. McClure, and Y. C. Tang, Phys. Rev. 185, (1969) 1351

3) R. E. Brown and Y. C. Tang, Phys. Rev. 176 (1968) 1235

4) K. Wildermuth, private communication

5) W. S. Chien, Ph.D. Thesis, U. Minn. 1974; W. S. Chien and R. E. Brown, to be pub.

6) T. A. Tombrello and L. S. Senhouse, Phys. Rev. 129 (1963) 2252; R. Nelson W. K. Jentchke, G. R. Briggs, R. O. Kerman, and J. N. Snyder, Phys. Rev. 109 (1958) 850; H. Werner and J. Zimmerer, Paris Conf. on Nuclear Physics, 1964, p. 241

P-P Final State Interaction in the $^2H(p,pn)p$ Reaction*

D. Bonbright, J.W. Watson, D.J. Roberts
Cyclotron Laboratory, University of Manitoba,
Winnipeg, Manitoba, R3T 2N2, Canada

Much work has been done on the energy dependence of "quasi-free" scattering from deuterium[1]. A similar study of the energy dependence of final state interactions should add further to our knowledge.

Within the energy range of the present experiment, 20-40 MeV, it has been emphasized that breakup of deuterium occurs via a complicated series of multiple scatterings[2]. It would seem that final state interactions, which are known to be sensitive only to the final interacting pair, would present noteable systematic behaviour. We have thus studied the behaviour of the p-p FSI from the reaction $^2H(p,pn)p$ as a function of bombarding energy. And, we have compared our results with a similar study of the n-p FSI as seen in the reaction $^2H(p,pp)n$[3].

We detect a proton and neutron in coincidence, using an experimental technique that has been described elsewhere[4]. A target of 9 mg/cm^2 CD_2 was used; its thickness was monitored before and after each run. Our neutron counter was kept at 70°, and our proton counter placed so as to allow a p-p FSI to occur. We present data taken at 22.1, 35.7 and 41.1 MeV.

Figure 1 shows a typical spectrum. All spectra show a pronounced minimum, due to Coulomb repulsion, where the protons have low relative energy. In contrast, a spectrum showing a n-p FSI in deuterium breakup would show strong peaking due to the strong n-p singlet state. In the case at hand, it is more difficult to infer that the dominant reaction mode is via FSI.

Figure 2 presents the energy behaviour of cross sections integrated over a FSI peak. For the present experiment, we have summed counts over

* Research supported by the Atomic Energy Control Board of Canada.

the region where the two protons have less than one MeV of relative energy. As a comparison, we give data taken for a study of the n-p FSI from deuterium[3], where one of the two protons was also detected at a fixed angle of 70°. The cross sections for the latter were integrated over the prominent part of the FSI peak. It is seen that the n-p FSI is about four times larger than the p-p FSI, an effect that was discussed above. Moreover, it is seen that both cross sections decrease with increasing energy in a remarkably similar manner. This may be taken as evidence that the production mechanism for both is the same. The solid line on this graph is a fit to the n-p FSI data given by a code due to W. Ebenhoeh[5]. It fits the data quite well. In the future, we hope to modify this code to fit the p-p FSI data by using a simple multiplicative factor applying Coulomb repulsion, as in a Watson-Migdal calculation.

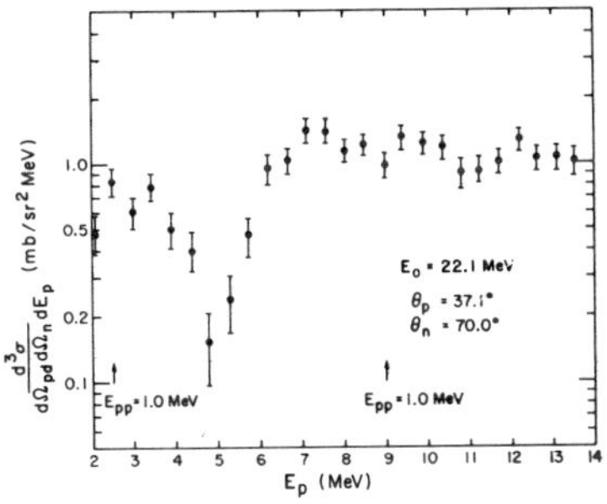

Fig. 1

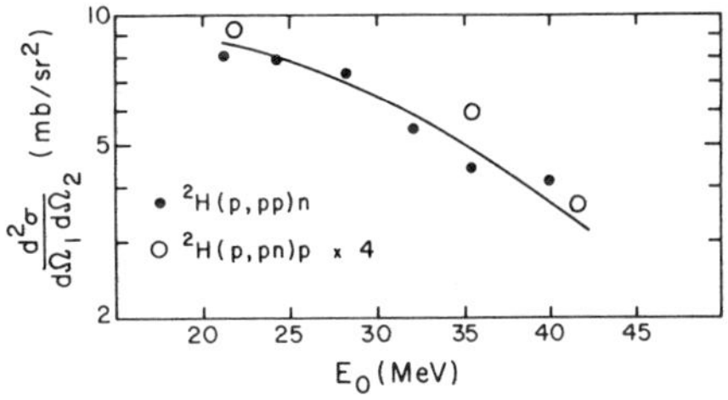

Fig. 2

REFERENCES

1) W.J. Braithwaite, J.R. Calarco, J.M. Cameron, and D.W. Storm, Nucl. Phys. A166 (1971) 515

2) R.T. Cahill and I.H. Sloan, Nucl. Phys. A165 (1971) 161

3) D.I. Bonbright, C.A. Miller, J.W. Watson, Bull. Am. Phys. Soc. 19 (1974) 505

4) C.A. Miller, J.W. Watson, D.I. Bonbright, F.J.S. Wilson, and D.O. Wells, Phys. Rev. Lett. 32 (1974) 684

5) W. Ebenhoeh, Nucl. Phys. A191 (1972) 97

Quasifree and Final State Processes in $^4He + ^3H$ at $E_{^4He} = 70$ MeV

R. G. Allas, L. A. Beach, R. O. Bondelid, E. M. Diener* and E. L. Petersen
Naval Research Laboratory, Washington, D. C. 20375

and

P. A. Treado†, J. M. Lambert,† R. A. Moyle and L. T. Myers†
Georgetown University, Washington, D. C. 20007 and Naval Research Laboratory

and

Ivo Slaus
Institute Ruder Bošković, Zagreb, Yugoslavia and Naval Research Laboratory

and

P. Tomas†
Institute Ruder Bošković, Zagreb, Yugoslavia and Georgetown University

A 70 MeV α-beam from the NRL cyclotron bombarded a ^{3}H-Ti self-supporting target, and kinematically complete data were obtained for the $^4He + ^3H$ reactions to final state systems with at least two charged particles. The reactions ^{3}H(α,αd) n, ^{3}H(α,αp) 2n, ^{3}H(α,ht) n, ^{3}H(α,tt) p, ^{3}H(α,dh) 2n, and ^{3}H(α,dt) d were investigated with kinematic conditions favoring quasifree scattering (QFS) and quasifree reaction (QFR) processes. Data was obtained for eight angle pairs. For each reaction, the momentum transfer region up to ~ 0.6 fm^{-1} was studied. The spectral data are frequently dominated by final state interaction (FSI) structure, but some QF enhancements have been observed. The reaction ^{3}H(α,αd) n has a pronounced QFS enhancement, and Table 1 describes the QF measurements.

Table 1

Particle 1	Particle 2	Spectator	Mechanism and Relative Strength
alpha	deuteron	neutron	QFS, α + d → α + d; strong
alpha	proton	dineutron	QFS, α + p → α + p; very weak
helion	triton	neutron	QFR, α + d → t + h; weak
helion	deuteron	dineutron	QFR, α + p → h + d; not observed
triton	deuteron	deuteron	QFR, α + n → t + d; not observed
triton	triton	proton	QFR, α + 2n → t + t; not observed

These data are analyzed in the PWIA in the same manner as described previously,[1] and the extracted square of the momentum form of the wave function for the strong QFS reaction is shown in Fig. 1.

FSI spectral structure is observed from the reactions ^{3}H(α,αd) n, ^{3}H(α,αp) 2n, ^{3}H(α,tt) p, ^{3}H(α,ht) n, ^{3}H(α,dd) t, ^{3}H(α,dt) d, and ^{3}H(α,pt) t. In the αd spectra, the ^{5}He ground state is present for the only angle pair used where it is kinematically allowed, the ^{5}He 16.7 MeV state is apparent in the five angle pairs where it is allowed, and there is some indication of previously reported structure in the ^{5}He excitation region of 18 - 20 MeV.[2] In the pα spectra, which are generally not clean due to the possibility of various relative energies of the 2n system, the ^{5}Li ground state is quite strong and there is an indication of the 1.8 MeV state of ^{6}He when it is kinematically allowed and no distinct indication of a broad 15 MeV ^{6}He state is present, although it was kinematically allowed for most angle pairs used. The ^{5}He states at 16.7 MeV and

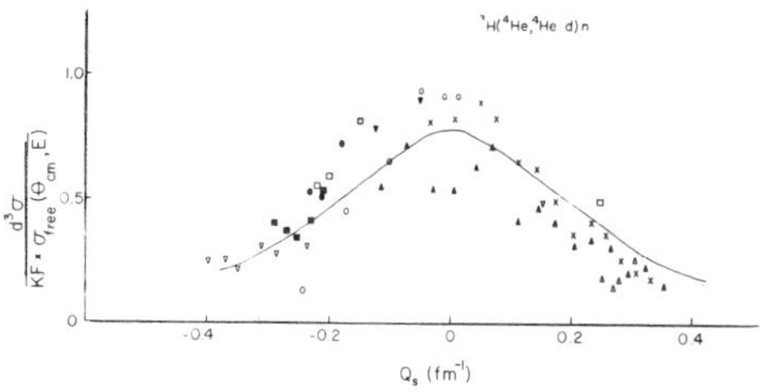

Figure 1. Data points are the extracted square of the momentum form of the wave function for $^3H(^4He,\alpha d)n$. Curve is (PWIA - Fourier transform) x 0.2.

at 18 - 20 MeV are observed in the dd and dt spectra. No evidence for the 28.4 MeV state in ^4He is present. In the dt spectra, the 20 MeV state of ^4He is quite strong. In the tt and pt spectra, the 20, 21.4 and 22.4 MeV states of ^4He are seen whenever they are kinematically allowed. There is some indication of a broad 15 MeV level in ^6He, but better statistics are required. Again, there is no indication of a 28.4 MeV level in ^4He. The paucity of data for the ht spectra limits the data analysis. At one angle pair, the 22.4 MeV state of ^4He is seen and at another angle pair, there exists minimal structure in the region where the 28 MeV state of ^4He would be. Figures 2 and 3 depict some of the FSI data.

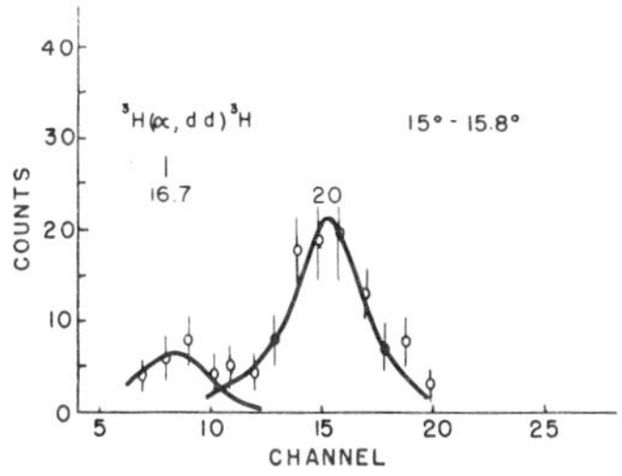

Figure 2. Data and possible ^5He excited state structure in a dd spectrum for one angle pair.

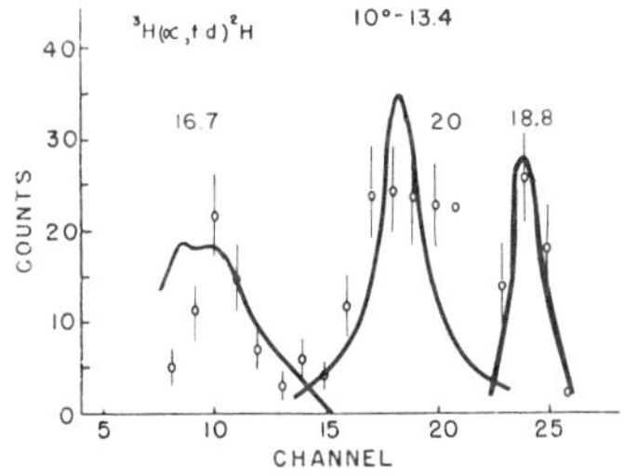

Figure 3. Data and possible ^5He excited state structure in a td spectrum for one angle pair.

*National Research Council - Naval Research Laboratory Research Associate

†Work Partially Supported by the National Science Foundation

References

1. I. Slaus et. al., Phys. Rev. C8, 444 (1973) and R. G. Allas et. al., Phys. Rev. C9, 787 (1974).
2. P. A. Treado et. al., Phys. Rev. C7, 1742 (1973).

Overlapping FSI and QFS in N-d Break-up

R.T. Cahill

Physics, Flinders University, Bedford Park 5042, Australia.

In the N-d break-up final state if one N-N pair is in the fsi region, then another pair may be in the qfs region: the fsi-qfs overlap. In kinematically incomplete spectra this overlap is often unavoidable, as in forward d(n,p)2n. The analysis of this overlap is essential to understanding the n-n fsi. Here we combine the general theories [1-3] of fsi and qfs to obtain the consequences of the overlap and to obtain a simple wave function description. The results explain significant features of n-n fsi in d(n,p)2n and also of elastic N-d scattering.

From either the Faddeev equations or S-matrix theory we may derive [1] the general structure of fsi. In this the fsi-pair ℓ^{th}-wave projection of the coherent fsi amplitude is [1]

$$F_n^{(\ell,m)}(s,z) = \frac{1}{2iq} e^{i\delta_n^{(\ell)}(s)} \left[e^{i\delta_n^{(\ell)}(s)} R_n^{(\ell,m)}(s,z) \right]_{disc.}, \tag{1}$$

where both $\delta(s)$ and $R(s,z)$ have a square root branch point at $s = 0$. Here $[....]_{disc.}$ is the branch cut discontinuity, $z = \hat{k}_p \cdot \hat{k}_n$, $s = q^2$ is the fsi pair subenergy and n labels the fsi. Eqn (1) is complimented by the discontinuity constraint [1] for R,

$$\left[R_n^{(\ell,m)}(s,z) \right]_{disc.} = \sqrt{4\pi} \frac{iq}{2\pi} \sum_{n'} \int d\hat{q} Y_{\ell m}(\hat{q}) \chi_{nn'} N_{n'}(s',z',\hat{q}'). \tag{2}$$

Here s', z' and $\hat{q}'$ are functions [1] of s, z and $\hat{q}$ by usual three particle kinematics. Now if $z \simeq +1$ and $s \in$ fsi region ($s \simeq 0$), then $s', z' \in$ qfs region. Hence qfs strongly influences the fsi. That is, the amplitude in the qfs region determines $[R]_{disc.}$ in the fsi region. Here $N_n(s,z,\hat{q})$ are the usual Faddeev amplitudes [1].

From the Faddeev equations we have derived [3] a general qfs theory which gives, for qfs region,

$$\sum_{n'} \chi_{nn'} N_{n'}(s',z',\hat{q}') \simeq c\Phi(\vec{q} - \tfrac{1}{2}\vec{\Delta}; \vec{k}), \tag{3}$$

where $\Phi(\vec{K};\vec{k}) = \int d^3r\, e^{i\vec{K}\cdot\vec{r}} \psi(r) D(\vec{r};\vec{k})$. Here $\psi(r)$ is the deuteron wave function, $\vec{\Delta}$ is the momentum transfer, say $\vec{k}_p - \vec{k}_n$ in n+d → n+n+p ($\vec{k}_n = \vec{k}$), and $D(\vec{r};\vec{k})$ is the smooth cut off function [3]. $D(\vec{r};\vec{k})$ causes $\Phi(\vec{K};\vec{k})$ - an "effective" deuteron wave function - to be smaller and sharper than the normal $\phi(K)$. D describes the strong effect on qfs of all higher order multiple scatterings.

Using simple dispersion theory we construct the simplest function that follows from (2) and (3),

$$R_n^{(\ell,m)}(s,z) \simeq \sqrt{4\pi}\,\frac{c}{2\pi^2}\int d^3q'\,\frac{\Phi(\vec{q}\,'-\frac{1}{2}\vec{\Delta};\,\vec{k})}{s'-s}\,Y_{\ell m}(\hat{q}'), \tag{4}$$

where we could add a non-singular (at s = 0) function. Then (1) is, say for $\ell = 0$,

$$F_n^{(o)}(s,z) \simeq \frac{c}{4\pi}\int d\hat{q}'\Phi(\vec{q}\,'-\frac{1}{2}\vec{\Delta};\vec{k}) + \frac{c}{2\pi^2}f_n^{(o)}(s)\int d^3q'\,\frac{\Phi(\vec{q}\,'-\frac{1}{2}\vec{\Delta};\vec{k})}{s'-s-i\varepsilon},$$

which gives the approximate fsi amplitude [3)]

$$F_n(s,z,\hat{q}) \simeq c\int d^3r\psi_n^{(-)*}(\vec{q},\vec{r})\,e^{i\vec{\Delta}.\vec{r}/2}\psi(r)D(\vec{r};\vec{k}). \tag{5}$$

This is similar to the usual fsi impulse approximation, and differs by the qfs cut off function. D occurs because of the fsi-qfs overlap. The influence of $D(\vec{r};\vec{k})$ on d(n,p)2n is to narrow the n-n fsi peak. Hence $D(\vec{r};\vec{k})$ accounts for the "dilation" effect seen [1)] in exact numerical solutions of the Faddeev equations in the fsi region: the "effective" deuteron wave function $\Psi(\vec{r};\vec{k}) = \psi(r)D(\vec{r};\vec{k})$ in (5) is more extended than $\psi(r)$.

If n is the $^3S_1$ n-p interaction then the residue of F_n at the deuteron pole ($s = -e_d$) is the corresponding n-d elastic amplitude, F(z). Whence (with $\vec{\Delta} \to \vec{k}_n' - \vec{k}_n$)

$$F(z) \sim \int d^3r\psi(r)\,e^{i\vec{\Delta}.\vec{r}/2}\psi(r)D(\vec{r};\vec{k}). \tag{6}$$

This describes forward n-d scattering. Essentially F(z) is a modified target form factor. The influence of D here will be to sharpen the forward peak and perhaps introduce diffraction minima, compared with the usual form factor. A similar expression arises in the fixed scatterer approximation [4)], but there the prescription for $D(\vec{r};\vec{k})$ differs from that derived [3)] from the Faddeev equations.

References

1) R.T. Cahill, Phys. Rev. C9 (1974) 473

2) S. Shirato *et al* Nucl. Phys. A125 (1973) 277

3) R.T. Cahill, Phys. Lett. B49 (1974) 150

4) R.T. Cahill and M.S. Poole, "Multiple Scattering Effects in a Three Particle System", Flinders preprint 1974.

AN APPLICATION OF THE CLUSTER MODEL REACTION THEORY TO THE d(α,α)n p BREAK UP PROCESS

P. Heiss and H.H. Hackenbroich

Institut fuer Theoretische Physik der Universitaet Koeln, Germany

In the last few years considerable progress has been made in the exact treatment of three-particle processes. However, to date no realistic microscopic calculation has actually been performed for three-body break up reactions in nuclear systems with A>3. Hence it seems desirable, to investigate the feasibility of quasi-two-particle methods for this type of reactions.

Recently a multichannel cluster model calculation has been performed for the ^{6}Li scattering system. As discussed in ref. [1]), the results for the elastic ^{4}He-d scattering data are in excellent agreement with experiment. In addition to ^{4}He-d and ^{3}He-t channels, this calculation also includes ^{5}He-p and ^{5}Li-n structures with the five nucleon systems in their lowest $J^{\pi} = 3/2^-$ state. These five nucleon states are approximately represented by quasi-bound wave functions. Nevertheless we assume, that the calculated ^{4}He-d$\rightarrow$(5)-(1) S-matrix elements are realistic, because these quantities only depend on the behaviour of the five nucleon functions in the interaction region.

As explained in ref.[2]), the three-particle cross section $d^2\sigma / d\Omega_1 d\Omega_s$ can be derived from the ^{4}He-d$\rightarrow$(5)-(1) S-matrix elements, when a sequential decay mechanism is assumed. In the present case we obtain

$$d^2\sigma / d\Omega_1 d\Omega_s \sim \frac{1}{4\pi} T_{00}(\Omega_1) + \frac{1}{\sqrt{4\pi}} \left[T_{20}(\Omega_1) Y_{20}(\Omega_s) + 2\,\mathrm{Re}\, T_{21}(\Omega_1) Y_{21}(\Omega_s) + 2\,\mathrm{Re}\, T_{22}(\Omega_1) Y_{22}(\Omega_s) \right]$$

Here Ω_1 is the c.m. angle in the first step (i.e. the angle of the excited recoil nucleus) and Ω_s is the decay angle relative to the direction of the recoil nucleus. The tensormoments $T_{lm}(\Omega_1)$ of the recoil nucleus in the first step can be calculated from the ^{4}He-d$\rightarrow$(5)-(1) S-matrix elements according to the definition of Welton.

In fig. 1 our results for the ^{4}He-d $\rightarrow$ ^{5}He-p $\rightarrow$ ^{4}He-n-p cross sections are compared with recent measurements of Prescher et al. [4]). For the three curves one common normalization factor has been used. The experimental points are obtained by integrating the three-particle energy spectra over the width of the ^{5}He resonance.

The results are in reasonable agreement for proton foreward angles. However, there are considerable discrepancies for ^{5}He foreward angles. This can be explained by the fact that for these angles other reaction-mechanisms (as the ^{4}He-p QFS) play an important role. Hence strong interference between the different reaction-mechanisms can not be excluded.

A further calculation, taking into account the energy dependence in the second step, is in progress. Finally we want to emphasize, that the presented calculation was not separately performed for the analysis of the d(α,α)n p break up, but is a part of a consistent and successful description of the ^{6}Li scattering system.

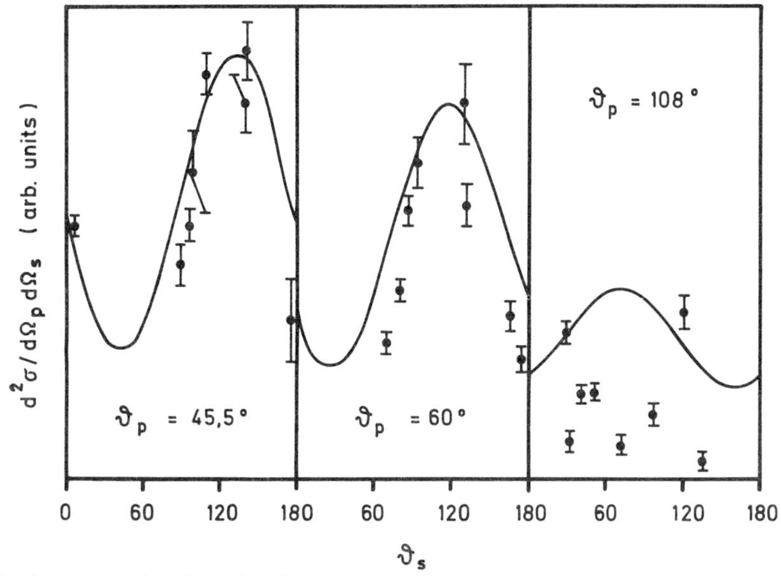

Fig. 1 Cross section for d(α,α)n p

References

1) H.H. Hackenbroich, P. Heiss and Le Chi Niem Nucl. Phys. A221 (1974) 461

2) H. Eichner, H. Stehle and P. Heiss Nucl. Phys A205 (1973) 249

3) T.A. Welton Fast Neutron Physics, vol. 2 (Interscience,New York,1973) p. 1317

4) K. Prescher et al. to be published

THREE-ALPHA BREAKUP OF THE 11.83-MeV STATE OF THE ^{12}C NUCLEUS

B. Antolković

Institute "Rudjer Bošković", Zagreb, Yugoslavia

The ^{12}C-3 alpha breakup has been extensively investigated to obtain data on ^{12}C excited states. The aim of the present work was to contribute more information on the spin and parity value of the 11.83-MeV state of ^{12}C, since the data obtained so far are controversial and assign either the 1^- or the 2^- value to this state (ref. 1 and references therein). Because of the large positive Q-values, ^{12}C states reached through the $p+^{11}B$, $d+^{10}B$ or $^{3}He+^{9}Be$ reactions have excitations higher than 15.96 MeV. To investigate the states lower than 15.96 MeV, one has to explore reactions with four particles in the final state.

Recently a detailed study of the $^{12}C(11.83)$-3 alpha decay was performed by means of the $^{12}C(n,n)3\alpha$ reaction[1)]. The three-alpha correlation was measured by the nuclear emulsion technique in the whole momentum space. The results obtained show that a satisfactory fit with the theoretical density distribution can be achieved with the spin and parity value 1^- assigned to the 11.83-MeV state of ^{12}C. The experiment was, however, performed at an incident neutron energy of 14.4 MeV, at which the kinematical conditions lead to a certain overlap of the momentum spaces where the reaction chains $^{12}C(n,n)^{12}C(\alpha)^{8}Be(2\alpha)$ and $^{12}(n,\alpha)^{9}Be_{2.43}(\alpha)^{5}He(n,\alpha)$ take place. Although the latter process is involved only with 10% in the total cross section, and its influence on the $^{12}C_{11.83}$-3α analysis was removed by eliminating all events for which one α-α-n group has an excitation $E(^{9}Be_{2.43})$ of 2.43 ± 0.3 MeV, some ambiguity in such a procedure is still present.

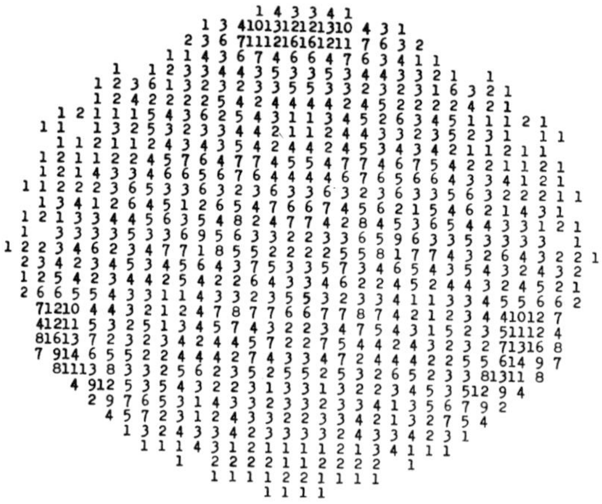

FIGURE 1

To avoid the overlap of the two competing processes going via

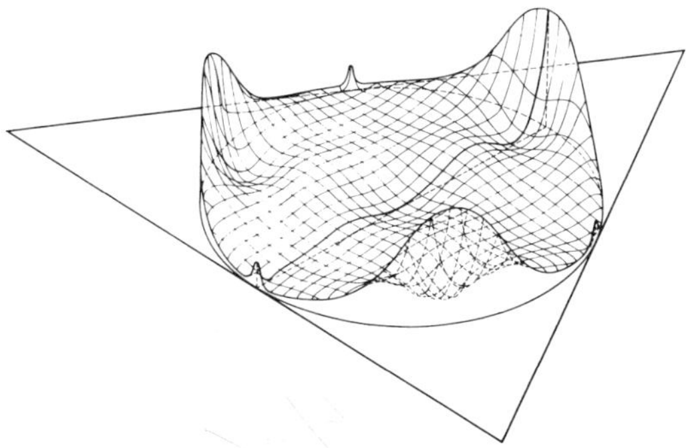

FIGURE 2

the $^{12}C_{11.83}$ and $^{9}Be_{2.43}$ intermediate states, the present experiment was performed at an incident energy of 18.2 MeV. The density distribution of the $^{12}C_{11.83}$-3 alpha breakup is shown in Fig. 1 in the frame of the Dalitz plot. The main feature of the alpha-alpha correlation spectrum is the clustering of the experimental points at the maximum energies of the three alpha particles.

The theoretical density distribution for the ^{12}C-3α breakup was calculated by an expression that takes into account the antisymmetrization of the exit channel with respect to the three alpha particles (see refs. 2 and 3) and assuming a sequential decay via the $^{8}Be_{2.9}(2^{+})$ state. A good fit was obtained for the spin and parity value 1^{-} assigned to the 11.83-MeV state of ^{12}C, and the angular momentum of the first emitted particle with respect to the center of mass of ^{8}Be formed by the other two alpha particles being $\ell=3$ (fig. 2). Assuming the unnatural spin and parity value 2^{-} for the $^{12}C_{11.83\ MeV}$ state, the theoretical density distribution cannot reproduce the high-energy peaking of the experimental data, since it decreases to zero when approaching the contour of the Dalitz plot, and the transition via the $^{8}Be_{g.s.}$ is not allowed. The results thus obtained confirm the data of ref. 1) and add some more evidence for the 1^{-} assignment to the 11.83-MeV state of the ^{12}C nucleus.

REFERENCES

1) B. Antolković, J. Hudomalj, to be published, and B. Antolković, Proc. Int. Conf. on Few Particle Problems in the Nuclear Interaction, Los Angeles 1972, p. 695.

2) K. Schäfer, Nucl. Phys. A140 (1970) 9

3) C. Jacquot, Y. Sahamoto, M. Jung and L. Girardin, Nucl. Phys. A201 (1973) 247

INTERACTION DANS L'ETAT FINAL DANS LES REACTIONS A TROIS NUCLEONS A 156 MeV.

T. Yuasa

Institut de Physique Nucléaire, Université de Paris-Sud, 91406 Orsay (France).

L'expérience cinématiquement complète effectuée à Orsay sur le "break-up" du deuton par le faisceau de protons de 156 MeV[1] a confirmé l'existence des interactions dans l'état final (FSI) p-n et p-p pour les conditions cinématiques qui correspondent à leur énergie relative presque zéro. Ces interactions FSI ont été observées dans les spectres de sections efficaces différentielles, $d^3\sigma/d\Omega_1\, d\Omega_2\, dE_1$, obtenus pour deux paires d'angles de diffusion ($\theta_1 = 45^o$, $\theta_2 = -57^o$) et ($\theta_1 = 48^o$, $\theta_2 = -55^o$) (les angles de diffusion et l'énergie cinétique d'un proton sont respectivement θ_1, θ_2 et E_1). Le premier spectre passe par un point correspondant à l'énergie relative entre un proton et le nucléon spectateur E^{rel}_{1-3} égale à 0.16 MeV et le deuxième passe par un point correspondant à E^{rel}_{1-3} égale à 0.0045 MeV.

Ces deux spectres pour (p, 2p) sont présentés dans les Figs. 1 et 2, tandis que pour (p, pn), le seul spectre pour ($\theta_1 = 48^o$, $\theta_2 = -55^o$) est présentés dans la Fig. 3.

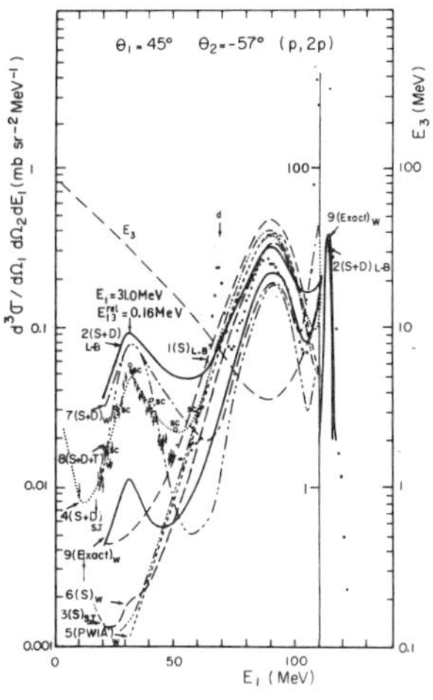

Fig. 1

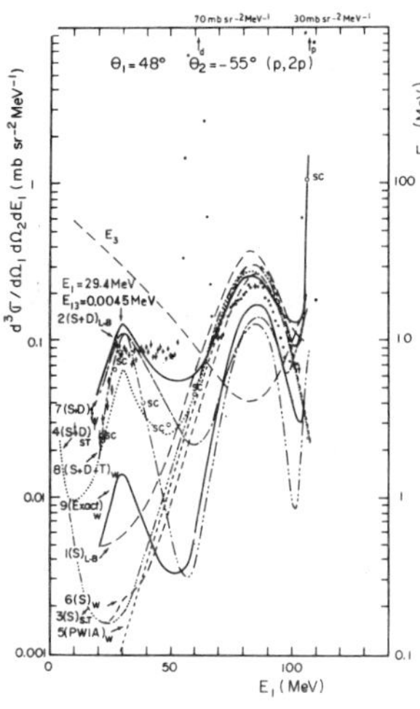

Fig. 2

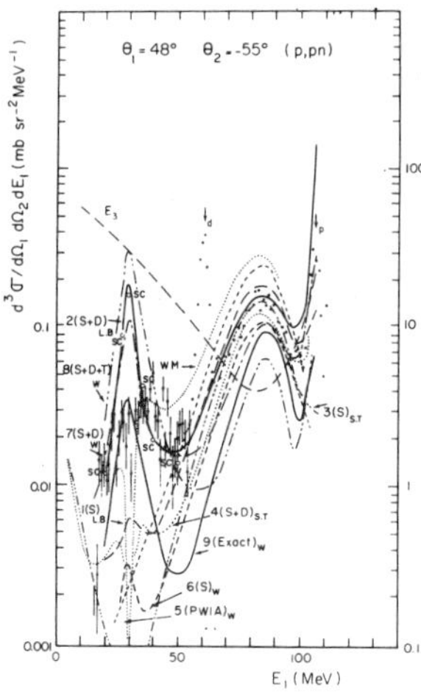

Fig. 3

D'après ces résultats expérimentaux, on peut remarquer que les FSI p-n et p-p sont très sensibles à leur énergie relative et que, d'autre part, la FSI p-n manifeste un pic, tandis que la FSI p-p présente un minimum et deux maxima en indiquant l'interaction Coulombienne comme dans la diffusion de deux protons libres.

Nous avons comparé nos spectres expérimentaux avec certains calculs théoriques : ceux de L'Huillier et Ballot [2], de Sakamoto et Takeutchi [3] et de Wallace [4] désignés respectivement L-B, S-T et W dans les figures 1, 2 et 3. La première méthode, celle des calculs de L-B et S-T tronque au deuxième ordre la diffusion multiple, mais utilise l'interaction réaliste de Hamada-Johnston, soit hors-couche de l'énergie (L-B), soit sur-couche de l'énergie avec l'interaction Coulombienne (S-T), tandis que la deuxième méthode, celle de Wallace calcule exactement la diffusion multiple, mais avec une interaction simplifiée. Les $d^3\sigma$ calculées, limitées aux ordres, premier, deuxième, troisième et de l'infini sont désignées respectivement dans les figures comme (S), (S+D), (S+D+T) et (exact), le calcul fait par DWIA par Wallace ainsi que le calcul de modèle spectateur sont aussi présentés comme $(DWIA)_W$ et SM. Les calculs $(S+D)_{L-B}$ semblent expliquer le mieux les résultats expérimentaux. D'autre part, les sections efficaces SM multipliées par le facteur de renforcement de Watson-Migdal sont présentées comme W-M. L'interaction Coulombienne n'est pas encore correctement calculée pour une interaction à trois nucléons.

Les interactions FSI p-n et p-p sont aussi confirmées dans les réactions ^{3}He (p, 2p) d et ^{3}He (p, pn) 2p [5].

Références :

[1] J.P. Didelez, I.D. Goldman, E. Hourany, H. Nakamura, F. Reide et T. Yuasa voir les références dans une autre communication à cette conférence [5]; Phys. Rev. (sous presse).

[2] M. L'Huillier, Nucl. Phys. A.122, 667 (1968) ; Thèse Orsay (1974).

[3] F. Takeutchi, Y. Sakamoto et T. Yuasa, Phys. Lett. 35B, 498 (1971) et Int. Symposium Budapest (1971), p. A6 ; F. Takeutchi et Y. Sakamoto, Proc. Int. Conf. Los Angeles (1972), p. 347 ; F. Takeutchi et Y. Sakamoto, Nucl. Phys. A185, 366 (1972).

[4] J.M. Wallace, Phys. Rev. C5, 609 (1972), C7, 10 (1973), C8, 1275 (1973).

[5] J.P. Didelez, R. Frascaria, N. Fujiwara, I.D. Goldman, E. Hourany, H. Nakamura, F. Reide et T. Yuasa, communication à cette Conférence.

The Watson factorization model and the exact three body theory in the case of neutron-proton final state interaction in the reaction pd - ppn.

W. Kluge, R. Schlüfter and W. Ebenhöh[+)]

Institut für Experimentelle Kernphysik der Universität Karlsruhe, Germany

This paper is a new analysis of a recent angular distribution measurement of np-FSI with 52 MeV deuterons at the Karlsruhe isochronous cyclotron. The aim of the analysis is

1) to compare the experimental data with the exact three-body-model using spindependent separable potentials for relative S-states of two nucleons.
2) to obtain information on the relation between the exact three-body theory and the Watson factorization ansatz.

The separable potential model is very successful in calculating the cross-sections of the break-up reaction. In the past a formfactor of the Yamaguchi type was used for the NN-vertex[1,2]. Recently also other types of formfactors have been discussed representing better approximations to the high energy behaviour of the nucleon-nucleon ^1S_o and $^3S_1$ phases than the Yamaguchi formfactor[3,4]. Fig.1 shows the measured angular distribution of np-FSI at zero relative energy[5]. The solid curve in Fig.1 was calculated with the

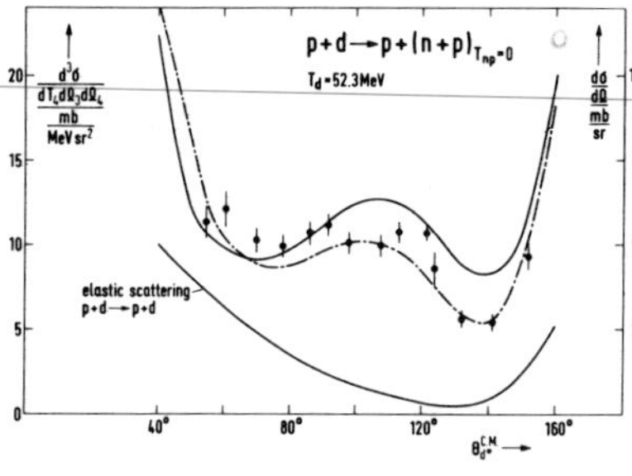

Fig.1

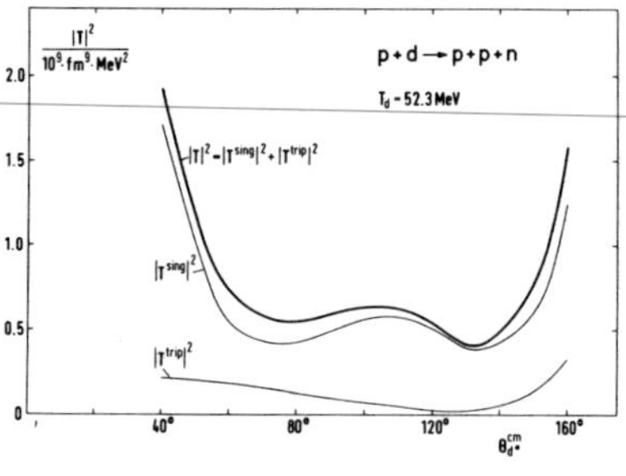

Fig.2

separable potential model using the Yamaguchi formfactor, while the dashed-dotted one was obtained with a quadratic formfactor representing a better fit to the S-phases of the nucleon-nucleon-scattering[4]. No normalization factor has been applied to the calculation. For comparison with the break-up reaction the angular distribution for elastic scattering of deuterons on protons representing scattering in the $^3S_1$ only is also shown in Fig.1. There is an obvious difference in the shape of the two angular distributions, which is interpreted as a predominant contribution of ^1S_o scattering to the break-up reaction. This can be seen clearly if one plots separately the squared transition matrix-elements for scattering into the ^1S_o and $^3S_1$ states $|T^{sing}|^2$ and $|T^{trip}|^2$ at T_{np}=0 calculated by the separable potential model versus the scattering angle as has been done in Fig.2. The triplet contribution is small and shows the same shape as the elastic deuteron-proton scattering. It can be concluded that the shape of the measured angular distribution of the break-up reaction is actually characterized by the dominant excitation of the ^1S_o state. The same conclusion was already drawn from a former analysis

+) Institut für Theoretische Physik der Universität Heidelberg, Germany using the Watson model[5]. The results of the analyses with the separable potential model and with the Watson model lead to a very interesting conclusion on the relation between the two models. The quantities X^s_{np} and X^t_{np} of the Watson model defined in ref. 5) are proportional to the squared transition matrixelements for production of neutron proton pairs in the singlet or triplet state. The absolute magnitudes of the parameters X^s_{np} and X^t_{np} must be determined by least square fits to the experimental data. The ratio X^s_{np}/X^t_{np} obtained by such an analysis can be used to check the applicability of the Watson factorization ansatz. This ratio must be equal to the ratio $|T^{sing}|^2/|T^{trip}|^2$ at $T_{np}=0$ if the Watson model shall make sense. A comparison of the ratio X^s_{np}/X^t_{np} with the ratio $|T^{sing}|^2/|T^{trip}|^2$ is shown in Fig.3. No normalization factor has been used in Fig.3. The agreement between these ratios is surprising. A possible explanation is given in ref. 2,6). The analysis presented in this paper and further investigations[7,8] show that the Watson factorization ansatz is a useful tool not only to analyse the shape of the coincidence spectra of np- or nn-FSI as demonstrated earlier[5] but also to analyse the angular distribution of the np-FSI.

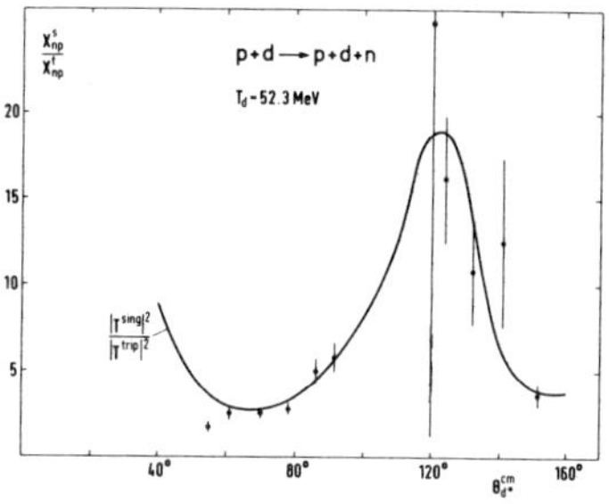

Fig. 3

The applicability of the Watson model under carefully chosen kinematical conditions should be helpful for the analyses of the np- or nn-FSI in the dd - dpn reaction, where exact models does not exist so far.

References

1) I.H. Sloan and R.T. Cahill: Nucl. Phys. A165 (1971) 161
2) W. Ebenhöh: Nucl. Phys. A191 (1972) 97
3) J.M. Wallace: Phys. Rev. C7 (1973) 10
4) J. Bruinsma et al.: 1974 submitted to Nucl. Phys.
5) H. Brückmann et al.: Nucl. Phys. A157 (1970) 209
6) W. Kluge, R. Schlüfter and W. Ebenhöh: 1974 submitted to Nucl. Phys.
7) H. Brückmann et al.: 1974 submitted to Nucl. Phys.
8) W. Kluge: submitted to Fortschr. der Physik

UNE RECHERCHE SUR LA DESINTEGRATION BETA DU DIPROTON DANS $^{3}He(\tau, {}^{2}He)^{4}He$

R. Pigeon, U. von Moellendorff†, C.R. Lamontagne, M. Irshad et R.J. Slobodrian

La synthèse des éléments dans les étoiles commence par la fusion de deux protons qui produit $e^{+} + {}^{2}H_{1} + \nu$. La très faible section efficace calculée pour cette réaction ne permettait même pas de tenter sa mesure[1-2]. L'objet du travail rapporté ici était d'observer les deutons de cette désintégration à la suite de la réaction $^{3}He(\tau, {}^{2}He)^{4}He$. La désintégration a alors une plus grande probabilité de se produire puisque les protons doivent sortir de la barrière coulombienne plutôt que d'y entrer.

L'expérience a été réalisée à l'aide d'un faisceau d'$^{3}He^{++}$ de 12.56 MeV produit par l'accélérateur van de Graaff CN de l'Université Laval. L'analyse électronique des impulsions comprenait un identificateur de particule et pour chaque détecteur un circuit qui rejette les empilements de deux impulsions quasi simultanées.

La forme de la distribution angulaire est en accord avec celle de la réaction $^{3}He(\tau, {}^{2}He)^{4}He$[3] et la section mesurée à 42°CM doit être inférieure à 4 nb/sr. Les particules observées (fig. 1) avec une identification correspondant à celle des deutons se groupent en deux régions placées de part et d'autre du domaine d'énergie attendu pour ces deutons. Le spectre d'identification de ces deux régions montre clairement qu'un groupe de particules d'identification deuton (B) se distinguent de l'entourage (fig. 2). L'accumulation à 28° du CM avec de l'air comme cible produit un spectre d'intensité presque nulle dans la région d'énergie de ces deux groupes.

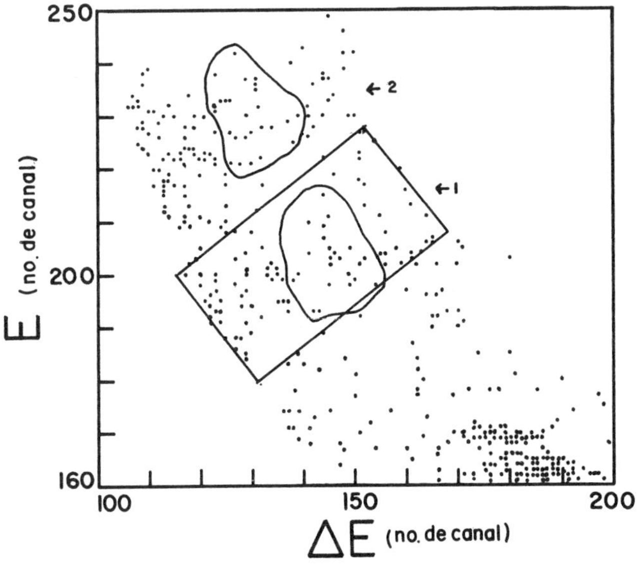

Figure 1

Les groupes 1 et 2 sont entourés

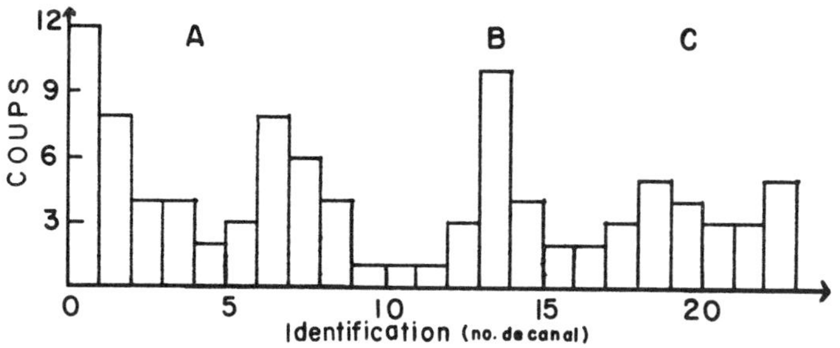

Figure 2

Spectre d'identification autour du groupe 1 (encadrement dans la fig. 1)

En plus des quasi-paraboles correspondant aux diverses identifications, les spectres bidimentionnels présentent également une structure de bandes presque perpendiculaires à ces courbes. La possibilité que deux particules puissent être détectées simultanément pour produire les énergies de ces groupes et une identification semblable à celle du deuton a été étudiée à l'aide de l'équation $I = (E_{tot})^{1.76} - (E_{résiduelle})^{1.76}$. L'empilement peut être constitué d'une particule qui s'arrête dans le détecteur ΔE et d'une autre complétement détectée. Parmi les empilements pouvant produire l'énergie du groupe 1, seule la combinaison de deux protons de certaines énergies peut simuler l'identification du deuton. Les deux groupes ont un comportement cinématique semblable à celui de la réaction $^3He(\tau, p)^5Li$. Pour les différents angles mesurés l'équation d'identification nous permet de croire que le groupe 2 est le résultat de l'empilement du proton accompagnant le ^5Li avec un proton du continuum de la réaction à 3 corps $^3He(\tau, 2p)^4He$. La possibilité pour le groupe 1 d'attribuer sa structure de bande à 45° à l'empilement du proton provenant d'une impureté de ^{15}N avec le continuum de proton ne respecte pas la cinématique qui devrait lui correspondre.

La section efficace de cette désintégration bêta est estimée inférieure à 1 nanobarn, et nous nous proposons de continuer les mesures à des angles plus grands.

Nous remercions le personnel de l'accélérateur pour sa collaboration et la Commission de Contrôle de l'Energie Atomique du Canada pour son appui financier.

BIBLIOGRAPHIE

1) E.M. Burbidge, G.R. Burbidge, W.A. Fowler, F. Hoyle, Revs. of Mod. Phys. 29 (1957) 547.

2) R.J. Blin-Stoyle and S. Papageorgiou, Nucl. Phys., 64 (1965) 1.

3) M.L.V.L. de Slobodrian, B. Cujec, R. Roy, R.J. Slobodrian, Nucl. Phys. A194 (1972) 577.

† adresse actuelle : IASR, Kernforschungszentrum, C.P. 3640, D-75 Karlsruhe, Allemagne

FINAL STATE INTERACTION IN THREE NUCLEON SYSTEM*

F. N. Rad

Lawrence Berkeley Laboratory, University of California, Berkeley, California 94720

D. P. Saylor

39 Otsego, Worcester, Massachusetts 01609

and

Mahavir Jain†

Texas A & M University, College Station, Texas 77843

The dynamics of three-nucleon system, as well as the interaction between the particles on and off the energy shell can be studied by the processes leading to three nucleons in the final state. There are several approximate descriptions of the three-nucleon breakup process which are often used. Each is useful in providing an understanding of and insight into the qualitative features of the breakup process in a limited part of phase space. The final state interaction (FSI) mechanism is appropriate for understanding that part of the breakup spectrum in which one of the three outgoing pairs of nucleons has a low relative energy. Typically, the cross sections are enhanced. The FSI theory of Watson and Migdal is often used to describe these enhancements. Theoretical calculations[1)] based on separable S-wave potentials give results in good agreement with the Watson-Migdal form of the final state peak. These calculations also predict a rather complex angular dependence of the FSI which may prove to be sensitive to the energy shell behavior assumed in the model.

Several attempts[2-5)] have been made to estimate the size of model dependent effects in N-d breakup reactions. The recent results of Brayshaw[4)] and of Haftel and Peterson[5)] (HP) are of particular interest because these estimates were based on comparison of models which gave identical on-shell two-nucleon scattering. For $E_{cm} \leqslant 20$ MeV Brayshaw and HP have found that significant model dependent variations were restricted to three-body ^{2}S-waves. They also found that if it is required that the N-d doublet scattering length, a_2, is held fixed the remaining variability is much reduced. Brayshaw believes that his analysis indicates that the two nucleon observables and a_2 determine the low energy trinucleon reactions to high precision for any plausible interaction. It would be significant if this expectation were to be experimentally verified. The conclusions of HP are different from that of Brayshaw while they provide a basis for selecting experiments. HP have found substantial variability in the ^{2}S partial wave even with a fixed doublet scattering length. The sensitive region is concentrated where singlet FSI dominates. The variability in the amplitudes results in variations of the FSI angular distributions and spectra which are of sufficient magnitude to be significant experimentally. In regard to these HP expectations the experiment and analysis of Brückman et al.[6)] (BKMSW) should be particularly valuable. They found that the breakup cross section in the np FSI region could be fitted very accurately by

$$\frac{d\sigma}{dSd\Omega_3 d\Omega_4} = \left[X^s_{np}(\theta_3)F^s_{np}(E_{45}) + X^t_{np}(\theta_3)F^t_{np}(E_{45})\right]\rho_S(E_3,E_4)$$

where F^s_{np} and F^t_{np} are the singlet and triplet Watson-Migdal enhancement factors. Empirically BKMSW have found that the triplet FSI angular distribution, $X^t_{np}(\theta_3)$, was the same as the experimental d-p elastic angular distribution and that the magnitude of the triplet FSI cross section could be successfully related to the elastic cross section. Their application of the Watson-Migdal analysis shows internal consistency with the np low-energy parameters. Other experiments which should be particularly valuable in isolating the np singlet FSI are the measurements of the vector to vector spin transfer parameters $K_y^{y'}$ in $^1H(\vec{d},\vec{p})$ and $^2H(\vec{p},\vec{p})$ for slightly inelastic p-d scattering. These experiments when combined with kinematically incomplete cross section measurements should allow one to extract $X^s_{np}(\theta_3)$. Assuming the validity of the Watson-Migdal analysis,

$$K_y^{y'} = \frac{{}^sK_y^{y'}\, X^s_{np}(\theta_3)\, F^s_{np}(E_{45}) + {}^tK_y^{y'}\, X^t_{np}(\theta_3)\, F^t_{np}(E_{45})}{X^s_{np}(\theta_3)\, F^s_{np}(E_{45}) + X^t_{np}(\theta_3)\, F^t_{np}(E_{45})}$$

where ${}^sK_y^{y'}$ and ${}^tK_y^{y'}$ are the spin-transfer parameters for the singlet and triplet components respectively. Using the simplified structure of the breakup amplitudes,[7] characteristic of the Amado model, it is easy to show that

$${}^sK_y^{y'} = 1 \qquad \text{for } {}^1H(\vec{d},\vec{p})$$

$${}^sK_y^{y'} = -\frac{1}{3} \qquad \text{for } {}^2H(\vec{p},\vec{p})$$

${}^tK_y^{y'}$ will depend on angle and energy but an implication of BKMSW is that ${}^tK_y^{y'}$ is nearly the same as the $K_y^{y'}$ parameters for elastic scattering. However in terms of the breakup amplitudes[7])

$${}^tK_y^{y'} = \frac{\frac{5}{3}|q|^2 - \frac{1}{12}|d_1|^2 - \frac{1}{4}|d_2|^2 - \frac{2}{3}\,\mathrm{Re}q^*d_1 + \sqrt{\frac{4}{3}}\,\mathrm{Re}q^*d_2 + \sqrt{\frac{1}{12}}\,\mathrm{Re}d_1^*d_2}{2|q|^2 + \frac{1}{4}|d_1|^2 + \frac{3}{4}|d_2|^2 - \sqrt{\frac{3}{4}}\,\mathrm{Re}d_1^*d_2} \qquad \text{for } {}^1H(\vec{d},\vec{p})$$

$${}^tK_y^{y'} = \frac{\frac{1}{9}|q|^2 + \frac{1}{36}|d_1|^2 + \frac{1}{12}|d_2|^2 + \frac{8}{9}\,\mathrm{Re}q^*d_1 - \sqrt{\frac{64}{27}}\,\mathrm{Re}q^*d_2 - \sqrt{\frac{1}{108}}\,\mathrm{Re}d_1^*d_2}{3|q|^2 + \frac{1}{4}|d_1|^2 + \frac{3}{4}|d_2|^2 - \sqrt{\frac{3}{4}}\,\mathrm{Re}d_1^*d_2} \qquad \text{for } {}^2H(\vec{p},\vec{p}),$$

where q is the amplitude for breakup in the quartet state ($S = 3/2$), and d_1 and d_2 are the doublet-state ($S = 1/2$) amplitudes in which the two identical nucleons are coupled to spin 1 or 0, respectively.

To a first approximation the slightly inelastic $K_y^{y'}$ parameters can be computed on the basis of a decomposition of the breakup cross section similar to that of BKMSW and a measurement of the corresponding $K_y^{y'}$ for elastic scattering. Thus a measurement of the $K_y^{y'}$ parameters have additional interest in that they provide additional checks of the final state interaction theory. We are currently working on the theoretical predictions using a computer code which solves the three particle Faddeev equation for separable spin dependent s-wave nucleon nucleon interactions.[8])

REFERENCES

*Work supported by the U. S. Atomic Energy Commission.
†Guest scientist, Los Alamos Scientific Laboratory, University of California, Los Alamos, New Mexico 87544.

1. R. T. Cahil and I. H. Sloan, Nucl. Phys. A165 (1971) 161; W. Ebenhöh, Proceedings of the International Conference on Few Particle Problems in the Nuclear Interactions, Los Angeles, California, 1972, edited by Slaus et al. (North-Holland, Amsterdam, 1972), pp. 325-328.
2. W. M. Kloet and J. A. Tjon, Nucl. Phys. A210 (1973) 380.
3. Mahavir Jain, J. G. Roger, and D. P. Saylor, Phys. Rev. Letters 31 (1973) 838.
4. D. D. Brayshaw, Phys. Rev. Letters 32 (1974) 382.
5. M. I. Haftel and E. L. Peterson, Bull. Am. Phys. Soc. 19 (1974) 505.
6. H. Brückman, W. Kluge, H. Mattäy, L. Schänzler and K. Wick, Nucl. Phys. A157 (1970) 209.
7. D. P. Saylor and F. N. Rad, Phys. Rev. C8 (1973) 507
8. Mahavir Jain and Gary D. Doolen, Phys. Rev. C8 (1973) 124.

$^{9}Be(\tau,\alpha)\alpha\alpha$ ET $^{10}B(d,\alpha)\alpha\alpha$: MECANISME DE REACTION ET LARGEUR DES ETATS DU ^{8}Be *

R. Roy †, J. Birchall ‡, B. Frois ††, U. von Moellendorff †††, C.R. Lamontagne et R.J. Slobodrian, Université Laval, Département de Physique, Laboratoire de Physique Nucléaire, Québec, P.Q. Canada G1K 7P4.

Les ondes planes et distordues dans le cadre de l'approximation de Born utilisées de façon extensive à l'étude des niveaux non-liés ont parfois conduit à des conclusions intéressantes sur certaines réactions avec les noyaux légers, comme par exemple, les réactions $^{7}Li(\tau,d)\,^{8}Be$ [1] et $^{11}B(p,\alpha)^{8}Be$ [2]. Dans la présente étude, les distributions angulaires théoriques des réactions $^{9}Be(\tau,\alpha)^{8}Be$ et $^{10}B(d,\alpha)^{8}Be$ à basse énergie pour les transitions à l'état fondamental et au premier état excité du ^{8}Be ont été calculées à l'aide du programme DWBA [3]. La figure 1 compare les résultats expérimentaux (Ex=0.0) à E_τ=9 et 10 MeV. Le potentiel optique de la voie élastique $\tau+^{9}Be$ a été déduit d'une analyse de la diffusion d'un faisceau d'^{3}He polarisé [4] à 18 MeV (courbes en pointillés); à ces paramètres a été ajouté un potentiel imaginaire de volume [5] (courbes continues). Les paramètres du potentiel de la voie $\alpha+^{8}Be$ sont ceux qui ont été utilisés pour le cas $\alpha+^{6}Li$ dans une étude [6] de la réaction $^{7}Li(\tau,\alpha)\,^{6}Li$. La correction de portée finie améliore appréciablement la qualité de la représentation . En particulier, les calculs ne peuvent autrement reproduire le premier maximum à $\Theta cm \approx 20^{o}$. Toutefois, l'importance de la portée finie est moins évidente pour le premier état excité à 3 MeV (figure 2). Les paramètres optiques sont les mêmes

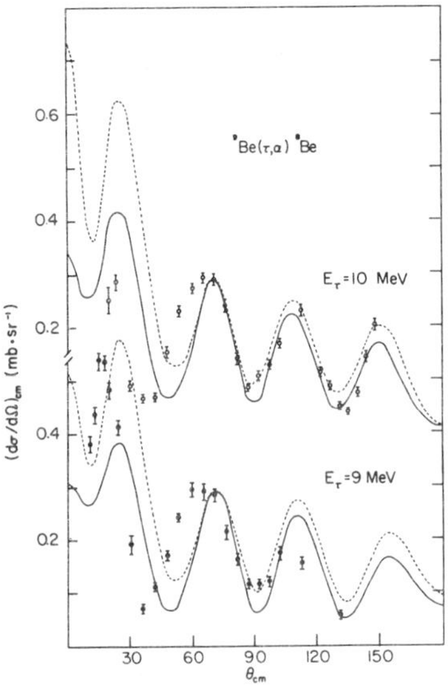

Fig.1 (gauche)

Les courbes théoriques sont normalisées aux points expérimentaux.

Fig.2 (droite)

Les points expérimentaux sont normalisés aux courbes théoriques.

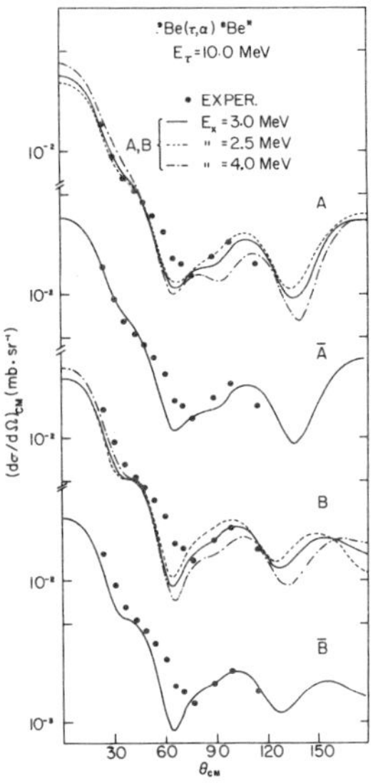

que ceux utilisés pour l'état fondamental, excepté dans la voie $\alpha+{}^{8}Be$ où la partie réelle du potentiel est plus étroite et à contour moins diffus. La dominance du terme $j=3/2$ sur le terme $j=1/2$ pour le nucléon transféré est en accord avec les facteurs spectroscopiques théoriques [7]. Les trois différentes courbes (A et B), calculées pour les énergies d'excitation indiquées, mettent en évidence un point intéressant. A cause de l'interaction directe, la section efficace varie sensiblement avec l'énergie d'excitation du noyau résiduel, ce qui implique une variation de la largeur apparente d'un niveau large tel que celui étudié. Les valeurs d'énergie d'excitation choisies, 2.5 et 4.0 MeV, correspondent approximativement aux points de demi-hauteur de la résonance. On peut observer cet effet expérimentalement, comme l'illustre la figure 3.

Fig.3 (voir texte)

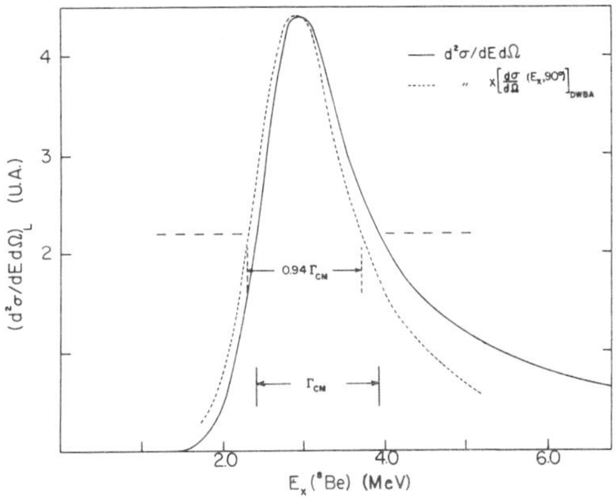

La courbe continue est le résultat d'une analyse [8] des spectres continus en énergie de l'état final à trois particules α et celle en pointillé est la même courbe corrigée pour la variation de la section efficace (voir figure 2) en fonction de l'énergie d'excitation, à $\Theta cm=90^{o}$.

Les mêmes remarques peuvent être faites à propos des calculs pour la réaction ${}^{10}B(d,\alpha_1){}^{8}Be^{*}$ (3MeV) à $E_d=5$ MeV. Toutefois un processus direct semble insuffisant pour reproduire les distributions expérimentales des deux premiers états du ${}^{8}Be$. Les courbes d'excitation [9] confirment cette remarque. Ainsi demeure-t-il difficile d'évaluer le formalisme du double transfert (d,α) à cette énergie. Toutefois, la méthode des ondes distordues semble adéquate pour le transfert d'un nucléon, malgré l'incertitude sur les paramètres optiques de la voie $\alpha+{}^{8}Be$. Comme il est prévisible pour les noyaux légers, la correction de portée finie pour la fonction d'onde du groupe transféré a un effet positif important, surtout pour le niveau fondamental.

Le transfert direct à un état résonnant a été le sujet de nombreux travaux récents. En particulier, un formalisme [10], restreint à des états plus étroits que celui étudié ici, a été développé dans ce cadre. Néanmoins il est intéressant de noter que la section efficace, calculée seulement au maximum de la résonance mais pondérée en énergie par une formule de résonance un peu différente de la présente formulation [8], a un effet net à peu près équivalent à celui obtenu dans la présente expérience à 90°c.m. Toutefois, ceci pourrait être différent pour d'autres angles (voir figure 2). Si l'effet prédit dans le cadre du présent travail est correct, il devrait être observable: une analyse [8] précise de la résonance et du continuum à de plus hautes énergies incidentes devrait permettre d'éclaircir la situation aux angles arrières. Il est possible qu'on puisse trouver là une partie de la solution au problème soulevé par Berkowitz [11] au sujet de la variation de la largeur apparente de l'état résonnant 2^+ du ^8Be.

---

* Travaux subventionnés en partie par la Commission de l'énergie atomique et par le Conseil national de recherches du Canada.

† A présent au Lawrence Berkeley Laboratory, Berkeley, California 94720, U.S.A.

‡ A présent au Physikalisches Institut der Universität Basel, Klingelbergstrasse 82, 4056 Basel, Switzerland.

†† A présent au Service de Physique Nucléaire à Haute Energie, CEN Saclay, B.p. No 2, 91190 Gif-sur-Yvette, France.

††† A présent au IASR Kernforschungszentrum, 75 Karlsruhe, Postfach 3640, West Germany.

Références

1. R.L. Dixon et R.D. Edge, Nucl. Phys. A156 (1970) 33.
2. J.F. Cavaignac, N. Longequeue et T. Honda, Nucl. Phys. A167 (1971) 207.
3. J.M. Nelson et B.E.F. Macefield, A General DWBA Computer Program for Particle Transfer Reaction (Nuclear Physics Laboratory, Oxford); J.M. Nelson, communication privée.
4. W.S. McEver, T.B. Clegg, J.M. Joyce, E.J. Ludwig et R.L. Walter, Nucl. Phys. A178 (1972) 529.
5. F.C. Young et A.R. Knudson, Nucl. Phys. A184 (1972) 563.
6. A.R. Zander, K.W. Kemper et N.P. Fletcher, Nucl. Phys. A175 (1972) 374.
7. S. Cohen et D. Kurath, Nucl. Phys. A101 (1967) 1.
8. R. Roy, J. Birchall, B.Frois et R.J. Slobodrian, Lett. al Nuovo Cim. 6 (1973) 374; et à publier.
9. C. Lemeille, D. Manesse, L. Marquez et N. Saunier, Jour de Phys. 26 (1965) 1.
10. E.I. Dolinsky et A.M. Mukhamedzhanov, Phys. Lett. 48B (1974) 95.
11. E.H. Berkowitz, Nucl. Phys. 60 (1964) 555.

COMPLETE MEASUREMENTS OF THE DEUTERON BREAKUP BY PROTONS AT E_p=8.5 MeV

B. Kühn, H. Kumpf, J. Mösner, W. Neubert and G. Schmidt
Zentralinstitut für Kernforschung der AdW der DDR, Rossendorf,
Bereich 2, DDR 8051 Dresden, Postfach 19

Although a great number of experimental investigations of the deuteron breakup by protons for incident particle energies from threshold up to several hundred MeV has been published, accuracy and resolution of most measurements leave much to be desired.

Indeed customary models such as the spectator model or pole-graph approximation [1)] gave deviations in absolute cross-sections from experiment by an order of magnitude. Therefore it appeared obsolete to take pains for e.g. a precise calibration of the cross-section scale.

But recently emerging computer codes for solving the exact three body integral equations by rigorous approximation methods permitted a much more realistic description of experimental data. A first survey of the existing body of experimental information revealed a rough agreement with calculations based on a separable Yamaguchi N-N-potential [2)]. Meanwhile Ebenhöh et al. [3)] succeeded in demonstrating a certain dependence of the calculated breakup cross-sections on the used formfactor in the separable N-N-interaction. The various formfactors were adjusted such a way as to guarantee fixed low energy scattering parameters, different high energy behaviour being allowed.

As with 10 MeV incident energy all the relative energies of two nucleon subsystems are well within the limits of validity of the effective range approximation, to our opinion the formfactor dependence of breakup quantities should be interpreted as reflecting a certain sensitivity to off-shell behaviour. Differences amounting to at most 20 % we considered precision measurements of double - differential cross-sections to be worth while.

Measurements were performed at the EGP-10 Tandem accelerator preferentially with 8.5 MeV protons. Two Si-SB detectors for the emerging protons of D(p,2p)n reaction were mounted on movable discs inside a reaction chamber providing for easy and precise adjustment [4)]. A computer-analyzer corrected the time signal for flight-time differences and rise-time shifts before recognizing an event as coincident and storing it in a 64 x 64 channels array according to the energies E_1 and E_2 of the two protons. In a second 64 x 64 array non-coincident events were collected for reliable background subtraction. For reducing deuterium loss during irradiation the deuterated polyethylen target (150 μgcm^{-2}) was rotated. In order to avoid the difficulties of absolute cross-section calibrations the protons elastically scattered by deuterium were monitored during the runs and the cross-section ref. [5] was taken as reference. Counting losses by dead-times were carefully determined.

Spectra were taken for the coplanar angular settings $\Theta_1 = 35^\circ$, $\Theta_2 = -19^\circ$, $+10^\circ$, $+20^\circ$, $+34.2^\circ$, $+40^\circ$, $+48.7^\circ$ and $+62.8^\circ$. Noncoplanar correlation spectra have been measured for 3 pairs of angles, namely $\Theta_1 = 47.7^\circ$ and $\Theta_2 = 47.7^\circ$, $\varphi = 120^\circ$; $\Theta_2 = 35.2^\circ$, $\varphi = 120^\circ$ and $\Theta_2 = 40^\circ$, $\varphi = 129^\circ$, respectively.

In fig. 1 a projected spectrum together with the theoretical cross section, calculated with Ebenhöh's computer code [2] using the value $a_{pp} = -7.8$ fm is shown. The finite target thickness and angular resolution were taken into account by a folding procedure. The error of the extracted cross sections is estimated to be ± 6 %.

References

1) K. Möller, ZfK-Report - 221, Sept. 1971
2) W. Ebenhöh, Nucl. Phys. A191 (1972) 97
3) J. Bruinsma, W. Ebenhöh, J. H. Stuivenberg and R. van Wageningen, Naturkunding Laboratorium der Vrije Universiteit Amsterdam 1974
4) J. Mösner, G. Schmidt, T. Stiehler, ZfK Annual Report 1973, p. 211
5) D. C. Kocher and T. B. Clegg, Nucl. Phys. A132 (1969) 455

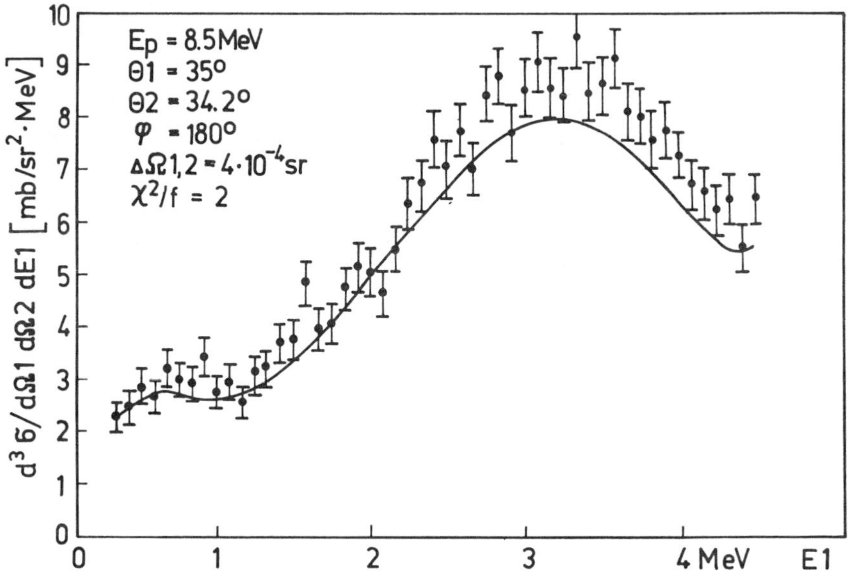

Fig. 1

Neutron Spectra from Proton Bombardment of Deuterium at 647 and 800 MeV

C. W. Bjork, P. J. Riley, U. of Tex.*; B. E. Bonner, J. E. Simmons, K.D. Williamson, Jr., LASL*; D. W. Werren, U. Geneva and SIN*; H. C. Bryant, C. G. Cassapakis, S. Cohen, B. D. Dieterle, C. P. Leavitt, D. M. Wolfe, U. of New Mex.*; M. L. Evans, G. Glass, J. C. Hiebert, Mahavir Jain, L. C. Northcliffe, Tex. A&M U.*

As part of the planning for neutron-proton scattering experiments at the Clinton P. Anderson Meson Physics Facility (LAMPF) of the Los Alamos Scientific Laboratory it was proposed to generate neutron beams by proton bombardment of deuterium. The emission of the highest energy neutrons from the D(p,n)2p reaction near zero degrees via quasifree scattering is enhanced and the spectrum narrowed by strong final state interaction of the two protons in a low relative energy state. Such considerations were given by Gluckstern and Bethe[1] in 1951, who also suggested that use of liquid deuterium for this purpose would provide a superior neutron beam. Larsen[2] used this method at 710 MeV in the external beam of the Berkeley S.C. for studies of neutron-proton charge exchange scattering. There have also been detailed measurements near 150 MeV by Langsford, *et al.*[3] and by Measday.[4]

The present measurements were made in the Nucleon Physics Lab. at LAMPF. The incident proton beam passed through a 10.8 cm long liquid deuterium (LD_2) cell, and was deflected through a 60° bend into a shielded beam stop. The beam intensity varied in the range 1/2 to one micro-ampere average and was measured by a toroid current transformer. The target cell and circulating cryogenic loop were cooled by means of ^{4}He gas flowing through a heat exchanger. The cold ^{4}He was provided by a turbine-type refrigerator capable of extracting 100 W of beam power from the LD_2. Neutrons emitted from the target at 0° passed through a steel collimator 2.54 cm diam. by 3.66 m long. Our recently developed medium energy neutron spectrometer[5] is located at the end of the collimator as shown in Fig. 1. The neutrons pass through a neutron monitor, a sweep magnet (M1) and impinge upon a liquid hydrogen scattering target. Here protons are produced via charge exchange scattering and by associated pion production as well as deuterons via the $^1H(n,d)\pi^0$ reaction. The momenta and velocities of the outgoing charged particles are measured by the spectrometer. The anti-coincidence requirement and time information are obtained by the scintillators, S0, S1, and S2, while position information is provided by multi-wire proportional chambers W1, W2, W3, and W4 which have 2 mm signal wire spacing. Data are collected by an on-line PDP-11 computer and acquisition program.[6] Off-line analysis on a large computer is required to obtain particle momenta and other relevant information from the measured wire chamber addresses by numerical integration of the particle trajectories through the measured magnetic field of M2.

Figure 2 shows a representation of the neutron spectrum from the reaction D(p,n) at 0° neutron angle and 800 MeV incident energy. Shown there is the measured proton recoil spectrum for proton polar angles less than 0.04 radians. This spectrum has not been corrected for variation of the n-p elastic charge exchange scattering cross-section with differing incident neutron momenta (which may be as much as a factor of 1.5 over the energy range shown). Contamination of the lower energy yield from inelastic events producing protons in the LH_2 radiator (mainly single pion production) is also present and may account for 20% of the broad low energy bump. Notice the strong, yet narrow, quasi-elastic peak with approximately 13 MeV FWHM, which includes experimental resolution. The ratio of the yield of the quasi-elastic peak integrated over all momenta contributing to that peak vs. the yield from an equal momentum bite in the valley is about 30:1. We estimate the peak differential cross section by using n-p charge exchange differential cross sections from Bizard, et al.[7] together with approximate measurement of the incoming beam intensity as follows: peak $d\sigma/d\Omega$ for D(p,n) at 0°: 22.5 mb/sr and 16.1 mb/sr at 800 and 647 MeV incident energy, respectively. These figures may have errors as large as ± 15%. At 800 MeV, the neutron beam intensity is approximately 6×10^5 neutrons/sec-μa of proton beam and the neutron beam radius is approximately 2 cm at 1/10 maximum intensity. The spectrum of neutrons at 647 MeV is very similar to that of 800 MeV except that the low energy bump is relatively less intense.

Space does not allow for comparison with other neutron beams. The measurements given above describe a very useful beam which is now being used for studies of n-p scattering.

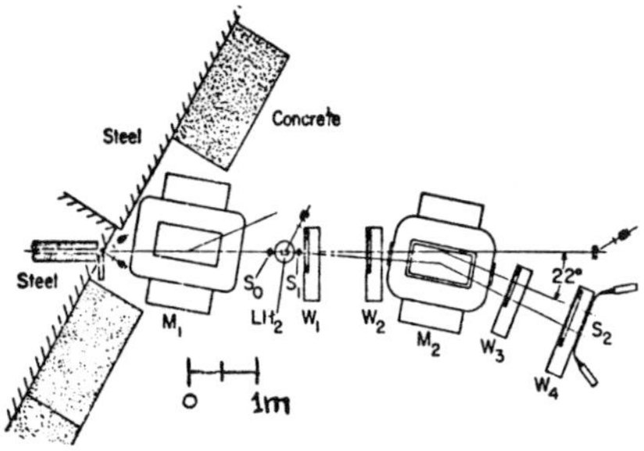

Figure 1

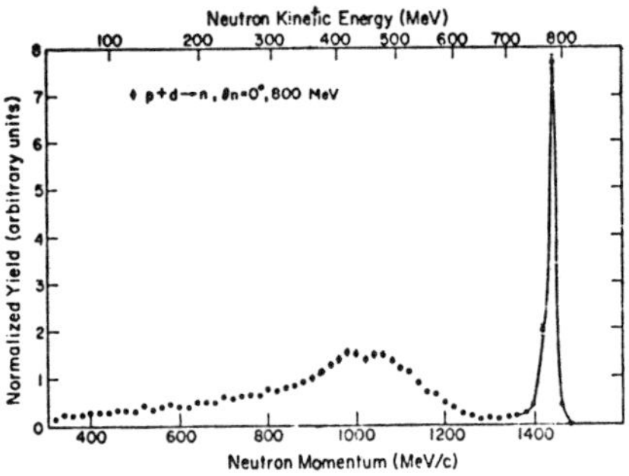

Figure 2

*Supported by the U.S. Atomic Energy Commission.

1. R. L. Gluckstern and H. A. Bethe, Phys. Rev. 81 (1951) 761.
2. R. R. Larsen, Nuovo Cimento 18 (1960) 1039.
3. A. Langsford, et al., Nucl. Phys. A99 (1967) 246.
4. D. F. Measday, Nucl. Inst. and Meth. 40 (1966) 213.
5. D. W. Werren, et al., LASL Report LA-5396-MS (1973).
6. J. C. Hiebert and A. Niethammer, LASL Report LA-5609-MS (1974).
7. G. Bizard, et al., Contribution to the IInd Aix-en-Provence International Conference on Elementary Particles (1973).

VI

ETATS LIES ET DE DIFFUSION
FORCES A PLUSIEURS CORPS

BOUND AND SCATTERING STATES
FEW-BODY FORCES

THREE PARTICLE SYSTEMS: CURRENT PROBLEMS AND PROGRESS

L. M. Delves†

Department of Mathematics, University of Victoria

and

Department of Computational and Statistical Science, University of Liverpool

Abstract

This talk attempts to summarise the current computational state-of-the-art in the nuclear three body problem. There are three main questions in this field:

(i) Can we do the calculations?

(ii) What does a comparison with experiment tell us about the two-nucleon potential?

(iii) What can we tell about mesonic, relativistic, and three-body potential corrections?

The answer to the first of these is still only: "more or less". However, we appear to have moved to the start of an exciting period when questions two and three are yielding non-null responses.

†Permanent address: University of Liverpool.

Introduction

In this talk I shall attempt to summarise recent progress in the three-body problem in nuclear physics. Because I am a theoretician I shall concentrate on the theory; and because it is the answers which matter I shall discuss numerical results and their comparison with experiment and each other, rather than the techniques for getting the results†.

Section I deals with three-body bound states; this is the segment of the problem which has attracted most work recently, and where some interesting new conclusions are starting to emerge. Section II is on three-body scattering states; here ones feelings are mixed - calculational techniques are improving, allowing reasonable comparisons with experiment, but there is currently doubt that any fundamental results will come from this work. Finally, section III briefly touches on progress in calculations in the four-body problem.

I. Three-body Bound States

A large amount of effort has gone into calculations of the ^{3}H and ^{3}He bound state systems, in the past few years. This effort has been aimed at getting information on:

(a) The shape of the two-nucleon potential, especially in the repulsive core region.

(b) Off-energy shell ambiguities in the two-nucleon potential.

(c) The presence and strength of any three-body potential.

(d) Mesonic effects on, for example, the charge and magnetic form factors of ^{3}H and ^{3}He.

(e) Relativistic effects on the form factors and binding energy.

We review these in turn; but it is first necessary to talk about methodology.

I.1. How well can we do the calculations

It is no good searching for 0.5-1 MeV contributions to the ^{3}H binding energy, or 3 fm shifts in the position of the diffraction minimum in the ^{3}He form factor, if the calculations are not more accurate than that. Unfortunately it is still the case that results from different groups for a given potential differ by as much as the above interesting numbers.

---

†But because I am a numerical analyst you may (or you may not) forgive a short section comparing results with themselves.

I.1.1. Calculations with the Reid potential: The binding energy

The situation is worst with the Reid soft core (RSC) potential, partly because this has attracted the most calculations, but mainly because the Yukawa singularity at the origin makes life hard - harder than with an infinite repulsive core, for example, and infinitely harder (as Schneider (23) has pointed out) than a super-soft-core (finite at the origin) and infinitely differentiable potential such as GPT or $SSC_{A,B,C}$ (see Table 3 for notation).

Table I reports the current computational situation with respect to the binding energy; the last column contains the figures which should be compared, and we see that even for the simplest truncated Reid potential the agreement is not good; for the full potential results still range from 6.8 to 7.75 MeV. Where do these large variations come from? I believe, from no special causes other than the innate form of this potential being unkind to some of our implicit (or even explicit) assumptions. Any calculation of the binding energy has two stages at which error is introduced:

Stage I. Algebraic simplification. We assume that certain potential components, or certain wave function components, will contribute negligibly, and drop them from the calculation completely. This stage defines the equations (whether Faddeev or Schrodinger) which we try to solve, and yields a truncation error.

Stage 2. Numerical Approximations. The equations from stage I are solved approximately, with a further unavoidable error: step-lengths are finite, Gauss quadrature rules have a maximum number of points, wavefunction expansions are truncated. Hopefully, we try to estimate these effects, and perhaps to extrapolate to yield a best value. The remaining error we term extrapolation error, although in practise this is more appropriate for the variational calculations. Now, for the Reid potential, we have all found out painfully that both types of error can be larger than we expected a priori. A number of authors have remarked that

> "small components in the three-body wave function can contribute significantly to the binding energy".

For explicit examples, see ref. (7). This means that we must be very careful which states are termed negligible in stage I, whether in a Faddeev or a variational calculation. For a Faddeev calculation, these decisions probably yield most of the error; note that they include decisions on how large a spectator angular momentum must be considered.

Variational calculations on the other hand usually find it easy to include all states (in principle) until stage 2; but it is no good having them in unless the expansion used represents them adequately. The problem is then to extrapolate reliably to an infinite number of states. Because the problem is multi-dimensional, so is the extrapolation; it seems worth commenting separately on the two main classes of basis.

(I) Bases using inter-particle distances.

These include the calculations of my own group (6) and shell model calculations (8,12). Here, all coordinates are treated equally, and there is

TABLE I

CALCULATIONS OF THE REID S.C. ^3H BINDING ENERGY

Authors	Ref.	Method	Best directly calculated E	Extra-polated values	Comments
		A: $^1S_0$ $^3S_2$ - $^3D_2$			
Malfliet & Tjon	(1)	Faddeev	6.9	6.9±0.2	4 components
Bruinsma, Van Vageningen	(2)	Hyperspherical basis; variational	6.05	6.6±0.2 6.8–7.5	Authors re-analysis
Harper et al	(3)	Faddeev, iterative	6.7	6.7±0.07	5 components
Laverne & Gignoux	(4)	r-space Faddeev, relaxation	7.0	7.0	
		B: A + $^1D_2$ $^3D_2$			
Gignoux & Laverne	(5)	r-space Faddeev	6.64	6.64	
Bruinsma, Van Vageningen	(2)	Hyperspherical basis	6.26	6.8±0.2 7.0–7.7	authors re-analysis
		C,D: ~ complete potential			
Hennell, Delves	(6)	variational; damped poly. basis	7.0	7.75±0.5	extrapolation appears unreliable
Demin et al	(7)	Hyperspherical basis	6.62	6.8	our extrapolation
Strayer and Sauer	(8)	Shell model basis	6.7	7.3±0.2	
Laverne et al	(4)	r-space Faddeev and perturbation	7.27	7.27	

TABLE I

CALCULATIONS OF THE REID S.C. ^{3}H BINDING ENERGY

Authors	Ref.	Method	Best directly calculated E	Extra-polated values	Comments
Bruinsma et al	(2)	Hyperspherical basis	6.26	6.8 ± 0.2 7.0–7.7	Author's extrapolation our extrapolation
Brandenburg & Kim	(9)	Faddeev	6.9	6.9	

...ailable a single parameter (the number of excitons for a shell model basis) to be extrapolated to infinity. The difficulty lies in carrying out the extrapolation sufficiently well. We illustrate these difficulties below; to cast the first stone in an appropriate direction, let us consider the value E = 7.75 ± 0.5 MeV given by my own group.

We note first that this calculation differs from all others in including potential contributions from all states; the result should therefore differ from those for "Potential C" by an amount representing the contributions from two-body states with J>2. The magnitude of these contributions is not known; a straightforward subtraction of our result from the best available value (E = 7.3 MeV - see below) yields 0.45 MeV, which seems rather high; but this of course ignores the error bars (± 0.5 MeV) on our value. These large quoted errors stem almost wholly from the extrapolation procedure, and should not be ignored; let us recall the procedure involved.

Figure (1) illustrates the energy obtained in ref (6) as a function of the number of terms included. A convergence analysis (see ref (24)) predicts that for large Q these should take the form

$$E_Q = E_\infty + AQ^{-\gamma}$$

and a fit to this form yields the value E_∞ = 7.75 MeV. The quoted error represents the range obtainable by fitting in various ways. If the contribution from states with J⩾2 should prove to be significantly less than 0.5 MeV, we would have to conclude that this procedure has over-extrapolated; that is, our calculation was just starting to converge significantly faster than the initial portion indicated, and we were therefore not in the asymptotic region demanded by the extrapolation form above.

(2) Hyperspherical bases.

With a hyperspherical basis there are two pitfalls for the unwary.

(a) As the maximum generalized angular momentum index K increases, the associated radial functions $f_K(r)$ become more and more illbehaved. Whether they are computed by solving a set of ordinary differential equations, or via a truncated complete set $\{h_i(r),\ i = 1,\dots,N_K\}$ we must ensure that they have sufficient accuracy for the inclusion of that K value to be meaningful.

It is possible to have a calculation converge smoothly (in K) to the wrong answer, by paying insufficient attention to $f_K(r)$ for large K.

(b) Convergence in K is at best as K^{-4} (ref. 23) and calculations performed so far seem to achieve slower convergence than this (why?). Summing a slowly convergent series is not easy, especially in the presence of numerical errors.

We illustrate remarks (b) and (a) by considering two recent calculations (2), (7).

Table 2 lists the results obtained with a hyperspherical basis by Bruinsma and van Vageningen. These were also extrapolated by the authors by fitting to a formula of the form given above and they conclude $\gamma \sim 3$ and $E_\infty = 6.8 \pm 0.2$. My own conclusions would go as follows:

(1) The first and second differences ΔE and $\Delta^2 E$ strongly indicate something fishy about the K = 20 result. Either the calculation is suddenly converging faster (in which case we cannot reliably extrapolate it from these results) - or this number contains quite large numerical errors and should be ignored (which is not unreasonable; most authors probably add terms until the calculation starts to go unstable).

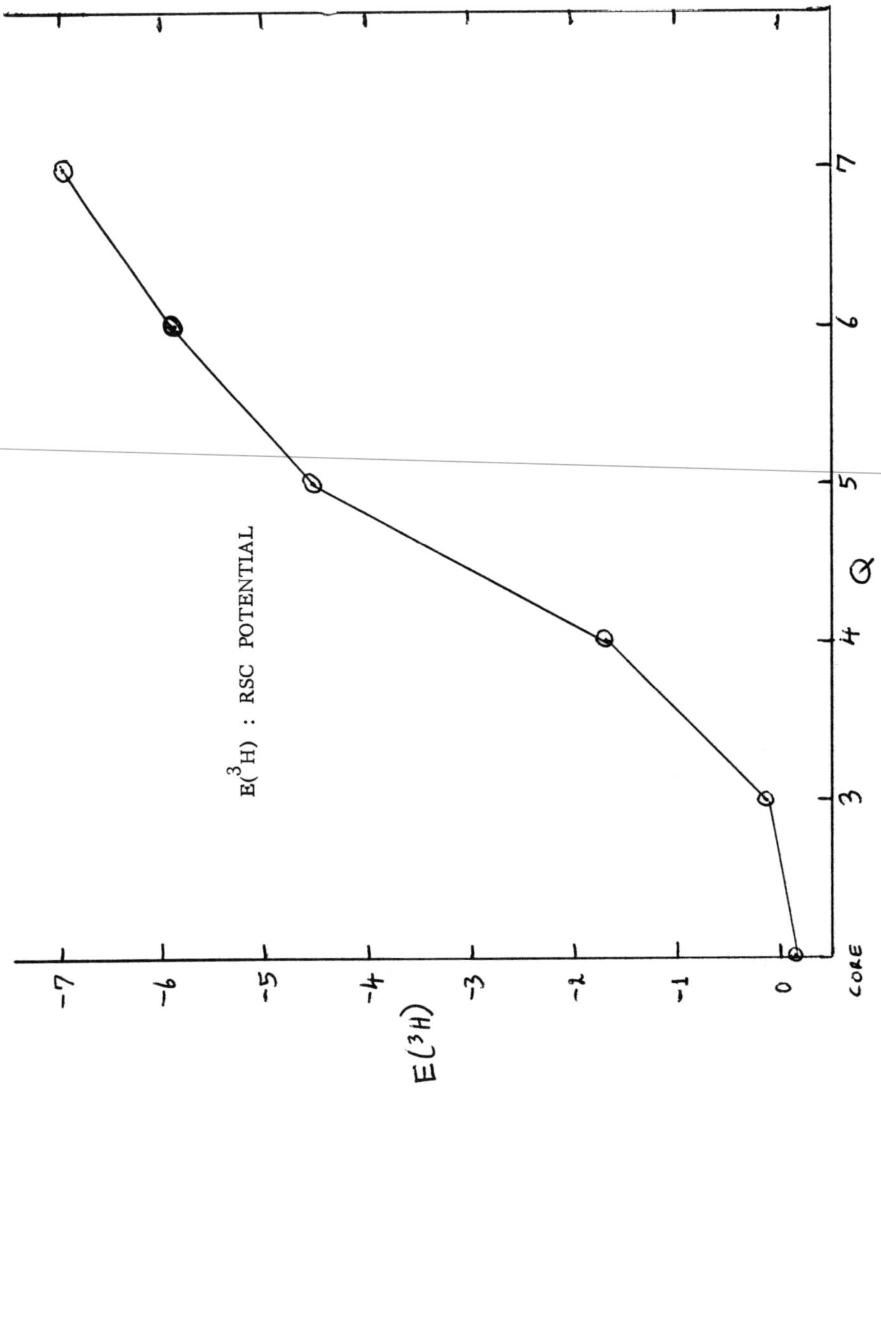

FIG. 1.

TABLE II

CONVERGENCE ANALYSIS FOR THE CALCULATION OF

BRUINSMA AND VAN VAGENINGEN (ref.2)

K	E	ΔE	$\Delta^2 E$	$\Delta E_{pred}, \gamma=2$	$\Delta E_{pred}, \gamma=3$
10	-2.325				
		1.982		1.57	2.44
12	-4.307		1.184		
		0.798		0.946	1.40
14	-5.105		0.205		
		0.593		0.614	0.70
16	-5.698		0.172		
		0.421		0.421	0.421
18	-6.119		0.282		
		0.139			
20	-6.258				

$$E_K = E_\infty + AK^{-\gamma} \quad ; \quad \gamma \leq 2$$

$$E_{16} \text{ and } E_{18} \rightarrow E_\infty = 7.70 \quad , \quad \gamma = 2$$

$$\rightarrow E_\infty = 7.11 \quad , \quad \gamma = 3$$

(2) If we ignore E_{20}, a fit to E_{16} and E_{18} assuming $\gamma = 3$ yields $E_\infty = 7.11$, but values of ΔE which state categorically that γ is not this large. Taking $\gamma = 2$ gives a rather better fit to ΔE ($\gamma<2$ would in fact seem indicated) and yields $E_\infty = 7.70$ MeV.

This large range of "possible extrapolated values" again indicates the difficulties in extrapolating such a calculation. It is not intended as a criticism of the calculation, which I believe to have been competently and carefully carried out.

As a third example, we consider the, we might think very similar, calculation of Demin and Pokrovsky (7). These authors used a hyperspherical basis in which the terms were split into "potential" terms - which should be important - and "nonpotential" terms - which they hope are not. Without the "nonpotential" terms, the basis is not complete; but both types of term were included. Their results, for K up to 28, are given in Table III. There is no major indication of instability; and the predicted differences show that certainly $\gamma>2$. With $\gamma=3$, we get essentially identical values of $E_\infty = 6.78$ using the pairs, (E_{28}, E_{26}) or (E_{26}, E_{24}): the sign of a smooth extrapolation. Nonetheless, this extrapolated value is, I believe, badly wrong. It seems certain that for RSC, $E > 7.0$; and I believe that the best available value is the result of Strayer and Sauer (8), namely

$$E(^3H) = 7.3 \pm 0.2 \text{ MeV}$$

while a recent survey by Levinger (25) quoted $E(^3H) = 7.1$ MeV as the "best" value. It would seem possible that Demin and Pokrovsky have either ignored significant states with $K \leq 28$, or that they have fallen into trap (2a) above: keeping too few examples of each angular momentum state. Presumably similar explanations apply to the Faddeev calculations which yield $E(^3H) < 7.0$ MeV.

Moral

We can draw a moral from these examples, and it is this. Estimating the effects of truncation and other calculational errors on our calculations is difficult. Correcting for them (extrapolating) is even more so. But that is no reason for shirking the job. No experimental paper would be accepted for publication without a discussion of error, both statistical and systematic, in the results. There seems no reason why similar standards should not apply to numerical calculations, and every reason why they should. Every such paper should discuss the numerical procedure used; its possible sources of error; ways in which the magnitude of the error was checked; final estimated results with error estimates where possible; and sufficient directly computed numbers to demonstrate how these estimates were arrived at. In short, I would like to plead for "full disclosure" in theoretical papers.

1.1.2 Calculations with the Reid potential: The Ground State Wavefunction

Similar discrepancies appear in other bound state parameters. The results of various authors are given in Table IV, and we draw the following conclusions:

(a) D state probability. Generally reasonable agreement at around 9%.

(b) S′ probability: A terrible situation with values ranging from 0.37 to 1.8%, but a clustering at around 1.6% suggesting this as a plausible value. Note however that the Faddeev calculations seem to yield a consistently higher value of P(S′) than do the variational - it is not clear why.

(c) Charge radius and charge form factor. Here again there are two groups, characterised by the position (~ 13 or ~ 18 fm^{-2}) of the diffraction minimum in Fch(^{3}He). These go with, respectively, a relatively small (1.9 fm) or large (≥ 2.0 fm) charge radius, but are not strictly drawn into variational or

TABLE III

CONVERGENCE ANALYSIS FOR THE CALCULATION OF

DEMIN AND POKROVSKY (ref. 7)

K	E	ΔE	$\Delta^2 E$	$\Delta E_{pred}, \gamma=2$	$\Delta E_{pred}, \gamma=3$
18	6.072				
		0.215		0.101	0.143
20	6.287		0.046		
		0.169		0.075	0.096
22	6.456		0.067		
		0.102		0.057	0.066
24	6.558		0.053		
		0.049		0.044	0.048
26	6.607		0.014 ?		
		0.035		0.035	0.035
28	6.642				

E_{28} and $E_{26} \rightarrow E_{\infty} = 6.94$ $(\gamma=2)$; $E_{\infty} = 6.78$, $(\gamma=3)$

E_{26} and $E_{24} \rightarrow E_{\infty} = 6.86$ $(\gamma=2)$; $E_{\infty} = 6.79$ $(\gamma=3)$

TABLE IV

3-NUCLEON PROPERTIES WITH THE R.S.C. POTENTIAL

Potential: $A = {}^1S_0\,,\ {}^3S_2 - {}^3D_2$; $B = A + {}^1D_2\,,\ {}^3D_2$; C = complete, $j \leq 2$; D = complete, all j .

Probably $C \approx D$ closely.

Authors	ref.	Potential	$E(^3H)$	E_c	P(D)%	P(S')%	$^3H, \langle R_{ch}\ ^3H\rangle^{\frac{1}{2}}$	$^3He, R_{ch}\ ^3He$	dip' $F_{ch}(^3He)$	Sec. max	height x 10^3
Malfliet & Tjon	1	A	6.9	0.63	6.47	1.6	1.8	2.05	17.0	-	-
Demin et al	7,10	C	6.8	0.66	8.87	0.37	1.77	1.90	13.8	19.0	1.93
Hennell & Delves	6	D	> 7.0	0.60	9.5±0.5	~1	1.78	1.90	12.8±0.3	19	2.0
Brandenburg & Kim	9	C	6.9	-	8.8	1.6	-	2.25	18	24	~ 0.6
Laverne & Gignoux	4	D	7.18	0.58	9.0	1.8	1.65	1.90	14.0	19	1.4
Harper, Kim, Tubis	11	C	6.7	-	8.56	1.6	-	1.96	15.5	-	-
Jackson et al	12	C	6.5	-	8.9	0.52	-	-	13.3	18	2.0
Strayer & Sauer	8	C	7.3		8.8	1.4	1.85	2.07	12.6	17.5	2.5

Faddeev camps. I believe that

The weight of the evidence suggests a value of around 13 fm^{-2} for the diffraction minimum in $F_{ch}(^3He)$.

Presumably calculations yielding a higher minimum are failing to treat adequately the high momentum components of the wavefunction. These are the parts affected by the Yukawa repulsive core.

I.1.3. Calculations with other "Realistic" potentials.

With other potentials, discrepancies between the results of different authors are smaller but not always insignificant. For the CPT super soft core potential there is on the whole good agreement between the groups of Gignoux and Laverne, and Hennell; but for the equally soft Spring-de Tourreil SSC potentials there is less good agreement between Demin and Laverne & Gignoux. Table V summarises "best available" results for a number of potentials; further results are to be reported at this conference by Afnan and Read.

I.2. What can we tell about the two-nucleon potential?

The calculations may not be perfect; but they are probably by now adequate to extract from the "best" values what information we can on the two-nucleon potential. And some interesting conclusions are starting to emerge.

I.2.1. Calculations with phase-equivalent families.

A phase equivalent family of potentials is a set of potentials which all yield the same two-nucleon phaseshifts (in all states, and for all energies) and the same deuteron binding energy. Sets of calculations with different "realistic" potentials, such as those listed in Table V, automatically form an approximate phase equivalent family, since they have all been fitted roughly to roughly the same data; but the degree of approximation is not always very high. It is an interesting question to ask: how much variation in the three-body output can we generate from a family of strictly phase-equivalent potentials. Note that this is not the same question as: "how much ambiguity in the three-body data does the two-body data leave (due to off shell freedom)" because other two-body data (the deuteron form factors, D-state probability, and quadrupole moment, for example) are measurable (maybe) but not constant within such a family.

A number of authors (refs. 26-32) have carried out calculations with phase equivalent potentials, and we can summarise their findings as follows (most alarming conclusion first, most relevant last).

(1) Phase equivalent central potentials can be found with variations in $E(^3H)$ of up to 6 MeV (ref. 26).

(2) If we insist on a fit to the deuteron form factors too, this variation drops to 1 MeV (ref. 26,28).

(3) The triton binding energy appears not to depend strongly on the 2-nucleon phaseshift above 50 MeV (perhaps even above 20 MeV (ref. 32)). It does depend strongly on the deuteron wavefunction.

Conclusion (2) suggests that variations in the off shell two-body t matrix, if tied by the deuteron form factor, do not lead to an alarmingly large range of three-body predictions. Conclusion (3) suggests that phase equivalent families are not really very relevant at all.

TABLE V

SUMMARY OF BEST AVAILABLE RESULTS
FOR VARIOUS "REALISTIC" POTENTIALS

Authors	Ref.	Potential	Ref.	$E(^3H)$	E_c	P(D)%	P(S')%	$R_{ch}(^3H)$	$R_{ch}(^3He)$	dip. $F_{ch}(^3He)$	Sec. Max	height x 10^3	$^2A_{nd}$
Delves & Hennell	13	HJ	14	6.5 ± 0.2	0.55	9.0	1.8	1.85	1.99	12.5 ± 0.3	17	1.0	2.5 ± 0.3
Delves & Hennell	13	GB	15	7.75 ± 0.2	0.60	6.7	1.8	1.65	1.85	15 ± 0.3	22	1.6	2.0 ± 0.3
CONCENSUS	-	Reid SC	16	$7.3^{+0.1}_{-0.2}$	0.60	9 ± 0.25	1.7 ± 0.1	1.7	1.9	13.0 ± 0.5	19	2	2.0
Gignoux, Laverne; Hennell Demin et al	18 19 20	CPT	17	8.6 ± 0.2	0.65	4.3 ± 0.2	1.2 ± 0.2	1.75	1.85	15.5	?	?	2.8
Laverne & Gignoux Demin et al	4 7,10	SSC_A	21	7.5 ± 0.15	0.61	6.3 ± 0.2	0.6-1.7	1.75 ± 0.1	1.9 ± 0.1	14.7	20	1.4	-
"	"	SSC_B	21	7.6 ± 0.15	0.62	6.2 ± 0.2	0.8-1.4	1.75 ± 0.1	1.9 ± 0.1	14.8	20	1.3	-
Laverne & Gignoux	4	SSC_C	21	7.34	0.60	8.0	1.4	1.62	1.83	14.4	-	-	-
Demin et al	7,10	BKR	22	6.2	0.63	8.8	0.5	1.82	1.97	13.2	18.5	1.9	-

With the inclusion of tensor potentials in the phase equivalent family (refs. (29),(30)) we find the following:

(4) For a family which fits the two-body phaseshift (moderately) and also E deut and the quadrupole moment Q(deut) the ^3H binding energy $E(^3H)$ is correlated strongly with the deuteron D state probability P_D(deut) (ref. 29). This result has been apparent for some time for the various "realistic" potentials, and these are shown together with the calculations of ref. (29), in Figure 2. The points are quite well fitted by the straight lines:

$$P_D(^3H) = \frac{9}{7} P_D(^2H)$$
$$E(^3H) = 11.27 - 6.3\, P_D(^2H)$$

The first of these is interesting since, as pointed out by T.A. Green (private communication), a first order perturbation treatment of the tensor potential in both ^2H and ^3H yields

$$P_D(^3H) = \frac{3}{2} P_D(^2H)$$

The second is perhaps even more interesting; its apparent accuracy is remarkably high. All of the results lie close to an unique line, suggesting that:

"The low energy (< 50 MeV) phaseshifts, and the deuteron parameters E(deut), Q_D(deut), determine the binding energy $E(^3H)$ essentially uniquely."

This suggestion is reinforced by the results of de Groot and Boersma (ref. 30), who perform a series of calculations for an approximately phase equivalent family with fixed P_D(deut), E(deut), Q_D(deut) and find a variation in $E(^3H)$ of only about 0.3 MeV.

I.2.2. Effect on other 3-body observables.

As well as the binding energy, we should look at the sensitivity of other 3-nucleon observables to allowable variations in the two-body potential. A start has been made by Harper, Kim & Tubis, who computed the charge and magnetic form factors for ^3He, for a family of phase-equivalent potentials based on the Reid potential RSC. The binding energy $E(^3H)$ varied by only 0.19 MeV among members of this family; but there were significant differences in the form factors at high momentum transfers, and in particular the diffraction minimum varied by up to 4 fm^{-2}. The results are shown in Figure 3.

I.2.3. Can we fit the three-body experiment?

The conclusions above suggest that families of potentials of the type used can probably fit the ^3He form factor data without requiring an explicit three-body potential. But the limited variation in $E(^3H)$ makes it unlikely that one can bridge the gap between around 7 MeV and 8.5 MeV with potentials of this type. Does this give strong evidence for the existence of a genuine three-body potential? The answer must be: not yet, since two alternative suggestions have been made. The first (ref. 13) is that an effective repulsive core of about 0.7 fm is required, such as is given by the boundary condition model of Feshbach and Lomon. The second (ref. 31, this conference) is that nonlocality of the OBEP type is needed (and is sufficient) to give the extra 1.5 MeV.

I.2.4. Relativistic and Mesonic Corrections.

Apart from three-body potentials, we expect corrections to the binding energy and to the form factors from mesonic exchange corrections, and from kinematic relativistic effects (we should perhaps not here enquire too closely about the distinction between all three effects; but in the next few years making such a

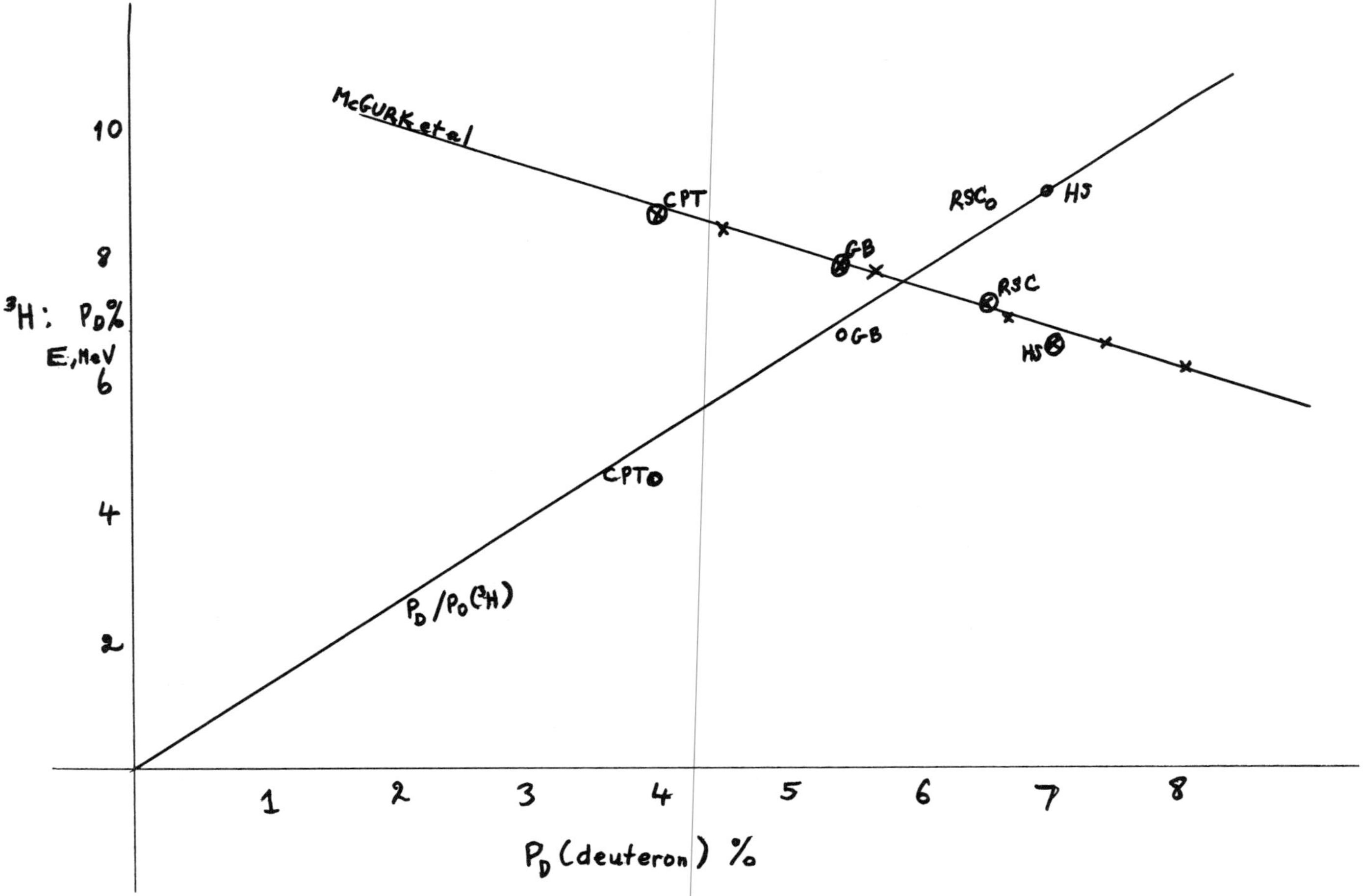
McGURK et al
CPT
GB
RSC
HS
RSC₀
HS
OGB
CPT
P_D/P_D(³H)
10
8
³H: P_D%
E, MeV
6
4
2
1
2
3
4
5
6
7
8
P_D (deuteron) %

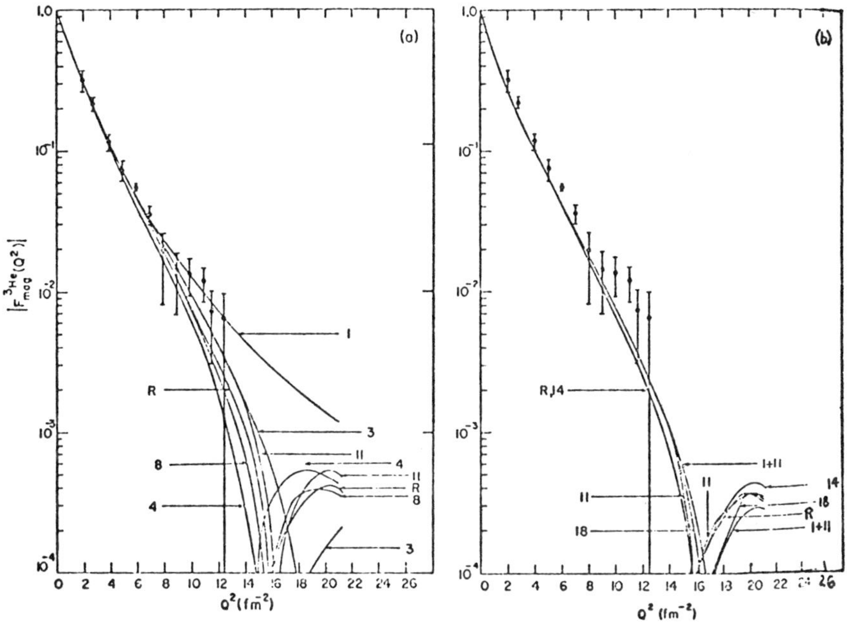

FIG. 2. (a), (b) ^{3}He magnetic form factor calculated with phase-equivalent interactions as indicated in Table II. The experimental points are taken from Ref. 24.

FIGURE 3.

Taken from Harper, Kim and Tubis (ref. 27)

distinction, and avoiding "double counting", will be crucial to progress in this field). We would particularly like to know whether plausible corrections exist which would enhance the charge form factor at high momentum transfer (to increase the calculated secondary maximum to somewhere near the experimental value) and to move the diffraction minimum to lower values of q^2. In papers to be presented at this conference, Kloet & Tjon (ref. 34) and Brandenburg et al (33) report calculations on these points. Their results are both based on RSC as a parent interaction; and show interestingly, that relativistic corrections move the diffraction minimum towards higher q^2 values by about 2 fm^{-2}, while mesonic corrections of the form considered in ref. (34) move it down by about 3 fm^{-2}, and enhance the secondary maximum considerably, bringing this more or less into line with experiment. Clearly, further studies of these effects should be carried out.

II. Scattering States of the Three-Body System.

We now pass on to the scattering states; and again the questions are:

Can we do the calculations?
How is the comparison with experiment?
Do we learn anything new about the nucleon-nucleon interaction?

II.1 The neutron-deuteron scattering lengths.

The simplest scattering calculations are those of the doublet and quartet scattering lengths, which experimentally are measured as

$$^4a = 6.3 \text{ fm}$$

$$^2a = 0.6 \text{ fm}$$

The quartet state at low energies contains little dynamic information, due to the Pauli principle; and any potential will yield a result around 6.3 fm. Interest has therefore centered on the doublet scattering length, and a number of calculations have been performed for realistic potentials; see Table V. Apart from the technical challenge, the main result of these calculations has been a realisation that ^2a is essentially uniquely determined by $E(^3H)$ (that is, the points lie on a "Phillips line") and that this line goes more or less adequately through the "experimental" point (E = 8.5 MeV, ^2a = 0.6 fm). This implies that, unless a particular potential deviates significantly from this line, we learn nothing new by calculating ^2a.

II.2 Elastic n-d scattering.

At low energies, steady progress is being made in systematising the study of n-d elastic scattering. We briefly report on this progress:

II.2.1 Generalised effective range expansions

The striking behaviour of the S-state $k\cot\delta$, (see ref. 35), with a sharp knee close to zero energy, has as its well understood origin (36) a pole at a small negative energy. In a paper presented to this conference, Sasakawa (37) has used the existence of this pole as the basis of a generalised effective range expansion, which fits the "experimental" phaseshifts rather well, with few parameters, up to a CM energy of at least 4 MeV.

In a similar vein, Fuda (38) has parametised the high-angular-momentum scattering state in a simple manner. These states should depend on only the tail of the deuteron wavefunction, which in turn depends only on the two-body effective range parameters. For $L \geq 3$ Fuda's results yield a method of computing the n+d phase-shifts at trivial cost below the breakup threshold.

II.2.2 Phaseshift analyses of the data.

The first phaseshift analysis of the n-d data was that of ref. (35): since then steady progress has been made on including splitting and coupling between states. More recently (Fayard et al, 39) comparisons with the result of separable calculations have been used to aid the analysis. Unfortunately, the analyses of different authors do not yet agree at all well.

II.2.3 Elastic scattering and breakup calculations.

Calculations with simple separable potentials showed very early on quite good agreement with experiment, considering the crudity of the potential models used. These calculations have undergone a steady refinement. For the elastic scattering channel we quote recent results of Fayard et al (39,40) using two-term separable potentials for both the ^1So and $^3S_1$ - $^3D_1$ two-nucleon states. Such potentials can make a reasonable stab at fitting the observed S_1 mixture parameter in this latter state. Figure 4 shows Fayard's result for the tensor analysing power of the deuteron at 10.85 MeV. The agreement with experiment is really very good. However, it is fair to say that single-term separable potentials (with a tensor component) also give a fairly good fit to both the vector and tensor polarising powers (41). We return to this point in (2.3) below.

As an example of the similar trend toward sophistication in breakup calculations, we note the breakup calculation using charge-dependent potentials to be reported here by Strivenberg et al (42). This calculation is of interest in showing that charge dependent effects may be significant (up to 20% in the breakup cross section at large production angles) and hence should be considered if a precise fit to experiment is looked for; an appropriate formalism has been given by Cahill & Sloan (44).

II.2.4 Photodisintegration of ^3H and ^3He.

It has been known for some time that a fairly detailed treatment of the final state (i.e.: not plane waves) is needed for photodisintegration calculations to be very meaningful, and that two-body and three-body breakup must be considered together since the threebody breakup is significant. There is nothing particularly difficult in this, since final state and bound state wavefunctions with separable potentials have been available. However, the effort of producing both together for a given potential, and in a form suitable for evaluating the matrix elements, seems to have been enough to delay consistent breakup calculations by some years. Even the recent papers (46) (47) by Barbour & Phillips, and by Hendry & Phillips, have not remedied this; these authors use an analytic form for the bound state wavefunctions albeit of a type suggested by the potential used. The latter includes a tensor interaction, and these calculations have been compared with previous and new experimental results by Ticcioni et al (45). The agreement for the 90 degree cross section is shown in Figure 5, and it is apparent that for small E_γ the agreement is not good. Recent work with central potentials by Matthews et al, and by Gibson et al, presented at this conference, shows clearly that this discrepancy stems largely if not wholly from the approximation used to the bound state function. It would now be interesting to repeat these latter, self consistent calculations including a tensor potential. However, current agreement is good enough, and the latest experimental data vary smoothly enough to remove all suggested evidence (48, 49) of a possible excited state in ^3He.

II.3 Do These States contain New Information on the Nucleon-Nucleon Interaction?

The ease with wich relatively good fits can be obtained to angular distributions and breakup spectra and now to polarization data, has always suggested that we might

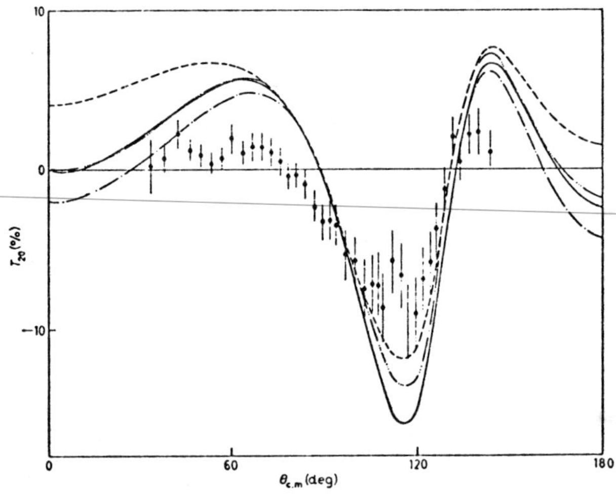

Fig. 1. – Tensor analysing power of the deuteron at 10.85 MeV. The experimental points are from ref. (19). ——— M5 (1.4 %), —·—·— M2 (1.1 %), — — — A.S. (7 %), —·····—·····— A.S. (2 %).

FIGURE 4

Taken from Fayard et al (ref. 40)

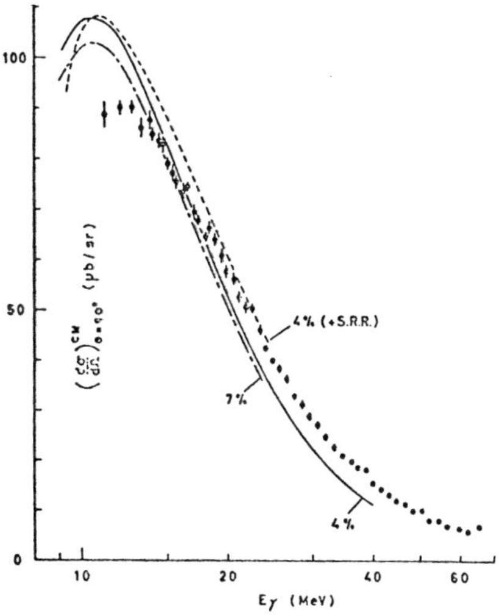

Fig. 5. Taken from Ticcioni et al (ref. 45)

find little new information on the two-nucleon force from scattering calculations, other than that already available from the bound state and doublet scattering lengths. How well founded is this suggestion?

A recent attempt to answer this question is given by Brayshaw (50). We require in principle to repeat the scattering calculations for as many families of phase equivalent potentials (which fit also the deuteron bound state data) as ingenuity will devise. Rather than this straightforward approach, which would be extremely expensive, Brayshaw (51) inputs the two-body data in terms of a set of boundary conditions on the three-body wavefunction in various asymptotic regions. Then, for any two-body potential yielding these limits, differences in the three-body behaviour must stem from the manner in which the wavefunction approaches these limits.

Applying the boundary conditions to an explicit representation for the three particle wavefunction, Brayshaw obtains a one-variable integral equation of the form

$$X = g + (K_1 + AK_2)X$$

where g, K_1, K_2 are uniquely specified by the two-body phaseshifts, X determines the three-body scattering; and A is an arbitrary real operator.

Brayshaw then chooses a particular two-body interaction to fix g, K_1, K_2 and carries out calculations for a range of operators A, with the constraint that the scattering lengths ^4a and ^2a be fitted. He finds that with these constraints, the scattering data are very insensitive to the form of A. This is illustrated in Figure 6, which shows the n-d breakup spectra at 14.4 MeV for one set of angles. The lower part of the figure shows the variation found in a previous comparison by Kloet & Tjon (52), while the upper half shows the maximum variation observed by Brayshaw. It is evident that constraining the scattering lengths to fixed values has drastically reduced the variation, and Brayshaw finds this is generally true. He concludes that (we quote):

> "The low energy trinucleon reactions can provide no new information, being essentially determined by the two-nucleon observables and the value of ^2a. It thus appears highly unlikely that this system will enhance our understanding of the short range nuclear force."

III. The Four-Body Problem.

Finally, we mention briefly progress in extending the definition of "few" past N = 3, up to N = 4. Here progress is being made, and we may be at the start of another game.

III.1 Analyses of p-^{3}He and n-T scattering.

The optical model has proven surprisingly good at correlating the experimental data for these reactions (53-55). This model of course tells us nothing fundamental about the dynamics; but in the absence of genuine four-body calculations is the best achieved so far. See however the paper by Shirato et al (56), which describes a Faddeev calculation treating n-T as (n+n+d). Their success in fitting the data is perhaps better than could be expected!

III.2 Bound state calculations with realistic forces.

There have appeared in the literature many calculations of the binding energy of ^{4}He using simple central potentials. More recently these have been extended to "realistic" potentials by Demin et al (ref. 20) using a hyperspherical basis. So far, "realistic" does not include the RSC potential; but super soft potentials (GPT)

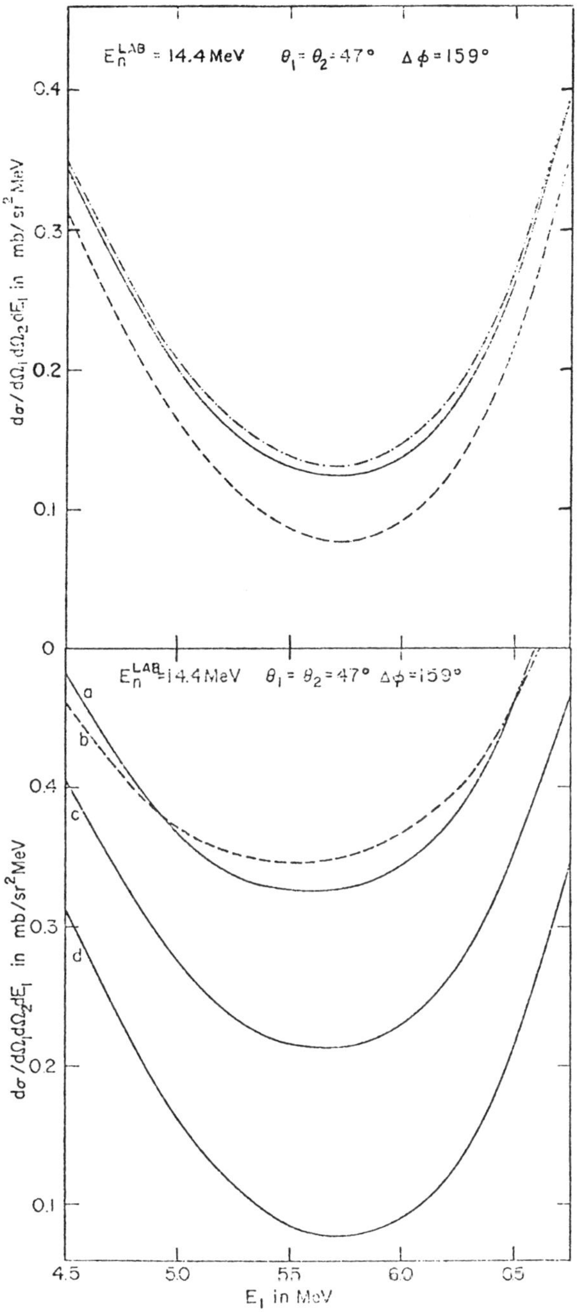

Brayshaw
Figure :5

appear amenable to current techniques. The calculations of (20) are not extensive enough to draw conclusions on the potentials; but they certainly demonstrate feasibility, and we can expect many more to follow.

Acknowledgement

This work was carried out during a period of sabbatical leave from the University of Liverpool, at the University of Victoria, and was supported by NRC research grant No. A5251. I am grateful to the Department of Mathematics, University of Victoria, for the facilities provided.

References

(1) R. A. Malfliet & J. A. Tjon, Proc. Intern. Conf. on Few Particle Problems, ed. I. Slaus et al (North Holland Publ. Company, 1973), p. 441.

(2) J. Bruinsma & R. Van Vageningen, Phys. Letters 44B (1973), 221.

(3) E. P. Harper, Y. E. Kim & A. Tubis, Phys. Rev. Letters 28 (1972), 1533.

(4) A. Laverne & C. Gignoux, Nuc. Phys. A205 (1973), 597.

(5) C. Gignoux & A. Laverne, Phys. Rev. Letters 29 (1972), 436.

(6) M. A. Hennell & L. M. Delves, Phys. Letters 40B (1972) 20.
Proc. Intern. Conf. on Few Particle Problems, ed. I. Slaus et al (1973), 419.

(7) V. F. Demin & Yu. E. Pokrovsky, Physics Letters 47B (1973), 394.

(8) M. R. Strayer and P. U. Sauer, Preprint IPNO/TH 74-16, Manchester (1974).

(9) R. A. Brandenburg and Y. E. Kim, Phys. Letters 49B (1974), 205.

(10) V. F. Demin, S. N. Ezkov, Yu K. Rogov & I. E. Zakharchenko, Phys. Letters 49B (1974), 217.

(11) E. D. Harper, Y. E. Kim & A. Tubis, Phys. Rev. Lett. 28B (1972), 1533.

(12) A. D. Jackson, A. Lande & P. U. Sauer, Phys. Lett. 35B (1971), 365.
S. N. Yang & A. D. Jackson, Phys. Letters 36B (1971), 1.

(13) L. M. Delves and M. A. Hennell, Nuc. Phys. A168 (1971), 347.

(14) T. Hamada & I. D. Johnston, Nuc. Phys. 34 (1962), 382.
T. Hamada, Prog. Theor. Phys. 33 (1965), 769.

(15) K. A. Brueckner & J. L. Gammel, Phys. Rev. 109 (1958), 1023.

(16) R. Reid, Ann. Phys. 50 (1968), 411.

(17) D. Cogny, P. Pires & R. de Tourreil, Phys. Lett. 32B (1970), 591.

(18) C. Gignoux and A. Laverne, Proc. Conf. on Few Particle Problems, ed. I. Slaus et al, (1973), 411.

(19) M. A. Hennell, this conference.

(20) V. F. Demin, Yu E. Pokrovsky and V. D. Efros, Phys. Lett. 44B (1973), 227.

(21) R. de Tourreil & D. W. L. Sprung, Nuc. Phys., A201 (1973), 193.

(22) C.N. Bressel, A.K. Kerman & B. Rouben, Nuclear Physics A124 (1969), 264.

(23) T. R. Schneider, Phys. Letters 40B (1972), 439.

(24) L. M. Delves, Advances in Nuclear Physics 5 (1973), p. 1; see p. 116.

(25) J. S. Levinger. The Two and Three Body Problem, Springer Tracts in Modern Physics (1974).

(26) M. I. Haftel, Phys. Rev. C7 (1973), 80.

(27) E. P. Harper, Y. E. Kim & A. Tubis, Phys. Rev. C6 (1972), 1601.

(28) J. P. Lavine & G. J. Stephenson, Phys. Rev. C9 (1974), 2095.

(29) N. J. McGurk, H. Fiedeldey, H. de Groot & H. J. Boersma, Phys. Letters, 49B (1974), 13.

(30) H. de Groot & H. J. Boersma, this conference.

(31) I. R. Afnan & J. M. Read, this conference.

(32) H. Fiedeldey, Letters al Nuovo Cimento, 9 (1974), 301.

(33) R. A. Brandenberg, Y. E. Kim, & A. Tubis, this conference.

(34) W. M. Kloet & J. A. Tjon, this conference.

(35) W. T. H. Van-Oers & J. D. Seagrave, Phys. Lett. 24B (1967, 560.

(36) G. Barton & A. C. Phillips, Nuc. Phys. A132 (1969),97.

(37) T. Sasakawa, this conference.

(38) M. G. Fuda, Phys. Letters 47B (1973), 288.

(39) C. Fayard; G.-h. Lamot & J. Pigeon, Lettere al Nuovo Cimento, 5 (1972), 647.

(40) C. Fayard, G.-H. Lamot & E. Elbaz, Lettere al Nuovo Cimento, 7 (1973), 423.

(41) S. C. Pieper, Nuc. Phys. A193 (1972), 529; Phys. Rev. C6 (1972), 1157.

(42) J. H. Stuivenberg, J. Bruinsma & R. M. Wageningen, this conference.

(43) J. Arvieux, R. Beurtey, J. Goudergues, R. Mayer, A. Papineau & J. Thirion, Nuc. Phys. A102 (1967), 503.

(44) R. T. Cahill & I. H. Sloan, Nuc. Phys. A165 (1971), 161.

(45) G. Ticcioni, S. N. Gardiner, J. L. Matthews & R. O. Owens, Phys. Letters 46B (1973), 369.

(46) I. Barbour & A. C. Phillips, Phys. Rev. Cl (1970), 165.

(47) J. A. Hendry & A. C. Phillips, preprint, University of Liverpool (1973), to be published.

(48) A. Van der Woude, M. L. Halbert, C. R. Bingham & B. D. Belt, Phys. Rev. Lett. 24 (1970), 1120; 26 (1971), 909.

(49) C. C. Chang, E. M. Diener & E. Ventura, Phys. Rev. Lett. 29 (1972), 307.

(50) D. D. Brayshaw, Technical report 74-037, University of Maryland, Number 1973, and this conference.

(51) D. D. Brayshaw, Phys.Rev. D8 (1973), 952.

(52) W.M. Kloet & J. A. Tjon, Nuc. Phys. A210 (1973), 380.

(53) B. S. Podmore & H. S. Sherif, this conference.

(54) R. M. Demes, J. L. Perrenoud & I. Slaus, Nuc. Phys. A1 88 (1972), 449.

(55) H. S. Sherif & B. S. Podmore, in Few Particle Problems in the Nuclear Instructions, ed. I. Slaus et al, North Holland (1972); p. 691 .

(56) S. Shimto, Y. Suda, S. Touruta and S. Oryu, this conference.

Figure Captions

Figure 1.

The convergence of the binding energy for the Reid soft core potential given by the variational calculation of ref (6) for increasing numbers of terms.

Figure 2.

The triton binding energy and D-state probability for a number of "realistic" potentials (see table 5) and for the phase equivalent family of Melynuk et al (ref. 29). There appears to be a causal relation in each case with the deuteron D-state probability P_D(deuteron).

Figure 3.

^{3}He charge form factors calculated for a family of phase equivalent potentials based on RSC (ref. 27).

Figure 4.

Tensor analysing power of the deuteron at 10.85 MeV, taken from ref. (40). The calculations assume two-term separable potentials in the ^1So and $^3S_1$ - $^3D_1$ channels. The experimental points are from ref. (43).

Figure 5.

The 90° photodisintegration correction for ^{3}He(γ,d)^{1}H. Taken from ref. (45). The theoretical calculations are those of Hendry & Phillips (47) for separable potentials yielding 4% - 17% deuteron D-state probability; the dotted curve contains a judge factor to simulate a short range repulsion.

Figure 6.

Differential corrections for n-d breakup vs the energy of the continuing neutron. Taken from Brayshaw (50). The lower set of curves arise from models which produce identical behaviour in the low energy two-nucleon system: (a) Malfliet-Tjon I-IV; (b) a separable potential; (c) I-III potential; (d) A = 0 in the analysis of Brayshaw. The standard curve in the upper half of the figure is the same as (d); the solid and dash-dot arrows represent the maximum deviation found by Brayshaw when ^2a and ^4a were continued to not vary.

DISCUSSION

CONZETT (Berkeley): I can't disagree with Brayshaw's and your statement that no off-shell information on the nucleon-nucleon interaction is likely to be gained through the three-nucleon studies. However, as I suggested in my talk yesterday, there is the possibility of deducing specific on-shell information which is not yet available from the nucleon-nucleon investigations themselves. That is the $^3S_1$ - $^3D_1$ mixing parameter ε_1 is poorly determined at 50 MeV, and the suggested appropriate experiments in np scattering, i.e. measurements of polarization transfer or spin correlation, are very difficult, particularly to the accuracy required. It now seems possible to deduce a better value of ε through much simpler polarization experiment in N-d scattering.

DELVES: I cannot possibly disagree with that statement. But Brayshaw's result says, and I believe it, that if you fix the two-body phase shifts then there is nothing to be learnt from the three-body break-up. Of course if you cannot fix the two-body phase shifts other than by doing the three-body experiments, which is what you are saying, then there is useful information from the three-body system.

RISKA (New York State): I am surprised that you claim that there is no evidence for the presence of meson exchange currents in the three-body system. After all, it has been known for 25 years that the magnetic moments of the three-body nuclei cannot be explained without taking into account the meson exchange currents. It has also been known for 3 years that taking into account the pionic range exchange currents, the magnetic moments can be explained. Another point I wish to make is that the magnetic form factors are probably more interesting quantities than the charge form factors. This is because the magnetic spin current operator can connect the S and D states of the wave function and this leads to a large destructive interference effect. In fact, we have seen in Phys.Rev.Letters this summer a calculation of the Purdue group showing that this interference effect brings in the predicted minimum in the magnetic form factor to a low value in momentum transfer where no minimum is seen experimentally. It has also been demonstrated by Hadjimichael and Barroso that the exchange current effects are large enough to push the minimum back out again.

DELVES: When I said that there was no evidence for any mesonic correction to the triton binding energy, I was playing a historical game, the game being: can one achieve 8.5 MeV by inputting a two-nucleon potential that fits the two-body data into a non-relativistic three-body calculation, and that is a game which is purely technical, and I immediately said afterwards that, of course, there are corrections and they should be calculated. In answer to your first point, I was making a technical remark about the game which I think is now over. The second remark, concerning the magnetic form factor: of course they are interesting, of course they are dominated by mesonic exchange corrections and it is for that reason that I have got into the habit of not even discussing them because one does not even have a zeroth order of approximation to discuss. That is a game in which the exchange corrections are sure to dominate and until we understand them better it is rather difficult to discuss the results one gets. One first has to understand the exchange corrections to the charge form factor where there is a reasonable chance of zeroth order approximation and then maybe one can understand the magnetic properties of the triton.

MITRA (Delhi): I have two comments. The first concerns the role of the ^1S-states. I completely disagree with Prof. Delves that it should play a qualitative role in deciding the position of the minimum in the He^3 form factor. The physical reason for this is, of course, that at least one pair of nucleons is antisymmetric in this state and this should bring about a nodal structure. An extreme example of this mechanism is represented by e.m. form factor of the nucleon as a three-quark composite. This argument played a not inconsiderable role in bringing people together to the view that the orbital wave function of the nucleon should be symmetric.

The second question concerns the role of the relativistic correction. There are two aspects to this problem. If one has to talk about full relativistic dynamics at the two and three-body levels, I am not aware if such calculations exist so far. On the other hand, there is another limited aspect to the relativistic correction which has to do with a second order correction to the potential (the Shirokov correction). The latter is an off-centre-of-mass effect for the two-body system and unfortunately gives a correction of the negative sign to the triton binding energy (i.e. tends to reduce the value). This is not to say that corrections to the kinetic energy could not give an opposite effect (i.e. positive correction) but such calculations would require a fully relativistic dynamical framework at the two and three body levels. I wonder if Prof. Kim would tell us about the nature of the relativistic corrections he considered for his calculations of $F_{ch}(He^3)$.

BRANDENBURG (Purdue): We used our three-body wave function to calculate matrix elements of the charge operator as given by J.Friar for example. The relativistic corrections were from those terms of order $\frac{1}{M^2}$.

SCHUHL (Saclay): I want to ask you two questions: (i) You showed us a $^3He(\gamma, pd)$ result at low energy. Do you think the same type of experimental measurement around 300 MeV is important to solve the three-nucleon problem? (ii) Do you think that the three-body breakup, the $^3He(\gamma,ppn)$ reaction, will help in the understanding of the three-nucleon calculation?

DELVES: Concerning the 300 MeV results, that is such a high energy that I think I don't know anything about that energy region; the two-body and three-body breakup are of course important experimentally and they are linked together theoretically. One cannot do a theoretical calculation of one without including the other. The sizes of the cross section you get are fairly sensitive to the details of the bound state wave function which you are using and in that sense photodisintegration experiments give us a link to the ground state wave function and may be useful in pinning down things like the ^1S probability for example. Although until the calculations have been performed with better potential, it is impossible, I think, to tell whether this will be the case. But they may be useful in this respect.

NOYES (Stanford): I would like to make an exception to this statement that we can learn nothing from experiments in the scattering region; you have already agreed with Conzett that they can be used to measure two-nucleon phase shifts which are not easily accessible in free two-nucleon experiments. But we saw earlier in this Conference that there are regions which are sensitive to effects that go beyond the two-nucleon phase shifts. They contain genuine three-body information coming from mesonic degrees of freedom. They can be extracted from the data by available techniques, as I remarked yesterday. I would not like to discourage experimentalists from going after this precious information.

DELVES: Let me reply to that. Maybe what Brayshaw shows is that these effects are going to be small, and if they are small, you may or may not view them as fundamental. At this stage of the game I think that they are going to be so small as not to be viewed as fundamental, because there are bigger effects one can see easily by looking at the ground state wave function. It is certainly true that the discrepancies won't be non-zero, and if people are interested in finding where they are going to be most non-zero, where the biggest ε is, then there is certainly a game to play there. I did not intend to convey the point that there is nothing there. But what is there is going to be unfortunately small and hard to extract. Now, whether that is good or bad, depends on how long you want to be working in the field.

SASAKAWA (Tohoku): I would remark, as I remarked in another session, that in the calculations in the coordinate representation, the use of the wave function which shows the correct asymptotic behaviour is very important for having a rapid convergence. It is important to note that both, the n-d channel and the break-up channel do exist in triton as the closed channel. The n-d channel behaves aymptotically as all of you know, and the break-up channel behaves asymptotically as $e^{-\sqrt{ER}}/R^{5/2}$ (R: the hyper radius). In the variational calculation, the break-up asymtotic behaviour is not taken into account, whereas in the hyperspherical calculations, the n-d asymtotic behaviour is not taken into account.

AMADO (Pennsylvania): You say that the phase equivalent potentials, that also leave the deuteron features the same, give at most 0.2 MeV variation in triton binding energy. Yet there is a hope for 2 MeV variation between, say, Hamada-Johnston calculations and Feshbach-Lomon. How do you justify these two statements? If, for example, the deuteron D-State is the most important single controlling parameter, are not triton binding calculations a very complicated way to determine it (assuming it has a well defined meaning) ?

DELVES: I did say both these things, didn't I, and they do not sound consistent. What

I should have said is that a family of calculations were carried out fixing those parameters, which are mentioned, the deuteron binding energy, quadrupole moment and D-state probability, and a given set of two-body phase shifts and a family of phase equivalent potentials, based on a particular set of these parameters, led to only 0.2 MeV difference in triton binding energy. But that was a particular family, with a particular deuteron D-state, and that is the fact which I did not make clear: that the variation, started at 6 MeV for a given family, and for a different family is reduced, when you put these extra constraints in to 0.2 MeV. But that does not mean that there are not other families which will get you to different D-state probabilities for the deuteron.

AMADO: No, but you did show that graph which seems to show that D-state probability meant more that any other parameter.

DELVES: That is right.

AMADO: Triton binding energy calculations are a very hard way of determining the deuteron D-state probability, if that probability has any meaning at all.

DELVES*: Nevertheless, the results show, that to a good approximation, that is what the triton energy depends on. Of course, we generate also a wave function; so far no such striking correlations have been found for its other equivalent parameter.

SPRUNG: This is just a short comment on the Reid soft core potential. Although it was too late to be included in his article in Ann. Phys. 50:411 (1968), Reid did fit a $^3D_3$ - $^3G_3$ potential. In nuclear matter this is the most important of the higher partial waves, giving 0.6 MeV/A attraction rather than 0.6 MeV/A repulsion as does OPEP. I suspect that in the triton this is also the most important of the higher waves. Did you use this $^3D_3$ state potential of Reid in your calculation?

DELVES: No, we did not. We used the published Reid states for j less than or equal to 2 to fit them to a local potential with a structure as in Hamada-Johnston, by just making the operator Schrödinger equations equivalent in the various two-body states and that automatically generates the higher j effective potential. We did not use the direct Reid potential for these higher states.

* Comment added later

WHAT DO WE KNOW ABOUT THE DEUTERON AND THE ANTI-DEUTERON?

Donald W. L. Sprung, Department of Physics, McMaster University
Hamilton, Ontario, Canada L8S 4M1

To a large extent, the nuclear force is a two body force, so the two body problem is the central problem of nuclear physics. As the only two nucleon bound state, the deuteron has been the object of intense experimental and theoretical interest for four decades. Yet, our knowledge of this simplest of all nuclei is far from satisfactory. Many of its properties have been measured to great precision, as shown in Table 1. To date, our understanding of its structure has remained much less precise.

Of the antideuteron, very little is known. Antideuterons have been observed at Brookhaven, during an experiment to look for quarks (Dor+ 65), and also at CERN (Mas+ 65). It has negative charge, a mass equal to M_D within 3% error, and a lifetime of order 10^{-7} sec. Here one can say that theory is in the lead.

The deuteron is a state of J=1, positive parity. The standard view is that it consists most of the time of a bound neutron-proton pair. The nonrelativistic wave function may be written

$$|\Psi^M\rangle = \frac{1}{r\sqrt{4\pi}}\{u(r) + \frac{w(r)}{\sqrt{8}} S_{12}\}|\chi_1^M\rangle \tag{1.1}$$

where χ_1^M is a triplet spin state, and S_{12} is the usual tensor operator. The S-wave u(r) and D-wave w(r) radial wave functions satisfy coupled Schroedinger equations

$$\begin{pmatrix} \frac{d^2}{dr^2} - \alpha^2 - v_{00} & - v_{02} \\ - v_{20} & \frac{d^2}{dr^2} - \alpha^2 + \frac{6}{r^2} - v_{22} \end{pmatrix}\begin{pmatrix} u \\ w \end{pmatrix} = 0 \tag{1.2}$$

with $v = m/\hbar^2\, V$ and $E = -\hbar^2/m\,\alpha^2 = -B$ the energy eigenvalue. In case we write the potential as

$$V = V_c(r) + V_{LS}(r)\underset{\sim}{L}\cdot\underset{\sim}{S} + V_T(r)S_{12} + V_{LL}(r)\underset{\sim}{L}^2 \quad ,$$

$$v_{00} = v_c \qquad v_{20} = v_{02} = \sqrt{8}\, v_T \tag{1.3}$$

$$v_{22} = v_c - 2v_T - 3v_{LS} + 6v_{LL} \quad .$$

The normalization is $\int_0^\infty (u^2 + w^2)dr = p_S + p_D = 1$. (1.4)

The D-state probability p_D is believed to lie between 4 and 9%.

All serious potential models of the N-N interaction attempt to reproduce the properties of the deuteron. The longest ranged component of the force is due to one pion exchange. This OPEP force is

$$V_{OPE} = f^2 \frac{\mu c^2}{3} \underset{\sim}{\sigma}^1\cdot\underset{\sim}{\sigma}^2\ \underset{\sim}{\tau}_1\cdot\underset{\sim}{\tau}_2\ [\frac{e^{-x}}{x} + k_2(x)S_{12}] \tag{1.5}$$

with $f^2 \approx 0.08$, $x = \frac{\mu c}{\hbar} r \simeq 0.7r$ and $k_2(x) = (1 + \frac{3}{x} + \frac{3}{x^2})e^{-x}/x$. For states of S=1, T=0 the appropriate pion mass is

$$\mu = (2m_{\pi+} + m_{\pi 0})/3 \quad .$$

The OPEP tail on the force is a fairly powerful constraint on the low energy properties, as was shown by Glendenning & Kramer (GK 62). A large enough r, the nuclear potential will be negligible, and the deuteron wave function will take its asymptotic form

$$\begin{aligned} u(r) &\to Ne^{-x} \\ w(r) &\to \eta N x k_2(x) \end{aligned} \tag{1.6}$$

with $x = \alpha r$, N a normalization constant and η the "asymptotic D/S ratio". In effective range theory, one finds (Wil 63)

$$N^2 = 2\alpha/[1 - \alpha\rho(-B, -B)] \tag{1.7}$$

with $\rho(-B, -B)$ being the effective range evaluated at the deuteron binding energy:

$$\rho(-B, -B) = 2\int_0^\infty [e^{-2\alpha r} - (u^2 + w^2)]dr \quad . \tag{1.8}$$

Thus the measured effective range parameters constrain the asymptotic normalization. D. Wong (Won 59) found that imposing the OPEP tail on the potential leads to $\eta \sim 0.029$, which is indeed found in most potential models. Recently it has been found that the tensor polarization of sub-Coulomb barrier deuterons scattered from nuclei, is sensitive to η. Knutson, Stephenson and Haeberli (KSH 74) have developed a classical model which gives a very physical explanation of this.

The wave functions of several well known potentials are shown in Fig. 1 and the calculated properties listed in Table 2. The general characteristics of these potential models include a weak attractive central force and an attractive tensor force, both long ranged due to their OPEP tails. The spin orbit interaction is required, to make v_{22} independent of v_{00} (= v_c) and v_{02} (= $\sqrt{8}\, v_T$). It is invariably found to be repulsive. In OBEP models this comes about because the spin orbit force is associated with the ρ meson (and is attractive in the triplet odd parity states). With only these components, the $^3D_2$ force would turn out to be too attractive, with the 300 MeV phase shift nearly double the experimental value. For this reason an additional, quadratic spin orbit force is usually invoked. The simplest possibility is simply $\underset{\sim}{L}^2$, as in Eq. (1.3). All potentials include some repulsive core, though the amount of repulsion used is quite arbitrary. In our work we have favoured very weak repulsion, balanced by weaker attractive forces at long range (TS 73, TSR 74).

It is the existence of the quadrupole moment Q which requires the tensor force and resulting D-state admixture. Although its value is small compared to the deuteron radius, nonetheless people fitting local potential models have difficulty reproducing so large a value. The well known expression (BW 52) is

$$Q = \frac{1}{\sqrt{50}}\int_0^\infty r^2 w(r)[u(r) - \frac{w(r)}{\sqrt{8}}]dr \quad . \tag{1.9}$$

The problem seems to be that Q is especially sensitive to u, w at large r, but it is just in this region that we have little freedom to vary the wave function. Recently Klarsfeld (Kla 74) has used the assumption that beyond some r=a, the force is OPEP, in order to derive a lower bound on the D-state probability $p_D \gtrsim 3.3\%$. Since his argument is rather pretty, I will repeat a simple version of it here. Beyond r=a we take u, w in the form Eq. (1.6). We have

$$p_D = \int_0^a w^2 dr + \int_a^\infty w^2 dr \quad ,$$

the latter now fixed. For the former we use the Schwartz inequality over the interval (0,a), following Levinger (Lev 69)

$$|w^2|\cdot|(r^2u)^2| \geq |r^2uw|^2 \simeq [\sqrt{50}\ Q - \int_a^\infty r^2uwdr]^2 = 50Q_I^2 \quad . \qquad (1.10)$$

Here we neglect the second term in Eq. (9) for Q, arguing that it amounts to about a 10% correction. This gives

$$p_D \gtrsim \frac{50Q_I^2}{\int_0^a r^4u^2dr} + \int_a^\infty w^2dr = F(a,\eta) \quad . \qquad (1.11)$$

So far, the asymptotic D/S ratio η is a free parameter. If we vary η, the normalization of u^2, w^2 will alter, and this has to be taken into account. The denominator involves r^4u^2, so it is not very sensitive to the values chosen for u - the asymptotic form $Ne^{-\alpha r}$ will be too large and this only reinforces the inequality. Klarsfeld then finds that F(a,η) is quadratic in η for fixed a, so choosing η at the minimum gives a lower bound on p_D. If one chooses a $\simeq$ 3 fm., $p_D \gtrsim$ 2%. In his more refined calculation, u and w are actual solutions in the OPEP potential beyond r=a. For a = 2 fm., this gives $p_D \gtrsim 3.3\%$, irrespective of what potential occurs inside a. This establishes that the low values of p_D achieved in separable potential models, are due to the long range deviations from OPEP.

The simplest model for the magnetic moment is

$$\mu_{op} = \mu_p\sigma_p + \mu_n\sigma_n + \frac{1}{2}L$$

$$= (\mu_p + \mu_n)J - (\mu_p + \mu_n - \frac{1}{2})L + \frac{1}{2}(\mu_p - \mu_n)(\sigma_p - \sigma_n) \quad . \qquad (2.1)$$

The deuteron magnetic moment is defined as the expectation value of μ_{op} in the state M=1. The last term does not contribute in a spin 1 state, so

$$\mu_D = \mu_p + \mu_n - (\mu_p + \mu_n - \frac{1}{2})J\cdot L/2$$

$$= \mu_p + \mu_n - \frac{3}{2}(\mu_p + \mu_n - \frac{1}{2})p_D \quad . \qquad (2.2)$$

Using

μ_p = 2.792 845 6 (11)

μ_n = -1.913 148 (66) ,

$\mu_p + \mu_n$ = 0.879 696 (67)

and μ_D = 0.857 406 (1) ,

all in units of $e\hbar/2m_pc$, the nuclear magneton. If the entire discrepancy between μ_D

and $(\mu_p + \mu_n)$ is ascribed to the D-state, 0.0223 = .5361 p_D, we have the well known estimate p_D = 4%. Almost all local potential models which fit the deuteron quadrupole moment, however, give values of p_D of order 7%, so the above estimate is never taken very seriously. In any case, the above model for the magnetic moment operator is subject to a large number of well known corrections, any one of which is capable of explaining the discrepancy induced by a 6 or 8% p_D.

One of these corrections is that induced by the momentum dependence of the spin orbit force. This was introduced by Feshbach (Fes 57), as an argument against such forces.

$$V_{SL}\underset{\sim}{S}\cdot\underset{\sim}{L} = \frac{1}{2\hbar} V_{LS}(r)(\underset{\sim}{S} \times \underset{\sim}{r})\cdot(\underset{\sim}{p}_1 - \underset{\sim}{p}_2) \quad . \qquad (2.3)$$

If particle 1 is the proton, in an e.m. field $\underset{\sim}{p}_1 \to \underset{\sim}{p}_1 - \frac{e}{c}\underset{\sim}{A}$ giving an interaction

$$\Delta H_{em} = - \frac{e}{2\hbar c} V_{LS}(r)(\underset{\sim}{S} \times \underset{\sim}{r})\cdot\underset{\sim}{A} \quad . \qquad (2.4)$$

In a uniform magnetic field $\underset{\sim}{A} = -\frac{1}{2}(\underset{\sim}{r}_1 \times \underset{\sim}{\mathcal{H}})$ leading to a magnetic moment correction

$$(\Delta\mu)_{LS} = - \frac{e}{16\hbar c} \langle\psi^M|V_{LS}\big((\underset{\sim}{S}\cdot\underset{\sim}{r})^2 - r^2\underset{\sim}{S}\cdot\underset{\sim}{J}\big)|\psi^M\rangle \quad . \qquad (2.5)$$

Inserting the deuteron wave function we find

$$(\Delta\mu)_{LS} = \frac{1}{6}\left[\langle u|r^2v_{LS}|u\rangle - \frac{1}{\sqrt{2}}\langle u|r^2v_{LS}|w\rangle - \langle w|r^2v_{LS}|w\rangle\right] \qquad (2.6)$$

which differs from Feshbach's result in the coefficient of the final term. [This has been noticed independently by Bergstrom (Ber 74).] The cross term and D-state term reduce the result significantly. Values of $(\Delta\mu)_{LS}$ for a number of modern potentials (before and after correcting the expression) are shown at the right of Table 2. The positive sign is due to the repulsive nature of the spin orbit force, a consistent feature of triplet even potentials already remarked upon. This correction allows p_D to increase from 4% to about 6%, so it makes local potential models more consistent with μ_D.

Relativistic corrections to the magnetic moment were discussed by Breit (Bre 47); they are small, of order $(\langle T\rangle/6mc^2)\cdot(\mu_p + \mu_n - 1)$, and difficult to compute reliably without a relativistic wave equation.

Meson currents are the most widely invoked explanation for the magnetic moment discrepancy. Starting from the pseudovector coupling Lagrangian

$$\mathcal{L}_{\pi NN} = \frac{f}{\mu}\bar{\psi}(x)\underset{\sim}{\sigma}\cdot\underset{\sim}{\nabla}\,\underset{\sim}{\tau}\cdot\underset{\sim}{\phi}\,\psi(x)$$

and invoking the usual e.m. coupling $\underset{\sim}{\nabla} \to \underset{\sim}{\nabla} - \frac{ie}{c}\underset{\sim}{A}$ induces an interaction which seems to be called the 'seagull term', shown in Fig. 2(a). One finds that the isospin coupling for this process goes like $(\underset{\sim}{\tau}_1 \times \underset{\sim}{\tau}_2)_z$, so is an isovector. Such a one pion exchange process cannot contribute in the T=0 deuteron ground state. A more general argument based on G-parity ($G = e^{i\pi T_y}C$) shows that all N=even pion states are forbidden to contribute to the isoscalar current (which has odd G-parity). Thus the lowest order meson current corrections involve three pions. Adler and Drell (AD 65, Adl 66) have estimated the process shown in Fig. 2(b), where two of the pions are taken as forming a ρ meson. Their result depends on the magnitude and the unknown sign of the coupling constant $g_{\rho\pi\gamma}$. If $g^2_{\rho\pi\gamma/4\pi} = 0.02$, corresponding to a width $\Gamma_{\rho\pi\gamma} = 0.5$ MeV, for ρ decaying into πγ, then the correction to Δμ could be as large as ±0.02 nuclear magnetons. It seems likely that $g^2_{\rho\pi\gamma}$ is considerably smaller, but the correction should not be ignored. More recently Horikawa et al. (HFY 72) have calculated this correction by dividing the correction into long range and short range effects. The former can be estimated using the OPEP force.

Blankenbecler and Gunion (BG 71) have proposed a two-body correction to the magnetic moment operator. Figure 2(c) illustrates a process that contributes to the nucleon form factor. When a second nucleon is present, it is possible for the ω-meson to scatter from the first nucleon before being absorbed on the second. This is a nonlinear contribution, which they estimated using a Glauber type approximation. Their result is proportional to the difference of the isoscalar and isovector moments, $(\mu^s - \mu^v)$ and to $|\psi(r=0)|^2$. They estimate the value as $\Delta\mu_{2B} \simeq 0.017$ n.m., quite comparable with other mechanisms.

In a contribution to this conference, F. Gross (Gro 74b) has computed a correction of 0.015 n.m., on the basis of a relativistic wave equation. I will say more about this in a few moments. The general conclusion is that the magnetic moment shows the inadequacy of the simple two body model of the deuteron, without favouring any of the various proposed remedies. Surely this is a problem which should soon be resolved. At the present time, there is no real evidence against any value of p_D from 3.3% to 9%. Remler and Miller (RM 74) have found in an analysis of 30-300 MeV nucleon deuteron scattering, a preference for p_D between 7 and 9%. Michael and Wilkin (MW 69) have made similar claims for GeV energy π-D scattering. The problem with these analyses is that they put in certain deuteron wave functions, which can very well influence the conclusions.

Electron scattering from the deuteron has two aims, to discover the structure of the deuteron wave function at small r, and to determine the neutron form factor. Due to the small mass and charge of D, the Born approximation should be valid. It is generally assumed that the deuteron structure is adequately described by the nonrelativistic wave function, Eq. (1.1). For point nucleons the cross section formula was deduced by Z. Jankus (Jan 56). M. Gourdin generalised these results in an impressive sequence of papers (Gou 63, 64a, 64b, 65a, 65b, GP 64) to include nucleon form factors, relativistic contributions to the single nucleon currents, and polarization in e-D scattering. Gross was the first to calculate all the relativistic corrections of order $1/m^2$, and $1/m_D 2$ (Gro 64, 66a, 66b, CG 67). Recently Friar (Fri 73a, 73b) has confirmed and beautifully simplified Gross' results.

Following Gourdin, we can write the unpolarized cross section in the form

$$\left(\frac{d\sigma}{d\Omega}\right) = \left(\frac{d\sigma}{d\Omega}\right)_{MOTT} \cdot \{A(q^2) + B(q^2)[1 + 2(1+\nu)\tan^2\theta/2]\} \quad (4.1)$$

where

$$\left(\frac{d\sigma}{d\Omega}\right)_{MOTT} = Z^2\alpha^2 \cos^2\theta/2 \Big/ 4E^2 \sin^4\theta/2\left(1 + \frac{2E}{M_D}\sin^2\theta/2\right) \tag{4.2}$$

$\nu = q^2/4M_D^2$, and $q^2 = \tilde{q}^2 - \tilde{q}_0^2$ is the four momentum transfer. Gourdin worked in the Breit frame where $q^2 = \tilde{q}^2$.

$$A(q^2) = G_0^2 + G_2^2 \tag{4.3}$$

$$B(q^2) = \frac{2}{3}\nu G_M^2 \tag{4.4}$$

while

$$G_0 = 2G_{ES}C_E \quad , \quad G_2 = 2G_{ES}C_Q \quad , \quad G_M = \frac{M_D}{m_p}(2G_{MS}C_S + G_{ES}C_L) \quad .$$

The nucleon isoscalar form factors are

$$2G_{ES} = (G_{Ep} + G_{En}) \quad , \quad 2G_{MS} = (G_{Mp} + G_{Mn}) \tag{4.5}$$

and the deuteron structure enters through the structure factors

$$\begin{aligned} C_E &= \int [u^2 + w^2] j_0(qr/2)dr & &\rightarrow 1 \\ C_Q &= 2\int w(r)[u(r) - \frac{w(r)}{\sqrt{8}}] j_2(qr/2)dr & &\rightarrow \frac{q^2}{3\sqrt{2}} Q \\ C_L &= \frac{3}{2}\int w^2(r)[j_0(qr/2) + j_2(qr/2)]dr & &\rightarrow \frac{3}{2} P_D \\ C_S &= \int (u^2 - \frac{1}{2}w^2) j_0(qr/2)dr + \frac{1}{\sqrt{2}} \quad w(r)[u(r) + \frac{w(r)}{\sqrt{2}}] j_2(qr/2)dr & &\rightarrow (1 - \frac{3}{2}P_D) \quad . \end{aligned} \tag{4.6}$$

The limiting values as $q \rightarrow 0$ are shown at the right. The limiting value of the magnetic form factor is

$$G_M(0) = \frac{M_D}{m_p}\left[(\mu_n + \mu_p)(1 - \frac{3}{2}P_D) + \frac{3}{4}P_D\right] \tag{4.7}$$

which is seen to agree with the deuteron magnetic moment, Eq. (2.2). Actually the electrostatic contribution to the magnetic form factor was not included in Gourdin's earlier papers, but was added (Gou 65a) in a paper where the relativistic corrections to the nucleon currents were considered (Darwin-Foldy and spin orbit corrections). In the same level of approximation there is a spin-orbital (magnetic) correction to the quadrupole form factor. In the limit of zero q^2, this gives a correction to Q:

$$Q \rightarrow Q_0 + \frac{3}{20}P_D(\mu_n + \mu_p - \frac{1}{2})\left(\frac{\hbar}{M_D c}\right)^2 \quad . \tag{4.8}$$

This is, however, completely negligible because of the small (0.11 fm) compton wave length of the deuteron.

The relativistic corrections (Fri 73b) to the deuteron form factors G_i are of two types. The first class, of order q^2/m^2, arise from using the nucleon current

$$\rho(\tilde{P},\tilde{q}) = (1 - \frac{q^2}{8m^2})G_E - i\tilde{q}\cdot(\tilde{\sigma} \times \tilde{P})/4m^2\cdot(2G_M - G_E)$$

which exhibits both the Darwin-Foldy correction and the spin-orbit term. The former is taken into account by using amended form factors

$$\tilde{G}_{\substack{E\\M}}(q) = G_{\substack{E\\M}}(q)/\sqrt{1 + q^2/4m^2} \quad , \tag{4.10}$$

This factor has been many times proposed for inclusion in the definition of the nucleon form factors (YLR 57, Gou 65a). To this same order, the Rosenbluth cross section can be written (DW 64)

$$\left(\frac{d\sigma}{d\Omega}\right)_{LAB} = \left(\frac{d\sigma}{d\Omega}\right)_{MOTT} \{\tilde{G}_E^2 + \nu\cdot\tilde{G}_M^2[1 + 2(1+\nu)\tan^2\theta/2]\} \quad . \tag{4.11}$$

The spin orbital contribution was shown by Bertozzi et al. (BFHN 72) to make an important contribution, even for heavy nuclei, with $S \neq 0$.

The second class of corrections, of order q/M_D^2, arise from imposing approximate Lorentz invariance on the collection of nucleons comprising the nucleus. Friar has shown that for G_0, G_2, these corrections have the effect of changing the argument

$$\text{from} \quad G_i(q^2) \quad \text{to} \quad G_i(q^2/(1 + q^2/4M_D^2)) \quad . \tag{4.12}$$

For the deuteron, then, one has the simple prescription

$$\begin{aligned} G_0(q^2) &\to 2\tilde{G}_{ES}(q^2)C_E(q^2/(1+\nu)) + 2(2\tilde{G}_{MS} - \tilde{G}_{ES})F^0_{SO}(q^2) \\ G_2(q^2) &\to 2\tilde{G}_{ES}(q^2)C_Q(q^2/(1+\nu)) + 2(2\tilde{G}_{MS} - \tilde{G}_{ES})F^2_{SO}(q^2) \end{aligned} \tag{4.13}$$

with

$$\begin{aligned} F^0_{SO}(q^2) &= \frac{q^2}{8m^2}\int_0^\infty \frac{3j_1(qr/2)}{(qr/2)} w^2(r)dr \\ F^2_{SO}(q^2) &= \frac{3}{m^2}\int_0^\infty j_2(qr/2)w(r)\{(\tfrac{u}{r})' - \frac{1}{\sqrt{2}}\frac{1}{r}w'\}dr \quad . \end{aligned} \tag{4.14}$$

Gourdin (Gou 65a) has given rather different formulae for these corrections, which may be equivalent. Since the spin orbit form factors are already q^2/m^2 corrections, their arguments could be replaced by $q^2/(1+\nu)$ for consistency.

The conclusion is that we may continue to use the simple formula (4.1-6) for the electron cross section, providing the relativistic form factors are evaluated according to Eq. (4.13), when we compare to theoretical calculations.

As is well known, the 'electric' scattering $A(q^2)$ may be separated from the magnetic term $B(q^2)$ by comparing two cross sections at the same momentum transfer q^2 but different angles. However, the monopole and quadrupole form factors cannot be separated by differential cross sections alone. This is illustrated in Fig. 3, copied from Galster et al. (Gal+ 71), as a typical case, showing a smooth function $\tilde{A}(q^2)$ (= $A(q^2) + B(q^2)$) composed mainly of G_0^2 at small q^2 and mainly G_2^2 at large q^2. Levinger et al. (BTL 72, Lev 74) have proposed to use a polarized deuteron target to make this separation. Following the Madison Convention (Dar 70) the analysing power T_{20} is defined as

$$T_{20} = Tr(M\tau_{20}M^\dagger)/Tr(MM^\dagger) \tag{5.1}$$

with

$$\tau_{20} = \frac{1}{\sqrt{2}} (3S_z^2 - 2) \quad . \tag{5.2}$$

Levinger sets

$$P_e = - T_{20} = \frac{\sqrt{2}}{3} \frac{(\sigma_0 - \sigma_1)}{\sigma_{un}} \tag{5.3}$$

where σ_M is the cross section for scattering from aligned deuterons whose spins have projection M along the axis $\hat{\tilde{q}}$. The unpolarized cross section $\sigma_{un} = \frac{1}{3}(\sigma_0 + 2\sigma_1)$ is given by Eq. (4.1). Gourdin (Gou 64b, GP 64) has given the values of σ_0, σ_1 separately, from which we have

$$P_e = (2G_0G_2 + \frac{1}{\sqrt{2}} G_2^2)/(G_0^2 + G_2^2) = (2x + x^2/\sqrt{2})/(1 + x^2) \tag{5.4}$$

in terms of the ratio $x = G_2/G_0$. The subscript "e" implies that we have 'extrapolated' to $\tan^2\theta/2 = -\frac{1}{2}/(1+\nu)$, to remove the magnetic scattering. The function $P_e(x)$ is shown in Fig. 4; it has a maximum value $\sqrt{2}$ at $x = \sqrt{2}$; a minimum $-\frac{1}{\sqrt{2}}$ at $x = -\frac{1}{\sqrt{2}}$, and tends to the value $1/\sqrt{2}$ as $x \to \pm\infty$. Looking now at Fig. 3 we can see how the values of $P_e(q^2)$ will vary as a function of q^2. It starts off rising rapidly for small q^2, and at the point where G_0 has fallen to equal G_2, $P(q_a^2) = 1.35$. As G_0 continues to fall, P rises to a maximum value $\sqrt{2}$ and then falls sharply to $P(q_b^2) = 0.7$ at the diffraction minimum of $G_0(q^2)$. Since the fall is very sharp, q_b^2 should be quite accurately determined. P_e will continue to fall through zero, pass through a minimum, and rise to zero at the point where $G_2(q_c^2) = 0$. Unfortunately, this would be at a very high momentum transfer where other corrections are likely to be important. Calculations of $P_e(q^2)$ by Levinger et al., and by Moravcsik and Ghosh (MG 74) bear out this qualitative behaviour. They did not, however, emphasize the fact that P falling through the value 0.7 is a measuring device for the diffraction minimum of G_0, crucial information for determining its Fourier transform. As pointed out by Levinger, for the unrealistic (?) Hulthén wave function, the rise of P is much less rapid, implying a broad region where x is constant, $\simeq 1$.

If relativistic corrections were negligible, $P(q^2)$ would be independent of the nucleon form factors, as is clear from Eq. (4.13). Unfortunately, q_b^2 will probably occur in the region $q \simeq 5 - 8\ \text{fm}^{-1}$, where this is not the case. This will complicate the interpretation of the experiment, and means that the existing calculations should be refined.

I will just say a few words about photodisintegration and its inverse reaction, radiative capture. The capture cross section for reactor neutrons is 334.2 ± 0.5 mb (CWC 65), a figure of enormous economic importance to Canadians, because it leads to the use of heavy water as moderator in our power reactors. The magnetic dipole transition dominates, and gives a theoretical (Noy 65) value 302.5 ± 4. mb. Riska and Brown (RB 72) have accounted for the 10% discrepancy as an exchange current effect involving the D-state component of the deuteron. Since the transition can go from a T=1 initial state to the T=0 final state, a single pion can contribute to the exchange current, unlike the case of the magnetic moment. Transitions from an initial $^1S_0$ state to both the $^3S_1$ and $^3D_1$ final states are important. This result has been confirmed by Gari and Huffman (GH 73). It appears then that radiative capture is reasonably well understood.

For photodisintegration, the situation is confused. The nonrelativistic theory was worked out in great detail by Partovi (Par 64) who carried out calculations using the Hamada Johnston (HJ 62) potential. These calculations seem to be the accepted standard of reference. Breit and Rustgi (BR 68) have discussed the limitations and refinements to the standard theory.

Experiments in the 15 - 60 MeV photon energy range (WS 71, Bag+ 73) have achieved an accuracy of 6-7%. These results are in discrepancy with Partovi's calculations in a number of ways. The variation of cross section wthh energy is not right, the predicted asymmetry in the differential cross section is too small, and the predicted isotropic component is too large by about 20%. These discrepancies indicate the need for better calculations.

Some contrary indication is provided, however, by electrodisintegration experiments in Virginia (PW 74) at 4 - 11 MeV and Saskatchewan (Cho+ 74) at 30 - 106 MeV. Using the virtual photon theory

$$\sigma_e(E) = \sum_{\ell} \int N_{\ell}(E - E_f)\sigma_{\gamma,\ell}(E - E_f)dE_f$$

and a Hulthén wave function, they find very good agreement with theory. It will be important to improve these calculations, to see whether the agreement persists. There is, however, a definite discrepancy between the electro and photodisintegration experiments, which must be sorted out.

There are many calculations of meson current corrections to the disintegration cross section. Hockert et al. (Hoc+ 73) have shown that near threshold, and at large momentum transfer, the impulse approximation gives a result very much less than experiment. By adding in the simplest meson current corrections, shown in Fig. 5 [(a) pair term, (b) meson current, (c) intermediate N*], the backward angle values are significantly raised. They point out that the 'Jankus theory' (Jan 56) as usually employed neglects the $^3D_1$ component, and this is not justified at large angles. Hadjmichael (Had 73) has considered the meson current corrections to photodisintegration at energies up to 25 MeV. He finds the biggest effect from the intermediate isobar process Fig. 5(c), with 5(a), 5(b) following. Figure 2(b), the Adler-Drell process, is negligible in this context, which may rule out its being the only important correction to the magnetic moment.

In general, meson current corrections cannot be too large because they affect mainly the M1 transitions. Except just at threshold the E1 transition, for which Seigert's theorem holds, is dominant. This is illustrated in Fig. 6, taken from the paper of Phenneger and Winter (PW 74).

One way to improve on the standard, nonrelativistic theory outlined in Sec. 1, is to introduce a relativistic wave equation. Of course, such a theory in the form of the Bethe-Salpeter equation (SB 51) has been available for over twenty years. However, the complications of applying this theory have ruled out its practical application, so far. What is wanted is a way to combine relativity with the fruitful phenomenology of the nonrelativistic theory. An approach which promises to provide this has been proposed by F. Gross (Gro 69). Gross starts with the Bethe-Salpeter equation, but restricts one particle to lie on the mass shell. Such an approach has striking advantages. It leads to a covariant, but three dimensional integral equation for the two body problem. The wave functions obtained, with one particle off-shell, are precisely those required to describe the deuteron in its interaction with external probes, in the impulse approximation (when one nucleon is a spectator).

Gross' equation has the general form

$$M(pp'W) = G(pp'W) + \frac{i}{(2\pi)^4} \int d^4k \; G(pkW)\Delta(W,k)M(kp'W) \quad . \tag{6.1}$$

The original Bethe-Salpeter equation has

$$\Delta^{SB}(W,k) = [M^2 - (\tfrac{1}{2} W + k)^2 - i\varepsilon]^{-1}[M^2 - (\tfrac{1}{2} W - k)^2 - i\varepsilon]^{-1} \tag{6.2}$$

while Gross has

$$\Delta(W,k) = [M^2 - (\tfrac{1}{2} W - k)^2]^{-1} \delta(M^2 - (\tfrac{1}{2} W + k)^2]\theta(\tfrac{1}{2} W + k_0) \quad . \tag{6.3}$$

For comparison, the more familiar Blankenbecler-Sugar choice, which takes only the two particle unitary cut into account, uses

$$\Delta^{BS}(W,k) = 2\pi \int_{4M^2}^{\infty} \frac{ds}{s - W^2} \delta(M^2 - (\tfrac{1}{2} \sqrt{s} + k)^2)\delta(M^2 - (\tfrac{1}{2} \sqrt{s} - k)^2) \quad . \tag{6.4}$$

Gross proposes to construct the kernel G(pk,W) by adding up the one and two meson exchange contributions. The amplitude M then represents the sum of iterated one and two meson processes. The restriction of one particle to the mass shell determines the fourth component of the 4-momentum, so that G and M depend only on the three momentum, but this restriction is done in a completely covariant way. When applied to the electron-proton system, his method produces the Grotch-Yennie equation (GY 69). This is the Dirac equation for the electron, containing an effective potential, good to order $1/M^2$.

In a series of recent papers, Gross (Gro 74a, 74b; HG 73, DG 74) has applied his method to the nuclear force problem. He considers the exchange of $\pi\rho\omega$ and σ mesons. The latter is introduced to account for two pion exchange. In the framework of Gross theory, there are good arguments for treating one and two pion exchange independently, which relate to the pseudoscalar nature of the pion. For the deuteron, Gross' theory reduces in coordinate space to a set of coupled integro-differential equations:

$$\begin{aligned} (\frac{\hbar^2\nabla^2}{M} + \varepsilon)\psi^+ &= V^{++}(x)\psi^+(x) + V^{+-}\psi^-(x) \\ 2M\psi^-(x) &= V^{-+}(x)\psi^+(x) + V^{--}\psi^-(x) \quad . \end{aligned} \tag{6.5}$$

The positive energy components $\psi^+(x)$ are precisely of the form, Eq. (1.1). The negative energy components

$$\psi^-(r) = -i\sqrt{\frac{3}{4\pi}}\frac{1}{r}\left[\frac{v_t(r)}{\sqrt{2}}\,\underset{\sim}{S}\cdot\hat{\underset{\sim}{r}} + v_s\,\underset{\sim}{D}\cdot\hat{\underset{\sim}{r}}\right] \quad ,$$

$$\underset{\sim}{D} = \frac{1}{2}(\underset{\sim}{\sigma}^1 - \underset{\sim}{\sigma}^2) \qquad (6.6)$$

are new. v_t, v_s are respectively $^3P_1$ and $^1P_1$ components, which do not occur in the usual theory. The potentials $V^{\pm\pm}$ contain both local and nonlocal terms. The wave function is normalised to

$$\int d^3r[|\psi^+(r)|^2 + |\psi^-(r)|^2] = \int_0^\infty dr(u^2 + w^2 + v_t^2 + v_s^2) = 1 \quad . \qquad (6.7)$$

By playing with the meson coupling constants and masses, Gross can make his effective potential for ψ^+ look very much like the Reid Soft Core potential (Rei 68). Assuming that with proper care one would end up with a sensible deuteron wave function, Hornstein and Gross (HG 73) take $\psi^+(r)$ to be essentially Reid's wave function; corrected for velocity dependence. $\psi^-(r)$ can then be calculated from Eq. (6.5) above. The P-state components go like $(\text{const.}\,m_\pi/M)^2\,\frac{e^{-x}}{x}\,u(r)$. The amounts of S, D, ^3P and ^1P state are calculated to be 93.08%, 6.38%, 0.53% and 0.01%.

In a contribution to this conference, Gross (Gro 74b) evaluates the correction to the magnetic moment due to the P-wave admixtures. He finds a correction to Eq. (2.2),

$$\Delta\mu_G = \int_0^\infty dr\,\frac{Mr}{\sqrt{3}}\{v_t(\frac{u}{\sqrt{2}} - w) - v_s(u + \frac{1}{\sqrt{2}}w)\} + \ldots$$

$$\approx 0.015 \qquad (6.8)$$

neglecting terms of order v_t^2, p^2/M^2, etc. This correction is larger than the spin-orbital correction of Feshbach, and so is very significant.

In another contribution, Dressler and Gross (DG 74) show that the large pair term (Fig. 5(a)) in the meson current corrections to radiative capture, can be formulated in the language of Gross' theory. It is likely that the presence of these odd parity components of the off-shell deuteron wave function will be important to our understanding of the deuteron.

Since the subject of nucleon-deuteron interactions has been covered in so many other sessions of this conference, it will not be necessary for me to say very much about this. I would like to comment on just three points which bear on our knowledge of the deuteron. The scattering matrix formalism for N-D scattering is given in the classic paper by Seyler (Sey 69). This is one of those lucky instances where a good formalism has been available to guide the presentation of the present crop of phase shift analyses, which are becoming more and more detailed (Arv 74, Sch+ 73).

The phase shift analysis can be simplified if the higher partial waves phase shifts are taken from a peripheral collision model, i.e., the neutron exchange pole in p-D scattering shown in Fig. 7(a). This is the analogue of the one pion exchange pole, used to great advantage in the phase shift analysis of N-N scattering by Cziffra,

MacGregor, Moravcsik and Stapp (Czi+ 59). In our case, the pole is very close to the backward direction and makes its presence felt in the angular distribution. Following Amado (Ama 63) the pole term is

$$T_{ex} \sim |N|^2/[B + \frac{p^2}{4m} + \frac{p^2}{m}(1 + \cos\theta)] \qquad (7.1)$$

with

$$N = \lim_{q^2 \to -B/2} [(2q^2 + B)\phi(q^2)] = \lim_{r\to\infty} e^{\alpha r}\phi(r) \quad , \qquad (7.2)$$

the asymptotic normalization of the deuteron wave function. Fuda (Fud 73) has worked out the pole term phase shift for neutron-deuteron scattering:

$$\delta_L(p) = (-)^L(\frac{4\Lambda N^2}{3})Q_L(\frac{\frac{5}{4}p^2 + B}{p^2}) \qquad (7.3)$$

where Λ is -1 or $\frac{1}{2}$ for the $S = \frac{3}{2}, \frac{1}{2}$ cases respectively. The constant N should be very well given by the effective range theory.

For proton-deuteron scattering, coulomb distortion effects can be taken into account following the suggestion of Bolsterli and Hale (BH 72) (actually made for the case of p-^{3}He scattering, Fig. 7(b)). They note that the phase shift, Eq. (7.3), is just that due to a Yukawa potential,

$$V_{BH}(r) = (-)^L K e^{-mr}/r \qquad (7.4)$$

with

$$K = \frac{3}{2}N^2\Lambda \quad \text{and} \quad m^2 = (2mB + \frac{p^2}{2}) \quad . \qquad (7.5)$$

The $(-)^L$ comes in because it is an exchange pole. If the coulomb force is added to V_{BH}, and the Schroedinger equation solved, the nuclear-coulomb interference effects are included. As suggested by Bolsterli and Hale, and also by Kim and Tubis (KT 72), the modified phase shift analysis can be carried out treating N^2 as a variable to be optimised, providing an alternative measurement.

Also on the subject of coulomb distortion effects, I would like to suggest that in place of the Blatt-Beidenharn phase shift convention suggested by Seyler, one should consider using the Stapp convention. Stapp expresses the S-matrix in the form

$$S = e^{i\Delta}Ee^{i\Delta} \qquad (8.1)$$

where $\Delta = \begin{pmatrix} \delta_1 & 0 & \cdots \\ 0 & \delta_2 & \cdots \\ \vdots & \vdots & \end{pmatrix}$ is a diagonal phase shift matrix and $E = e^{2iC}$ with

$$C = \begin{pmatrix} 0 & \alpha & \beta & \cdots \\ \alpha & 0 & \gamma & \cdots \\ \beta & \gamma & 0 & \cdots \\ \vdots & \vdots & \vdots & \end{pmatrix}$$

is a symmetric unitary matrix. For two coupled channels the result is

$$S = \begin{pmatrix} \sqrt{1-\rho^2}\, e^{2i\delta_1} & i\rho\, e^{i(\delta_1+\delta_2)} \\ i\rho\, e^{i(\delta_1+\delta_2)} & \sqrt{1-\rho^2}\, e^{2i\delta_2} \end{pmatrix} \quad (8.2)$$

but for three channels I have not seen the explicit form worked out. Here ρ is $\sin 2\alpha$ and α (or ρ) is called a coupling parameter. When inelasticity is present, S is no longer unitary. A very unsatisfactory way to allow for inelasticity is to allow the phase parameters ρ, δ_i to become complex. This is because it leads to complicated expressions and the physical significance of the parameters so introduced is not at all clear. In the case of nucleon-nucleon scattering at high energies, an elegant parametrization has been found by Ueda, Nack and Green (UNG 73). They write

$$S = \begin{pmatrix} (1 - \frac{r_1}{r_2}\rho^2)^{1/2}\, r_1 e^{2i\delta_1} , & i\rho\sqrt{r_1 r_2}\, e^{i(\delta_1+\delta_2+\phi)} \\ i\rho\sqrt{r_1 r_2}\, e^{i(\delta_1+\delta_2+\phi)} , & (1 - \frac{r_2}{r_1}\rho^2)^{1/2}\, r_2 e^{2i\delta_2} \end{pmatrix} \quad (8.3)$$

The inelastic cross section can then be expressed very simply, as a sum of terms $\frac{\pi}{4k^2}(1 - r_i^2)$ in terms of the 'reflection parameters' r_i. These are a simple generalization of the uncoupled case (limit $\rho \to 0$). It would be very useful if a similar expression could be found for the three channel case of interest in N-D scattering.

Although it is conventional to think of the deuteron as a bound n,p pair, quantum mechanics allows the deuteron wave function to possess components in other channels. Figuratively, we can imagine it as containing part of the time, nucleon isobars. Such components seem to have been first suggested by Arenhövel, Danos and Williams, and Kisslinger and Kerman (ADW 70, Kis 69). Kisslinger has emphasized the role of the N*(1688), with $J = \frac{5}{2}$, as forming additional D-wave components in the deuteron, which can contribute to the magnetic moment. In the reference cited, Arenhövel et al. considered the Δ(1236), N*(1470), N*(1550) components as well. About half the magnetic moment discrepancy could be accounted for by such admixtures. Kerman and Kisslinger (KK 69) invoked the N*(1688) admixture to account for the backward peak in high energy p-d scattering. That this is necessary has been disputed, however. More recently Miller and Arenhövel (MA 73) have calculated the effect on deuteron photodisintegration. At lower energies they find a decrease in σ_{TOT} due to reduced probability of the normal components. At high energies, however, σ_{TOT} is enhanced (5% at 120 MeV), and $d\sigma/d\Omega$ is more forward peaked. These trends are favoured by experiment.

Undoubtedly such resonance components should exist; and more work will be directed towards finding situations where their presence can be unambiguously detected. A recent experiment by Horne et al. (Hor+ 74) has looked for Δ-Δ components by scattering pions in a deuterium bubble chamber. The evidence is not conclusive, due to a background from N* production in π-N collisions, but a $P_\Delta \lesssim 0.7\%$ can be deduced under certain assumptions. The N*(1688) may be a more favoured component because, having $T = \frac{1}{2}$, it can exist singly in the T=0 ground state.

So far, most calculations of isobar admixtures have used perturbation theory. To the extent that one believes that a large part of the intermediate range NN force is due to virtual isobar states, one should attempt a coupled channels formalism, where there are NN and NN* components for the two nucleon wave function. The potential becomes a matrix

$$V \rightarrow \begin{pmatrix} V_{NN} & V_{NN^*} \\ V_{NN^*} & V_{N^*N^*} \end{pmatrix} . \qquad (9.1)$$

The usual two body force is then the effective NN-channel interaction

$$V_{eff} \sim V_{NN} + |V_{NN^*}|^2 \frac{1}{H_{ON^*} - V_{N^*N^*}} . \qquad (9.2)$$

The coupling potential can be taken from perturbation theory, and V_{NN} can be adapted from current phenomenological potential models. Such an approach has been investigated by Jena (Jen 73).

The deuteron continues to attract vast amounts of attention. There are no lack of ideas of how to solve the many puzzling discrepancies between theory and experiment. Let us hope that enough persistent effort will go into pursuing these, that the next few years will bring a sharp increase in our understanding of the deuteron.

AD 65 R.J. Adler and S.D. Drell Phys. Rev. Lett. 13 (1964) 349

Adl 66 R.J. Adler Phys. Rev. 141 (1966) 1499

ADW 70 H. Arenhövel, M. Danos and H.T. Williams Phys. Lett. 31B (1970) 109

Ama 63 R.D. Amado Phys. Rev. 132 (1963) 485

Arv 74 J. Arvieux Nucl. Phys. A221 (1974) 253

Bag+ 73 J.E.E. Baglin, R.W. Carr, E.J. Bentz and C.P. Wu Nucl. Phys. A201 (1973) 593

Ber+ 73 R.W. Berard, F.R. Buskirk, E.B. Dally, J.N. Dyer, X.K. Maruyama, R.L. Topping and T.J. Traverso Phys. Lett. 47B (1973) 355

Ber 74 J.C. Bergstrom Phys. Rev. C 9 (1974) 2435

BFHN 72 W. Bertozzi, J.L. Friar, J. Heisenberg and J.W. Negele Phys. Lett. 41B (1972) 408

BG 71 R. Blankenbecler and R.J. Gunion Phys. Rev. D 4 (1971) 718

BH 72 M. Bolsterli and G. Hale Phys. Rev. Lett. 28 (1972) 1285

BR 68 G. Breit and M.L. Rustgi Phys. Rev. 165 (1968) 1075

Bre 47 G. Breit Phys. Rev. 71 (1947) 400

BTL 72 T.J. Brady, E.L. Tomusiak and J.S. Levinger Bull. Am. Phys. Soc. 17 (1972) 438

BW 52 J.M. Blatt and V.F. Weisskopf Theoretical Nuclear Physics, Wiley, N.Y. 1952

Cho+ 74 K.F. Chong, M.C. Phenneger, Y.M. Shin, D.M. Skopik and E.L. Tomusiak Nucl. Phys. A218 (1974) 43

CWC 65 A.E. Cox, S.A.R. Wynchank and C.H. Collie Nucl. Phys. 74 (1965) 497

Czi+ 59 Criffra, MacGregor, Moravcsik and Stapp Phys. Rev. 114 (1959) 880

Dar 70 S.E. Darden, in "Polarization Phenomena in Nuclear Reactions", editors H.H. Barschall and W. Haeberli, Univ. Wisconsin Press (1970) 39

DG 74 E. Dressler and F. Gross Contribution to this conference

Dor+ 65 Dorfan, Eades, Lederman, Lee and Ting Phys. Rev. Lett. 14 (1965) 310

DW 64 S.D. Drell and J.D. Walecka Ann. Phys. (N.Y.) 28 (1964) 18

Fes 57 H. Feshbach Phys. Rev. 107 (1957) 1626

Fri 73a J.L. Friar Phys. Lett. 43B (1973) 108

Fri 73b J.L. Friar Ann. Phys. (N.Y.) 81 (1973) 332

Fud 73 M.G. Fuda Phys. Lett. 47B (1973) 288

Gal+ 71 Galster, Klein, Moritz, Schmidt and Wegener Nucl. Phys. B32 (1971) 221

GB 66 R.C. Greenwood and W.W. Blake Phys. Lett. 21 (1966) 702

GC 67 F. Gross and B.M. Casper Phys. Rev. 155 (1967) 1607

GH 73 M. Gari and A.H. Huffmann Phys. Rev. C 7 (1973) 994

GK 62 N.K. Glendenning and G. Kramer Phys. Rev. 126 (1962) 2159

Gou 63 M. Gourdin Nuovo Cimento 28 (1963) 533

Gou 64a M. Gourdin Nuovo Cimento 32 (1964) 493E

Gou 64b M. Gourdin Nuovo Cimento 33 (1964) 1391

Gou 65a M. Gourdin Nuovo Cimento 35 (1965) 1105

Gou 65b M. Gourdin Nuovo Cimento 36 (1965) 129

GP 64 M. Gourdin and C.A. Pecketty Nuovo Cimento 32 (1964) 1137

Gro 64 F. Gross Phys. Rev. 136 (1964) B146

Gro 66a F. Gross Phys. Rev. 142 (1966) 1025

Gro 66b F. Gross Phys. Rev. 152 (1966) 1517E

Gro 69 F. Gross Phys. Rev. 186 (1969) 1448

Gro 74a F. Gross Phys. Rev. D 10 (1974) issue No 1

Gro 74b F. Gross Contribution to this conference

GY 69 H. Grotch and D.W. Yennie Rev. Mod. Phys. 41 (1969) 350

Had 73 E. Hadjmichael Phys. Lett. 46B (1973) 147

HFY 72 Y. Horikawa, T. Fujita and K. Yazaki Phys. Lett. 42B (1972) 173

HG 73 J. Hornskin and F. Gross Phys. Lett. 47B (1973) 205

HJ 62 T. Hamada and I.D. Johnston Nucl. Phys. 34 (1962) 382

Hoc+ 73 J. Hockert, D.O. Riska, M. Gari and A. Huffmann Nucl. Phys. A217 (1973) 14

Hor+ 74 C.P. Horne and 18 others Phys. Rev. Lett. 33 (1974) 380

HW 67 T. Houk and R. Wilson Rev. Mod. Phys. 39 (1967) 547

Jan 56 Z. Jankus Phys. Rev. 102 (1956) 1586

Jen 73 S. Jena Thesis Carnegie Mellon Univ. (Univ. Microfilms Order #73-22869)

Kis 69 L.S. Kisslinger Phys. Lett. 29B (1969) 21

Kla 74 S. Klarsfeld Orsay preprint IPNO/TH 74-5 (1974)

KSH 74 L.D.Knutson, E.J.Stephenson and W.Haeberli, Phys. Rev. Lett.32 (1974) 690

KK 69 A.K. Kerman and L.S. Kisslinger Phys. Rev. 180 (1969) 1483

KT 72 Y.E. Kim and A. Tubis Proceedings Los Angeles Conf. on Few Particle Problems, North Holland (1972) 671

Lev 69 J.S. Levinger Phys. Lett. 29B (1969) 216

Lev 74 J.S. Levinger, in "Two-Nucleon and Three-Nucleon Systems", Springer, Berlin (1974) Ch. 2

Lin 65 I. Lindgren, in K. Seigbahn's "Alpha, Beta and Gamma Ray Spectroscopy", p. 1623

MA 73 H.G. Miller and H. Arenhövel Phys. Rev. C 7 (1973) 1003

Mas+ 65 Massam, Muller, Righini, Schneegans and Zichichi Nuovo Cimento 39 (1965) 10

MG 74 M.J. Moravcsik and P. Ghosh Phys. Rev. Lett. 32 (1974) 321

MW 69 C. Michael and C. Wilkin Nucl. Phys. B11 (1969) 99

Noy 65 H.P. Noyes Nucl. Phys. 74 (1965) 508

Noy 72 H.P. Noyes Ann. Rev. Nuc. Sci. 22 (1972) 465

Par 64 F. Partovi Ann. Phys. 27 (1964) 79

Pre+ 65 Prestwich, Kennett, Hughes and Fiedler Can. J. Phys. 43 (1965) 2086

PW 74 M.C. Phenneger and R.G. Winter Nucl. Phys. A219 (1974) 589

Rei 68 R.V. Reid Ann. Phys. 50 (1968) 411

RM 74 E.A. Remler and R.A. Miller Ann. Phys. (N.Y.) 82 (1974) 189

RV 72 R.V. Reid and M.L. Vaida Phys. Rev. Lett. 29 (1972) 494

SB 51 E.E. Salpeter and H.A. Bethe Phys. Rev. 84 (1951) 1232

Sch+ 72 P.A. Schmelzbach, W. Gruebler, R.E. White, V. König, R. Risler and P. Marmier Nucl. Phys. A197 (1972) 273

Sey 69 R.G. Seyler Nucl. Phys. A124 (1969) 253

TS 73 R. de Tourreil and D.W.L. Sprung Nucl. Phys. A201 (1973) 193

TSR 74 R. de Tourreil, D.W.L. Sprung and B. Rouben Contribution to this conference

UNG 73 T. Ueda, M.L. Nack and A.E.S. Green Phys. Rev. C 8 (1973) 2061

Wil 63 R. Wilson, "The Nucleon-Nucleon Interaction", Interscience, N.Y. (1963) 13

Won 59 D.Y. Wong Phys. Rev. Lett. 2 (1959) 406

WS 71 B. Weissman and H.L. Shultz Nucl. Phys. A174 (1971) 129

YLR 57 D.R. Yennie, M.M. Levy and D.G. Ravenhall Rev. Mod. Phys. 29 (1957) 144

TABLE 1

MEASURED DEUTERON PROPERTIES

$E_D = -2.22461(7)$ MeV (Pre+ 65, GB 66)

$\mu_D = 0.8577406(1) \frac{e\hbar}{2M_pc}$ (Lin 65)

$Q = 0.2875(20)$ e fm^2 (RV 72)

$r_c = 2.0952(60)$ fm (Ber+ 73)

$r_D = 1.9635(45)$ fm (Ber+ 73)

$a_t = 5.411(4)^{(a)}, 5.413(5)^{(b)}$ fm

$\rho(-B, 0) = 1.747(4)^{(a)}, 1.748(6)^{(b)}$ fm

$\rho(-B,-B) = 1.764(6)^{(b)}$

(a) HW 67 : (b) Noy 72

$P_D = 3.3 - 9\%$

TABLE 2

DEUTERON PROPERTIES CALCULATED FOR SOME WELL KNOWN PHENOMENOLOGICAL POTENTIALS

	E_D	p_D	A_D/A_S	Q	$\Delta\mu_{LS}$	$\Delta\mu_{LS}^{(F)}$
Reid S.C.	-2.225	6.471	.0263	.280	.0066	.0080
Reid SCA	-2.225	6.218	.0260	.276	.0093	.0123
Reid HC	-2.225	6.498	.0259	.277	.0073	.0106
Yale	-2.196	6.961	.0252	.276	.0066	.0123
deTS	-2.225	5.452	.0255	.279	.0076	.0096
deTSR	-2.225	5.802	.0260	.279	.0089	.0116
Ham-Jo (a)	-2.253	7.002	.0285	.285	.0024	.0034
Ham-Jo (b)	-2.224	6.951	.0264	.285	.0024	.0033

[a] using Hamada Johnston values μc^2=139.4 MeV, $\hbar/\mu c$=1.415 fm, M/μ=6.73

[b] using Humberston's values $\hbar c$=197.32857 MeV fm, $\hbar^2/M$=41.47 MeV fm^2, $\mu c/\hbar$=.7082 fm^{-1}

Fig. 1: Deuteron Radial Wave Functions for Several Well Known Potential Models

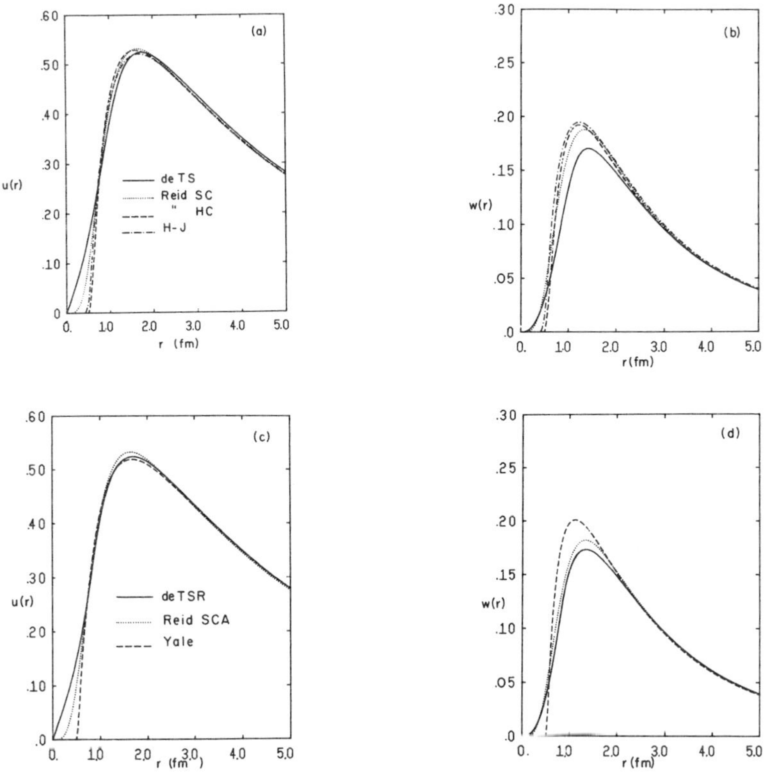

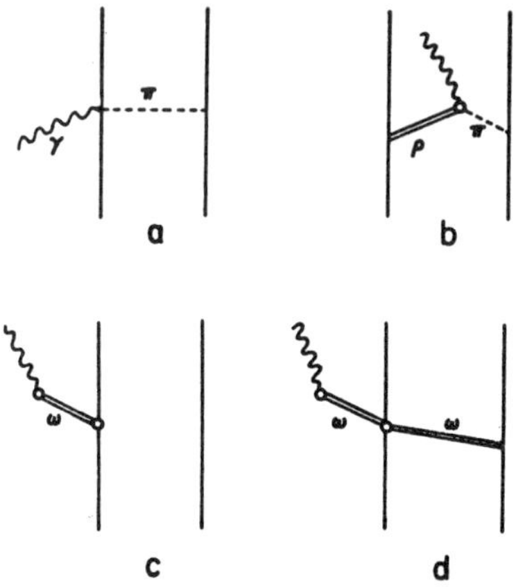

Fig. 2: Processes Contributing to Deuteron Magnetic Moment, Described in Text

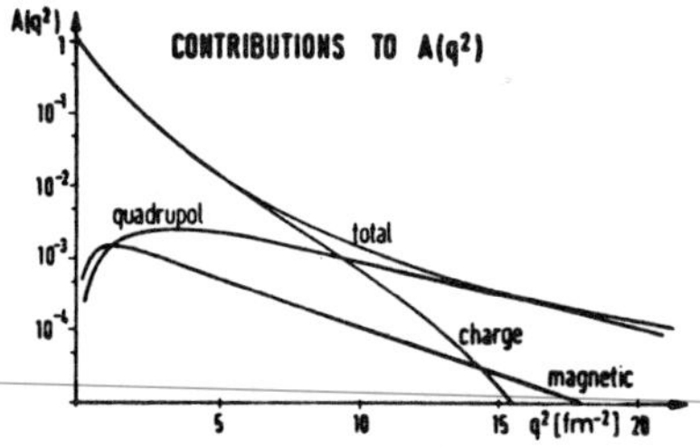

Fig. 3: Contributions to $\tilde{A}(q^2) = A(q^2) + B(q^2)$, from Gal+ 71

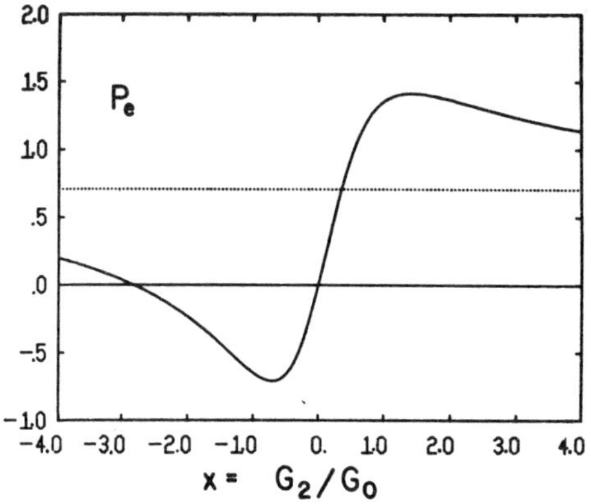

Fig. 4: Behaviour of Deuteron Polarization P_e as a Function of $x = G_2/G_0$

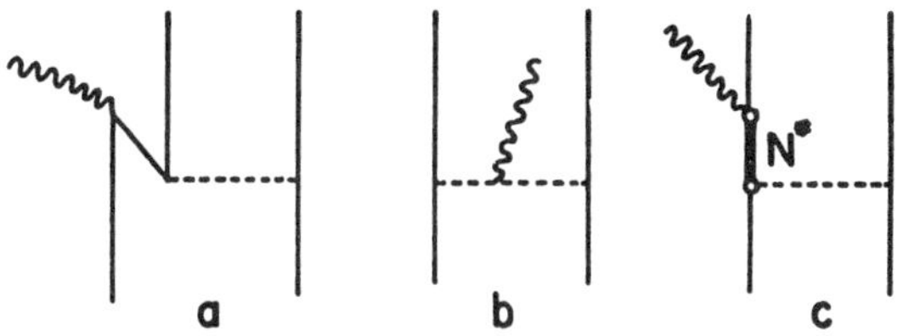

Fig. 5: Processes Contributing to Deuteron Photodisintegration

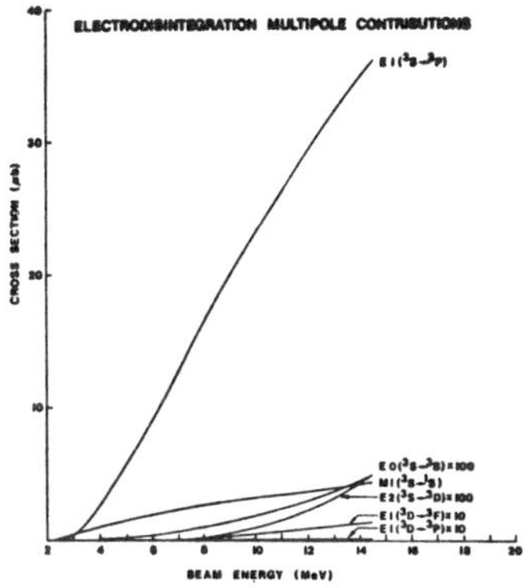

Fig. The multipole contributions to the ^{2}H(e, e'n) cross section, integrated over angles and over final electron energy.

Fig. 6: Multipole Contributions to Deuteron Electrodisintegration

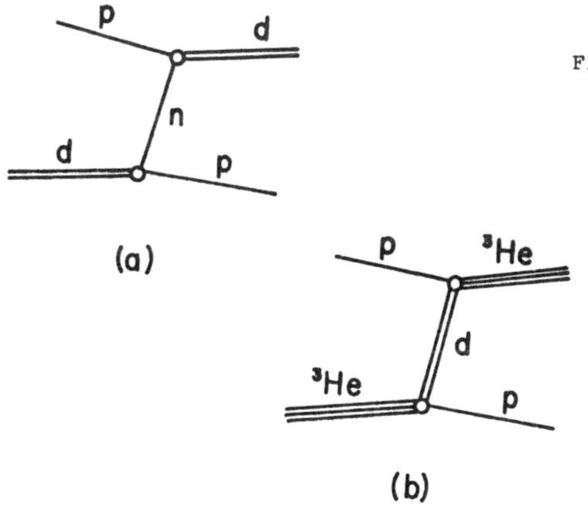

Fig. 7: Pole Terms Contributing to Backward (a) p-d, (b) p-^{3}He Scattering

DISCUSSION

MATTHEWS(MIT) I'd like to correct a slightly misleading impression you gave regarding the deuteron photo-disintegration problem. In addition to the Saskatchewan electro-disintegration results, there are two recent accurate photo-disintegration measurements: one of Ahrens et.al., at Mainz, for E = 10-30 MeV for which the data points lie right on Partovi's curve; and also an experiment from Saclay which, if anything, is slightly higher than Partovi, but within systematic errors, probably agrees. So not all photon workers have retreated in disarray.

SPRUNG: It should be pointed out that the written text of my talk differs at this point from what I said orally, namely that it now seems to be agreed that the photo-disintegration experiment of Buglin et al. must be in error. This resolves the apparent discrepancy discussed in the written text. I am certainly grateful to you for pointing out these additional experiments which further establish the consistency of electro- and photo-disintegration.

NOYES(Stanford): Two comments: as Wong showed a long time ago one can get the S/D asymptotic ratio quite accurately from the π exchange pole by a dispersion calculation; this would presumably improve the lower limit on the deuteron quadrupole moment if included in the calculation. Is that correct ?

SPRUNG: Yes, but unfortunately that minimum falls at just about 0.029 anyway. Maybe the lower limit on P_D would be 3.5 %.

NOYES: Oh, that does not help much. The other comment: since my value of 305 mb, for n-p → γd was quoted, I would like to retract it. This is the correct value given by the Bethe-Longmire shape independent formula. If one believes one should calculate using only free particles and scattering, I would argue that the Bosco calculation, which gives 321 mb, is the correct non relativistic approximation. The correction due to a free π was calculated by Skolnick, and applied correctly, then gives the experimental value of 333 mb.

SPRUNG: Of course I was quoting your earlier paper which is part of the folklore. Thanks for the precision.

CHERTOK(SLAC): We would like to bring to the attention of this Conference our experiment which is in progress at the Stanford Linear Accelerator Centre, to measure electron-deuteron elastic scattering out to high momentum transfer well beyond $q^2 \approx 25\ fm^{-2}$. This is a precision experiment which will be exploring high q^2 to clarify the physics of the deuteron.

SPRUNG: Thank you.

LEHAR(Saclay): The quadrupole moment of the deuteron is connected with the parameters ε_1 of N-N scattering for T = 0. This parameter is negative at -2.2 MeV, it is zero for zero energy and so it will be positive for small energies of N-N scattering. On the other hand, most of the phase shift analyses give a negative sign for ε_1 at low energies. Do you have an explanation for this effect ?

SPRUNG: Well I think so. Dr. Conzett gave us the explanation that the phase shift analyses do not have a sharp minimum for ε_1, they just have a broad through from -12 to +10 degrees, from the lack of experimental data to determine that parameter in the phase shift analyses. The phase shift analyses are certainly consistent with the positive values and, theoretically, I agree with you that it has to be positive.

SRIVINASA (Madras); In the case of $\gamma d \to d\pi$ we find that the ORSAY differential cross section data at fixed momentum transfers (250 MeV/c, and 350 MeV/c) are in good agreement with theory, when we make use of deuteron wave functions which have about 7 % D-state admixture. (ref: Prog.Theo.Phys., 52(no.1)1974).

SPRUNG: I think there is no doubt that you can fit things with large D-state percentages, but, I am not sure that proves that you have to have one.

MILJANIC(Rice): In respect to the n-p capture, I got the impression that you understand everything at thermal energies. As an experimentalist, I would like to know whether that means that we cannot learn anything new from the experiments at higher energies.

SPRUNG: Well, personally I don't understand everything, but I think that in general it seems that the theory is in good accord with experiments for this type of processes up to certainly 100 MeV. Maybe Dr. Shin has a comment on some of these questions.

SHIN(Saskatchewan): I don't have a comment on this question, but I would like to make a remark to theoreticians here concerning the photo-disintegration of the deuteron. The photo-disintegration cross section is quite sensitive to the electromagnetic structure of nucleons, at say $\geqslant$ 100 MeV. Hence before introducing effects such as the Δ-resonance, exchange currents, etc, one should include the well-known structure effects; only then introduce dynamic effects.

THOMAS(British Columbia): I have one comment and a question. First, I am somewhat surprised that little mention was made of the OBEP works which seem to produce lower values of p_D. It is well-known that a somewhat lower value of this parameter can produce significant increases in nuclear binding energies.

My question concerns the work by Michael and Wilkin which you mentioned. I recall that they showed a 'canonical' D-state probability of about 6% could fill in the minimum. Do you know of any calculation of this cross section using a lower value of p_D ?

SPRUNG: No, I don't. As for the first point I was not discussing any potential models. Our own supersoft core OBEP force gives p_D 5.8%, not much different from other forces.

RISKA(New York State): I would like to answer this question. At higher energies for np-dγ, the electric transitions, which are pretty well understood, dominate the reaction. Since $d\gamma \to np$ is well understood at higher energies, $np \to d\gamma$ should be so too. Let me also draw some attention to the richer reaction $ed \to epn$, which has one more variable under control, i.e. the photon 4-momentum. In this reaction one can switch off the electric transitions completely by going to the backward direction. The magnetic transitions are very much affected by the exchange currents. And at large momentum transfers the exchange current effects dominate. It is much easier to look at exchange current effects here than in the complicated three body-system.

SPRUNG: I think, I skipped over that part of my talk.

MATTHEWS: I'd like to answer the question as to whether higher energy n-p capture experiments are useful from an experimental viewpoint: that such data would be very useful in checking the photo-disintegration measurements which are in even greater disarray in the 100 MeV region. The trouble with photon measurements is that you must rely on theoretical bremsstrahlung spectra to determine your incident flux. Some workers at Birmingham University are planning an experiment on $^1H(n,d)\gamma$: they detect the recoil deuterons and normalize the capture cross section to the well known elastic n-p scattering cross-section. I think such measurements will be very valuable in the few-nucleon photo-disintegration problems.

SPRUNG: Thank you very much for that explanation.

SHIN: Concerning the importance of exchange current effects, measurements of polarization of photo-nucleons in the photo-disintegration of the deuteron is quite sensitive to the M1 transitions, hence the exchange current contributions seen in the thermal n-p capture can be seen at high energies. If there is an intense source of polarized neutrons, an n-p capture experiment with the polarized source would give a solution to this problem.

Note added in Proof:

Prof. F. Coester has kindly pointed out that the paragraph preceding eq. (4.1) ends on a misleading note. It is clear from reading Friar's paper that he has obtained all but one of Gross' relativistic corrections. The missing term involves $V(r)/mc^2$ and is called I_v. It will contribute terms similar to the spin-orbital ones eq. (4.14). According to Coester's calculations, this is the largest part of the relativistic correction for $q^2 \sim 10$ fm^{-2}.

DETERMINATION OF THE COUPLING CONSTANT FOR THE VERTEX (tdn) FROM THE SOLUTION OF THE FADDEEV EQUATIONS

V.B.Belyaev, B.F.Irgasiev*, Yu.V.Orlov**

We report the result of calculation the coupling constant G^2, which expressed by the amplitude of the virtual decay $t \rightleftarrows d+n$ and relates to C^2 - the coefficient in asymptotic part of the radial wave function of the d-n relative motion: $G^2=3C^2\pi\kappa(h/\mu c)^2$, where $\kappa=\sqrt{2\mu(\epsilon_t-\epsilon_d)}/h$, μ -the reduced mass, ϵ_i the binding energy of a nucleus i. As soon as the quantity G^2 is determined[1] from experimental data ($G^2 \approx 1.02$-1.13 fm) for different nuclear reactions, one should consider G^2 as an independent physical characteristic of the three-nucleon system. Some time ago G^2 was calculated with the use of the Faddeev equations solution by Kim and Tubis[2] for the Reid potential ($G^2=1.02$ fm) and by the authors[3] of this report (O.Yu.V. and V.B.B.) for the Malfliet-Tjon potential[4] ($G^2 \simeq 1.6$ fm) and the Darewich-Green potential[5] ($G^2 \simeq 1.4$ fm). Bearing in mind the sensitivity of the G^2 to the potential form we calculated the vertex functions $W_{1,2}(Q^2)$ (see fig.2) and function $G(q,Q) = (Q^2+\kappa^2)v(q,Q)/\phi_d(q)$ (see fig.1), with the modified version[7] of the Bressel-Kerman-Rouben (BKR) potential (here $v(q,Q)$ [6] and $\phi_d(q)$ are the tritium and deuteron space wave functions). For this potential we found the value $G^2 \simeq 1$ fm from the equation $G=W_{1,2}(Q^2=-\kappa^2)=G(q,Q=i\kappa)$ with the direct use of Faddeev equations for the unphysical region ($Q^2<0$) .

The polynomial extrapolation of the two-dimensional function $G(q,Q)$, based on the numerical calculations in the physical region only gives too large value $G^2 \simeq 1.6$ fm for the BKR potential due to nonmonotonic behaviour of $G(q,Q)$ for $Q^2 \leq 0$ (see fig.1).

So it seems reasonable to revise the results[2,3], obtained by such a kind of extrapolation.

The authors would like to thank Dr. H.Schulz for help. Two of us (I.B.F. and O.Yu.V.) gratefully acknowledge Prof. V.G.Soloviev for the kind hospitality.

REFERENCES

1) M.P.Locher, Nucl.Phys., B23, 116 (1970); E.I.Dolinsky et al. Nucl. Phys. A202 (1973) 97; L.S.Kisslinger, Phys.Rev.Lett., 47B (1973)93; I.Borbely, Preprint JINR E4-7993, Dubna 1974.
2) Y.E.Kim, A.Tubis, Phys.Rev.Lett., 29, 1017 (1972).
3) Yu.V.Orlov, V.B.Belyaev, JETP Lett., 17, 385 (1973).

4) R.A.Malfliet, J.A.Tjon, Nucl.Phys.A127, 161 (1969).
5) G.Darewich, A.E.S.Green., Phys.Rev., 164, 1324 (1967).
6) A.G.Sitenko, V.F.Kharchenko, N.M.Petrov, Phys.Lett., 28B, 308 (1968).
7) V.B.Belyaev, H.Schulz, Preprint JINR, E4-6353, Dubna 1972.

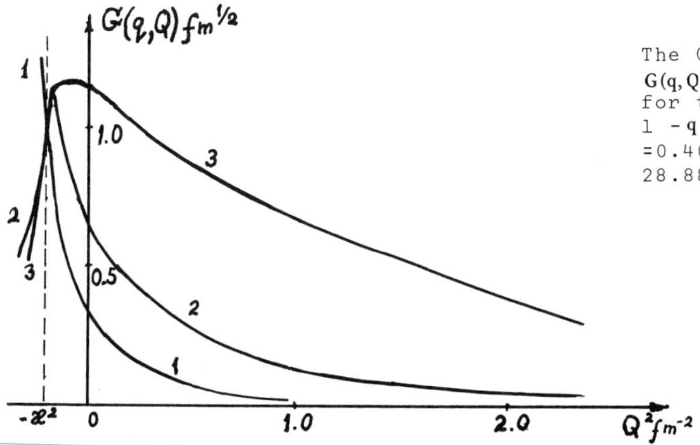

The Q^2- dependence of the G(q,Q) for different q for the BKR-potential. 1 - q =0.078fm^{-1}, 2 - q = =0.405fm^{-1}, 3 - q =2.553, 28.887 f^{-1}.

Fig.1

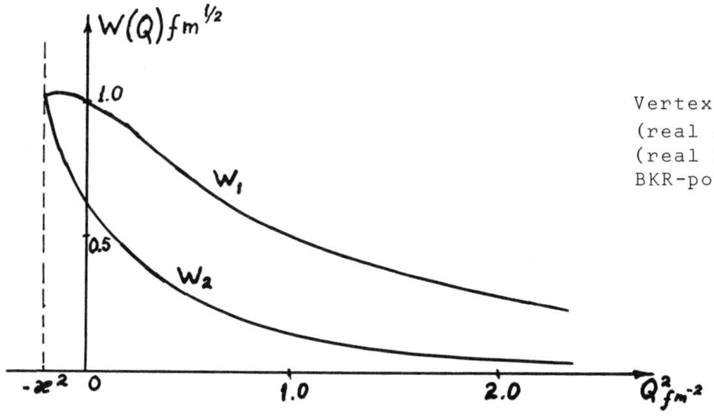

Vertex form factors $W_1(Q^2)$ (real deuteron) and $W_2(Q^2)$ (real neutron) for the BKR-potential.

Fig.2

FOUR BODY FORCES IN NUCLEAR MATTER

D. W. E. Blatt and R. G. L. Hewitt, Department of Theoretical Physics, University of Sydney, N.S.W. 2006, Australia and B. H. J. McKellar, School of Physics, University of Melbourne, Parkville, Victoria 3052, Australia.

The contribution of the three body force to the binding energy of nuclear matter has been calculated by Loiseau, Nogami and Ross[1)] (LNR) using an effective potential method. Similar calculations[2)] have been made using realistic internucleon correlations instead of the simple cutoffs used by LNR. It was found that using the correlations[3)] obtained by solving the Bethe-Goldstone equation with the Reid hard core potential gave a binding energy contribution of -1.7 MeV. This result could be duplicated by using LNR type cutoffs of 0.8 fm. A surprisingly large binding energy contribution of -6 MeV was however obtained if correlations calculated from the Reid soft core potential were used[4)].

The above method has been extended to give the binding energy contribution from the four body force. A four body potential $W(\underset{\sim}{r}_1\ \underset{\sim}{r}_2\ \underset{\sim}{r}_3\ \underset{\sim}{r}_4)$ can be defined to reproduce the 4 body scattering matrix given by the Feynman diagram of figure 1.

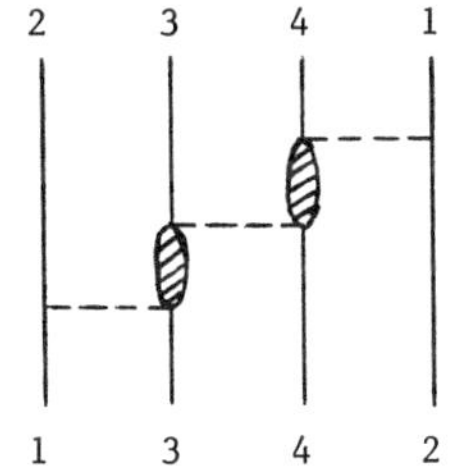

FIGURE 1

Feynman diagram defining the four body force. The blobs represent the πN scattering amplitude with the forward propagating nucleon lines removed.

An effective two body potential is defined by a weighted integration over the "spectator nucleons" 3 and 4:

$$V_{3PE}\ (r_{12}) = \mathcal{E} \iint W\ \tilde{\rho}(\underset{\sim}{r}_1\ \underset{\sim}{r}_2\ \underset{\sim}{r}_3\ \underset{\sim}{r}_4)\ d\underset{\sim}{r}_3\ d\underset{\sim}{r}_4 \quad . \tag{1}$$

The weighting function $\tilde{\rho}$ is the density distribution of the four nucleons given in the cutoff approximation by

$$\tilde{\rho}(\underset{\sim}{r}_1\ \underset{\sim}{r}_2\ \underset{\sim}{r}_3\ \underset{\sim}{r}_4) = \rho^2 \prod_{i<j} \theta(|\underset{\sim}{r}_i - \underset{\sim}{r}_j| - c) \quad . \tag{2}$$

The effect of exchanges among the spectator nucleons, a process which has no analogue in the three body case, has also been computed; it can be taken into account approximately by including an overall multiplicative factor $\mathcal{E} \approx 3/4$.

V_{3PE} is of OPEP form with a central and tensor part :

$$V_{3PE} = \underset{\sim}{\tau}_1 \cdot \underset{\sim}{\tau}_2 \, [f_c \, (r_{12}) \, \underset{\sim}{\sigma}_1 \cdot \underset{\sim}{\sigma}_2 + f_T \, (r_{12}) \, S_{12} \, (\underset{\sim}{r}_{12})] \tag{3}$$

so the binding energy calculations can be done in the usual manner. The results are given in table I for form factor II defined in LNR. $\Delta E^{(1)}$ and $\Delta E_2^{(2)}$ are the first and second order contributions and $\Delta E_1^{(2)}$ is the second order cross term between the two body potential and the effective four body potential. The total for a cutoff of 0.8 fm is approximately 4% of the corresponding three body force contribution.

	3 body	4 body	
Cutoff (fm)	0.8	0.6	0.8
$\Delta E^{(1)}$	1.0	0.26	0.16
$\Delta E_1^{(2)}$	-2.2	-0.68	-0.22
$\Delta E_2^{(2)}$	-0.1	-0.03	-0.01
Total	-1.3	-0.45	-0.06

TABLE I
Binding energy contributions from the three and four body forces (MeV).

The method has also been extended to give the effective potential for the n-body force. This effective potential also separates into a central and tensor part as in (3). Without exchanges of spectator nucleons a convergence ratio of about 0.5 is obtained. Exchanges however become increasingly important and give almost exact cancellation for $n \gtrsim 6$.

REFERENCES

1. B. A. Loiseau, Y. Nogami and C. K. Ross, Nucl. Phys. A165, 601 (1971).
2. D. W. E. Blatt, Ph.D. Thesis, University of Sydney, Australia (1974).
3. A. M. Green and T. H. Schucan, Nucl. Phys. A188, 289 (1972).
   Note correlations in only one internucleon distance were included in this work.
4. D. W. E. Blatt and B. H. J. McKellar, Phys. Lett. (to be published).

This work was supported in part by the Science Foundation for Physics within the University of Sydney.

COO-3235-44

UNITARITY MODIFICATIONS OF THE COUPLED CHANNEL T AND K OPERATOR EQUATIONS FOR MANY-BODY SCATTERING

D.J. Kouri*, Max-Planck-Institut für Strömungsforschung, 3400 Göttingen, Federal Republic of Germany, and F.S. Levin†, Physics Department, Brown University, Providence, R.I. 02912

We have recently developed a new method for treating the n-body, N-channel scattering problem[1,2,3]. The method involves the introduction of an N x N channel-coupling matrix W that links the various (arrangement) channels together and whose seemingly arbitrary elements are fixed by requiring the iterated kernel of the coupled equations for the transition (and reactance) operators to be connected[2]. Properties of these equations, their connection with other formalisms, and application to $e^- + H$ scattering have been discussed in detail elsewhere[1,3]. In this report, we show how the basic equations are modified by examining the unitarity relation obeyed by the transition operators.

The coupled equations for the transition operators $T_{jk}(Z)$ are, in matrix form,

$$T(Z) = \mathcal{V} + \mathcal{V}\mathcal{G}(Z)T(Z) \quad , \tag{1}$$

where T_{jk} is the operator describing transitions from channel k to channel j; $\mathcal{V}_{jk} = V_j W_{\ell k}$, where V_j is the perturbation in channel j which goes to zero when the separation between all the stable subsystems in channel j becomes large, and $W_{\ell k}$ is an element of the channel coupling array W [ℓ is arbitrary and may be chosen at our convenience] obeying $\sum_\ell W_{\ell k} = 1 = \sum_k W_{\ell k}$; $\mathcal{G}_{jk}(Z) = \delta_{jk}\, G_k(Z)$, with $G_k(Z) = (Z - H_k)^{-1}$, the kth channel resolvent operator [$H_k = H - V_k$ is the unperturbed channel Hamiltonian and H is the total n-body Hamiltonian]; and Z is a complex energy parameter.

The matrix of reactance operators K(E) is defined by

$$K(E) = \mathcal{V} + \mathcal{V}\mathcal{G}^P(E + i0)K(E) \, , \tag{2}$$

where $\mathcal{G}^P(E + i0) = \mathrm{Re}\,\mathcal{G}(E + i0)$ is the matrix of standing wave Green's functions and E is the total energy of the system. We also define the matrix $\Delta(E)$ as $\mathcal{G}(E + i0) = \mathcal{G}^P(E) - i\pi\Delta(E)$. Then K(E) and T(E + i0) are related by the matrix damping equation[2]

$$T(E + i0) = K(E) - i\pi K(E)\,\Delta(E)\,T(E + i0) \quad . \tag{3}$$

In the one-body scattering case, (3) is an expression of unitarity. In the n-body case, this seems not to be the case because breakup states in an n-body channel (m<n) can be counted more than once, e.g., as asymptotic states in an (m + 1)-body channel. [We define channels in the same manner as Eckstein[4].] Furthermore (3) does not lead to Lovelace's unitarity relation[5] for n = 3.

The key to understanding this is that the resolvent matrix $\mathcal{G}(Z)$ is analytic everywhere in Z but on the right hand cuts, where its boundary value $\mathcal{G}(E + i0)$ needs to be more carefully determined than merely writing $\lim_{\varepsilon \to 0^+} \mathcal{G}_{ij}(E \pm i\varepsilon) \to G_j(E + i0)\delta_{ij}$. We have examined the discontinuity of $\mathcal{G}(Z)$, $D\mathcal{G}(Z)$, across the real axis using two different methods, one of them Lovelace's[6)], and with each have found $D\mathcal{G}(Z) = 2i\, \mathrm{Im}\, \mathcal{G}_{ij}(Z) = 2i\delta_{ij}\tilde{G}_j(Z)$, where $\tilde{G}_j(Z) \equiv G_j(Z)P(j)$ and $P(j)$ is the projection operator onto the set of all internal bound states in channel j[6)]. That is, $P(j)$ includes no breakup (i.e. continuum) states in channel j. In particular, this means that the operator $\Delta(E)$ of (3), whose elements are $\Delta_{ij}(E) = \delta_{ij}\delta(E - H_j)$, is changed to $\tilde{\Delta}(E)$, where $\tilde{\Delta}_{ij}(E) = \Delta_{ij}(E)P(j)$. Such a replacement guarantees that the contribution from any unitarity cut to the discontinuity is not counted more than once.

This analysis shows that $\mathrm{Im}\, \mathcal{G}(Z) = -\pi\tilde{\Delta}(Z)$. The real part of $\mathcal{G}(Z)$ can be found using Cauchy's theorem. By construction, there are no n-body bound state poles associated with $\mathcal{G}(Z)$, since H_i contains no interactions that could bind the various stable subsystems forming the channels. The only singularities are the set of unitarity cuts along the real axis. We may thus use the contour Γ and Cauchy's theorem in the form $\mathcal{G}(Z) = (2\pi i)^{-1} \int_\Gamma \mathcal{G}(\zeta)/(\zeta - Z)$ to establish[6)] that $\mathrm{Re}\mathcal{G}_{ij}(Z) = [\mathrm{Re}\, \tilde{G}_j(Z)]\delta_{ij}$. Hence, we find that by examining the discontinuity of $\mathcal{G}(Z)$ carefully and using Cauchy's theorem, $\mathcal{G}(Z)$ should be replaced by $\tilde{\mathcal{G}}(Z)$, where $\tilde{\mathcal{G}}_{ij}(Z) = \tilde{G}_j(Z)\delta_{ij}$. The effect of this is that in all channels but the n-body, free-particle one [labelled j = 0], only bound internal states occur in the expansion of each resolvent $G_j(Z)$. A similar result holds in the differential form of our equations, a result compatible with the effective orthogonality condition found by Hahn[7)] to hold in multichannel scattering.

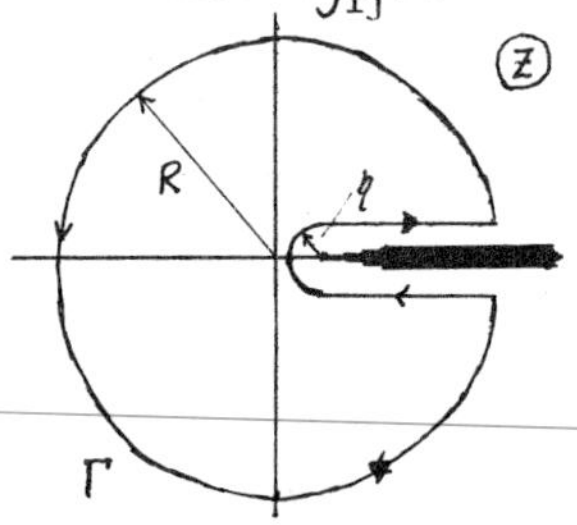

References

1. D. J. Kouri and F.S. Levin, A New Method in the Theory of Many Body Scattering, these Conference proceedings.
2. D.J. Kouri and F.S. Levin, Bull. Am. Phys. Soc. 19(1974)489; Physics Letters (in press).
3. D.J. Kouri and F.S. Levin, submitted for publication.
4. H. Eckstein, Phys. Rev. 101 (1956) 880.
5. C. Lovelace, in Strong Interactions and High Energy Physics, ed. by R. Moorhouse (Oliver and Boyd, 1964).
6. D.J. Kouri and F.S. Levin, submitted for publication.
7. Y. Hahn, Physics Letters 30B (1969) 595; Phys. Rev. C3 (1971) 1432.

---

*Present address: Departments of Chemistry and Physics, University of Houston, Houston, Texas 77004

†Work supported in part by the U.S. Atomic Energy Commission

THEORETICAL PHASE-SHIFT ANALYSIS IN THE (n-d) ELASTIC SCATTERING

C. Fayard, G.-H. Lamot and E. Elbaz

Institut de Physique Nucléaire, Université Claude Bernard Lyon-I and IN2P3
43, Bd du 11 novembre 1918, 69621 Villeurbanne, France

Many progress have been done recently in (n-d) elastic scattering phase-shift analysis. Arvieux [1)] (denoted RV hereafter) neglect the splitting and the coupling, while Schmelzbach et al [1)] (denoted SCH), analysing the available results between 3 and 6 MeV, take into account all degrees of freedom up to J = 3/2, thus giving the splitted phase shifts and the corresponding mixing coefficients.

We have compared these analyses with theoretical results yielded by solving the Lovelace equations with different non local separable interactions [1)], including the ^1S_o, $^3S_1$ - $^3D_1$ and P two-body partial waves [2)].

For ^1S_o wave, the interaction is of Mongan's type, denoted M 5, and for the P waves we use the potential of ref. 2). In the $^3S_1$ - $^3D_1$ channel, we use a one-term rank-two Serduke's interaction denoted ACS 4 and ACS 7 for P_D = 4 and 7 %, and the two-term Mongan IV parametrization with P_D = 1.4 %.

Our calculations, performed between 3 and 15 MeV, evaluate with a minimization method the eigen phase shifts and mixing coefficients up to J = 5/2 with Seyler's parametrization of the S-matrix.

Doublet phase shifts - We note (Fig. 1) that the $^2S_{1/2}$ phase shift decreases when P_D increases, the best result being obtained for P_D = 4 % with ACS 4. At 3 MeV appears a pathological behaviour of this phase shift. Fig. 2 shows the $^2P_{1/2}$, $^2P_{3/2}$ and ^2P phase shifts. Our results are in complete disagreement with SCH (as for the ^2D phase shifts) and we can note the strong dependence on the two-body P-waves. Our ^2P phase shift agrees with RV data only if the P-waves are neglected.

Quartet phase shifts - Even if SCH results cannot be reproduced completely, a global agreement is obtained with RV results, especially for $^4S_{3/2}$ (Fig. 1).

Mixing parameters - Fig. 2 shows the very good agreement with SCH results for $\varepsilon_{1/2}(^2P_{1/2} - {}^4P_{1/2})$ and $\varepsilon_1^{3/2}(^2P_{3/2} - {}^4P_{3/2})$. The other mixing parameters are not so good, especially the $\eta_{1/2}(^2S_{1/2} - {}^4D_{1/2})$ which is considerably smaller than the value predicted by SCH.

Figs. 4 and 5 show P_N neutron vector polarization and T_{20} deuteron tensor polarization calculated at 10.85 MeV. These two figures exhibit the strong influence of the coupling between odd waves.

REFERENCES

1) G.H. Lamot, C. Fayard and E. Elbaz, Lettere al Nuovo Cimento 9, No 16 (1974) 653 (In this paper are given all references concerning experimental results and used interactions).

2) J.P. Meunier, W. Plessas, J. Pigeon, R. Nahabetian, Phys. Rev. (to be publ.).

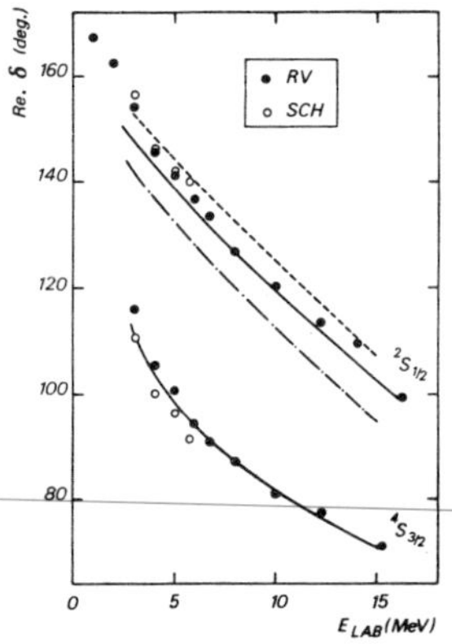

Figure 1 - S phase shifts with P-waves. (---) M IV (—) ACS 4 and (-·-) ACS 7

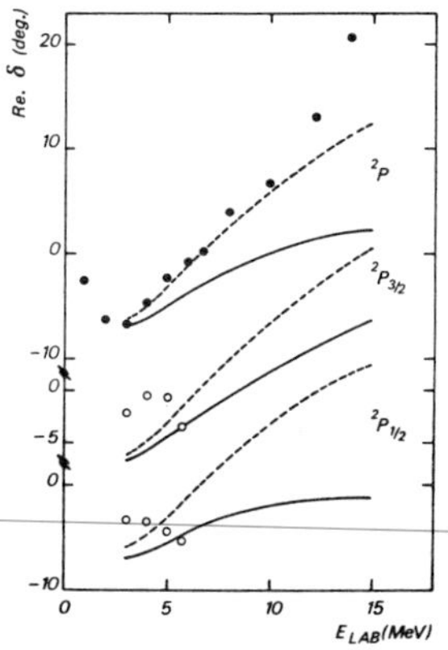

Figure 2 - P phase shifts

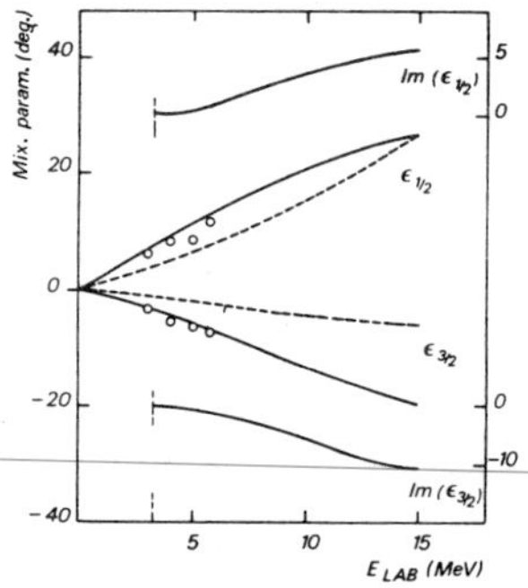

Figure 3 - Mixing param. with ACS 4 - M 5 parametrization (---) without P-waves (—) P-waves included

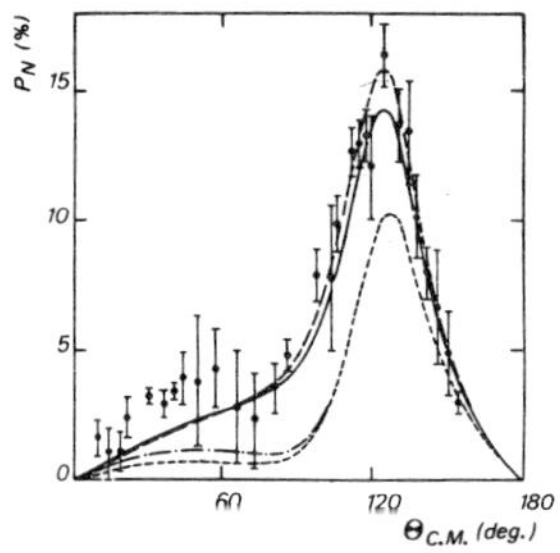

Figure 4 - Neutron vector polarization
(—) "exact" calculation
(---) all coupling neglected
(-·-) even coupling only
(—–) odd coupling only

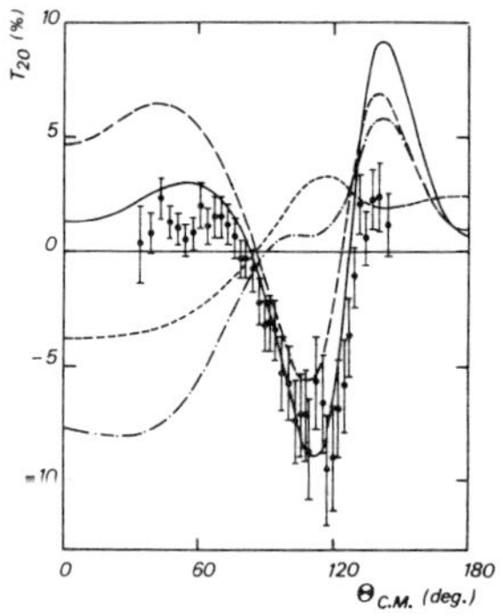

Figure 5 - Deuteron tensor polarization T_{20}. The lines are the same as in Fig. 4.

NUCLEON-DEUTERON BREAKUP CALCULATIONS WITH CHARGE DEPENDENT POTENTIALS

J.H. Stuivenberg, J. Bruinsma and R. van Wageningen
Natuurkundig Laboratorium der Vrije Universiteit,
Amsterdam, the Netherlands

During the last few years many analyses[1-6] of nucleon-deuteron breakup experiments have been performed with the Amado model using separable Yamaguchi interactions. Elsewhere we have reported the effect of taking separable interactions with other formfactors[7], which reproduce better the experimental high energy phase shifts. In accurate comparisons between theory and experiment it is evident that also the charge dependence of the nuclear forces will have to be taken into account. This means that for the doublet state of the system one has to solve a system of three coupled integral equations instead of two for charge independent potentials. In the formalism of Cahill and Sloan[8] this is called the YYY model. The breakup program of Ebenhöh[9] also contains charge dependence, but only in an approximate way. He followed a suggestion of Aaron and Amado[10], that it is a reasonable approximation to use isospin invariant interactions in the kernels of the integral equations and charge dependent ones in the inhomogenuous terms only. Effectively this means that only the important final rescattering of the particles is taken to be charge dependent. The advantage of this procedure is that in the doublet state only two coupled equations have to be solved. For the singlet interaction in the kernels a potential is taken which is fitted to the averaged singlet scattering parameters

$$a_s^{-1} = {}^2/3\ a_{np}^{-1} + {}^1/3\ a_{NN}^{-1} \qquad N = n \text{ or } p$$

$$r_{o_s} = {}^2/3\ r_{o_{np}} + {}^1/3\ r_{o_{NN}}$$

To check the reliability of this procedure which we call "hybrid", we compared the results of these calculations with exact charge dependent calculations. Therefore we extended the Ebenhöh code to three coupled equations.

As an example we consider the 16 MeV p-d breakup experiment of Klein[2], which originally has been analysed with the hybrid approach. In figs. 1a and 1b two typical situations are shown. From these figures it can be seen that in the quasi-free scattering region the approximation is not too bad (differences of about 4%). But in the np final state interaction peaks considerable differences may occur, which turn out to be strongly dependent on the production angle of the fsi pair. At large production angles (140° in fig. 1a) they can be of the order of 20%. For comparison also the charge independent calculations are shown in the figures. At other energies differences of the same order of magnitude are found.

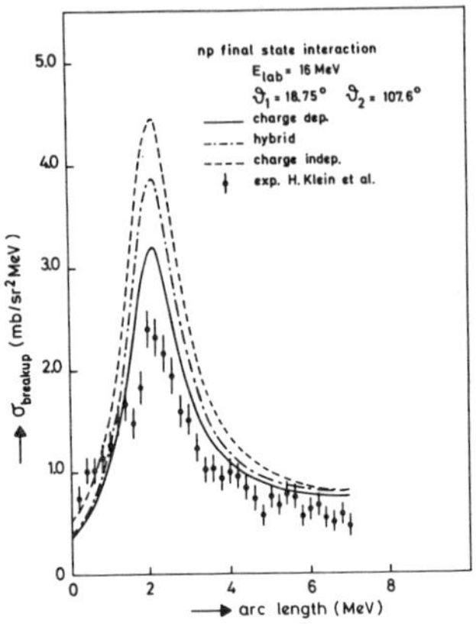

fig. 1a

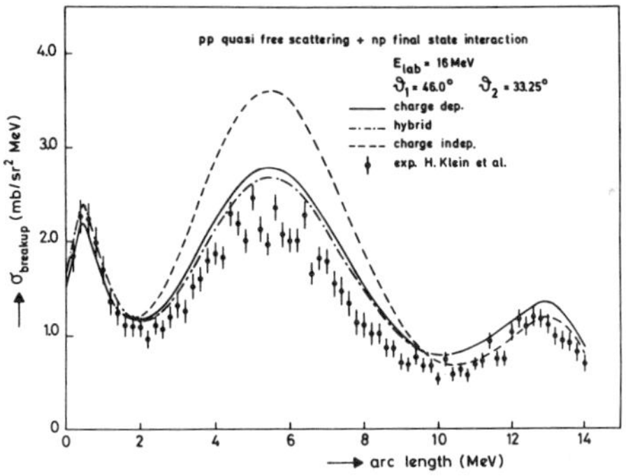

fig. 1b

It is to be expected that the hybrid approach will work somewhat better for n-d calculations since here the differences between the np singlet (a = -23.68 fm) and nn singlet (a = -16 fm) is not so large as for the p-d case (a_{pp} = -7.76 fm). Still it will be necessary to reanalyse the n-d experiments of Zeitnitz[5] with exact charge dependence. A preliminary analysis showed already that the value of $r_{o_{nn}}$, determined from this experiment, will be influenced considerably[11].

Our conclusion is that they hybrid approach is not in all cases a valid approximation. So one has to be careful in applying it, especially if one is looking for effects of other parts of the nucleon-nucleon interaction like e.g. higher partial waves or tensor interaction.

References

1) R. Bouchez, S. Desreumaux, J.C. Gondrand, C. Perrin, P. Perrin, R.T. Cahill, Nucl. Phys. A185(1972)166

2) H. Klein, H. Eichner, H.J. Helten, H. Kretzer, K. Prescher, H. Stehle, W.W. Wohlfarth, Nucl. Phys. A199(1973)169

3) G. Anzelon, I. Slaus, D.J. Margaziotis, M.B. Epstein, Nucl. Phys. A202(1973)593

4) E.L. Petersen, M.I. Haftel, R.G. Allas, L.A. Beach, R.O. Bondelid, P.A. Treado, J.M. Lambert, M. Jain, J.M. Wallace, Phys. Rev. C9(1974)508

5) B. Zeitnitz, R. Maschuw, P. Suhr, W. Ebenhöh, Proc. Intern. Conf. on Few Particle Problems in the Nuclear Interaction, Ivo Slaus et al. eds., North-Holland Publ. Comp., Amsterdam 1973, p. 117

6) W.H. Breunlich, S. Tagesen, W. Bertl, A. Chalupka, Nucl. Phys. A221(1974)269

7) J. Bruinsma, W. Ebenhöh, J.H. Stuivenberg, R. van Wageningen, to be published

8) R.T. Cahill, I.H. Sloan, Nucl. Phys. A165(1971)161

9) W. Ebenhöh, Nucl. Phys. A191(1972)97

10) R. Aaron, R.D. Amado, Phys. Rev. 150(1966)857

11) B. Zeitnitz, private communication

TRITON BINDING ENERGY AND THE DEUTERON WAVE FUNCTION

H. de Groot and H.J. Boersma

Natuurkundig Laboratorium, Vrije Universiteit
Amsterdam, The Netherlands

Recent investigations (1,2) with phase-equivalent spin-independent S-wave interactions indicate that the triton binding energy (E_T) is mainly sensitive to changes in the deuteron wave function. The results of Fiedeldey (1) show that even modifications of the two-body phaseshifts down to 20 MeV hardly influence E_T, once the deuteron wave function has been fixed. Afnan and Serduke (3) found a similar dependence with central forces in the $^3S_1$ state, having fixed the $^1S_0$ interaction. All these calculations neglect the tensor force which couples the $^3S_1$ and $^3D_1$ states. Here we consider this dependence in the more realistic case, when a tensor force is present and the deuteron properties are well reproduced. For this purpose we have constructed separable tensor potentials with the same deuteron wave function as realistic phase-equivalent potentials. These have been generated from the Reid soft-core potential by Haftel and Tabakin using the method of unitary transformations (4). We consider two sets of potentials. In the first set only the L=0 component of the deuteron wave function differs from Reid's, while in the second set only the L=2 component differs from Reid's. The form factors of the separable potentials, with the conventions of Tabakin (5), are then given by

$$g_0(p) = \lambda \left\{C(p) + \mu\frac{(p^2+k_D^2)}{(p^2+\alpha^2)^2}\left((\alpha+\beta) - \frac{4\alpha^2\beta}{p^2+\alpha^2}\right)\right\} \qquad g_2(p) = -\lambda T(p) \qquad \text{(first set)}$$

$$g_0(p) = \lambda C(p) \qquad g_2(p) = -\lambda\left\{T(p) + \mu p^2\frac{(p^2+k_D^2)}{(p^2+\alpha^2)^3}\left(1 - \frac{6\alpha\beta}{p^2+\alpha^2}\right)\right\} \qquad \text{(second set)}$$

C(p) and T(p) describe the separable potential with the deuteron wave function of the Reid potential, as given in ref. 6 (table 2). The additional terms follow from the specific unitary transformation applied to the Reid wave function; k_D is the deuteron wave number, 0.23161. The other parameters are given in table 1. We have labeled the potentials as the corresponding potentials of ref. 4. The chosen potentials give the strongest variation of the deuteron wave function. All potentials give the same D-state percentage in the deuteron, P_D = 6.57%, nearly the same electric quadrupole moment and, within experimental error, the same electric formfactor. We note that the usual separable potentials with simple analytical formfactors do not reproduce the electric formfactor (7). The scattering parameters for the potentials generally differ less than 1% at 10 MeV, and still less than 10% at 100 MeV.

The values obtained for E_T with these potentials are given in table 1. In the $^1S_0$ state we employed the unitary pole approximation to the Reid soft-core potential (8). We observe a variation in E_T of 0.88 MeV, but only 0.27 MeV if we exclude the unrealistic case 23. Two studies of the triton with some of the potentials of Haftel and Tabakin can be found in the literature. However, in ref. 9 very rough extrapolations for the E_T values had to be made. Harper, Kim and Tubis (10) obtained an accuracy of 0.3 MeV in E_T and found a variation of 0.32 MeV with a set of four tensor potentials. Our E_T values show a comparable variation, despite the fact, that the separable potentials are *not* phase-equivalent above 100 MeV. This suggests that also in the presence of tensor forces the triton binding energy is not sensitive to variations in the two-nucleon T-matrix at high positive energies, but mainly to changes in the deuteron wave function. The constraint of a realistic deuteron wave function seems to allow only a small variation of E_T, when P_D has been fixed. We note finally that the realistic deuteron properties and approximate phase-equivalence below 100 MeV of our potentials make a more general study in the three-nucleon system worthwhile.

Table 1

Parameters for the tensor potentials and triton binding energy results

	potential	$\alpha(fm^{-1})$	$\beta(fm^{-1})$	λ	μ	$\|E_T\|$ (MeV)
	R	-	-	1.63686	0.0	7.42
set 1	8	2.4	0.8	1.60336	-0.27844	7.34
	18	4.0	1.3	1.51784	3.60543	7.55
	21	2.7	0.91	1.70744	1.10755	7.61
	23	3.0	1.2	0.98696	7.90774	6.73
set 2	10	3.0	0.85	1.58668	-5.82518	7.40
	13	3.6	1.0	1.43342	-21.1780	7.34
	15	3.0	0.78	1.60386	9.49136	7.42

References.

(1) H. Fiedeldey, Lettere al Nuovo Cimento 9 (1974) 301.
(2) M.I. Haftel, Phys. Rev. C7 (1973) 80.
(3) I.R. Afnan and F.J.D. Serduke, Phys. Lett. 44B (1973) 143.
(4) M.I. Haftel and F. Tabakin, Phys. Rev. C3 (1971) 921.
(5) F. Tabakin, Ann. of Phys. 30 (1964) 51.
(6) S.C. Bhatt, J.S. Levinger and E. Harms, Phys. Lett. 40B (1972) 23.
(7) H.C. Pradhan and Y. Singh, Can. J. Phys. 51 (1973) 343.
(8) E. Harms and L. Laroze, Nucl. Phys. A160 (1971) 449.
(9) E. Hadjimichael and A.D. Jackson, Nucl. Phys. A180 (1972) 217.
(10) E. Harper, Y. Kim and A. Tubis, Phys. Rev. C6 (1972) 1601.

CRITERION FOR THE VALIDITY OF THE UNITARY POLE APPROXIMATION AND CALCULATION OF CORRECTION TERMS

B.L.G. Bakker† and W. Sandhas††

† Natuurkundig Laboratorium, Vrije Universiteit, Amsterdam, The Netherlands

†† Physikalisches Institut, Universität Bonn, GFR

Model calculations with potentials containing repulsion have shown, that the unitary pole approximation (UPA), where only the deuteron bound state and the anti-bound state are incorporated in the Faddeev kernel, yields already remarkably good results. In the following we want to explain the origin of this fact.

One way to study this problem is provided by introducing additional separable terms besides the pole terms, i.e., to expand the two-body transition operators into a series of the form (γ denotes the subsystem under consideration)

$$T_\gamma(z) \;=\; \sum_{n,m} |g_{\gamma n}\rangle \, \Delta_{\gamma,nm}(z) \langle g_{\gamma m}| \;. \qquad (1)$$

The form factors $|g_{\gamma n}\rangle$ occurring in this unitary pole expansion (UPE) are eigenfunctions of the Lippmann-Schwinger kernel $V_\gamma G_o(E)$ belonging to the eigenvalues $\eta_{\gamma n}$. If z is near to E, eq. (1) reads

$$T_\gamma(z) \sim \sum_n |\, g_{\gamma n}\rangle \frac{-\eta_{\gamma n}}{1-\eta_{\gamma n}} \langle g_{\gamma n}| \qquad (2)$$

In the case of the Malfliet-Tjon potentials I (singlet) and III (triplet)[1], the first six eigenvalues are given in the following table[2]

n	1	2	3	4	5	6
η_n^{I}	0.927	-2.717	-0.660	-0.291	-0.163	+0.115
η_n^{III}	1.000	-2.005	-0.496	-0.220	+0.159	-0.123

Proceeding in the spirit of Weinberg's argumentation[3] at least $|g_{\gamma 1}\rangle$ and $|g_{\gamma 2}\rangle$ have to be taken into account. One also expects $|g_{\gamma 3}\rangle$ to yield a significant contribution. However, we read off from the next table[2] that the effect of $|g_{\gamma 2}\rangle$ and $|g_{\gamma 3}\rangle$ on the triton binding energy E_T and the doublet scattering length ${}^2a$ is very small, a result know also from other calculations[4].

n	1	2	3
E_T	-8.63	-8.50	-8.48
${}^2a$	0.94	1.07	1.09

A suitable tool to investigate the origin of this phenomenon is provided by the quasi-Born approach[5]. Here only the pole term is extracted from T_γ,

$$T_\gamma = |\gamma 1> t_{\gamma 1} <\gamma 1| + T_\gamma' \quad , \tag{3}$$

with $|\gamma 1> = (1 + T_\gamma' G_o)|g_{\gamma 1}>$. Inserting this in the Faddeev equations we end up with an effective two-body formulation as in the UPA case. However, the effective potentials are modified now. In the first quasi-Born approximation they read, for $\beta \neq \alpha$,

$$V_{\beta\alpha} = (1-\delta_{\beta\alpha})<g_{\beta 1}|G_o|g_{\alpha 1}> + \sum_\gamma <g_{\beta 1}|G_o T_\gamma' G_o| g_{\alpha 1}> \tag{4}$$

I.e., as compared to the UPA result an additive term, linear in T_γ' occurs. A similar modification holds for the effective free Green's functions.

Expanding T_γ', in analogy to (1), into a series of separable terms which now, however, starts from n = 2, we have for the correction term

$$<g_{\beta 1}|G_o T_\gamma' G_o|g_{\alpha 1}> = \sum_{n=2}^{\infty} \sum_{m=2}^{\infty} <g_{\beta 1}|G_o|g_{\gamma n}> \Delta'_{nm} <g_{\gamma m}|G_o|g_{\alpha 1}> \tag{5}$$

For $n \geq 2$ we have $\eta_n << 1$ and consequently $|\Delta'_{nm}| < 1$. But even if Δ'_{nm} is near unity, which in our case happens for n=m=2, we have a small correction term if

$$|<g_{\beta 1}|G_o|g_{\gamma n}> <g_{\gamma m}|G_o|g_{\alpha 1}>| \quad << 1$$

The smallness of these overlap integrals, therefore, represents the decisive criterion for the validity of the UPA.

For the Malfliet-Tjon potentials I and III it turns out that only some of these expressions are of the order of few percents. In addition they partly cancel due to different signs. This explains the success of the UPA in the present case.

Taking into account only the biggest correction terms on the right-hand side of (5), and solving the effective two-body equations represents a new method for calculating three body observables. We have studied in this way the contribution due to $|g_2>$ obtaining E_T=-8.49 MeV and ^2a = 1.08 fm, in excellent agreement with column 2 of the second table. We found the same agreement when incorporating further terms.

References

1) R. A. Malfliet and J. A. Tjon, Nucl. Phys. A127, 161 (1969)
2) B. L. G. Bakker, Thesis,Vrije Universiteit, Amsterdam, 1970
3) S. Weinberg, Phys. Rev. 131, 440 (1963)
4) E. Harms and V. Newton, Phys. Rev. C2, 1214 (1970); compare also I. R. Afnan and J. M. Read, Phys. Rev. C8, 1294 (1973)
5) E. O. Alt, P. Grassberger and W. Sandhas, Nucl. Phys. B2, 167 (1967)

INFLUENCE OF THE HIGHER PARTIAL WAVES OF LOCAL POTENTIALS ON THREE-NUCLEON OBSERVABLES

E.O. Alt† and W. Sandhas††

† Institut für Physik, Universität Mainz, GFR

†† Physikalisches Institut, Universität Bonn, GFR

The unitary pole approximation (UPA) recently constructed for the Malfliet - Tjon (MT) potential has been shown to reproduce excellently the three-nucleon Padé calculations below the break-up threshold[1-4]. One therefore expects that the UPA represents a good starting point for quasiparticle calculations[5] which automatically incorporate all higher partial wave contributions of the two-body potentials.

In this paper we report results obtained in 1. quasi-Born approximation (QBA) for the local MT potential. The triton binding energy B_t and the doublet scattering length ^2a, calculated in 1. QBA, are compared in the following table with 0. QBA results (= UPA to the MT potential) and with various other calculations.

	0.QBA[2]	1.QBA	Padé[1,4]	UPE[2]	ref.6	ref.8	Yamaguchi[3]
B_t (MeV)	8.63	8.78	8.3	8.58	8.56	8.77	10.16
^2a (fm)	0.94	0.67	0.9	0.90	1.2	-	-0.41

The remarkable difference between the two separable potential results ("0.QBA" and "Yamaguchi") demonstrates the importance of adequate form factors in the driving terms. In comparison only small corrections to the 0.QBA are obtained when incorporating via unitary pole expansion (UPE) the neglected S-wave contributions of the two-body potentials. (The reason for this phenomenon is explained in ref. 7). The 1. QBA contains in addition the contribution of all higher subsystem partial waves. Their effect is evidently somewhat more important, but still small (see column "1.QBA"). In this context we also mention the similar conclusion drawn from hyperspherical methods[8].

The doublet contribution to the elastic nd-cross-section, calculated in 0. and 1. QBA, is shown in the figure. Furthermore, results obtained for the S-wave parts of the MT potentials[4] and for separable potentials of the Yamaguchi type[3] are given. Again the improvement resulting from the use of appropriate form factors is considerable (curves "Yamaguchi" and "0.QBA"). Smaller corrections originate from

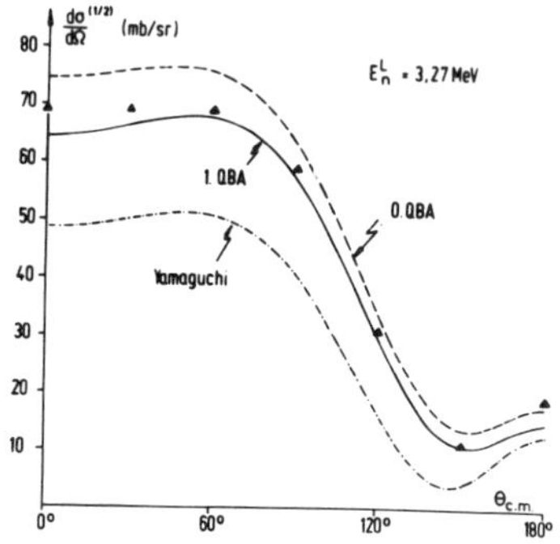

Fig. Doublet contribution to the differential cross section for the scattering of 3.27 MeV neutrons off deuterons. The triangles represent selected values of ref. 4

the inclusion of the higher partial waves of the MT potentials (curve "1.QBA").

The above demonstrated smallness of the first order correction of the quasi-Born series has to be contrasted with what has been found in a quasiparticle calculation for purely attractive Yukawa potentials[9]. There the differences between 0.QBA and 1.QBA turned out to be rather drastic for all doublet quantities. I.e., due to the repulsion occurring in the present more realistic model, much less sensitivity against the inclusion of higher subsystem partial waves is found.

References

1) R. A. Malfliet and J. A. Tjon, Nucl. Phys. A127 (1969) 161
2) B. L. G. Bakker, Thesis, Vrije Universiteit, Amsterdam, 1973
3) E. O. Alt and B. L. G. Bakker, to be published
4) W. M. Kloet and J. A. Tjon, Ann. Phys. (N.Y.) 79 (1973) 407
W. M. Kloet, Thesis, Rijksuniversiteit Utrecht, 1973
5) E. O. Alt, P. Grassberger and W. Sandhas, Nucl. Phys. B2 (1967) 167
6) V. B. Belyaev, J. Wrzecionko and A. L. Zubarev, Soviet J. Nucl. Phys. 12 (1971) 503
7) B. L. G. Bakker and W. Sandhas, contribution to this conference
8) R. van Wageningen and G. Erens, Phys. Letters 34B (1971) 184
9) E. O. Alt, Acta Physica Austriaca, Suppl. IX (1972) 139;
E. O. Alt and W. Sandhas, in Few Particle Problems in the Nuclear Interaction, edited by I. Slaus, S. A. Moszkowsky, R. P. Haddock and W. T. H. van Oers (North-Holland, Amsterdam, 1972)

LA VOIE ELASTIQUE $^{3}He + {}^{3}He$ ET LE MODELE OPTIQUE

R. Pigeon et R.J. Slobodrian
Université Laval, Département de Physique, Laboratoire
de Physique Nucléaire, Québec G1K 7P4, Canada

La réaction $^{3}He(^{3}He\ ^{4}He)2p$ et son miroir $^{3}H(^{3}He\ ^{4}He)2n$ ont une importance évidente pour la détermination des paramètres de l'interaction neutron-neutron. Les résultats expérimentaux indiquent jusqu'à présent[1] que la région des spectres des particules α peuplées par les configurations du système à trois corps correspondant à la paire p-p dans l'état $^{1}S_{o}$, proviennent d'un mécanisme direct. Une analyse de la réaction dans ce contexte en utilisant la théorie des ondes distordues de Born est peut-être la plus simple sans avoir recours à un calcul à plusieurs corps. Le premier pas d'une telle approche est la mesure des sections efficaces pour la diffusion élastique. Nous avons mesuré la section efficace pour la diffusion élastique à 13.55 MeV d'énergie du laboratoire, de 32° à 66° dans le système du centre de masse, à l'aide de l'accélérateur Van de Graaff CN de l'Université Laval. L'analyse des résultats a été faite avec le programme de recherche automatique Magali [2]. Nous avons fait simultanément une recherche pour les mesures de Bacher et al. [3]. Une analyse semblable du système $^{3}He - {}^{3}H$ a également produit un mauvais accord du calcul et de l'expérience [4]. Pour le système $^{3}He - {}^{3}He$ la situation est analogue et les résultats sont indiqués sur les figures ci-jointes.

Bacher et al. calculé avec les meilleurs paramètres

Nos données avec les meilleurs paramètres

Nos données avec les paramètres $^{3}He - {}^{3}H$ de Roberts et al.

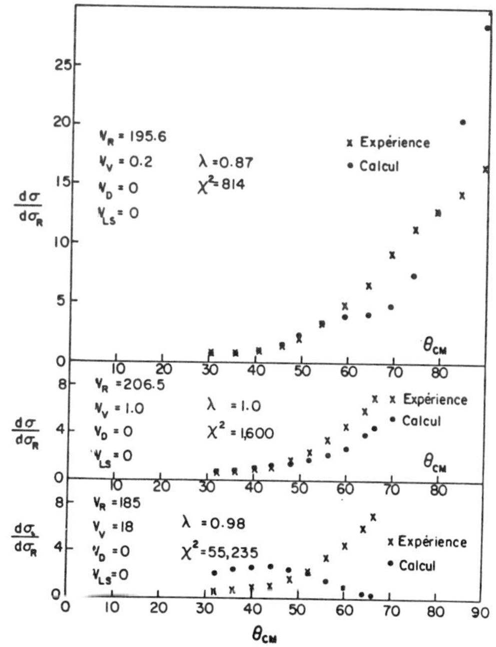

Le modèle optique est peut-être justifiable même pour des systèmes aussi légers que les noyaux d' ^{3}He , surtout si on considère que les résultats correspondent à une moyenne sur un grand nombre de collisions. Néanmoins le formalisme habituel du modèle optique considère seulement les interactions spin-orbite pour les projectils possédant un spin, et néglige les interactions spin-spin. Par ailleurs l'identité du projectil et de la cible devrait requérir dans le cas du système ^{3}He - ^{3}He l'antisymétrisation de la fonction d'onde. Ces considérations n'ont pas été respectées jusqu'à date dans les analyses des systèmes ^{3}He - ^{3}H et ^{3}He - ^{3}He, par exemple celles de Roberts et al. [4] faites dans le but de comparer les sections efficaces des réactions $^{3}H(^{3}He, {}^{4}He)^{2}H$ et $^{3}He(^{3}He, {}^{4}He)2p$ en regard du théorème de Barshay-Temmer [5]. Vu le mauvais accord de ce genre de paramétérisation avec l'expérience nous proposons d'incorporer l'espace de spin approprié avec les interactions spin-spin et l'antisymétrisation des amplitudes de diffusion. En correspondance l'analyse des réactions avec les ondes distordues devraient être modifiées pour tenir compte des nouveaux paramètres optiques et des interactions présentes dans la voie élastique.

BIBLIOGRAPHIE

1) M.L.V.L. de Slobodrian, B. Cujec, R. Roy, R.J. Slobodrian, Nuc. Phys. A 194 (1972) 577.

2) J. Raynal, 1969 D. PhT/69, 42 (CEN Saclay).

3) A.D. Bacher, Spiger, T.A. Tombrello, Nuc. Phys. A 119 (1968) 481.

4) W.V. Roberts, E.E. Gross, E. Newman, Phys. Rev. C9(1974) 149.

5) S. Barshay, G.M. Temmer, Phys. Rev. Let., 12 (1964) 728.

ELECTROMAGNETIC FORM FACTORS OF ^{3}He*

R. A. Brandenburg† and Y. E. Kim,† Los Alamos Scientific Laboratory, Los Alamos, New Mexico 87544, and A. Tubis, Purdue University, West Lafayette, Indiana 47907, U.S.A.

Calculations of bound-state properties of the three-nucleon system are a useful means for studying the nucleon-nucleon interaction, and of particular interest are the electromagnetic form factors. We have previously reported[1)] on a complete solution of the Faddeev equations for ^{3}He using the entire Reid soft-core potential. In this report we use the resulting wave function to estimate the relativistic corrections to the order $(1/M^2)$ for the ^{3}He charge form factor, F_{CH}, and to calculate the ^{3}He magnetic form factor, F_{MAG}, including off-diagonal S-D contributions and the nucleon convection currents.

In calculating F_{CH}, we use the charge density operator (to the order $1/M^2$)[2)],

$$\rho(\vec{Q}) = \sum_{j=1}^{3} \left[e_j - \frac{Q^2}{8M^2} e_j - \frac{i}{4M^2} (\vec{Q} \cdot \vec{\sigma}_j \times \vec{k}_j)\, \tilde{\mu}_j \right] \tag{1}$$

where $\vec{Q}$ is the electron momentum transfer, $\vec{\sigma}_j$ and $\vec{k}_j$ are the Pauli spin operator and the momentum for the jth nucleon, respectively, and

$$e_j = f^p_{CH}(Q^2)\tfrac{1}{2}(1 + \tau_z(j)) + f^n_{CH}(Q^2)\tfrac{1}{2}(1 - \tau_z(j)), \tag{2}$$

$$\tilde{\mu}_j = \left[2(2.793)f^p_{MAG}(Q^2) - f^p_{CH}(Q^2)\right]\tfrac{1}{2}(1 + \tau_z(j)) + \left[2(-1.913)f^n_{MAG}(Q^2) - f^n_{CH}(Q^2)\right] \times \times \tfrac{1}{2}(1 - \tau_z(j)) \,. \tag{3}$$

$f_{CH}(Q^2)$ and $f_{MAG}(Q^2)$ are respectively the charge and magnetic form factors for the appropriate nucleon, which we take to be the analytic forms of Janssens, et al.[3)] In Fig. 1, the solid curve gives $|F_{CH}(Q^2)|$ calculated from the first term of Eq. (1), the dashed curve from the first and second terms, and the dot-dashed curve from all three terms of Eq. (1).

In calculating F_{MAG}, we use the usual impulse approximation and nonrelativistic reduction of the nucleon-current matrix elements[4,5)]. Contributions from meson-exchange currents are neglected. A convenient expression for F_{MAG} is obtained if we: a) use the fact (following from charge-current conservation) that only the component of the nucleon current perpendicular to the electron momentum transfer $\vec{Q}$, contributes, b) orient our coordinate system so that $\vec{Q}$ lies in the x-direction, and c) choose the z-axis as the quantization axis for magnetic quantum numbers. We then find

$$F_{MAG}(Q^2) = \frac{3}{Q} \int d\vec{p}_1 \int d\vec{q}_1\, \psi^*(\vec{p}_1, \vec{q}_1 - \frac{\vec{Q}}{\sqrt{3}}) J_y^{(1)}\, \psi(\vec{p}_1, \vec{q}_1) \tag{4}$$

where $\psi(\vec{p}_1, \vec{q}_1)$ is the ^{3}He center-of-mass momentum-space wave function with $J = J_z = \frac{1}{2}$. (Spin and isospin variables are not indicated.) The momenta, $\vec{p}_1$ and $\vec{q}_1$, are given in terms of the nucleon momenta $\vec{k}_i$ by

$$\vec{p}_1 = (\vec{k}_2 - \vec{k}_3)/2, \quad \vec{q}_1 = (\vec{k}_2 + \vec{k}_3 - 2\vec{k}_1)/2\sqrt{3} \,. \tag{5}$$

$J_y^{(1)}$ is the y component of $\vec{J}^{(1)}$, the momentum matrix-element of the current operator for nucleon 1, consisting of convection and magnetic-spin-current contributions[4,5).

$$\vec{J}^{(1)} = \vec{J}_C^{(1)} + \vec{J}_M^{(1)} = -\left(\frac{4}{\sqrt{3}}\vec{q}_1 + \vec{Q}\right)e_1 + i(\vec{\sigma}_1 \times \vec{Q})\mu_1 \quad , \tag{6}$$

$$\mu_1 = 2.793\ f^p_{MAG}(Q^2)\ \frac{1}{2}\ (1 + \tau_z(1)) - 1.913\ f^n_{MAG}(Q^2)\ \frac{1}{2}\ (1 - \tau_z(1)) \quad . \tag{7}$$

In Fig. 2, the solid curve gives $|F_{MAG}(Q^2)|$ calculated with the total $\vec{J}^{(1)}$. The experimental data are from reference 6 (solid-circle) and reference 7 (open-circle). The separate contributions from the spin and convection-current terms are plotted as dot-dash and dashed curves respectively. All three curves are renormalized so that $|F_{MAG}(Q^2 = 0)| = 1$. The effect of the convection current is to shift the minimum from 7.8 to 7.1 fm^{-2}.

REFERENCES

1) R. A. Brandenburg, Y. E. Kim and A. Tubis, Phys. Lett. 49B, 205 (1974).
2) W. Bertozzi, et al., Phys. Lett. 41B, 408 (1972).
3) T. Janssens, et al., Phys. Rev. 142, 922 (1966).
4) K. W. McVoy and L. Van Hove, Phys. Rev. 125, 1034 (1962).
5) T. de Forest, Jr., and J. D. Walecka, Advances in Physics 15, 1 (1966).
6) J. S. McCarthy, et al., Phys. Rev. Lett. 25, 884 (1970).
7) M. Bernheim, et al., Lett. Al. Nuovo Cimento 5, 431 (1972).

*Research supported in part by the National Science Foundation and by the U. S. Atomic Energy Commission.

†On leave from Purdue University, West Lafayette, Indiana, U. S. A.

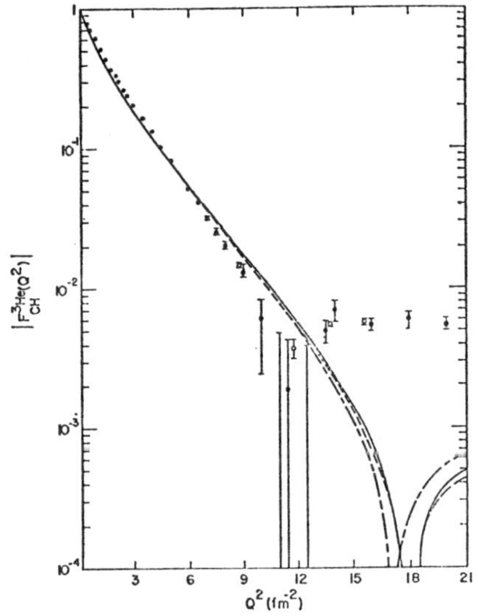

Figure 1

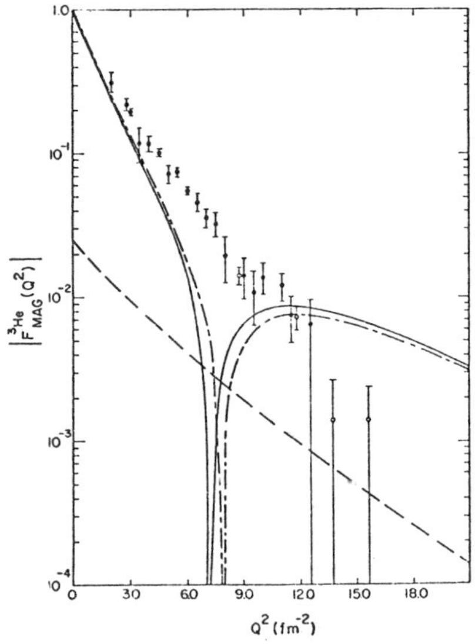

Figure 2

OPTICAL MODEL ANALYSIS OF p-^{3}He SCATTERING*

B.S. Podmore and H.S. Sherif
Nuclear Research Centre, The University of Alberta, Edmonton, Alberta, Canada

The recent success [1,2] of the optical model in describing n-T elastic scattering has encouraged us to carry out a similar investigation for the p-^{3}He system. This system should provide a more stringent test of the model since much more data are available, in particular *both* cross section and polarization measurements have recently been performed at several energies. Ten sets of data, for incident proton energies ranging from 5.51 to 156 MeV, were analysed using a conventional optical model potential [2]. The well geometries were fixed at their values found in the n-T analysis and only the well depths were allowed to vary. As was the case with the n-T system, a general preference for volume absorption prevailed. Table 1 gives the best fit parameters determined using the search code SNOOPY [3]. Note that the real well depth decreases with energy as expected. In the energy range considered, its energy dependence can be represented by

$$V_0 = 53.75 - 0.869\,E + 0.004\,E^2 \qquad (1)$$

The fits are generally in good agreement with experiment, but are consistently poor at backward angles for cross sections. An example is given by the dashed curves in figure 1 for the scattering of 30 MeV protons [4].

Calculations using the resonating group method have indicated the importance of

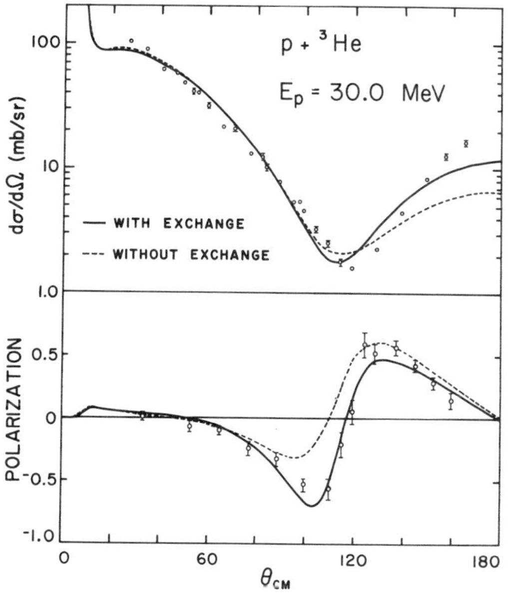

FIGURE 1

Differential cross section and polarization for 30 MeV protons scattered by ^{3}He. The data points are from ref. 4). The dashed curves are optical model best fits using the parameters listed in Table 1. The solid curves are those obtained with exchange terms added using the parameters of eq. 3.

exchange processes in light systems[5]. This effect may be taken into account in the optical model by including Majorana exchange terms in the potential[6]. The following substitutions:

$$V_0 \rightarrow V_0\ [1 + C_0(-)^{\ell}] \qquad W_V \rightarrow W_V\ [1 + C_V(-)^{\ell}] \qquad V_{so} \rightarrow V_{so}\ [1 + C_{so}(-)^{\ell}] \tag{2}$$

were made in the potential and a search over the exchange parameters C_0, C_V and C_{so} was performed for the 30 MeV data. The best fit obtained with the exchange terms is shown by the solid curves in figure 1. The values for the parameters are

$$C_0 = -0.065 \qquad C_V = -0.4 \quad \text{and} \quad C_{so} = 0.06 \tag{3}$$

It is evident from the figure that the inclusion of the Majorana exchange terms improves the fit to backward angle cross sections. In addition, the polarization fit is substantially improved. Investigation of the effect of the exchange terms for the other data sets is presently in progress.

TABLE 1

OPTICAL MODEL PARAMETERS

$r = 1.488$ fm $\quad r_i = 1.501$ fm $\quad r_{so} = 1.049$ fm $\quad r_c = 1.3$ fm

$a_0 = 0.144$ fm $\quad a_i = 0.378$ fm $\quad a_{so} = 0.289$ fm

E_p (MeV)	V_0 (MeV)	W_V (MeV)	W_D (MeV)	V_{SO} (MeV)	χ^2_σ	χ^2_p
5.51	50.25	0.0	0.0	0.72	112.3	7.4
6.82	48.26	0.0	0.0	0.86	149.9	12.9
8.82	46.63	1.20	0.0	1.31	18.7	8.0
10.77	44.61	1.37	0.0	1.43	13.6	9.9
13.6	41.63	1.18	0.0	1.47	304.7	40.2
16.2	38.99	0.94	0.0	1.60	320.3	50.0
19.4	38.33	3.55	0.0	3.21	286.4	277.3
30.0	32.20	6.05	0.0	7.34	34.6	7.3
49.5	17.44	12.64	0.0	12.64	60.0	21.0
156.0	9.86	8.61	0.0	4.26	275.4	20.3

REFERENCES

1. R.M. Devries, J.L. Perrenooud and I. Slaus, Nucl. Phys. A188 (1972) 449
2. H.S. Sherif and B.S. Podmore in *Few Particle Problems in the Nuclear Interactions*, I. Slaus, *et al.*, ed. (North-Holland, 1972) p.691
3. P. Schwandt, University of Colorado Report (1969), unpublished
4. J.R. Morales-Pena, Thesis, University of California, Davis (1970), unpublished; S.A. Harbison, *et al.*, Nucl. Phys. A112 (1968) 137
5. D.R. Thompson and Y.C. Tang, Phys. Rev. C4 (1971) 306; R.E. Brown, Y.C. Tang and D.R. Thompson in *Few Particle Problems in the Nuclear Interaction*, I. Slaus, *et al.*, ed. (North-Holland, 1972) p.703
6. G.W. Greenlees and Y.C. Tang, Phys. Lett. 34B (1971) 359

*Work supported in part by the Atomic Energy Control Board of Canada.

Effective Three-Nucleon Interaction

I. Lovas*

Institut für Kernphysik, Kernforschungsanlage Jülich, BRD.

The Hilbert space of the three-baryon system is divided into two orthogonal subspaces denoted by P and Q. The P subspace is spanned by the three-nucleon states, while the states containing excited baryons /resonances/ and the states corresponding to simultaneous two - or more - boson exchange processes belong to the Q subspace. The bound states of the three-baryon system $\Psi = \Psi_1 + \Psi_2 + \Psi_3$ can be obtained from the Faddeev equations given by

$$\Psi_\alpha = G_o \sum_\beta T_\alpha (1 - \delta_{\alpha\beta}) \Psi_\beta ,$$

where $G_o = (E-H_o)^{-1}$ and T_α satisfies the two-body Lippmann-Schwinger equation:

$$T_\alpha = V_\alpha + V_\alpha G_o T_\alpha .$$

By the help of the projection operators P and Q we can get a coupled system of equations for the projected components of the wave function. Eliminating the components Ψ_α^Q the following equations can be derived:

$$\Psi_\alpha^P = G_o^{PP} \sum_{\beta\gamma} R_{\alpha\gamma}^{PP} (1 - \delta_{\gamma\beta}) \Psi_\beta^P$$

where

$$R_{\alpha\gamma}^{PP} = T_\alpha^{PP} \delta_{\alpha\gamma} + T_\alpha^{PQ} G_{\alpha\gamma}^{QQ} T_\gamma^{QP}$$

and $G_{\alpha\gamma}^{QQ}$ are the solutions of the following Faddeev-type equations:

$$G_{\alpha\gamma}^{QQ} = (1 - \delta_{\alpha\gamma}) G_o^{QQ} + \sum_\mu (1 - \delta_{\alpha\mu}) G_o^{QQ} T_\mu^{QQ} G_{\mu\gamma}^{QQ} .$$

Using the iterative solution of these equations, the second term of the modified kernel $R_{\alpha\gamma}^{PP}$ can be represented by the following diagrams

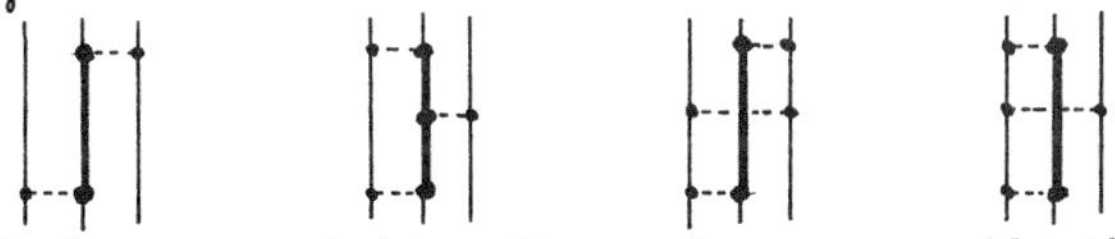

From this graphical representation it can be seen easily that the second term of the modified kernel corresponds to an effective three-body interaction. The appearence of the three-body interaction is a natural consequence of the elimination of the degrees of freedom associated with the Q subspace.

---

*On leave of absence from the Central Research Institute for Physics, Budapest, Hungary.

The most relevant element of the Q subspace can be identified with the NNΔ state where the resonance characterised by T=3/2, S=3/2, m=1236 MeV and Γ=120 MeV is denoted by Δ. For the sake of simplicity only this single state of the subspace Q will be treated here explicitely. In this case the operator T^{PQ} describes the NN-NΔ transition. By an appropriate choice of V^{PQ} it is possible, however, to include contributions coming from other states lying at about 300 MeV above the three-nucleon states. At low energies it is enough to consider the S=0, L=0, J=0, T=1 ($^1S_0$) state of the NN system which is coupled to the S=2, L=2, J=0, T=1 ($^5D_0$) state of the NΔ system. The t-operators being present in the modified kernel $R^{PP}_{\alpha\gamma}$ satisfy the following type of equations:

$$T^{AB}_{\alpha} = V^{AB}_{\alpha} + V^{AP}_{\alpha} G^{PP}_{0} T^{PB}_{\alpha} + V^{AQ}_{\alpha} G^{QQ}_{0} T^{QB}_{\alpha}; \quad \{A,B\} = \{P,Q\} .$$

We approximate the solution of these equations by separable expressions:

$$\langle \vec{p}_\alpha \vec{q}_\alpha \xi_\alpha \sigma_\alpha | T^{AB} | \vec{p}_\alpha' \vec{q}_\alpha' \xi_\alpha' \sigma_\alpha' \rangle = \sum_{m_\alpha m_\alpha'} g_{m_\alpha}(\vec{p}_\alpha \xi_\alpha)\, t^{AB}_{m_\alpha m_\alpha'}\, g^{*}_{m_\alpha'}(\vec{p}_\alpha' \xi_\alpha')$$

where $$t^{AB}_{m_\alpha m_\alpha'} = \delta(\vec{q}_\alpha - \vec{q}_\alpha')\, \delta_{\sigma_\alpha \sigma_\alpha'}\, h^{AB}_{m_\alpha m_\alpha'}(E-q_\alpha^2) \quad .$$

Here the usual momentum and spin variables are denoted by $\vec{p}_\alpha, \vec{q}_\alpha$ and $\xi_\alpha, \sigma_\alpha$, respectively. The quantity $h^{AB}_{m_\alpha m_\alpha'}$ can be obtained either by the Unitary Pole Approximation or by means of nonlocal, separable interactions. For the sake of brevity we approximate $G^{QQ}_{\alpha\gamma}$ by its first iteration $(1-\delta_{\alpha\gamma})\, G^{QQ}_{0}$. Using these approximations we can express the matrix elements of the modified kernel in the following form:

$$\sum_{m_\alpha m_\gamma} g_{m_\alpha}(\vec{p}_\alpha \xi_\alpha) \left[t^{PP}_{m_\alpha m_\gamma} \delta_{\alpha\gamma} + f^{PP}_{m_\alpha m_\gamma}(1-\delta_{\alpha\gamma}) \right] g^{*}_{m_\gamma}(\vec{p}_\gamma \xi_\gamma)$$

where $$f^{PP}_{m_\alpha m_\gamma} = \sum_{m_\alpha' m_\gamma'} h^{PQ}_{m_\alpha m_\alpha'}(E-q_\alpha^2)\, z^{QQ}_{m_\alpha' m_\gamma'}\, h^{QP}_{m_\gamma' m_\gamma}(E-q_\gamma^2)$$

and $$z^{QQ}_{m_\alpha m_\gamma}(\vec{q}_\alpha \sigma_\alpha \vec{q}_\gamma \sigma_\gamma) = \sum_{\xi_\alpha \xi_\gamma} \int g^{*}_{m_\alpha}(\vec{p}_\alpha \xi_\alpha) \langle \vec{p}_\alpha \vec{q}_\alpha \xi_\alpha \sigma_\alpha | \vec{p}_\gamma \vec{q}_\gamma \xi_\gamma \sigma_\gamma \rangle g_{m_\gamma}(\vec{p}_\gamma \xi_\gamma) (E - p_\alpha^2 - q_\alpha^2 - \varepsilon_Q)^{-1}\, d\vec{p}_\alpha d\vec{p}_\gamma$$

As it is seen this representation of the effective three-nucleon interaction preserves all the advantages of the Separable Approximation consequently it can be incorporated easily into the conventional three-nucleon calculations.

The $\frac{1}{2}^{+}$ Three Nucleon states for a Super Soft-Core Potential

M. A. Hennell
Dept. of Computational & Statistical Science, University of Liverpool, Liverpool 3, U.K.

In recent times a number of commendable efforts have been made to produce nucleon-nucleon potentials which, whilst reproducing the two-body data to a high degree, do not give rise to computational difficulties when utilised in computations of more complex systems. Such a potential is that of Cogny, Pires and De Tourreil[1)] (CPT) which is a super-soft core potential being finite at the origin. In this paper results are presented for the $J = \frac{1}{2}^{+}$ three-nucleon system for this potential.

The potential which is local has the usual spin singlet/triplet, parity even/odd components. Like the Hamada-Johnston[2)] (HJ) potential it has central, tensor, L.S. and L.L. components. The radial part of these components consists of a sum of Gaussian terms so that:

$$V(r) = \sum_{i=1}^{n} a_i e^{-\alpha r^2} .$$

The Gaussian form was chosen to allow certain many body problems to be simplified. The potential was fitted both to the two-body data and also to the spherical nucleii properties.

The calculation was performed variationally, the main details of which are to be found in ref.[5)]. The only changes made to the program suite used for the calculation with the Reid potential [6)] was that a small routine which calculates the potential values, was replaced. Thus the trial function for this potential is identical to that used for the Reid potential.

In Table 1 are the principal parameters of the $J = \frac{1}{2}^{+}$ state where the predictions for the CPT potential are compared with those of the HJ, Gammel-Bruckner[8)] (GB) and Reid potentials. It may be seen from Table 1 that the CPT potential gives a Triton binding energy which is close to the experimental value and is somewhat greater than the values for the other potentials.

The Coulomb energy predictions for CPT are also higher than those for the other potentials. This is consistent with the increased binding energy. The discrepancy with the experimental value observed with other potentials holds for this potential also.

From Table 1 it can also be seen that the predicted minimum of the ^3H_e electron scattering charge form-factor occurs at a substantially higher value of the momentum transfer q than is the case for the Reid and HJ potentials but is close to the G.B. value. The magnitude of the secondary maximum is substantially less for CPT than for the other potentials being a factor of ten less than experiment.

The neutron-deuteron doublet scattering length ^2a can also be seen to be rather larger than the value for the other three potentials.

Also included in Table 1 are the two body predictions for the four potentials. It may be seen that the CPT potential does not fit the two-body bound state parameters. This factor may enhance the binding energy of the three body bound state whilst not decreasing appreciably the doublet scattering length. That the binding energy is close to experiment may be attributed to the fitting of the potential to the spherical nuclei binding energies. The two poor predictions are then the doublet scattering length and the form factors. These together with the overbound deuteron suggest that the potential is too strong in the intermediate region and possibly not strong enough at very small distances.

Finally, the predictions for the CPT potential presented here, are in agreement with the results obtained by Gignoux and Laverne[7], who predict $\bar{E}_T \sim -8.3$ Mev and $q^2 = 15.6$ for the $|F_{ch}|$ minimum. The discrepancy of 0.4 Mev in the binding energy can be explained by the inclusion of all two-body J states in this calculation and the inclusion of only the low J states in [8].

Property	H-J	G-B	Reid S.C.	Expt	CPT S.S.
Binding Energy Mev	-6.5 ± 0.2	-7.75 ± 0.2	-7.75 ± 0.5	-8.5	-8.7 ± 0.2
E_c(point)Mev	0.562	0.620	0.617	-	0.66
E_c(finite) Mev	0.547	0.595	0.593	0.764	0.62
$<r^2 ch>^{\frac{1}{2}}$ fm	1.85	1.65	1.78	1.70	1.65
$<r^2 mass>^{\frac{1}{2}}$fm	1.90	1.74	1.86	1.87	1.74
P(S) %	89.1	91.5	89.5 ± 0.5	~90	94.5
P(S') %	1.8	1.8	~1	?	1
P(D) %	9.0	6.7	9.5 ± 0.5	4-10	4.5
Diff.MIN $F_{ch}(^3He)$	12.5 ± 0.3	15.0 ± 0.3	12.8 ± 0.3	11.8	15.5 ± 0.3
$^2a(n-d)$	2.5 ± 0.3	2.0 ± 0.3	~	0.1 -0.7	2.8
$E_{Deuteron}$	2.216	2.297	2.225	2.226	2.650
P(D)%(Deuteron)	6.96	5.37	6.47	3.85	4-6

Table 1.

References

1. D. Cogny, P. Pires and R. De Tourreil. Phys. Letts. 32B, 1970, 591.
2. T. Hamada and I. D. Johnston. Nucl. Phys. 34(1972)382.
3. R. V. Reid. Ann. of Phys. 50(1968) 411.
4. K. A. Bruckner and J. M. Gammel. Phys. Rev. 109 (1958) 1023.
5. L. M. Delves and M. A. Hennell. Nucl. Phys. '168 (1971) 347.
6. M. A. Hennell and L. M. Delves. Proc. Conf. on Few particle problems in the nuclear interaction. L.A., 1972, North Holland, p.419.
7. C. Gignoux and A. Laverne. Ibid, p.411.
8. M. A. Hennell and L. M. Delves. Nucl. Phys. A204 (1973) 552.

THE MESONIC EXCHANGE CHARGE FORMFACTORS OF ^{3}He AND ^{3}H

W.M. Kloet, University of Maryland, Department of Physics and Astronomy
College Park, Maryland 20742, U.S.A.
J.A. Tjon, University of Utrecht, Institute for Theoretical Physics
Utrecht, Netherlands

The effect of meson exchange on the charge formfactor (CFF) of the tri-nucleon system is calculated. The diagrams considered are the same which give the dominant exchange effects in the case of the magnetic formfactor[1]. They are shown in Fig. 1.

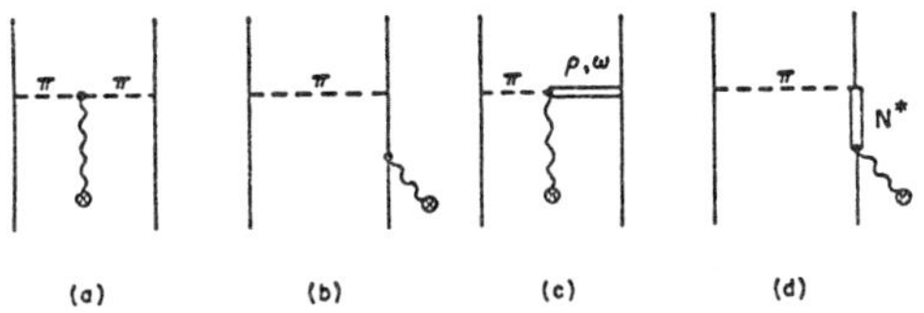

Fig.1.

For the three-body wavefunction we used the solution of the Faddeev equation for the Reid soft core potential as described in Ref. 2.

The calculation for diagrams 1a, 1b and 1c is straightforward and was published before[3]. The effect of these three diagrams on the CFF of ^{3}He is that the dip is moved towards lower q^2 by about 3 fm^{-2} and the amplitude of the secundary maximum is increased by a factor three.

The isobar diagram 1d is evaluated with the help of a low energy theorem for photopionproduction which relates the non-Born amplitude of this process to the p-wave pion-nucleon scattering phaseshifts. It is the same method as discussed by Chemtob and Rho[1] for the magnetic formfactor, but adjusted to the case of the CFF. As in Ref. 1. the isoscalar part is neglected.

The result for diagram 1d enhances the effect of the first three diagrams in the case of ^{3}He. The dip moves to 10 fm^{-2}, the secundary maximum occurs now at q^2=15 fm^{-2}, and it has an amplitude of .54 x 10^{-2}. The total exchange effect and the corrected CFF for ^{3}He are shown in Fig. 2. In the range of q^2=8 fm^{-2} the theoretical curve is about 40 % lower than experiment. This discrepancy is considerably reduced if more partial waves in the two-nucleon interaction are included[5] and if the effect of Δ(1236) in the one-body formfactor is taken into account[6]. Also the fact that we are considering a three-nucleon system which has too little binding (6.5 MeV compared to 8.5 MeV experimentally) may be partly responsible for a CFF that falls off too rapidly. In the region of the secundary maximum the exchange effect is the dominant contribution and it is encouraging that the result is compatible with experiment. However one must bear in mind that the isobar contribution is model dependent and also that relativistic effects probably cannot be ignored at these values of q^2. Nevertheless it is clear from this calculation that exchange effects are very important already at not too large momentum transfer. This may well imply that also for heavier nuclei one cannot ignore exchange effects, and that they must be taken into account before crucial information about the wavefunction is subtracted from the charge formfactor.

In Fig. 3. the exchange effect on the CFF of ^3H is shown. In this case the dip in the one-body formfactor is completely filled in by the exchange correction.

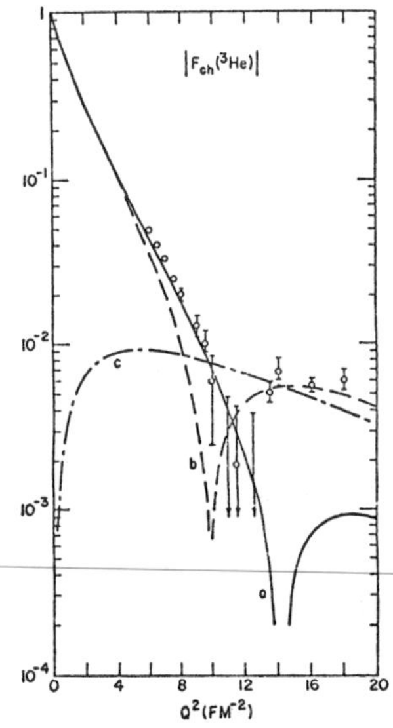

Fig.2.

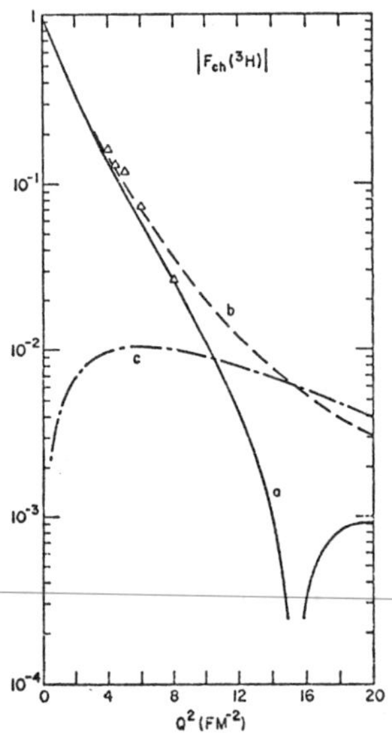

Fig.3.

In Figs. 2,3 curves a, b, c represent respectively the one-body charge formfactor, the corrected formfactor and the exchange correction. Experimental points are from Ref. 4.

BIBLIOGRAFIE

1. M. Chemtob, M. Rho, Nucl.Phys. A163, 1 (1971).
E.P. Harper, Y.E. Kim, A. Tubis, M. Rho, Phys.Lett. 40B, 533 (1972).
E. Hadjimichael, A. Barroso, Phys.Lett. 47B, 103 (1973).
2. R.A. Malfliet, J.A. Tjon, Ann.Phys. 61, 425 (1970).
3. W.M. Kloet, J.A. Tjon, to be published in Phys.Lett.
4. J.S. McCarthy, I. Sick, R.R. Whitney, M.R. Yearian, Phys.Rev.Lett. 25, 884 (1970).
H. Collard, R. Hofstadter, E.B. Hughes, A. Johansson, M.R. Yearian, R.B. Day, R.T. Wagner, Phys.Rev. 138, B57 (1965).
5. R.A. Brandenburg, Y.E. Kim, A. Tubis, Phys.Lett. 49B, 205 (1974).
6. A.J. Kallio, P. Toropainen, A.M. Green, T. Kouki, preprint Helsinki (1974).

Description of High-Energy Scattering on Composite Targets Without Eikonal Approximations

S.A. Gurvitz and A.S. Rinat (Reiner)
Weizmann Institute of Science, Rehovot, Israel

In the following we report results on calculations of the scattering of high energy projectiles from composite targets using the multiple-scattering formalism in the fixed scatterers approximation (FSA) [1] with non-overlapping potentials (NOP) [2]. The FSA lies also at the basis of the standard Glauber theory[3], but contrary to the latter, we did not invoke any eikonal approximation. Our aim is then to asses the accuracy of the Glauber theory and the interpretation of reported agreement with data.

Our results can be summarized as follows:

1) Amplitudes describing the scattering on n different nucleons in a target can be derived for a restricted geometry resulting from the NOP assumption. When overlap corrections are small these amplitudes give an accurate rendition of multiple scattering.

2) The amplitudes above can be expressed in terms of the product of n scattering amplitudes f_{xN} and (n-1) functions g_i, describing effective radial-wave propagation.

3) When decomposing each g_i, portions coming from Im g_i come close (but are not identical) to on-shell Glauber amplitudes of corresponding order.

4) Off-shell contributions in the non-eikonal NOP theory do not vanish nor cancel and thus contribute genuine corrections to the Glauber amplitude. This result should be contrasted with the proven result that the summed off-shell contributions cancel in the eikonal approximation[4] and actually vanish in each order if in addition potentials do not overlap[5].

5) For nearly purely imaginary f_{xN}, on-shell and off-shell parts have opposite phase and may strongly affect regions in differential cross sections, where amplitudes of different multiplicity interfere.

Our results are easiest applied if elementary amplitudes have a Gaussian dependence on momentum transfer q. Averaged (effective) elementary amplitudes are roughly of such a form up to medium q. Also lack of information on the dependence of

$$\rho_{xN}(q) = \frac{\text{Re } f_{xN}(Eq)}{\text{Im } f_{xN}(Eq)} \sim \rho_{xN}(0)$$ makes a tentative test only meaningful for a restricted q region.

In Figs. 1,2,3 we give experimental results for proton scattering (E_p = 0.6, 1.0 GeV) on ^3He and ^4He [6)-8)] together with the Glauber and NOP predictions. For details, see [9].

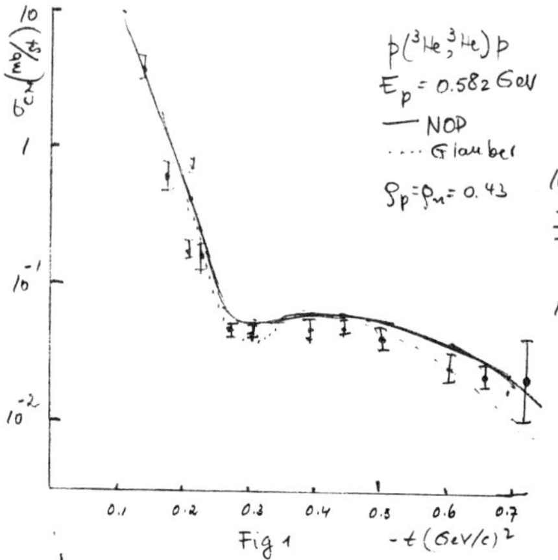

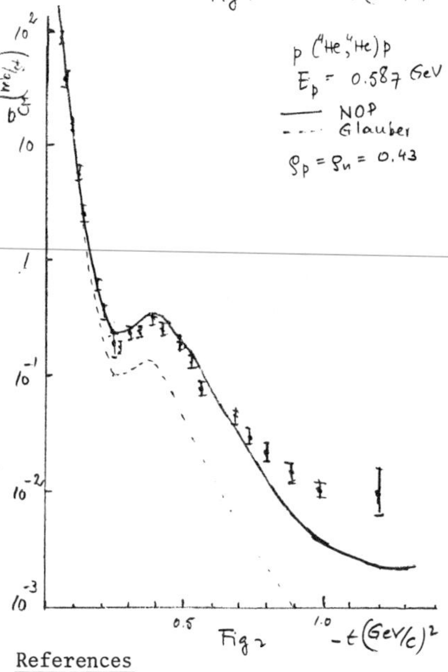

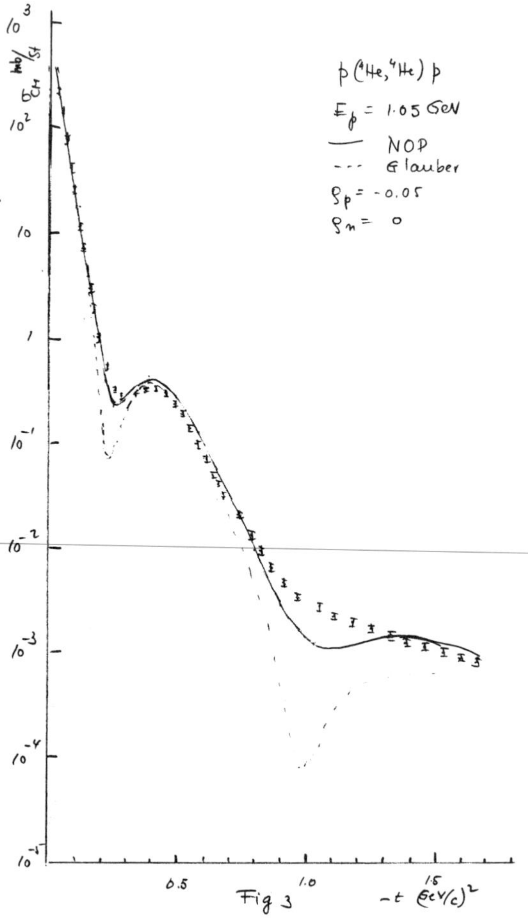

<u>References</u>

1) L.L. Foldy and J.D. Walecka, Ann. of Phys. (NY) <u>54</u> (1969) 447.
2) D. Agassi and A. Gal, Ann. of Phys. <u>75</u> (1973) 56.
3) R.J. Glauber, Lecures on Theoretical Physics, Vol. 1, Interscience N.Y., 1959.
4) D.R. Harrington, Phys. Rev. <u>184</u> (1969) 1745; J.M. Eisenberg, Ann. of Phys. (N.Y.) <u>71</u> (1972) 542.
5) D.R. Harrington, to be published.
6) E.T. Boschitz et al., Phys. Rev. <u>C6</u> (1972) 457.
7) L. Meritet <u>et al</u>., Journ. de Phys. Suppl. <u>34</u> (1973) 146 Paper III.28
8) S.D. Baker et al., Phys. Rev. Lett. <u>32</u> (1974) 839.
9) S.A. Gurvitz, A.S. Rinat (Reiner) and Y. Alexander, to be published.

How good is the adiabatic assumption for scattering on a composite target?

Y. Avishai, Ben-Gurion University of the Negev, Beer Sheva, Israel
A.S. Rinat (Reiner), Weizmann Institute, Israel

We discuss adiabatic approaches to intermediate and high-energy scattering. In it, the amplitude is first computed for a static fixed configuration of scatterers and is subsequently averaged over the target density.

Kujawski and Lambert[1)] performed detailed comparisons between several possible versions and approximations. Particular stress has been paid to the non-uniquely determined kinematics in the elementary collisions between a projectile and constituent nucleons. These collisions could for instance be described in the CM of nucleon and target (fixed scatterers, FS, kinematics) or alternatively in the impulse kinematics (IA). The reported calculations[1)] only stress differences in approach but do not, nor can, decide the absolute quality of either approximation.

In order to settle that question we compared versions of the adiabatic approximation to exact solutions of the Fadde'ev equations for the scattering of alternatively, a fictitious projectile and a nucleon on a D target[2)]. Our results for a nucleon projectile are presented in Fig. 1a, b for E_p = 100 and 500 MeV respectively. More extensive results are given and discussed in Ref. 3.

From it one infers that the trend of small angle scattering is reproduced in either kinematics. As expected the 'impulse' kinematics is favored for increasing energies, but both fail completely in the backward hemisphere. In the case of nucleon scattering, one can point at the neglect of nucleon exchange present in the exact theory and which is impossible in any adiabatic approach. No such mechanism exists if the projectile is not a nucleon in which case and reliable treatment of recoil becomes of primary importance.

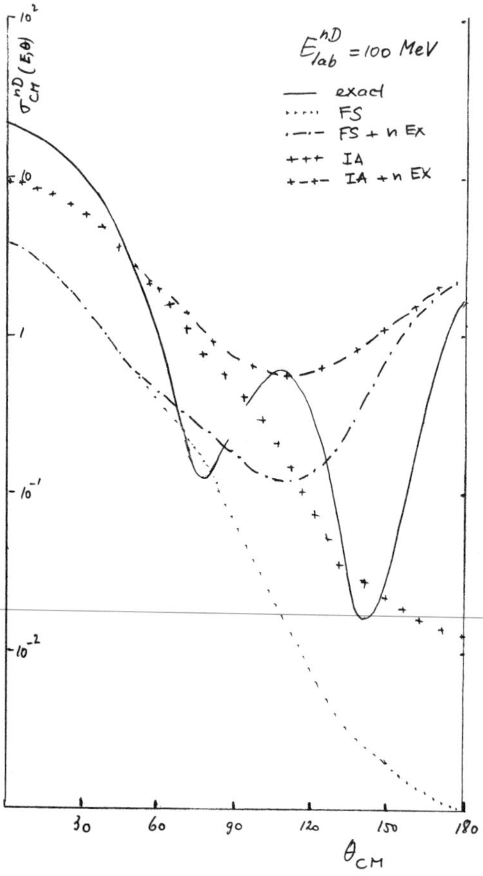

Fig 1a

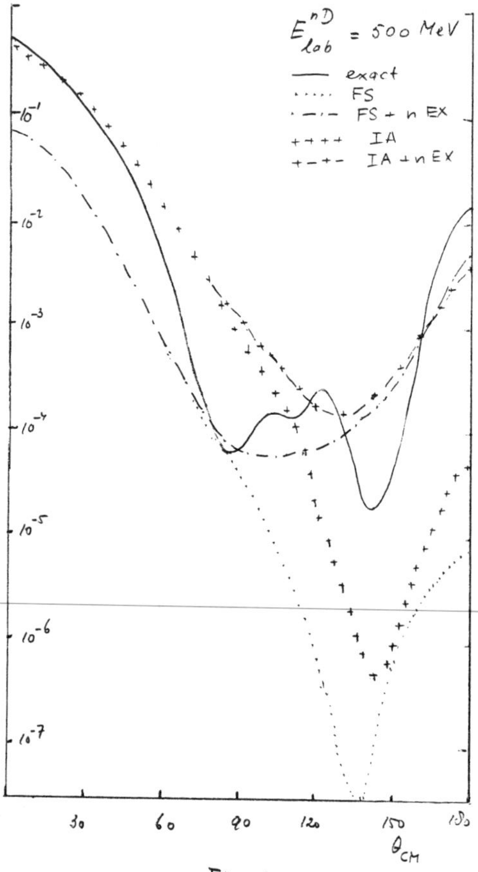

Fig 1b

References

1. E. Kujawski and E. Lambert, Ann. of Phys. (NY) 81 (1973) 591.
2. L.D. Fadde'ev, Sov. Phys. JETP 12 (1961) 1014.
3. Y. Avishai and A.S. Rinat, to be published.

Application of Partial Wave Dispersion Relations to a Treatment of the Coupled $n+{}^4He$, $d+{}^3H$ Channels

M. Stingl, University of Münster, W. Germany
A.S. Rinat (Reiner), Weizmann Institute, Israel

Encouraged by initial successes in the applications of the ND^{-1} method for solving non-relativistic scattering problems[1)2)], we applied the formalism[3)] to the nuclear 5-body problem. The following points motivated our choice:

a) Possibility to study a reaction ($n+{}^4He$ $n+{}^4He$) which is elastic over a wide energy range and has a simple spin-isospin structure.

b) Study of resonances in that channel.

c) Tracing of observed spin-orbit splitting to nucleon-nucleon force.

It has been assumed that the analytic structure of amplitudes can in principle been found from a non-relativistic dynamical theory involving only nucleonic degrees of freedom. In the absence of an exact theory the following additional assumptions have been made

i) Only two-cluster channels are retained.

ii) ND^{-1} equations are driven by single and double cluster exchange.

iii) Anomalous thresholds may be neglected. Together with these the input consists of parameters of vertex functions[4)]. No adjustable parameters appear.

Qualitative agreement has been found for the following observables on which we comment

1°) $n({}^4He,{}^4He)n$
- a) an s wave phase shift, which decreases too strongly
- b) resonating $p_{3/2}$, $p_{1/2}$ phases
- c) an insufficient spin-orbit splitting, due to the tensor component in the NN force
- d) a sharp resonance in $d_{3/2}$ phase just below the $d+{}^3H$ threshold

2°) $d({}^3H,{}^4He)n$ — Calculations qualitatively reproduce observed cross sections except for observed secondary maximum around $\theta=100^o$.

3°) $d({}^3H,{}^3H)d$ — About same measure of agreement is obtained as for stripping reaction.

It can be argued that at least part of the observed disagreements may be due to the neglect of anomalous thresholds[7)].

An extensive report will be published elsewhere[8)].

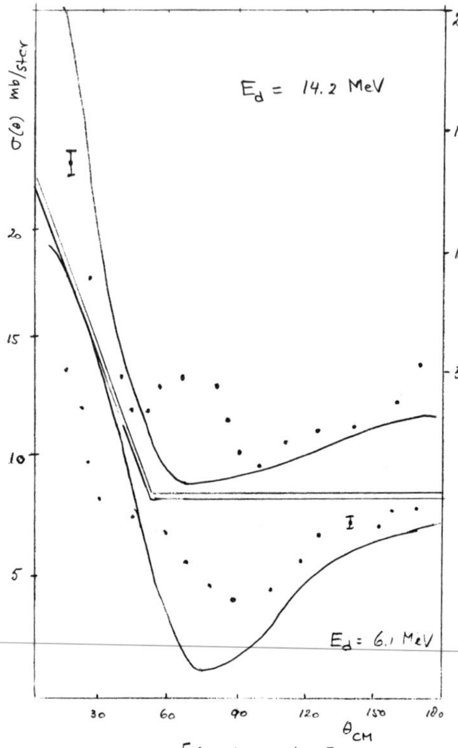

σ [$^3H(d,n)^4He$]

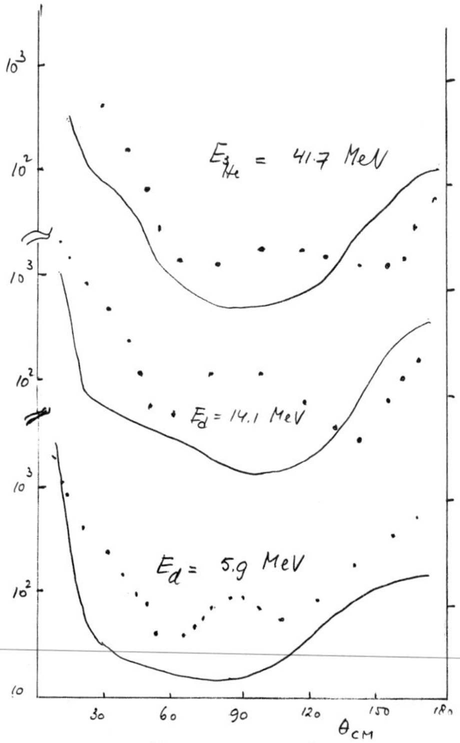

σ [$^3H(d,d)^3H$]

References

1) Y. Avishai, W. Ebenhöh and A.S. Rinat (Reiner), Ann. Phys. NY 55 (1969) 341; Phys. Lett. 29B (1969) 638.
2) G. Barton and A.C. Phillips, Nucl. Phys. A132 (1969) 97.
3) A.S. Rinat (Reiner) and M. Stingl, Ann. Phys. NY 65 (1971) 141.
4) A.S. Rinat (Reiner), L.P. Kok and M. Stingl, Nucl. Phys. A190 (1972) 328.
5) L. Stewart, J.E. Brolley, Jr. and L. Rosen, Phys. Rev. 119 (1960) 1649.
6) J.E. Brolley et al., Phys. Rev. 117 (1960) 1307; T.R. King and R. Smythe, Nucl. Phys. A183 (1972) 657.
7) L.P. Kok and L. Grebben, private communication.
8) M. Stingl and A.S. Rinat (Reiner), Phys. Rev., to be published.

Search for 3-Body Forces With $^1H + ^2H$ at E_p = 23 MeV

J. M. Lambert* and P. A. Treado*
Georgetown University, Washington, D. C. 20007 and Naval Research Laboratory
and
R. G. Allas, L. A. Beach, R. O. Bondelid and E. M. Diener†
Naval Research Laboratory, Washington, D. C. 20375

In few nucleon reactions, the dominant processes are usually strong only in restricted regions of phase space. In kinematically complete experiments on the $^2H(^1H,2p)n$ reaction, the usually observed processes are quasifree scattering (QFS) and the two-nucleon final state interaction (FSI). In both cases, if one detector is moved 15 degrees away from the kinematic condition for maximum cross section, the amplitudes for these reactions drop substantially. In the CM system, angular ranges of this order for one detector cover only a few percent of the available phase space.

In this experiment, a small region of phase space is investigated which corresponds to another distinct kinematic condition; namely, one in which the neutron is at rest in the total CM system. This condition produces a configuration in which the three nucleons are colinear. For momentum to be transferred in such a way as to produce this condition, the three nucleons have to be close together, hence, three-body effects and off-energy shell might be relatively important. This particular geometry also has a unique feature in that the colinearity of the three outgoing particles could be thought of as leading to a shielding effect of three-body forces transmitted by meson exchange. At appropriate pairs of angles, off-energy shell effects would show up along the entire locus, while "shielding" effects would be very localized, possibly producing a narrow peak or valley. The $^1H + ^2H$ reaction was chosen because it is the one three-body final state system for which the Faddeev equations have been solved so that the experiment can be examined in terms of deviations from these calculations.

Angle pairs for which the neutron is at rest in the total CM system were calculated and theoretical predictions from the Haftel modified Ebenhöh code were examined to optimize the experimental cross sections while avoiding spectral structure due to other effects in the phase space region of interest. This condition usually occurs near a minimum in the cross section. For angle pairs 57.8°-57.8° and 50.0°-66.1°, data with better than 7% statistics in the region of interest were obtained. Standard coincidence electronics with particle identification and background measurement were used.

For purposes of comparison, each point of the experimental spectra was divided by theoretical predictions and each spectrum was normalized at the FSI peak. The theoretical predictions are from the potentials (labeled HA1-8 and HA2-8, Haftel and Petersen, this conference) which give extreme peak to valley ratios for reasonable potentials. Figure 1 shows the two predictions for $\theta_1 = \theta_2 = 57.8^\circ$. Figure 2 shows the ratios of experiment to theory for these potentials at the two angle pairs. The comparison shows statistically significant deviations from a straight line, centered around the expected energy, and having a width of about 150 keV in the neutron CM energy. Although it is very tempting to speculate that a three-body effect has been observed, it is necessary to consider other possible interpretations. The deviations which appear in this region could be associated with on-shell changes in a separable potential or could be caused by the tail of the quasifree scattering process. It will be necessary to do the experiment at some angles that are quite close to those used but that do not include the kinematic condition of the neutron being at rest in the CM system to determine if the peak structure in the comparison disappears. This should provide evidence to indicate whether a three-body force has been seen, or whether this is just a particularly sensitive condition for observing the effects of the potentials used in the Faddeev analysis.

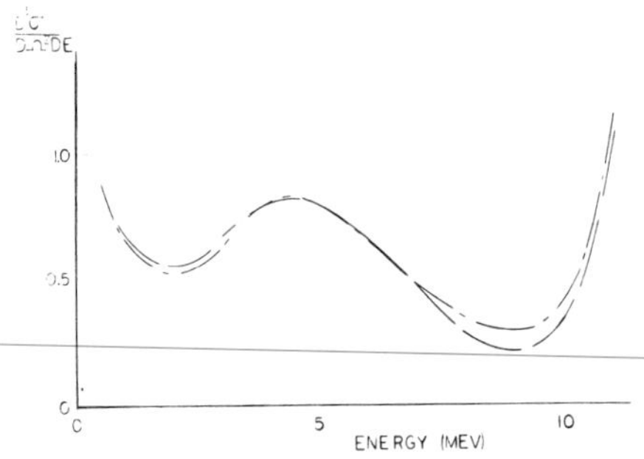

Figure 1. This shows the results of two potentials (see text) produced by the Ebenhöh code for $\theta_1 = \theta_2 = 57.8^o$.

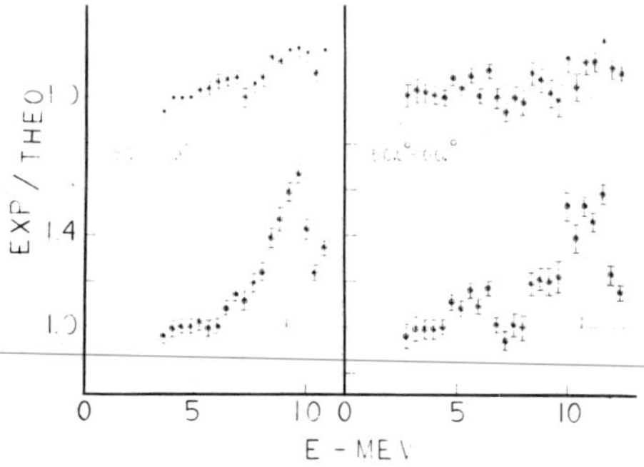

Figure 2. The ratio of experiment to theory at two angle pairs for the two potentials shown in Fig. 1. The lines under the peaks indicate the position of $E_n(CM) = 0$.

*Work Partially Supported by the National Science Foundation

†National Research Council - Naval Research Laboratory Research Associate

SCATTERING-EQUIVALENT MULTICHANNEL SYSTEMS AND MANY-BODY FORCES

A. W. Sáenz and W. W. Zachary*

Naval Research Laboratory, Washington, D. C. 20375 and

Department of Physics, Princeton University, Princeton, N. J. 08540

Consider a system of N particles governed by the usual type of nonrelativistic Hamiltonian H acting in an appropriate Hilbert space $\mathcal{H}$. We have derived rigorously necessary and sufficient conditions for this "original" system to yield the same scattering amplitudes as a "transformed" system with Hamiltonian $\widetilde{H}=W^{\dagger}HW$, W being a unitary operator on $\mathcal{H}$. Generally $\widetilde{H}$ has nonlocal many-body interactions, even if H has only local two-body ones. This work encompasses the presence of spin and of identical particles and generalizes our earlier studies[1] for the single-channel case.

Let us first take the N particles to be distinguishable, for simplicity. For each cluster decomposition D of these particles,[2] let W_D be a unitary operator on $\mathcal{H}$ which, in an appropriate sense, preserves this cluster structure and is asymptotic to W for large intercluster separations. The channels of the transformed system corresponding to the channels $\alpha, \beta, \ldots$ of the original one are denoted by $\widetilde{\alpha}, \widetilde{\beta}, \ldots$ Consider a fixed cluster decomposition D and an arbitrary channel α consistent with D. The respective subspaces of channel states for α and $\widetilde{\alpha}$ are $\mathcal{H}_\alpha$ and $\widetilde{\mathcal{H}}_\alpha = W_D^{\dagger}\mathcal{H}_\alpha$, and the respective channel Hamiltonians are H_α and $\widetilde{H}_\alpha = W_\alpha^{\dagger} H_\alpha W_\alpha$, W_α being the restriction of W_D to $\widetilde{\mathcal{H}}_\alpha$.

For given α, β, the original and transformed systems are said to be scattering-equivalent for the scattering processes $\alpha \to \beta$ and $\widetilde{\alpha} \to \widetilde{\beta}$ if the amplitudes $(\psi_\beta, S_{\beta\alpha}\varphi_\alpha)$ and $(\widetilde{\psi}_\beta, \widetilde{S}_{\beta\alpha}\widetilde{\varphi}_\alpha)$ are equal for all $\varphi_\alpha \in \mathcal{H}_\alpha$ and $\psi_\beta \in \mathcal{H}_\beta$. Here $\widetilde{\varphi}_\alpha \equiv W_\alpha^{\dagger}\varphi_\alpha \in \widetilde{\mathcal{H}}_\alpha$, $\widetilde{\psi}_\beta \equiv W_\beta^{\dagger}\psi_\beta \in \widetilde{\mathcal{H}}_\beta$, and $S_{\beta\alpha}$ and $\widetilde{S}_{\beta\alpha}$ are the partial S operators for these respective processes. Under very weak assumptions, we have proved that this equivalence holds for given α, β if the norm of $(W-W_\gamma)\exp(-it\widetilde{H}_\gamma)\chi_\gamma$ approaches zero as $t \to \pm\infty$ for each $\chi_\gamma \in \widetilde{\mathcal{H}}_\gamma$ when $\gamma = \alpha, \beta$. This condition is very natural physically. Examples of operators W and W_D for which the latter condition (which is entailed by our more general necessary and sufficient ones) is obeyed for all channels have been constructed

*On leave from the Naval Research Laboratory

for arbitrary $N \geq 2$.

The above theory, with due inclusion of the Pauli principle, yields a practical phenomenological way of studying many-body nuclear force effects on few-nucleon electromagnetic form factors, by varying these form factors via a suitable W while keeping strictly constant all scattering amplitudes.[3] We illustrate this for the (p,p,n) system. Let H_3 be a Hamiltonian for this system involving only two-body interactions and adequately predicting the relevant p-d elastic scattering and breakup data. $\tilde{H}_3 \equiv W^\dagger H_3 W$ is the corresponding transformed Hamiltonian, involving both two and three-body interactions, where now W is a unitary operator (e.g. as given in the above examples) for which the original and transformed systems are equivalent for the pertinent channels. The deuteron and ^{3}He wave functions $\tilde{\psi}_d$ and $\tilde{\psi}_h$ for the transformed system are, of course, specified in terms of the corresponding wave functions ψ_d and ψ_h of the original one via W and the appropriate W_D. The idea is to compare the form factors predicted by $\tilde{\psi}_d$ and $\tilde{\psi}_h$ with those entailed by ψ_d and ψ_h, to determine whether the observed ^{3}He charge form factor can be significantly better fit by including three-body forces, without spoiling the agreement with the measured deuteron charge form factor.

REFERENCES

1. A. W. Sáenz and W. W. Zachary, "On the Equality of S Operators Corresponding to Unitarily Equivalent Hamiltonians in Single-Channel Scattering" (to be published).

2. Our terminology is essentially that of W. Hunziker, in Lectures in Theoretical Physics, edited by A. O. Barut and W. E. Brittin (Gordon and Breach, New York, 1968), Vol. XA, p. 1.

3. M. I. Haftel has performed detailed computations of this kind for ^{3}He and reports on them in these proceedings.

Unitary Three Body Transformations and Three Body Forces

Michael I. Haftel

Naval Research Laboratory, Washington, D. C. 20375

Many investigators of two-nucleon off-shell effects have employed two-body unitary transformation methods.[1] This paper proposes a three-body unitary transformation technique that should prove useful in the study of three-body forces. We apply a special case to the calculation of the ^{3}He charge form factor ($F_{ch}(q^2)$) since the disagreement between theory and experiment here has suggested to some workers the existence of three-nucleon forces.[2,3]

We start with a Hamiltonian for the three-body system $H = \sum_{i=1}^{3} T_i + \sum_{j>i=1}^{3} V_{ij}$, where V_{ij} is the interaction between pairs. This Hamiltonian yields the bound state wave function(s) Ψ_B and energy (energies) E_B. We now introduce a unitary operator U that may have two and three-body parts. By the unitary property the "transformed" Hamiltonian $\tilde{H} = UHU^\dagger$ gives the same spectrum as H, but yields altered bound-state eigenfunctions $\tilde{\Psi}_B = U\,\Psi_B$. Under certain conditions[4] H and $\tilde{H}$ will yield the same scattering predictions in all channels. In these cases $\tilde{H} = \sum_{i=1}^{3} T_i + \sum_{j>i=1}^{3} \hat{V}_{ij} + H^{(3)}$, where $\hat{V}_{ij}$ and V_{ij} are phase-shift-equivalent two-body potentials, and $H^{(3)}$ represents all three-body force terms. As in the two-body unitary transformation method we view U as a device of manufacturing three-body Hamiltonians that leave spectra and scattering properties unaltered but have altered off-shell properties and three-body forces. Of course U does alter wave-functions and, for example, could alter three-body bound-state form factors.

To date Faddeev calculations have been unable to simultaneously fit the binding energy (E_B) and the position of the diffraction minimum of ^{3}He with two-nucleon forces that fit the two-body data. Even reasonable (and some unreasonable) off-shell variations of "realistic" potentials fail to fit both E_B and $F_{ch}(q^2)$ simultaneously.[2] One could conceivably use the two-body off-shell freedom to fit E_B and then use U to fit $F_{ch}(q^2)$.

This paper applies to the ^{3}He problem the rank-two separable unitary operator

$$U = 1 - (1-\cos\theta)\,(|g_1\rangle\langle g_1| + |g_2\rangle\langle g_2|) + \sin\theta\,(|g_1\rangle\langle g_2| - |g_2\rangle\langle g_1|) \quad (1)$$

where $\langle \vec{r}_1\vec{\rho}_1|g_n\rangle = g_n(R) = Ne^{-\alpha_n R}(\beta_{on} + \beta_{1n}R + \beta_{2n}R^2)$, $\langle g_n|g_m\rangle = \delta_{nm}$, $\vec{r}_i = \vec{x}_j - \vec{x}_k$, $\vec{\rho}_i = (3)^{-\frac{1}{2}}(\vec{x}_j + \vec{x}_k - 2\vec{x}_i)$ (ijk cyclic) and $R = \sqrt{r_i^2 + \rho_i^2}$. The choice of V_{ij} is the Malfliet-Tjon V potential (MTV)[5]. While MTV is a S-wave spin-independent potential, it yields an E_B of 7.65 MeV, in close agreement to the ^{3}He experimental binding energy (7.8 MeV), and yields a $F_{ch}(q^2)$ similar to that predicted by "realistic" potentials.

Fig. 1 summarizes the results of the calculation. Each curve illustrates the values of $\langle H^{(3)}\rangle \equiv \langle\tilde{\Psi}_B|H^{(3)}|\tilde{\Psi}_B\rangle$ and the position of the diffraction minimum of $F_{ch}(q^2)$ for certain fixed values of $F_{ch}(q^2{=}6\,\mathrm{fm}^{-2})$. Results for these curves are for $\alpha_1 = \alpha_2$, $\beta_{01} = \beta_{02} = 0$, $\beta_{11} = \beta_{22} = -1$, $\beta_{21} \approx -\alpha_1/7$ ($\langle g_1|g_2\rangle = 0$ fixes β_{12}); however, points calculated from alternate choices of the parameters fall very close to the plotted curves. The values of α_1 range from 1.0 fm^{-1} (which corresponds to the rightmost part of each curve) to approximately 2.4 fm^{-1} (the topmost part of each curve). This figure indicates that there are limits to the minimum value of q^2 for the diffraction minimum without drastically changing the values of $F_{ch}(q^2)$ at lower q^2, at least for the type of transformation considered. For instance, the experimental value of $F_{ch}(q^2 = 6fm^{-2})$ is consistent with diffraction minima ranging into q^2=13.6 fm^{-2} (experiment $q^2 \approx 11.8\ fm^{-2}$) but with a cost of $\langle H^{(3)} \rangle$ = -6MeV. Figure 2 illustrates $F_{ch}(q^2)$ for one of the best overall fits (at least to $q^2 \leq 11\ fm^{-2}$). Here $\langle H^{(3)} \rangle$ = = -2.31 MeV, a value consistent with meson-theoretic estimates.[6]

Meson exchange corrections to $F_{ch}(q^2)$ are important near the diffraction minimum.[7] Curves like Fig. 1 should be useful in future calculations when one has a better idea of what the "experimental" values of $F_{ch}(q^2)$ are <u>before</u> exchange corrections. Also, more general transformations should be explored, especially cases where $\hat{V}_{ij} \neq V_{ij}$ (for Eq. (1) $\hat{V}_{ij} = V_{ij}$). Then one could see in addition what range of three-body results one should expect from two-body forces.

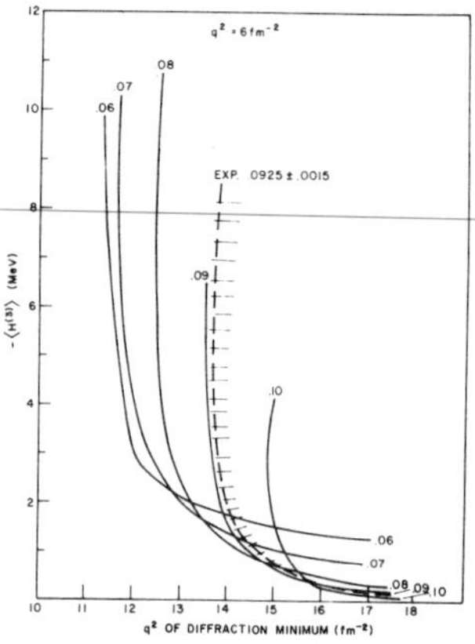

Fig. 1 $\langle H^3 \rangle$ and the position of the diffraction minimum for fixed values of $F_{ch}(q^2)/(G_{Ep}+5G_{Ep})$ for q^2=6fm^{-2}.

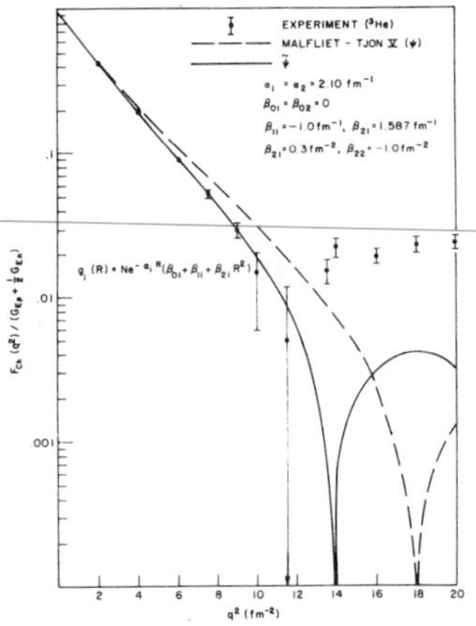

Fig. 2 $F_{ch}(q^2)$ for a particular choice of U (θ=5°). G_{Ep} and G_{En} are the proton and neutron electric form factors.

REFERENCES

1. F. Coester, S. Cohen, B. Day and C. M. Vincent, Phys. Rev. <u>C1</u>, 769 (1970).
2. M. I. Haftel, Phys. Rev. <u>C7</u>, 80 (1973); E. P. Harper, Y. E. Kim, and A. Tubis, Phys. Rev. <u>C6</u>, 1601 (1972).
3. D. D. Brayshaw, Phys. Rev. <u>C7</u>, 1731 (1973).
4. See the contribution of A. W. Sáenz and W. W. Zachary in these proceedings.
5. R. A. Malfliet and J. A. Tjon, Nucl. Phys. <u>A127</u>, 161 (1969).
6. S. N. Yang, Brookhaven preprint.
7. W. M. Kloet and J. A. Tjon, University of Maryland Technical Report 74-043.

TOTAL REACTION CROSS SECTIONS FOR p-^{3}He AND p-^{4}He

A. M. Sourkes, N. E. Davison, S. A. Elbakr, J. L. Horton*, A. Houdayer†, and W.T.H. van Oers, Cyclotron Laboratory, Department of Physics, University of Manitoba, Winnipeg, Manitoba R3T 2N2

and

R. F. Carlson, Department of Physics, University of Redlands, Redlands, California 92373

For the p-^{3}He system the total lack of σ_R information up to now has hampered theoretical calculations in properly taking into account the inelastic channels. The small amount of accurate p-^{4}He σ_R data to date has not permitted an adequate evaluation of the existing theoretical predictions. Thus, using protons from the University of Manitoba Cyclotron, total reaction cross sections for the p-^{3}He and p-^{4}He systems have been measured at ten and sixteen energies respectively, between 15 and 50 MeV, employing a beam attenuation technique [1].

The results for the p-^{3}He system are indicated in Fig. 1. Shown also for comparison are the 30 and 50 MeV results deduced from the work of Griffiths et al. [2] and the sum of the 3- and 4-body breakup reaction channel cross sections for the n-^{3}He system reported by Drosg et al. [3]. The present results enable an extension of resonating group calculations [4] to include an imaginary potential for the p-^{3}He system, accounting for the inelastic channels [5]. As well, extension of phase shift analyses (such as [6]) to include imaginary phases becomes more feasible.

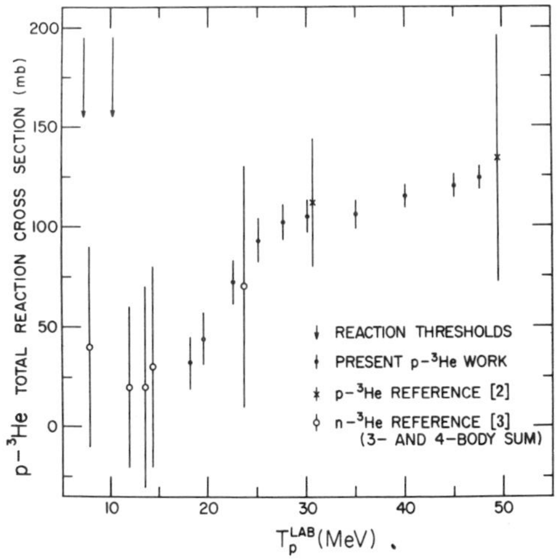

Figure 1

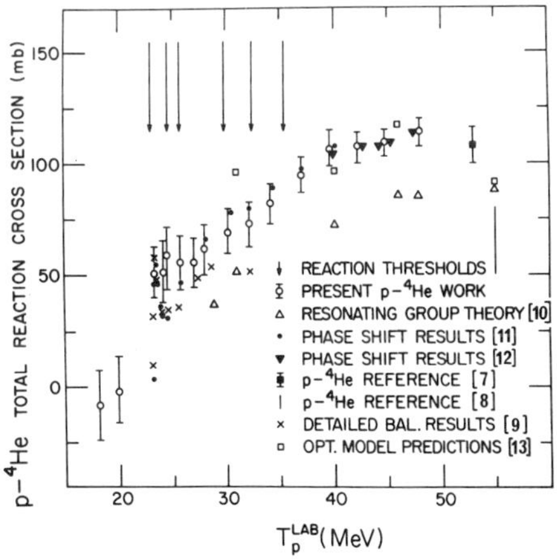

Figure 2

The p-^{4}He σ_R results are shown in Fig. 2 along with the 53 MeV cloud chamber measurement of Cairns et al. [7] and the 55 MeV estimate of Hayakawa et al. [8]. Detailed balance values from the ^{3}He(d,p)^{4}He work of Yarnell et al. [9] are also plotted for comparison and show good agreement with the present work between 23 and 25 MeV where ^{4}He(p,d)^{3}He is the dominant reaction channel. The resonating group predictions [10] exhibits a reasonable energy dependence, but are uniformly low. Results from the phase shift analysis of Plattner et al. [11] between 23 and 40 MeV and from the extension of this analysis to 50 MeV by Houdayer and van Oers [12] are in good agreement with experiment. The latter work used the σ_R measurements to select among final phase shift solutions. Thompson, Epstein and Sawada [13] in their 30 to 55 MeV p-^{4}He optical model analysis got σ_R values of about 150 mb. Only at a sacrifice in the elastic scattering fit were they able to obtain σ_R values near 95 mb (Fig. 2).

In summary, the p-^{3}He σ_R results provide information enabling extension of the theoretical treatment of this system. The p-^{4}He σ_R results indicate the need for refinement in the model dependent calculations.

REFERENCES

[1] R. F. Carlson, W. F. McGill, T. H. Short, J. M. Cameron, J. R. Richardson, W.T.H. van Oers, J. W. Verba, P. Doherty, and D. J. Margaziotis, to be published.

[2] R. J. Griffiths, and S. A. Harbison, Astr. J. 158, (1969) 711.

[3] M. Drosg, D. K. McDaniels, J. C. Hopkins, J. D. Seagrave, R. H. Sherman, and E. C. Kerr, Phys. Rev. C9, (1974) 179.

[4] I. Reichstein, D. R. Thompson, and Y. C. Tang, Phys. Rev. C3, (1971) 2139.

[5] R. E. Brown, Private Communication.

[6] J. R. Morales-Pena, Ph.D. Thesis, University of California at Davis (1970) unpublished.

[7] D. J. Cairns, T. C. Griffith, G. J. Lush, A. J. Metheringham, and R. H. Thomas, Nucl. Phys. 60, (1964) 369.

[8] S. Hayakawa, N. Horikawa, R. Kajikawa, K. Kikuchi, H. Kobayakawa, K. Matsuda, S. Nagata, and Y. Sumi, J. Phys. Soc. Jap. 19, (1964) 2004.

[9] J. L. Yarnell R. H. Lovberg, and W. R. Stratton, Phys. Rev. 90, (1953) 292.

[10] D. R. Thompson, Y. C. Tang, and R. E. Brown, Phys. Rev. C5 (1972) 1939.

[11] G. R. Plattner, A. D. Bacher, and H. E. Conzett, Phys. Rev. C5, (1972) 1158.

[12] A. Houdayer, and W.T.H. van Oers, Private Communication.

[13] G. E. Thompson, M. B. Epstein, and T. Sawada, Nucl. Phys. A142, (1970) 571.

Work supported in part by the Atomic Energy Control Board of Canada and by the Research Corporation.

---

* Present address: Radiation Laboratory, Aberdeen Proving Grounds, Aberdeen, Maryland.

† Present address: Foster Radiation Laboratory, McGill University, Montreal, Quebec.

DIFFERENTIAL CROSS SECTION MEASUREMENTS AND PHASE SHIFT ANALYSIS FOR p-^{4}He ELASTIC SCATTERING BETWEEN 20 AND 55 MeV[†]

N. E. Davison, S. A. Elbakr, A. Houdayer[††], A. M. Sourkes and W.T.H. van Oers
Cyclotron Laboratory, Department of Physics, University of Manitoba
Winnipeg, Manitoba R3T 2N2
and
A. D. Bacher, Department of Physics, Indiana University, Bloomington, Indiana 47401

Recently Bacher et al. [1] reported the measurement of a set of p + ^{4}He polarization angular distributions of improved accuracy. Subsequently, a phase shift analysis was made by Plattner et al. [2] of these polarization data together with available differential cross section information in the energy range 20-40 MeV.

In order to extend the p + ^{4}He phase shift analysis of Plattner et al. [2] to higher energies and to calculate the elastic attenuation correction in p + ^{4}He total reaction cross section measurements [3], differential cross section angular distributions of accuracy comparable to that of the polarization data were measured using the proton beam of the University of Manitoba sector-focussed cyclotron.

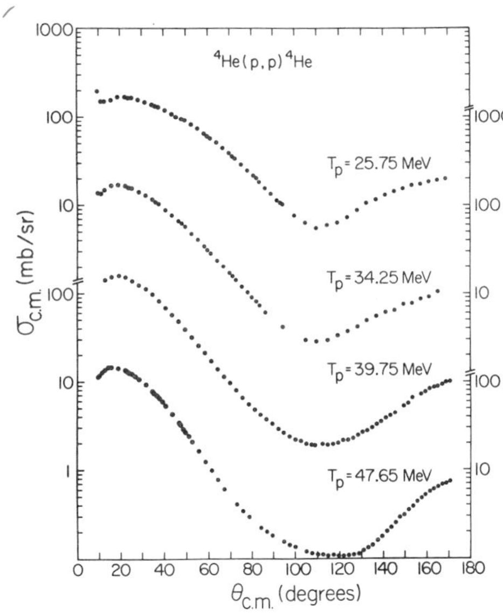

Figure 1

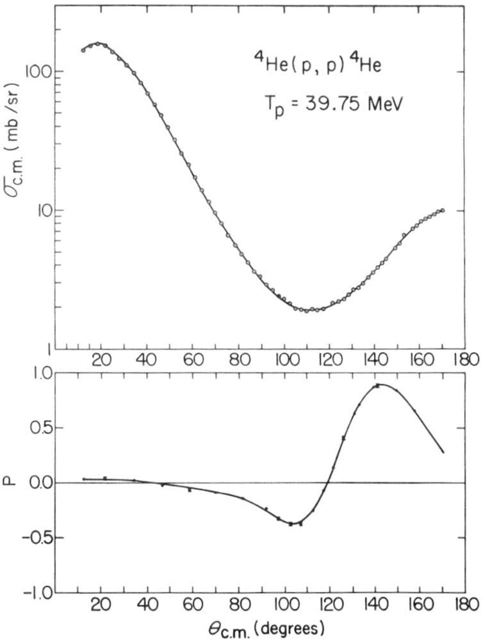

Figure 2

Angular distributions covering the range 10° to 170° cm in steps of 2.5° or 5.0° were measured at incident energies of 21.85, 23.90, 25.75, 28.10, 30.35, 32.15, 34.25, 36.90, 39.75, 42.45, 44.05, 44.95 and 47.65 MeV. The uncertainty in the incident energy was less than 150 keV while the beam energy spread was estimated to be 150 keV (FWHM). The measurements were made using a 10 cm diameter gas cell covered with a 25 μ thick kapton-H foil and filled with helium gas of 99.99% purity to a pressure slightly in excess of 1 atmosphere. Scattered particles were detected in an array of eight equally spaced detectors with an appropriate gas scattering collimation system. The typical relative error of the data points is 2%; the error in the absolute scale of the measurements is less than 3%. Four representative differential cross section angular distributions are shown in Fig. 1.

For the energy range 20-55 MeV a phase shift analysis has been made of the present differential cross section data together with selected polarization (e.g. those of ref. [2]) and differential cross section data from the literature [4]. Also included in the analysis were total reaction cross section data [3], which provide an important constraint on the imaginary phases, as well as an angular distribution of the spin-rotation parameter at 48 MeV [5]. Typical fits obtained to the differential cross section and polarization angular distributions are shown in Fig. 2. Except for the $^2D_{3/2}$ resonance the phase shifts show a smooth energy dependence. The higher energies analysed required the inclusion of small G-wave phases.

REFERENCES

† Work supported in part by the Atomic Energy Control Board of Canada

†† Present address: Foster Radiation Laboratory, McGill University, Montreal, Quebec

[1] A. D. Bacher, G. R. Plattner, H. E. Conzett, D. J. Clark, H. Grunder, and W. F. Tivol, Phys. Rev. C5, 1147 (1972).

[2] G. R. Plattner, A. D. Bacher, and H. E. Conzett, Phys. Rev. C5, 1158 (1973).

[3] A. M. Sourkes, N. E. Davison, S. A. Elbakr, J. L. Horton, A. Houdayer, and W.T.H. van Oers, to be published.

[4] E. T. Boschitz, M. Chabre, H. E. Conzett, E. Shield, and R. J. Slobodrian, in Proceedings of the 2nd International Symposium on Polarization Phenomena of Nucleons, edited by P. Huber and H. Schopper (Birkhauser Verlag, Basel, 1966). p.328;
M. K. Craddock, R. C. Hanna, L. P. Robertson, and B. W. Davies, Phys. Letters 5, 335 (1963);
B. W. Davies, M. K. Craddock, R. C. Hanna, Z. J. Moroz, and L. P. Robertson, Nucl. Phys. A97, 241 (1967;
S. Hayakawa, N. Horikawa, R. Kajikawa, K. Kikuchi, H. Kobayakawa, K. Matsuda, S. Nagata, and Y. Suma, J. Phys. Soc. Japan 19, 2004 (1964).

[5] T. C. Griffith, D. C. Imrie, G. J. Lush, and L. A. Robbins, Phys. Rev. 146, 626 (1966).

SUPPRESSION OF $V_{\Sigma\Sigma}$ IN A SEPARABLE $^3S_1-^3D_1$ ΛN POTENTIAL

L. H. Schick, Department of Physics and Astronomy, University of Wyoming, Laramie, Wyo. 82070 USA.

Most of our knowledge of the low-energy lambda-nucleon (ΛN) interaction comes from the bound states of light hypernuclei. In at least one of these ($^5_\Lambda He$) suppression of the reaction $\Lambda N \to \Sigma N$ (i.e. suppression of the closed Σ channel) plays a crucial role, especially in the use of this hypernucleus to determine the spin triplet ΛN scattering length and effective range. Therefore, an investigation of suppression effects in the determination of phenomenological ΛN potentials has been undertaken. Since, of course, no direct data is available on the ΛN interactions with various parts suppressed, the results predicted by a meson-theoretic-potential (MTP), adjusted to give the experimental values of the free Λp scattering length and effective range, are used as input. Because of its ease of applicability to multiparticle problems a nonlocal separable ΛN potential was chosen for the phenomenological interaction.

In the present work it was found that the simplest form of two-channel $^3S_1-^3D_1$ ΛN separable potential could not be made "suppression equivalent" to the MTP of Brown, Downs, and Iddings.[1)]

The hyperon-nucleon potential energy operator was taken to be a 2x2 matrix operator, each of whose matrix elements could be written in a relative-momentum-space representation as $\langle \vec{k}_x | V_{xy} | \vec{k}_y \rangle = p_{xy} \lambda_{xy} v_x(\vec{k}_x)\, v_y(\vec{k}_y)$, where x, y = Λ or Σ, $v_x(\vec{k}_x) = C_x(k_x) + T_x(k_x)\, S_{12}(\vec{k}_x)/\sqrt{8}$, $p_{\Lambda\Lambda} = 1$, $p_{\Lambda\Sigma}=\varepsilon$, and $p_{\Sigma\Sigma}=\gamma$. Here C_x and T_x are the central and tensor parts of v_x, and S_{12} is the usual tensor operator, The parameter ε is the Σ-channel suppression parameter; i.e. for ε=1 the Σ channel is unsuppressed, for ε=0 this channel is fully suppressed, and for $0<\varepsilon<1$ this channel is partially suppressed. Similarly, γ is the Σ-potential suppression parameter; i.e. $\gamma \to 0$ suppresses $V_{\Sigma\Sigma}$. In previous work [2)] the dependence of the low-energy ΛN interaction on ε and the tensor force were studied. Here it is the γ-dependence of this interaction which was of prime interest.

With this potential it was a simple matter to calculate the values of the S=1 ΛN inverse scattering length $\alpha(\varepsilon,\gamma) \equiv 1/a(\varepsilon,\gamma)$ and effective range $r_0(\varepsilon,\gamma)$ as functions of ε and γ.

Four relations among the potential parameters were obtained by fitting $\alpha(0,\gamma)$, $r_0(0,\gamma)$ - both of which are actually independent of γ - $\alpha(1,\gamma)$, and $r_0(1,\gamma)$ to the BD12 values for Λp scattering. One of these relations had the form $\lambda^2_{\Lambda\Sigma} = A_\Lambda(\gamma)\, B_\Sigma$, where $A_\Lambda(\gamma)$ depends on γ and v_Λ, whereas B_Σ is independent of γ, is dependent on v_Σ, and satisfies $B_\Sigma>0$ for any C_Σ and T_Σ.

The specific requirements that with both the Σ channel and the Σ potential unsuppressed (ε=1, γ=1) α and r_0 yield the values of BD12, and that with ε=1 but $V_{\Sigma\Sigma}$ at least partially suppressed ($\gamma=\nu<1$) α and r_0 yield the BD12 values, then result in

$A_\Lambda(1) = A_\Lambda(\nu)$. With u defined by $u(k_\Lambda) \equiv C_\Lambda{}^2(k_\Lambda) + T_\Lambda{}^2(k_\Lambda) = u_0[1 + u_r k_\Lambda{}^2 + 0(k_\Lambda{}^4) + \cdot\cdot\cdot]$, this last relation determines u_r in terms of the BDI2 values for $\alpha(0,1)$, $\alpha(1,1)$, $\alpha(1, \nu)$ $r_0(0,1)$ and $r_0(1,\nu)$.

However, the BDI2 values for α and r_0 satisfy $\alpha(\varepsilon,\gamma)<0, [\alpha(1,\gamma) - \alpha(0,\gamma)]>0$, $[r_0\ (0, \gamma)-r\ (1, \gamma)] >0$ for all ε, γ. Under these conditions it is easy to show from the form of $A_\Lambda(\gamma)$ that $\lambda^2_{\Lambda\Sigma}>0$ holds for all γ only if $u_r>-0.5[r_0(0,1)-r_0(1,\gamma)]/[\alpha(1,\gamma) - \alpha(0,1)]$ for all γ. But no matter what value of ν is used to determine u_r this inequality is violated.

In sum then, independent of the shapes C_Λ, T_Λ, C_Σ, and T_Σ, the simple $^3S_1$-$^3D_1$ separable ΛN potential used here cannot be made to fit for any fixed $\nu<1$ the ($\varepsilon=0$, γ), ($\varepsilon=1$, $\gamma=1$), and ($\varepsilon=1,\gamma=\nu$) values of the S=1 Λp scattering length and effective range as given by the MTP of BDI2.

On the basis of simplicity it might be tempting to fit the form of separable potential discussed above, using a specific shape for each C_x and each T_x, to the $^3S_1$-$^3D_1$ ΛN phase shift over a range of Λ energies and use the resulting potential in a hypernuclear bound state calculation. The results presented here show that for hypernuclei where suppression is important this temptation should be resisted.

REFERENCES

1) J. T. Brown, B. W. Downs, and C. K. Iddings, Nuclear Physics B47 (1972) 138. Hereafter referred to as BDI2.

2) See R. Russell and L. H. Schick, Phys. Rev. D5 (1972) 2804 and references cited there.

THE PHILLIPS PLOT FOR PHASE EQUIVALENT SEPARABLE POTENTIALS

N.J. McGurk and H. Fiedeldey
Physics Department, University of South Africa
P.O. Box 392, Pretoria, South Africa

Using the phase equivalent separable potentials constructed in ref. 1 by direct inversion of the constrained (n,p) energy-dependent scattering data of MacGregor, Arndt and Wright (MAW)[2] we have calculated the doublet n-d scattering length ^2a in addition to the triton binding energy E_T. In the $^3S_1$ - $^3D_1$ channel, however, we had to replace the experimental mixing parameter $\varepsilon_1(k)$ by a single analytical form $\varepsilon_1(k) = \arctan\,(0.026k^2/k_D^2)$, with $E_D = -\hbar^2k_D^2/m$ for reasons explained in ref. 1. To investigate the off-shell sensitivity of ^2a and E_T we also performed additional calculations with potentials having a different off-shell behaviour in the singlet channel. In table 1 we list the off-shell parameters ($\gamma_0,\beta_0,\gamma_1,\beta_1$) using the same notation as in ref. 1 (γ_0,β_0 are the singlet parameters). All these potentials (except no. 1) produce underbinding in the triton like realistic local potentials.

Table 1

	$\gamma_0(fm^{-3/2})$	$\beta_0(fm^{-1})$	$\gamma_1(fm^{-3/2})$	$\beta_1(fm^{-1})$	P_D	E_T(MeV)	^2a(fm)
1)	1.9	1.2	4	1.2	4.45	8.4±0.2	0.77
2)	1.9	1.2	6	1.2	5.65	7.7±0.2	1.32
3)	1.9	1.2	8	1.2	6.64	7.0±0.2	1.82
4)	1.9	1.2	10	1.2	7.41	6.6±0.2	2.27
5)	1.9	1.2	12	1.2	8.01	6.2±0.2	2.66
6)	6.0	1.2	4	1.2	4.45	7.3±0.2	1.58
7)	10.0	1.2	4	1.2	4.45	7.0±0.2	2.01

The estimated error of our doublet scattering calculations is less than 10% of the calculated value of ^2a.

In fig. 1 E_T is plotted against ^2a, the so-called Phillips plot.

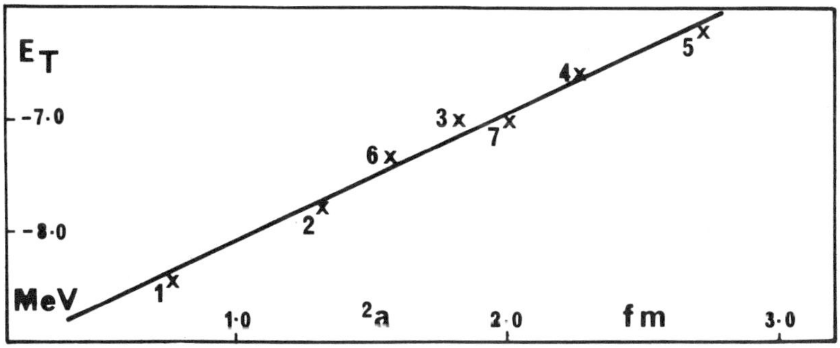

FIGURE 1 Phillips plot for the potentials of table 1.

All the calculated points lie within the quoted accuracies on a straight line, the Phillips line, with the equation

$$^{2}a = 0.87(E_T + 8.5) + 0.62$$

This equation should be compared with the one given by Harms[3] for a large number of separable potentials, i.e. $^{2}a = 0.75(E_T + 8.5) + 0.75$. Those potentials are not phase equivalent, however, and consequently the points are scattered on a wider band around the line. It has been shown by van Wageningen[4] and Bakker[5], that for potentials which are not phase equivalent, the points are scattered in a broad band around the Phillips line. McGurk et al[6] showed that the same conclusions hold for purely central potentials. It should be noted that eq. 1 is in good agreement with the experimental result of Dilg et al $^{2}a = 0.65 \pm 0.04$ fm[7]. The results of table 1 (potentials 1 to 5) also imply that E_T and ^{2}a are both linearly related to P_D for a fixed singlet potential, a result also found by Brady[8]. The linear relations are of the same accuracy as in fig. 1. The lines are shifted if we change the singlet interaction only, (potentials 1, 6 and 7), but in such a way that the same linear relation between E_T and ^{2}a still holds, independent of the $^{1}S_0$ potential and apparently generally independent of off-shell effects.

From table 1 it is seen that potential 1 nearly reproduces the experimental results for E_T and ^{2}a. However, this potential has a rather low value of P_D (=4.45%). Although this still lies within the range of acceptable values of P_D (4-8%), present indications favour a higher P_D value of 6-8%. We also note that off-shell effects due to variations in the $^{1}S_0$ potential (potentials 1, 6 and 7) of the order of 1.5 MeV in E_T and 1.2 fm in ^{2}a are obtained, with a fixed triplet potential (P_D = 4.45%).

Finally, we point out that potential 3 with a P_D and E_T closely corresponding to the values for the Reid potential produces a significantly lower value of ^{2}a, than the value of 2.4 ± 0.2 fm[9] for the Reid potential.

One of the authors (HF) acknowledges assistance by the Council for Scientific and Industrial Research, Pretoria.

REFERENCES

1) N.J. McGurk, H. de Groot, H. Fiedeldey and H.J. Boersma, Phys. Lett. 49B (1974) 13
2) M.H. MacGregor, R.A. Arndt and R.M. Wright, Phys. Rev. 182 (1969) 1714
3) E. Harms, "XIV Latin American School of Physics". Caracas, 1972A, Reidel, Holland
4) R. van Wageningen, Symposium on "Present Status and Novel Developments in the Nuclear Many-Body Problem", Rome, 1972
5) B.L.G. Bakker, Ph.D. thesis, Vrije Universiteit, Amsterdam, Netherlands, 1973 (unpublished)
6) N.J. McGurk and H. de Groot, to be published
7) W. Dilg, L. Koester and J.D. Seagrave, Phys. Lett. 36B (1971) 208
8) T. Brady, Phys. Lett. 32B (1970) 85
9) R.A. Malfliet and J.A. Tjon, Ann. Phys. 61 (1970) 425

UPA AT POSITIVE ENERGIES FOR HARD CORE POTENTIALS

A.V.Lagu, C.Maheshwari and V.S.Mathur
Department of Physics, Banaras Hindu University,
Varanasi-221005,India

A separable potential or a separable approximation to a local potential brings the two-body t-matrix into a separable form. Such a separable t-matrix, in turn, reduces the two-dimensional coupled Faddeev integral equations of three-body problem to one-dimensional equations thus reducing computational problems considerably[1]. Consequently several separable expansions and approximations to the t-matrix have been proposed in recent years[2]. Amongst these unitary pole expansion of Harms[3] and its one term version unitary pole approximation (UPA)[4] have been extensively studied at both negative and positive energies and their goodness tested through two and three body calculations[5]. In general UPA has been found to be adequately satisfactory except for some results by Bakker and Ruig[6] and van Haeringen and Kok[6]. We, however, find that no direct calculations have been reported at positive energies for local potentials containing hard core although Brady et al.[5] report on Tabakin's rank-2 separable potential containing hard shell repulsion. Since we had already tested the UPA at negative energies for hard core potentials[7], we have carried the calculations to positive energies as well and report them here. Moreover, these calculations are also required as input in a scheme we are pursuing in connection with off-shell continuation of two-body t-matrix with bound states.

Since we can go uniquely from the half off shell t-matrix to full off shell t-matrix through Low equation, instead of comparing the full off shell t-matrix elements at positive energies (through their real and imaginary parts), we have compared rather the half shell functions defined by Baranger et al. (BGMS)[8] and the phase shifts. Moreover, the determination of either of half shell function or the phase shift automatically gives the other.

For our calculations we have chosen the Herzfeld potential (hard core with square well outside)[9] with parameters V_0(well depth)=1.54 fm^{-2}, a (hard core radius)=0.4 fm, b=1.737 fm and where the potential binds the two body with energy B=0.435 MeV. Using the method outlined by Laughlin and Scott[10], we determine the exact S-wave half shell function (the expression is not given here being lengthy) and phase shift as

$$\delta_0(k) = \tan^{-1}\left\{ \frac{k}{\beta} \tan \beta (b-a)\right\} - kb$$

where $\beta^2 = k^2 + V_o$. It may be stated that the definition of on shell t-matrix element followed by us confirms to reference (10) and not to BGMS. We have checked that the half shell function $\varphi_{exact}(k,k')$ and $\delta_o(k)$ are consistent. Moreover, $\delta_o(k)$ can also be obtained from the definition of S-matrix[2].

For the calculation of UPA half shell function and phase shifts, we make use of the expressions already given for Herzfeld potential[9] and obtain UPA half shell function $\varphi_{UPA}(k,k')$ and phase shift $\delta_{UPA}(k)$ in terms of the form factors[3] $\Psi_1(q,-B)$ and expressions involving principal value integrals. We have again checked that φ_{UPA} and δ_{UPA} are consistent. It may be stated that the expressions for φ_{UPA} and δ_{UPA} are valid irrespective of the potential. For calculation of principal value integrals, we have followed the method suggested by BGMS.

We have calculated φ_{exact}, φ_{UPA}, δ_o and δ_{UPA} at various values of k. We find that UPA phase shifts are neither in very good or bad agreement with exact phase shifts. The exact phase shift changes sign at about k^2=2.6 fm^{-2} while UPA phase shift does not. Regarding half shell functions, we find that at low momenta, the agreement is poor but at higher momenta, it is much better. This behavior is similar to the one reported by us[7] earlier at negative energies for a potential containing hard core. Thus at least for the potential under considera-tion, UPA does not seem particularly good at positive energies. We also find that Levinson theorem holds in case of this potential even though it contains a hard core.

AVL and CM are grateful to CSIR & DAE respectively for financial help.

REFERENCES

1) T.A.Osborn,SLAC Report 79(1967), unpublished
2) A.V.Lagu,Ph.D.Thesis (unpublished), B.H.U.(1972) & references cited
3) E.Harms,Phys.Rev.C1(1970)1667
4) C.Lovelace,Phys.Rev.135(1964)B1225; M.Fuda,Nucl.Phys.A116(1968)83
5) J.S.Levinger and J.O'Donoghue,Preprint (1972) & references cited, & other papers of Prof.Levinger's group;T.Brady et al.,Phys.Rev. 186(1969)1069.
6) B.L.G.Bakker and P.Ruig,Proc.Int.Conf.Few Part.Prob.Nucl.Inter.(1972) Ed.I.Slaus et al.(North-Holland),p.357;van Haeringen & Kok,Ibid,p.351
7) V.S.Mathur et al.,Nucl.Phys.A178(1972)365
8) M.Baranger et al., Nucl.Phys. A138(1969)1
9) C.Maheshwari et al.,Nucl.Phys. & Sol.State Phys.(India)14B(1972)223.
10) R.Laughlin and B.L.Scott, Phys.Rev. 171(1968)1196.

DIFFERENTIAL CROSS SECTIONS FOR ELASTIC AND INELASTIC SCATTERING OF 14.1 MEV NEUTRONS BY TRITONS

Shoji Shirato, Yasuhiko Suda and Shizuo Tsuruta (Department of Physics, St. Paul's (Rikkyo) University, Nishi-Ikebukuro, Tokyo, Japan) and Shinsho Oryu (Department of Physics, Science University of Tokyo, Noda, Chibaken, Japan)

The n-t elastic data have been well reproduced by the optical model.[1] At 14 MeV, some discrepancy is found among the data.[2-6] Data on the spectrum and angular distribution for the $^3H(n,d)2n$ reaction are very few. It is a problem how well both the (n,n) and (n,d) cases are explained by the Faddeev approach as well as the optical model.

We obtained the differential cross sections for n-t elastic and inelastic scattering at 14.1 MeV by measuring t, d and p with a counter telescope. The apparatus and arrangement are the same those used in our n-p and n-d experiments.[7] The measurements were done at θ_o = 0°, 10°, 20°, 30° and 40°. A ^3H-Ti-Cu target (12.3 Ci of ^3H in 2.77 mg/cm^2 Ti) was used together with a Ti-Cu target for background measurement. The particle identification for p, d and t was almost complete by two dimensional analysis of ΔE and E signals from 76 μm and 1.5 mm Si(Li) detectors. The results are shown in figs. 1-3.

The optical model calculation through the program ELIESE-3[8] was done using the following parameters obtained by interpolation from those of Sherif and Podmore[1]: r_0=1.488 fm, a_0=0.144 fm, V_0 =44.88 MeV; r_i=1.501 fm, a_i=0.378 fm, W_I=3.30 MeV; r_{so}=1.049 fm, a_{so}=0.289 fm, V_{so}=2.89 MeV. The calculated total cross section is 1079 mb. A least squares fit to our data and Kootsey's data[3] at Θ_n < 90° gives $\sigma_{el}(\Theta_n) = 76.0(1-1.48P_1(x)+1.22P_2(x)-0.028P_3(x)+0.034P_4(x)+0.068P_5(x))$ mb/sr, where $x = \cos\Theta_n$ and P_ℓ is Legendre polynomials, with the 10 % error in the number of ^3H atoms. This gives the total elastic cross section of 955 mb, which is in agreement with the calculation (992 mb for singlet 2n and 965 mb for triplet 2n). As seen in fig. 2, the optical model calculation as well as the Born approximation calculation of Gammel and MacKellar[9] could not reproduce our data for the $^3H(n,d)2n$ reaction.

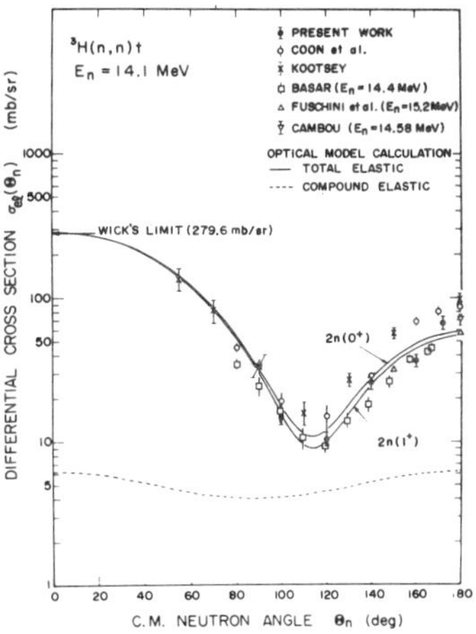

Fig. 1. Angular distribution for n-t elastic scattering.

We made the Faddeev approach using the separable potentials under the assumption of the 3-body system (n+n+d). First, by means of χ^2-minimization to the elastic data we obtained 3 complex parameters a,b,c, to be multiplied to the pick-up, knock-on and remained black-box amplitudes in order to produce the whole elastic amplitude. Here the black box was assumed to be independent of angle. By using these elastic amplitudes, the breakup-deuteron energy differential cross section was calculated taking into account the n-n final-state interaction of the singlet scattering length a_{nn} = -17 fm and the effective range r_{nn} = 2.84 fm. The quantity $\tilde{\gamma}$ related to the triton form factor was adjusted. Details of the theory are described elsewhere.[10] The calculated spectra smeared with the experimental resolution of 390 keV in FWHM are shown in fig. 3. The shape of the calculated spectra is satisfactory, although the cross sections are too large in the very forward direction. The result of integration over the spectrum is shown in fig. 2.

It is concluded that the triton breakup can be explained to a considerable extent by our frozen deuteron model and the observed shape of the breakup deuteron spectrum is reproduced by taking into account the n-n final-state interaction.

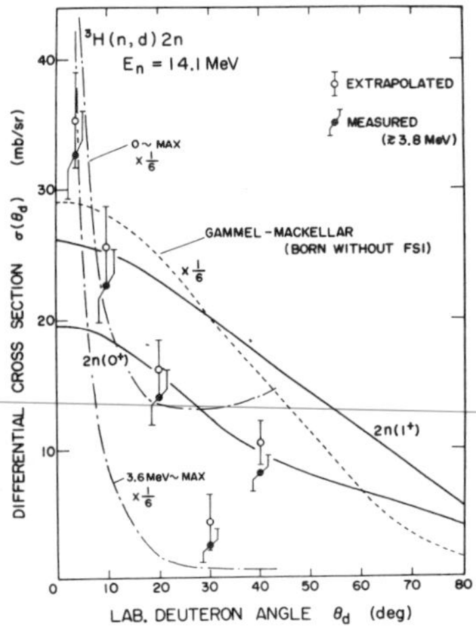

Fig. 2. Angular distribution for ^{3}H(n,d)2n. The solid and the solid-dot curves represent the optical model and our model calculations respectively.

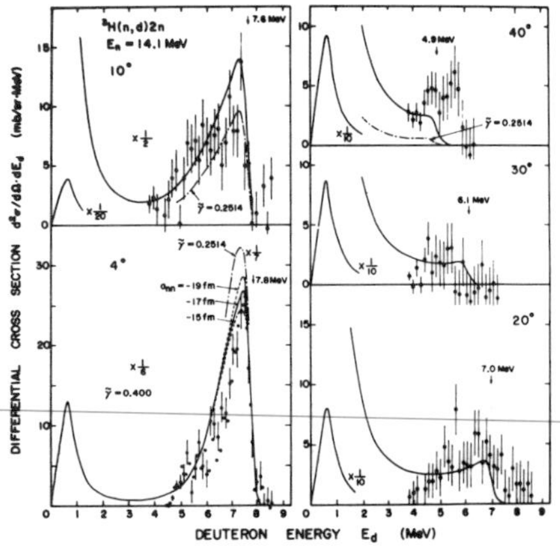

Fig. 3. Energy spectra of breakup deuterons. The solid curves represent our model calculation in the case $\tilde{\gamma}$ = 0.400 (weakly bounded n-d), and the solid-dot curves in the case $\tilde{\gamma}$ = 0.2514 (n -d bound state). a_{nn} = -17 fm in both cases.

REFERENCES

1) H.S. Sherif and B.S. Podmore, Few Particle Problems in the Nuclear Interaction, ed. I. Slaus et al. (North-Holland, Amsterdam, 1972) p. 691.

2) J.H. Coon, C.K. Bockelman and H.H. Barschall, Phys. Rev. 81 (1951) 33.

3) J.M. Kootsey, Nucl. Phys. A113 (1968) 65.

4) I. Basar, Few Body Problems, Light Nuclei, and Nuclear Interactions, ed. G. Paic and I. Slaus (Gordon and Breach, New York, 1968) p. 867.

5) E. Fuschini, C. Maroni, A. Uguzzoni, E. Verondini and A. Vitale, Nuovo Cim. 48B (1967) 190.

6) F. Cambou, CEA/R-2002 (1961).

7) S. Shirato and K. Saitoh, J. Phys. Soc. Japan 36 (1974) 331.
S. Shirato, K. Saitoh, N. Koori and R.T. Cahill, Nucl. Phys. A215 (1973) 277.

8) S. Igarasi, JAERI-1224 (1972).

9) J.L. Gammel and A.D. MacKellar, Phys. Rev. 133 (1964) B1476.

10) S. Oryu, Contribution in this conference.

LOW ENERGY FORMULA FOR n-d SCATTERING

T.Sasakawa

Tohoku University, Department of Physics,

980-Sendai, Japan

So far, the n-d low energy behavior[1] has been theoretically treated on the basis of the dispersion relation of coupled two-body channels with the exchange process[2),3)]. In the present paper, the Faddeev equation is formulated and solved in the coordinate space. The Jost theory of the potential scattering is extended to the three-body system. The corresponding operator is now 3×3 matrix, if we keep only s-wave. This matrix is designated by J. The matrix J obeys the Alt-Grassberger-Sandhas-type equation[4], but with the Green's function which vanishes at large distances from the center of mass of the three-body system.

Except for the difference in the Green's function which is employed, similar manipulations as in the K-matrix theory of the three-body problem[5] yields the n-d scattering amplitude in a simple form

$$T = T_{\alpha\alpha} + T_{\beta\alpha} + T_{\gamma\alpha} = \mathrm{Im}F/F \qquad (1)$$

for $E < 0$, k real, where $E = -|E_d| + \dfrac{3\hbar^2}{4m}k^2$, (2)

and

$$F = 1 + A + 2A' \quad , \qquad (3)$$

$$A = \langle \alpha | J | \alpha'' \rangle \quad , \quad A' = \langle \beta | J | \alpha'' \rangle \qquad (4)$$

with

$$|\alpha\rangle = u_\alpha \frac{\sin k_\alpha \rho_\alpha}{\sqrt{k_\alpha}\,\rho_\alpha} \frac{1}{\sqrt{4\pi}} \chi_\alpha ,$$

$$|\alpha''\rangle = u_\alpha \frac{e^{ik_\alpha \rho_\alpha}}{\sqrt{k_\alpha}\,\rho_\alpha} \frac{1}{\sqrt{4\pi}} \chi_\alpha , \qquad (5)$$

u_α; the wave function of deuteron.

χ_α; the spin-isospin function.

After examined the analyticity of A and A', we found that F is approximately parametrized as

$$F \simeq \left\{1 - \frac{g}{z}\left[\tan^{-1}\frac{3}{2}z - \tan^{-1}\frac{z}{2} + \frac{i}{2}\ln\left((1 + 9z^2/4)/(1 + z^2/4)\right)\right] + az^2 + i(bz + cz^3)\right\}\cdot\left[d + ifzE^2(\ln E - i\pi)/\pi\right]. \quad (6) \quad \left(z = k/\alpha ,\ \alpha = \sqrt{(m/\hbar^2)|E_d|}\right)$$

This function satisfies the three-body unitarity. This function is an entire function on the upper half of the complex k-plane, thus satisfying the causality. The bound state is at the zero of the first factor of (6).

If we use eqs.(1) and (6), and assume that f is small the phase shift reads

$$k\cot\delta = \frac{1 - (g/z)\,(\tan^{-1}(3/2z) - \tan^{-1}(z/2)) + az^2}{-\,(g/2z^2)\,\ln((1 + 9z^2/4)/(1 + z^2/4)) + b + cz^2} \,. \qquad (7)$$

The parameters given in Table I yield k cotδ as shown in Fig. 1 (Fig.2) for the doublet (quartet) scattering[6] and also the correct binding energy of triton in the case of the doublet scattering.

	g	a	b	c
doublet	0.537	-0.102	0.607	0.0451
quartet	-1.075	-0.125	1.975	1.147

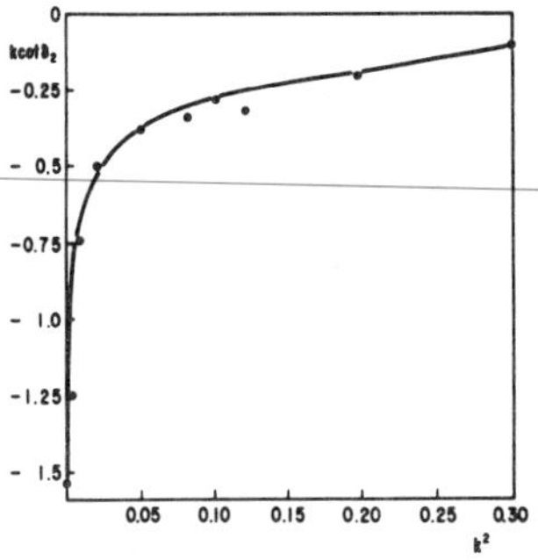

Fig. 1

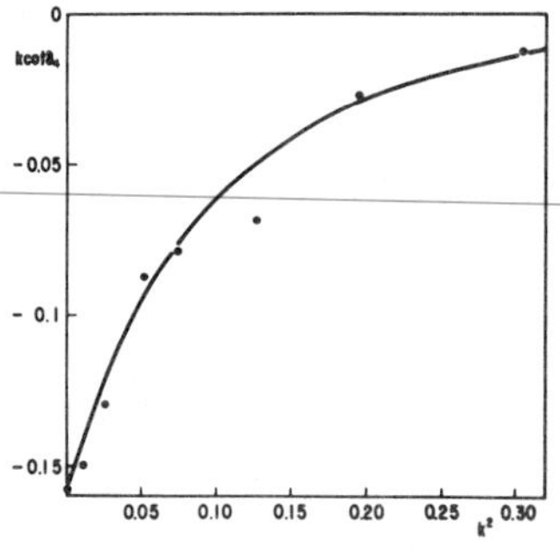

Fig. 2

References

1) W.T.H.van-Oers and J.D.Seagrave, Phys. Lett. 24B(1967)562.
2) G.Barton and A.C.Phillips. Nucle. Phys. A132(1969)97.
3) A.S.Rinat and M.Stingel, Ann. Phys. 65(1971)141.
4) E.O.Alt, P.Glassberger and W.Sandhas Nucle. Phys. B2(1967)167.
5) T.Sasakawa, Nucle. Phys. A203(1973)496.
6) W.Dilg, L.Koester and W.Nistler, Phys. Lett. 36B(1971)208.

This work was financially supported by the fund for research of science of the Japanese Ministry of Education.

THE DEPENDENCE OF THE TRITON BINDING ENERGY ON THE ASSUMED TWO-NUCLEON CORE

I.R. Afnan and J.M. Read, School of Physical Sciences, The Flinders University of South Australia, Bedford Park. 5042. Australia.

Recent calculations of the triton binding energy (E_T) using the Reid soft-core potential have been in close agreement and the value $E_T = 7.1$ MeV is now generally accepted.[1)] From the results of many authors, it has also become clear that E_T is almost exclusively dependent on the two-nucleon bound state in the $^3S_1$-$^3D_1$ channel, and the antibound state in the $^1S_0$ channel.[1)] Using the unitary pole approximation (UPA) for the two-nucleon T-matrix, when correctly constructed in the $^3S_1$-$^3D_1$ channel, gives a value of $E_T = 7.15$ MeV for the Reid potential.[2)] This accuracy suggests that the UPA provides a simple method of investigating the dependence of the trinucleon binding energy on the assumed form of the two-nucleon interaction.[3)]

The available local phenomenological potentials agree exactly at large distances due to the universal acceptance of the one-pion-exchange tail (OPET), but disagree appreciably at small distances. This difference is most easily categorized by the core used to provide the short-range repulsion thought characteristic of the inter-nucleon force. The various interactions may be divided into the three classes:

(i) Soft-core (SC) - those for which $V(r) = \frac{\text{const.}}{r}$ as $r \to 0$.

(ii) Hard-core (HC) - those for which $V(r) = \infty$ for $r \lesssim r_c$.

(iii) Super soft-core (SSC) - those for which $V(r) = \text{const.}$ as $r \to 0$.

In this paper we investigate the dependence of the triton binding energy E_T and n-d doublet scattering length on the type of core used in the two-nucleon interaction. To determine E_T and ^2a, we use the UPA for the two-nucleon T-matrix in the Faddeev equations. In Table 1 we compare the UPA values of E_T with calculations of other authors, for potentials of the above form, to ensure the accuracy of this scheme. These results show that the UPA is certainly accurate enough for the purpose of this study.

In Table 2 we present the values of E_T and ^2a for various SC, HC, and SSC potentials. We have also included the values of the deuteron binding energy E_d and quadrupole moment Q_d to ensure that these potentials correctly predict the deuteron bound state properties. It is clear that the HC potentials predict $E_T \lesssim 7.0$ MeV ($^2a \simeq 2$F) and are in disagreement with the experimental value of $E_T = 8.48$ MeV ($^2a = .65\pm.04$F). The SC potentials, the RSC, and RSCA are also nearly 1.5 MeV below the experimental value. The difference in E_T for these two potentials is due to the difference in their deuteron D-state probabilities which in turn reflects the RSCA potential's lower quadrupole moment. The SSC potentials seem to increase E_T and lower ^2a relative to the HC and SC models. However, the TSA and TSB have an unrealistically low value of Q_d ($Q_d = 0.261$ F^{-2}). This is partially corrected in the TSC model ($Q_d = 0.278$ F^{-2}), but then both E_T and ^2a move away from the experimental results. If the quadrupole moment was further increased to agree with the latest value of $Q_d = 0.287\ F^{-2}$, E_T would be depressed to around 7 MeV.

We may conclude that the local phenomenological potentials constrained to have the OPET and to give a correct value of Q_d underbind the triton by some 1.5 MeV, *irrespective of the type of core assumed.* Thus, unless one is prepared to introduce a very large contribution from three-body forces, it is not possible to reconcile the calculated values of E_T with experiment, using local potentials.

TABLE 1

Potential	Type	E_T(UPA)	$E_T(^1S_0\ \&\ ^3S_1-^3D_1)$	E_T†
Reid Soft-Core	SC	7.15 MeV	7.02 MeV [2]	7.1 MeV [1]
Gammel-Brüeckner	HC	7.61 MeV	-	7.75 ± 0.2 MeV [4]
Tourriel-Sprung	SSC	7.42 MeV	7.46 MeV [5]	7.34 MeV [5]

†The higher partial waves are included in these calculations.

TABLE 2

Potential	Type	E_d(MeV)	P_d(%)	$Q_d(F^{-2})$	E_T(MeV)	^2a(F)	Reference
Reid Hard-Core	HC	2.2246	6.50	0.277	6.96	1.97	6
Hamada-Johnston	HC	2.2689	7.02	0.284	6.96	2.04	7
Yale	HC	2.1888	6.95	0.276	6.73	2.10	8
Reid Soft-Core (RSC)	SC	2.2246	6.47	0.280	7.15	1.80	6
Reid Soft-Core Alternate (RSCA)	SC	2.2246	6.22	0.276	7.32	1.68	6
Tourriel-Sprung							9
Model A (TSA)	SSC	2.224	4.43	0.261	7.52	1.46	
Model B (TSB)	SSC	2.228	4.25	0.261	7.62	1.38	
Model C (TSC)	SSC	2.224	5.45	0.278	7.42	1.52	

REFERENCES

1) J.S. Levinger, *The Two and Three-Body Problem*, Springer Tracts in Modern Physics (1974).

2) I.R. Afnan and J.M. Read, Phys. Rev. C8, 1294 (1973).

3) J.M. Read, Ph.D. Dissertation, The Flinders University of S.A., March 1974.

4) M.A. Hennell and L.M. Delves, Nucl. Phys. A204, 552 (1973).

5) A. Laverne and C. Gignoux, Nucl. Phys. A203, 597 (1973).

6) R.V. Reid, Ann. Phys. (N.Y.) 50, 411 (1968).

7) T. Hamada and I.D. Johnston, Nucl. Phys. 34, 382 (1962).

8) K.E. Lassila, *et al.* Phys. Rev. 126, 881 (1962).

9) R. Tourriel and D.W.L. Sprung, Nucl. Phys. A201, 193 (1973).

TRINUCLEON RESULTS FOR SEVERAL ONE-BOSON EXCHANGE POTENTIALS

I.R. Afnan and J.M. Read, School of Physical Sciences, The Flinders University of South Australia, Bedford Park. 5042. Australia.

Recent calculations have shown that the phenomenological interaction models, as typified by the Reid soft-core and Hamada-Johnston potentials underbind the triton by some 1.5 MeV.[1)] This deficit has been attributed to contributions from three-body forces, relativistic corrections, or to an incorrect off-shell behaviour of the two-nucleon T-matrix. The features common to these potentials are their one-pion-exchange tail (OPET) and their locality in each partial wave. If the quadrupole moment of the deuteron (Q_d) is to be correctly predicted, then the D-state probability P_d is constrained to approximately 7%.[2)] This large value of P_d leads to underbinding in ^{3}H.

For the one-boson-exchange potentials (OBEP), the non-locality introduced by the exchange of scalar and vector mesons permits a P_d of 4-5½%, yet still retaining the OPET and allowing a correct value of the quadrupole moment. By comparing the values of P_d and Q_d given in the table for the Bryan-Scott and Bryan-Gersten potentials, we see that any constraint between these quantities is removed upon the introduction of explicit non-locality. Both these potentials represent a similar fit to the two-nucleon scattering data. A low value of P_d is also attractive because it gives a reasonable value of the deuteron magnetic moment, without the introduction of excessively large corrections.

We have tabulated the triton binding energy E_T and the doublet n-d scattering length ^2a for a number of realistic OBEP. These values were obtained by solving the Faddeev equations, using the unitary pole approximation (UPA) for the two-body T-matrix in the $^1S_0$ and $^3S_1$-$^3D_1$ channels.[3)] This approximation has been shown to give an accuracy of approximately .1 MeV for the Reid soft-core potential,[4)] and by directly comparing the two-body T-matrix elements, we conclude that it is even more accurate for the OBEP. Included also in the table are the deuteron binding energy E_d, D-state probability P_d, and quadrupole moment Q_d.

These results show that the OBEP do, in fact, lead to a closer agreement with experiment. The only exception is the Ueda-Green potential which overbinds the triton. This is due to its excessively large deuteron binding energy. The variation in E_T and ^2a for the other three OBEP also reflects the variation in the deuteron observables. We feel that most of the objections that are used against the OBEP are due to the impreciseness of the fit to the two-nucleon bound state parameters. Future potentials of this type must correctly predict E_d, Q_d, and the effective range parameters, if they are to give meaningful results in a many-particle context.

Finally, we see that the discrepancy between the experimental value of E_T = 8.48 MeV and the values of E_T predicted by the OBEP may be explained by assuming very reasonable contributions from the higher partial waves (≈ 0.2 MeV),[1] and from relativistic corrections (≈ 0.25 MeV).[1] Given that the non-locality predicted by these potentials is substantially correct, it is not necessary to introduce any contribution from three-body forces in order to reconcile theory and experiment. By comparing the results described in this paper with a similar calculation for the phenomenological potentials,[5] it is evident that the triton binding energy is very sensitive to the degree of non-locality present in the two-nucleon interaction. It is to be hoped that the triton electric form factor can be used to give more information on this sensitivity.

TABLE

Potential	E_d(MeV)	P_d(%)	$Q_d(F^{-2})$	E_T(MeV)	^2a(F)	Reference
Bryan-Scott	2.1977	5.47	0.258	7.94	1.10	6
Bryan-Gersten	2.3006	5.06	0.275	7.81	1.43	7
Stagat *et al.*	2.3344	4.56	0.270	8.31	1.15	8
Ueda-Green	2.7105	5.60	0.259	8.83	1.59	9
Experiment	2.2246	4-8	0.287	8.48	0.65±0.04	

REFERENCES

1) For an excellent discussion of the *status quo* concerning realistic three-nucleon calculations, see J.S. Levinger, *The Two and Three-Body Problem*, Springer Tracts in Modern Physics (1974).

2) P. Signell, Proc. Int. Conf. on Few Particle Problems in the Nuclear Interaction, Los Angeles (1972) (North-Holland 1973).

3) The details of this calculation are given in J.M. Read's Ph.D. Dissertation, The Flinders University of South Australia (March 1974).

4) I.R. Afnan and J.M. Read, Phys. Rev. C8, 1294 (1973).

5) I.R. Afnan and J.M. Read, *The Dependence of the Triton Binding Energy on the Assumed Two-Nucleon Core*, This conference.

6) R.A. Bryan and B.L. Scott, Phys. Rev. 177, 1435 (1969).

7) R.A. Bryan and A. Gersten, Phys. Rev. D6, 341 (1972).

8) R.W. Stagat, F. Riewe, and A.E.S. Green, Nucl. Phys. B10, 289 (1969).

9) T. Ueda and A.E.S. Green, Nucl. Phys. B10, 289 (1969).

LOW ENERGY FORMULA FOR n-d SCATTERING

T.Sasakawa
Tohoku University, Department of Physics,
980-Sendai, Japan

So far, the n-d low energy behavior[1] has been theoretically treated on the basis of the dispersion relation of coupled two-body channels with the exchange process[2),3)]. In the present paper, the Faddeev equation is formulated and solved in the coordinate space. The Jost theory of the potential scattering is extended to the three-body system. The corresponding operator is now 3×3 matrix, if we keep only s-wave. This matrix is designated by J. The matrix J obeys the Alt-Grassberger-Sandhas-type equation[4], but with the Green's function which vanishes at large distances from the center of mass of the three-body system.

Except for the difference in the Green's function which is employed, similar manipulations as in the K-matrix theory of the three-body problem[5] yields the n-d scattering amplitude in a simple form

$$T = T_{\alpha\alpha} + T_{\beta\alpha} + T_{\gamma\alpha} = \mathrm{Im}F/F \tag{1}$$

$$\text{for } E < 0,\ k \text{ real, where } E = -|E_d| + \frac{3\hbar^2}{4m}k^2 , \tag{2}$$

and

$$F = 1 + A + 2A' , \tag{3}$$

$$A = \langle \alpha | J | \alpha'' \rangle , \quad A' = \langle \beta | J | \alpha'' \rangle \tag{4}$$

with

$$|\alpha\rangle = u_\alpha \frac{\sin k_\alpha \rho_\alpha}{\sqrt{k_\alpha}\,\rho_\alpha} \frac{1}{\sqrt{4\pi}} \chi_\alpha ,$$

$$|\alpha''\rangle = u_\alpha \frac{e^{ik_\alpha \rho_\alpha}}{\sqrt{k_\alpha}\,\rho_\alpha} \frac{1}{\sqrt{4\pi}} \chi_\alpha ,$$

[u_α; the wave function of deuteron.

χ_α; the spin-isospin function.]

After examined the analyticity of A and A', we found that F is approximately parametrized as

$$F \simeq \Big\{1 - \frac{g}{z}\Big[\tan^{-1}\frac{3}{2}z - \tan^{-1}\frac{z}{2} + \frac{i}{2}\ln((1 + 9z^2/4)/(1 + z^2/4))\Big] + az^2 + i(bz + cz^3)\Big\}\cdot[d + ifzE^2(\ln E - i\pi)/\pi]. \tag{6}$$

$$(z = k/\alpha,\ \alpha = \sqrt{(m/\hbar^2)|E_d|})$$

This function satisfies the three-body unitarity. This function is an entire function on the upper half of the complex k-plane, thus satisfying the causality. The bound state is at the zero of the first factor of (6).

If we use eqs.(1) and (6), and assume that f is small the phase shift reads

$$k \cot \delta = - \frac{1 - (g/z)\left(\tan^{-1}(3/2z) - \tan^{-1}(z/2)\right) + az^2}{-(g/2z^2)\ \ln((1 + 9z^2/4)/(1 + z^2/4)) + b + cz^2} . \quad (7)$$

The parameters given in Talbe I yield k cotδ as shown in Fig. 1 (Fig.2) for the doublet (quartet) scattering[6] and also the correct binding energy of triton in the case of the doublet scattering.

	g	a	b	c
doublet	0.537	-0.102	0.607	0.0451
quartet	-1.075	-0.125	1.975	1.147

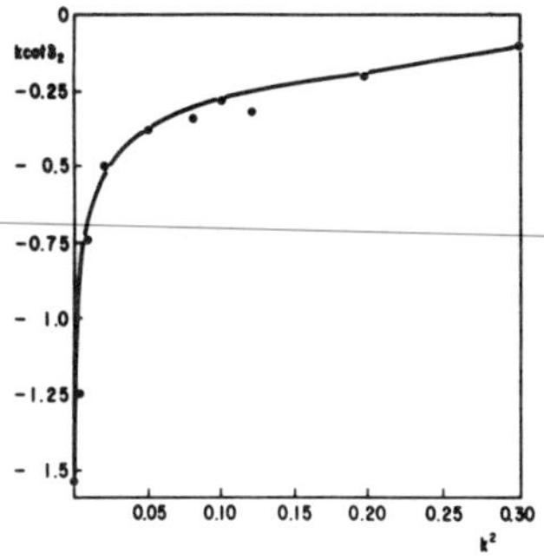

Fig. 1

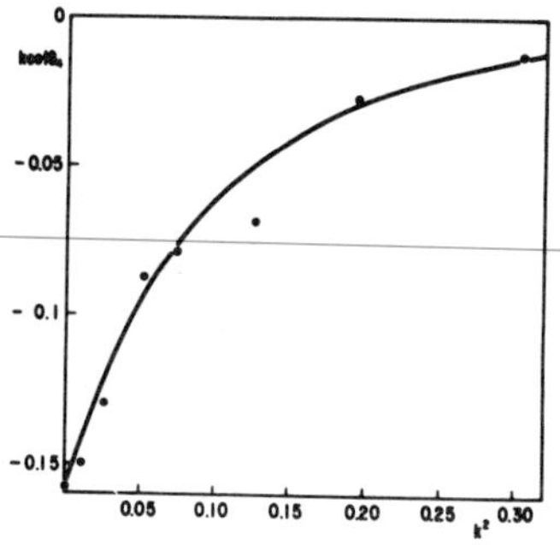

Fig. 2

References

1) W.T.H.van-Oers and J.D.Seagrave, Phys. Lett. 24B(1967)562.
2) G.Barton and A.C.Phillips. Nucle. Phys. A132(1969)97.
3) A.S.Riant and M.Stingel, Ann. Phys. 65(1971)141.
4) E.O.Alt, P.Glassberger and W.Sandhas Nucle. Phys. B2(1967) 167.
5) T.Sasakawa, Nucle. Phys. A203(1973)496.
6) W.Dilg, L.Koester and W.Nistler, Phys. Lett. 36B(1971)208.

This work was financially supported by the fund for research of science of the Japanese Ministry of Education.

PRECISE MEASUREMENT OF THE ELASTIC N-D DIFFERENTIAL CROSS SECTION AT LOW ENERGIES

L Amtén, L Gönczi, A Johansson, L Nilsson and B Sundqvist
Tandem Accelerator Laboratory, University of Uppsala, Sweden

Scattering phenomena in three-nucleon systems have been studied extensively in recent years both experimentally and theoretically. The general conclusion at the present time seems to be that no significant discrepancies exist between calculated and measured data within the theoretical and experimental uncertainties, which are of the order of 10 %. It has, for instance, not been possible to identify effects due to different off-shell behaviour of the two-nucleon t-matrix, to three-body forces or to the present lack of complete knowledge of the on-shell two-nucleon t-matrix. The dynamics of the reactions seems to be well understood and in order to reveal something new about the validity of the two-nucleon input data it appears necessary that the precision of both calculations and measurements is improved to the 1 % level or better.

Such detailed comparisons between theory and experiment are, however, difficult to make. The ppn system can be handled experimentally with great accuracy (although few attempts have been made to push the accuracy to its limit) but the theoretical treatment is complicated due to Coulomb effects. The nnp system, on the other hand, is easier to calculate but normally hard to study from the experimental point of view due to the problems associated with neutron beams and neutron detection. In the present paper a method to obtain accurate differential cross sections for elastic n-d scattering is discussed. The method can in principle be extended to breakup reactions, such as d(n,pn)n.

The method consists of using as a target a foil of a known mixture of $(CH_2)_n$ and $(CD_2)_n$ and measuring the ratio between the number of recoiling deuterons and protons into a ΔE-E telescope. At low energies the n-p differential cross sections are known to between .25 and .5 % and the desired n-d differential cross section is simply obtained by multiplying the measured ratio with the n-p cross section as calculated from the prescription given by Lomon and Wilson [1].

In this way most of the difficulties with neutron experiments are avoided; e g there is no need to measure the incident neutron flux, the absolute number of deuterium atoms in the target foil or neutron detection efficiencies and a precise knowledge of solid angles and neutron beam optics is not required.

This method has been used to measure the 180° n-d elastic cross section at 8.10 MeV using neutrons produced in the d-d reaction with deuterons from the Uppsala tandem accelerator. The ΔE detector was kept very close to the target foil, which reduced the background protons below 1 %. A clean separation between protons, deuterons and background events was achieved (Fig 1). Two $(CH_2)_n$-$(CD_2)_n$ foils were used with the mixing ratios differing by a factor of two. The measured ratios between the recorded numbers of deuterons and protons divided by the relative content of deuterium and hydrogen atoms in the foils, were 2.364 ± 0.040 and 2.331 ± 0.023 with the average

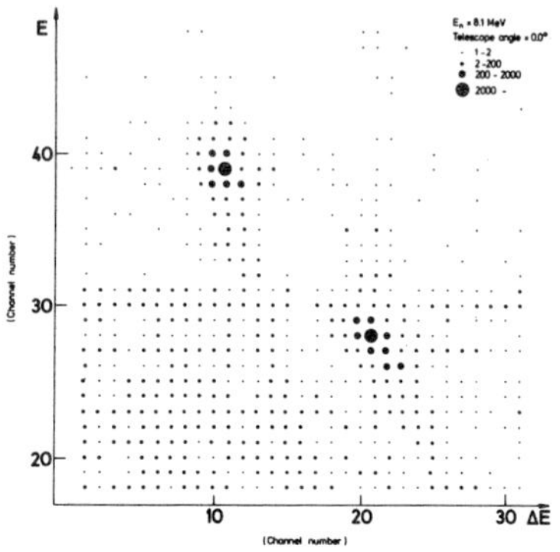

Fig 1 Two-parameter ΔE-E spectrum showing clean proton-deuteron separation

value 2.340 ± 0.020. Background protons and deuterons were measured without target foil and subtracted. The n-p cross section was calculated to be 92.1 ± 0.3 mb/sr [1]. To obtain the n-d cross section at 180°, its shape in the backward direction was used to correct the measured ratio for the finite angular resolution (mean laboratory scattering angle 5.3°). Two different shapes were used, one calculated with separable potentials and Yamaguchi form factors with a programme due to Ebenhöh [2], the other taken from values calculated by Kloet and Tjon [3] with local potentials. The 180° n-d cross sections obtained were 244.6 ± 2.1 and 246.9 ± 2.1 mb/sr, respectively, where the errors are due to counting statistics only. Other errors are estimated to be small, the largest one (.4 %) being due to imperfect knowledge of the angular resolution function.

The elastic n-d scattering is known to be dominated at low energies by the quartet scattering states and in the backward region the exchange process is expected to be mainly responsible. Thus the low-energy n-d cross section at 180° should predominantly be determined by the n-p triplet parameters, which are well-known from n-p scattering and the deuteron properties. (The percentage of D state in the deuteron ground state is probably the least well known quantity of relevance.) One could thus expect little parameter dependence of the calculated 180° cross section and any accurate three-nucleon model should reproduce the experimental values very closely. The Ebenhöh [2] calculations give a 180° n-d cross section of 216.3 mb/sr, which is significantly smaller than our measured value. The calculations by Kloet and Tjon [3] with local potentials yield approximately 10 % larger cross sections and are thus favoured by our measurements.

REFERENCES

1) E Lomon and Richard Wilson, Phys Rev C9 (1974) 1329
2) W Ebenhöh, Nucl Phys A191 (1972) 97 and private communication
3) W M Kloet and J A Tjon, Proc Int Conf on Few Particle Problems (ed Slaus et al) North Holland 1972, p 376

SCATTERING FROM A TWO-BODY BOUND STATE AT HIGH AND INTERMEDIATE ENERGY

J.J. BENAYOUN, Institut des Sciences Nucléaires, BP n° 257 38044 Grenoble,France

In the study of elastic scattering of hadrons at high and intermediate energy, several approximations are usually employed, such as the so-called "Fixed ScattererApproximation" (FSA). The use of these approximations has been usually justified by their successful fits to experimental data.

To better understand and to determine the limits of applicability of these approximations, we have chosen an exactly soluble model on which we could make different approximations and so control the effects by comparison with the exact solution.

The "deuteron" target consists of two identical particles of mass M interacting via a s - wave Yamaguchi potential ; the interaction between the projectile of mass m and the target particles (in s - wave only) are identical and can be different from the interaction which binds the "deuteron". The exact solution of the corresponding Faddeev equations is easily obtained.

Following Foldy and Walecka (1) we have performed closure on the target eigenstates assuming that they are completely degenerate. In the case of non local potentials this approximation, called here approximation FW, does not correspond to the scattering from two fixed scatterers.

To determine the region of applicability of the approximation FW we have first varied the projectile kinetic energy from 50 MeV to 500 MeV. The difference between exact and approximate results decreases when the kinetic energy increases, as one would expect.

In varying the mass ratio R = m/M to approach the limit of very massive, fixed scatterers for the Faddeev equations, we obtained the somewhat surprising result that the limit does not correspond exactly to the FW results (although this difference is small, mainly at high energies ; see fig.). It seems that this is due to the incorrect asymptotic behaviour of the FW wave function ; we do not find the "deuteron" in the scattering amplitude. Finally, we employed this model to study the effect of variation of the binding energy and the target density. In the limit of very strong binding, our exact results correspond to those of the FW model.

The calculations performed by Kujawski and Lambert (2) assuming fixed scatterers for the same model and for three identical particles lead to results very different from ours. It is not surprising since assuming fixed scatterers with non-local interactions leads to the non-conservation of the center-of-mass momentum.

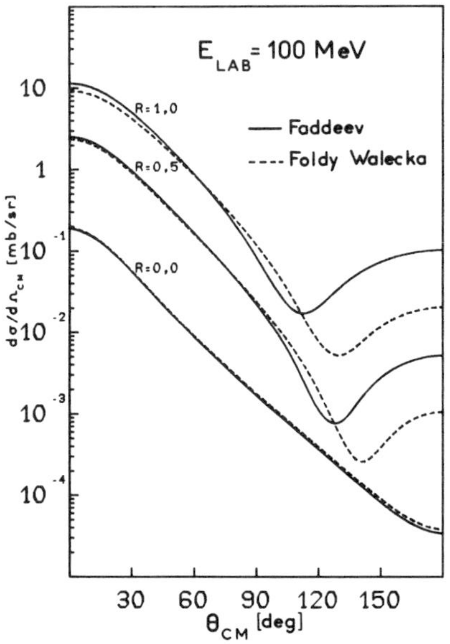

Angular distribution for hadron-"deuteron" elastic scattering at 100 MeV. R is the ratio of projectile mass of the target.

(1) Foldy, L.L., Walecka, J.D., Ann. Phys. 54 (1969) 447.

(2) Kujawski, E., Lambert, E., Ann. Phys. 81 (1973) 591.

CLUSTER LEVELS APPEARED AS EXCITED STATES OF TRITON

T.Sasakawa

Tohoku University, Department of Physics,
980-Sendai, Japan

If the three-body system consisting of deuteron and other particle of partial wave ℓ subjects to the exchange process, the Jost function F_ℓ is parametrized as [1)]

$$F_\ell = 1 + H_\ell(g'_\ell) + \lambda_\ell + a'_\ell Z^2 + iZ^{2\ell+1}(b'_\ell + c'_\ell Z^2), \quad (1)$$

$Z = k/\alpha$ (α; the wave number of deuteron; $0.2315\times 10^{13}\mathrm{cm}^{-1}$), Here $H_\ell(g'_\ell)$ denotes the exchange singularity: For S-wave

$$H_0(g'_0) = -(g'_0/Z)[(\tan^{-1}(3Z/2) - \tan^{-1}(Z/2)) + (1/2)i\,\ln((1 + 9Z^2/4)/(1 + Z^2/4))] \quad (2)$$

and for P-wave

$$H_{\bar{1}}(g'_1) = -(g'_1/Z)[\{(1/Z^2)(1 + 5Z^2/4)(\tan^{-1}(3Z/2) - \tan^{-1}(Z/2))-1/Z\} + i\{(1/2Z^2)(1 + 5Z^2/4)\ln((1 + 9Z^2/4)/(1 + Z^2/4))-1\}]. \quad (3)$$

with the renormalization

$$g_\ell = g'_\ell/(1 + \lambda_\ell), \quad a_\ell = a'_\ell/(1 + \lambda_\ell) \text{ etc.} \quad (4)$$

We can express the phase shift formual as

$$k^{2\ell+1}\cot\delta_\ell = -\alpha^{2\ell+1}\frac{1 + \mathrm{Re}\,H_\ell(g_\ell) + a_\ell Z^2}{\frac{1}{Z^{2\ell+1}}\mathrm{Im}\,H_\ell(g_\ell) + b_\ell + c_\ell Z^2}. \quad (5)$$

From the phase shift analysis (and the binding energy of triton for the ^{2}S-state), we determine the parameters in the formula (5). The result is listed in Table 1. The big values of parameter g_1 for ^{4}P-state is due to the renormalization (4). It is plausible that λ_1 for the ^{4}P-state is very close to -1. The realistic two-body force suggests that g'_1 which is effective for ^{4}P-state is very small.

If the equation

$$F(k) = 0 \qquad (6)$$

is satisfied, this value of k is the "wave number" of the three-body bound state (anti-bound state, resonance) for $k = i|k|$, ($k = -i|k|$ and $k = k_0 - i\gamma$). We found the solution of eq.(6). See, Table 2. Since the exchange singularity is predominant in eq.(6), these states are the cluster state consisting of deuteron and another particle.

Table 1

Parameters of eq.(5)

State	g	a	b	c
^2S	0.5374	-0.1015	0.6070	0.0451
^4P	4.0447	3.6275	-1.6806	0.1229

Table 2

Roots of eq.(6) on the second energy sheet. The energy is measured from the n-d threshold.

State	$k(10^{13}cm^{-1})$	E_{CM}(Mev)
^2S	-0.1256i	-0.49
^4P	0.0326 - 0.0909i	-0.336 - 0.277i

Reference

1) T.Sasakawa, Tohoku University preprint (1974), Tu/74/121.

Rescattering effects in the inelastic proton-deuteron scattering at proton energies up to 100 MeV

K. Möller
Joint Institute for Nuclear Research, Lab. for Theor. Physics, Dubna, USSR

In the last years various methods and computer codes have been developed to solve Faddeev equations for scattering states. At the same time to describe proton-deuteron scattering simple models remain to be of interest. As compared with sophisticated calculations models have the advantage of giving a more obvious picture of the physical process going on. For example the final state interaction model (FSI) and the quasi free scattering model (QFS) have proved very useful to interpret the deuteron break-up by protons, although they give only qualitative agreement with the experiment. The FSI model and the QFS model represent the first approximation (pole graph) of the Faddeev equations. To investigate whether we can achieve an improvement of the theoretical description of the experiments in the present paper we calculated the next approximation (rectangular graph) of the Faddeev equations. The presented calculations are based on the papers [1), 2), 3)].

The conditions of the calculations are the following:

1. The exact Faddeev equations have been used including spin, isospin and correct antisymmetrization of the matrix elements.
2. Coulomb and three body forces have been neglected.
3. The two body forces have been assumed to be charge independent.
4. The two body potential between particles i and k has been choosen in the form:

$$V(ik) = V_1 P_\sigma^{(t)} P_\tau^{(s)} + V_2 P_\sigma^{(s)} P_\tau^{(t)} + V_3 P_\sigma^{(t)} P_\tau^{(t)} + V_4 P_\sigma^{(s)} P_\tau^{(s)}$$

(The operators $P_{\sigma/\tau}^{(t/s)}$ are spin and isospin operators projecting on triplet and singlet states correspondingly)

5. In the two body potential we restricted ourselves to S-waves. Then it follows that $V_{tt}=V_{ss}=0$.

The remaining coefficients V_{ts} and V_{st} have been choosen to correspond to the following two body t-matrix:

$$t_\beta(\vec{k}',\vec{k},\eta) = \frac{4\pi\hbar^2/m}{a_\beta^{-1} + j\sqrt{\eta} - \frac{r_{0\beta}}{2}\eta} \quad \text{with}$$

$\sqrt{\eta} = j\sqrt{|\eta|}$ for $\eta < 0$, $\eta = \frac{m}{\hbar^2}(E + j0)$

E= energy in the two-particle system

ß= triplet, singlet

a_β= scattering length

$r_{0\beta}$= effective radius

7. The Faddeev equations can be decomposed into a series of graphs. In schematic representation we get:

In the present paper we investigated the first two graphs. The results are given in fig.1.

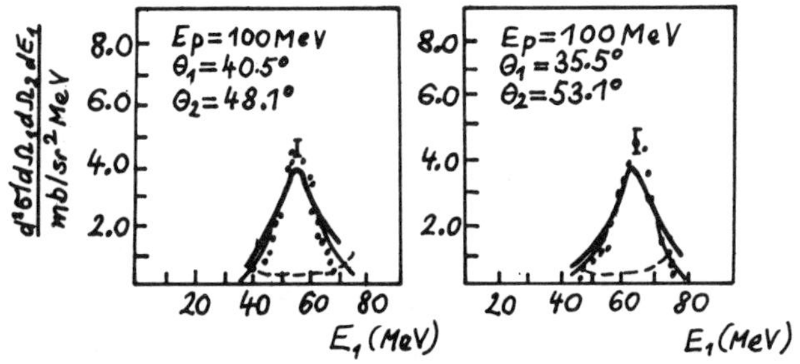

Fig. 1 Energy spectrum of the (pp)-quasi free scattering
Experimental data taken from [4].
——— graph (1) ---- graph (2) ▬▬ graph (1)+(2)

From fig.1 it can be seen that for 100 MeV the pole graph agrees approximately with the experiment in the form of the spectrum as well as in the absolute value. In the considered kinematic region the contribution of the rectangular graph is small and does not improve the agreement with the experiment. Calculations have been performed also for lower energies. It turns out that at lower energies the pole graph overestimates the absolute value of the spectrum. The rectangular graph gives no improvement. In the contrary it worsens the situation especially in the region of the final state interaction. So we can conclude that under the given assumptions for the two-particle potential the rectangular graph gives no improvement of the theoretical description of the deuteron break-up.

Literature:

1) K.Möller, ZfK-221
2) B.Kühn et al, Nucl. Phys. A120 (1968) 285
3) K.Möller, ZfK-262, Jahresbericht 1973
4) P.J.Pan et al, Nucl. Phys. A150 (1970) 216

VII

PHENOMENES DE POLARISATION

POLARIZATION PHENOMENA

POLARIZATION PHENOMENA IN THE THREE-NUCLEON SYSTEM

H. E. Conzett

Lawrence Berkeley Laboratory
University of California
Berkeley, California 94720

1. Introduction

There has been an impressive quantity and quality of polarization data acquired during the past few years on the mass 3 to mass 6 systems, essentially with beams of polarized protons and deuterons. I have chosen, in this paper, to limit the discussion of polarization effects to those of the three-nucleon system. I do this for two reasons: 1) There is now a rather extensive variety of experimental results on this system, and the ever more detailed three-body calculations of these polarization effects have been remarkably successful. Thus, a description of the past developments and present status of this research, in the detail that is warranted by this substantial progress, will take my allotted time. 2) Certainly a central role in this and in the past few-body conferences has been that of the "exact" three-body theory, which calculates the three-nucleon observables from the two-nucleon interaction. Thus, there is, so far, a natural separation between descriptions of three-nucleon data and those of mass 4 and higher. Polarization results are certainly important in the latter systems, but the appropriate theoretical descriptions are generally those of R-matrix analysis, and the direct connection to the nucleon-nucleon force is not made.

2. Terminology

Since the community of experimentalists involved in investigations of polarization phenomena is rather small, let me begin with some definitions and terminology.[1] Fig. 1 summarizes everything I will use in this discussion. For the analyzing powers I list an "experimental" definition along with its theoretical counterpart. We follow the Madison convention[2] throughout; the incident beam is along the z axis and the y axis is taken along $\vec{k}_i \times \vec{k}_f$, i.e. perpendicular to the reaction plane. In direct analogy with the spin 1/2 case, the fully vector-polarized deuteron beam (indicated in the top figure by the single-headed arrow) has zero population of the m = -1 deuteron magnetic substate and twice the normal population of the m = +1 state, where normal means the population in an unpolarized beam. The m = 0 state, which is not indicated, has its normal population. The vector analyzing power A_y (cartesian form) or

iT_{11} (spherical form) is given as shown, where σ_y and σ_{-y} are differential cross sections with the vector polarization oriented in the +y and -y directions, respectively, and σ_0 is the spin-averaged, or unpolarized, cross section. In the theoretical expression for A_y, M is the transition matrix and $\mathcal{P}_y$ is a basic spin-one 3 x 3 matrix operator. A fully tensor polarized beam is one with zero population of the m = 0 substate and normal populations of the m = ±1 substates. The vector polarization is zero, and this tensor polarization corresponds to an alignment of the deuteron indicated by the double-headed arrows in Fig. 1. The tensor analyzing powers A_{zz}, A_{xz}, and $A_{xx}-A_{yy}$ are given then by the differences in cross-sections for deuterons with the alignment axis oriented in the different directions shown with respect to the beam direction. The $\mathcal{P}_{jk}$ are the corresponding matrix operators.

Deuteron analyzing powers

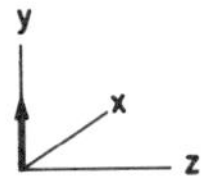

$$\frac{(\sigma_y-\sigma_{-y})}{2\sigma_0} = A_y = \frac{2}{\sqrt{3}}\, iT_{11} \qquad A_y = \frac{Tr(M\mathcal{P}_y M^\dagger)}{Tr(MM^\dagger)}$$

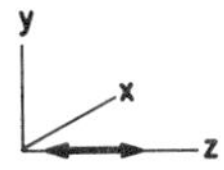

$$\frac{2(\sigma_{zz}-\sigma_0)}{\sigma_0} = A_{zz} = \sqrt{2}\, T_{20}$$

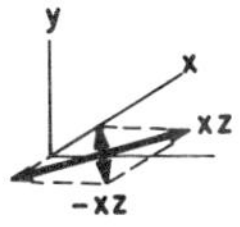

$$\frac{(\sigma_{xz}-\sigma_{-xz})}{\sigma_0} = A_{xz} = -\sqrt{3}\, T_{21} \qquad A_{jk} = \frac{Tr(M\mathcal{P}_{jk}M^\dagger)}{Tr(MM^\dagger)}$$

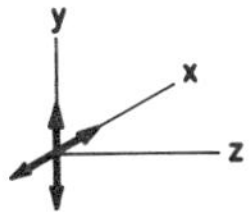

$$\frac{2(\sigma_{xx}-\sigma_{yy})}{\sigma_0} = A_{xx}-A_{yy} = 2\sqrt{3}\, T_{22}$$

Polarization transfer coefficients ($\vec{p}$ to $\vec{d}$)

$$p_{i'} = (P_{i'} + \sum_k K_k^{i'} p_k)\, I_0/I \qquad K_k^{i'} = \frac{Tr(M\sigma_k M^\dagger \mathcal{P}_{i'})}{Tr(MM^\dagger)}$$

$$p_{i'j'} = (P_{i'j'} + \sum_k K_k^{i'j'} p_k)\, I_0/I \qquad K_k^{i'j'} = \frac{Tr(M\sigma_k M^\dagger \mathcal{P}_{i'j'})}{Tr(MM^\dagger)}$$

XBL748-3966

Fig. 1

The polarization transfer coefficients for the case of proton-to-deuteron polarization transfer are defined in the bottom two lines. In the first expression, $\mathcal{P}_{i'}$, is the deuteron vector polarization resulting from N-d scattering with an initial nucleon beam of polarization components p_k, $\mathbf{P}_{i'}$ is the deuteron vector polarization produced in the scattering of an unpolarized beam, I_o and I are the cross sections for scattering of unpolarized and polarized nucleons, respectively, and $K_k^{i'}$ are the vector-to-vector polarization transfer coefficients. For example, with $\mathbf{P}_{i'} = 0$, $I = I_o$, and for a nucleon beam polarized along the y-axis, $p_{y'} = K_y^{y'} p_y$. As shown in the second expression, the vector-to-tensor polarization transfer coefficients $K_k^{i'j'}$ are defined in the same manner. Here $p_{i'j'}$ are the tensor polarizations of the scattered deuterons and $P_{i'j'}$ are the tensor polarizations that occur in the scattering with unpolarized nucleons.

3. Analyzing Powers in Nucleon-Deuteron Elastic Scattering

The first comparison, some 10 years ago, between the calculated[3] and experimental[4] proton analyzing power in N-d scattering below 100 MeV is shown in Fig. 2.

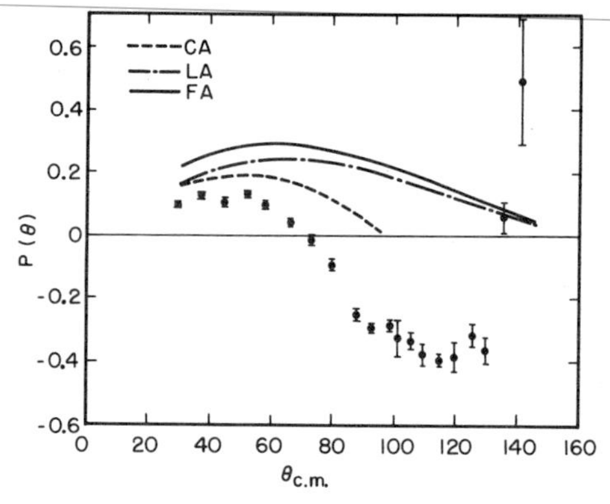

Fig. 2

MUB-2813

This impulse-approximation calculation was quite inadequate to explain the data. In fact, the discrepancy between the experimental and calculated results increased in going to the more complete versions of the calculation. Soon thereafter the early three-nucleon calculations, based on the Faddeev equations with simple S-wave nucleon-nucleon potentials, were very successful in fitting the elastic N-d differential cross section data up to about 50 MeV. Since only S-wave forces were used, those calculations could not provide the observed polarizations. There was already a substantial amount of nucleon analyzing-power data up to 50 MeV[5] and a few measurements of the deuteron vector and tensor analyzing powers at lower energies.[6] Only within the past three years have more realistic N-N potentials been used in efforts to fit the polarization data. Aarons and Sloan[7] used a two-body force with separable terms corresponding to both the $^1S_0$ and the coupled $^3S_1$-$^3D_1$ (tensor) interactions. This calculation gave deuteron tensor polarizations in qualitative agreement with experiment over the range E_N = 3-11 MeV. Fig. 3 shows the comparison near 11 MeV. The solid and dashed curves are for the percentage D-state in the deuteron of 7% and 2%, respectively. In contrast to this agreement, the calculated nucleon and deuteron vector polarizations were much too small, as shown in Fig. 4. Soon thereafter Pieper[8] and Doleschall[9] independently included S- and P-wave interactions, and the improvement over the previous calculations of the vector polarizations was dramatic. The nucleon polarizations were in excellent agreement with the experimental data up to 14 MeV, and qualitative agreement was achieved beyond that to 40 MeV. As an example, Fig. 5 shows Piepers' comparison with the nucleon polarization near 23 MeV. Sets C and E correspond to different P-wave potentials, and set D adds D-wave forces to set C. The agreement with the available deuteron vector analyzing-power data was not as good, as is shown in Fig. 6 near E_d = 20 MeV. It soon proved, however, that the absolute normalization of these data was in error by about a factor of two, due to an uncertainty in the value of the beam polarization. These were some of the first measured deuteron

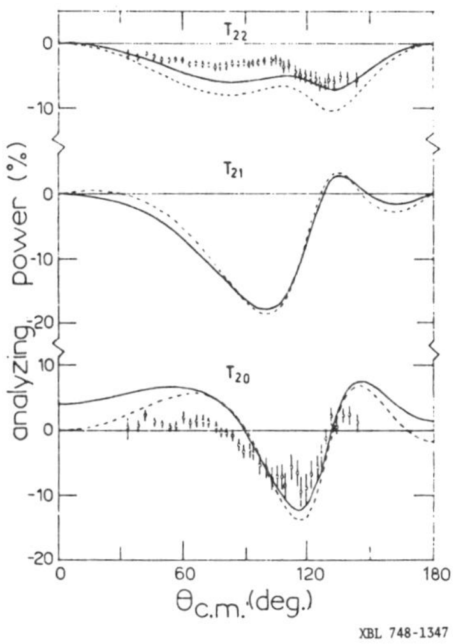

Fig. 3

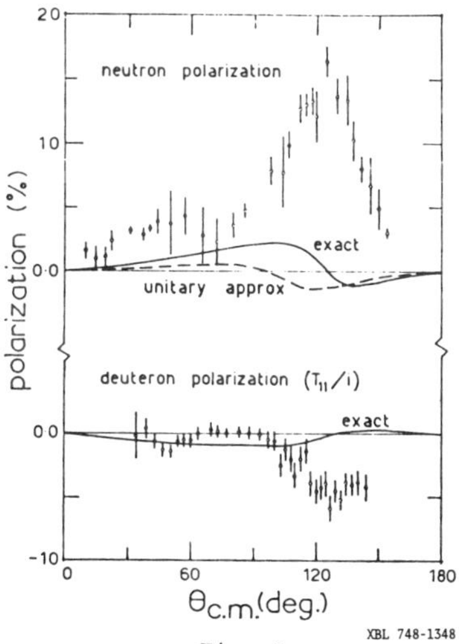

Fig. 4

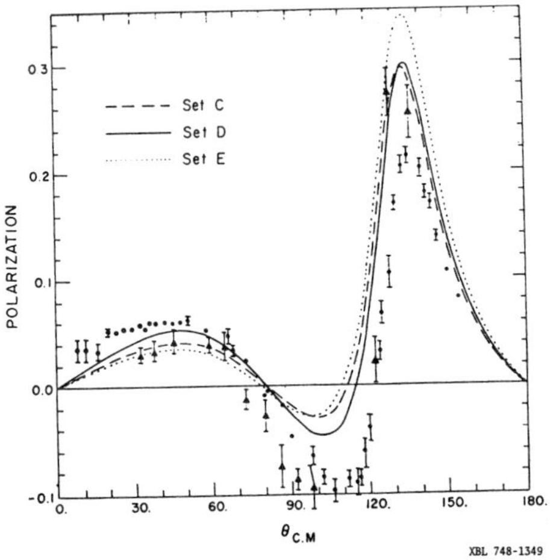

Fig. 5

analyzing powers, and there had not yet been established any standard analyzer with which to determine and to monitor deuteron beam polarizations. Almost concurrently with these calculations, more precise determinations were made of the deuteron vector analyzing powers at E_d = 20 and 30 MeV,[10] and these are shown in Figs. 7 and 8. The dashed line in Fig. 7 represents the data of Fig. 6. The solid curve is Pieper's prediction. In Fig. 8 the curves are the calculated results at E_d = 28.2 MeV. Doleschall's calculation (dashed line) was an exact one with separable N-N potentials for the $^1S_0$, $^3S_1$-$^3D_1$, and P-wave components of the two-nucleon interaction. Pieper's calculation (solid line) included D-waves, but not the $^3S_1$-$^3D_1$ tensor force. Also, his calculation treated the S-wave part exactly in the Faddeev equations, while the part containing the P- and D-wave N-N input information was treated in first-order perturbation theory. Thus, one could not conclude that the

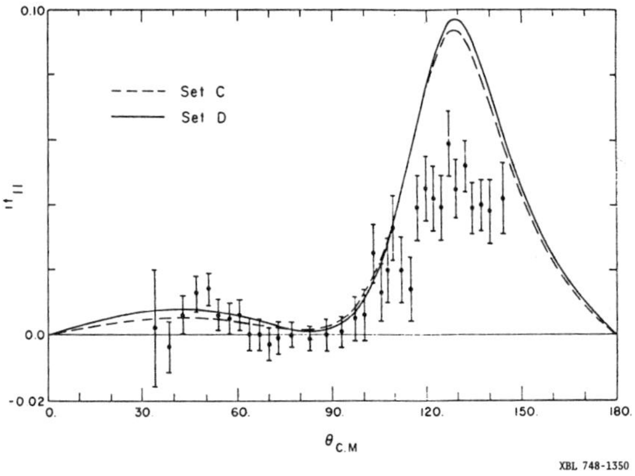

Fig. 6

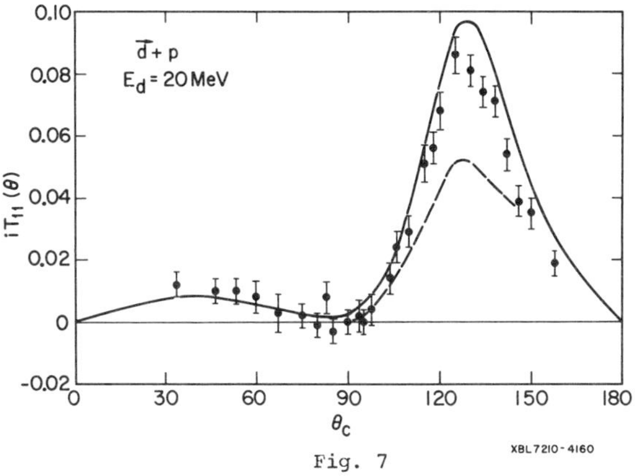

Fig. 7

differences seen in Fig. 8 were due mainly to the additional D-wave interactions in Pieper's calculation. In fact, in a subsequent paper Pieper[11] compared his perturba-

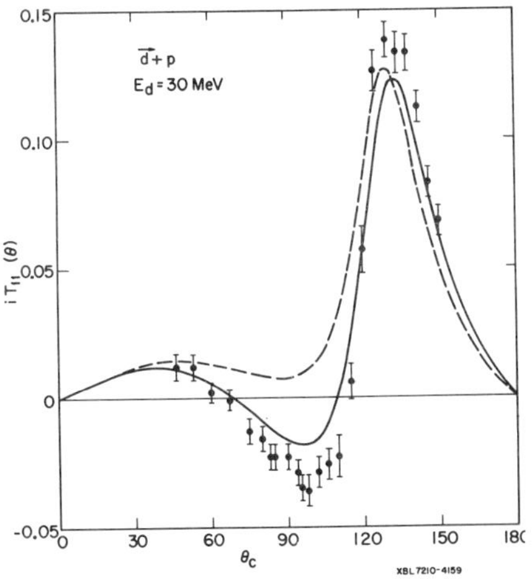

Fig. 8

tive calculation with the exact result of Doleschall for the <u>same</u> N-N input interactions, and he found that significant differences, as shown in Fig. 9 for the nucleon polarization near 23 MeV, were indeed due to the different methods of calculation.

The dotted curve is the perturbative calculation, so the fact that it agrees better with the data than does the exact calculation appears to be a fortuitous result.

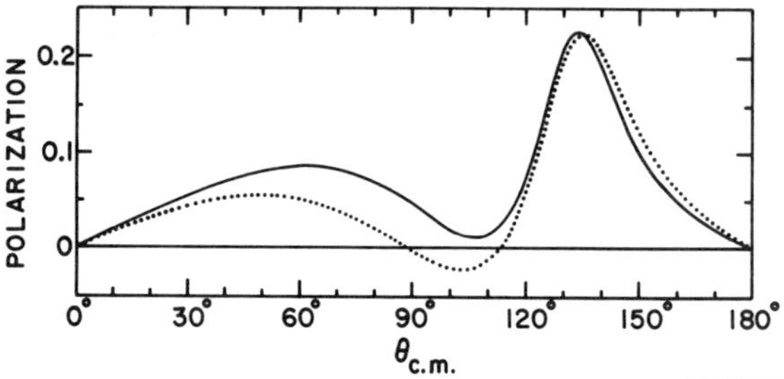

Fig. 9

These calculations have also provided very good fits to the recently measured tensor analyzing powers.[12] Examples near E_d = 25 MeV are shown in Fig. 10. The predictions, which preceded the data, are for E_d = 28.2 MeV. The dashed line is Doleschall's result, and the solid line is that of Pieper[13], which now includes the ($^3S_1$-$^3D_1$) tensor force. Q and R are linear combinations of T_{20} and T_{22} [$Q = (1/2\sqrt{2})(T_{20} + \sqrt{6}\ T_{22})$, $R = (1/2\sqrt{2})(T_{20} - \sqrt{6}\ T_{22})$].

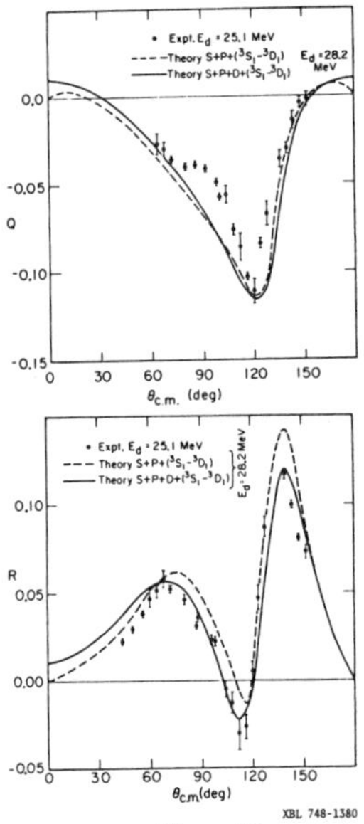

Fig. 10

At this point, then, these three-nucleon calculations had shown very considerable success in fitting the several available N-d elastic scattering observables. It was, also, clear that the polarization data required the use of the more realistic, i.e. more detailed and more complicated, N-N interactions, and thus provided the more sensitive and significant tests of the calculations. The conclusions then were that 1) the N-N P-wave interactions were chiefly responsible for the observed vector polarizations, whereas 2) the $^3S_1$-$^3D_1$ tensor force was the source of the N-d tensor polarizations.

The stage was now ready for an examination of the sensitivity of the calculated N-d observables to changes in the N-N input interactions. Certainly, the first goal of the three-body theory has been to reproduce three-nucleon data with calculations that use two-nucleon forces. Clearly, a second goal is to pursue the possibility of deducing, from

N-d scattering and the three-body calculations, information on the N-N interaction which has not been available from N-N scattering itself. Very recent investigations have suggested that this possibility exists. In contrast to the two quite definite conclusions noted above, there have been conflicting opinions concerning the effect of the N-N tensor force on the N-d vector polarizations. Pieper[13] reported only slight changes with the addition of the tensor force, and he suggested[11] that changes in the $^3S_1$-$^3D_1$ potential would have little effect on the nucleon polarization. This conjecture was based on Sloan and Aarons result,[14] which demonstrated that none of the N-d polarizations were very sensitive to reasonable changes in the tensor interaction. However, that calculation did not include P-waves, so the vector polarizations were unrealistically small. Doleschall's earlier calculation[9] showed a substantial change in the vector polarizations with the addition of the tensor force to the S- and P-wave interactions, and his most recent calculation[15] demonstrates that the vector polarizations are quite sensitive to the details of the $^3S_1$-$^3D_1$ potential used. First of all, as shown in Fig. 11, his P-wave interactions were improved to give a better representa-

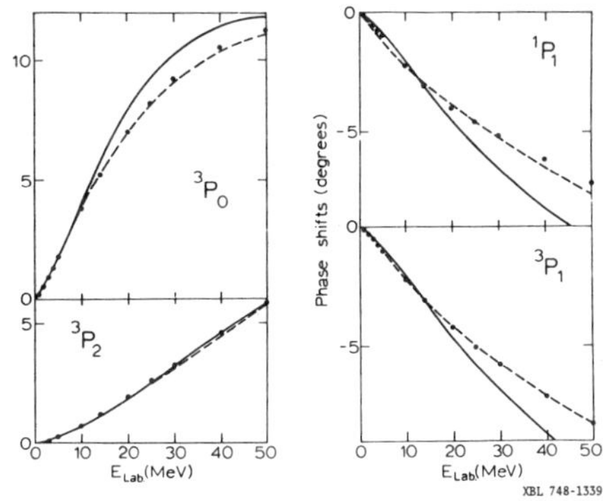

Fig. 11

tion of the N-N P-wave phase shifts, which are shown as the dots. This had the effect shown in Fig. 12, where the solid and dashed curves correspond to the solid and dashed P-wave curves of Fig. 11. Also, he constructed rank-2 tensor interactions in an attempt to simultaneously reproduce the $^3S_1$ and $^3D_1$ phase shifts, the corresponding mixing parameter ε_1, and the deuteron properties. As seen in Figs. 13 and 14, it was not possible to find a single rank-2 tensor force which satisfied all of these criteria, so two such potentials were used. One, the T4D force, reproduced the low energy ($\leqslant$ 100 MeV) $^3D_1$ phase shifts but gave larger values of ε_1 than have been deduced from n-p

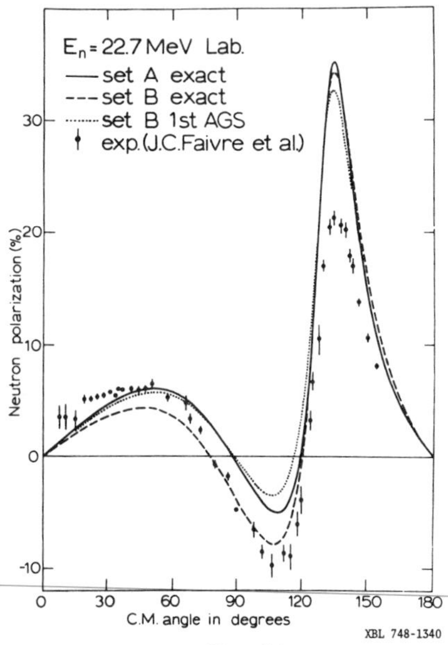

Fig. 12

scattering.[16] The other, the T4M force, reproduced the low energy ε_1 behavior but not that of the $^3D_1$ phase shift. Fig. 15 shows the separate effects of these two different tensor interactions on the nucleon polarization in N-d scattering at 22.7 MeV. It is clear (compare with Fig. 11) 1) that the tensor interaction has a substantial effect on the (vector) polarization, and 2) that the addition of the T4M force gives an excellent fit to the data at angles larger than $\theta_c = 80°$. In a further effort to improve the agreement with the data, Doleschall also included a $^3D_2$ interaction. Computational limitations precluded the addition of a complete set of D-wave interactions. The result is shown in Fig. 16 as the solid curve, which shows some improvement in the fit to the forward angle data. Doleschall also calculated deuteron vector polarizations at the same energy, so we very recently at Berkeley made measurements[17] of the deuteron vector analyzing power, iT_{11}, in $\vec{d}$-p scattering at E_d = 45.4 MeV for direct comparison

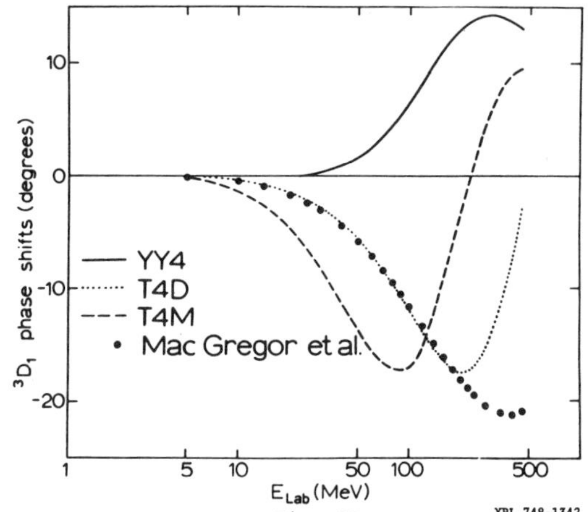

Fig. 13

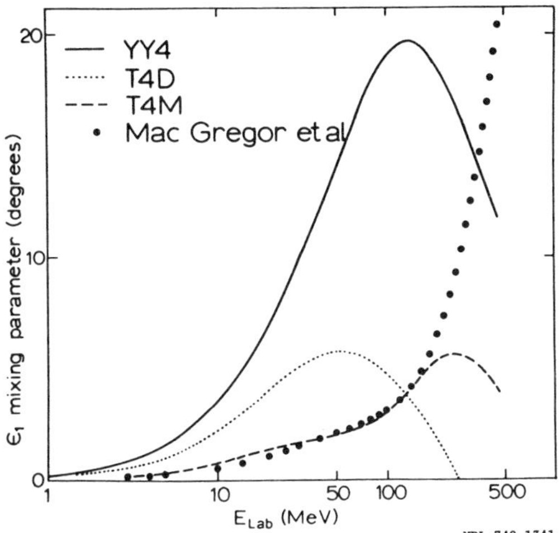

Fig. 14

with the calculated vector polarization at the equivalent nucleon energy of 22.7 MeV. Our data are compared in Fig. 17 with the calculated results for the different N-N interactions. The dotted curve is the result with S- and P-waves plus the T4D tensor potential; the dashed line, with T4M in place of T4D; and the solid line, with the addition of the $^3D_2$ terms to the T4M case. Just as with the nucleon polarization, this final step gives improved agreement with the forward angle data at the expense of a slightly poorer fit in the region $\theta_c = 85^\circ$ to 115°.

The angular region forward of $\theta_c = 120^\circ$, wherein the remaining discrepancies between experiment and theory exist, is just the region of greatest sensitivity to the details of the $^3S_1$-$^3D_1$ tensor interaction. Clearly, it would be most interesting and useful to do the calculation with a tensor force which simultaneously reproduces the N-N $^3D_1$

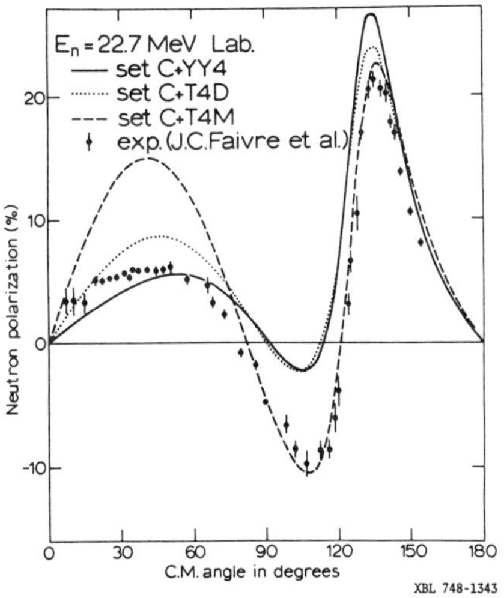

Fig. 15

phase shift and the mixing parameter ε_1. For example, the rank-4 potential recently constructed by Pieper[18] does just that.

I must now digress momentarily in order to connect these results with a problem of rather long standing in n-p scattering. MacGregor et al., in their phase shift analyses of n-p data five years ago, found that the $^1P_1$ phase shift $\delta(^1P_1)$ and the mixing parameter ε_1 were strongly correlated and poorly determined below 80 MeV.[16] Neither, in-fact, was near the theoretical expectation. Arndt, Binstock, and Bryan[19] have recently examined this problem in considerable detail near 50 MeV, including in their analyses some more recently available n-p data. They did not find a strong ε_1 - $\delta(^1P_1)$ correlation. Concerning $\delta(^1P_1)$, they reemphasize that the differential cross section is the observable most sensitive to that phase shift. They conclude that the existing forward angle data are suspect, and they recommend that forward angle absolute $d\sigma/d\Omega$ measurements, accurate to ±1%, be made in order to pin down $\delta(^1P_1)$. Concerning ε_1, they show (Fig. 18) that the present n-p data (σ_{TOT}, $d\sigma/d\Omega$, $P(\theta)$) near 50 MeV leave ε_1 undetermined between -10° to $+3^\circ$. They also examine the sensitivity of other experimental observables to ε_1, and they find that the neutron-to-proton polarization transfer coefficient D_t (our $K_y^{y'}$) combines fairly high sensitivity with reasonable experimental feasibility. Fig. 18 also shows the effectiveness of adding to the data set presumed values of D_t, which were taken from a calculation with ε_1 fixed at a theoretical expectation near 2.8°. One sees that ε_1 is determined to about $\pm 1^\circ$ in each case. Note, also, the assumed

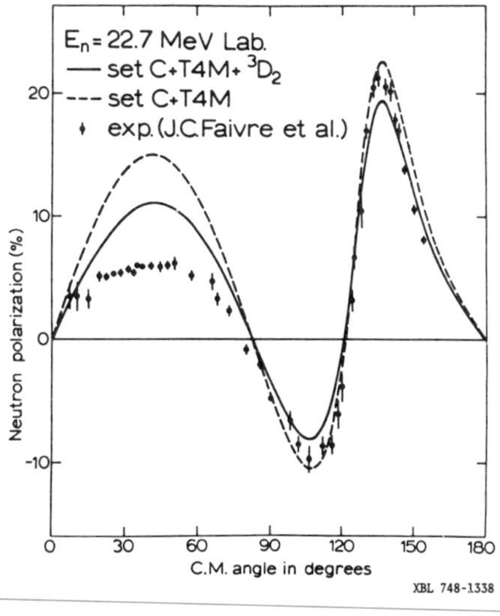

Fig. 16

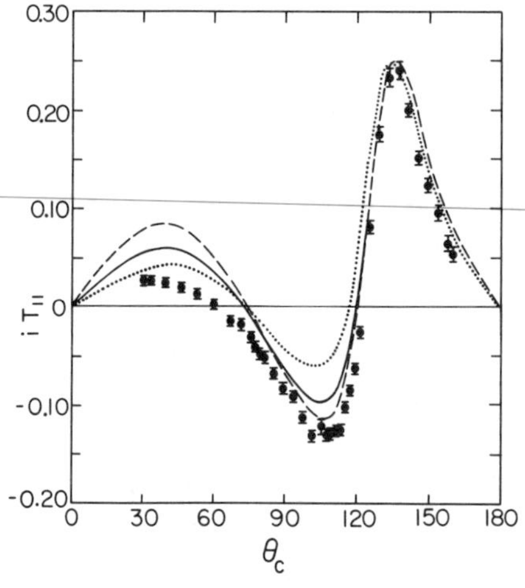

Fig. 17

±0.01 absolute erron on D_t, and be assured that such an experiment, if reasonable, is difficult.

Let us return now to the three-nucleon calculation. In view of the demonstrated sensitivity of the vector polarizations to the N-N tensor interaction, it should be possible, with a tensor force which simultaneously reproduces the N-N $\delta(^3D_1)$ and ε_1, to vary ε_1 in a search for improved fits to the N-d vector analyzing power data. It seems possible that this procedure could more easily provide a better determination of the low energy values of ε_1 than is possible via the more difficult measurement of D_t. If this should prove to be so, one would indeed have deduced from the three-nucleon investigations specific information about the two-nucleon interaction that has not yet been attainable.

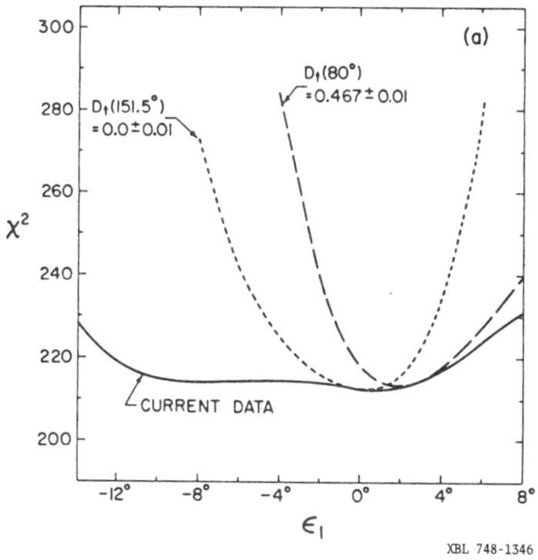

Fig. 18

4. Polarization Transfer in Nucleon-Deuteron Elastic Scattering

The first of the other N-d spin observables measured at energies below 100 MeV were the Wolfenstein spin rotation parameters[20] D, R, and A (our $K_y^{y'}$, $K_x^{x'}$, $K_z^{x'}$). These describe the reorientation of the projectile nucleon's spin in the scattering process, and are included in the more general class (Fig. 1) of polarization-transfer coefficients. Measurements had been made of them near 50 MeV,[21] and Pieper's perturbative three-body claculation[8] of these spin rotation parameters were in qualitative agreement with the data. The experimental errors of 10% or larger did not encourage any attempt to fit the data in greater detail. At the Los Angeles Few Particle Conference two years ago Ohlsen[22] pointed out that polarized beams and experimental techniques have been sufficiently perfected that these more difficult measurements of polarization transfer coefficients could be made. At the same time, the Los Alamos group reported the first determinations of vector-to-tensor polarization transfer coefficients in p-d elastic scattering.[23] Their measurements were made at two angles at energies between 5 and 9 MeV, and the values were consistent with zero within a typical error of ±0.05.

Pieper's later three-body calculation,[11] an example of which is shown in Fig. 19, gave predicted values of these coefficients to be less than 0.1 for energies below 10 MeV, in essential agreement with the data. Of particular interest was his result that the

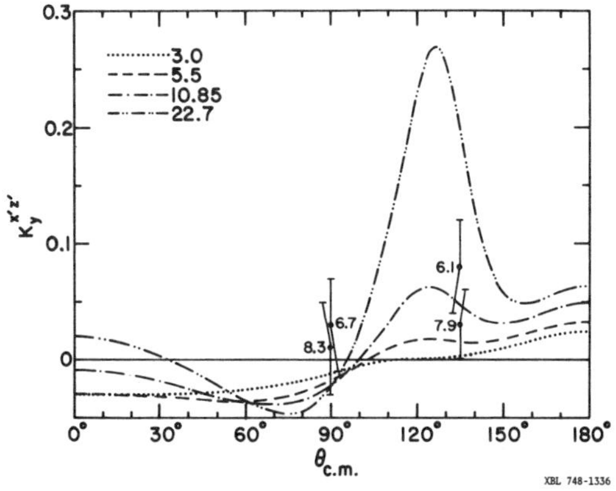

Fig. 19

calculated vector-to-vector transfer coefficients differed more substantially from zero, so his suggestion was that measurements of those could be a fruitful area for experiments. A single measured value of $K_x^{x'}$ at 9 MeV[24] was two standard deviations from the calculated value, but it was clear that more data were required for a significant comparison with the predictions. Now, in a contribution to this conference, the Berkeley group reports on measurements of $K_y^{y'}(\theta)$ at E_p = 22.7 MeV.[25] Fig. 20 shows these data along with Piepers predicted curve. The agreement is certainly very good. Since this provides the first significant comparison between experimental and calculated vector-to-vector polarization transfer coefficients, this agreement represents yet another substantial succes of the three-body calculations.

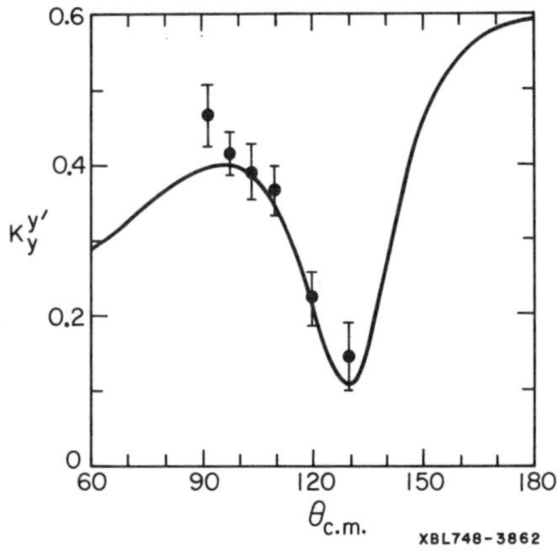

Fig. 20

We have noted the considerable progress that has been made during the past two years in both the experimental and theoretical determinations of polarization observables in N-d elastic scattering. It seems to me that an important undertaking now is to specifically examine the sensitivities of the various analyzing powers and transfer coefficients to the details of the input two-body interactions; for example, sensitivity to variations of $\delta(^1P_1)$ and ε_1, and to the addition of a complete set of D-wave interactions in the exact calculation. This would provide invaluable guidance in the choice of further experiments, since it is clear that many of the polarization observables can now be measured to just about whatever accuracy is required for specific and detailed comparison with predictions.

5. Polarization effects in the N-d Breakup Reaction

The present status of studies of polarization phenomena in the N-d breakup transition to three-nucleon final states is comparable to that which existed for the elastic channel almost ten years ago. That is, only a few experiments have been done which even show the presence of polarization effects, and theoretical interpretation and predictions via exact three-body calculations have not, as yet, been made. Such calculations[26] have been successful in fitting N-d breakup cross-sections, but they have so far been limited to S-wave N-N input interactions. It appears that experimental evidence of significant polarization effects in the breakup channel are required in order to encourage, or even compel, the addition of the tensor force and P-wave contributions to these calculations.

Perhaps the first polarization effects seen in the breakup reaction below 100 MeV were those observed by Arvieux et al. in the reaction $D(\vec{p}, 2p)n$ at 10.5 MeV.[27] Their results are shown in Fig. 21. The open circles are their measurements of the proton analyzing power for the transition to the np final-state-interaction region of the 3-body continuum spectrum, in their case selected to be the region of relative np energies $E_{np} \leqslant 0.5$ MeV. They noted, for comparison, the similarity of the trend of these data to that of the elastic channel analyzing power at 11 MeV, as shown by the solid dots connected by the dashed line. In a contribution to this conference, Blyth et al.[28] report an investigation of the deuteron vector analyzing power at several angles in the same reaction $H(\vec{d}, 2p)n$, but now induced with a beam of 12.2 MeV vector polarized deuterons. Their reported values are all consistent with zero, within errors of ±0.01 to ±0.03, but it should be noted that in this case the elastic channel analyzing power at the nearby deuteron energy of 11.5 MeV has a maximum value of less than 0.05.[29] We have very recently obtained at Berkeley some results for the deuteron vector analyzing power in the same inelastic transition at $E_d = 45.4$ MeV.[30] These are shown in Fig. 22. Again, for comparison, the elastic scattering analyzing power is

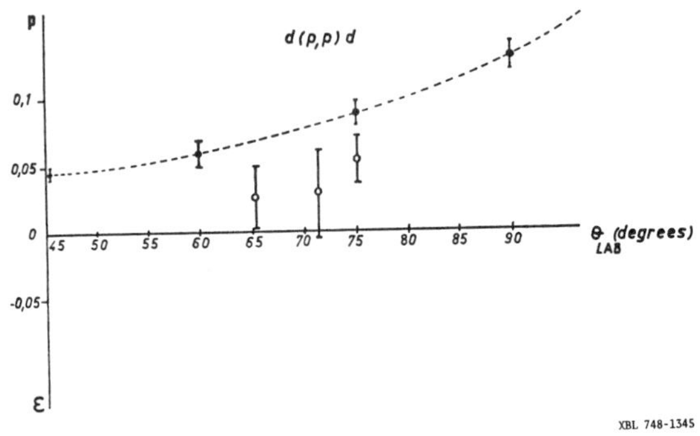

Fig. 21

shown as the smooth curve. Here the similarity between the inelastic and elastic analyzing powers is quite definite. This similarity is rather unexpected in view of the results that were reported by Brückmann et al., in their analysis of cross section data

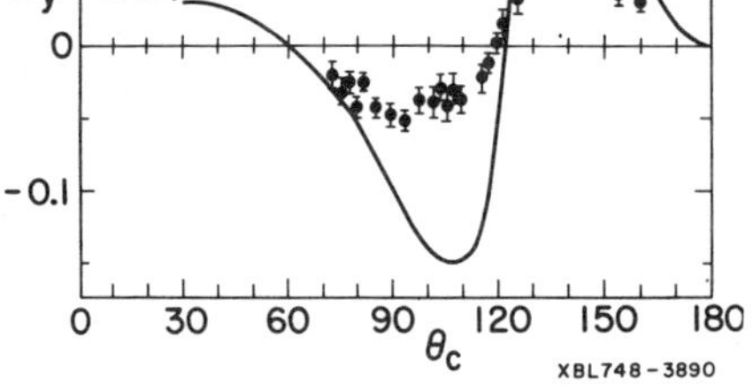

Fig 22

in this reaction at the slightly higher energy E_d = 52.3 MeV.[31] Their findings are displayed in Fig. 23. In their analysis they determined the separate contributions of n-p singlet and triplet pairs to the observed final-state-interaction peak at the relative n-p energy E_{np} = 0. These separate contributions are shown in the figure. The solid curve, which is in excellent agreement with the cross section for production of n-p triplet pairs, represents a Born approximation calculation in which the final state n-p wave function used was effectively that of a deuteron with binding energy E_B = 0. If triplet n-p production were the major contribution to the cross section, we could expect the similarity between inelastic and elastic vector ana-

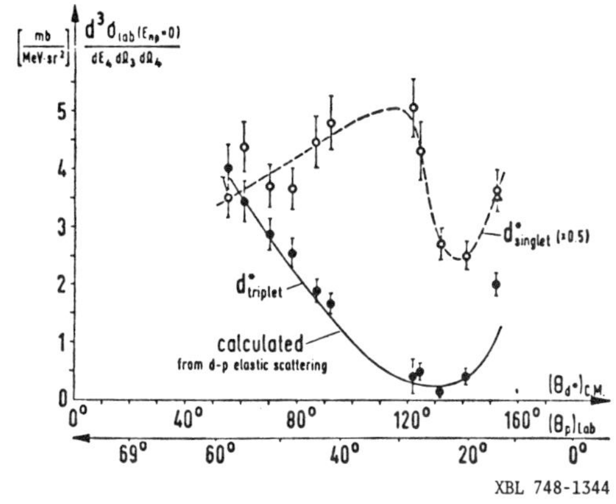

Fig. 23

lyzing powers. However, in just the backward angular region of maximum analyzing-powers, Fig. 23 shows that the major cross-section contribution is the production of n-p singlet pairs. Thus, one is led to the conclusion that the contribution to the inelastic analyzing power from the production of n-p singlet pairs has an angular distribution similar to that of the elastic analyzing power; and that is surprising. Since Ebenhöh's three-body calculation[26] quite successfully reproduces the form of the singlet n-p contribution to the cross section shown in Fig. 21, it would be of considerable interest to add to such a calculation the N-N spin dependent interactions that are necessary for the calculation of these analyzing powers.

Another study of polarization effects in the breakup reaction was that of Rad et al.[32] in their measurement of the neutron polarization in p + d breakup at E_p = 21.4 MeV and $\theta_L = 18^\circ$. Fig. 24 shows some of their results, here d on p at E_d = 42.8 MeV. The n-p quasifree scattering (QFS) contribution occurs in the region of the maximum polarization values of about -0.05. Indeed, the corresponding polarization in n-p scattering has essentially the same magnitude but is opposite in sign. Their conclusion then is that predictions based on QFS near this energy can not be reliable. So here again is a polarization result which awaits a three-body calculation for its explanation.

In a contribution to this conference Walter et al.,[33] report measurements of the proton-to-neutron polarization transfer in the $^2H(\vec{p}, \vec{n})2p$ breakup reaction at $\theta = 0^\circ$ and at several energies between 10.5 and 15 MeV. They found $K_y^{y'}(0^\circ)$ to be near 0.2 for the production of p-p singlet pairs of relative energy near zero. Associated with the

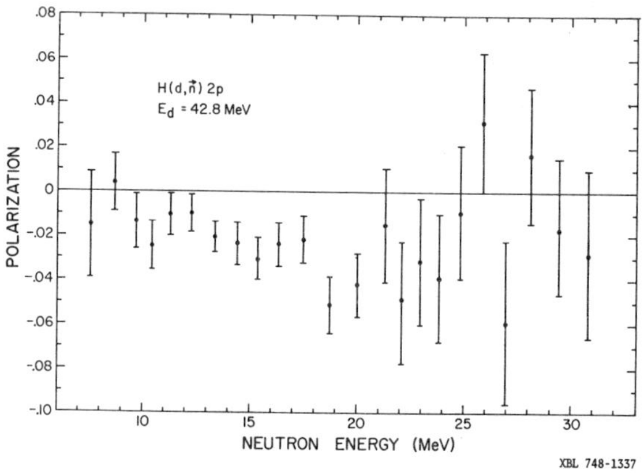

Fig. 24

production of p-p pairs of several MeV relative energy they found $K_y^{y'}(0^\circ)$ reaching values of +0.5. The negative values are qualitatively understood in a simple knockout process, because the singlet p-p pairs would have been formed from the incident polarized proton and protons of opposite spin from the target deuteron; then the correlated neutron spin would be opposite to that of the incident proton. If one includes the deuteron m = 0 magnetic substate, this qualitative argument[34] yields $K_y^{y'}(0^\circ) = -1/3$. This description neglects the deuteron D-state and nucleon-nucleon spin exchange terms; thus, a three-body calculation is again indicated in order to examine the connection between the behavior of this observable and the specific components of the nucleon-nucleon force.

6. Summary

The three-body calculations have been impressively successful in reproducing and predicting the cross-sections and spin-observables in elastic N-d scattering below 50 MeV. We have reached the point where the distinct possibility exists that specific information on the two-nucleon, i.e. on the mixing parameter ε_1, may be obtained from these three-nucleon studies. We clearly need calculations which investigate the sensitivities of the various N-d observables to the details and to the complexity of the input N-N interaction; the $^3S_1$-$^3D_1$ separable potential used should reproduce all of the parameters of that channel. I can only imagine the difficulty, and cost, of such calculations. However, the theorists' suggestion to measure D_t in n-p scattering at 50

MeV to an accuracy of ±0.01 allows me considerable latitude in my suggestion.

The situation concerning the N-d breakup to three-nucleon final states is reminiscent of that which existed with respect to the elastic channel several years ago. As we have seen here, there are now several experimental determinations of spin observables in this breakup reaction, and an appropriate three-body calculation is required for their quantitative interpretation. When the more complicated spin-dependent N-N interactions are included and spin observables are calculated, be assured that more such data will become available.

REFERENCES

1. See, for example, S. E. Darden, Polarization Phenomena in Nuclear Reactions, edited by H. H. Barschall and W. Haeberli (Univ. of Wisc. Press, Madison, 1971) p. 39; G. G. Ohlsen, Rep. Prog. Phys. 35 (1972) 717.
2. Ibid., p. xxv.
3. K. L. Kowalski and D. Feldman, Phys. Rev. 130 (1963) 276.
4. H. E. Conzett, H. S. Goldberg, E. Shield, R. J. Slobodrian, and S. Yamabe, Phys. Lett. 11 (1964) 68.
5. W. Haeberli, Proc. Int. Conf. on Three-Body Problem, edited by J. S. C. McKee and P. M. Rolph (North-Holland, Amsterdam, 1970) p. 188.
6. P. Extermann, Nucl. Phys. A95 (1967) 615; R. E. White, W. Grüebler, V. König, R. Risler, A. Ruh, P. S. Schmelzbach, and P. Marmier, Nucl. Phys. A180 (1972) 593; J. Arvieux, R. Beurtey, J. Goudergues, B. Mayer, A. Papineau, and J. Thirion, Nucl. Phys. A102 (1967) 503.
7. J. C. Aarons and I. H. Sloan, Nucl. Phys. A182 (1972) 369.
8. S. C. Pieper, Nucl. Phys. A193 (1972) 529.
9. P. Doleschall, Nucl. Phys. A201 (1973) 264.
10. J. S. C. McKee, H. E. Conzett, R. M. Larimer, and Ch. Leemann, Phys. Rev. Lett. 29 (1972) 1613.
11. S. C. Pieper, Phys. Rev. C8 (1973) 1702.
12. A. Fiore, J. Arvieux, Nguyen Van Sen, G. Perrin, F. Merchez, J. C. Gondrand, C. Perrin, J. L. Durand, and R. Darves-Blanc, Phys. Rev. C8 (1973) 2019.
13. S. C. Pieper, Phys. Rev. C6 (1972) 1157.
14. I. H. Sloan and J. C. Aarons, Nucl. Phys. A198 (1972) 321.
15. P. Doleschall, Nucl. Phys. A220 (1974) 491.
16. M. H. MacGregor, R. A. Arndt, and R. M. Wright, Phys. Rev. 182 (1969) 956.
17. F. N. Rad, J. Birchall, H. E. Conzett, S. Chintalapudi, R. M. Larimer, and R. Roy, Bull. Amer. Phys. Soc. 19 (1974) 476, and to be published.
18. S. C. Pieper, Phys. Rev. C9 (1974) 883.
19. R. A. Arndt, J. Binstock, and R. Bryan, Phys. Rev. D8 (1973) 1397; J. Binstock and R. Bryan, Phys. Rev. D9 (1974) 2528.
20. L. Wolfenstein, Ann. Rev. Nucl. Sci. 6 (1956) 43.
21. N. M. Stewart and W. R. Gibson, Polarization Phenomena in Nuclear Reactions, edited by H. H. Barschall and W. Haeberli (Univ. of Wisc. Press, Madison, 1971) p. 455; N. M. Stewart, W. R. Gibson and F. G. Kingston, Nucl. Phys. A171 (1971) 225; N. M. Stewart, W. R. Gibson, and J. H. P. C. Megaw, Nucl. Phys. A174 (1971) 338.
22. G. G. Ohlsen, Few Particle Problems in the Nuclear Interaction, edited by I. Slaus, S. A. Moszkowski, R. P. Haddock, and W. T. H. van Oers (North-Holland,

Amsterdam, 1973), p. 585.

23. G. G. Ohlsen, C. K. Mitchell, and G. C. Slazmann, ibid., p. 460.

24. G. G. Ohlsen, C. K. Mitchell, G. C. Slazmann, J. L. Gammel, D. C. Dodder, and K. Witte, Bull. Am. Phys. Soc. 17 (1972) 922.

25. F. N. Rad, J. Birchall, H. E. Conzett, and R. Roy, These Proceedings.

26. R. T. Cahill and I. H. Sloan, Nucl. Phys. A165 (1971) 161; W. Ebenhöh, Nucl. Phys. A191 (1972) 97.

27. J. Arvieux, J. L. Durand, J. C. Faivre, D. Garreta, A. Papineau, J. Sura, and A. Tarrats, Nucl. Phys. A150 (1970) 75.

28. C. O. Blyth, N. T. Okumasoglu, N. Berovic, and J. S. C. McKee, These Proceedings.

29. R. E. White et al., reference 6.

30. J. Birchall, H. E. Conzett, F. N. Rad, and R. Roy, to be published.

31. H. Brückmann, W. Kluge, H. Matthäy, L. Schänzler, and K. Wick, Nucl. Phys. A157 (1970) 209.

32. F. N. Rad, L. C. Northcliffe, D. P. Saylor, J. G. Rogers, and R. G. Graves, Phys. Rev. C8 (1973) 1248.

33. R. L. Walter, P. W. Lisowski, R. C. Byrd, and T. B. Clegg, These Proceedings.

34. G. G. Ohlsen, reference 1.

DISCUSSION

NOYES(SLAC): This beautiful data emphasizes again a point which has already been made in other ways at this Conference and which needs restating in this context. In order to do justice to the data, all three-nucleon calculations should be constructed in such a way that the on-shell (N-N phase shift) input be separated from the off-shell input. Only the latter comes from specific interactions, and should always be normalized to identical (over several hundred MeV) N-N phase shifts. Once this is done, data sensitive to off-shell effects can be used to explore them. Both are important. I urge my theoretical colleagues to follow this procedure in the future, using any of several methods now available. Only in this way can we do justice to the data.

CONZETT: I should remark that what I have said about the possibility of getting information about ε_1 contradicts in no way what Brayshaw said last night. He was talking about off-shell information. There is still some information, we believe, that can be obtained about the specific neutron-proton phase parameter ε_1 through three-body studies.

ROGERS(Alberta): Regarding your proposal for determining ε_1 for n-p scattering by comparing exact three-body calculations with p + d vector analysing power, will the uncertainty in the calculation due to the omitted Coulomb force make the dependence of the calculation on the tensor force uncertain ? Since the major sensitivity of the calculations to the tensor force is in the angular region of the nuclear-Coulomb interference minimum in the p+d cross section, it seems that the Coulomb interaction might influence the p+d polarization also.

CONZETT: In all, save one, the analysing powers in n-d and p-d scattering at 16, 22, 35 and 50 MeV are the same within experimental errors. The experimental errors are fairly large, larger typically in the neutron case. There is one point at 35 MeV which disagrees, but that point stands alone.

BRAYSHAW(Maryland): I just wish to point out that my technique can be used very effectively for the purpose of probing parameters such as ε_1. The Coulomb effect can be put in exactly , the off-shell effects can be eliminated as background, and one deals only with an effective rank one calculation regardless of δ or ε_1.

CONZETT: Thank you, we will send you our data immediately.

MASTERSON(British Columbia): Have the Wisconsin n-p differential cross section data at 24 and 28 MeV been included in the most recent calculations of the ε_1 parameters ? These data extend to 19° (lab), have accuracies of 2 - 4 %, and were published in Physical Review articles by Masterson (1972) and Burrows (1973).

SIMMONS(LASL) There were data published about one and a half years ago. Are those the ones you are referring to ?

MASTERSON: 1972-1973.

POLARIZATION TRANSFER IN THE $D(\vec{p},\vec{n})pp$ REACTION AT $\theta = 0°$ FOR PROTON BOMBARDING ENERGIES FROM 10.5 TO 15 MeV

R. L. Walter, P. W. Lisowski, R. C. Byrd, Duke University and Triangle Universities Nuclear Laboratory + and T. B. Clegg, University of North Carolina and TUNL +

Although the emphasis in the theoretical studies of three body processes lies at intermediate energies, recently several successful calculations of the cross section for N-D breakup have been reported below 20 MeV.[1)] In order to provide experimental data of a unique type to more stringently compare the reliability of the different theoretical techniques which give similar results for cross sections in our energy range, we have measured the transverse polarization transfer coefficient $K_y^{y'}$ for the $D(\vec{p},\vec{n})pp$ process at a reaction angle of 0° for five proton energies from 8 to 15 MeV.

The proton beam used in this experiment was produced by the Lamb-shift source of the TUNL tandem accelerator laboratory and was about 80% polarized. The experimental technique for these measurements is the same as is described in other contributions to this conference. To provide a reasonable counting rate the deuterium target thickness was chosen to be approximately 750 keV thick.

The upper part of Fig. 1 shows an experimental spectrum generated by the continuous distribution of neutrons produced in the $D(\vec{p},\vec{n})pp$ reaction at our highest energy, E_p = 15 MeV. The highest energy neutrons detected at a reaction angle of 0° are associated with the two protons being emitted near 180° with low relative energies. The solid lines in Fig. 1a represent a simple attempt using a two-gaussian fit to differentiate the neutrons associated with the p-p final-state interaction from neutrons arising from other breakup processes. In Fig. 1b the experimentally observed asymmetry calculated for each energy channel is shown. Here the structure in the asymmetry as a function of neutron continuum energy is clearly evident. Figure 1c shows the polarization transfer coefficient calculated from a five-channel average of the asymmetry data. This illustrates the detailed shape of the transfer coefficient throughout the breakup distribution. It is obvious that the neutrons from the p-p final-state region have a negative polarization transfer coefficient, while the other neutrons have $K_y^{y'}$ values described by a positive-going structure. Qualitatively, the negative values at the highest energy in Fig. 1c are expected from the Pauli exclusion principle and occur because the p-p pairs would be in singlet states and must therefore have been formed from the incident polarized protons and protons of opposite spin direction from the target deuteron. Figure 2 shows the transfer coefficient for five other proton energies calculated for the neutron energies corresponding to maxima in the yield as shown in Fig. 1a. Not only did the opposite polarizations mentioned above for E_p = 15 MeV occur at all proton energies, but in fact, the resulting transfer coefficients show no energy dependence over the 8 to 15 MeV range.

+ Work supported in part by the U. S. Atomic Energy Commission.

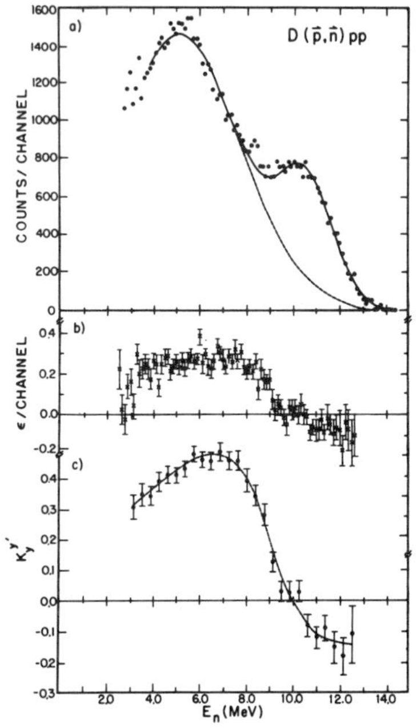

Figure 1

Experimental results for the D($\vec{p}$,$\vec{n}$)pp reaction at E_p = 15 MeV. The curve in part c is a guide to the eye.

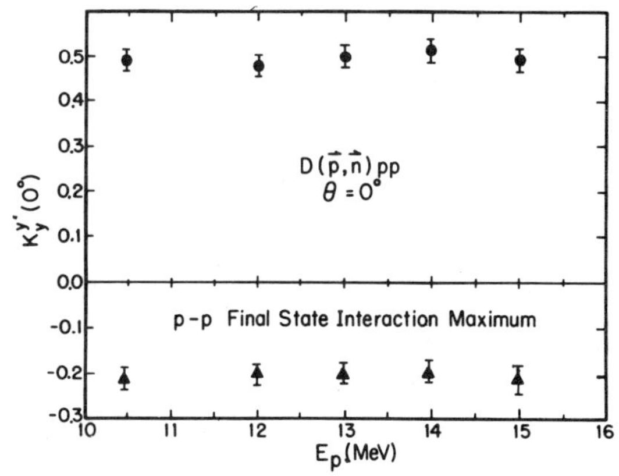

Figure 2

Polarization transfer coefficients extracted from two regions of the D($\vec{p}$,$\vec{n}$)pp spectra (as explained in the text) plotted as a function of proton energy.

Bibliography

1) J. M. Wallace, Phys. Rev. 9C (1974) 897 and references therein

POLARIZATION TRANSFER IN THE $D(\vec{d},\vec{n})^3He$ REACTION AT $\theta = 0°$

P. W. Lisowski, G. Mack, and R. L. Walter, Duke University and Triangle Universities Nuclear Laboratory+ and C. E. Busch and T. B. Clegg, University of North Carolina and TUNL+

The possibility of producing a polarized neutron beam by using a polarized charged particle initiated reaction has been investigated by several authors.[1] Above a deuteron energy of about 3 MeV, the $D(d,n)^3He$ reaction has a large 0° cross section and a high probability for spin polarization transfer from the incoming deuteron to the outgoing neutron. Initial measurements conducted at Los Alamos[2] and at Birmingham[3] showed that the 0° polarization transfer coefficient $K_y^{y'}$ is not only large, but is also nearly constant with deuteron energy. These features provide an experimental versatility not available when an unpolarized charged particle beam is used to produce polarized neutron beams through specific reactions at specific energies and angles.

In fact, the $D(\vec{d},\vec{n})^3He$ reaction has provided polarized neutrons for several elastic scattering polarization studies conducted at our laboratory.[4] Analysis of those data indicated a need for more detailed knowledge of the parameters necessary to determine the neutron polarization. In this paper we report the results of two experiments which were conducted to provide additional accurate information about the outgoing neutron polarization produced in the $D(\vec{d},\vec{n})^3He$ reaction at an angle of 0° as a function of deuteron energy. These measurements were carried out using a polarized deuteron beam which had equal vector and tensor polarizations of about 0.75 as measured by the quench-ratio method (see footnote 14 in Ref. 7). From 20 to 60 nA of beam was incident on a deuterium gas target of about 400 keV thickness.

The polarization $p_{y'}$ of the outgoing 0° neutrons is given by the equation:

$$p_{y'} = 3/2\, p_y\, K_y^{y'} / \left[1 + 1/2\, p_{zz} A_{zz}\right],$$

where $K_y^{y'}$ is the vector polarization transfer coefficient, A_{zz} is the zero-degree analyzing power, and p_y and p_{zz} are the vector and tensor beam moments. As indicated in the above formula, it is necessary to determine A_{zz} as well as $p_{y'}$ in order to calculate $K_y^{y'}$ in this experimental arrangement. The polarization of the neutrons from the deuterium target was analyzed by scattering from a high pressure ^4He cell into a left-right pair of plastic detectors positioned at $\theta_{lab} = 120°$. The general experimental technique is the same as that reported earlier.[5] Values of A_{zz} were deduced from ratios of the neutron intensity in the helium cell for incident deuteron beams with the spin oriented transversely to the intensity with the spin oriented longitudinally.

The results of these A_{zz} measurements and other values reported by Simmons et al.[2] and Salzman et al.[6] are shown in Fig. 1. These data show no structure above a deuteron energy of about 3 MeV, but increase in magnitude for lower deuteron energies. Values of $K_y^{y'}$ are shown in Fig. 2 along with the earlier results of Simmons et al.[2] and Blythe et al.[3]. Here the data also show a rapid increase from about 1 to 4 MeV. Above 4 MeV the data appear to be nearly constant, declining

---

+ Work supported in part by the U. S. Atomic Energy Commission.

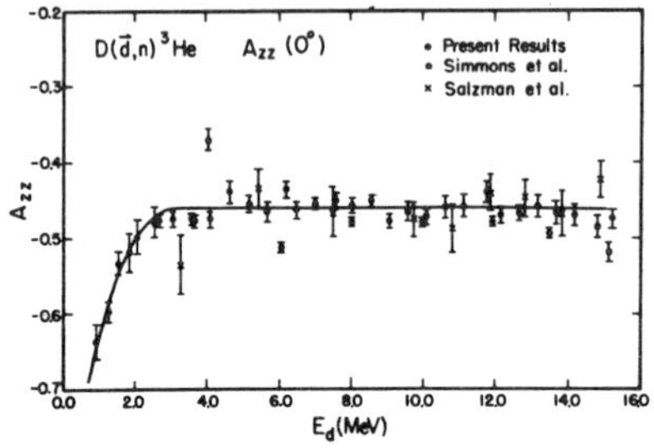

Figure 1

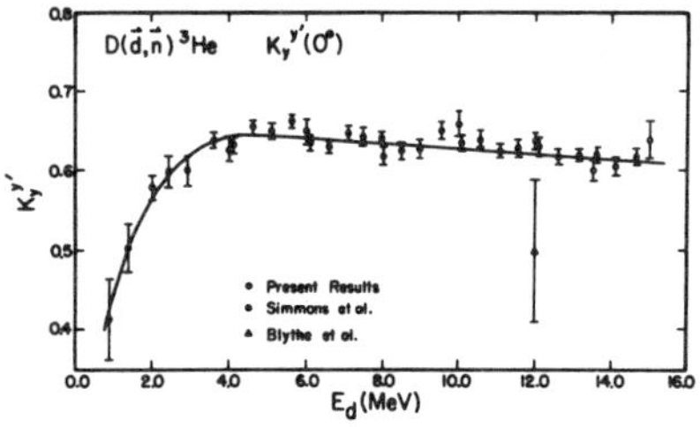

Figure 2

slightly with increasing energy. The data in the region above 4 MeV may be parameterized as $A_{zz}(0°) = -0.462$; $K_y^{y'}(0,E_d) = 0.662 - 0.003 E_d$. This simple energy dependence demonstrates that there is no structure which would be detrimental to the use of the reaction as a source of polarized neutrons.

Our experiment further confirms the previously reported[7] similarity of the values of $K_y^{y'}$ for the $D(\vec{d},\vec{p})T$ and $D(\vec{d},\vec{n})^3He$ mirror reactions. Figure 3 illustrates this feature by showing a comparison between the least squares parameterization of the present data and the $D(\vec{d},\vec{p})T$ results of Clegg et al.[7] The similarity between these reactions is further evidence for the charge symmetry of the nuclear force in the four body problem.

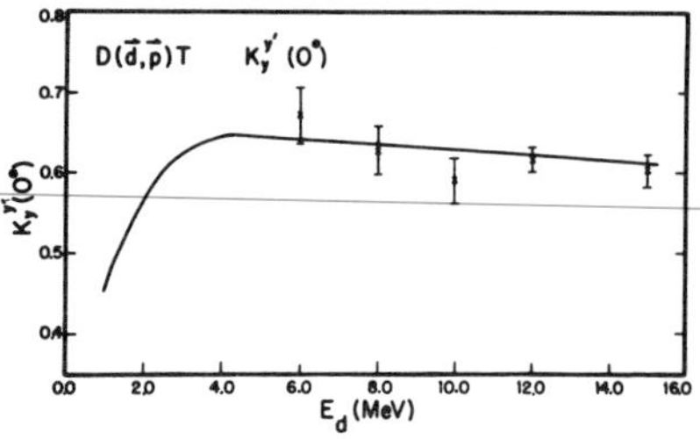

Figure 3

Bibliography

1) J. E. Simmons, W. B. Broste, T. R. Donoghue, R. C. Haight and J. C. Martin, Nucl. Instr. and Meth. 106 (1973) 477 and references therein

2) J. E. Simmons, W. B. Broste, G. P. Lawrence, J. L. McKibben and G. G. Ohlsen, Phys. Rev. Letters 27 (1971) 113

3) C. O. Blythe, P. B. Dunscombe, J. S. C. McKee, and C. Pope, Phys. Letters 33B (1970) 211

4) P. W. Lisowski, T. C. Rhea, R. L. Walter, T. B. Clegg and C. E. Busch, Bull. Am. Phys. Soc. 18 (1973) 687

5) G. Spalek, R. A. Hardekopf, J. Taylor, Jr., Th. Stammbach and R. L. Walter, Nucl. Phys. A191 (1972) 449 and references therein

6) G. C. Salzman, J. C. Martin, J. J. Jarmer, J. E. Simmons, G. G. Ohlsen and T. R. Donoghue, Phys. Letters 45B (1973) 123

7) T. B. Clegg, D. D. Armstrong, R. A. Hardekopf and P. W. Keaton, Jr., Phys. Rev. C8 (1973) 922

SCATTERING OF POLARIZED NEUTRONS FROM ^{3}He AT 8 AND 12 MeV

P. W. Lisowski, R. L. Walter, Duke University and Triangle Universities Nuclear Laboratory+
and C. E. Busch, and T. B. Clegg, University of North Carolina and TUNL +

Recently Hardekopf et al.[1] compared their new data on the polarization produced in the $T(\vec{p},p)T$ elastic scattering process to available $^3He(\vec{n},n)^3He$ polarization data.[2,3] The observed differences in these sets of data were inconsistent with the small differences one might expect for nucleon scattering from mirror nuclei of such low A values. That is, since the nuclear force is believed to be charge independent except for restrictions arising from the Pauli exclusion principle, the n-n, n-p and p-p interactions should be identical. One thus expects that there is little distinction between the four-nucleon interactions for n-^{3}He and p-T or for p-^{3}He and n-T, at least at energies somewhat above the Coulomb barrier. Because of experimental difficulties with sizeable tritium targets, data on the T(n,n)T reaction is scarce, limiting comparisons with the relatively abundant $^3He(p,p)^3He$ data. To provide data for an accurate comparison for the mass-4 system, we have remeasured the $^3He(\vec{n},n)^3He$ polarization function at 8 and 12 MeV with better techniques than have been employed previously.

Polarized neutrons were scattered from ^{3}He contained in a high-pressure scintillator cell, and the asymmetry for scattering into plastic side detectors was measured as a function of scattering angle. The incident polarized neutron beam was produced in a unique way by using the high polarization transfer capabilities of the $D(\vec{d},\vec{n})^3He$ reaction at $\theta = 0°$. Deuterons which originated in a Lamb-shift polarized ion source bombarded a deuterium-filled gas cell to produce a clean, highly polarized neutron beam emitted at a $D(\vec{d},\vec{n})^3He$ reaction angle of 0°. The deuteron beam polarization was deduced from frequent measurements in a separate target chamber using the known 0° analyzing power of the $^3He(\vec{d},p)^4He$ reaction[4].

The obvious differences between our $^3He(\vec{n},n)^3He$ data and the earlier results[2,3] are exhibited in Fig. 1. The new data are in much better agreement with the $T(\vec{p},p)T$ data[1,5] shown in Fig. 2. The solid line in Fig. 2 represents the polarization based on single energy phase shifts determined as part of a fit to the available n-^{3}He cross-section and polarization data for the region from 3 to 24 MeV. These phase shifts provide a good representation of the data and are similar to those obtained by Hardekopf et al. for T(p,p)T. However, there are not yet sufficient data to permit the determination of a unique set of phase shifts. The current differences seen in Fig. 2 in the polarization for these mirror reactions are indeed small and may be attributable to the known Coulomb energy differences in the two interactions. That is, there no longer is evidence for a strong violation of charge symmetry in these two cases.

+ Work supported in part by the U. S. Atomic Energy Commission.

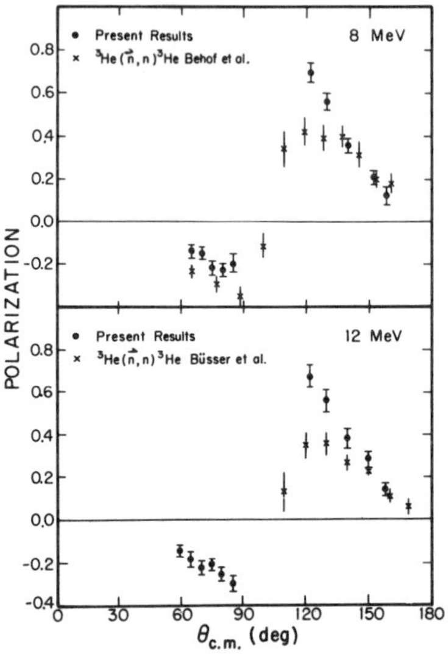

Figure 1

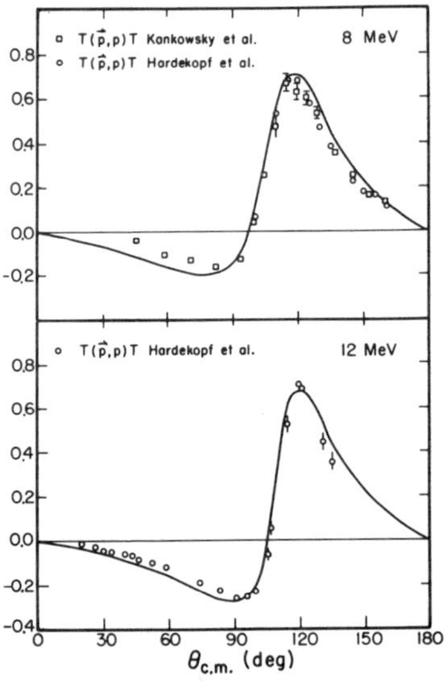

Figure 2

Fig. 1: Comparison of the present results with previous $^3He(\vec{n},n)^3He$ data.

Fig. 2: Comparison of $T(\vec{p},p)T$ results with polarizations calculated from newly obtained $^3He(n,n)^3He$ phase shifts.

Bibliography

1) R. A. Hardekopf, P. W. Lisowski, T. C. Rhea, R. L. Walter and T. B. Clegg, Nucl. Phys. A191 (1972) 481

2) A. F. Behof, J. M. Hevezi, and G. Spalek, Nucl. Phys. 84 (1966) 290

3) F. W. Büsser, H. Dubenkropp, F. Niebergall and K. Sunram, Nucl. Phys. A129 (1969) 666

4) W. Grüebler, V. König, A. Ruh, R. E. White, P. A. Schmelzbach, R. Risler and P. Marmier, Nucl. Phys. A 165 (1971) 505

5) R. Kankowsky, K. Killian, and D. Fick, private communication (1971)

POLARIZATION TRANSFER IN THE $(\vec{d},\vec{n})$ BREAKUP REACTION ON ^2H, ^4He and Ta

R. L. Walter, P. W. Lisowski, Duke University and Triangle Universities Nuclear Laboratory + and T. B. Clegg, University of North Carolina and TUNL +

With the advent of intense polarized beams, a variety of polarization transfer experiments have now become feasible without requiring excessive accelerator time. Using proton and deuteron beams produced in our Lamb-shift polarized ion source, we have investigated the energy dependence of the polarization transfer coefficient for $(\vec{p},\vec{n})$ and $(\vec{d},\vec{n})$ interactions at 0° reaction angle. The measurements have involved transverse vector-polarized incident beams and determination of the transverse polarization in the outgoing neutron beam. (This type of polarization transfer coefficient is similar to the Wolfenstein depolarization parameter D.) The transfer coefficient was very large in all the $(\vec{d},\vec{n})$ reactions investigated, suggesting that the influence of spin-dependent forces and compound nuclear disturbances has a small effect at 0°. Usually the reactions studied were limited to those cases where the residual nucleus was left in the ground state.

It was recognized in the $D(\vec{d},\vec{n})^3He_{gs}$ measurement reported in another contribution to this conference that it would be possible to obtain values of the transfer coefficient for neutrons produced in deuteron breakup reactions as a function of the outgoing neutron energy E_n, the neutron detection angle being 0°. To avoid measuring the tensor coefficient A_{zz}, this experiment was conducted with the polarized ion source adjusted to produce an incident beam with pure vector polarization. The magnitude of the neutron polarization was determined by scattering from ^4He into plastic side detectors at 120°. The energy of the ^4He recoil was recorded in coincidence with the detected neutrons. As the ^4He energy is proportional to E_n, analysis of the observed polarization as a function of ^4He recoil energy permits determinations of polarization of the neutrons which have a continuous distribution of energies E_n.

In Fig. 1 a spectrum of the coincident recoil ^4He pulses is illustrated for the $D(\vec{d},\vec{n})^3He_{gs}$ and $D(\vec{d},\vec{n})pD$ reactions for $E_d = 11.4$ MeV; the spread in energy for the $^3He_{gs}$ neutrons is due to the finite geometry and the electronic resolution of the analyzer system. Below the spectrum in Fig. 1, the extracted values for the polarization transfer coefficient $K_y^{y'}$ are plotted as a function of channel number, which is linearly related to E_n. It can be seen that over much of the breakup region $K_y^{y'}$ is nearly constant at about 0.5. This value should be viewed in comparison to $\sqrt{2}/2$, the maximum theoretical value for a deuteron-induced reaction. In Fig. 2, the same information is exhibited for the $^4He(\vec{d},\vec{n})p^4He$ reaction at $E_d = 11.4$ MeV; here, $K_y^{y'}$ ranges about 0.65 over the region where the breakup yield is large. Similarly high transfer coefficients were also observed in investigations of these reactions for deuteron energies of 8.7 and 14.4 MeV. For Ta + d reactions other spectra at $E_d = 14$ MeV showed no large maximum in the yield attributable to the breakup contribution but rather a distribution which decreased monotonically for increasing neutron energy. The $K_y^{y'}$ for this interaction was also large for all E_n although the quality of the data is probably too poor to conclude that $K_y^{y'}$ is as large as that for the ^4He target.

[+] Work supported in part by the U. S. Atomic Energy Commission.

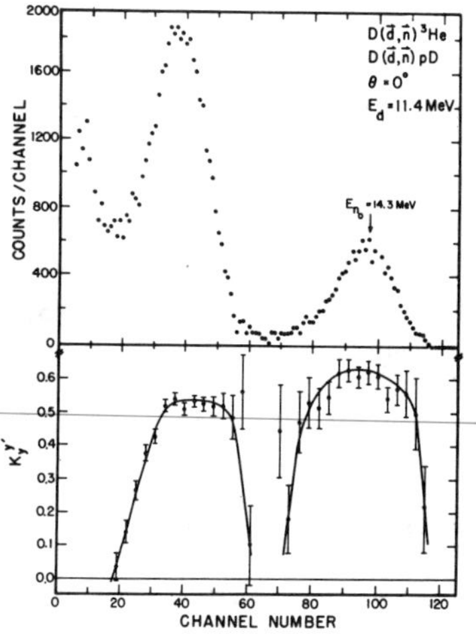

Figure 1

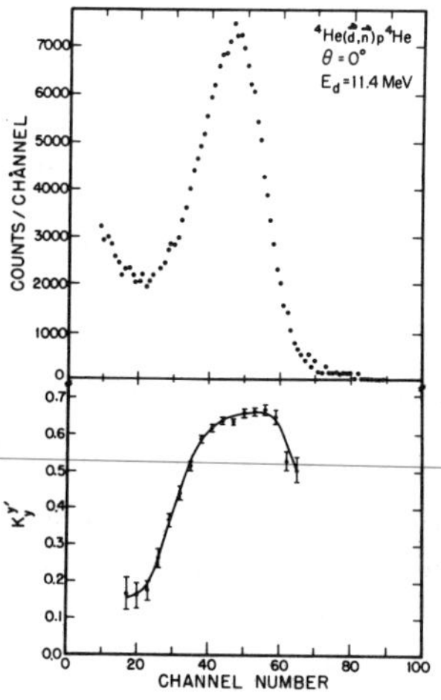

Figure 2

Fig. 1: Experimental results for the $D(d,n)^3He$ and $D(d,n)pD$ reactions.

Fig. 2: Experimental results for the $^4He(d,n)p^4He$ reaction.

LINEAR RELATIONS BETWEEN THE RANK-1 POLARIZATIONS OF A NUCLEAR REACTION

F. Seiler+
Physics Dept., University of Basel, CH-4056 Basel, Switzerland

The angular distributions of the various rank-1 polarizations in reactions between light nuclei show striking similarities. A typical example is the $^3He(d,p)^4He$ reaction which has been the object of considerable experimental [1] and theoretical [2] work. Numerous broad levels have been identified in light nuclei, inviting the investigation of the effect of an overlap between two large matrix elements $R_i = \langle \ell_i' s_i' J_i^\pi | R | \ell_i s_i J_i^\pi \rangle$ on rank-1 polarizations. For the two-body reaction a+A→a'+A' with spins i+I→ i'+I', the various vector polarizations have been computed and the Legendre polynomial coefficients compared, using convenient normalizations [3,4]. For the incident channel the relation

$$\frac{a_{11}(L)}{a_{0011}(L)} = \frac{d_{11}(L)}{d_{0011}(L)} = (-)^{s_1-s_2}\ \delta(s_1',s_2')\ \sqrt{\frac{2i+1}{2I+1}}\ \frac{\begin{Bmatrix} i & i & 1 \\ s_1 & s_2 & I \end{Bmatrix}}{\begin{Bmatrix} I & I & 1 \\ s_1 & s_2 & i \end{Bmatrix}} \tag{1}$$

results. This ratio depends solely on the channel spins s_i of the interfering elements R_1 and R_2. The formula for the outgoing channel is obtained by interchanging primed and unprimed parameters. An example is the behavior of the L=2 coefficients in the $^3He(d,p)^4He$ reaction below 1 MeV. The overlap seen in fig. 1 is due to the s-wave $J^\pi = \frac{3}{2}^+$ element, resonant at 430 keV, and a d-wave element, identical in all other quantum numbers. This fact, established in previous analyses [4], is demonstrated easily by the use of eq. (1). The relevant ratios and data are shown in fig. 1. Clearly both elements must have channel spin $s = \frac{3}{2}$.

Linear relations between polarizations in the incoming and outgoing channels have been widely discussed. For the overlap of two resonances

$$\frac{d_{11}(L)}{d_{00}^{11}(L)} = -2\sqrt{\frac{i'(i'+1)}{i(i+1)}}\ \frac{\delta(s_1',s_2')}{\delta(s_1,s_2)}\ \frac{i(i+1)+s(s+1)-I(I+1)}{i'(i'+1)+s'(s'+1)-I'(I'+1)}\ \frac{s'(s'+1)}{s(s+1)}A$$

with

$$A = \frac{\ell_1(\ell_1+1)-\ell_2(\ell_2+1)-J_1(J_1+1)+J_2(J_2+1)}{\ell_1'(\ell_1'+1)-\ell_2'(\ell_2'+1)-J_1(J_1+1)+J_2(J_2+1)} \tag{2}$$

is obtained. Similar ratios are found for the other observables [4]. All depend on the detailed spin configuration of the interfering elements, which must satisfy both δ-functions in order to appear in both observab-

---
+Present address: Lawrence Berkeley Laboratory, Berkeley, CA 94720, USA.

les. An example is the $^3H(p,n)^3He$ reaction [5], a rather special case, for which neutron polarization and proton analyzing power should be nearly equal [6]. For equality eq. (2) must yield $d_{11}(L) = -2d^{11}_{00}(L)$, for which A=1 is required. Generally this will hold only if both elements satisfy $\ell_i=\ell_i'$. Indeed the deviations in fig. 2 have been traced to terms involving elements with $\ell_i' = \ell_i+2$. The main effect, however, is due to an overlap of two resonant p-wave elements with $J = 0^-$ and 2^-, which clearly belong to the class for which A=1.

In the analysis of polarization data eqs. (1) and (2) can thus be used to restrict the element combinations responsible for structures of common origin in different rank-1 observables.

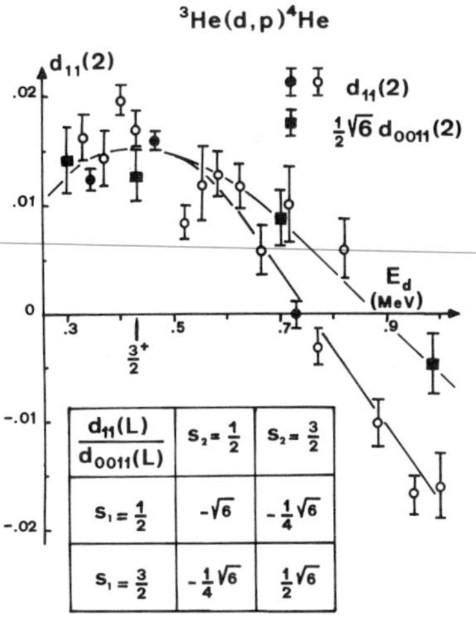

Fig.1

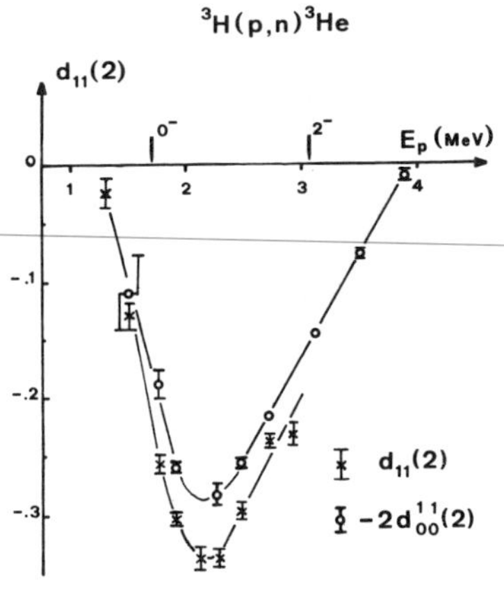

Fig.2

References

1) D.M. Hardy, S.D. Baker, D.H. McSherry and R.J. Spiger, Nucl. Phys. A160 (1971) 154
2) B. De Facio and R.K. Umerjee, Nucl. Phys. A169 (1971) 257
3) F. Seiler and E. Baumgartner, Nucl. Phys. A153 (1970) 193
4) F. Seiler, Nucl. Phys. A187 (1972) 379 and to be published
5) L.G. Arnold, R.G. Seyler, T.R. Donoghue, L. Brown and U. Rohrer, Phys. Rev. lett. 32 (1974) 310
6) H.E. Conzett, to be published

ASYMMETRY IN THE $^1H(d,2p)n$ REACTION STUDIED WITH 12.2MeV POLARISED DEUTERONS

C. O. Blyth, N. T. Okumusoglu, N. Berovic and J. S. C. McKee. Department of Physics, The University of Birmingham, B15 2TT, England.

Arvieux et al (1970) have reported small, but in one case significant asymmetries observed in the reaction $^2H(p,2p)n$ at an incident proton energy of 10.5 MeV. The relevant datum occurred at energies and angles where the neutron-proton final state interaction was significant.

Using the model described in their paper, which compares the $^2H(p,p^1)^2H^*$ reaction with the $^2H(p,p)^2H$ elastic scattering, they express surprise at the similarity observed in the experimental results, since the latter case requires a larger number of transition amplitudes for a full description of the process.

The object of the present work was to make similar measurements at another energy, in this case using 12.2MeV polarised deuterons incident on a hydrogen target. The two protons in the final state were detected in coincidence in the reaction plane. The overall experimental arrangement was similar to that of Arvieux, protons being selected by means of solid state detector telescopes with a fast coincidence requirement between them.

Accepted pulses were later displayed in two dimensions. A routing system kept track of the switching of the incident spin direction and tagged the detected events accordingly. As the Birmingham Radial Ridge Cyclotron is naturally a pulsed machine, random background is determined by the introduction of a delay equivalent to a separation of eight beam bursts into one of the fast proton signal lines.

The locus of coincident events was clearly apparent on the visual display, and various areas were summed both on and off the final state interaction peak.

The table below summarises the asymmetries and analysing powers observed for events in the peak, all of which are consistent with zero. In other cases the results are also consistent with zero, but with considerably poorer statistics.

Counter Geometry		Asymmetry	Analysing Power
θ_1	θ_2	$\varepsilon = \frac{N_{up} - N_{down}}{N_{up} + N_{down}}$	$iT_{11} = \frac{\varepsilon}{2it_{11}}$
22^0	30^0	0.0066 ± 0.017	6.6×10^{-3}
22^0	25^0	0.0083 ± 0.010	8.1×10^{-3}
22^0	20^0	0.003 ± 0.025	2.8×10^{-3}
22^0	15^0	0.0019 ± 0.011	1.8×10^{-3}

REFERENCES

1. J. ARVIEUX, J. L. DURAND, J. C. FAIVRE, D. GARRETA, A. PAPINEAU, J. SURA and A. TARRATS. Nucl. Phys. A150, 75 (1970).

POLARIZATION OF NEUTRONS FROM THE D(α,n) REACTION

H. D. Knox, R. G. Graves, and L. C. Northcliffe, Cyclotron Institute, Texas A & M University, College Station, Texas 77843

and

H. Nakamura and H. Noya, Department of Physics, Hosei University, Tokyo, Japan

The measurement of the polarization of neutrons emitted from the D(α,n) reaction is of interest for two reasons. First, earlier measurements[1)] of the differential cross section of neutrons from this reaction indicated a strong interaction between the neutron and the alpha particle in the final state. Because of this, one might speculate that this reaction could be a source of highly polarized neutron. Second, recent theoretical calculations[2,3)] using a modified impulse approximation (MIA) have been quite successful in predicting differential cross sections of this reaction. However, in these calculations cross sections are determined by the absolute values of amplitudes only, whereas the polarization depends on the relative phases of the amplitudes. Measurement of the polarization of neutrons from this reaction would, then, give a further check of the validity of the MIA.

A beam of 42.3 MeV alpha particles from the Texas A & M 88 inch cyclotron was incident on a high pressure deuterium gas target. The alpha beam was time tuned and had a pulse width of approximately 2.5 nsec. The target pressure was 5 atm and the resulting target thickness was 1.07 MeV. Havar windows at the entrance and exit of the gas cell caused an energy loss of 2.4 MeV in the alpha beam. Immediately after passing through the target, the alpha beam was magnetically deflected through 90° and collected in a heavily shielded Faraday cup. The polarization of neutrons emitted at 18° to the incident alpha beam was measured with a liquid helium polarimeter located 4.5 meters from the target. A pair of transverse field spin precession magnets in the 18° line were used to help eliminate false asymmetries.

The neutron polarization measured in the present experiment along with the differential cross section data (E_α = 38.8 MeV) obtained earlier[1)] at Texas A & M are shown in Figures 1 and 2, respectively. The cross at 19.2 MeV in the polarization results indicates the polarization of neutrons elastically scattered by alpha particles

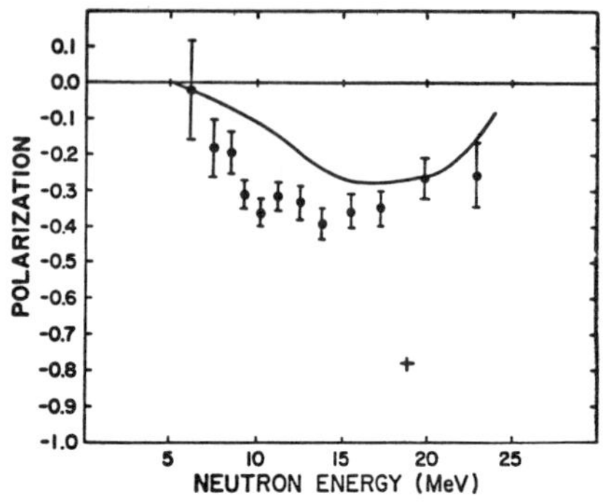

Figure 1

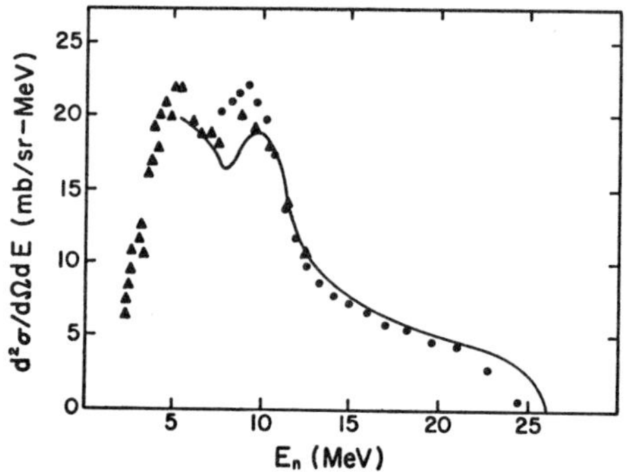

Figure 2

of the same incident energy as the present experiment. The fits to both the differential cross section and the polarization were obtained with the modified impulse approximation using the same methods and parameter set as in Reference 2 and assuming E_α = 42.3 MeV. Although the fits to the data are generally good, the present results indicate that some improvements must still be made in the theory.

REFERENCES

1). C. Lewis, D. Saylor, and L. Northcliffe, Phys. Rev. C3 (1971) 28.

2). H. Nakamura, Nucl, Phys. A208 (1973) 207.

3). H. Nakamura, Nucl. Phys. to be published.

POLARIZATION TRANSFER IN PROTON-DEUTERON ELASTIC SCATTERING*

F. N. Rad, J. Birchall, H. E. Conzett, and R. Roy

Lawrence Berkeley Laboratory, University of California, Berkeley, California 94720

Within the past two years considerable theoretical progress has been made in fitting the nucleon polarizations measured in nucleon-deuteron elastic scattering at energies below 50 MeV.[1,2] These three-body calculations, based on the Faddeev equations, have also provided predictions of the deuteron vector and tensor polarizations and, more recently, of polarization transfer coefficients[2] at energies below 25 MeV. Subsequent determinations of deuteron polarizations[3] were in good agreement with the calculations. These studies showed that the P-wave part of the two-body N-N interaction was chiefly responsible for the vector polarizations, whereas, in addition, the $^3S_1$-$^3D_1$ N-N tensor force was required in order to provide agreement with the observed tensor polarizations.

It is clear that the polarization results have provided the important and essential tests of the calculations and of the sensitivity of the N-d observables to the details of the input N-N potentials. As a further test of the theoretical predictions, we report here on measurements of the vector polarization of the recoil protons in d-p elastic scattering with incident 45.4 MeV vector polarized deuterons. These $^1H(\vec{d},\vec{p})^2H$ measurements provide determinations of the vector-to-vector polarization transfer coefficient, $K_y^{y'}(\theta)$, defined by the equation[4]

$$p_{y'} = (A_{y'} + \frac{3}{2} p_y K_y^{y'})/(1 + \frac{3}{2} p_y A_y) \quad ,$$

where p_y in the polarization of the purely vector polarized incident deuteron beam, $p_{y'}$ is the observed polarization of the outgoing (recoil) proton, and A_y and $A_{y'}$ are, respectively, the deuteron vector analyzing power and the proton analyzing power in N-d elastic scattering. The Madison convention[5] is followed in defining the cartesian forms of the polarization and analyzing power, and the y and y' axes are taken along $\vec{k}_d \times \vec{k}_p$. From time-reversal-invariance, the same coefficient describes the transfer of vector polarization in the inverse $^2H(\vec{p},\vec{d})^1H$ scattering process at the equivalent proton energy of 22.7 MeV[4]. Thus, our measurements can be compared directly with Pieper's[2] calculated $K_y^{y'}(\theta)$ at that energy. Previous measurements[6] of vector-to-tensor polarization-transfer coefficients in p-d scattering below 10 MeV were consistent with zero within the experimental error, while a single measurement[7] of $K_x^{x'}$ was found to be two-standard-deviations from the theoretical value.[2] It was clear that more data were required for a significant comparison with the calculated values, and that substantial values of $K_y^{y'}$ were expected at E_p = 22.7 MeV for the θ_{cm} that we found to be experimentally accessible. The polarized deuteron beam from the Berkeley 88-in. cyclotron, passed through a high-pressure liquid-nitrogen cooled hydrogen gas target, positioned in the 36-in-diam scattering chamber. The areal density of the gas was 12.8 mg/cm^2 at the operating pressure of 12.4 atm. and the mean-energy in the gas was 45.4 MeV. Two silicon polarimeters, similar to those described elsewhere,[8] were used in these measurements and were positioned at equal angles to the left and right of the beam. The angular resolution defined by tantalum collimators, was 0.8° (FWHM). The vector polarization of the beam, monitored continuously with a ^{4}He polarimeter placed downstream of the scattering chamber, was .815±.015 of the maximum possible value p_y = 2/3, with zero tensor components. For each $K_y^{y'}(\theta)$ determination, runs were taken with the spin vector of the beam oriented up and down with respect to the scattering plane. This mode of operation along with the left-right polarimeters provided several internal consistency checks. The silicon polarimeters were calibrated with polarized protons elastically scattered from ^{4}He and ^{12}C, for which $K_y^{y'}$ = 1.0.

Preliminary values of our experimental results are shown in fig. 1., where the errors indicated are purely statistical. The solid curve is Pieper's calculated result, and it is seen that the agreement with experiment is very good. Since this constitutes the first significant comparison between experimental and calculated polarization-transfer coefficients in N-d elastic scattering, this agreement represents yet another substantial success of the three-body calculations.

We are grateful to R. M. Larimer for her assistance during the course of the experiment.

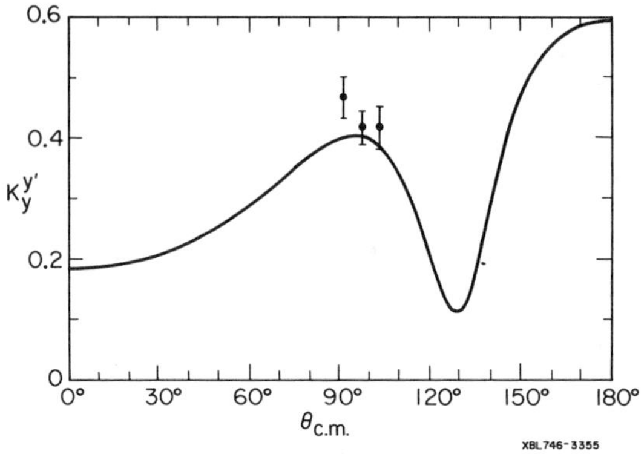

FIGURE 1

REFERENCES

1) S. C. Pieper, Nucl. Phys. A193 (1972) 529; P. Doleschall, Nucl. Phys. A201 (1973) 264, A220 (1974) 491
2) S. C. Pieper, Phys. Rev. C 8 (1973) 1702
3) J. S. C. McKee, H. E. Conzett, R. M. Larimer, and Ch. Leemann, Phys. Rev. Letters 29 (1972) 1613; A. Fiore, J. Arvieux, Nguyen Van Sen, G. Perrin, F. Merchez, J. C. Gondrand, C. Perrin, J. L. Durand, and R. Darves-Blanc, Phys. Rev. C 8 (1973) 2019
4) G. G. Ohlsen, Rep. Prog. Phys. 35 (1972) 717
5) Polarization Phenomena in Nuclear Reactions, edited by H. H. Barschall and W. Haeberli (Univ. of Wisconsin Press, Madison, 1971), p. xxv
6) G. G. Ohlsen, C. K. Mitchell, and G. C. Salzman, Few Particle Problems in the Nuclear Interaction, edited by I. Slaus, S. A. Moszkowski, R. P. Haddock, and W. T. H. van Oers (North-Holland, Amsterdam, 1973), p. 460
7) G. G. Ohlsen, C. K. Mitchell, G. C. Salzman, J. L. Gammel, D. C. Dodder, and K. Witte, Bull. Am. Phys. Soc. 17 (1972) 922
8) B. Frois, J. Birchall, R. Lamontagne, R. Roy, and R. J. Slobodrian, Nucl. Instr. Methods 96 (1971) 431

*Work performed under the auspices of the U. S. Atomic Energy Commission.

POLARISATION MEASUREMENTS IN NEUTRON-DEUTERON ELASTIC SCATTERING

R.C.Brown, J.A.Edgington, M.P.May and Y.Onel,Queen Mary College, London, England
I.M.Blair,AERE, Harwell, Oxfordshire, England
N.M.Stewart, Bedford College, London, England

There are few reported measurements of the polarisation P(θ) in n-d elastic scattering, to complement the extensive p-d data. What comparison of the two systems has been made, near 23MeV[1)] and at 35 MeV[2)], seems to indicate systematic discrepancies with P(θ) being significantly more negative in n-d than in p-d scattering, near angles corresponding to the minimum in the cross section. We report here some preliminary results of a measurement of P(θ) in n-d scattering at higher energies. The experiment was performed at the Harwell 160 MeV synchrocyclotron, and the results presented derive from about one quarter of the data taken.

A beam of polarised neutrons was obtained by deflecting the internal proton beam onto an aluminium target; neutral particles emerging at 47° travelled 50 m before striking a target of liquid deuterium. Outgoing charged particles were then detected in scintillation counter telescopes, protons and deuterons being distinguished by a combination of time of flight and pulse height analysis. A solenoid was used to precess the neutron spin, initially vertical, towards the horizontal. This horizontal component P_n (which varied with neutron energy) resulted in an up-down asymmetry, A, of the recoil deuteron. By reversing the field direction, the effects of small misalignments of the up and down counter telescopes were minimised. Incident neutron energies were measured by standard time of flight techniques, and the quantity $A = P_n P(\theta)$ was measured at energies between 50 and 120 MeV, and in the angular range $70^o < \theta_{cm} < 170^o$.

In a separate measurement P_n was found by Schwinger scattering from ^{238}U at an angle of 1°. As a check on the methods of data reduction the liquid deuterium was replaced by hydrogen, and the asymmetry in free n-p scattering found. The corresponding values of P(θ) are in excellent agreement with previous determinations[3)] of this quantity.

In the figure we show P(θ) in n-d elastic scattering, for three energy bins,each 10 MeV wide, at 70, 80 and 90 MeV. Errors are statistical, and ther is an additional normalisation error with an estimated s.d. of $\pm$ 7%. Also shown are lines representing the trend of the p-d measurements at 50 MeV[4)] and near 140 MeV[5)]. Future analysis will approximately double the number of angles, and include energies above and below those shown here. The implications of possible differences between n-d and p-d polarisations are discussed in another paper submitted to this Conference[6)]. There, the discrepancies at lower energies are ascribed mainly to the rapid variation of P(θ) with energy; the near-constancy of P(θ) between 50 and 140 MeV would therefore suggest that, between these energies, such differences would be small. Our results indicate that any differences that do exist lie within the present experimental errors.

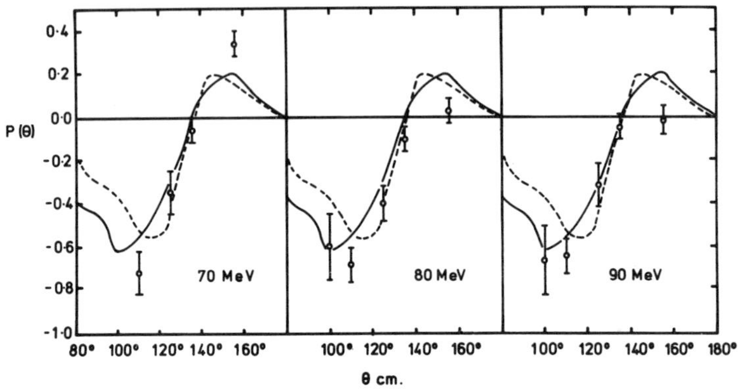

Polarisation P(θ) in n-d elastic scattering at 70, 80 and 90 MeV

———— Trend of p-d measurements near 140 MeV

- - - - - - Trend of p-d measurements near 50 MeV

Ⴔ Results of this experiment

REFERENCES

1) J.C.Faivre, D.Garetta, J.Jungerman, A.Papineau, J.Sura and A.Tarrats, Nucl. Phys. A127, 169 (1969)
J.J.Malanify, J.E.Simmons, R.B.Perkins and R.L.Walter, Phys. Rev. 146, 632 (1966)

2) S.N.Bunker, J.M.Cameron, R.F.Carlson, J.R.Richardson, P.Tomas, W.T.H. van Oers and J.Verba, Nucl. Phys. A113, 461 (1968)
J.Zamudio-Cristi, B.E.Bonner, F.P.Brady, J.A.Jungerman and J.Wang, Phys. Rev. Lett. 31, 1009 (1973)

3) A.Langsford, P.H.Bowen, G.C.Cox, G.B.Huxtable and R.A.J.Riddle, Nucl. Phys. 74, 241 (1965)

4) A.R.Johnstone, W.R.Gibson, E.A.McClatchie, J.H.P.C.Megaw, G.T.A.Squier, R.J.Griffiths and F.G.Kingston, Rutherford Laboratory Report RHEL/R 136 (1966), p.32

5) H.Postma and R.Wilson, Phys. Rev. 121, 1229 (1961)
M.Poulet, A.Michalowicz, Y.LeGuen, K.Kuroda,D.Cronenberger and G.Coignet, J. Phys. Rad. 26, 399 (1965)

6) J.A.Edgington and C.J.Oram, these Proceedings

POSSIBLE TESTS OF TIMEREVERSAL INVARIANCE IN POLARIZED $p\text{-}^3H$ OR $p\text{-}^3He$ SCATTERING

M. Simonius

Laboratorium für Kernphysik, Eidg. Techn. Hochschule

Zürich, Switzerland

Though K^o decay data are in agreement with superweak interaction predictions, experiments are not yet able to eliminate electromagnetic or millistrong T-violating interactions which both predict T-violation in non-weak nuclear processes of the order of 10^{-3} (ref.[1]). The object of this note is to point out and discuss the possibility for timereversal invariance tests in (low energy) polarized $p\text{-}^3H$ and $p\text{-}^3He$ scattering to an accuracy of this order. The quantities of interest are the analysing power $A \equiv A_y \equiv \sqrt{2}\, i\, T_{11}$ and the polarization $P = p_y = \sqrt{2}\, i\, t_{11}$ as defined e.g. in the Madison convention[2,3]. Timereversal invariance requires[1,2] $P = A$. The experimental possibilities for such measurements and the reason for choosing $p\text{-}^3H$ and $p\text{-}^3He$ scattering have been discussed in ref.[2]. The three main sources of difficulties are statistics, angle and energy definition, and calibration of the polarization since P and A are obtained in two different measurements. The calibration may be eliminated by considering the ratio $P(p\text{-}^3He)/P(p\text{-}^3H)$ and similarly for A since the two reactions have the same kinematics. Apart from nucleon-nucleon scattering, where higher energies are needet for sensible tests, $p\text{-}^3H$ and $p\text{-}^3He$ are the simplest systems with favorable spin structure[2].

The question which remains is the sensitivity of such tests to the timereversal violating nuclear scattering amplitudes and nucleon-nucleon interaction. Since $P - A$ is odd under timereversal it is due to interference between T-violating (TV) and T-conserving scattering amplitudes. In nucleon-nucleon or $p\text{-}^3He$ scattering below breakup threshold (or as long as breakup can be neglected) where only elastic scattering is possible the phase of the T-violating amplitude can be obtained from the strong phases and unitarity yielding for instance

$$\langle {}^3S_1 | S^{TV} | {}^3D_1 \rangle = -\langle {}^3D_1 | S^{TV} | {}^3S_1 \rangle = g_1\, e^{i(\delta_1+\delta_2)} \qquad (1)$$

where the δ_i are the eigenphases in the space spanned by $|{}^3S_1\rangle$ and $|{}^3D_1\rangle$ and g is real (positive or negative). In this case the interference term is uniquely given if the strong scattering amplitudes are known. For $p\text{-}^3He$ P-A may be obtained from the table on the next page and (θ is the center of mass scattering angle)

$$P-A = \left(|k|^2 \frac{d\sigma}{d\Omega}\right)^{-1} \sum_{L,\alpha} g_\alpha\, G_L^\alpha\, d_{10}^L(\theta) = \sum_\alpha g_\alpha\, K^\alpha(\theta) \qquad (2)$$

E_p	G_1^1	G_2^1	G_1^2	G_2^2	$\theta = 110^0$ K^1	K^2	$\theta = 124^0$ K^1	K^2
4.0	0.07	0.01	-0.08	0.78	-0.05	0.38	-0.02	0.36
5.5	-0.11	0.02	-0.07	0.94	0.09	0.44	0.06	0.44
6.9	-0.25	0.02	-0.07	1.00	0.24	0.61	0.16	0.54
8.8	-0.38	0.02	-0.05	1.01	0.46	0.76	0.29	0.71
10.8	-0.44	0.01	-0.03	0.99	0.63	0.88	0.39	0.86

The coefficients in eq.(2) for p-^{3}He scattering calculated using S-, P- and $^3D_1$-amplitudes from ref.[4]. E_p is the proton Lab. energy.

where g_2 is connected to $\langle ^1P_1|T^-|^3P_1\rangle$ as in (1). The strong amplitudes are taken from ref.[4] and k is the CM wave number. Similar results may be obtained also for p-^{3}H scattering where, however, the phases are not fixed from unitarity. These results give an estimate of the sensitivity of P-A on the two principal T-violating amplitudes at low energies. In spite of the fact that the 2 possible T-violating amplitudes considered compete with 11 T-conserving S-, P- and $^3D_1$-amplitudes one thus obtaines fairly good sensitivity. It should be mentioned that the result could be considerably larger without the phase restriction (1). Note also that the K^α are rising with energy leading one to expect still better sensitivity at somewhat higher energy.

Remains the question of the sensitivity of g_1 and g_2 to a T-violating N-N interaction. This depends naturally on the models chosen for the latter. If we consider meson exchange potentials[6] we find that ρ exchange can contribute to g_2 only and g_1 is due to a (second rank) tensor interaction caused by a 1^+ exchange (A_1) or Huffmanns π-γ exchange potential[5]. Calculations, which must use good p-^{3}He or p-^{3}H scattering states, have not been done yet. However, qualitative arguments show that P-A in these reactions is fairly sensitive to T-violation in the N-N-system.

I would like to thank D. Wyler for the numerical calculations.

References:

1) R.J.Blin-Stoyle, Fundamental Interactions and the Nucleus (North Holland, Amsterdam, 1973).

2) M.Simonius, Lectures presented at the Symposium on Nuclear Polarization Physics, Ebermannstadt, Oct.1973 , (Lecture Notes in Physics, in Press).

3) Madison Convention, in Polarization Phenomena in Nuclear Reactions (University of Wisconsin Press, Madison, 1971) p.XXV.

4) P.Heiss and H.H.Hackenbroich, Nucl.Phys.A182(1972)522.

5) A.H.Huffmann, Phys.Rev.D1(1970)882.

6) M.Simonius and D.Wyler, submitted to this conference.

STUDY OF ^{4}He WITH THE POLARIZED PROTON CAPTURE REACTION

H. F. Glavish, J. R. Calarco, G. King, D. G. Mavis, V. K. C. Cheng,
E. Kuhlmann and S. S. Hanna
Department of Physics, Stanford University, Stanford, California 94305 USA*

The lowest excited states of ^{4}He obtained from resonance analyses of (p,p) and (p,n) reactions have been summarized by Fiarman and Meyerhof[1]. The negative parity levels are consistent in spin and location with the 1p1s^{-1} states predicted by the shell model. Positive parity states are expected to be partly 1p-1h and partly 2p-2h excitations. For a self-conjugate nucleus, such as ^{4}He, E1 transitions to the ground state are allowed only from T = 1, 1^- levels. Hence the two T = 1, 1^- levels postulated at 27.4 and 30.5 MeV are expected to dominate the characteristics of the E1 radiation in nucleon capture reactions which lead to ^{4}He. The presence of third and fourth degree Legendre polynomials P_3, P_4 in the gamma angular distribution $d\sigma/d\Omega = A_o(1 + \sum a_k P_k(\cos\theta))$ is evidence for E2 radiation which is thought to reflect a T = 0, 2^+ state[1] in the region of 33 MeV in ^{4}He.

The unpolarized angular distribution measurements[2] show that for E1 radiation the S = 1 component is strongly suppressed. Crone and Werntz[3] have shown that this is strong evidence for dominant singlet strength in the lower T = 1, 1^- level. However this is contradicted by the $^3\vec{He} + \vec{p}$ elastic scattering measurements[4] which imply the dominant singlet strength is in the upper T = 1, 1^- level, at least in ^{4}Li. The key to resolving this problem is to determine the phases as well as the amplitudes of both the S = 0 and S = 1 components of the S-matrix elements for proton capture and to relate[5] these through the unitarity of the S-matrix to the S = 0 and S = 1 phase shifts and mixing parameter obtained from $^3H + \vec{p}$ elastic scattering measurements.[6]

In order to determine the phases as well as the amplitudes of the capture S-matrix elements, the reaction $^3H(\vec{p},\gamma)^4He$ has been studied at the energies E_p = 6.0, 9.8, 14.0 and 16.0 MeV. At each energy the unpolarized yield $d\sigma(\theta)/d\Omega$ and the analyzing power $A(\theta)$ were measured simultaneously at several angles using the polarized proton beam from the Stanford tandem Van de Graaff and detecting the gamma rays with a 24 cm × 24 cm NaI spectrometer. At the higher energies, sufficient angles were chosen to enable the E2 matrix elements to be extracted. The results for $d\sigma/d\Omega$ and $A d\sigma/d\Omega$ are shown in Fig. 1. The extracted amplitudes and phases obtained from a least squares fit are shown in Fig. 2. The quantities P_o and P_1 are the singlet and triplet P wave (E1) matrix elements while D_o and D_1 refer to the E2 radiation. The phases of P_1 and D_o relative to P_o are ϕ_1 and δ_o, respectively.

A preliminary analysis suggests that the capture S-matrix elements are consistent with the ^{3}H + p elastic scattering phase shifts extracted by Hardekopf et al.[6] with a large mixing parameter. The E2 amplitudes we have extracted from the data provide more quantitative information on the E2 strength in ^{4}He.

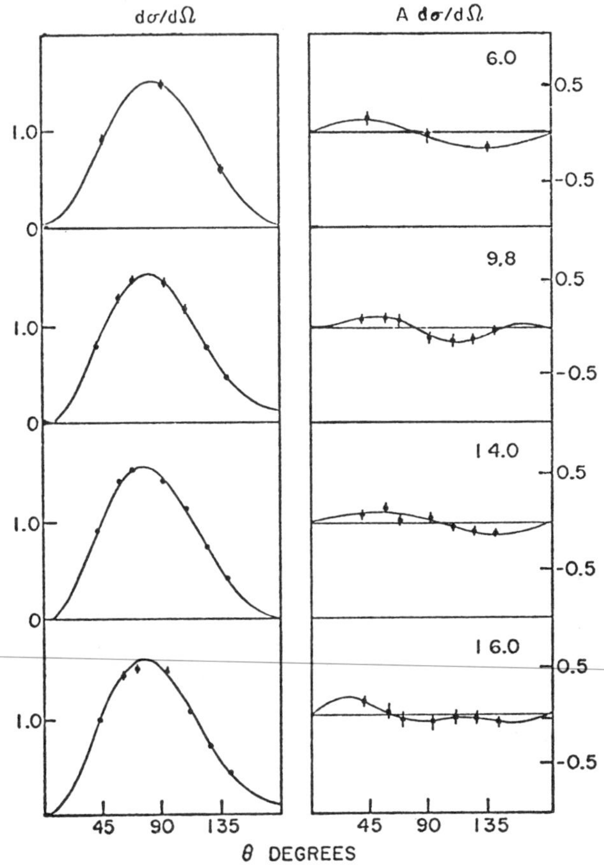

Fig. 1. Angular distributions of cross sections and analyzing powers measured in the reaction $^{3}H(\vec{p},\gamma)^{3}He$.

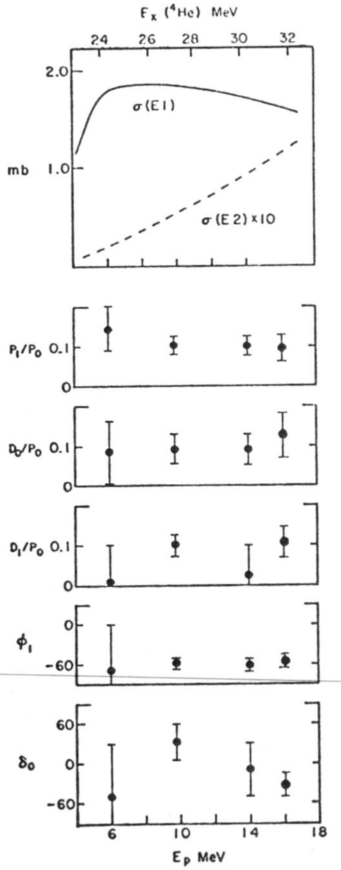

Fig. 2. Top: E1 and E2 cross sections from $^{3}H(p,\gamma)^{3}He$ (Ref. 2). Below: amplitudes and phases extracted from the data in Fig. 1.

REFERENCES

* Supported in part by the National Science Foundation.

1) S. Fiarman and W. E. Meyerhof, Nucl. Phys. A206 (1973) 1

2) W. E. Meyerhof, M. Suffert and W. Feldman, Nucl. Phys. A148 (1970) 211

3) L. Crone and C. Werntz, Nucl. Phys. A134 (1969) 161

4) D. H. McSherry and S. D. Baker, Phys. Rev. C1 (1970) 888

5) H. F. Glavish et al., Few Particle Problems in the Nuclear Interaction, I. Slaus, S. A. Moskowski, R. P. Haddock and W. T. H. Van Oers, eds. (North Holland, Amsterdam, 1972) p. 632

6) R. A. Hardekopf et al. Duke University, preprint

TRANSPARENCY TO VECTOR POLARIZATION TRANSFER IN DEUTERON STRIPPING REACTIONS

F. D. Santos

Laboratório de Física e Engenharia Nucleares, Sacavém, Portugal

Recently there has been increasing interest on polarization transfer for the (d,p) and (d,n) reactions[1]. The tensors $M^{ss'}_{kq}$, defined by Santos[2], are specially suitable for the discussion of polarization phenomena and in particular of polarization transfer due to: (a) their simple transformation properties under rotations of the coordinate system since they transform as a spherical harmonic Y^q_k; (b) the physical significance of the upperscripts s and s' which are spin transfer quantum numbers. When both projectile and outgoing particle have spin one of the allowed values of spin transfer tends to be strongly favoured, the other value(s) being dependent on less well known weaker interactions.

In deuteron stripping reactions $s = \frac{1}{2}$ and $\frac{3}{2}$, the lower value being considerably favoured. In the usual DWBA theory spin transfer $\frac{3}{2}$ arises exclusively from spin dependent distortion and the D-state component of the deuteron internal wave function[3]. Neglecting spin transfer $\frac{3}{2}$, assuming parity conservation, but without further approximations on the transition matrix there are only two non-vanishing tensor components[2], namely $M^{\frac{1}{2}\frac{1}{2}}_{00}$ and $M^{\frac{1}{2}\frac{1}{2}}_{11}$. In cartesian tensor form the outgoing nucleon polarization components are then given by

$$P_x(\Theta) = \frac{p_x + \frac{2}{\sqrt{3}}\, p_{xy}\, A(\Theta)}{1 + \sqrt{3}\, p_y\, A(\Theta)} \quad \text{(1a)}, \qquad P_y(\Theta) = \frac{p_y + \frac{1}{\sqrt{3}}\,(1 + 2p_{yy})\, A(\Theta)}{1 + \sqrt{3}\, p_y\, A(\Theta)} \quad \text{(1b)},$$

$$P_z(\Theta) = \frac{p_z + \frac{2}{\sqrt{3}}\, p_{yz}\, A(\Theta)}{1 + \sqrt{3}\, p_y\, A(\Theta)} \quad \text{(1c)}, \quad \text{where} \quad A(\Theta) = -i\sqrt{\frac{2}{3}}\,\frac{M^{\frac{1}{2}\frac{1}{2}}_{11}(\Theta)}{M^{\frac{1}{2}\frac{1}{2}}_{00}(\Theta)} \,. \quad (2)$$

Here Θ is the scattering angle (C.M. system) and p_i, p_{ij} (i,j = x, y, z) are the cartesian moments of the deuteron polarization referred, like P_i, to a coordinate system with z axis along $\vec{k}_{in}$ and y axis along $\vec{k}_{in} \times \vec{k}_{out}$. The form of eqs.(1) shows that the outgoing nucleon polarization components become independent of $A(\Theta)$ for particular values of the deuteron polarization. In fact, for

$$p_{xy} = \frac{3}{2}\, p_x p_y \,, \quad p_{yy} = \frac{1}{2}\,(3p_y^2 - 1) \,, \quad p_{yz} = \frac{3}{2}\, p_y p_z \qquad (3)$$

the eqs.(1) give, respectively,

$$P_x(\Theta) = p_x \,, \quad P_y(\Theta) = p_y \,, \quad P_z(\Theta) = p_z \,. \qquad (4)$$

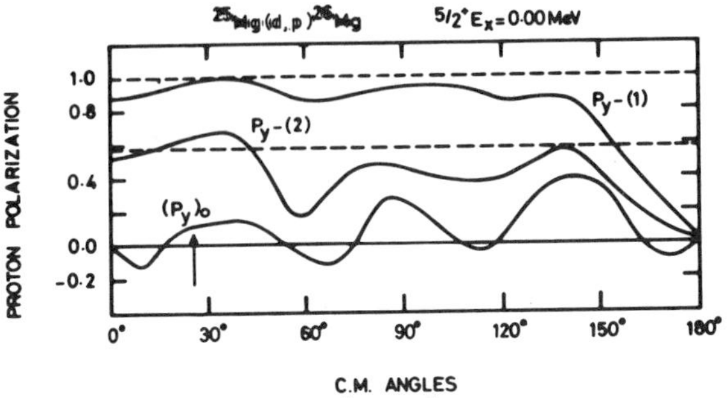

Figure 1

With these deuteron polarizations the reaction model gives perfect vector polarization transfer between the incoming and outgoing particles. The reaction is, as it were, transparent to vector polarization transfer and the polarization of the outgoing particle independent of θ and of the deuteron incident energy.

Fig.1 shows the result of DWBA calculations[3,4], including D-state effects and spin dependent distortion, of proton polarization angular distributions for the $^{24}Mg(d,p)^{25}Mg$ ground state transition at 12.3 MeV. The curves $(P_y)_o$ correspond to an unpolarized incident beam and those marked P_y to a polarized incident beam with the following moments satisfying eqs.(3): (1) $p_y = p_{yy} = 1$; (2) $p_y = 1/\sqrt{3}$, $p_{yy} = o$ (the other moments being zero) . The arrow indicates the position of the stripping cross section main peak and the broken lines the values of p_y . Analogous DWBA calculations for (d,p) reactions on ^{12}C, ^{28}Si and ^{40}Ca, at the same energy, have shown that the proton polarization P_y is close to p_y in the angular region of the stripping main peak, the difference being always smaller than 20%. Since this is, of course, the region of stronger flux polarization transfer under transparency conditions in (d,p) and (d,n) reactions may be used as a method of producing fast polarized neutrons (or protons) of known polarization provided a suitable calibrated source of polarized deuterons is available. The approximation leading to transparency in vector polarization transfer, namely neglect of spin transfer $\frac{3}{2}$, is independent of any particular reaction model such as the DWBA. Thus it is not restricted to the usual scope of that theory and may occur for deuteron stripping on very light targets[5], which presents experimental advantages.

REFERENCES

1) - R.A. Hardekopf, D.D. Armstrong, W. Gruebler, P.W. Keaton Jr. and U. Meyer--Berkhout, Phys. Rev. C8 (1973) 1692 and references therein.

2) - F.D. Santos, to be published.

3) - R.C. Johnson and F.D. Santos, Particles and Nuclei, 2 (1971) 285.

4) - R.C. Johnson, F.D. Santos, R.C. Brown, A.A. Debenham, G.W. Greenlees, J.A.R. Griffith, O. Karban, D.C. Kocher and S. Roman, Nucl. Phys. A208 (1973) 221.

5) - J.E. Simmons, W.B. Broste, G.P. Lawrence, J.L. Mckibben and G.G. Ohlsen, Phys. Rev. Lett. 27 (1971) 113.

POLARIZATION IN n-d ELASTIC SCATTERING FOR 50.0 MeV NEUTRONS †

A.L. Sagle, M.W. McNaughton, N.S.P. King, F.P. Brady, Crocker Nuclear Laboratory, University of California, Davis, and B.E. Bonner, P-DOR Los Alamos Scientific Laboratory.

There exist considerable data for p-d polarization up to 150 MeV[1]; however only limited data is available for the charge symmetric n-d reaction at neutron bombarding energies above 20 MeV. As a test for violation of charge symmetry in nuclear forces, the comparison of p-d and n-d polarizations is not as yet conclusive, particularly for angles near the negative minimum. Recent work by Morris et al.[2] indicates no discrepancy between elastic n-d and p-d polarizations below 22 MeV. However, the 35 MeV n-d polarization data of Zamudio-Cristi et al.[2] shows a possible difference at the large negative minimum near 120° c.m.

The motivation for this experiment is therefore to extend the comparison of n-d and p-d polarizations to an energy where coulomb effects are presumably reduced. In addition the higher energy simplifies the detection problem in the region of the polarization minimum.

We have recently developed an improved polarized neutron beam using the $T(d,\vec{n})\,^4He$ reaction collimating neutrons at a laboratory angle of 30°. Preliminary results from an absolute calibration using the reciprocal reaction $^4He(\vec{n},d)T$ at identical c.m. energy and angle indicate a polarization of 0.49 + 0.03 for 50.0 MeV neutrons [3]. This beam of neutrons was used to bombard a cooled high pressure deuterium gas target. Neutrons of lower energy were separated by reference to the neutron time of flight. Scattered deuterons were detected in four NE102A EΔE telescopes placed symmetrically on either side of the beam, defining an angular acceptance of 5°. A neutron spin rotation magnet in the beam was employed to eliminate false equipment asymmetries. Target empty data were obtained and subtracted to eliminate contributions from the aluminum walls of the gas target.

Preliminary data at three angles are shown in Figure 1 in relation to the 50 MeV p-d polarization data of reference[4]. Corrections have not been included for the angular spread in the present measurements, but crude calculations indicate a correction of -0.03 for the 120° data point, so that its error bars would overlap those of both p-d data points at this angle.

Unfortunately the existing p-d polarization data at 50 MeV are not as precise as lower energy measurements, however, within statistical uncertainties no discrepancy between p-d and n-d polarizations is evident.

---

†Supported by the National Science Foundation

FIGURE I: Angular distribution of polarization for p-d and n-d scattering at 50 MeV.

Open Circles : p-d using liquid deuterium target

Closed Circles: p-d using CD_2 target

Crosses : n-d present work

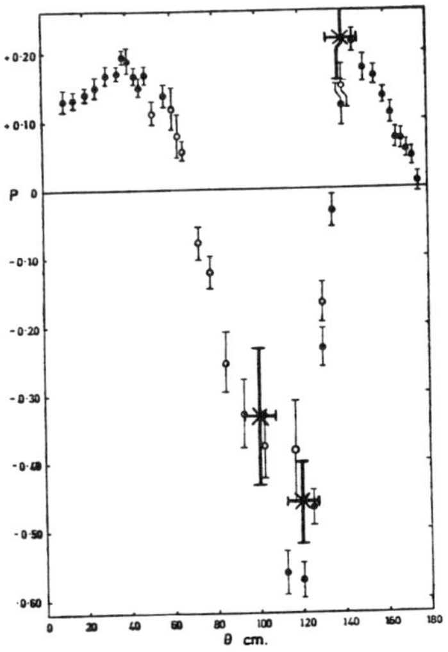

C.M. Angle	Polarization
100	-0.33 ± 0.10
120	-0.46 ± 0.06
140	+0.22 ± 0.06

TABLE I:

Polarization in n-d scattering at 50.0 MeV. Does not include finite angular corrections

REFERENCES

1) Three Body Problems In Nuclear and Particle Physics, ed. J.S.C. McKee and P.M. Rolph, North Holland 1969 and ref. therein.

2) G.L. Morris, et al, Phys. Rev. C9, 1687 (1974)
J.J. Malanify et al Phys. Rev. 146, 532 (1966)
R.L. Walter and C.A. Kelsey, Nucl. Phys. 46, 66 (1963)
J. Zamudio-Cristi, Phys. Rev. Lett. 31, 1009 (1973)

3) A.L. Sagle et al, this conference

4) W.R. Gibson et al, Proc. Intern. Nucl. Phys. Conf. (Academic Press, New York, 1967) p. 1016

MEASUREMENTS OF POLARIZATION AND SPIN ROTATION IN PROTON-PROTON ELASTIC SCATTERING AT 3,83 GeV/c.

J. DEREGEL, C. BRUNETON, J. BYSTRICKY, G. COZZIKA, Y. DUCROS, A. GAIDOT, F. LEHAR
A. de LESQUEN, J.P.MERLO, S. MIYASHITA, J. MOVCHET, J.C. RAOUL, L. VAN ROSSUM

DPhPE - CEN SACLAY, FRANCE.

The polarization P and the spin rotation parameter R have been measured for proton-proton elastic scattering at 3.83 GeV/c incident momentum, in the slow extracted beam of the SACLAY Proton Synchrotron SATURNE. The two measurements were carried out simultaneously using the same target polarized in the horizontal plane at an angle of about 90° with respect to the beam axis. The polarization parameter P_0 is measured by detecting scattering in the vertical plane, and the linear combination C = 0.999 R + 0.07 A of the spin rotation parameters R and A is measured by detecting scattering in the horizontal plane and measuring the polarization of the recoil proton. This polarization depends mainly on R, and only weakly on A, A', R', D and P_0. [1]

The part of the apparatus for the spin rotation measurements is similar to the one used in a previous experiment at the CERN PS [1] and has been described elsewhere [2] . A description of the present apparatus for simultaneous measurements of P and R will be published [3] . The SACLAY experiment used a butanol target of 1.5 cm diameter and of 8.5 cm length, in a magnetic field of 2.5 T. The average target polarization was 60%.

The external proton beam of SATURNE is collimated and focused into a spot of less than 1 cm in diameter. The intensity is reduced to several 10^6 particles per burst of 350 msec length, the momentum spread is very small (< 0.5%) and the central momentum is 3.83 ± 0.05 GeV/c. The horizontal and the vertical divergence is ± 3 mrad and ± 4 mrad respectively. The direction of the incident particle is measured by four scintillator hodoscopes H1 to H4 (Fig.1). A set of anticoincidence counters with 5 mm thick lead converters reduces the triggering on inelastic interactions.

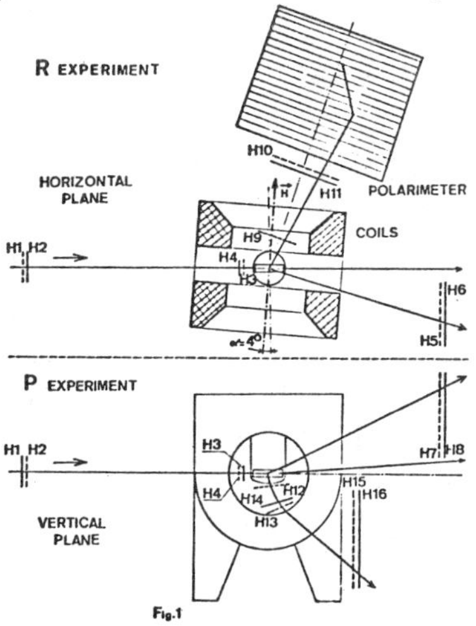

Fig.1

The P experiment.

The detectors are arranged in the vertical plane. The hodoscopes H7 H8 measure the direction of the fast proton. Two different positions are used depending on the region of four-momentum transfer. For measurements in the interval $0.35 \lesssim |t| \lesssim 1.63\ (GeV/c)^2$ the slow proton is detected by the hodoscopes H12, H13, H14 located below the target inside the coils of the target magnet. For measurements in the large angle region, $1.3 \lesssim |t| \lesssim 3.0\ (GeV/c)^2$, the slow proton is detected in the hodoscope H12, and in H15,H16 placed outside the coils of the target magnet. A total of 4.5×10^5 events with the butanol target and 2.3×10^5 events with the dummy target were registered.

The results of the analysis are shown on the Figure 2. Our data are in excellent agreement with the results of J.PARRY et al. [4] at 3.75 GeV/c. There is definite evidence for a minimum of P_0 at $-t \simeq 1.0\ (GeV/c)^2$ followed by a second maximum at $-t \simeq 1.5\ (GeV/c)^2$. A similar t-dependence with minima and maxima at about the same t- values appears in all data at higer energies, including the recent measurements at 45 GeV/c [5]. The errors shown on Fig. 2 represent statistical uncertainties only. The relative systematic error representing mostly the uncertainty in the effective target polarization is $\pm$ 5% for both experiments. At small four-momentum transfer, $0.35 \lesssim |t| \lesssim 0.5\ (GeV/c)^2$, we find a polarization of $P_0 \simeq 0.18$ which is somewhat lower than the results at 3.66 GeV/c of Neal and Longo [7] but much lower than the results of Grannis et al. [6] at 3.67 GeV/c.

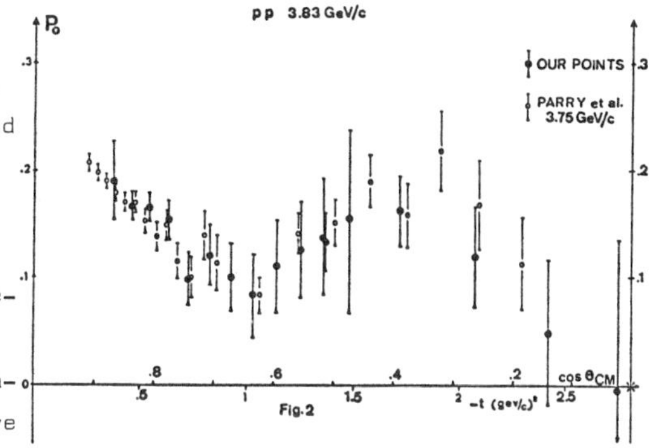

Fig.2

The R experiment.

The spin rotation of recoil particles is measured by selecting elastic scatterings in the plane containing the polarization of the target, and analyzing the polarization of the recoil protons by a second scattering in a carbon plate polarimeter with wire spark chambers. The principle, and the major part of the apparatus, are the same as in previous experiments [1], [8].

A total of 3.9×10^5 events with the butanol target and 0.5×10^5 events with the dummy target have been registred. The target background is of about 10%.

In order not to obstruct the path of the scattered proton by the coils of the target magnet, the direction of the magnetic field makes an angle of $\alpha = 86^\circ$, instead of 90°, with respect to the beam direction. In consequence, the experiment measures the combination $C = R \sin\alpha + A \cos\alpha$, instead of R only.

Because of the spin precession of the recoil proton between the target and the polarimeter and because of the finite azimutal acceptance for the first scattering, the measured recoil proton polarization depends also, however weakly, on the polarization P, the depolarization D and on the combination $C' = R' \sin\alpha + A' \cos\alpha$. The analysis determines the quantities C, C' and D from the data but only only C ($\simeq$ R) is obtained with a good precision (Fig. 3a and table 1). For the polarization P_0, the analysis uses the values measured independently in the P experiment, and the data from reference [7] . The errors on P_0, C' and D have a very weak influence on the error on the error for C.

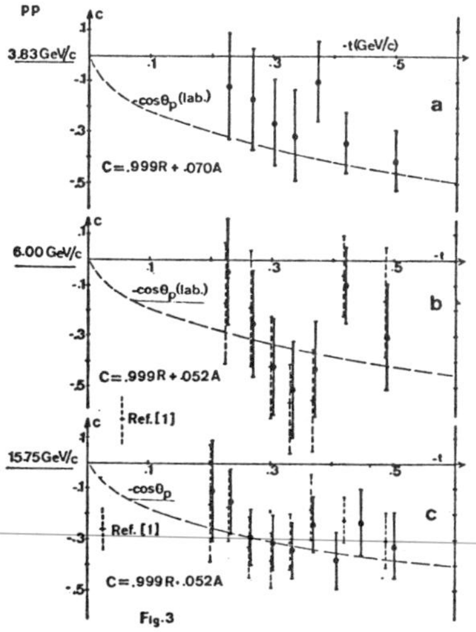

Fig.3

In order to study the energy dependence of the spin rotation parameter R ($\simeq$ C), our previously published data at 6 and 15.75 GeV/c,[1] have been reanalyzed by exactly the same method as used at 3.83 GeV/c. The new analysis represents some improvements in the method, but the differences in the results are negligible with respect to the statistical errors (Fig. 3b, 3c)

The values obtained at all three energies for C ($\simeq$ R), C' and D averaged over the entire interval of t are presented in Table 1.

TABLE 1

P GeV/c	(-t)	C (R)	C' (R')	D
3.83	0.18 - 0.57	-0.275 $\pm$ 0.058	0.82 $\pm$ 0.31	0.61 $\pm$ 0.52
6.00	0.18 - 0.54	-0.294 $\pm$ 0.073	0.41 $\pm$ 0.52	0.84 $\pm$ 0.59
15.75	0.17 - 0.54	-0.286 $\pm$ 0.040	0.72 $\pm$ 0.33	1.08 $\pm$ 0.33

No variation of C with energy, between 3.8 and 16 GeV/c, is apparent from the data. The t- dependence of C (Fig. 3), is similar at all three energies. The data are not far from the values $R = -\cos\theta_p$ where θ_p is the laboratory scattering angle of the slow proton. This would be the expected result if proton-proton elastic scattering at these energies were dominated by Pomeron exchange with no helicity flip in the s-channel.

REFERENCES.

1 J. DEREGEL et al : Phys. Letts 43B (1973) 338

2 P. AUTONES et al : Nucl. Insti. and Methods 103 (1972) 211

3 P. AUTONES et al : Apparatur for Simultameous Measurement ot the Polarization and Spin-Rotation Parameters in High Energy Elastic Scattering on Polarized Protons - to be published in Nucl. Instr, and Methods.

4 J. PARRY et al : Phys. Rev. D8 (1973) 45

5 A. GAIDOT et al : The XVII th Int. Conf on High Energy Physics, London 1974 paper 1019

6 P. GRANNIS et al : Phys. Rev. 148 (1966) 1297

7 H.A. Neal and M.J. LONGO : Phys. Rev. 161 (1967) 1374

8 A. de LESQUEN et al : Phys. Letts 40B (1972) 277

A THEORETICAL STUDY OF THE ANALYSING POWER A(Θ) IN ($\vec{p}$,γ) REACTIONS IN LIGHT NUCLEI

K. Ramavataram, C.L. Rao and S. Ramavataram
Université Laval, Laboratoire de Physique Nucléaire,
Département de Physique, Québec G1K 7P4, Canada.

The 1p-1h continuum model of the giant dipole resonance (GDR) has been successful[1,2] in interpreting the observed energy dependence of the photo-absorption cross section in light nuclei such as ^{12}C and ^{16}O. In the latter case when T-mixing is taken into account a satisfactory correspondence can be obtained between the 1^- states predicted by the 1p-1h bound sate calculations[3] and the resonances generated in the coupled-channel treatment[2]. Comparison of the theoretical (full curve; FIG. 1a) and experimental (dashed curve; FIG.1a) (p,γ_0) excitation functions[2,4] suggests that just below the GDR, the theory predicts one 1^- resonance in the region of E_p=7.0 to 8.0 MeV while experimentally there are two small peaks in this region. Since in the theoretical calculations the next lowest 1^- resonance is around E_p of 6.0 to 7.0 MeV and since there are other resonances in the experimental spectrum in this region it was assumed[2] that the experimentally observed 18.97 MeV resonance has no counter part in the 1p-1h continuum calculation. Recent measurements[5] of the analysing power A(Θ) in the ($\vec{p},\gamma_0$) reaction throw additional light on these weak 1^- resonances. While the magnitude of A(45°) is relatively constant (dashed curve: FIG.1b) in the region of the GDR, for proton energies of 7.0 to 8.0 MeV the analysing power goes through a minimum. In order to understand this behaviour, we calculated the theoretical excitation function of A(45°) (full curve : FIG.1b) in the frame-work of the 1p-1h coupled-channels method using the parameters of ref. 2. The remarkable feature is that theory also predicts a minimum for A(45°) in the right region of proton energies. On examining the parameter dependence of the theoretical calculation, we found that the minimum in the analysing power is associated with the interference between the T = 1, 1^- resonance at 19.5 MeV and the GDR. Analysis of the theoretical expression for A(Θ) in terms of angular functions[8] indicates that the zero in the analysing power is due to the fact that the phase difference (Δ_{ds}) between d and s-waves goes through π at this energy; further, Δ_{ds} has this value only if the resonance at 19.5 MeV is a T = 1,1^-

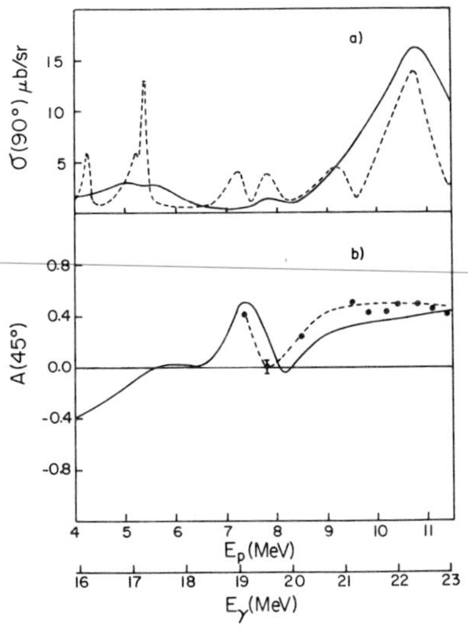

FIGURE 1. Comparison between theory and experiment for the reactions $^{15}N(p,\gamma_0)^{16}O$ and $^{15}N(\vec{p},\gamma_0)$ (see text for details).

with a predominantly $p_{1/2}^{-1}\ s_{1/2}$ configuration.

In the case of ^{12}C, the presence of broad single particle resonances prevents one from making a detailed comparison between the predictions of the 1p-1h bound-state calculations[3] and the coupled-channels calculation[1]. However, the results of the latter are in good agreement with experiment[6] over the main GDR component. Several smaller peaks observed in the $^{11}B(p,\gamma_0)^{12}C$ reaction cannot be reproduced in the simple 1p-1h model. Experimental data on the energy dependence of A(θ) in the $^{11}B(\vec{p},\gamma_0)^{12}C$ reaction have recently become available[7] and it is of interest to compare these results with predictions of the model of ref. 2 in the region of the main photo-peak. We therefore calculated the excitation function of A(45°) by the coupled-channels method and using the parameters of ref. 1. It can be seen that in the region of the main GDR (top half of FIG. 2), the predicted A(45°) has the right magnitude(bottom half of FIg.2) in comparison with experiment[7]. The experimental values of the analysing power change sign at 29 MeV, a trend not reproduced in the present calculations.

This feature is very likely due to the presence of a relatively strong 1^- resonance in the region of 26.0 to 27.0 MeV excitation in ^{12}C which is observed in the (p,γ) reaction and which cannot be reproduced in the 1p-1h model for the reaction. Suitable modifications to the formalism of ref. 2 are being considered in order to interpret satisfactorily the observed (p,γ) cross section and the associated energy dependence of the analysing power.

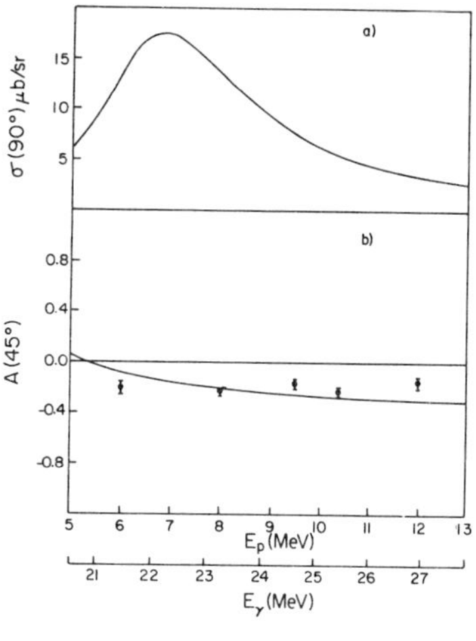

FIGURE 2. The calculated cross section for the $^{11}B(p,\gamma_0)^{12}C$ reaction and a comparison between theory and experiment for A(45°) in the $^{11}B(\vec{p},\gamma_0)^{12}C$ reaction.

REFERENCES

1) M. Marangoni and A.M. Sarius, Nucl. Phys. A132 (1969) 649.
2) B. Buck and A.D. Hill Nucl. Phys. A95 (1967) 271.
3) V. Gillet and N. Vinh-Mau, Nucl. Phys. 54 (1964) 321.
4) E.D. Earle and N.W. Tanner, Nucl. Phys. A95, (1967) 241.
5) S.S. Hanna et al. Phys. Lett. 40B, (1972) 631.
6) R.G. Allas et al Nucl. Phys. 58 (1964) 122.
7) H.F. Glavish et al Phys. Rev. Lett. 28 (1972) 766.
8) K. Ramavataram, C.L. Rao and S. Ramavataram, to be published.

POLARIZATION AT 50 MeV NEUTRONS FROM THE $T(d,n)^4He$ REACTION†

A.L. Sagle, M.W. McNaughton, N.S.P. King, F.P. Brady, Crocker Nuclear Laboratory, University of California, Davis, and B.E. Bonner, P-DOR, Los Alamos Scientific Laboratory.

An absolute measurement of neutron polarization at E_n = 50 MeV was made using a double scattering technique in which the first scattering was the $T(d,\vec{n})^4He$ reaction and the second was the $^4He(\vec{n},d)T$ reciprocal reaction. When the center of mass energies and angles are equated in these two scatterings, the neutron polarization (P) is equal to the deuteron analyzing power (A) because of the reciprocity of nuclear reactions.(1) The quantity measured is the deuteron left-right asymmetry e = PA which in this case is equal to P^2.

The equality of the center of mass energies and angles is a constraint which along with kinematics uniquely determines all energies and angles when either the incoming deuteron energy or the outgoing neutron energy is specified. Because Q = +17.35 MeV in the $T(d,\vec{n})^4He$ reaction, the neutron laboratory angle of the outgoing 50 MeV neutrons is 29.75°. The corresponding laboratory angle of the outgoing deuteron in the second scattering (for equal CM angles) is 25.0°. From previous measurements of the deuteron analysing powers in the charge symmetric reciprocal reaction $^4He(\vec{p},d)^3He$ at E_p = 52, 40, 50, and 52.5 MeV(2) it is seen (in Fig. 1) that σ_{lab} = 29.75° is very close to the maximum of P^2I, the figure of merit.

To measure the neutron polarization a 38 MeV deuteron beam passed through a liquid nitrogen cooled high pressure tritium gas cell. 50 MeV neutrons emerging at σ_{lab} = 29.75° passed through a spin rotation magnet and then through a liquid nitrogen cooled ^4He gas cell. Deuterons from the second reaction at σ_{lab} = 25° were detected with ΔE-E scintillators. The asymmetry was deduced from spin up and spin down right-left measurements made with the ^4He target filled and then evacuated. Figure 2 shows the polarization excitation function of the mirror reactions at σ_{lab} = 30°. The preliminary value for the neutron polarization is 0.49± .03 which is in agreement with the proton polarization in the charge symmetric reaction.

†Supported by the National Science Foundation

REFERENCES

1) H.H. Barschall, Helv. Phys. Acta 29, 149(1956).

2) A.L. Sagle, M.W. McNaughton, N.S.P. King, F.P. Brady, Crocker Nuclear Laboratory Progress Report UCD-CNL 175 (1973).

3) R.I. Brown and W. Haeberli, Phys. Rev. 130(1963)1163.

4) W.G. Weitkamp and W. Haeberli, Nucl. Phys. 83(1966)46.

5) E.T. Boschitz et al., UCRL-16218(1965).

6) G.S. Mutchler, W.B. Broste, and J.E. Simmons, Phys. Rev. C3, 1031(1971).

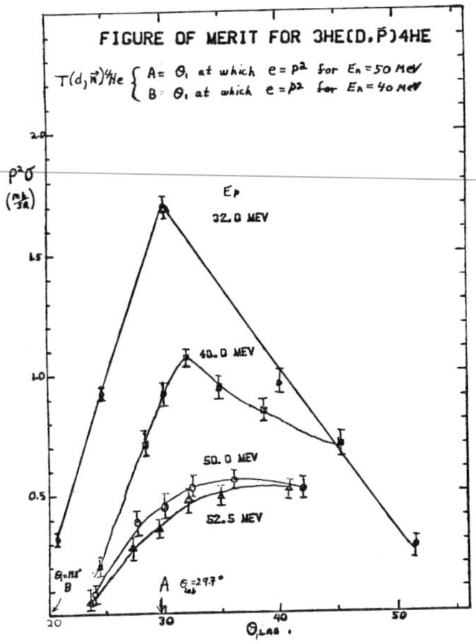

Figure 1

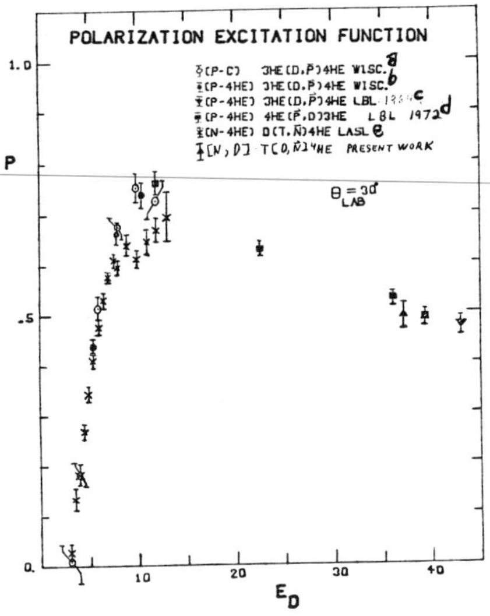

Figure 2

a - Ref. 3

b - Ref. 4

c - Ref. 5

d - Ref. 2

e - Ref. 6

PRELIMINARY MEASUREMENT OF THE PROJECTILE SPIN DEPENDENCE IN THE REACTIONS $\vec{d} + {}^3He \rightarrow p + t + p$ and $\vec{d} + {}^3He \rightarrow p + {}^3He + n$ by H. O. Meyer, G. G. Ohlsen, R. A. Hardekopf, R. V. Poore, and J. W. Sunier, Los Alamos Scientific Laboratory, Los Alamos, New Mexico

We have measured coincidence spectra for the following two breakup reactions induced by 15 MeV polarized deuterons incident on ^3He nuclei:

$$\vec{d} + {}^3He \rightarrow p + t + p - 1.46 \text{ MeV}$$

$$\vec{d} + {}^3He \rightarrow p + {}^3He + n - 2.22 \text{ MeV}$$

The chief purposes of the measurements were to establish the experimental feasibility of this class of measurement with polarized beams, and to draw the attention of theorists to the availability of this type of data for the study of breakup reactions.

The experimental procedure was as follows. A polarized deuteron beam with a vertical quantization axis was used to bombard a gaseous ^3He target (pressure 270 torr). ΔE-E solid state detector telescopes at 30° to the left and to the right of the beam, were used to detect and identify protons, tritons, and ^3He particles. Coincidences between the left and right telescopes triggered computer storage of the pulse heights from each of the four detectors as well as from a time-to-amplitude-converter which measured the time difference between the left and right signals. The observation angles of ± 30° were chosen because the elastic scattering of deuterons on a small hydrogen contaminent in the target as also leads to coincidences, thereby providing us with a check for systematic errors.

A data set consisted of three runs with equal integrated beam current: 1) with $m_I = 1$ polarization ($p_y = p_{yy} = f$), 2) with $m_I = 0$ polarization ($p_y = 0$, $p_{yy} = -2f$), and 3) with $m_I = -1$ polarization ($p_y = -f$, $p_{yy} = f$), where $f(\approx 0.8)$ is the fraction of the beam which is polarized in the particular ideal state. If N_1, N_0, and N_{-1} represent the corresponding counts observed in an arbitrary region on the kinematically allowed locus of a given breakup channel, where the protons are detected by the left telescope, the corresponding vector and tensor analyzing powers are defined by:

$$A_y = \frac{1}{f}\,\frac{N_1 - N_{-1}}{N_1 + N_0 + N_{-1}} \quad \text{and} \quad A_{yy} = \frac{1}{f}\,\frac{N_1 + N_{-1} - 2N_0}{N_1 + N_0 + N_{-1}} ,$$

in accordance with the Madison convention.[1)] Because of the symmetric angles chosen, the same parameters can also be deduced from the data which corresponds to protons going to the right detector.

Simultaneously with the left-right measurements, we also measured coincidences between a set of telescopes at ± 30° in the up down plane. In this case we obtain the quantities A_x (required to be zero by parity conservation) and A_{xx}, a second tensor analyzing power.

The results for the two breakup channels investigated are given in the figure, where A_y, A_{yy}, and A_{xx} are plotted as a function of proton energy. This is, all of the events have been projected onto their respective proton energy axes. Statistics are rather poor, but one can still make the following observations: 1) There are fairly large A_{xx} and A_{yy} effects, and small A_y effects; and 2) the data for the $p + {}^3He + n$ channel and for the p + t + p channel are consistent with each other for each of the observables.

In conclusion, we have demonstrated that analyzing power measurements in kinematically complete experiments are feasible, and in at least one case, yield non-zero results.

BIBLIOGRAPHY

1) Proc. 3rd Intern. Symp. on Polarization Phenomena in Nuclear Reactions, H. H. Barschall and W. Haeberli, Eds. (Univ. of Wisconsin Press, Madison, 1971) p. xxv.

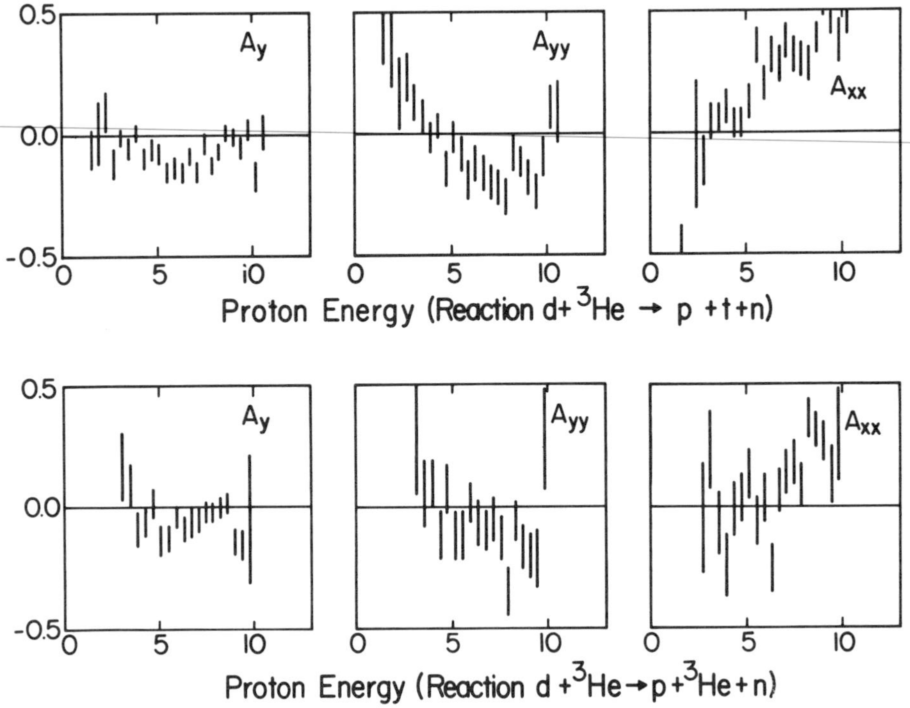

THE POLARIZATION OF 50 MeV NEUTRONS FROM THE $T(d,\vec{n})^4He$ REACTION†

A.L. Sagle, M.W. McNaughton, N.S.P. King, F.P. Brady, Crocker Nuclear Laboratory, University of California, Davis, and B.E. Bonner, P-DOR, Los Alamos Scientific Laboratory.

An absolute measurement of the neutron polarization in the $T(d,\vec{n})^4He$ reaction was made for E_n = 50 MeV. The technique used was first proposed by Barschall and the measured value is 0.45 ± 0.02 which is in disagreement with the deuteron analyzing power in the charge symmetric reciprocal reaction $^4He(\vec{p},d)^3He$.

†Supported by the National Science Foundation.

LA POLARIZATION DE 50 MeV NEUTRONS DE LA $T(d,\vec{n})^4He$ REACTION†

A.L. Sagle, M.W. McNaughton, N.S.P. King, F.P. Brady, Crocker Nuclear Laboratory, University of California, Davis, and B.E. Bonner, P-DOR, Los Alamos Scientific Laboratory.

On donne la mésure absolue de la polarization des neutrons de la reaction $T(d,\vec{n})^4He$ à E_n = 50 MeV. On employe la technique de Barschall. La rèsultat est 0.45 ± 0.02 que n'agrèe pas avec la deuton "analysing power" de la "charge-symmetric reciprocal reaction" $^4He(\vec{p},d)^3He$.

†Ce travail a été appuyé par la National Science Foundation.

THE POLARIZATION OF 50 MeV NEUTRONS FROM THE $T(d,\vec{n})^4He$ REACTION†

A.L. Sagle, M.W. McNaughton, N.S.P. King, F.P. Brady, Crocker Nuclear Laboratory, University of California, Davis, and B.E. Bonner, P-DOR, Los Alamos Scientific Laboratory.

An absolute measurement of neutron polarization in the $T(d,\vec{n})^4He$ reaction has been made via a double scattering technique in which the first scattering was the $T(d,\vec{n})^4He$ reaction and the second, the $^4He(\vec{n},d)T$ reciprocal reaction. When the center of mass energies and angles are equated in these two scatterings, the neutron polarization (P) is equal to the deuteron analyzing power (A) due to the reciprocity of nuclear reactions.[(1)] The quantity measured was the deuteron left-right asymmetry $e = PA$ which in this case is equal to P^2.

The equality of the center of mass energies and angles is a constraint which along with kinematics, uniquely determines all energies and angles when either the incoming deuteron energy or the outgoing neutron energy is specified. Because $Q = +17.35$ MeV for the $T(d,\vec{n})^4He$ reaction, the neutron laboratory angle required for outgoing 50 MeV neutrons is 29.75°. The corresponding laboratory angle of the outgoing deuteron in the second scattering (for equal CM angles) is 25.0°.

To measure the neutron polarization, a 38 MeV deuteron beam bombarded a liquid nitrogen cooled, high pressure, tritium gas cell. The resulting 50 MeV neutrons emerging at $\theta_{lab} = 29.75°$ passed through a spin rotation magnet and then through a liquid nitrogen cooled ^4He gas cell. Deuterons from the second reaction at $\theta_{lab}=25°$ were detected with ΔE-E scintillators and the asymmetry deduced from spin up and spin down right-left measurements made with the ^4He target filled and then evacuated. Figure 1 shows the polarization excitation function of the mirror reactions at $\theta_{lab} = 30°$. The preliminary value for the neutron polarization is $0.45 \pm .02$ which differs from the proton polarization in the charge symmetric reaction at 50 MeV. A comparison of the charge symmetric analysing power for the $^4He(\vec{p},d)^3He$ reaction[2)] with the present experiment is shown in Figure 2. The dotted line represents the $^4He(\vec{p},d)^3He$ data angle-averaged to compensate for the fact that no angular corrections have been applied to the $^4He(\vec{n},d)T$ results. Finite geometry effects would appear to be ruled out as the reason for the difference in analysing powers. A recent experiment by Ryu et.al[3)] at 40 MeV also shows a similar difference. However, the larger uncertainty in their results does not eliminate the possibility of agreement between the two reactions. In conclusion, the difference in analysing powers at $\theta_{lab} = 30°$ for the $^3He(d,\vec{p})^4He$ and $^3H(d,\vec{n})^4He$ reactions observed in earlier low energy experiments[4-7)] also seems to be observed in this experiment.

†Supported by the National Science Foundation

REFERENCES

1) H.H. Barschall, Helv. Phys. Acta 29, 149(1956).
2) Sagle et. al., LBL-1666(1972)p.58 and UCD-CNL 175(1973).
3) N. Ryu et. al., Nucl. Phys. A180(1972)657.
4) R.I. Brown and W. Haeberli, Phys. Rev. 130(1963)1163.
5) W.G. Weitkamp and W. Haeberli, Nucl. Phys. 83(1966)46.
6) E.T. Boschitz et. al., UCRL-16218(1965).
7) G.S. Mutchler, W.B. Broste, and J.E. Simmons, Phys. Rev. C3, 1031(1971).

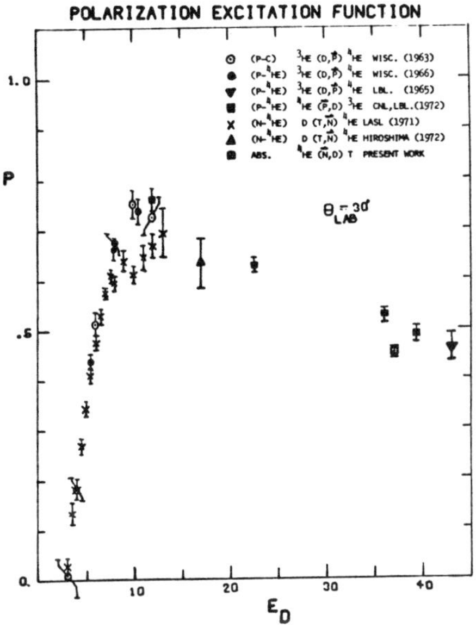

Figure 1

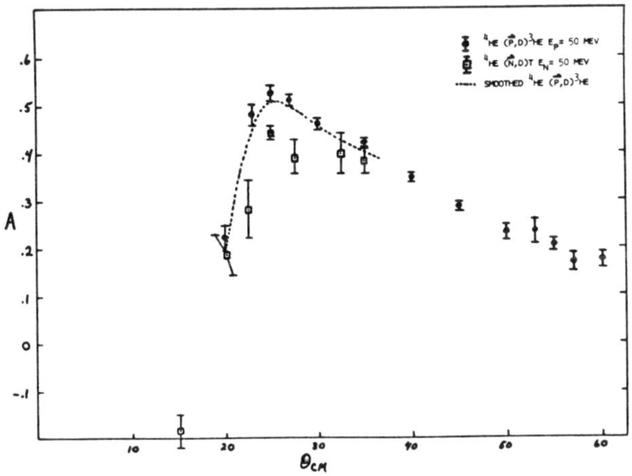

Figure 2

VIII

CORRELATIONS A COURTE PORTEE
DIFFUSION QUASI-LIBRE

SHORT RANGE CORRELATIONS
QUASI-FREE SCATTERING

SHORT-RANGE CORRELATIONS IN NUCLEAR MATTER

O. Kofoed Hansen
CERN, Geneva, Switzerland.

Abstract

It is shown that the trivial correlations in nuclear matter caused by the centre-of-mass constraint and the Pauli principle result in grave difficulties for the direct experimental observation of dynamical hard-core correlations. Both theory and experiment will have to undergo considerable improvement before we can hope to investigate hard-core effects in scattering experiments.

CORRELATIONS DE FAIBLE PORTEE DANS LA MATIERE NUCLEAIRE

O. Kofoed Hansen
CERN, Geneva, Switzerland.

Résumé

Nous montrons que les corrélations triviales causées dans la matière nucléaire par la contrainte du centre de masse et par le principe de Pauli ont pour conséquence de rendre très difficile l'observation directe en expérience des corrélations dynamiques dans le coeur dur des nucléons. Les recherches théoriques et expérimentales devront toutes deux faire des progrès considérables avant que nous puissions espérer étudier les effets du coeur dur dans les expériences de diffusion.

SHORT-RANGE CORRELATIONS IN NUCLEAR MATTER

O. Kofoed Hansen

CERN, Geneva, Switzerland.

INTRODUCTION

I would like to remind you of the fact that high-energy physics came into being because one believed that a complete knowledge of the two-body system would enable us to compute everything in nuclear physics. But nature was more complicated than we had expected. Now we have beautiful N-N data; we have what looks like an infinity of particles or particle-states; we have learned how to excite the nucleon; and we still don't really know how to compute the bound states of the three-nucleon system.

In talking to you about correlations in nuclear physics I shall endeavour to remind you of the fact that all fields of physics or other sciences which turn out to be complicated land themselves in studies of correlations. This has happened to statistical mechanics, to motion of stars in galaxies, to Brownian motion, to motion of fluids and gases, to studies of noise of any kind you can dream of, to diffusion, etc., and it has happened to nuclear physics and, during the last few years, to strong interactions. All these fields are so complicated that we have grave difficulties in describing them in detail. We consequently make our escape into statistics. With this in mind, and realizing how broad and unifying a subject statistics is, I will start with a brief reminder of basic statistics.

Next comes the nuclear physics part of my talk. Here I shall remind you of what I call the trivial correlations: the centre-of-mass (c.m.) correlation and the Pauli correlation. These two correlations make it very difficult to find genuine dynamical or non-trivial short-range correlations in nuclear physics. Other correlations are more easily seen like shape correlations or cluster correlations.

I shall finish my talk with a description of multiple scattering and there show how correlations enter into the Glauber formalism[1].

REMINDER OF BASIC NOTIONS IN STATISTICS[2]

If, for example, we are looking at a Fermi gas then we will see particles moving past the observation point in what will first appear to be a random fashion. We will observe some kind of signal $F(t-t_k)$ from the kth particle which passes our point of observation at time t_k and the signal evolves in time, being zero for large values of $|t-t_k|$. One nice example of such a system would be electrons in a resistance and our measurement could be the resulting current, in which case we are investigating what in old days was called the shot effect, i.e. the background noise in amplifiers. What we see is a current

$$I = \sum_k F(t - t_k) \tag{1}$$

and this current is a fluctuating current due to the random passage of the kth electron. There is a distribution function P(I) normalized according to

$$\int_{-\infty}^{\infty} P(I)\, dI = 1 \,. \tag{2}$$

As you know P(I) is completely determined by the moments of I, i.e. by

$$M_n = \int_{-\infty}^{\infty} I^n P(I)\, dI \,. \tag{3}$$

However, M_n is not, in general, very sensitive to details of P(I). It is exactly for this reason that instead of the second moment we always ask for the mean square fluctuation $\sigma^2 = M_2 - M_1^2$. In statistics this is generalized through the introduction of the so-called semi-invariants λ_n which are generated through the characteristic function

$$f(u) = \int_{-\infty}^{\infty} P(I)\, e^{iIu}\, dI \tag{4}$$

by

$$\log f(u) = \sum_{n=1} \frac{\lambda_n}{n!} (iu)^n \,. \tag{5}$$

There is, of course, a one-to-one correspondence between the set of moments M_n and the set of semi-invariants, but the set λ_n has several nice properties which make them more useful than the set M_n. One of these properties is that the λ_n is very close to the sensitive part of the nth moment over and above what was contained in the first n-1 moments. Thus

$$\begin{aligned}
\lambda_1 &= M_1 \\
\lambda_2 &= M_2 - M_1^2 = \langle (I - M_1)^2 \rangle \\
\lambda_3 &= M_3 - 3M_2M_1 + 2M_1^3 = \langle (I - M_1)^3 \rangle \\
\lambda_4 &= M_4 - 4M_3M_1 + 12M_2M_1^2 - 3M_2^2 - 7M_1^4 \\
&= \langle (I - M_1)^4 \rangle - 3\langle (I - M_1)^2 \rangle^2
\end{aligned} \tag{6}$$

characterize the relation of P(I) to a Gaussian. The Gaussian distribution has all $\lambda_n = 0$ for $n > 2$. In this expression λ_3 is called the skewness and λ_4 the excess.

rise to a series of what can be called inclusive densities defined by

$$\rho^{(1)}(\vec{r}) = \int \rho(\vec{r},\vec{r}_2,\vec{r}_3, \ldots \vec{r}_A)\, d^3r_2\, d^3r_3 \ldots d^3r_A$$

$$\rho^{(2)}(\vec{r}_1,\vec{r}_2) = \int \rho(\vec{r}_1,\vec{r}_2,\vec{r}_3, \ldots \vec{r}_A)\, d^3r_3 \ldots d^3r_A \qquad (8)$$

etc.

Each of these functions is symmetric and normalized to unity when integrated over all of its coordinates. Also in nuclear physics it is convenient to consider the uncorrelated case where $\rho^{(n)} = \rho^{(1)}(r_1) \cdot \rho^{(1)}(r_2) \ldots \rho^{(1)}(r_n)$.†) We now want to define correlations in such a manner that deviations from this single-particle picture give correlation functions different from zero. That is, we define

$$C^{(2)}(r_1,r_2) = \rho^{(2)}(r_1,r_2) - \rho^{(1)}(r_1)\rho^{(1)}(r_2)$$

$$C^{(3)}(r_1,r_2,r_3) = \rho^{(3)}(r_1,r_2,r_3) - \rho^{(2)}(r_1,r_2)\rho^{(1)}(r_3) - \rho^{(2)}(r_1,r_3)\rho^{(1)}(r_2) - \rho^{(2)}(r_2,r_3)\rho^{(1)}(r_1) + 2\rho^{(1)}(r_1)\rho^{(1)}(r_2)\rho^{(1)}(r_3)\,, \qquad (9)\ ^{*)}$$

etc.

There is an obvious parallel between these definitions and the definitions of the semi-invariants, and clearly $C^{(n)}$ vanishes for all $n \geq 2$ if the density factorizes.

It is often convenient to work in momentum space instead of configuration space. We shall use the same notation as in Eq. (9) changing only the variables. Thus

$$C^{(2)}(q_1,q_2) = \int e^{i\vec{q}_1\vec{r}_1+i\vec{q}_2\vec{r}_2}\, C^{(2)}(r_1,r_2)\, dr_1\, dr_2\ , \text{ etc.} \qquad (10)$$

A special case, namely $\bar{C}^{(2)}(q) = C^{(2)}(q, -q)$ is often called the correlation function. Its particular rôle in nuclear physics stems from the fact that owing to completeness of the possible set of states, $\bar{C}^{(2)}(q)$ can, in principle, be measured in reactions between an incoming particle and a single nucleon in the nucleus. Otherwise, it should be evident that correlations need a measurement of a reaction which can only be

†) Omitting also the CM-correlation and thereby using an unrealistic model

*) Generating function, see, for example, Reference (1).

A Poisson distribution has all $\lambda_n = M_1$. Thus $f_n = \lambda_n - M_1$ characterize the deviation of P(I) from a Poisson*). Those who are high-energy physicists, were prejudiced when they observed the first high-energy multiplicity distributions. They expected Poisson distributions and so for this reason always talked about $\langle(n-\langle n\rangle)^2\rangle - \langle n\rangle = \lambda_2 - M_1$ in their discussions. They have also taken this over in their definition of correlations, hence confusion.

We could already start to talk about correlations on the basis of Eq. (1). Such correlations

$$\lim_{T\to\infty} \frac{1}{T} \int_0^T I(t)I(t+\tau)\,dt$$

are called auto-correlations and are very useful, say, in hydrodynamics. I think we will skip them here because they are of a different type from those in which we are interested, namely correlations between two or more entities.

Thus let us consider two or more variables $I_1, I_2, \ldots, I_n$. Then clearly we have a distribution function $P(I_1, I_2, \ldots, I_n)$. I shall now digress in order to remind you of the main reason why λ_n or $\lambda_n - M_1$ are nice magnitudes in contrast to M_n. Suppose we will do an inclusive experiment and only measure

$$I = \sum_{i=1}^{n} I_i \qquad (7)$$

and suppose that the I_i's are uncorrelated, i.e. that $P(I_1, I_2, \ldots, I_n)$ factorizes into $P(I_1)P(I_2) \ldots P(I_n)$. Then one could fold the n distributions together with Eq. (7) as the content of a δ function factored into the product and after the n integrations obtain P(I). However, one could also remember that folding equals multiplication of Fourier transforms and that multiplication becomes addition in expressions exponentiated according to Eq. (5) and thus it follows that P(I) is obtained by adding the sets of λ_m's. In short, the semi-invariants (or $\lambda_m - M_1$) are additive magnitudes in statistics for sums of uncorrelated random variables. For this reason the λ's are forerunners of measures of correlations.

HOW TO DEFINE CORRELATIONS

Let us now change notation and let our physical magnitudes be nucleon positions in a nucleus. Thus our density distribution described by†) $\rho(\vec{r}_1, \vec{r}_2, \vec{r}_3 \ldots \vec{r}_A)$ can give

---

*) The definition of f_n in high-energy physics is slightly more complicated[3)].

†) We now work in the CM-system of the nucleus such that $\sum_{L=1}^{A} \vec{r}_L = 0$

described by a two- or more-particle operator. And, of course, a sum rule consisting of a sum of cross-sections describable by one- and many-particle operators permits the information on many-particle operators to be extracted by measuring all the cases where single-particle operators only play a rôle. Thus if we have measured elastic electron scattering and extracted $\rho^{(1)}(q)$ from that, then a measurement of all nuclear inelastic electron scattering gives information on correlations. The cross-section for the excitation of a final nuclear state f, without pion production, depends again only on a one-body operator:

$$\left(\frac{d\sigma}{d\Omega}\right)_{0\to f} \propto \left|\sum_i \left\langle f\left|e^{i\vec{q}\cdot\vec{r}_i}\right|0\right\rangle\right|^2 . \tag{11}$$

However, if we sum over all the final nuclear states (including the ground state) then:

$$\left(\frac{d\sigma}{d\Omega}\right)_{total} \propto \sum_f \sum_{i,j} \left\langle 0\left|e^{-i\vec{q}\cdot\vec{r}_j}\right|f\right\rangle \left\langle f\left|e^{i\vec{q}\cdot\vec{r}_i}\right|0\right\rangle . \tag{12}$$

For a high-energy electron and a not too large q^2, it is perhaps reasonable to forget that there is insufficient energy to excite some of the final states f and make use of closure

$$\sum_f |f\rangle\langle f| = I \tag{13}$$

to eliminate the explicit dependence on the final states. The resulting expression

$$\left(\frac{d\sigma}{d\Omega}\right)_{total} \propto \sum_{i,j} \left\langle 0\left|e^{i\vec{q}\cdot(\vec{r}_i-\vec{r}_j)}\right|0\right\rangle \tag{14}$$

for the total nuclear cross-section (i.e. elastic plus break-up) is proportional to the expectation value in the ground state of a two-body operator and can therefore lead to correlation phenomena. Performing the sums over i and j, and using our previous notation, we obtain

$$\left(\frac{d\sigma}{d\Omega}\right)_{total} \propto Z\left[1 + (Z-1)\left(\rho^{(1)}(q)\right)^2 + (Z-1)\bar{C}^{(2)}(q)\right] . \tag{15}$$

Given that $\rho^{(1)}(q)$ is known from elastic-electron scattering, in principle we can deduce $\bar{C}$ from the summed cross section. It should be noted that even in the absence of dynamical correlations, $\bar{C}$ will be non-zero because of the centre-of-mass constraint and the effect of the Pauli principle.

CENTRE-OF-MASS, PAULI CORRELATIONS AND CLUSTERS

It is not my intention to go very deeply into the effects of c.m. constraint, Pauli correlations or cluster formation, but a rough estimate is of considerable interest. Let us first disregard the Pauli principle and let us take the simplest density one can imagine

$$\rho(\vec{r}_1, \vec{r}_2, \dots \vec{r}_A) = N\, e^{-a(r_1^2+r_2^2+\dots r_A^2)}\, \delta(\vec{r}_1 + \vec{r}_2 + \vec{r}_A) \tag{16}$$

which contains the c.m. constraint. If the δ function is represented by (in 1 dimension)

$$\delta(x) = \lim_{B\to\infty} \sqrt{\frac{B}{\pi}}\, e^{-Bx^2} \tag{17}$$

everything in Eq. (17) is multi-dimensionally Gaussian, i.e. the exponential of a quadratic form. We can also consider $e^{i\vec{q}\vec{r}}$ as a quadratic form by counting $\vec{q}$ among our variables and hence we can use a general theorem for all integrations. The result is that in this simple case

$$(Z - 1)\bar{C}(q) = (Z - 1)\, e^{-q^2/2a}\, (1 - e^{q^2/2Za}) \tag{18}$$

i.e. a negative correlation which, of course, just tells us that the nucleons are pulled away from each other by the c.m. constraint. The maximum of the absolute value of Eq. (18) is:

$$(Z - 1)\, \mathrm{Max}\left(\mathrm{abs}\left[\bar{C}(q)\right]\right) = \left(\frac{Z - 1}{Z}\right)^Z \tag{19}$$

which goes towards 1/e for $Z \to \infty$. Thus we see that the c.m. correlation amounts as a maximum to something like $\sim$ 37%.

To estimate the magnitude of the Pauli correlation and the combination of c.m. and Pauli correlations is slightly more complicated. In the simplest possible approximation we can use harmonic oscillator wave-functions and the results for $\bar{C}(q)$ are shown in Fig. 1 for the doubly closed shell configuration of ^{16}O. For scattering of hadrons we must consider $(A-1)\bar{C}(q)$ instead of $(Z-1)\bar{C}(q)$. For this reason the value for $\bar{C}(q)$ equal to $1/[(A-1)e]$ is marked in the figure corresponding to our previous estimate of the maximum value of the c.m. correlation alone. We note three points:

i) correlations do not add linearly;
ii) the trivial correlations may amount to 50% for $(A-1)\bar{C}(q)$; and
iii) there are points where $\bar{C}(q)$ is independent of the c.m. constraint.

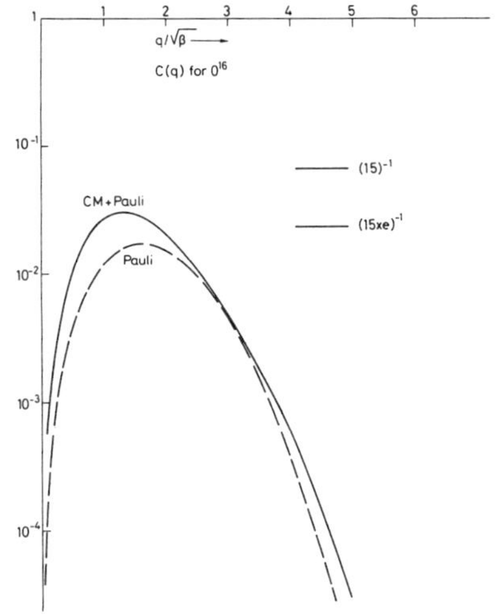

Fig. 1 : $\bar{C}(q)$ vs. q (q in arbitrary units) for the A = 16 harmonic oscillator model of ^{16}O showing $\bar{C}(q)$ as obtained when both the c.m. and the Pauli correlation is taken into account.

Here all of these statements are derived on the basis of a simple model but in reality they are quite general. This is trivial for the first two points; for point (iii) it follows from the fact that $\bar{\rho}^{(2)}(q)$ is independent of the c.m. coordinate, whereas $\rho^{(1)}(q)$ is quite sensitive to the c.m. constraint. However, $\rho^{(1)}(q)$ has zeros, and the heavier the nucleus is the more zeros there are. Thus at these zeros $\rho^{(2)}(q) - (\rho^{(1)}(q))^2$ has the same value and slope versus q c.m. constraint included or not. For heavy nuclei this leads to the fact that the two curves cannot be distinguished in praxis.

The computations leading to Fig. 1 were carried out on a code which generates the results for the harmonic oscillator Fermi gas to any number of closed shells. Although such wave-functions are quite unrealistic, the general trend of the results is quite indicative. The results confirm in detail points (i) to (iii) above.

Correlations of relatively long range are, of course, also present in nuclear matter. Recent experiments make it reasonable to talk about cluster formation, and there are papers on this subject at the present meeting. Again we can make a simple estimate of the possible order of magnitude of such effects.

Consider pairs of particles such that their position vectors $\vec{r}_1$ and $\vec{r}_2$ are close to a vector $\vec{\ell}_1$ which can fill out the nuclear volume, i.e. with a density distribution that can be approximated as

$$\rho(\vec{r}_1,\vec{r}_2) = N\, e^{-\beta\ell_1^2-\alpha(\vec{r}_1-\vec{\ell}_1)^2-\alpha(\vec{r}_2-\vec{\ell}_2)^2} \times \delta\left(\vec{\ell}_1 - \frac{\vec{r}_1+\vec{r}_2}{2}\right) \qquad (20)$$

Note that I have not introduced $\vec{\ell}_1$ as an extra variable; it is simply the centre of gravity of the pair in question. Note also that the over-all nuclear size is described by $2\beta\alpha/(\beta + 2\alpha)$. Assume that there are $\sim$ A/2 such correlated pairs but no other correlation. One then finds

$$(A-1)\bar{C}(q) = \exp\,(-\,q^2/2\alpha) - \exp\left(-\frac{q^2}{2}\cdot\frac{\beta + 2\alpha}{2\beta\alpha}\right) \tag{21}$$

i.e. a positive correlation with the maximum value

$$(A-1)\bar{C}(q)_{max} = x^{x/(1-x)}\,(1-x) \tag{22}$$

where

$$x = 2\beta/(\beta + 2\alpha) \tag{23}$$

and the maximum is at

$$q^2/2\alpha = -\frac{x}{1-x}\,\ell n\; x\;. \tag{24}$$

It is evident that also this type of correlation will mask hard-core correlations.

Whereas it is an easy matter to deal with harmonic oscillators almost anything else is insolvable. Consequently people discard the c.m. constraint right from the start and then other physicists write long articles about spurious states, lack of translational invariance, unexpected correlations, etc., ad nauseam. There is probably more nonsense in the literature on this point than on any other. I shall abstain from references.

HARD-CORE PROBLEMS

Instead I shall now turn to the hard-core of my talk. We have seen that Pauli, c.m. and cluster correlations give $(A-1)\bar{C}(q)$ values of between $\frac{1}{3}$ and $\frac{1}{2}$, that they (and other correlations) do not add linearly, that the harmonic oscillator is wonderful for illustrating where the problems are, but that more realistic models are almost impossible to deal with in exact computations and/or in good approximations. Also we have two experimental facts:

i) nuclear saturation, as demonstrated by the nuclear volume being proportional to nuclear-mass number; and
ii) the 1S phase-shift δ becomes negative at about 250 MeV lab. energy in pp scattering, which can be cleared up assuming a hard, or almost hard, core.

These facts are clear evidence of a repulsive component in the nucleon-nucleon interaction, but this does not necessitate a hard-core potential. Combinations of exchange forces, velocity-dependent potentials, three (or more) particle forces, non-local potentials and, of course, soft- and hard-core potentials might do as well. As far as I know there does not exist a thorough and comprehensive analysis of these complicated possibilities. It is interesting to see that, at this meeting, Wölfli et al. have experimental evidence from bremsstrahlung against Hamada-Johnston potentials, and that Epstein wants non-locality to fit parity violating γ-ray polarization in $n + p \to d + \gamma$ as claimed experimentally by Lobashev et al.

In this situation it would be extremely interesting to have direct experimental evidence for strong short-range correlations. Such evidence could be interpreted as an exclusion area in the two-particle density, i.e.

$$\rho^{(2)}(\vec{r}_1,\vec{r}_2) = 0 \quad \text{for} \quad |\vec{r}_1 - \vec{r}_2| \text{ small} . \tag{25}$$

Let us consider an infinitely extended bit of nuclear matter where there is no outer limitation on $\rho^{(2)}(r_1,r_2)$. In this case $\rho^{(2)}$ can only depend on $|\vec{r}_1 - \vec{r}_2|$ so we have a rather complete set of possibilities in the set

$$\rho_n^{(2)}(\vec{r}_1,\vec{r}_2) = N\left[1 - e^{-d_n(r_1-r_2)^2}\right]^n \tag{26}$$

from soft to hard exclusion volume. Also if we now make d_n such a function of n that the exclusion volume

$$V_{ex} = \int \left[1 - (1 - e^{-d_n r})^n\right] d^3r \tag{27}$$

is independent of n, we can compare the different types of correlation functions. In order to carry out a comparison, the simplest possible picture of the over-all density was used multiplying Eq. (26) by $e^{-\alpha(r_1^2+r_2^2)}$. Setting $\alpha/d \sim 0.16/(1.44\ A^{2/3})$, and varying A the results in Fig. 2 were obtained for the maximum value of $(1-A)\bar{C}(q)$ corresponding to this particular set of two-particle densities. Three points are evident:

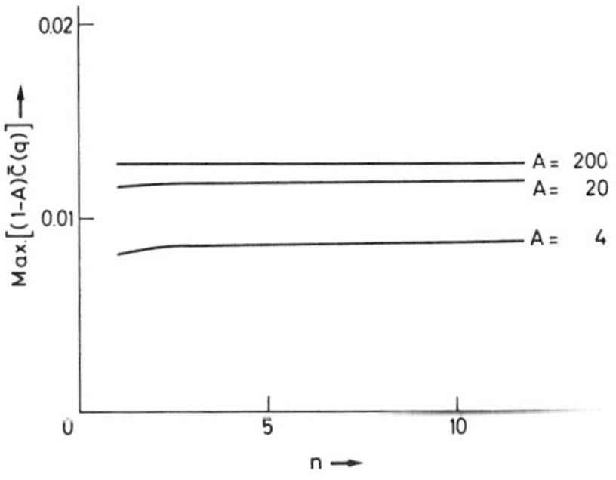

Fig. 2 : Maximum value of $(1-A)\bar{C}(q)$ for the particular set of two-particle densities based on Eq. (35) and with parameters as shown in the figure.

i) $\bar{C}(q)$ measures nothing else than the exclusion volume, the rest (like variation with n) gives higher order effects only;

ii) the variation with A soon reaches close to the asymptotic value (i.e. infinite nuclear matter); and

iii) the total effect is very small compared to the 0.3 - 0.5 value from the trivial correlations.

Now, admittedly, Eq. (26) multiplied by an over-all nuclear shape is a simplification, and although the literature as such is rather quiet on such topics, I have certainly witnessed discussions where many other proposals for hard-core correlations have been put forward. However, for computational purposes it is convenient to approximate by Gaussians as much as pssible. Thus let us look at the shape:

$$\rho_n^{(2)}(\vec{r}_1,\vec{r}_2) = N\, e^{-\alpha(r_1^2+r_2^2)} \left[1 - b_1\, e^{-d_1(\vec{r}_1-\vec{r}_2)^2} + b_2\, e^{-d_2(\vec{r}_1-\vec{r}_2)^2}\right]^n \qquad (28)$$

with the condition

$$1 - b_1 + b_2 = 0 \qquad (29)$$

ensuring a zero in $\rho_n^{(2)}$ for $|\vec{r}_1 - \vec{r}_2| = 0$. In order to ensure that Eq. (28) is positive-definite, we must demand $d_1 > d_2$. Equation (28) corresponds to a correlation where part of the density taken out of the hole lies as a small wall around the hole. It is quite difficult to make a parameter choice such that comparisons with Eq. (26) are meaningful.

Examples (curves B and C) based on unpublished computations by Wong for ^{16}O and parametrized according to Eq. (28) by Tanner[4)] are shown in Fig. 3. These cases (which do not seem realistic to me) give much larger values for $(A-1)\bar{C}(q)$ than what we have seen previously and if one wants to be very prudent then one would say that we know very, very little about what to expect as regards hard-core correlations. Another case which was obtained by searching for such values of b_1, d_1 and d_2 that the missing volume

$$V'_{ex} = \int \left[1 - (1 - b_1\, e^{-d_1 r^2} + b_2\, e^{-d_2 r^2})^n\right] d^3r \qquad (30)$$

equals that of Eq. (27) for both n = 1 and 2 and with the choice of d_n in Eq. (26) used previously. Such a case exists and results in the curves shown in Fig. 4. The resulting correlations are larger than for our standard example. However, if one defines the exclusion volume as

$$V''_{ex} = \int_0^{r_0} \left[1 - (1 - b_1\, e^{-d_1 r^2} + b_2\, e^{-d_2 r^2})^n\right] 4\pi\, r^2\, dr\ , \qquad (31)$$

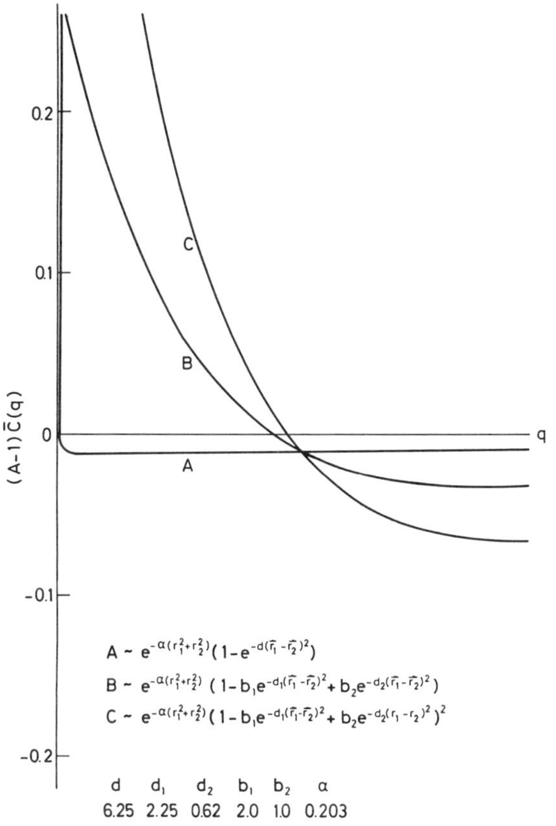

Fig. 3 : $(A-1)\bar{C}(q)$ vs. q for the model Eq. (36) with parameters as shown in the figure.

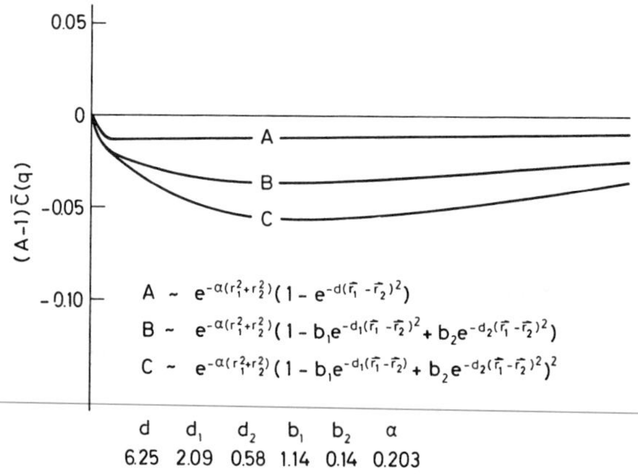

Fig. 4 : As Fig. 3, but with other parameters.

where r_0 is the value of r where the integrand in Eq. (31) is zero, then again one finds that the maximum value of $(1-A)\bar{C}(q)$ is merely a measure of the exclusion volume and roughly speaking linear in this volume. The linearity was checked in detail with the model of Eq. (26) and is almost perfect as long as V_{ex} is smaller than the volume per nucleon (10% deviation from linearity at this point and for heavy nuclei).

MULTIPLE SCATTERING

The possibility of using multiple scattering in order to learn something about hard-core correlations was investigated by Colin Wilkin and myself[5] a few years ago for scattering on ^{4}He.

The main point in our approach was to optimize the similarity between form factors for a c.m. correlated and a c.m. + hard-core correlated density. The necessity for this procedure is apparently difficult to make clear. Also at this meeting there is a paper talking about hard-core correlations seen in the nuclear form factor -- however, that is a model-dependent statement and will never constitute independent experimental proof of hard-core effects. In effect, we found that the two models, respectively, with or without hard-core correlations predict very small effects when they have practically identical form factors; everything is masked by the c.m. correlation. We investigated both elastic and inelastic, single and multiple scattering and for parameters covering all available projectiles.

The general theory of multiple scattering was worked out already by Glauber[1] in 1959. Defining particle densities through

$$n^{(m)}(\vec{r}_1,\vec{r}_2, \ldots \vec{r}_m) = \frac{N!}{(N-m)!}\,\rho^{(m)}(\vec{r}_1,\vec{r}_2, \ldots \vec{r}_m) \tag{32}$$

for the general N particle system he finds that the optical phase shift function is given by

$$\begin{aligned}\chi_{opt}(\vec{b}) &= i\int n^{(1)}(\vec{r})\Gamma(\vec{b}-\vec{s})\, d^3r \\ &- \frac{i}{2}\iint\left[n^{(2)}(\vec{r}_1,\vec{r}_2) - n^{(1)}(\vec{r}_1)n^{(1)}(\vec{r}_2)\right]\Gamma(\vec{b}-s_1)\Gamma(\vec{b}-s_2)\, d^3r_1\, d^3r_2 \\ &+ \frac{i}{3!}\int\left[n^{(3)}(\vec{r}_1,\vec{r}_2,\vec{r}_3) - 3n^{(2)}(\vec{r}_1,\vec{r}_2)n^{(1)}(\vec{r}_3) + \right. \\ &\qquad \left. + 2n^{(1)}(\vec{r}_1)n^{(1)}(\vec{r}_2)n^{(1)}(\vec{r}_3)\right] \\ &\times \Gamma(\vec{b}-\vec{s}_1)\Gamma(\vec{b}-\vec{s}_2)\Gamma(\vec{b}-\vec{s}_3)\, d^3r_1\, d^3r_2\, d^3r_3 + \ldots\ .\end{aligned} \tag{33}$$

We recognize the structure of the "semi-invariants" in Eq. (33). $\chi_{opt}(b)$ determine the elastic scattering since the scattering amplitude is essentially the Fourier transform of $[1 - \exp(i\chi_{opt})]$. The profile functions Γ are Fourier transforms of the hadron-nucleon amplitudes.

The structure of Eq. (33) is such that the main difference between two models with the same $\rho^{(1)}(q)$ but where one is correlated and the other uncorrelated is given by $\bar{C}(q)$. Much of the above is therefore still valid.

Outside the diffraction region inelastic scattering is described by the effective number of nucleons participating in the n-fold collision

$$N_n(A;\sigma) = \frac{1}{n!\sigma} \int e^{-\sigma T(b)} \, |\sigma T(b)|^n \, d^2b \tag{34}$$

with T(b) the projection of the nuclear density (normalized to A) into the plane of the impact parameter, i.e. the plan perpendicular to the direction of the incoming particle.

Also here, however, correlations play a rôle. The most important is the Pauli correlation[6]. The main consequence of the Pauli principle is to change the effective number of nucleons in Eq. (34). The results are shown in Fig. 5.

It is seen from the figure that the Pauli correction consists of two effects. One of these has to do with the omission of $1/N_s$ order terms when exponentiating. This effect suppresses $N_n(A;\sigma)$ for small values of A and large values of n. The second effect increases $N_n(A;\sigma)$ because the Pauli principle prevents particles from falling in line; thereby the attenuation is suppressed.

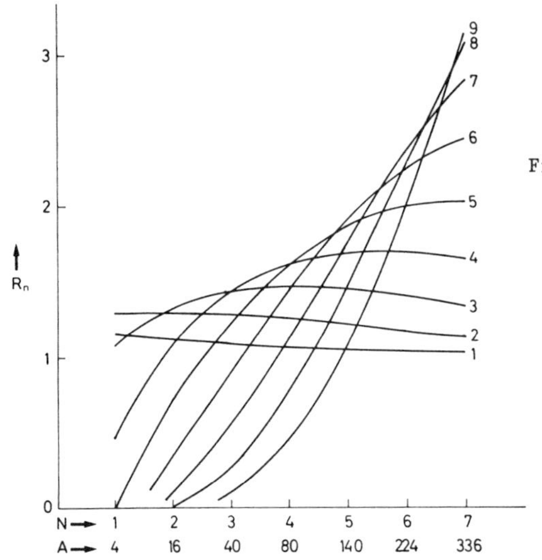

Fig. 5 : The ratio $R_n = \{N_n(A;\sigma)(\text{Pauli})/N_n(A;\sigma)[\text{Eq. (34)}]\}$ vs. number of four-fold closed shells or vs. mass number A. This ratio gives an estimate of the correction factor resulting from the Pauli principle. Computations are carried out for $\sigma = 40$ mb. Compressed harmonic oscillator densities.

COMPARISON WITH EXPERIMENT

We shall now finally compare the theoretical picture with experiments performed by the CERN-Rome collaboration[7]. We neglect all correlations but the c.m. correlation for deuterium and all correlations but the Pauli principle for nuclei heaver than helium. The basis for the comparison between theory and experiment is the elastic and hadronic inelastic p-p data obtained by the same group[8].

The results for deuterium are shown in Fig. 6 and for Be and Pb in Fig. 7. Finally the 67 mrad Pb spectrum is decomposed into its components in Fig. 8. Remember now that, for example, a 10% error in the p-p data will show up as a 50% theoretical uncertainty in predictions on five-fold scattering. Thus the results are nowhere in disagreement with prediction. Results for small scattering angles have also been obtained, they show very little structure and are not reproduced here. Including such results, however, one may conclude that there is agreement between prediction and measurement inside the experimental and theoretical errors over a $d\sigma/d\Omega\, dk'$ range from $\sim 2.10^{-22}$ cm²/(ster. GeV/c) in the forward-direction for Pb down to $\sim 2.10^{-31}$ cm²/(ster. GeV/c) for large angles and deuterium, i.e. over almost nine decades, over several GeV/c in k' and all of this including elastic, hadron-inelastic, and nuclear-inelastic scattering. In addition we have seen that the possible effects of hard-core correlations are likely to be small and we can thus conclude by the statement that considerable increase in precision both experimentally and theoretically is needed if we hope to measure effects of hard-core correlations.

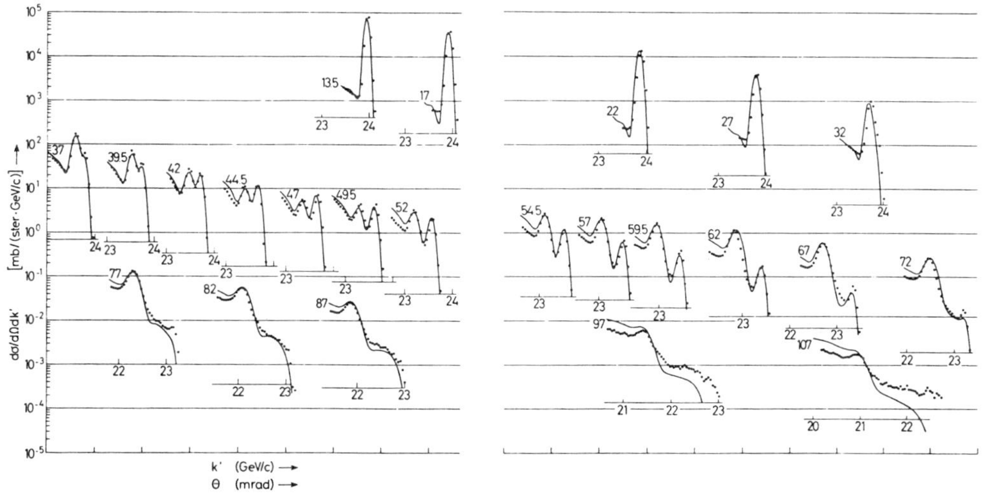

Fig. 6 : The figure compares measured[7] and computed double differential cross-sections $d^2\sigma/d\Omega dk'$ for 24 GeV/c protons on deuterium. Each spectrum is identified through the appropriate value of the nominal laboratory scattering angle θ in mrad.

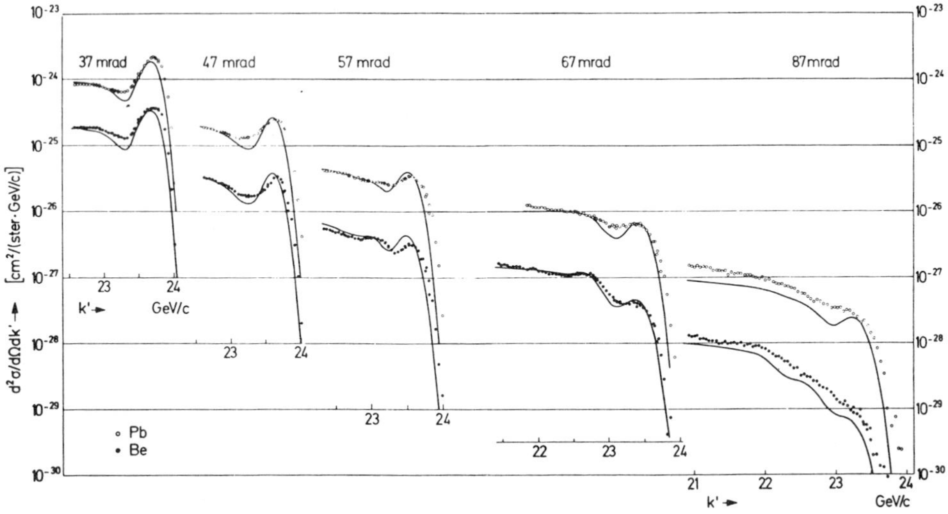

Fig. 7 : The figure compares measured[7] and computed double differential cross-sections $d^2\sigma/d\Omega dk'$ for 24 GeV/c protons on Pb (upper curves) and Be (lower curves). Spectra are identified as in Fig. 6.

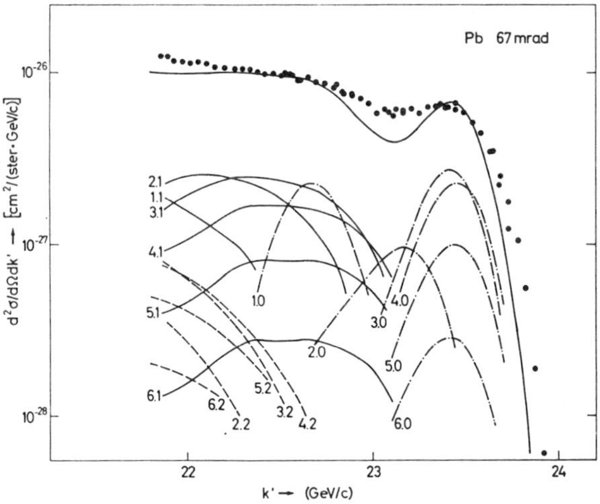

Fig. 8 : The detailed decomposition of the computed momentum-loss spectrum for $k = 24$ GeV/c, $\theta = 67$ mrad and Pb. The curves are indicated n, m with n the multiplicity of nuclear inelastic events out of which also m are hadronic inelastic.

REFERENCES

1) R.J. Glauber, Lectures in Theoretical Physics, edited by W.E. Brittin and L.G. Dunham, p. 315, Interscience Publishers, New York (1959).

2) See, for example: S.O. Rice, Selected papers on noise and stochastic processes, edited by N. Wax, p. 133, Dover Publications, New York (1954).

3) Z. Koba, Proceedings of the 1973 CERN-JINR School of Physics, edited by O. Kofoed-Hansen, p. 171, CERN Report (1973).

4) N. Tanner, private communication.

5) O. Kofoed-Hansen and C. Wilkin, Ann. Phys. 63, 309 (1971).

6) O. Kofoed-Hansen, Nuclear Phys. B54, 42 (1973).

7) A. Klovning, O. Kofoed-Hansen and K. Schlüpmann, Nuclear Phys. B54, 29 (1973); U. Amaldi, R. Biancastelli, C. Bosio, G. Matthiae, J.V. Allaby, A.N. Diddens, R.W. Dobinson, A. Klovning, J. Litt, L.S. Rochester, K. Schlüpmann and A.M. Wetherell, Nuclear Phys. B39, 39 (1972).

8) J.V. Allaby, A.N. Diddens, R.W. Dobinson, A. Klovning, J. Litt, L.S. Rochester, K. Schlüpmann, A.M. Wetherell, U. Amaldi, R. Biancastelli, C. Bosio and G. Matthiae, Phys. Letters 34B, 431 (1971); U. Amaldi, R. Biancastelli, C. Bosio, G. Matthiae, J.V. Allaby, A.N. Diddens, R.W. Dobinson, A. Klovning, J. Litt, L.S. Rochester, K. Schlüpmann and A.M. Wetherell, Phys. Letters 34B, 435 (1971).

DISCUSSION

SAUER(MIT): You parametrized the two-body density phenomenologically. Does the two-body density you use yield the correct experimental one-body density, once the second variable r_2 is integrated over ?

KOFOED-HANSEN: Several years ago Colin, Wilkin and I tried to do as well as we could in relation to the question, in the case of ^{4}He. We were a little bit more ambitious and we didn't talk only about two particle density. That one we can always deal with mathematically very easily. But in the ^{4}He we took two different nuclear densities, the first without correlations (except the C-M constraint),

$$\prod_{i=1}^{4}\left[e^{-\alpha r_i^2}\left(1-be^{-\beta r_i^2}\right)\right]\delta(\Sigma\, r_i)$$

and secondly

$$\prod_{i=1}^{4}\left[e^{-\alpha' r_i^2}\prod_{j>i}\left(1-b'\, e^{-\beta' r_{ij}^2}\right)\right]\delta(\Sigma\, r_i)$$

where we have genuine correlations. We searched on computer for values of α, b and β that would fit the ^{4}He form factor as well as possible. Then we searched for α', b' and β' which would do the same. Then we computed predictions for electron scattering cross section and multiple scattering in order to obtain both total and elastic differential scattering cross sections. In this search we have chosen some of the points where one could expect major deviations, like the position at the diffraction minimum and the height of the second diffraction hump. We have then compared such things with the aid of a computer, and obtained cross sections and slopes in the initial cross section for hadrons on nuclei over the whole region of all known hadrons, and we could only find the possibility of differences of the magnitude of 7%. Why is it not possible to do better than to obtain approximately identical form factors ? This is because there are simply different numbers of Gaussian terms in these two series and those cannot be made identical. But if you take two-particle densities, then we can match the two exactly and still handle the things with the computer.

LEVINGER(Rennselaer): I wonder why you have not discussed experiments which many years ago lead to the idea of correlations: e.g. the neutron-proton pairs following photon absorption, or pairs from absorption of negative π's. Is this relevant?

KOFOED-HANSEN: It is relevant to the topic, but it would have been more relevant if there had been better results available than those we have had so far. There is evidence for α-clusters and for pairing correlations, but that doesn't reflect the existence of a hard core. If you in particular think about π capture, there are no results giving definite information on hard core correlations. If you think about double charge exchange, the story is the same. We think that there ought to be hard core correlations since we know the saturation in the nuclear matter and because we know the S-wave in the p-p data changes sign when we pass 250 MeV. You know that better than I, and you deal with these things more than I do. You also know that you can try to interpret these things by non-local potentials, etc.

LEVIN(Brown): Is it the total cross section that yields information on correlations, the cross section determined by a transmission experiment, or the angle integrated differential cross-section ?

KOFOED-HANSEN: The elastic cross section here is the elastic differential cross section. The elastic cross section will not give any dependent information on short range correlations, but only on the mass (or charge) density in the CM-system of the nucleus. This density might have structure as for a crystal. Some people call that correlations. In a crystal it is extremely long range correlation. The total cross section I am talking about is the total differential cross section. It does contain information on hard core but the effects are small and are only a few percent for the hard core part. It contains about 50% from Pauli principle, centre of mass, etc.

RINAT(Weizmann): Some time ago, a Polish group reported calculations of the inclusive cross section with 'uncorrelated' and 'correlated' wave functions; if my memory does not deceive me, these authors reported an effect of the order ≃2%, not of 7%. Is my memory wrong, or is that the case with that calculation ?

KOFOED-HANSEN: I think you are right. However, I don't think that they had the same form factor for the correlated and the uncorrelated case. It is really a funny state of affairs caused by our having been accustomed to work with single particle wave functions for so many years. You see now and again, and ever so often, talks about lack of translational invariance, spurious states and a series of other statements which are just so obvious. If we don't take the CM-constraint into account, then of course these troubles are there.

EFFECT OF SHORT RANGE CORRELATIONS IN PION PHOTOPRODUCTION FROM NUCLEI

K.Srinivasa Rao, MATSCIENCE, The Institute of Mathematical Sciences, Madras-600020, India and V.Devanathan, Department of Nuclear Physics, University of Madras, Madras-600025, India.

It is usually accepted that the nucleon-nucleon interaction contains a hard core. But the short range correlations induced by the hard core are neglected in most shell model calculations even though the correlation corrections are expected to become increasingly important as one considers nuclear phenomena at high momentum transfers. Following Villars[1)], short range correlations can be introduced into the nuclear wave function using a unitary model operator e^{iS}, where S is a hermitian two-body operator. The wave function which contains short-range correlations induced by such a unitary operator can be written as:

$$\psi \doteq e^{iS}\Phi \quad , \quad S^+ = S \tag{1}$$

where Φ is the usual shell model wave function. Da Providencia and Shakin[2)] have shown that the presence of short range correlations causes a significant modification of electron scattering form factors at momentum transfer of about 3 fm^{-1}. Besides, there are many studies[3)] of the influence of short range dynamical nucleon-nucleon correlations on elastic electron and nucleon scattering cross sections and these provide a strong evidence for the existence of such correlations. Since the momentum transfer in pion photoproduction can be varied by varying the angle at which the pion is observed, it is interesting to study the effect of short range correlations in photopion production from nuclei. If the important assumption that the short range correlations are the same in the ground state and exicted states of the system is made, then we need only a single model operator to correlate the shell model wave functions. On the basis of this assumption, it becomes possible to calculate the physical quantities using uncorrelated shell model states if one uses a set of transformed operators, $\tilde{O}$, related to the transition operator O, by:

$$\tilde{O} = e^{-iS}\, O\, e^{iS} \tag{2}$$

$$= \tilde{O}^{(1)} + \tilde{O}^{(2)} + \cdots \tag{3}$$

where

$$\tilde{O}^{(1)} = \sum_{\alpha,\beta} a^+_\alpha \langle \alpha | O | \beta \rangle a_\beta \ , \tag{4}$$

$$\tilde{O}^{(2)} = \sum_{\alpha,\beta,\gamma,\delta} a^+_\alpha a^+_\beta \left[\langle \alpha\beta | e^{iS}(O_1 + O_2) e^{iS} | \gamma\delta \rangle - \langle \alpha\beta | (O_1 + O_2) | \gamma\delta \rangle \right] a_\delta a_\gamma , \tag{5}$$

in the notation of second quantization. The operator e^{iS} is assumed to have a simple representation in the relative and center-of-mass system and the two-body part of the model operator is:

$$[e^{iS}]^{(2)} = \sum_{\substack{nl,N\mathcal{L},LM\\ SS_z,TT_z}} |\widetilde{nl}, N\mathcal{L}, LM, SS_z, TT_z\rangle \langle nl, N\mathcal{L}, LM, SS_z, TT_z| , \qquad (6)$$

where the bra and ket are identical in all quantum numbers except nl which correspond to a state of relative motion governed by a wave function of the harmonic oscillator form. The state $\widetilde{nl}$ that replaces nl via the unitary transformation is the correlated radial wave function. The details regarding the construction of the two-body operator constructed from two one-body operators, and the explicit expressions for the one-body and two-body matrix elements M_1 and $\widetilde{M}_2$ are given in Devanathan[6]. The differential cross section, is given by

$$\frac{d\sigma}{d\Omega} = (2\pi)^{-2}\mu\mu_0 [M_1 M_1^* + M_1 \widetilde{M}_2^* + M_1^* \widetilde{M}_2 + \widetilde{M}_2 \widetilde{M}_2^*] . \qquad (7)$$

The calculation of the one-body part of the operator $\widetilde{O}$, viz. $\widetilde{O}^{(1)}$ leading to the one-body matrix element M_1, between the ground state of ^{16}O and the particle-hole states of ^{16}N has been carried out explicitly in refs.[6]. The numerical calculation involving $\widetilde{M}_2$ is rather tedious due to the large number of quantum numbers which arise as a consequence of the construction of the two-body operator and the transformation which takes us from motion of two particles about a common center to a description of the relative and center-of-mass motion of the two particles. One trestriction which we have introduced is to consider S-wave correlations only. The correlated harmonic oscillator wave functions are:

$$\widetilde{R}_{no}(r) = \frac{1}{\sqrt{N_{no}}} f(r)\, R_{no}(r) , \qquad (8)$$

where N_{no} is the normalization factor and $f(r)$ is the correlation function:

$$f(r) = \begin{cases} 0 & \text{for } r \le c, \\ 1 - \exp[-\beta\{(\frac{r}{c})^2 - 1\}] & \text{for } r > c, \end{cases} \qquad (9)$$

with β and c as parameters. This form of $f(r)$ was employed by Dabrowski[7] in his variational calculation on the binding energy of ^{16}O.

The 'partial' differential cross sections at 260 MeV incident photon energy with the ^{16}N states ($J^P = 2^-, 0^-, 1^-, 3^-$) described by the simple independent model configurations, carried out on CDC 3600 computer, reveal that the correlation corrections are not as yet the dominant feature in the study of $^{16}O(\gamma,\pi^+)^{16}N$. This is in conformity with the finding of Da Providencia and Shakin[2] that the short range correlations studied via the process of inelastic electron scattering, cause a significant modification of the form factor only for momentum transfers around 3 fm^{-1}. Since at 260 MeV, the maximum momentum transfer involved is only 1.9 fm^{-1}, we conjecture that for larger incident photon energies (say, around 400 MeV and above) these correlation corrections may become significant. In any case, a definite conclusion about the effect of short range correlations on pion photoproduction from nuclei can be drawn only when experimental angular distributions of the cross sections are available.

The authors wish to thank Professor Alladi Ramakrishnan for his interest in this work.

REFERENCES

1) F.Villars, 'Proc. of the Int. School of Physics', Enrico Fermi, Course 23, 1961
2) J.Da Providencia and C.M.Shakin, Ann. Phys. 30 (1964) 95, Nucl. Phys. 65 (1965) 54.
3) For example, see C. ciofi Degli Atti, ISS 68/24, Instituto Superiore di Sanita, Laboratori di Fisica, for a list of references.
4) G.F.Chew, M.L.Goldberger, F.E.Low and Y.Nambu, Phys. Rev. 106 (1957) 1345.
5) V.Devanathan, Symposia in Theoretical Physics and Mathematics, Ed. by Alladi Ramakrishnan, Plenum Press 10 (1970) 115. Also, K.Srinivasa Rao, Ph.D. thesis (1970), University of Madras (unpublished).
6) V.Devanathan, M.Rho, K.Srinivasa Rao and S.C.K.Nair, Nucl. Phys. B2 (1967) 329; K.Srinivasa Rao and V.Devanathan, Phys. Letts. 32B (1970).
7) J.Dabrowski, Proc. Phys. Soc. 71 (1958) 658, ibid. 72 (1958) 499.

MULTIPLICITY SCALING AND COMPOUND POISSON DISTRIBUTIONS

T.S.Santhanam and R.Sridhar, MATSCIENCE, Madras. India.

Recently it has been pointed out ~~that~~ by Le Bellac et al. ~~x~~ that a compound Poisson Distribution (CPD)

$$P(n,\bar{n}) = \int f\left(\frac{x}{\langle n\rangle}\right) \frac{x^n e^{-x}}{n!} \frac{dx}{\langle n\rangle} , \qquad 1)$$

for the multiplicity distribution taken purely as an emperical formula offers an interesting alternative to K.N%O Scaling. In this paper we point out that the assumption of CPD for the distribution function is indeed identical incontent to the assumption of scaling of the factorial moments

$$F_q = \langle n(n-1)\ldots(n-q+1)\rangle$$
$$= C_q \langle n\rangle^q \qquad 2)$$

with

$$C_q = \int_0^\infty f(u)\, u^q\, du \qquad 3)$$

It is, of course, known that to leading power in log s (s= square of the c.m. energy) Eq.2 leads to

$$\langle n^q\rangle \simeq C_q \langle n\rangle^q \qquad 4)$$

and hence to the scaling (supposedly assymptotic)

$$P(n) = \frac{\sigma_n}{\sigma_{inel}} = \frac{1}{\langle n\rangle} f\left(\frac{n}{\langle n\rangle}\right) \qquad 5)$$

where σ_n denotes the inclusive cross-section for n final particles. If $P(n)$ is the distribution function we define the generating fucntion $G(y)$ as

$$G(y) = \sum_{n=0}^{\infty} y^n P(n) \qquad 6)$$

From the definition it follows that

$$P(n) = \frac{1}{n!} \left.\frac{\partial^n G(y)}{\partial y^n}\right|_{y=0} \quad \ldots \qquad 7)$$

It is also clear from Eq.6 that

$$F_q = \langle n(n-1)\cdots(n-q+1)\rangle = \left.\frac{\partial^q G(y)}{\partial y^q}\right|_{y=1} \qquad 8)$$

so that one can write G as

$$G(y) = 1 + \sum_{r=1}^{\infty} \frac{(y-1)^r}{r!} F_r \qquad 9)$$

From Eqs.7and 9 one gets the inversion formula

$$P(n) = \frac{1}{n!} \sum_{m=0}^{\infty} \frac{(-1)^m}{m!} F_{m+n} \qquad 10)$$

In fact the inversioh formula Eq.10 has been recently used by many people in the recentl literature. The factorial moments carn be expressed as integrals over 'Product densities' of Ramakrishnan. It is clear from Eq.10 that when

$$F_q = (F_1)^q , \qquad F_1 = \langle n\rangle \qquad 11)$$

(expressing no correllation) one gets the usual Poisson distributions When

$$F_q = q!\,(F_1)^q \qquad 12)$$

one gets the Yule-Furry distribution . If,however,

$$F_q = c_q (F_1)^q, \qquad c_q = \int_0^{\infty} f(u)\, u^q\, du$$

one gets the CPD Eq.1 with $u F_1 = x$. This completes the proof. The Poisson and Y.F distributions are obtained as special cases when $f(u) = \delta(u-1)$ and $f(u) = e^{-u}$ respectively. The question, however,is whether Eq.1 can be understood from physical models. The authors of Ref.1 admit that the derivation of Eq.1 from physical considerations is far from being satisfactory. This,in fact, makes them to assume Eq.1 more as an emperical formaula than anything else.

References

1) M.Le Bellac et.al Nuclear Physics

2)Z.Koba,H.Nielson and P.Olesen, Nuclear Physics B40 (1972)317
This is referred in the text as K.N.O.

3) See,for instance, W.I.Weisberger,Phys.Rev.8D(1973)1388

4)Alldi Ramkrishnan,Handbuch Der Physik,Springer Verlag(1959)p.549.

Attenuation and p-p and p-n Quasifree Scattering in Deuteron Breakup

T. K. Lim and R. D. Haracz

Department of Physics and Atmospheric Science
Drexel University, Philadelphia, Pa. 19104

In a series of publications[1], we have shown that it is unnecessary to abandon PWIA as a viable instrument for low-energy QFS analyses provided one includes attenuation effects on the dominant impulse approximation mechanism. In the framework of our new attenuation model[2], the PWIA is modified by a factor which takes the expected absorption into account. Thus, the QFS angular distribution for the general breakup reaction $X(p_1,p_1p_2)S$ is given by

$$\left(\frac{d^3\sigma}{d\Omega_1 d\Omega_2 dE_1}\right)_{expt} = T \cdot \left(\frac{d^3\sigma}{d\Omega_1 d\Omega_2 dE_1}\right)_{PWIA} \quad (1)$$

where the transmission factor

$$\begin{aligned} T = &\int \psi^*\left(\vec{r}_{p_1S}\right) T\left(\sigma_{p_1S}, \vec{r}_{p_1S}\right) \psi\left(\vec{r}_{p_1S}\right) d\vec{r}_{p_1S} \\ &\times \int \phi^*\left(\vec{r}_{p_2S}\right) T\left(\sigma_{p_2S}, \vec{r}_{p_2S}\right) \phi\left(\vec{r}_{p_2S}\right) d\vec{r}_{p_2S} \end{aligned} \quad (2)$$

ψ and ϕ are the relative cluster functions of p_1 and p_2 with the spectator nucleus S, σ_{p_iS} is the total scattering cross section of p_i on S, and the transmision probability

$$\begin{aligned} T\left(\sigma_{p_iS}, \vec{r}_{p_iS}\right) &= \tfrac{1}{2}\left(1 + \sqrt{1 - \frac{\sigma_{p_iS}}{4\pi r^2_{p_iS}}}\right), \quad \sigma_{p_iS} < 4\pi r^2_{p_iS} \\ &= 0 \quad\quad , \quad \sigma_{p_iS} \geq 4\pi r^2_{p_iS} \end{aligned} \quad (3)$$

We have been encouraged by the model's success in describing (p,2p) quasifree scattering on ^2H, ^3He and ^4He to use it to explain the large difference in QFS cross sections for the two kinematically-equivalent modes of deuteron breakup under proton bombardment. In our model, we find that the (p,pn) QFS has two identical particles (protons) interacting in one of the exit channels and therefore one of the cluster functions to be used in eq.(2) must represent the p-p system. Now, since the p-p system differs from the p-n system in not forming a bound state, our model predicts a branching difference for (p,2p) and (p,pn) breakup on deuteron which will surface in our evaluation of the transmission factor. We have chosen to represent the p-p cluster function by the continuum wave function near threshold

$$\psi_{p-p} = \frac{a_s}{r} \left(1 - \frac{r}{a_s} - e^{-\xi r}\right) \quad (4)$$

and with it, have obtained the results displayed in Fig.1. We conclude from the model's success here and in ref. 2 that we have at our disposal a realistic method of QFS analysis which makes it unnecessary, at this stage, to discard PWIA in favor of more exact and laborious methods.

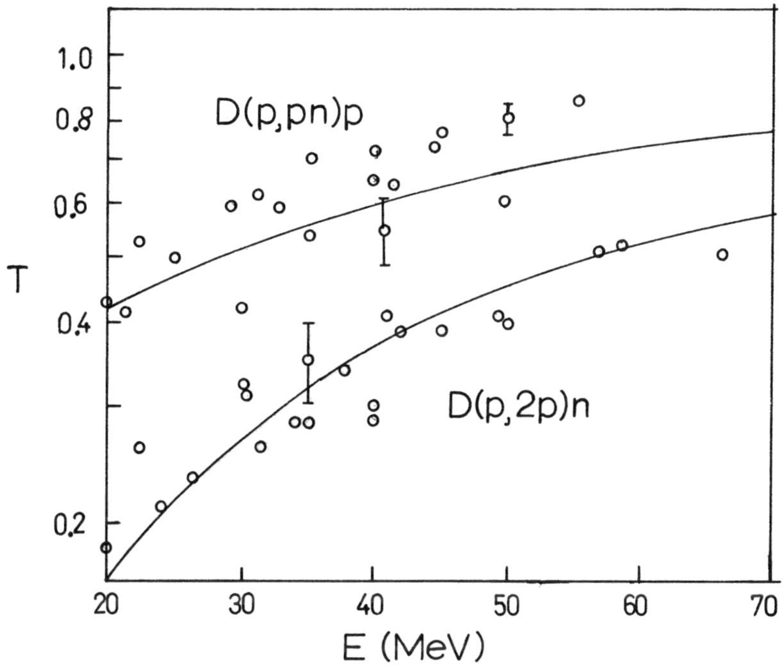

References:

1. T. K. Lim, Phys. Rev. Lett. 31, 1258 (1973); R. D. Haracz and T. K. Lim, Phys. Rev. Lett. 31, 1263 (1973); T. K. Lim, Phys. Lett. 47B, 397 (1973).

2. R. D. Haracz and T. K. Lim, Phys. Rev. (to be published).

3. E. L. Petersen, et al, Phys. Lett. 31B, 209 (1970); J. L. Durand, et al, Phys. Rev. C6, 393 (1972).

4. T. K. Lim, Nucl. Phys. 180, 668 (1972); L. Hulthen and M. Sugawara, Handbuch der Physik 39, 106 (1957).

REACTIONS QUASI ELASTIQUES p-p ET p-d SUR ^4He A 155 MeV

R. Frascaria, P.G. Roos[+], M. Morlet, N. Marty, V. Comparat, N. Fujiwara[++] et A. Willis

Institut de Physique Nucléaire, Orsay, France.

Les réactions $^4He(p, 2p)T$ (réaction A) et $^4He(p, pd)d$ (réaction B) ont été étudiées dans des cinématiques laissant la particule de recul avec un faible moment. La réaction A permet alors d'extraire le recouvrement de ^4He en p + T et la réaction B le recouvrement de ^4He en d + d. Cette information peut être obtenue si le mécanisme d'interaction est correctement décrit : la DWIA, pour la réaction A permet un tel traitement compte tenu du fait qu'à 155 MeV sur les noyaux légers les effets de distorsion sont faibles et que les termes d'échanges du nucléon incident avec les nucléons du coeur sont négligeables pour des petits moments k_p du proton dans $^4He(k_p < 0.7 Fm^{-1})$.

L'expérience utilise le faisceau de protons de 155 MeV du Synchrocyclotron d'Orsay et trois cibles liquides de ^4He, ^2H, ^1H géométriquement identiques. Deux télescopes T_1 et T_2 détectent les particules (p, d) émises aux angles θ_1 et θ_2. Un plastique scintillant à arrêt total placé sur chaque bras permet la mesure des énergies E_1 et E_2 des particules. Une jonction silicium de 300 µm placée sur une des voies discrimine leur nature. Une anticoïncidence élimine les protons de diffusion élastique sur chaque voie. L'écart en temps des deux voies Δt est mesuré par un convertisseur temps-amplitude, sa visualitation dans une bande de 100 ns (2 micropaquets de la machine) permet d'observer simultanément les coïncidences vraies ou fortuites. Les quatre informations E_1, E_2, ΔE_1, Δt permettent de classer l'événement en vrai ou fortuit et de le rattacher aux réactions A ou B. Chaque événement (E_1, E_2) est affecté d'un coefficient qui représente les efficacités des télescopes à ces énergies. Ces efficacités sont mesurées expérimentalement à l'aide des diffusions $^1_1H(p, p)$ et $^2_1H(p, p)$.

Les sections efficaces obtenues sont présentées sur la fig. 1 pour la diffusion quasi-élastique pp($\theta_1 = 40.2^o$ $\theta_2 = 40.4^o$, $E_3 = 0$ pour $E_1 = 68$ MeV) et sur la fig. 2 pour le quasi-élastique p-d($\theta_1 = 40^o$, $\theta_2 = 60^o$, $E_3 = 0$ pour $E_1 = 64$ MeV) en fonction de l'énergie E_1.

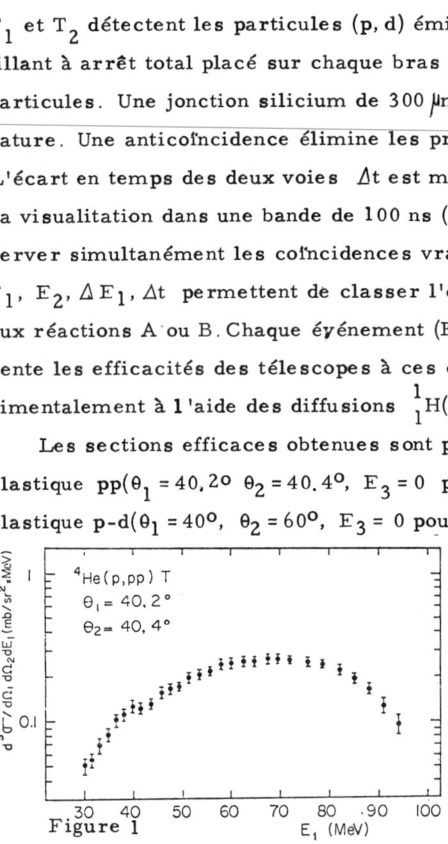

Figure 1

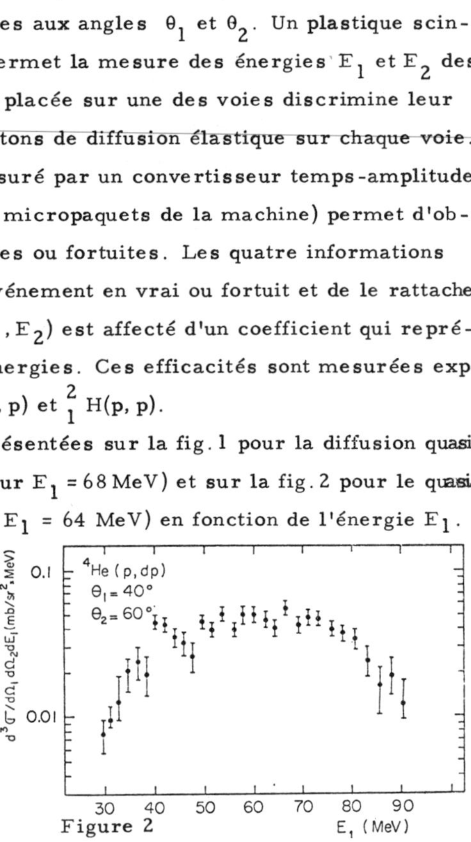

Figure 2

Les barres d'erreurs tiennent compte des erreurs statistiques et des incertitudes sur les efficacités. Il faut ajouter une erreur globale de 5% pour la réaction A et de 7% pour la réaction B, due aux valeurs des sections efficaces des références p-p[réf. 1)] et p-d[réf. 2)].

Les calculs en DWIA de la réaction A à différentes énergies[réf. 3, 4, 5)] ont été faits en utilisant le code de N.S. Chant[réf. 6)]. Les calculs ont été décrits ailleurs par Roos[réf. 7)]. Brièvement la section efficace s'écrit comme

$$\frac{d^3\sigma}{d\Omega_1 d\Omega_2 dE_1} = PSF \times \left(\frac{d\sigma}{d\Omega}\right)^{1/2}_{pp} \times \Phi^2_{DW} \times C$$

où PSF est le facteur cinématique ; $\left(\frac{d\sigma}{d\Omega}\right)^{1/2}_{pp}$ est la section efficace proton proton demi-couche[réf. 8)] pour une énergie incidente inférieure ou égale à 100 MeV et la section efficace expérimentale pp prise dans l'approximation de Stern Chamberlain aux énergies supérieures ; Φ_{DW} est la distribution en moment calculée en DWIA.

Pour décrire le proton dans l'^{4}He, on a utilisé la forme d'Eckart proposée par Lim[réf. 9)]. Les fonctions de distorsion ont été obtenues à partir de potentiels optiques décrivant la diffusion élastique. Sur la fig. 3 est présentée la distribution de moments expérimentale pour un recul nul (aux 6 énergies des réf. 3, 4 et 5). Nous présentons aussi les calculs DWIA pour C=2 qui suppose un facteur spectroscopique égal à 2. Si on prend C=1, ce que suggère le formalisme de la réf. 10, on obtient un excellent accord pour les 5 premiers points. Les effets de distorsion commencent à devenir importants en dessous de 100 MeV ce qui justifie le traitement DWIA au dessus de cette énergie.

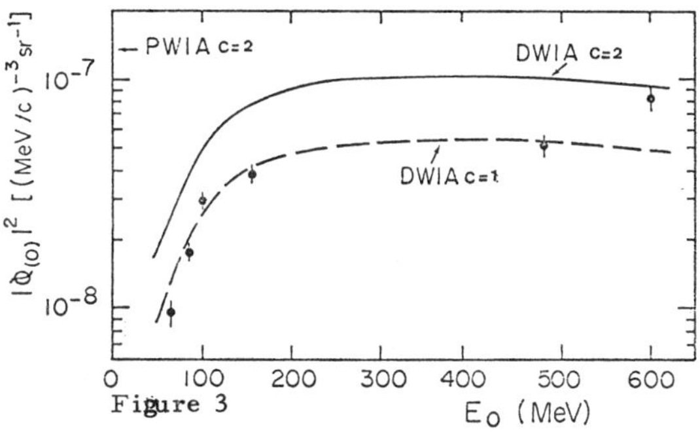

Figure 3

Un traitement plus élaboré des résultats (y compris la réaction B) utilisant des fonctions d'ondes plus sophistiquées est en cours.

BIBLIOGRAPHIE

1) C. Caverzasio, A. Michalowicz, J. Phys. Rad., 21(1960) 314
2) K. Kuroda, Thèse d'Etat, Orsay (1966)
3) M. Tyren et al. Nuclear Physics 79 (1966) 321
4) C. F. Perdrisat et al., Phys. Rev. 187, 1201(1969)
5) H. G. Pugh et al., Phys. Lett. 46B (1973) 192
6) N. S. Chant, private communication
7) P. G. Roos, Phys. Rev. C, to be published
8) E. F. Redish, G. J. Stephenson and G. M. Lerner, Phys. Rev. C2(1970)1665
9) T. K. Lim, Phys. Lett. 44B, 341(1973)
10) G. Jacob and TH. A J. Maris, Rev. of Mod. Phys. vol38, n°1(1966)121

† Adresse permanente: University of Maryland (U.S.A.)
†† Université de Kyoto (Japon)

HOW ACCURATELY CAN r_{nn} BE DETERMINED FROM QUASIFREE SCATTERING?

D. Vranić, I. Šlaus, G. Paić, P. Tomaš
Institute "Rudjer Bošković", Zagreb, Yugoslavia

Neutron-neutron effective-range parameters have been determined from FSI in multiparticle reactions: $a_{nn} = -16.4 \pm 0.9$ fm, $r_{nn} = 3.4 \pm 0.6$ fm[1]. It has been shown[2] that the extraction of nuclear effective-range parameters from the proton-proton interaction is subject to large uncertainties. Therefore, the determination of violation of charge independence depends on the knowledge of a_{nn} and r_{nn}. The difference between the binding energies of ^{3}H and ^{3}He indicates[3] that charge symmetry could also be violated. All this makes more accurate determination of a_{nn} and r_{nn} imperative. Therefore, the Ebenhöh code has been checked for sensitivity to variations of both a_{nn} and r_{nn}.

Comparisons of the sensitivity have been made in the region of the phase space usually referred to as the "quasifree region", namely, for kinematic conditions under which the proton has a very low momentum.

The calculation has proved that

i) the cross section in the mentioned region is only slightly influenced by the change of a_{nn}.

ii) The variation of r_{nn} has a much stronger influence on the cross section than the change of a_{nn}.

iii) It has been demonstrated that the QFS cross section is insensitive to off-energy-shell variations, while the FSI region could be more sensitive to it[4].

The predominance of the influence of r_{nn} over the region of quasifree scattering makes it possible to use that part of the spectrum for the determination of r_{nn}.

Figure 1 shows the variation of the breakup cross section and of the n-n elastic cross section with varying r_{nn}.

An interesting result is that $\Delta\sigma_{BU}/\Delta r_{nn} > 1$ beyond $E_{nn} = 7$ MeV. Besides, an amplification over elastic scattering is present, i.e., $\Delta\sigma_{BU}/\Delta\sigma_{nn} > 1$.

The fact that the sensitivity increases with increasing relative n-n energy cannot be fully exploited, because going higher than 10 MeV means running into important p-wave contributions, and higher-order terms in the effective-range expansion appear. The determination of the value of r_{nn} by comparison of theoretical and experimental breakup spectra may follow two different methods.

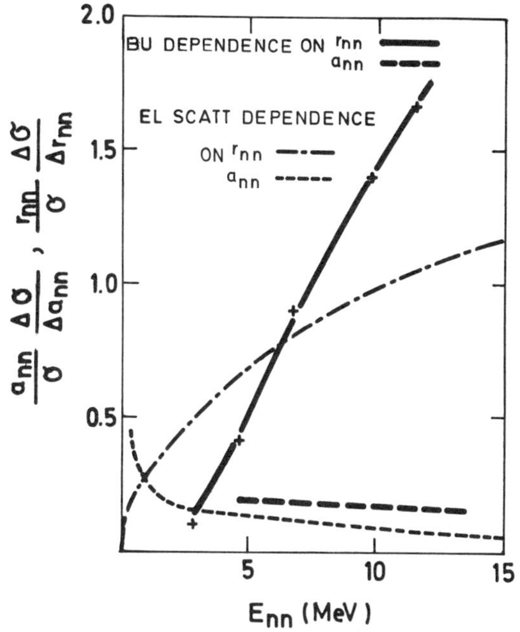

FIGURE 1. Sensitivity of the cross section to scattering parameters.

1) <u>The absolute cross section measurement</u>. In this case the relation between the error in the absolute cross section and the error in the determination of r_{nn} measured at E_n = = 30 MeV for the kinematic situation $\theta_3 = \theta_4 = 40^o$, is

$$\Delta r_{nn} = 1/1.5\ \Delta\sigma_{abs}$$

2) <u>The relative cross section measurement</u>. In this case the different dependence of the cross section for different E_{nn} makes possible the determination of r_{nn} using the quotient of the cross section measured under two different kinematical conditions, e.g., at E_{inc}= = 30 MeV for $\theta_3 = \theta_4 = 40^o$ and $\theta_3 =$ $= \theta_4 = 30^o$ the error in r_{nn} is essentially connected only with the statistical error in the measurement in the following way and thus it dampens the possible absolute error in the theory, which is $\approx$1-2%. The error in that case is:

$$\Delta r_{nn} \approx 2\ \Delta\sigma_{statistical}$$

A 2% determination of r_{nn} by the second method requires

1) Neutron energy 30 MeV, $E_{inc} \approx 1$ MeV.
2) Neutron flux on the target $10^7\ s^{-1} - 10^8\ s^{-1}$.
3) Angular resolution $\Delta\theta = 1^o$.
4) Efficiency of the counter 10-20%, $\Delta E_n \approx 1$ MeV.

The duration of the experiment under these conditions is estimated to be 1000 hours.

REFERENCES

1) In "Few Particle Problems in the Nuclear Interaction" (eds. I. Slaus et al.), North Holland Co., 1972
2) P. Sauer, Phys. Rev. Lett. 32 (1974) 626
3) K. Okamoto and C. Pask, Ann. of Phys. 68 (1971) 18
4) M. Haftel et al., to be published

MESURE DE LA COINCIDENCE NEUTRON-PROTON DANS LA REACTION p + D→p + p + n A BASSE ENERGIE ET SA RELATION A LA DIFFUSION QUASI-LIBRE

F. Coçu, G. Ambrosino, D. Guerreau[+], Service de Physique Nucléaire,
Centre d'Etudes de Bruyères-le-Châtel, B.P n° 61, 92120 Montrouge, France

Des mesures récentes[1,2,3] sur la rupture du deutéron par un faisceau de protons à des énergies incidentes de l'ordre de 10 MeV ont été effectuées avec un dispositif expérimental déterminant complètement la cinématique des trois nucléons libérés . Ces expériences ont souvent été interprétées à l'aide de l'approximation d'impulsion qui reproduit l'aspect général du phénomène, bien que dans le domaine des petites énergies, son application ne soit pas justifiée . Nous nous sommes proposés d'étudier le mécanisme de la réaction à trois corps p + D→p + p + n dans une géométrie coplanaire, le proton étant détecté dans une direction faisant 19°5 avec l'axe d'incidence , l'angle pour le neutron étant de -30° . Les coïncidences proton-neutron étaient enregistrées sous la forme d'un spectre de temps de vol déclenché par l'impulsion associée au proton . Nous avons eu l'avantage d'employer une cible gazeuse . L'énergie incidente a varié de 6,8 à 11,8 MeV par pas de 1 MeV . Les résultats expérimentaux sont reproduits sur la figure 1 , ils représentent en fonction de l'énergie du proton détecté, le nombre de coïncidences avec le neutron exprimés en section efficace différentielle .

Nous nous sommes efforcés de retrouver cette section efficace par un calcul dans lequel nous avons utilisé l'approximation de Born au premier ordre, en ondes planes, le potentiel coulombien ayant été négligé . Le potentiel nucléon-nucléon est représenté par : $V_1 f(r) + V_2 f(r) \vec{\sigma}_i \vec{\sigma}_j$ ou la fonction f (r) est prise sous la forme d'une gaussienne . Les différents coefficients sont choisis afin de reproduire les longueurs de diffusion des deux états de spin . Nous avons imposé en outre la conservation de l'isospin et respecté l'antisymétrie complète des fonctions d'ondes initiales et finales. De ce calcul nous tirons quatre éléments de matrice qui décrivent l'ensemble de l'interaction . Physiquement deux des éléments de matrice correspondent aux cas où le neutron est spectateur, et les deux autres représentent le cas où le proton est spectateur . Les probabilités de transition pour chacun des trois états de spin du système final sont représentées sur la figure 2 . Les sections efficaces obtenues par cette méthode, multipliées par un facteur de normalisation voisin de 1 sont indiquées sur la figure 1 et reproduisent convenablement les résultats expérimentaux, particulièrement dans la région centrale . On remarquera la valeur du coefficient de correction voisin de 1 qui paraît indiquer que la description utilisée représente l'essentiel du phénomène. Le processus dominant est donc la diffusion quasi-libre de deux nucléons. Parmi les épreuves auxquelles a été soumise cette description, nous distinguons l'application que nous en avons faite lorsqu'on compare dans la même géométrie les coïncidences proton-proton et les coïncidences neutron-proton . D'après Valkovic[1] le rapport expérimental de la section efficace n p au pic à celle de la section efficace p p au pic est voisin de 2 dans une gamme d'énergie de 6 à 13 MeV, pour un couple d'angles de détection symétriques. Notre calcul donne avec une bonne approximation la valeur de ce rapport $\left(\frac{\sigma_{np}}{\sigma_{pp}}\right)_{pic} \simeq \frac{3 V_t^2 + V_s^2}{4 V_s^2}$ Pour les valeurs $V_t = -60$ MeV $V_s = -41$ MeV qui reproduisent avec une bonne approximation les sections efficaces expérimentales ce rapport est voisin de 1,85 . Compte tenu de l'incertitude sur les longueurs de diffusions le résultat est très satisfaisant .

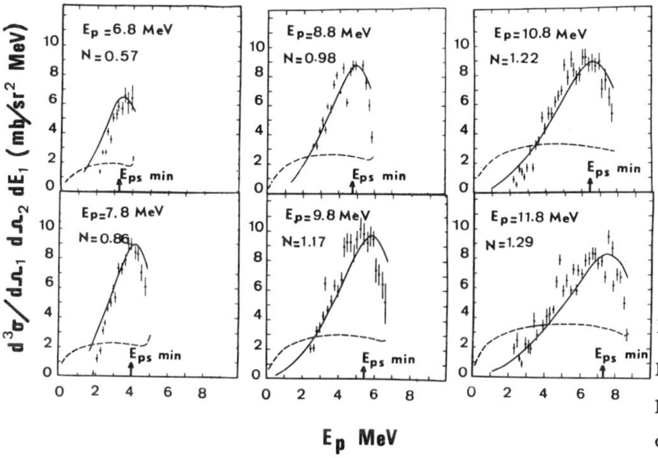

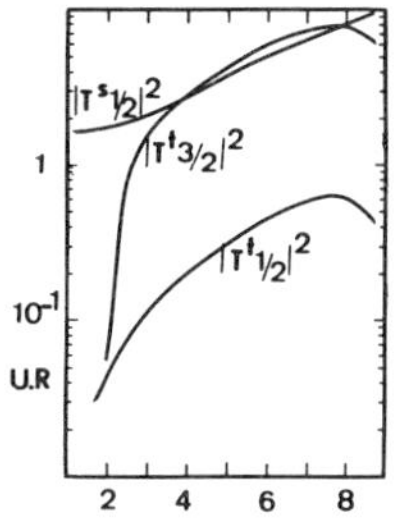

FIGURE 2
Evaluation des carrés des probabilités de transitions pour les trois possibilités différentes de spin

FIGURE 1
Spectres de coïncidence proton-neutron pour E_p=6,8 - 11,8 MeV . Le facteur de normalisation N est défini comme le rapport de la section efficace calculée à la section efficace expérimentale au maximum du spectre .

BIBLIOGRAPHIE

1) V. Valkovic , D. Rendic, V.A. Otte, W. Von Witsch et G.C. Phillips,Nucl. Phys. A 166 (1971) 547

2) V. Valkovic, I. Duck, W.E. Sweeney et G.C. Phillips,Nucl. Phys. A 183 (1972) 126

3) E.L. Petersen, R.G. Allas, R.O. Bondelid, A.G. Pieper et R.G. Theus, Phys. Lett. 31 B (1972) 209

Ce travail a été fait au Centre d'Etudes de Bruyères-le-Châtel B.P n° 61, 92120 Montrouge, France .

CASSURE DU DEUTERON DANS LA REACTION d+D → n+p+d

F. Coçu, G. Ambrosino, D. Guerreau[+], Service de Physique Nucléaire,
Centre d'Etudes de Bruyères-le-Châtel, B.P n° 61, 92120 Montrouge, France

Le processus de quasi diffusion libre est étudié dans la réaction d+d → n+p+d pour des énergies de deuterons incidents allant de 6,7 MeV à 11,7 MeV par pas de 1 MeV. La détection simultanée du proton et du deuteron a été faite en coïncidence avec le neutron associé pour le couple d'angles $\Theta_p = \Theta_d = 19°,5$, $\Theta_n = -30°$.

Le faisceau de deuterons produit par l'accélérateur Van de Graaff 12 MeV de Bruyères-le-Châtel avec une intensité de 6 nanoampères bombarde une cible gazeuse de deuterium. La mesure simultanée de la différence de temps de vol Δt entre le neutron et la particule chargée détectée et de l'énergie de cette dernière nous permet de séparer ^1_1H et ^2_1H par analyse biparamétrique. L'estimation relative du nombre de protons et de deuterons enregistrés par bandes d'énergie d'environ 0,2 MeV a été effectuée par un lissage des deux pics enregistrés à l'aide de deux gaussiennes de largeur peu différente.

Les résultats expérimentaux sont présentés ci-contre et une analyse théorique basée sur l'approximation de Born est actuellement en cours.

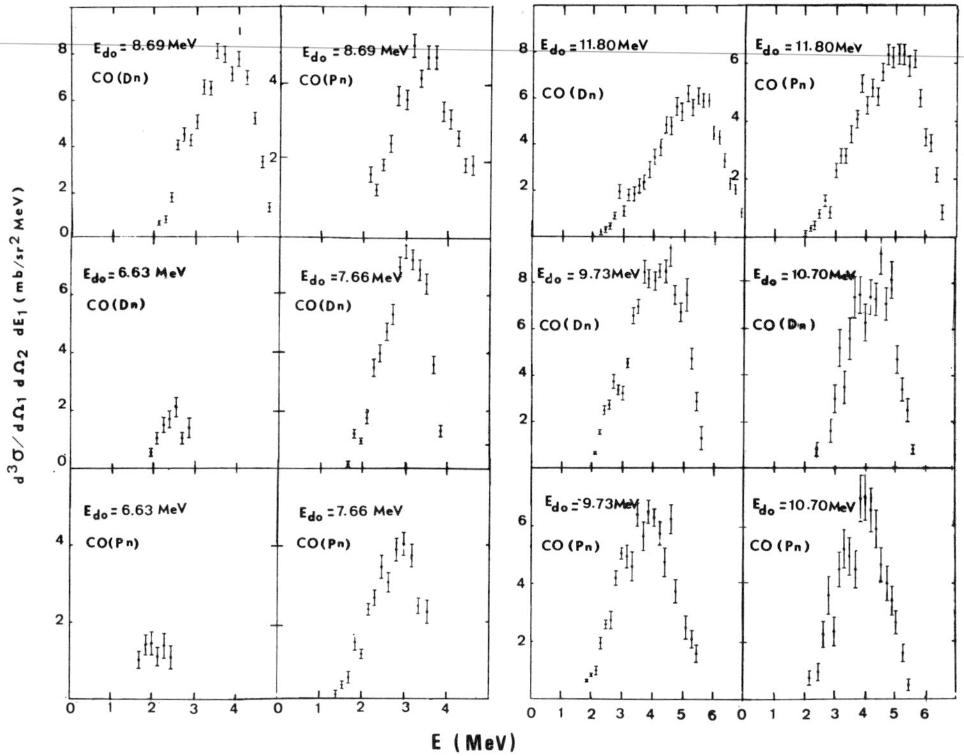

QUASI-FREE SCATTERING IN THE $^{6}Li(e,e'\alpha)^{2}H$ AND $^{6}Li(e,e'd)^{4}He$ REACTIONS

M. Toyama
Department of Physics, Kyoto Sangyo University, Kyoto, Japan
Y. Sakamoto
Institüt fur Kernphysik der Universität Frankfurt am Main
Frankfurt am Main, Germany

In the quasi-free scattering (QFS) model, the cross sections for the e-α and e-d QFS are given

$$d^3\sigma/d\Omega_e d\Omega_{\alpha(d)} dE_e = K\, n_{\alpha(d)}\, (d\sigma/d\Omega)_{e-\alpha(d)}\, |\phi_{\alpha-d}(\Delta k)|^2, \qquad (1)$$

where n_α and n_d are the so called effective numbers of α and d clusters in the nucleus, and $\phi_{\alpha-d}(\Delta k)$ is the overlap integral between the α-d intercluster and scattering-state wave functions. The plane wave QFS model predicts the ratio of $n_d/n_\alpha = 1.0$, in contrast with that of 0.53[1] extracted from the observed cross sections for the e-α and e-d QFS on ^{6}Li.

Kinematical conditions chosen[1] for the e-α and e-d QFS on ^{6}Li give different relative energies between the α particle and deuteron in the final state. The ratio different from unity may be explained with the QFS model only if the final-state interactions act to hinder the overlap integral in the cross section for the e-d QFS much stronger than that for the e-α QFS. However, the intercluster wave function corresponding to 2S-state relative motion has an oscillatory structure in the α-d configuration of ^{6}Li. This oscillatory structure cancels the contribution to the overlap integral from the interior region for the kinematical conditions corresponding to QFS. Hence, the optical absorptions due to the final-state interactions do not strongly affect the cross sections.

In the microscopic model, the factors n_α and n_d depend on the intracluster wave functions[2],

$$n_\alpha = |\gamma_d\, F_\alpha(q)/F^f_\alpha(q)|^2 \quad \text{and} \quad n_d = |\gamma_\alpha\, F_d(q)/F^f_d(q)|^2, \qquad (2)$$

where

$$\gamma_d = \langle \psi_d | \phi_d \rangle, \quad F_\alpha(q) = \langle \psi_\alpha | e^{i\mathbf{q}\cdot\mathbf{r}_{\alpha j}} | \phi_\alpha \rangle \quad \text{and} \quad F^f_\alpha(q) = \langle \psi_\alpha | e^{i\mathbf{q}\cdot\mathbf{r}_{\alpha j}} | \psi_\alpha \rangle$$

for n_α, and similar overlaps for n_d. Here ψ_α and ψ_d are free α-particle and deuteron wave functions; ϕ_α and ϕ_d are α and d intracluster wave functions in ^{6}Li, and q is the momentum transferred to scattered electrons.

When the α and d intracluster wave functions are constructed respectively with S- and P-state single-particle harmonic-oscillator functions whose parameters are fixed to reproduce[3] the form factor of ^{6}Li obtained by analysing electron scattering data, the ratio predicted with eq.(2) is

$$n_d\Big|_{\substack{q=1.33\ \mathrm{fm}^{-1}\\ \Delta k=0}} \Big/ \; n_\alpha\Big|_{\substack{q=1.49\ \mathrm{fm}^{-1}\\ \Delta k=0}} = 1.2.$$

To reproduce the observed ratio, it should be assumed that the intracluster wave functions are considerably different from corresponding free particle wave functions, and also that the structure of the α intracluster wave function differs completely from that of the d cluster. For example, the ratio

$$n_d\Big|_{\substack{q=1.33\ \mathrm{fm}^{-1}\\ \Delta k=0}} \Big/ \; n_\alpha\Big|_{\substack{q=1.49\ \mathrm{fm}^{-1}\\ \Delta k=0}} = 0.54$$

is obtained when the parameter for the P-shell nucleons, i.e. for the d cluster is changed so that its r.m.s. radius is about two times as large as that found by analysing electron scattering data, while the parameter for the S-shell, i.e. for the α cluster is kept to be the same as that found by analysing electron scattering data.

The intercluster wave function constructed with the harmonic-oscillator functions is matched with a correct asymptotic form. The 2S-state square-well function is also used for describing the intercluster motion. The radius of square-well potential is taken to be 3.5 fm. The calculated r.m.s. radius of ^{6}Li is 2.97 and 2.67 fm for the modified harmonic-oscillator and square-well functions, together with the intracluster wave functions which predict the above ratio of 0.54.

FIGURE 1

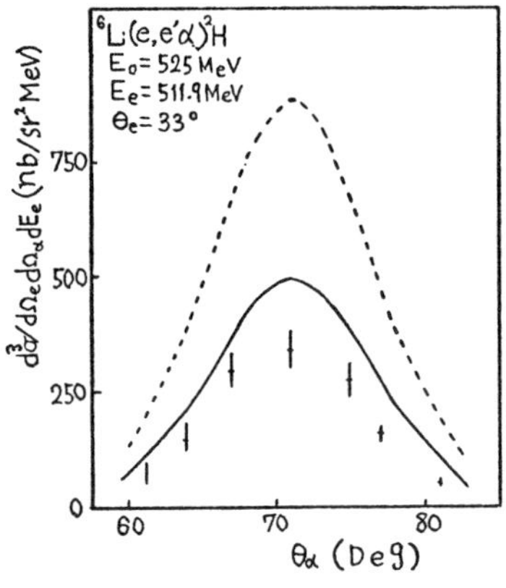

FIGURE 2

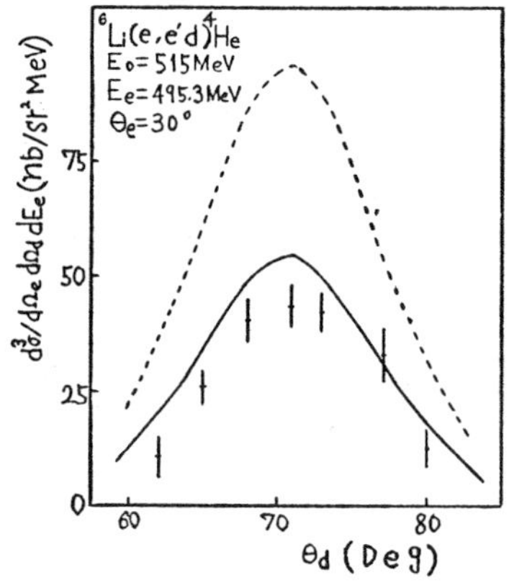

The cross sections for the QFS in the ^{6}Li(e,e'α)^{2}H and ^{6}Li(e,e'd) ^{4}He reactions are shown in Figures 1 and 2, where the solid and dotted curves are calculated with the square-well and modified harmonic-oscillator intercluster wave functions.

REFERENCES

1) J. Julien et al., to be published
2) Y. Sakamoto, Phys. Rev. 134 (1964) B1211
3) C. Ciofi Degli Atti, Phys. Rev. 175 (1978) 1256

PROTON INDUCED ^{3}He BREAK-UP AT 156 MeV

J. P. Didelez, R. Frascaria, N. Fujiwara, I. D. Goldman [+], E. Hourany [++],
H. Nakamura-Yokota [+++], F. Reide and T. Yuasa.
Institut de Physique Nucléaire, Université de Paris-Sud, 91406 Orsay, France.

A few years ago, the simultaneous measurement of the D(p, 2p)n and D(p, pn)p cross sections were performed at Orsay, with the 156 MeV proton beam. It allowed a direct comparison between p-p and p-n data in the quasi-free scattering (QFS) and in the final state interaction (FSI) kinematic regions and the cross sections obtained in this study were compared with some theoretical calculations [1]. We desired to extend such studies to the four-nucleon system by means of proton induced ^{3}He break-up. Recently, ^{3}He (p, 2p)d and d^{*} (the $^{1}S_0$ virtual state of the deuteron) and ^{3}He (p, pd)p reactions at 156 MeV have been studied at Orsay for small recoil momentum of the spectator particle [2]. It was interesting to investigate the QFS in the ^{3}He (p, pn)2p reaction leading to a p-p quasi-bound state and to compare the data with the ^{3}He (p, 2p)d and d^{*} reaction data in a wide range of the recoil momentum of the spectator particle. In addition, we have attempted to detect a possible virtual three-nucleon scattering state arising from FSI between three-nucleons.

A 156 MeV proton beam and a liquid ^{3}He target were used. The two emerging particles were detected in coincidence using counter telescopes. Typical spectra are shown in Figs. 1a, 1b and 2. The cross sections for the (p, 2p), (p, pn) and (p, pd) reactions are determined using the p-H and p-D cross sections obtained in coincidence. Figs. 1a and 1b show the energy-sharing spectra, $d^3\sigma/d\Omega_1 d\Omega_2 dE_1$ for the ^{3}He (p, 2p) d and d^{*} reactions and for the ^{3}He (p, pn)2p reaction at $\theta_{p1} = 40^{o}$ and θ_{p2} (or θ_n) = -44°. This angular pair was chosen to observe p-p and p-n QFS process. The solid curves E_3, E_{13} and E_{23} correspond respectively to the kinematic energy of the recoil particle (or the quasi-particle) and the relative energies between a detected particle and the recoil particle (or the quasi-particle). The data for the ^{3}He (p, pn)2p reaction indicate a dominant knockout mechanism when the relative energy between the two residual nucleons is small as for the ^{3}He (p, 2p) d^{*} reaction [2]. The shapes of the (p, 2p) and (p, pn) QFS spectra almost coincide with each other in the wide energy region of the recoil particle ($0 \sim 10$ MeV). The ratio of these two QFS peak cross sections divided by each free two-nucleon cross section $(d^3\sigma(p,2p)/d\sigma_{pp})/(d^3\sigma(p,pn)/d\sigma_{pn})$ is found to be 2.0 ± 0.1. This ratio is equal to the ratio of spin-isospin factors of p + pn ($S = {}^{1}_{0}$ $T = {}^{0}_{1}$) and n + 2p ($S = 0$ $T=1$)

configurations in the ^{3}He nucleus. The dotted curve in Fig. 1a is obtained by DWIA calculation using a Fabre wave-function for the ^{3}He [2] and corrected at the peak, with the ratio $(d + d^* / d)_{exp}$.

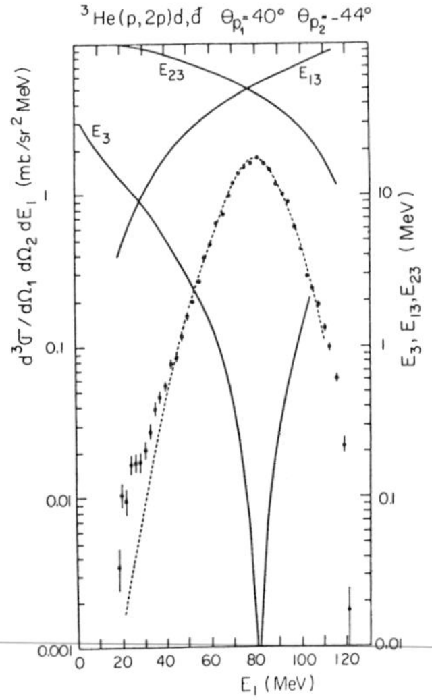

The (p, 2p) QFS

Fig. 1a

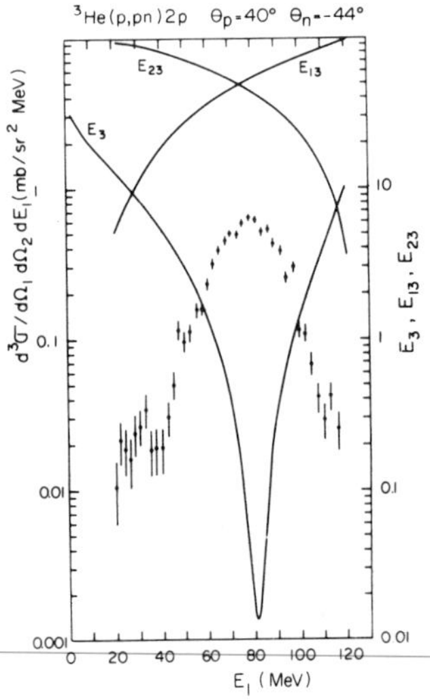

The (p, pn)QFS

Fig. 1b

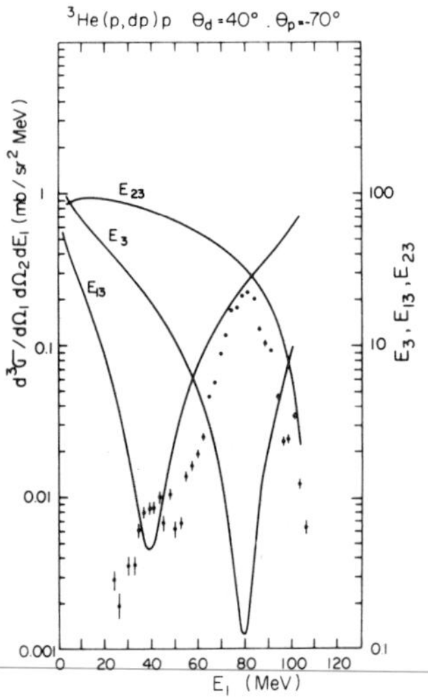

The (p, dp)QFS and d-p FSI

Fig. 2

Fig. 2 shows the energy-sharing spectrum for the ^{3}He (p, dp)p reaction at $\theta_d = 40^\circ$ and $\theta_p = -70^\circ$. This angular pair was chosen to observe p-d QFS process and possible p-d FSI process. The peak around the minimum kinematical energy of the recoil proton arises from a p-d QFS process. One notices that the spectrum has some sort of a bump on both sides of this QFS peak. The peak around $E_d \sim 38$ MeV, $E_{dp} \sim 0.5$ MeV is probably due to a contribution from the p-d FSI process.

References :

[1] J. P. Didelez, I. D. Goldman, E. Hourany, H. Nakamura, F. Reide and T. Yuasa (Proc. Europ. Conf. Phys.) J. de Phys. 33, 129 (1972) ;
Proc. Int. Conf. Los Angeles (North-Holland, Amsterdam, 1973) p. 483 ;
Proc. Int. Conf. Munich (North-Holland, Amsterdam, 1973) vol. 1, p. 400 ;
Phys. Rev. (to be published)

[2] R. Frascaria, V. Comparat, N. Marty, M. Morlet, A. Willis and N. Willis, Nucl. Phys. 178, 307 (1971)
R. Frascaria, Thesis Orsay 1973.

Quasi-Free Contributions in the p, d, and ^{16}O Induced Reactions on ^{6}Li

Đ. Miljanić*, T. Zabel and G. C. Phillips
T. W. Bonner Nuclear Laboratories**, Rice University, Houston, Texas 77001, USA

Prominent mechanisms of three-body reactions are quasi two body processes, namely final state interactions (FSI) and quasi-free reactions (QFR). At low bombarding energies FSI predominate. However, as we have recently shown[1)], if kinematical conditions are properly chosen, QFR could be observed even at energies a few MeV's above the reaction threshold. In order to get further insight into the significance of QFR at low energies we studied three body reactions on ^{6}Li using p, d, and ^{16}O beams. The ^{6}Li nucleus was chosen because of its low binding energy, its prominent d-α cluster structure and because of the S-state of d-α relative motion, a consequence of which is the peaking of the QFR cross section for spectator momentum $q_S = 0$.

Experiments were performed using isotopically separated ^{6}Li targets and p, d, and $^{16}O(6^+)$ beams from the Rice University tandem Van de Graaff accelerator. ΔE-E and/or Et^2 methods were used for identification of the coincident particles. Standard electronics together with an IBM 1800 computer analyzer system were used for collection, storage and analysis of the data. Measurements were performed for kinematical conditions which emphasized at least one of the following QFR: pd-pd, pα-pα, dd-dd, dα-dα, dd-pt, $^{16}O\alpha$-$^{16}O\alpha$, and ^{16}Od-$^{14}N\alpha$. Data were also taken simultaneously for kinematic conditions where contributions from these processes should be negligible. Absolute cross sections were determined for the p and d reactions by monitoring elastic scattering and normalizing to the known cross section[2,3)].

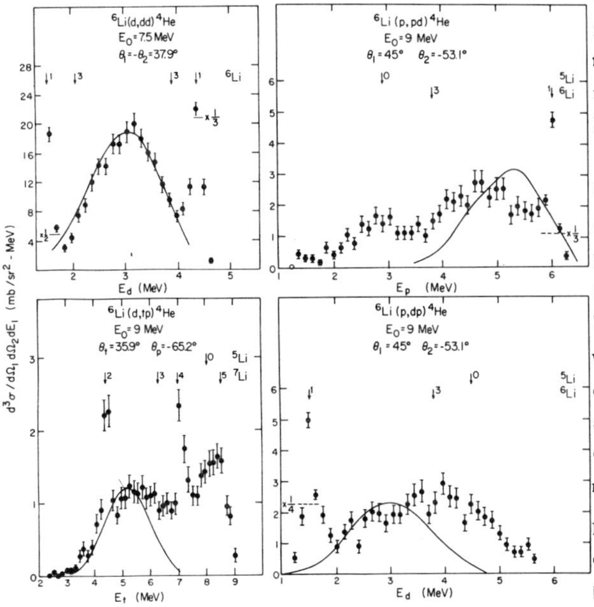

fig. 1

Typical coincidence spectra from proton and deuteron induced reactions are shown in fig. 1 as projections on the energy axis of the particle detected at angle θ_1. The vertical arrows show positions where the contributions from sequential reactions through known states of the nuclei involved are expected. For all these cases the minimum energy of the undetected particle is below 10 keV. The curves represent plane wave impulse approximation calculations (PWIA). In these calculations the cluster wave function ψ_C from ref. 4 and two body cross section data from refs. 5, 6, and 7 were used. It was necessary to introduce a radial cutoff R = 10 fm in the cluster wave function so that good agreement could be achieved between the shapes of the theoretical curve and experimental results for dd-dd QFR. The same cutoff radius was then used for all other calculations. The theoretical curves from fig. 1 were normalized to the experimental data by a factor of 2 in the deuteron induced reactions and by a factor of 1 in p induced reactions. Shifts of the QFR peaks from the expected positions, as seen in fig. 1 for proton induced QFR, are also seen in other cases for both deuteron and proton induced QFR. pα-pα and dα-dα QFR were also observed.

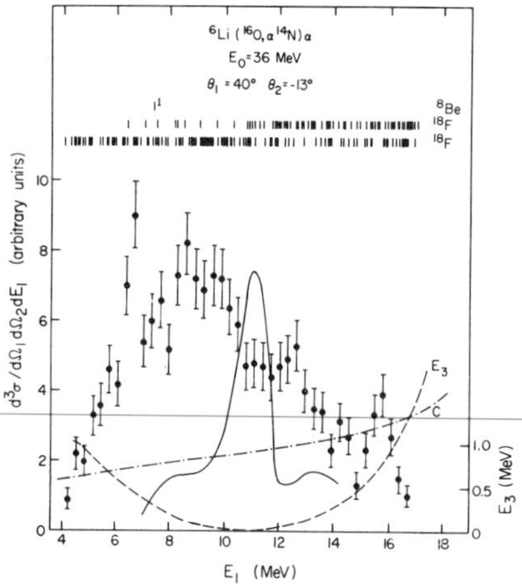

fig. 2

Fig. 2 displays typical data from an ^{16}O induced reaction. The unlabeled curve represents the PWIA calculation, which was calculated in the same way as in the previous case, using cross section data for two body processes from ref. 8. It is hard to establish the presence of QFR in this case, due to the many possible contributions from states of ^{18}F and also the possible shift of the QFR peak. However from a comparison of all 15 measured spectra it was noticed that integrated cross sections for the cases which favor QFR were always bigger than those measured for the conditions where these contributions should be negligible. This possibly could indicate the presence of QFR even in ^{16}O reactions.

From these results it appears that the QFR mechanism has significant contributions even at low energies. The large and non-constant normalization factors, the need for the introduction of a cutoff, and shifts of QFR peaks obviously all join to show that a much more elaborate theory than PWIA should be developed to describe these processes properly.

*On leave from Institute: "Ruđer Bošković," Zagreb, Yugoslavia

**Work supported in part by the U. S. Atomic Energy Commission.

References

1. Đ. Miljanić et al., Phys. Lett., to be published
2. H. G. Bingham et al., Nucl. Phys. A173 (1971) 265
3. W. D. Harrison, Nucl. Phys. A92 (1967) 253, 260
4. Đ. Miljanić et al., Nucl. Phys. A215 (1973) 221
5. A. S. Wilson et al., Nucl. Phys. A126 (1969) 193
6. A. S. Wilson et al., Nucl. Phys. A130 (1969) 624
7. J. E. Brolley et al., Phys. Rev. 107 (1957) 820
8. O. Dietzsch et al., Nucl..Phys. A114 (1968) 330

Quasifree Processes in the Reactions $^3H + {}^3He$ for $E_{^3He} = 50, 65,$ and 78 MeV

R. G. Allas, L. A. Beach, R. O. Bondelid, E. M. Diener* and E. L. Petersen
Naval Research Laboratory, Washington, D. C. 20375

and

P. A. Treado,† J. M. Lambert† and R. A. Moyle
Naval Research Laboratory and
Georgetown University, Washington, D. C. 20007

and

Ivo Slaus
Institute Ruder Bošković, Zagreb, Yugoslavia and
Naval Research Laboratory

In the interaction between ^3H and ^3He, there are a number of three-particle breakup combinations possible. We have performed a series of kinematically complete experiments in which charged particle pairs are detected and conditions are chosen favoring the observation of quasifree processes. Results on the quasifree scattering (QFS) processes $^3H(^3He,d^3He)n$, $^3H(^3He,p^3He)2n$ and the quasifree reaction (QFR) processes $^3H(^3He,dd)d$, $^3H(^3He,pt)d$, $^3H(^3He,dt)p$ and $^3H(^3He,p\alpha)n$ at a ^3He bombarding energy of 50 MeV have been previously reported.[1] Additional data have been obtained for some of these reactions at ^3He bombarding energies of 65 and 78 MeV.

The data for a given reaction are analyzed in terms of the Plane-Wave-Impulse-Approximation (PWIA) and the square of the nucleon-deuteron or nucleon-dinucleon pair cluster momentum wave function $|\Psi(Q_s)|^2$ is extracted. When this extracted square of the tritium momentum wave function is plotted as a function of spectator momentum Q_s, there is consistency between the data obtained from the various angle pairs used in the experiment, and a pronounced enhancement is observed at $Q_s = 0$, thus establishing the importance of the quasifree processes. We observe that the ratio of various quasifree processes involving the same transferred particle and the same virtual decomposition vertex are equal to the ratio of the corresponding free cross sections, thus suggesting that the effects of final state interactions are reasonably small. Observed quasifree enhancements are illustrated in Figs. 1 and 2. In Table 1, we show ratios of the experimental and theoretical cross sections for some of the processes at ^3He bombarding energies of 50, 65 and 78 MeV. The lack of dependence of this ratio as a function of bombarding energy for QFS processes and a decrease in this ratio for the QFR processes suggest the possibility of new outgoing channels available to the final state particles as the beam energy increases.

Table 1

Reaction	Experiment/Theory		
	50 MeV	65 MeV	78 MeV
QFS $^3H(^3He,dh)n$	.24	.20	.20
QFS $^3H(^3He,ph)2n$	.12	.07	.075
QFR $^3H(^3He,dd)d$	.52	.50	.20
QFR $^3H(^3He,dt)p$	1.7	1.0	.56

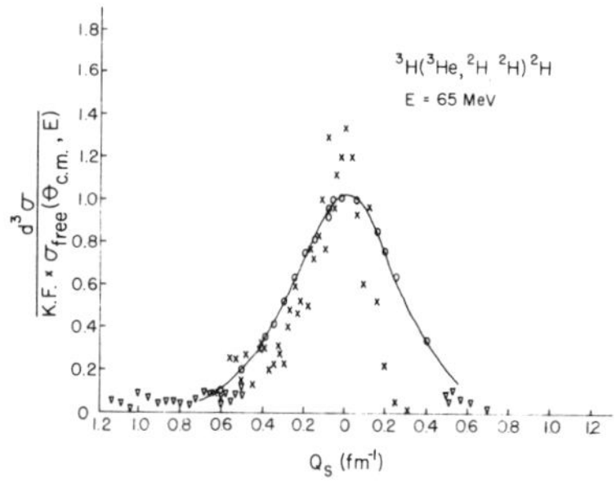

Figure 1. The cluster momentum wave function extracted from the $^3H(^3He,dd)d$ data. The solid curve is the calculated Fourier transform normalized by N = 0.5.

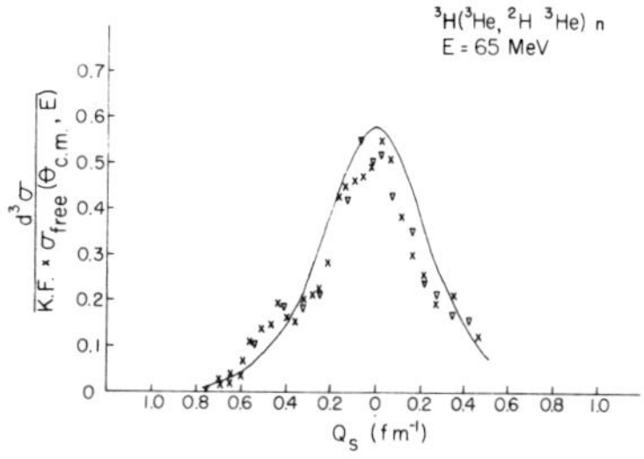

Figure 2. The cluster momentum wave function extracted from the $^3H(^3He,dh)n$ data. The solid curve is the calculated Fourier transform normalized by N = 0.2.

The shapes of the experimental spectra are reasonably well fit with the PWIA using either the overlap between the Irving-Gunn tritium wave function and the Hulthén deuteron wave function, or describing the nucleon-deuteron or nucleon-dinucleon pair relative motion by the expression $u(r) = (e^{-gr} - e^{-br})/r$ where $b = 1.202\ fm^{-1}$, and $g = 0.4478\ fm^{-1}$ corresponding to the neutron separation energy, or $g = 0.5225\ fm^{-1}$ corresponding to the tritium binding energy. The free cross sections change significantly with energy and angle and play an important role in the predicted cross section. It is still possible that comparison of the reactions involving $t \to nd$, $t \to p(2n)$ and $t \to n(np)$ vertices as a function of bombarding energy can provide some insight into the structure of the tritium ground state wave function.

*National Research Council - Naval Research Laboratory Research Associate.

†Work partially supported by National Science Foundation.

1. I. Slaus, R. G. Allas, L. A. Beach, R. O. Bondelid, E. L. Petersen, J. M. Lambert and D. L. Shannon, Phys. Rev. C8, 444 (1973) and R. G. Allas, L. A. Beach, R. O. Bondelid, E. M. Diener, E. L. Petersen, J. M. Lambert, P. A. Treado and I. Slaus, Phys. Rev. C9, 787 (1974).

THE (p,2p) AND (p,pd) REACTIONS ON ^3H AND ^3He AT 45 MeV*

D. I. Bonbright, S. A. Elbakr, A. Houdayer†, C. A. Miller††, D. J. Roberts, E.S.Y. Tin†††, W.T.H. van Oers and J. W. Watson
Cyclotron Laboratory, Department of Physics, University of Manitoba, Winnipeg, Manitoba R3T 2N2

The (p,2p) and (p,pd) reactions on ^3H and ^3He have been studied at 45 MeV in a symmetric and asymmetric coplanar geometry at many pairs of angles favoring quasi-free p-p and p-d scattering. The targets consisted of tritium absorbed in titanium on a gold backing and helium-3 gas of 99.9% purity, respectively. Charged reaction products were observed using two counter telescopes each consisting of a 200μ surface barrier ΔE detector, a 1 mm surface barrier first E detector, a 5 mm Si(Li) second E detector, and a VETO detector.

Momentum distributions of the d and [nn] systems in ^3H and the d and d* systems in ^3He have been extracted from the data (Fig. 1) within the framework of the plane-wave impulse approximation. With [nn] and d* are denoted a n-n pair and a n-p pair with low energy for their relative motion, respectively. The following observations can be made:

1) Good agreement can be obtained for the shape of the theoretical momentum distributions if the overlap integral is calculated using an Irving-Gunn wave function for the trion. However, a cut-off radius is found to be necessary for the $^3H(p,pd)n$ and $^3He(p,2p)d$ reactions. The cut-off radius for the former reaction (4.0 fm) is physically barely acceptable indicating the extreme importance of multiple scattering effects in the low energy region.

2) The PWIA prediction over estimates the differential cross section by a large factor. This factor decreases monotonically for increasing incident energy [1].

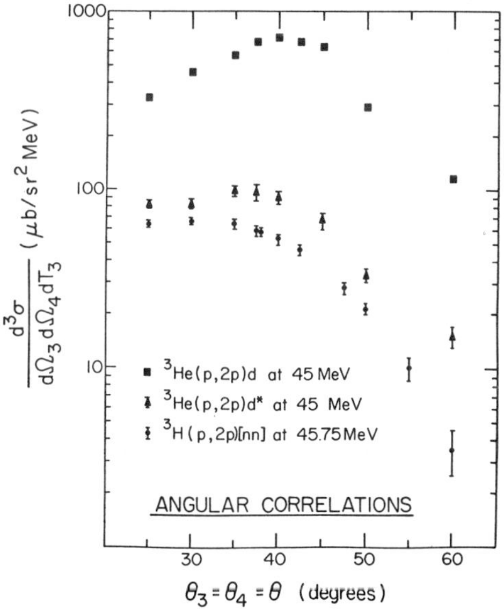

Fig. 1

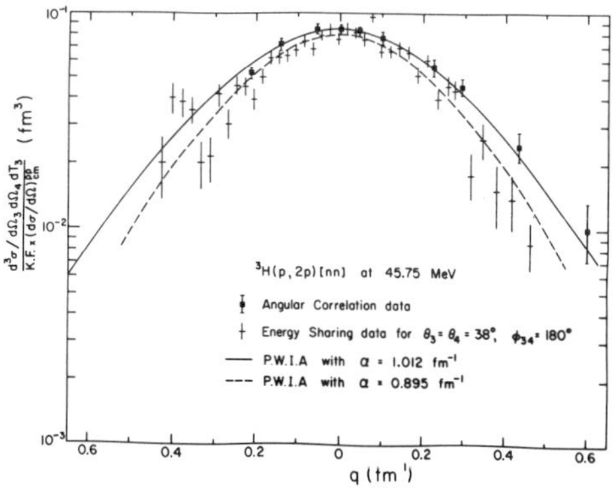

Fig. 2

3. There are noticeable differences between the angular correlation and energy sharing distributions for the larger spectator momenta (Fig. 2), requiring further studies at higher incident energies.
4. Energy shifts are observed in the positions of the quasi-free scattering peaks for the ^{3}He(p,pd)p and ^{3}H(p,pd)n reactions. Similar shifts were observed for the ^{3}He(p,pd)p reaction at 35 MeV [2]. These shifts can be understood qualitatively as due to a contribution from the neutron pick-up reaction [3].
5. The Watson-Migdal model fails to explain the final-state interaction between the unobserved pair of nucleons in both the ^{3}H(p,2p)[nn] and ^{3}He(p,2p)d* reactions.
6. Comparison of the continua for the ^{3}He(p,2p)np and ^{3}H(p,2p)nn reactions with four-body differential phase space distributions [4] shows no evidence for a resonance in the three-nucleon system. However, the ^{3}H(p,2p)nn continua reveal evidence for hitherto unobserved pseudo two-body processes, namely the p-d* quasi-free scattering process and the p + t → d* + d* reaction.

References

[1] H. G. Pugh, P. G. Roos, A. A. Cowley, V.K.C. Cheng, and R. Woody III, Phys. Letters 46B, 192 (1973);
A. A. Cowley, P. G. Roos, H. G. Pugh, V.K.C. Cheng, and R. Woody III, Nucl. Phys. A220, 429 (1974);
R. Frascaria, V. Comparat, N. Marty, M. Morlet, A. Willis and N. Willis, Nucl. Phys. A178, 307 (1971);
R. Frascaria, V. Comparat, N. Marty, M. Morlet and A. Willis in Few Particle Problems in the Nuclear Interaction, edited by I. Slaus, S. A. Moszkowski, R. P. Haddock and W.T.H. van Oers (North-Holland Publishing Company, 1972) p. 628.

[2] I. Slaus, M. B. Epstein, G. Paić, J. R. Richardson, D. L. Shannon, J. W. Verba, H. H. Forster, C. C. Kim, D. Y. Park and L. C. Welch, Phys. Rev. Letters 27, 751 (1971).

[3] M. B. Epstein, I. Slaus, D. L. Shannon, J. R. Richardson, J. W. Verba, H. H. Forster, C. C. Kim, and D. Y. Park, Phys. Letters 36B, 305 (1972).

[4] W.T.H. van Oers, M. Jain, and S. N. Bunker, Nucl. Instr. 112, 405 (1973).

* Work supported in part by the Atomic Energy Control Board of Canada.

† Present address: Foster Radiation Laboratory, McGill University, Montréal, Québec.

†† Present address: Nuclear Research Center, University of Alberta, Edmonton, Alberta T6G 2J1.

††† Present address: Power Projects, Atomic Energy of Canada Limited, Mississauga, Ontario L5K 1B2.

IX

MODELES HADRONIQUES COMPOSÉS
STRUCTURE SUBNUCLEONIQUE

COMPOSITE HADRON MODELS
SUBNUCLEON STRUCTURE

COMPOSITE MODELS OF HADRONS AND LARGE TRANSVERSE PROCESSES*

MODELES COMPOSES DES HADRONS ET PROCESSUS A UNE GRANDE IMPULSION TRANSVERSALE

Stanley J. Brodsky

Stanford Linear Accelerator Center, Stanford University, Stanford, California 94305

I. INTRODUCTION

Throughout the history of physics, our knowledge of the underlying structure of matter at smaller and smaller distances has progressed by the scattering of particles to ever larger momentum transfer. Now we are at the stage where the internal structure of the hadrons is being resolved through measurements of deep inelastic scattering of electrons, muons, and neutrinos, and, most recently, through large transverse momentum processes involving the collisions of hadrons. The most recent measurements, at the CERN-ISR p-p storage rings[1] and the proton accelerator at FNAL[2] have involved particle production at transverse momentums $p_\perp$ up to 9 GeV, corresponding to a resolving length $\ell \sim p_\perp^{-1}$ of order 10^{-15} cm. These measurements have shown that the cross section for hadron production at large transverse momentum is much larger than what would be extrapolated from the fast exponential fall-off of low momentum transfer reactions. Moreover, a striking pattern of scaling laws observed in both exclusive and inclusive processes at large $p_\perp$ has given support to the quark-parton field-theoretic models of composite hadrons, and some basic properties of the underlying dynamics of the constituents are now beginning to emerge.

II. THE QUARK-PARTON MODEL

The evidence for a composite model of the hadrons based on SU(3)-quark degrees of freedom has been steadily accumulating. The quarks of Gell-Mann and Zweig seem to be the underlying common denominator of a whole range of hadronic phenomena:

(1) The SU(6) spectroscopy - which associates baryon states with bound states of three spin 1/2 quarks and mesonic states with $q\bar{q}$ bound states - is extraordinarily successful - although not quite on par with the Mendeléeff table of the elements. There do remain classes of missing baryonic resonance states, and some disarray among the mesons.

(2) The quantitative features and symmetry properties of the electromagnetic and weak interactions of the hadrons, their decays and production amplitudes, are beautifully summarized by Gell-Mann's "current algebra"; the weak and electromagnetic currents are constructed exactly as if the fundamental carriers of the charges within the hadrons are elementary Dirac-spin 1/2 quarks. The vector nature of the electromagnetic current and vector-axial nature of the weak interactions leads to a closed SU(6) × SU(6) symmetry algebra among the currents. Recently,

* Work supported by the U. S. Atomic Energy Commission.

extremely elegant field theories (developed by Weinberg and Salam) which can incorporate the quark current algebra, have been shown to lead to a renormalizable unified theory of the weak and electromagnetic interactions of the leptons and hadrons. Thus, for the first time we can contemplate real theories of the weak interactions which are a natural extension of quantum electrodynamics.

(3) The most dramatic evidence for an elementary-constituent structure of the hadron comes from the scale-independent behavior of deep inelastic electron (and muon) nucleon scattering[3] measured at SLAC and NAL. The scaling of the inclusive cross section is of the form ($s \gg M^2$, $t/s, \mathcal{M}^2/s$ fixed)

$$\frac{d\sigma}{d^3p/E}(ep \to eX) = \frac{1}{s^N} f(t/s, \mathcal{M}^2/s) \tag{1}$$

where f is a function of ratios of invariants [$t = q^2 = (p_e - p'_e)^2$ is the invariant momentum transfer squared, $s = (p_e + p)^2 = E^2_{cm}$ and $\mathcal{M}^2 = (p+q)^2$ is the invariant missing mass-squared]. The measured value, $N \cong 2$, which was predicted from current algebra by Bjorken,[4] indicates there is no internal length scale [larger than the resolving distance of $\sim 10^{-15}$ cm]. To see what this dramatic absence of a length scale means in more detail, we can use the parton model of Feynman[5] and Bjorken and Paschos,[6] and use a Fock-space representation of the proton wave function (which is an eigenstate of the total Hamiltonian) in terms of quark states (which are eigenstates of the free Hamiltonian). Deep inelastic scattering then resembles electron disintegration in nuclear physics (see Figure 1).

It is convenient to choose the Lorentz frame so that $|\vec{p}| = p_z$ is very large. The constituents of the n-particle state have momenta $\vec{p}_i = x_i\vec{p} + \vec{k}_{i\perp}$, $\sum_{i=1}^{n} x_i = 1$ $\sum_{i=1}^{n} \vec{k}_{i\perp} = 0$. If a quark can recoil from the electron and stay close to the mass shell, then it must have $x = -q^2/2p\cdot q$ [ignoring corrections of order $((m^2 + k_\perp^2)/q^2)$]. We thus have, in agreement with Eq. (1),

$$\frac{d\sigma}{dt\,dx} \sim \frac{4\pi\alpha^2}{t^2} \sum_a \lambda_a^2 f_a(x)\Big|_{x = -\frac{q^2}{2p\cdot q}} \tag{2}$$

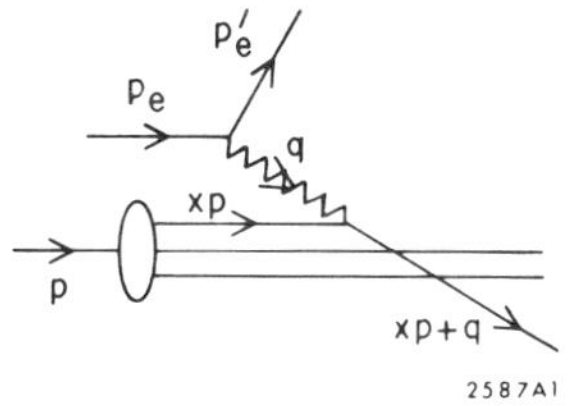

FIGURE 1

where $4\pi\alpha^2/t^2$ represents elementary electron-quark scattering, and $f_a(x)$ is the probability that the quark of type a (and charge λ_a) has longitudinal momentum fraction x (we ignore inessential spin complications here). Thus, in the parton picture, the absence of a length scale corresponds directly to point-like electron-constituent scattering. In the analogous case of $\nu p \to \mu X$, the predictions of the simplest quark model using an elementary V-A Fermi interaction at the quark level have been recently confirmed by NAL and CERN experiments. In particular, the angular dependence, predicted from the Dirac-like structure of the quark, has been found to be, within rough errors, consistent with experiment.[7]

The quantity $F_2(x) = \sum_a \lambda_a x f_a(x)$, $0 < x < 1$ is referred to as the structure function of the proton. Note that if the hadronic binding were turned off we would have $f_a(x) \rightarrow \delta(x_a - \frac{m_a}{M_{tot}})$ since $p_a \rightarrow (m_a/M_{tot})p$ in the rest frame. Thus $f_a(x)$ includes the effects of Fermi motion of the quarks in the proton, but within a relativistic model. I will discuss further applications of this approach below.

Of course, from the above description there is no apparent reason why quarks could not be easily knocked out of the nucleon. A number of attempts are now under way to construct consistent quark-containment models in which the large-distance binding effects are separable from the short-distance features which yield scale independence. In the approach discussed by Weinberg, Casher, Kogut and Susskind, Bjorken, and others, a gauge theory for quark interactions of the type used to unify weak and electromagnetic interactions is constructed.[8] At short distances the quark interactions become weak ("asymptotic freedom") and Bjorken scaling (modulo logarithmic corrections) results.[9] However at large distances the strength of the quark-quark interaction apparently grows so fast that meson production is more favored than quark separation. In other approaches (the "bag" models) boundary conditions[10] or specific scalar field potentials[11] are introduced to achieve the same goals. The quark model is thus an extraordinary type of bound state theory! Hadron physics seems to be at the stage that nuclear physics was just before the shell model.

III. EXCLUSIVE PROCESSES AT LARGE TRANSVERSE MOMENTUM

Since point-like constituents carry a finite fraction of the hadronic momentum, one expects that hadrons should be able to scatter to large transverse momentum by the hard, large-angle scattering of their constituents. This leads in general to power-law behavior for cross sections at large transverse momentum. Recently, it has been found that the ansatz of scale-invariance at short distances leads to dimensional-counting scaling laws of the form[12-14]

$$\frac{d\sigma}{dt}(A + B \rightarrow C + D) \rightarrow \frac{1}{s^{n-2}} f(t/s) \tag{3}$$

i.e.,

$$\frac{d\sigma}{d\Omega_{cm}}(A + B \rightarrow C + D) \rightarrow \frac{1}{s^{n-3}} f(\Theta_{cm})$$

for the asymptotic behavior of fixed center-of-mass angle scattering processes. The integer n is given by the minimum total number of elementary lepton, photon, or quark fields carrying a finite fraction of the momentum in the particles A, B, C, D; $n = n_A + n_B + n_C + n_D$. These scaling laws represent, in the simplest possible manner, a connection between the degree of complexity of a hadron and its dynamical behavior.

We can derive these counting rules in the following heuristic way:[12,13] Assuming limited hadronic binding, the computation of the hadronic amplitudes $M_{A+B\rightarrow C+D}$ is asymptotically identical to the computation of the n-particle scattering amplitude M_n obtained by replacing each hadron by a collection of quarks with the appropriate spin, each constituent carrying a finite fraction of the hadronic momentum. Note that M_n has dimension $[\text{Length}]^{n-4}$. If there is no internal scale - as is characteristic of Born graphs in renormalizable perturbation theory (due to dimensionless coupling constants) - then we expect $M_n \to 0\ [\sqrt{s}]^{4-n}$ since (barring infrared corrections) only $s^{-1/2}$ sets the length scale at fixed angle. Eq. (3) then follows from $d\sigma/dt \sim s^{-2}|M|^2$.

Applied to hadron scattering, the scaling laws (3) using quark counting seem to fare well, being consistent with a range of experiments for meson-baryon scattering, and more accurately verified for photoproduction $\gamma p \to \pi p$ and $pp \to pp$. We predict $d\sigma/dt \sim s^{-7}$ and s^{-10}, respectively, at fixed cm angle; experiment gives $s^{-7.3\pm0.4}$ and $s^{-9.7\pm0.5}$, respectively (see Figure 2). In general, using quark-counting, we have for any exclusive cross section integrated over a fixed cm region [12] (fixed invariant ratios):

$$\Delta\sigma \sim s^{-1-N_M-N_B} \tag{4}$$

where N_M (N_B) is the total number of mesons (baryons) in the initial and final states. Note that for processes involving only leptons and photons, $\Delta\sigma$ is scale invariant; each meson (baryon) introduces one (two) extra constrained fields which suppress the cross section.

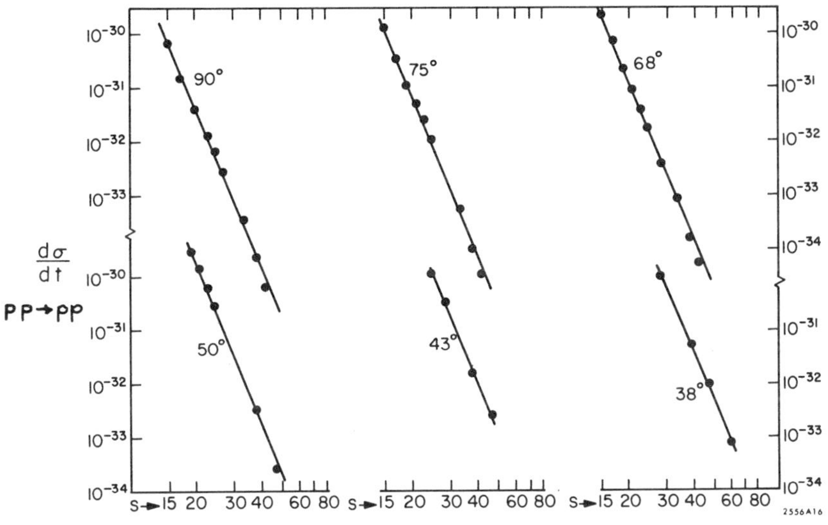

FIGURE 2

IV. FORM FACTORS AT LARGE t

One of the most important consequences of Eq. (3) is its application to elastic electron-hadron scattering; the rule immediately connects the asymptotic dependence of the spin-averaged form factor to the minimum number of fields in the hadron:[12,13,15]

$$F_M(t) \sim t^{1-n_M} \tag{5}$$

Thus one pays the penalty of a factor of t^{-1} for each additional quark whose direction must be changed from p to p+q. Using the quark model, we predict $F(t) \sim t^{-1}$ for mesons and $F_1(t) \sim t^{-2}$ for baryons. We also find in field-theory models $F_2 \sim t^{-3}$ and thus $G_E \sim G_M \sim t^{-2}$ scaling.[12] These results are consistent with the dependence indicated by present experiments, although the data for the meson form factors (which come from timelike t in $e^+e^- \to \pi^+\pi^-$) are not at all conclusive. A plot of $t^2 G_M(t)$ is shown in Figure 3.

A simple illustration of how the dimensional counting rule arises in the Bethe-Salpeter computation of the meson form factor is illustrated in Figure 4.[12]

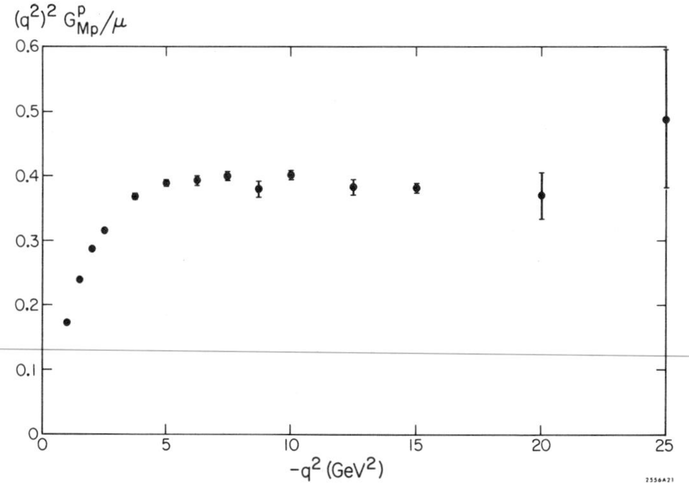

FIGURE 3

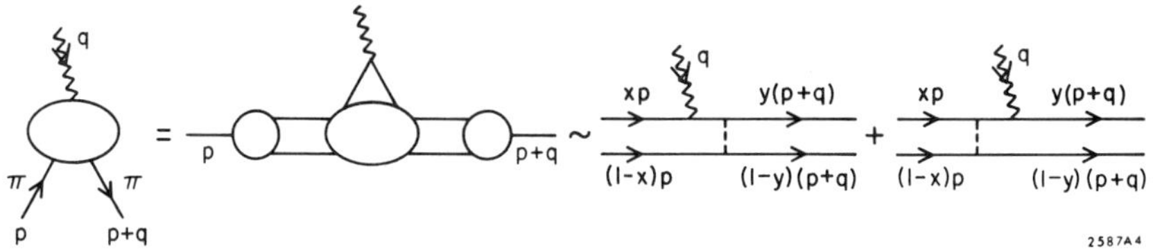

FIGURE 4

If we assume sufficient fall-off in the dependence of the Bethe-Salpeter wavefunctions at large relative momentum corresponding to a finite wavefunction in coordinate space, then the leading contribution to the asymptotic form factor comes from the iteration of the Bethe-Salpeter kernel wherever large relative momentum is required, as indicated in the figure. A simple computation then gives $F_M(t) \sim t^{-1}$ (modulo a logarithm) assuming a scale-invariant kernel. The inverse factor of t^{-1} comes from the off-shell quark line. In the n-body state, n-1 quark lines are off-shell, giving the result (5). Note that the minimum field contribution gives the leading asymptotic behavior.

If we are bold, we can apply the dimensional counting rule to the deuteron form factor (see Figure 5):

$$F_D(t) \sim t^{-5}$$

$$\frac{d\sigma}{dt}(eD \to eD) \sim \frac{4\pi\alpha^2}{t^2} F_D^{\,2}(t) \qquad (s \gg t \gg M^2) \qquad (6)$$

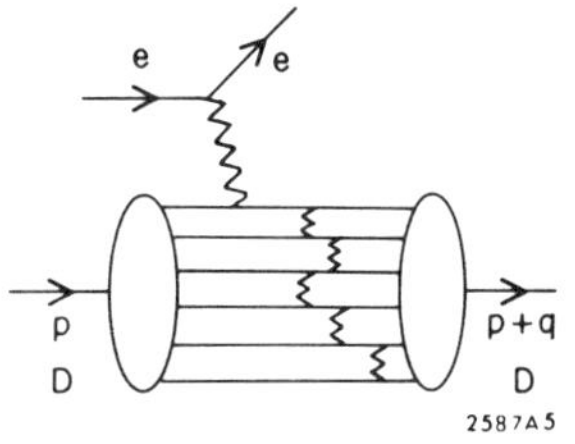

FIGURE 5

counting six quarks at short distance. A measurement of large momentum transfer elastic electron-deuteron scattering is now in progress at SLAC by a group headed by N. Chertok. The asymptotic prediction, however, is probably not useful until each of the off-shell quark lines is off-shell by $\gtrsim \frac{1}{2}$ GeV, i.e., for $|t| \gtrsim 10$ GeV^2. A convenient phenomenological approach which should be applicable at smaller $|t|$ is to plot $F_D(t)/F_{1p}^2(t/4)$ to see if a $(1-t/m^2)^{-1}$ behavior is observed.

The dimensional counting rule for form factors has recently been looked at more carefully within the context of specific renormalizable field theories.[12,16] In the case of asymptotic freedom theories, Appelquist and Poggio[17] have shown that the asymptotic behavior of the full Bethe-Salpeter kernel is effectively one logarithm more convergent at large momentum transfers than indicated by ladder approximation. They can then show that the quark-antiquark Bethe-Salpeter wavefunction falls with the required asymptotic dependence when both legs go off-shell at a constant rate; the coordinate space wavefunction is finite at the origin up to a calculable logarithm. Assuming that the wavefunction has no anomalous infrared behavior (which would be inconsistent with a bound system of finite size) when one leg goes on-shell, one predicts $F_\pi(t) \sim t^{-1}$ (modulo logarithms). A similar proof holds when a renormalizable theory has a small anomalous dimension. In the case of QED, the full Bethe-Salpeter kernel again clearly falls faster than indicated by simple ladder approximation, leading to a finite wavefunction at the origin. The true asymptotic dependence of the kernel to all orders in perturbation theory is not known, but to any finite order in perturbation theory the form factor of positronium obeys the dimensional counting rule - modulo powers of log (-t). [It should be noted that the singular behavior of the Bethe-Salpeter <u>ladder</u> approximation, where the wavefunction singularity at $x_\mu \to 0$ depends on the coupling constant (which in turn is restricted ad hoc by hermiticity), is misleading for the analysis of asymptotic behavior.] Schierholz and Alabiso[18] have shown explicitly how the dimensional counting analyses[19] go through in the case of three-body bound states.

V. THE CONSTITUENT INTERCHANGE MODEL

The angular dependence of the scaling cross section is of great fundamental interest itself. Note that for $s \gg M^2$, the function

$$f_{A+B\to C+D}(\theta_{cm}) = s^{n-2}\frac{d\sigma}{dt}(A+B\to C+D) \qquad (7)$$

becomes a "universal" function independent of s, no matter how large! The possibility of making asymptotic predictions such as this of course stems from the ultimate point-like structure of the constituents. It would not be surprising to see logarithmic modifications to the scaling laws (as expected in asymptotic freedom theories, for example).

Models that have been proposed for understanding the form of $f(\theta_{cm})$ are of two general categories: (1) gluon exchange and (2) quark interchange. In the first case, which has evolved from the early work of Wu and Yang,[20] hadron-hadron scattering derives from an elementary Yukawa interaction between quarks, and one predicts (for vector-gluon exchange)

$$\frac{d\sigma}{dt} \sim \left| F_{AC}(t) \frac{1}{t-\lambda^2} F_{BD}(t) + \text{u-channel terms} \right|^2 \tag{8}$$

Although this can be an excellent description of (diffractive) low momentum transfer, high energy hadron scattering, there are several difficulties which become critical at large $p_\perp$: First, the effective trajectory $\alpha_{eff}(t)$, defined via the parametrization $\frac{d\sigma}{dt} = \beta^2(t)\, s^{2\alpha(t)-2}$ for $s >> t$, stays at $\alpha \sim 1$ (i.e., energy independence at fixed t) even at large t, whereas the data indicate α becomes negative at large t (i.e., strong energy dependence at fixed t). Second, scale-invariant quark-quark interactions lead to an essential conflict with the observed large $p_\perp$ hadron production data, as we discuss in the next section. Third, as shown by Landshoff,[21] multiple Glauber-like gluon exchange between on-shell quarks actually will violate the dimensional counting rules because of a linear infrared behavior in the quark mass. One finds

$$M_{had} \sim \left(\frac{i\sqrt{s}}{m}\right)^{L-1} \left(\frac{1}{\sqrt{s}}\right)^{n-4} \tag{9}$$

where L is the number of on-shell quark pair interactions. This yields $d\sigma/dt \sim s^{-8}$ for p-p scattering (since L = 3), contrary to the data. Coleman[22] has shown that the only graphs with infrared behavior worse than logarithmic are these multiple, scale-invariant, on-shell quark scattering contributions. Thus, on empirical grounds, it appears that gluon exchange is not an important scattering mechanism at large transverse momentum.

Regardless of the importance of gluon exchange, composite states can always scatter by the interchange of their common constituents. In atomic physics, atoms can scatter via electron rearrangement, and in nuclear physics, deuteron-deuteron scattering can occur via nucleon rearrangement. In the case of p-p scattering, the nucleons can scatter via the interchange of a common p or n quark. The simplest example is $K^+p \to K^+p$ scattering which goes by p-quark interchange (see Figure 6).

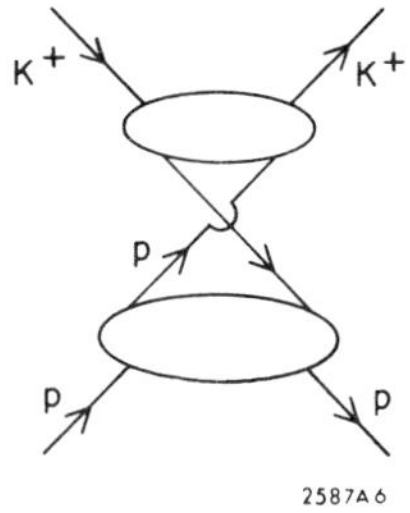

FIGURE 6

A simple calculation gives $(z = \cos\theta_{cm})$[14,23]

$$\frac{d\sigma}{dt} \sim \frac{1}{s^2}\frac{1}{u^2}\frac{1}{t^4} \propto \frac{1}{s^8}\frac{1}{(1+z)^2}\frac{1}{(1-z)^4} \tag{10}$$

and $\alpha_{eff}(t \to -\infty) = -1$. A comparison with K^+p data shows excellent agreement. Other cases assume further model dependence, but the general features of the angular distributions of the constituent interchange model (CIM), including the prediction of negative values of $\alpha_{eff}(-\infty)$, seem to be confirmed. Further, the CIM diagrams are consistent with the dimensional counting rules, assuming the two-body Bethe-Salpeter kernel for the quark binding is scale-invariant. We will discuss applications to inclusive processes in the next section.

VI. THE IMPULSE APPROXIMATION IN HADRON PHYSICS[24]

One of the crucial differences between particle physics and nuclear physics is the presence of Regge behavior, and the resulting general inapplicability of the impulse approximation in high energy hadronic reactions. Because of the unlimited number of particles possible in a virtual intermediate state, the probability function f(x) diverges as $x \to 0$: $f(x) \sim x^{-\alpha}$ (with $\alpha \sim 1$); i.e., there is a very large probability for a particle to have a very small fraction of the target hadron momentum. Thus a typical forward hadronic interaction occurs on a constituent at an effective energy $s_{eff} = xs$. Integration over the spectrum gives

$$M \sim \int_0^\infty \frac{dx}{x} f(x)\, M_0(xs) \sim s^\alpha \tag{11}$$

and thus $\sigma_{TOT} \sim s^{-1}\,\mathrm{Im}\,M \sim$ const. (for $\alpha \sim 1$) with most of the scattering occurring at the minimum possible s_{eff}. Thus high energy is <u>not sufficient</u> to study short distances in hadronic processes since the interactions can occur between particles with only a small fraction of the available energy. In contrast, when high transverse momentum particle production is required we must have $s_{eff} > 4p_\perp^2$, and a true high energy collision is involved. Further, in a large t exclusive process, coherence of the reaction allows only the fewest particle number states to contribute, α_{eff} is negative, and s_{eff} is of order s_{eff}.

It is interesting to note that at large momentum transfer, Compton scattering $\gamma p \to \gamma p$ and $\gamma p \to \gamma X$ is dominated by Thomson scattering off the point-like constituents of the hadron - exactly in analogy to high energy Compton scattering on atoms.[25] The calculations are extremely simple using the infinite momentum frame. For intermediate states with a finite number of particles, the high energy spin-averaged forward elastic Compton scattering amplitude is given by

$$M_{\gamma p \to \gamma p} \to -\sum_a \lambda_a^2 \frac{e^2}{m} \int_0^1 \frac{f_a(x)}{x}\, dx = -\sum_a \lambda_a^2 \frac{e^2}{m_{eff}} \tag{12}$$

where $m_{eff}^{-1} = m^{-1}\langle x^{-1}\rangle$ plays the role of an effective mass of the quark in the hadron. This can be compared with the atomic physics calculations of Goldberger and Low.[25] The calculations which account for hadronic Regge behavior entail a Regge subtraction consistent with analytic continuation in the trajectory α. The infinite-momentum frame analyses serve as a natural extension of the Schroedinger many-body theory to the relativistic domain, and have many applications to nuclear physics problems. Some time-ordered perturbation calculations for quantum electrodynamics using the $P \to \infty$ method are discussed in Ref. 26.

VII. INCLUSIVE REACTIONS AT LARGE TRANSVERSE MOMENTUM

One of the most exciting areas of particle physics is the study of inclusive large transverse momentum reactions $A+B \to C+X$. For example, at the CERN intersecting storage rings, measurements of the reaction $pp \to \pi^0 X$ have been made at a center of mass energy of 52.7 GeV (equivalent to a lab energy of ~ 1300 GeV) where the π^0 is detected at $\theta_{cm} \sim 90^0$ with $p_\perp$ up to 9 GeV.[1] In terms of $t = (p - p_{\pi^0})^2$ this is $|t| \sim 475\ \mathrm{GeV}^2 \sim 12000\ f^{-2}$. Over the range of the experiment, the results are consistent with a scaling law

$$\frac{d\sigma}{d^3p/E}(pp \to \pi^0 X) \sim \frac{1}{(p_\perp^2)^N} f(\theta_{cm}, \frac{\mathcal{M}^2}{s}) \tag{13}$$

where $2N = 8.24 \pm 0.05$ (± 0.70 including systematic errors). Here $\mathcal{M}^2/s$ measures the distance from the edge of phase space: $\mathcal{M}^2/s \cong 1 - p_{cm}/p_{max}$. Thus, as in the case of exclusive processes, the cross section has a power law scaling in $p_\perp^2$ (or s) at fixed ratios of invariants. For the ISR data $0.1 < p_{cm}/p_{max} < 0.4$.

In order to see what is involved physically let us imagine an experiment where two muonium atoms collide in the cm system and a high energy photon is detected at 90^0. The process is clearly due to the hard (Mott) scattering of the constituent leptons followed by the bremsstrahlung from a scattered lepton (see Figure 7). Since the electron only carries a fraction $x_e \sim m_e/(m_\mu + m_e)$ of the atomic momentum, it is evident that only muon scattering can produce a photon at an energy which is a sizable fraction of E_{cm}. Thus we expect ($x_\gamma = p_\gamma/\frac{1}{2}E_{cm} = 1 - \mathcal{M}^2/s$)

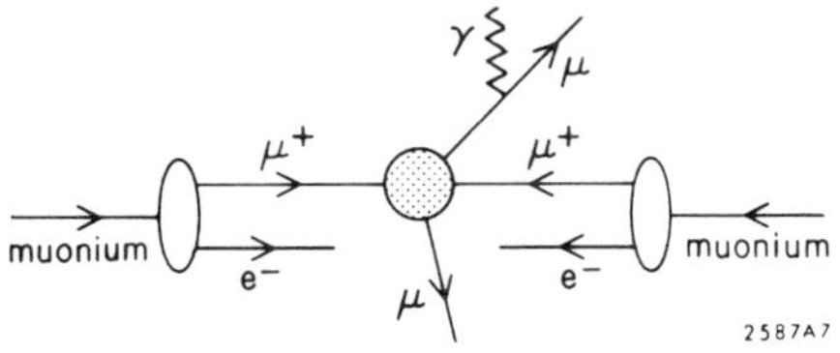

FIGURE 7

$$\frac{d\sigma}{d^3p/E} \sim \frac{\alpha^3}{p_\perp^4} \log\frac{p_\perp^2}{m_\mu^2} f(x_\gamma, \theta_{cm}) \tag{14}$$

where $f(x_\gamma)$ decreases with a bremsstrahlung spectrum from the maximum value $x_\mu = m_\mu/(m_e + m_\mu)$. More generally, we could include the Fermi motion due to atomic binding. Note that the scale-invariant $p_\perp^{-4}$ behavior is derived from the scale-invariant $\mu^+ - \mu^+$ scattering.

The fact that we do not see $p_\perp^{-4}$ behavior in $pp \to \pi^0 X^-$ is thus strong evidence against scale-invariant quark-quark scattering[27] being responsible for this process. Presumably the scattering of particles with form factors[28] is involved. This suggests that perhaps quark-hadron scattering and even hadron-hadron scattering are the important underlying hard scattering processes. In order to allow for all of the possible parton model processes we can use a description based on Figure 8, where particle C is produced by the fragmentation of particle C after a hard collision of a+b → c+d: Using dimensional counting,[12] the cross section has the form of Eq. (13) with a power law $(p_\perp^2)^{2-n_{active}}$ where $n_{active} = n_a + n_b + n_c + n_d$ are the number of active fields involved in the large angle collision. In fact, we can go further: for $\mathcal{M}^2/s$ small, i.e., where $p_\perp^c$ is a large fraction of the available momentum, we have[29]

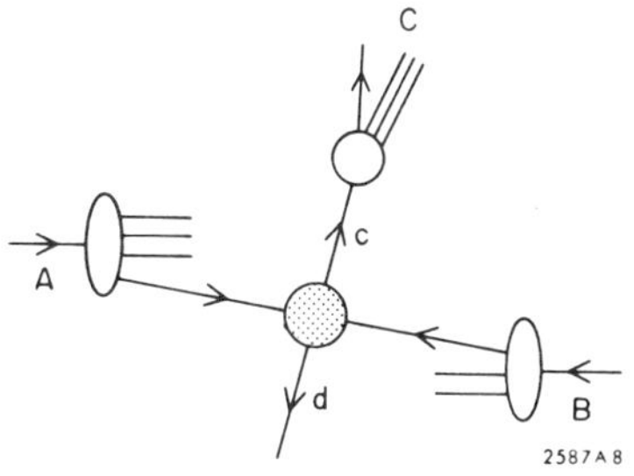

FIGURE 8

$$f(\theta_{cm}, \frac{\mathcal{M}^2}{s}) \sim f(\theta_{cm})\left(\frac{\mathcal{M}^2}{s}\right)^{2n_{passive}-1} \tag{15}$$

where $n_{passive}$ is the number of passive spectators of A, B, and C not involved in the production process - but which use up the available phase space. This rule accounts for the distribution of momentum of quarks in the hadrons, and can be easily derived from simple graphs in renormalizable perturbation theory. As a special case we can rederive the expression (2) for deep inelastic electron scattering and predict the threshold behavior

$$F_{2H}(x) \sim (1-x)^{2n_{spec}-1} \tag{16}$$

for the Bjorken structure functions; here $n_{spec} = n_H - 1$ is the number of non-interacting quarks in deep inelastic scattering. The reader can check that these results are consistent with the Drell-Yan-West[30] relation since $F_H(t) \sim t^{1-n_H}$. As a special case of Eqs. (14) and (15), we recover the dimensional counting rules for A+B → C+D using $n_{passive} = 0$.

The fact that 2N ~ 8 suggests that the dominant hard processes for the ISR measurements involve $n_{active} = 6$, perhaps processes such as q+M → q+M or q+q → p+$\bar{q}$. The dependence of the data on $\mathcal{M}^2/s$ given by Eq. (15) with $n_{spec} = 5$ is also consistent with this interpretation. Much more work relating quantum number characteristics, correlations, and multiparticle features of the cross section will be needed before we can become confident that this is indeed a correct explanation. Note that the dominance of quark-hadron scattering is the central hypothesis of the CIM model.[14]

The physics become much more complicated when the FNAL data for hadron production at large $p_\perp$ from proton-heavy target collisions are taken into account.[2] The unraveling of the nuclear physics effects is a fascinating problem here, especially in regard to the particle ratios. The π-production data, which cover $p_{cm}/p_{max} > 0.4$, are outside the ISR range, and yield 2N ~ 11 for $p_\perp > 4$ GeV. Assuming that nuclear effects do not influence the value of N appreciably, we can conclude that either the parton model scaling law is invalid, or two terms ~ $p_\perp^{-8}$ and $p_\perp^{-12}$ are operating. In fact, as p_{cm}/p_{max} increases, we should expect that more active quarks are required in order that they have a sufficiently large fraction of the available energy. A $p_\perp^{-12}$ ($n_{active} = 8$) process like $q + (qq) \to M + B^*$ (with $n_{spec} = 3$) has just this feature; because we are close to the edge of phase space, its contribution is expected to be of comparable importance to the $p_\perp^{-8}$ terms.

Quite good fits[31] to the 90° NAL, ISR, and low energy ($p_{lab} \sim 24$ GeV/c) data have been made using the predicted form[29]

$$\frac{d\sigma}{d^3p/E} = \frac{A}{(p_\perp^2 + m_8^2)^4}\left(1 - \frac{p}{p_{max}}\right)^9 + \frac{B}{(p_\perp^2 + m_{12}^2)^6}\left(1 - \frac{p}{p_{max}}\right)^5 \qquad (17)$$

with m_8^2, $m_{12}^2 \sim 1\ \mathrm{GeV}^2$. The effective power of $p_\perp$ in Eq. (13) thus changes from $p_\perp^{-8}$ to $p_\perp^{-12}$ over the range of the data.

VIII. CONCLUSIONS

Evidence from many different areas - particularly, deep inelastic scattering and large transverse momentum processes - supports the hypothesis that quarks determine the essential degrees of freedom of hadronic physics. The dimensional counting rules and the constituent interchange model seem to provide a convenient, simple, description of the scaling behavior of exclusive and inclusive processes at large transverse momentum. Despite the well-known present difficulties with the interpretation of the $e^+e^- \to$ hadron results of CEA and SLAC, it does seem that the quark model is the common denominator of many diverse hadronic phenomena.

As we have seen, there are some intriguing consequences for nuclear physics, e.g., the predicted t^{-5} asymptotic behavior of the spin-averaged deuteron form factor. A general consequence of the counting rules discussed above is the prediction[29] $G_{A/B}(x) \sim (1-x)^{2n-1}$ where $G_{A/B}$ is the probability of finding hadron A with a fraction x of the momentum of a fast-moving hadron B, and $n = n(\bar{A}B)$ is the number of quarks left behind. This gives a number of consequences for hadron physics - particularly in the triple Regge region of inclusive reactions. For deuterons we have $G_{p/d}(x) \sim (1-x)^5$ for the tail of the Fermi distribution of the nucleon momentum. This enters into the impulse calculations on a deuteron target, as well as being directly measurable in a diffractive breakup of a fast-moving deuteron. We also might remark here that since quarks carry the electromagnetic current, the Z-graph contribution to the meson exchange current amplitude in $e+p \to e'+p+$meson required in the nuclear form factors involves a quark pair rather than a $p\bar{p}$ pair.

It is now clearly important to establish theorems on which aspects of nuclear physics are indeed independent of nucleon substructure - e.g., low energy theorems, such as the Kroll-Ruderman theorem, and rigorous results valid for large nucleon mass which can establish the validity of the effective potential approach.[32] It seems clear that results requiring infinite sums over resonance states, subtraction constants, or questions of asymptotic behavior will require input from a theory of the underlying nucleon structure.

REFERENCES

1. F.W. Büsser et al., Phys. Letters 46B, 471 (1973).
2. J.W. Cronin et al., Phys. Rev. Letters 31, 1426 (1973); also "Production of Hadrons at Large Transverse Momentum at 200, 300 and 400 GeV," EFI preprint, University of Chicago (1974).
3. W.K.H. Panofsky, rapporteur's talk in Proceedings of the Fourteenth International Conference on High Energy Physics, Vienna, Austria (CERN Scientific Information Service, Geneva, Switzerland, 1968), p. 23.
4. J.D. Bjorken, Phys. Rev. 179, 1547 (1969).
5. R.P. Feynman, Phys. Rev. Letters 23, 1415 (1969); and R.P. Feynman, Photon-Hadron Interactions (W.A. Benjamin, Reading, Mass., 1972).
6. J.D. Bjorken and E.A. Paschos, Phys. Rev. 185, 1975 (1969).
7. For an excellent review, see F.J. Gilman, SLAC-PUB-1455 (1974), rapporteur's talk, XVII International Conference on High Energy Physics, London, England (1974).
8. See, e.g., A. Casher, J. Kogut, and L. Susskind, Phys. Rev. D 10, 732 (1974).
9. D.J. Gross and F. Wilczek, Phys. Rev. Letters 30, 1343 (1973); G. 't Hooft (unpublished); H.D. Politzer, Phys. Rev. Letters 30, 1346 (1973).
10. K. Johnson et al., MIT preprint (1974). See also L.I. Schiff, Ann. Phys. (N.Y.) 63, 248-260 (1971), and references therein.
11. S. Drell et al., "Heavy Quarks and Strong Binding," SLAC-PUB (1974) (to be published).
12. S. Brodsky and G. Farrar, Phys. Rev. Letters 31, 1153 (1973), and SLAC-PUB-1473 (1974) (to be published).
13. V. Matveev, R. Muradyan, and A. Tavkhelidze, Nuovo Cimento Letters 7, 719 (1973), and Joint Institute for Nuclear Research Report No. E2-8048 (1974).
14. Scaling laws of this form were first proposed in the constituent interchange model by R. Blankenbecler, S.J. Brodsky, and J.F. Gunion, Phys. Rev. D 6, 2652 (1972), Phys. Letters 42B, 461 (1973), Phys. Letters 39B, 461 (1973), Phys. Letters 39B, 649 (1972), Phys. Rev. D 8, 287 (1973); and also by P.V. Landshoff and J.C. Polkinghorne, Phys. Rev. D 8, 4159 (1973), Phys. Letters 45B, 361 (1973) and preprint DAMTP 73/31 (1973), Phys. Rev. D 8, 927 (1973).
15. A related non-relativistic version of this result is given by J. Stack, Phys. Rev. 164, 1904 (1967).
16. Z.F. Ezawa, preprint DAMTP 74/5 (1974) (to be published in Nuovo Cimento).
17. T. Appelquist and E. Poggio, Harvard preprint (1974).
18. C. Alabiso and G. Schierholz, SLAC-PUB-1395 (1974).
19. For related work see also M. Creutz and L. Wang, BNL preprint (1974), and W.R. Theis, Phys. Letters 42B, 246 (1972).
20. T.T. Wu and C.N. Yang, Phys. Rev. 137, 3708 (1965); H. Abarbanel, S.D. Drell, and F. Gilman, Phys. Rev. Letters 20, 280 (1968), and Phys. Rev. 177, 2458 (1969).
21. P.V. Landshoff, preprint DAMTP 73/56 (1973) (to be published in Phys. Rev.). See also references 16, 12, and P. Cvitanovic, Phys. Rev. D 10, 338 (1974).

22. S. Coleman, private communication.

23. See also P. G. O. Freund and S. Nandi, University of Chicago preprint EFI 74/33 (1974); K. Kinoshita and Y. Myozyo, Kyushu University preprint (1974); K. Kinoshita and H. Noda, Kyushu University preprint KYUSHU-74-HE-7 (1974); and K. Kinoshita, Y. Kinoshita, Y. Myozyo, and H. Noda, Kyushu University preprint KYUSHU-74-HE-11 (1974).

24. R. Blankenbecler, S. Brodsky, J. Gunion, and R. Savit, SLAC-PUB-1378 (1974) (to be published in Phys. Rev.).

25. S. Brodsky, J. Gunion, and F. Close, Phys. Rev. D 5, 1384 (1972); D 6, 177 (1972); and D 8, 3678 (1973); M. Goldberger and F. Low, Phys. Rev. 176, 1778 (1968).

26. S. Brodsky, R. Roskies, and R. Suaya, Phys. Rev. D 8, 4573 (1973).

27. S.M. Berman, J.D. Bjorken, and J. Kogut, Phys. Rev. D 4, 3388 (1971); J.D. Bjorken, Phys. Rev. D 8, 4098 (1973); D. Cline, F. Halzen, and H. Waldrop, Nucl. Phys. B55, 157 (1973); S.D. Ellis and M.B. Kislinger, Phys. Rev. D 9, 2027 (1974).

28. See reference 14; S.D. Ellis, Phys. Letters 49B, 189 (1974); M. Bander, R.M. Barnett, and D. Silverman, Phys. Letters 48B, 243 (1974); G. Preparata, CERN preprint TH.1859 (1974).

29. R. Blankenbecler and S. Brodsky, SLAC-PUB-1430 (1974); J. Gunion, Phys. Rev. D 10, 242 (1974).

30. S.D. Drell and T.M. Yan, Phys. Rev. Letters 24, 181 (1970).

31. See R. Blankenbecler, preprint SLAC-PUB-1438 (1974), talk presented at the IXth Balaton Symposium on Particle Physics, Hungary (1974); J. Cronin, review presented to SLAC Summer Institute (1974). For other recent reviews of this topic, see S.D. Ellis and R. Thun, IXe Rencontre de Moriond (1974), CERN preprint TH.1874 (1974); S.D. Ellis, P.V. Landshoff, and J. Gunion, review talks at the XVII International Conference on High Energy Physics, London (1974); S.J. Brodsky, Proceedings of the Stony Brook Conference on High Energy Collisions (1973) (SLAC-PUB-1329). J. Bjorken, Aix Conference (1974).

32. See F. Gross, Phys. Rev. D 10, 223 (1974), and references therein.

DISCUSSION

BOYMOND (Princeton): Concerning the production at large transverse momentum, can you make some predictions on the particle-antiparticle ratio?

BRODSKY: Oh yes, that's right, I didn't include this subject in this talk, but Blankenbecler, Gunion and I have benn working this out. In the interchange model you certainly can make specific predictions; i.e., for $\sigma(K^+)/\sigma(K^-)$, the K^+'s are more favored by a power of $(1-x_\perp)^2$ as you go to the exclusive boundary. This requires using the Bjorken, Kogut correspondence principle in order to make predictions. In fact it's somewhat encouraging that there is a difference between K^- and K^+ in the data and in this model; in contrast, this might not be true in a simple statistical model. You do retain in these models information from the quark structure of the incident beams to a great extent. It also points out that the meson-baryon scattering at large transverse momentum is going to be a crucial test, because predictions are much simpler for that case and the quantum number characteristics will be dramatic.

SLOBODRIAN(Laval): I am of course a layman in most of the things that have been explained this morning . However, what should be the binding energies of the quarks? Is there any hope of pulling them out or are they pointlike massless charged and neutral entities that 'exist' only within the nucleons?

BRODSKY: That's an excellent question. There are many answers that people have been giving. I can just try to give some of them. The most conservative argument is that there are no quarks at all so that the question does not have to be answered: in fact they are just 'mathematical' objects. In some models there is an infinite barrier for a quark to come out -the 'bag' model from M.I.T. does this by imposing a boundary condition. Preparata's model does this by postulating that the quark mass is effectively infinite outside the hadronic boundary. In any of these pictures, including the SLAC scalar field model, inside the boundaries of the wave functions, the quarks act as free, almost massless, particles. In the case of the asymptotic freedom theories where it's a complicated field theoretic problem, the quarks again are massless in the interior and as you pull them out further and further they acquire a virtual mass which becomes effectively infinite. But in fact, it's our tangle of language which entails interpreting the mathematics here. But I might say, it would be fascinating (maybe for the next Few body conference or the one after) to invite people to speak on this subject if there is additional progress in this field. I think that a natural extension of the boundaries of this Conference would be exacly the few body quark problem.

SASAKAWA (Tohoku): Does the angular distribution of the high energy elastic π p scattering have a backward peak? I am thinking of the u-channel amplitude of exchanging one quark.

BRODSKY: In the case of K^- -p at asymptotic energies it has not, but in the case of π^+ -p, because of both the cross graph plus the box graph, it has both peaks, and thus the backward peak comes in a natural way. I should say that there are currently many people that follow the same scheme as the C.I.M of trying to predict the angular distribution with all the right features. This includes a new paper by Freund and Nandy; there is a whole series of papers by various Japanese groups, especially those of Kinoshita (Japan) and co-workers; also the Russian group of Muradyan et al. have begun making predictions on the angular distribution based on the structure of the basic duality graphs.

PUGH (Maryland): My question is half nuclear physics. I believe the high P_t data from FNAL which showed a P_t dependence of P_t^{-11} were obtained with nuclei as targets and by making the assumption that the interaction occured with a single nucleon in the nucleus. If this is the case, how reliable are the scaling results? Also, how can you test the six-quark structure for the deuteron if the complicated structure of the target is important in these results?

BRODSKY: In the inclusive reactions one is dealing only with a relatively small number of quarks in the target so it doesn't matter how they are counted, Your first point is excellent. The data from FNAL is extracted primarily from proton scattering on a tungsten target. There is some data (at just one energy) from beryllium. From the beryllium data and from some copper data the Chicago-Princeton group can extract the A dependence of the large transverse momentum reactions. The results are rather odd, at first glance anyway, because for the meson case the A dependence at low transverse momentum is, as you expect, like $A^{2/3}$, but at large transverse momentum it saturates at $A^{1.1}$, so in fact there are some secondary processes that must be occuring. In fact, all the conclusions based on the data, especially the particle production ratio data must be done with some caution because the A-dependence effects could be complicating the physics. It is a fascinating nuclear physics problem to understand the impulse approximation at large transverse momentum particle production.

CHERTOK (SLAC): The charge form factor of the neutron, G_{En}, has an interesting t-dependence in that it rises as t out to about $200(MeV/c)^2$ and falls slowly, almost t-independent out to $1000(MeV/c)^2$. Does this interesting t-dependence final an explanation in your dimensional scaling picture of form factors?

BRODSKY: Certainly, it had better be that the G_{En} falls asymptotically as $1/t^2$ or the dimensional counting rules fail. But, of course, at t near 0 one has the coherent sum of the quarks which add up to total charge equal zero. Thus one has in fact a complicated coherent problem in the small t region which is more sensitive to the details of the model than the large t region. This could perhaps be profitably studied in these models.

OLSEN (Rochester): Does your model predict the production of μ pairs from hadron reactions?

BRODSKY: Yes, it agrees with the Drell-Yal prediction of scale invariance. The description of the experiment that Dr. Olsen is referring to is this: A quark in the incoming proton can annihilate with an anti-quark in the target. The annihilation goes through a virtual photon which produces a μ-pair with large invariant mass, or one could look at one muon at large transverse momentum. The prediction is that $n_{active} = 4$ so you get $1/p_\perp^4$, i.e.: scale invariance. But, in fact, thus far, the data has not even been checked at different energies to see whether it is scale invariant. The simplest prediction of the parton model is that the scale invariance has to be there, and if it isn't evident, there could be reasons why it is presumably just below other competing processes which are complicating the problem. One characteristic of quark-model theorists, as you know, is that there is always an excuse.

MEDIUM ENERGY PHENOMENOLOGY OF COMPOSITE HADRONS

D. D. Brayshaw †
University of Maryland
Dept. of Physics & Astronomy
College Park, Maryland 20742 U.S.A.

In view of the impressive array of evidence suggesting an underlying composite structure for the hadrons, it is of some interest to examine rather different phenomena in this context. Important steps in this direction have been taken by R. Tamagaki et al. and V. G. Neudatchin et al. in articles linking the short distance behavior of the N-N interaction to a composite model of the nucleon. [1)] In this article we extend the work of Neudatchin by establishing a connection between composite structure and an energy-independent boundary-condition on the wave function. This interpretation yields a number of predictions for N-N and π-π scattering which appear to be satisfied empirically.

The basic idea is to assume that the N-N wave function is not extinguished in the core region, but has an interior maximum with an energy-independent node at $r \lesssim .5F$. This picture is suggested by microscopic analyses of α-α scattering, in which such interior nodes arise as a consequence of Pauli exclusion. [2)] The position of the node would "explain" the success of hard-core models; observable consequences of the interior structure are discussed by Neudatchin. To connect this with scattering, let us consider the following simple model:

$$\begin{aligned} \psi_\ell(r) &= F_\ell(r)\sin\kappa_\ell r, & r \le r_\ell; \\ &= j_\ell(\kappa r) + i\sin\delta_\ell e^{i\delta_\ell} h_\ell(\kappa r), & r \le r_\ell. \end{aligned} \tag{1}$$

For the s-wave case we take $\kappa_o R_c = \pi$, where R_c is the hard-core radius, and ask for a suitable radius r_o at which the two representations may be matched for momenta κ sufficiently above the effective range region. Empirically, $\kappa_o \sim 1.2$ GeV/c, for N-N, which implies that a smooth matching can be achieved at energies of several hundred MeV only at the <u>first exterior maximum</u>. We are thus led to choose $\kappa_o r_o \simeq 3\pi/2$, or $r_o \simeq 3/2\, R_c$. This relation between the hard core and boundary-condition radii is satisfied to within a few percent in the N-N problem; to the author's knowledge this is the only model which attempts to relate them.

In the case of p-waves, we assume a canonical retardation of phase by π/2 relative to the s-wave as we approach the exterior. Thus $\kappa_1 R_c = \pi/2$, and the smooth matching condition leads to $r_1 = 2r_o$. Applied to s- and p-wave π-π scattering, we come to the

---

†Alfred P. Sloan Foundation Fellow.

following prediction; for sufficiently large κ,

$$(r\psi_0'/\psi_0)_{r=r_0} \quad = \quad (r\psi_1'/\psi_1)_{r=2r_0} \quad = \quad 0. \qquad (2)$$

Empirically, by fitting energy-dependent derivatives ψ'/ψ to the data in the range 500-900 MeV, [3] and extrapolating to large κ, we obtain r_0 = .32F, r_1 = .67F. The corresponding derivatives [Eq. (2)] are 0.28 and 0.39, respectively, The fit suggests that this description should pertain to $M\pi\pi \gtrsim 1.4$ GeV; present data is suggestive, but is not adequate to provide a decisive test at present.

Although naive, the arguments presented appear physical, and lead to a number of predictions which are empirically correct. Equally important, they suggest a certain consistency between processes as disparate as α-α, N-N, and π-π scattering. Naturally, the crucial question is whether or not a composite structure consistent with the high energy data will yield the required nodal behavior. One would expect such behavior to be intimately connected to the statistics of the constituents, which may provide an important new clue in attempts to construct more sophisticated composite models.

REFERENCES

1) R. Tamagaki, Rev. Mod. Phys. 39(1967) 629; V. G. Neudatchin, V. I. Kukulin, V. L. Korotkich and V. P. Korennoy, Phys. Lett. 34B(1971) 581.

2) I. Reichstein and Y. C. Tang, Nucl. Phys. A139(1969) 144. Qualitative features of the wave functions assumed below are observed in the α-α system.

3) J. P. Baton, G. Laurens and J. Reignier, Phys. Lett. 33B(1970) 528.

CALCULATION OF THE ω IN A RELATIVISTIC 3π MODEL

D. D. Brayshaw †
University of Maryland
Dept. of Physics & Astronomy
College Park, Maryland 20742 U.S.A.

Although there are many interesting three-body problems in elementary-particle physics, the covariant dynamical formalism proposed by several authors [1] has found scant employment. In view of the fact that numerical results, such as Basdevant's calculations in the 3π system [2], were quite discouraging, this is not too surprising. However, the results described below indicate that a rather different approach may prove quite valuable in classifying and interpreting the spectroscopic zoo of high energy physics.

The principal idea is to abandon a dynamical description in favor of the boundary-condition (BC) formalism proposed previously by this author.[3] In this way one can probe the consequences of two-particle "on-shell" input (such as the ρ) without postulating a complete hamiltonian framework. Although the equations are derived in terms of a covariant asymptotic representation of three on-shell particles, the resulting equations differ from the non-relativistic version only in the kinematical relationships between energy and momentum. As in 3), the two-particle input is given in terms of an energy-dependent BC:

$$(r\psi_\ell'/\psi_\ell)_{r=a} = \alpha + \gamma/(\kappa^2 - \beta^2). \qquad (1)$$

For the π-π p-wave a=.67F, α=0.39, $\beta = 2.2\ F^{-1}$ and $\gamma = 3.21\ F^{-2}$. In addition to this "minimal" input, off-shell corrections suggested by meson exchange were simulated via an additive term

$$K_{off}(q',q) = \frac{\lambda\ g(q')g(q)\ (\kappa^2+\mu^2)}{\kappa\ D_1^{(0)}(\kappa)}\ \rho_1(q) \qquad (2)$$

to the kernel; for notation see 3). The scale λ is fixed in terms of the on-shell input at 0.8, while $g(q) = q\ (q^2 + \mu_1{}^2)^{-1}\ (q^2 + \mu_2{}^2)^{-1}$. On general grounds one would expect $\mu = \mu_1 \simeq 2m$, $4m < \mu_2 < 5m$, where m is the π mass; the effect on the 1^-, I=0 3π resonance produced by varying μ_2 is shown in Table I. The entries A and B in that table illustrate the dependence on the ρ mass and width.[4]

†Alfred P. Sloan Foundation Fellow.

TABLE I.			
Type	μ_2 (m)	M (MeV)	Γ (MeV)
N	4.0	808	120
N	4.3	782	95
N	4.6	752	65
N	4.7	735	45
A	4.0	826	155
B	4.0	757	95

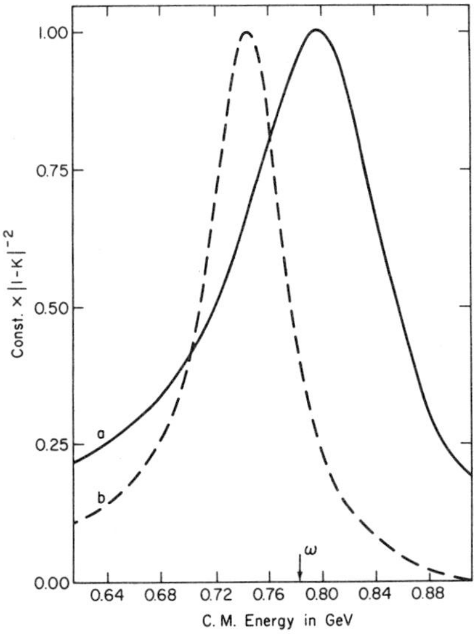

Fig. 1. Inverse $|1-K|^2$ vs. C.M. energy normalized to unit height. Solid line is N(4.0), dashed is N(4.6) of Table I.

The first and third entries in Table I. are plotted in Fig. 1, in which $|1-K|^{-2}$ is used to indicate the rapidly varying factor in an appropriate cross section. Although the precise position and width depend on the off-shell parameter μ_2, the essential location of the effect is specified by the on-shell (ρ) input. In order to reproduce the small experimental width (Γ=10 MeV) of the ω (784) more sophisticated off-shell input is apparently required. It is important to note that, in contrast to Basdevant's calculations, no spurious effects were found in any of the 0^- or 1^- isospin states.

REFERENCES

1) V. A. Alessandrini and R. L. Omnes, Phys. Rev. 139 (1965) B167; D. Z. Freedman, C. Lovelace and J. M. Namyslowski, Nuovo Cimento 48 (1966) 258.
2) J. Basdevant and R. E. Kreps, Phys. Rev. 141 (1966) 1409.
3) D. D. Brayshaw, Phys. Rev. D8 (1973) 952.
4) Model A corresponds to a "ρ" of width 60 MeV, model B has the ρ width but a mass of 660 MeV.

RELATIONS DE BASSE ENERGIE SUR LA BRISURE DE LA SYMETRIE CHIRALE ET L'ALGEBRE DES COURANTS.

Jean-Yves Harnois
University of Toronto, Physics Department,
Toronto, Ontario, Canada M5S 1A7

Dans cette analyse, nous faisons un examen critique des contraintes de l'algèbre des courants et de la brisure de la symétrie chirale sur les amplitudes de diffusion kaon-nucléon. Pour obtenir ces théorèmes de basse énergie, les trois hypothèses suivantes sont nécessaires:

(i) le commutateur des courants quadrivecteurs axiaux obéit à la relation

$$[A_a^o(\vec{r},t),A_b^\mu(\vec{r}',t)] = if_{abc}\ V_c^\mu(\vec{r}',t)\delta(\vec{r}-\vec{r}')(a,b,c=1,\ \dots\ 8)$$

(ii) il est possible de définir un champ scalaire local $\Sigma_{ab}(x)$ par le commutateur

$$[A_a^o(\vec{r},t),\phi_b(\vec{r}',t)] = i\Sigma_{ab}(\vec{r}',t)\delta(\vec{r}-\vec{r}')$$

où $\phi_b(x)$ dénote la composante b de l'octet des mésons.

(iii) le terme qui brise la symétrie chirale dans l'hamiltonien des interactions fortes est local. Conséquemment, nous avons

$$\partial_\mu A_a^\mu(\vec{r},t) = -i[Q_a^5(t),H_{S.B.}(\vec{r},t)]$$

où $Q_a^5(t) = \int d r A_a^o(\vec{r},t)$ sont les générateurs des transformations chirales.

Considérons la diffusion[1] $K(q) + N(p) \to K(q') + N(p')$. Il est possible, à partir des relations de commutation données ci-haut et de la définition du champ du kaon

$$\partial_\mu A_a^\mu(x) = -\mu^2 F_K \phi_a(x) \qquad (a = 4,5)$$

où μ et F_K désignent respectivement la masse et la constante de désintégration du kaon, d'obtenir une identité de Ward. D'après cette identité, les combinaisons paires et impaires des amplitudes sous symétrie de croisement $\{T^{(\pm)} \equiv \frac{1}{2}[T(K\text{-}N) \pm T(K\text{+}N)]\}$ sont contraintes par le terme sigma et l'algèbre des courants respectivement.

Les relations intéressantes à considérer sont évaluées à l'énergie de seuil, c'est-à-dire $\nu = \mu, \nu_B = -\frac{\mu^2}{2m}$ où $\nu = -(q'+q)\cdot(p'+p)/4m, \nu_B = q',q/2m = (t-2\mu^2)/4m$ et m est la masse du nucléon.

Après avoir explicité les termes de Born, pôles Λ et Σ, ainsi que les résonnances $\Lambda'(1405)$ et $\Sigma'(1385)$ dans l'approximation du pôle, nous obtenons, entre autres, la relation d'Adler-Weisberger mais sans que les constantes g_Λ^Λ et $g_A^{\Sigma^o}$ n'apparaissent explicitement,

$$G(\mu,-\frac{\mu^2}{2m}\,;\,K^\pm p) = \frac{4\pi}{\mu}(1+\frac{\mu}{m})\frac{1}{2}(\bar{a}_s-a_s) = \mu^2 \sum_{y=\Lambda,\Sigma^o}(\frac{g_{K-py}}{m_y+m})^2[(4m^2-\mu^2)-(m_y+m)(3m-m_y)]\cdot$$

$$\frac{1}{[(m+\mu)^2-m_y{}^2][(m-\mu)^2-m_y{}^2]} + (\text{termes semblables pour } \Lambda' \text{ avec } m_y \to -m_{\Lambda'})$$

$$+\ G_{\Sigma'}(\text{seuil}) + \frac{1}{F_K{}^2}\{1+(\frac{\mu}{m})^2\frac{1}{2}(d-e+f)\}$$

où $G(\nu,\nu_B) = \nu^{-1}(A^{(-)}-\nu B^{(-)})$, $G_{\Sigma'}$ indique la contribution de $\Sigma'(1385)$ et le reste (dernier terme) est exprimé par les fonctions complètement régulières et sans dimension d, e et f. Les paramètres $\bar{a}_s = \frac{1}{2}(\bar{a}_s{}^{(0)} + \bar{a}_s{}^{(1)})$ sont les parties réelles des longueurs de diffusion dans l'onde s du système K^-p avec étrangeté $S = -1$ et isospin $I = 0,1$. Semblablement, $a_s = a_s{}^{(1)}$ est la longueur de diffusion du système K^+p avec $S = 1$, $I = 1$. Les calculs démontrent[2,3] qu'en ce point le reste est complètement négligeable et que les contributions des résonnances de spin $\frac{3}{2}$ sont nulles. Nous obtenons $a_s{}^{(1)} =$ -0.13 f alors que les diverses determinations se situent entre -0.19 f et -0.05 f, -0.13 f étant la valeur obtenue par Von Hippel et J.K. Kim.[4]

Parmi les amplitudes paires, nous obtenons

$$A^{(+)}(\text{seuil};K^\pm n) = \frac{\mu^2}{F_K}\Sigma(t=0) + \frac{1}{F_K}\{\Sigma_{\text{pôle}} + \Sigma'_{\text{res}} + (\frac{\mu}{m})^4 a + (\frac{\mu}{m})^2(-b+c)\}$$

Dans cette expression, les termes à l'intérieur des crochets sont sans dimension et réguliers. Les calculs donnent $\Sigma_{\text{pôle}} \approx 0.32$, $\Sigma'_{\text{res}} \approx -14.12$, reste ≈ -0.05, indiquant que $\Sigma(t=0)$ est déterminé en bonne approximation par la résonnance $\Sigma'(1385)$ uniquement. De ces valeurs numériques nous obtenons $\mu^2 F_K \Sigma(t=0) = 1160$ MeV.

En se servant de l'hypothèse additionnelle que la brisure de la symétrie chirale dans le lagrangien d'interactions fortes appartient à la représentation $(3,3^*) \oplus (3^*,3)$ du groupe $SU(3) \otimes SU(3)$[5], c'est-à-dire,

$$L_{S.B.} = -u_o - cu_8,$$

nous pouvons calculer $<N|\mu_o|N>$,

$$<N|\mu_o|N> \approx 1228 \text{ MeV},$$

et partant,déterminer le terme sigma du système pion-nucléon: $\sigma^{\pi\pi}_{NN}$ = 104 MeV. Ce qui correspond à la valeur trouvée par Cheng et Dashen[6] à t=0.[7]

L'auteur est heureux de remercier le Dr. M.A. Gonzàlez pour les fructueuses discussions qui l'ont conduit à cette analyse.

BIBLIOGRAPHIE

1. Une analyse similaire du système pion-nucléon a été faite par L.S. Brown, W.J. Pardee et R.D. Peccei, Phys. Rev. D4, 2801 (1971).

2. Nous utilisons les determinations suivantes: $g^2_{KN\Lambda}/4\pi = 13.5$, $g^2_{KN\Sigma}/4\pi = 0.2$, $g^2_{KN\Lambda'}/4\pi = 0.32$, $g^2_{KN\Sigma'}/4\pi = 4.5 \text{ GeV}^{-2}$, cette dernière valeur correspondant à un couplage égal aux 3/4 de la valeur trouvée par SU(3). Voir R.F. Dashen, Y. Dothan, S.C. Frautschi et D.H. Sharp, Phys. Rev. 143, 1185 (1966): B.R. Martin, Springer Tracts in Modern Physics, édité par G. Höhler (Springer, New York 1970). Vol. 55, p. 73.

3. Nous utilisons les longueurs de diffusion consistentes avec les constantes de couplage ci-haut mentionnées. Voir G. Ebel et al., Nucl. Phys. B33, 317 (1971).

4. F. Von Hippel et J.K. Kim, Phys. Rev. Letters 20, 1303 (1968).

5. M. Gell-Mann, R. Oakes et B. Renner, Phys. Rev. 175, 2195 (1968); S. Glashow et S. Weinberg, Phys. Rev. Letters 20, 224 (1968).

6. T.P. Cheng et R. Dashen, Phys. Rev. Letters 26, 594 (1971).

7. H. Pagels et W.J. Pardee, Phys. Rev. D4, 3335 (1971).

Ce travail a été appuyé en partie par le Conseil national de recherche du Canada.

X

PROCESSUS DE PRODUCTION COHERENTE
ET PROCESSUS A PLUSIEURS PARTICULES
REACTIONS DE CASSURE
ELECTRODESINTEGRATION

MULTIPARTICLE AND COHERENT
PRODUCTION PROCESSES
BREAK-UP REACTIONS
ELECTRODISINTEGRATION

STATISTICAL TREATMENT OF COHERENT AND OTHER REACTIONS

B. MARGOLIS, Physics Department, McGill University

Introduction

Statistical concepts in particle physics go back at least twenty-five years when they were studied by Enrico Fermi[(1)]. Fermi proposed that reaction rates for producing strongly interacting particles are determined essentially by the total amount of phase space available to them. Fermi studied mainly pion production at the time. This proposal as we shall see requires some modification due to the rich spectrum of hadronic states and due to effects of competition. Nevertheless phase space considerations are central to the statistical picture to be presented here.

The work to be presented here has general interest in strong interaction physics. As we shall see we will present a theory and basis for the calculation of branching ratios for the production of hadrons in both reactions induced by hadrons and real and virtual photons. The targets for these reactions may be protons, complex nuclei and even electrons or positrons. For the case of complex nuclear targets we shall be especially interested in coherent reactions where the amplitudes for production of particles on the different nucleons are more or less in phase with one another. The production cross sections in this case yield beautiful diffractive patterns as a function of momentum transfer or angle.

An interesting proposal made by Hagedorn[(2)] and developed further by Frautschi[(3)] was responsible for reviving statistical considerations in elementary particle physics. Hagedorn proposes that hadronic matter can be considered as a superposition of resonances, fireballs or simply states which have a density as a function of mass and other quantum numbers which we write as $\rho(m, \ldots)$. This density of states is to be derived from a now well known bootstrap condition to be described further below.

Given this density of states we can test the theory by simply counting the number of resonances as a function of mass and other quantum numbers. This has been done by Hagedorn up to masses where the resonances are no longer resolvable. A strong exponential growth with increasing mass is observed. It is difficult to think of other direct tests of this statistical bootstrap picture. However, as we shall see, other less direct tests are possible. Perhaps our knowledge of $\rho(M, \ldots)$ can be used to calculate reaction rates, decay rates for processes involving hadrons in either the initial or final states. The reactions may be induced by leptons in fact as we shall see. For example one can make predictions for the branching ratios of reactions induced by e^+e^- collision. In principle then we can say something about electromagnetic form factors.

The Calculation of Reaction Rates

All the reactions we will consider here can be grouped into two classes as follows:

I. $a + b \rightarrow M \rightarrow c + d$

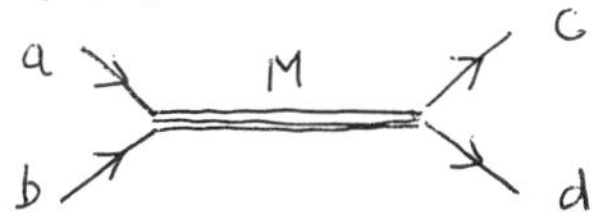

Here the intermediate s-channel states of mass M have a density of states given by $\rho(M)$. According to duality principles the amplitude above is, at energies which are high enough, ($\gtrsim$ a few Gev) dual to Regge type exchange diagrams of the type:

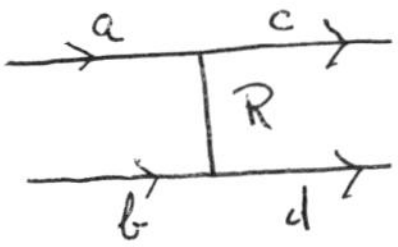

There is perhaps not too much value in adding up great numbers of resonances s-channel wise if duality is a sound principle. Why not use the conventional Regge description?

The fact of the matter is that we shall be interested in a badly neglected energy range for reaction I. That above the discrete resonance region but below where the t-channel Regge picture may be safely expected to work. We shall see some surprises here in the data which we will explain using a statistical picture of resonance superposition Again we shall see that we will be able to discuss a class of electromagnetically induced hadron production processes using such a picture.

We consider now an extension of the above to a second class of reactions.

II. $a + b \rightarrow M + c$

$\quad\quad\quad\quad\quad\quad \hookrightarrow d + e + \ldots$

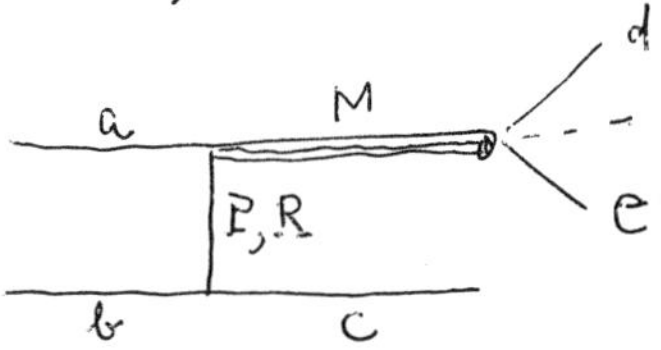

Here we shall be more interested in high energy situations for the reactions as much as intermediate energies. This is especially so if the exchanged trajectory is the pomeron.

Let us return to the reactions of Class I. We write the amplitude for this process in the forward direction, suppressing angular momentum labels as

$$A \propto \sum_i \frac{\gamma^i_{ab}\,\gamma^i_{cd}}{\sqrt{s} - m_i + \frac{1}{2} i \Gamma_i} \tag{1}$$

For elastic scattering

$$A_{el} \sim \sum_i \frac{(\gamma^i_{ab})^2}{\sqrt{s} - m_i + \frac{1}{2} i \Gamma_i} \tag{1a}$$

Given many resonances described by average widths and a level density, given by the statistical bootstrap say, we can convert the sum to an integral.

Looking at the imaginary part

$$\mathrm{Im}\, A_{el} \sim \int \frac{\Gamma_{el}\, \Gamma/2}{(\sqrt{s} - m)^2 + \frac{1}{4}\Gamma^2}\, \rho(m)\, dm \tag{2}$$

Then, assuming the integral is strongly peaked at the peak in the Breit-Wigner component we can write

$$\mathrm{Im}\, A_{el} \sim \Gamma_{el}(\sqrt{s})\, \rho(\sqrt{s}) \cdot \pi \tag{2a}$$

However according to duality, at high enough energies

$$\mathrm{Im}\, A_{el} \sim s^{\alpha(0)} \tag{2b}$$

where $\alpha(t)$ is the t-channel exchanged Regge trajectory. It follows that

$$\Gamma_{el}(\sqrt{s}) \underset{\sim}{\propto} 1/\rho(\sqrt{s}) \tag{3}$$

i.e. the partial width in any channel such as ab decreases exponentially with increasing energy.

We can get a more precise expression for partial widths as follows. The total width is the sum of the partial widths. Then working in the C.M. frame we have that the total width

$$\Gamma(M) = \sum_{k=2}^{\infty} \frac{1}{k!} \prod_{\ell=1}^{k} \int \frac{d^3p_\ell}{2E_\ell}\, dM_\ell\, \rho(M_\ell)\, \delta(M - \sum_\ell E_\ell)\, \delta^3(\sum_\ell \vec{p}_\ell)\, \Gamma(p_1, M_1, \dots p_k, M_k) \tag{4}$$

where $\Gamma(p_1, M_1, \dots\dots p_k, M_k)$ is the partial width for emitting particles with masses $M_1 \dots M_k$ and momenta $p_1 \dots p_k$. The factorial assures that each state is counted but once. Comparing expression (4) with the statistical bootstrap equation for the density of fireball states[(2)(3)]

$$\rho(M) = \sum_{n=2}^{\infty} \left[\frac{V}{(2\pi\hbar)^3}\right]^{n-1} \frac{1}{n!} \prod_{i=1}^{n} dm_i\, \rho(m_i) \int d^3p_i\, \delta(M - \sum_{i=1}^{n} E_i)\, \delta^3(\sum \vec{p}_i) \tag{5}$$

we find the expression for the partial widths

$$\Gamma(p_1, M_1, \dots, p_k, M_k) = \left[\frac{V}{(2\pi\hbar)^3}\right]^{k-1} \frac{\Gamma(M)}{\rho(M)} \prod_{\ell=1}^{k} (2E_\ell)_{CM} \tag{6}$$

so that for the two body channel

$$\Gamma_{ab}(M, m_a, m_b) = \frac{V}{(2\pi\hbar)^3} \frac{\Gamma(M)}{\rho(M)} \frac{M^4 - (m_a^2 - m_b^2)^2}{4M^2} \tag{6a}$$

The quantity V is the hadronic colume containing the constituents of the fireballs.

We see then that if the total width is slowly varying with mass M each partial width is a slowly varying function of M times $\rho^{-1}(M)$. There is then a connection between the number of channels and the density of resonances. Provided the total width is essentially constant (or slowly varying) each channel has a strength $\propto 1/\rho$ so that the number of channels $\sim \rho(M)$.

We return now to the non-elastic reaction

$$a + b \rightarrow \mathbf{M} \rightarrow c + d$$

Neglecting spin we can write the amplitude for this reaction in the form

$$f_{ab,cd}(E,\Theta) = -\frac{1}{2k}\sum_{\ell=0}^{\infty}(2\ell+1)P_\ell(\cos\Theta)\sum_j \frac{\gamma_{ab}^{j\ell}\gamma_{cd}^{j\ell}}{E-E_{j\ell}+\frac{i}{2}\Gamma^{j\ell}} \qquad (7)$$

where $\gamma_{ab}^{j\ell}$ is the square root of the width for elastic scattering of a by b for the resonance of mass $E_{j\ell}$ and angular momentum ℓ. The square roots of the partial widths, $\gamma_{ab}^{j\ell}$, being coupling constants may be positive or negative. We have then for the angle integrated cross section

$$\begin{aligned}\sigma_{ab,cd} &= \int |f_{abcd}|^2 d\Omega \\ &= \frac{\pi}{k^2}\sum_\ell (2\ell+1)\left\{\sum_j \frac{\Gamma_{ab}^{j\ell}\Gamma_{cd}^{j\ell}}{(E-E_{j\ell})^2+\frac{1}{4}(\Gamma^{j\ell})^2}\right. \\ &\left.+\sum_{j\neq j'}^{j}\sum^{j'} \frac{\gamma_{ab}^{j\ell}\gamma_{cd}^{j\ell}\gamma_{ab}^{j'\ell}\gamma_{cd}^{j'\ell}}{(E-E_{j\ell}+\frac{1}{2}i\Gamma^{j\ell})(E-E_{j'\ell}-\frac{1}{2}i\Gamma^{j'\ell})}\right\} \qquad (8)\end{aligned}$$

Here $\Gamma_{ab}^{j\ell} = (\gamma_{ab}^{j\ell})^2$.

For rapid oscillation or random sign of the coupling constants the sum of the terms with $j \neq j'$ will not contribute. We have, in this case, converting the remaining sum to an integral and ignoring orbital angular momentum

$$\sigma_{ab,cd} \approx \frac{\pi}{k^2} \cdot \frac{2\pi}{\Gamma(E)} \Gamma_{ab}(E) \Gamma_{cd}(E) \rho(E) \tag{9}$$

Since each partial width goes like $\rho^{-1}(E)$ we have that the angle integrated cross section $\sigma_{ab,cd} \tilde{\propto} \rho^{-1}(E)$. Therefore since $\rho^{-1}(E)$ falls exponentially with increasing E so does (9). This is as opposed to the resonant part of elastic scattering which goes, as we saw above as

$$\sigma^{Res}_{ab,ab} \tilde{\propto} E^{-\text{Neg. Power}}$$

In fact for a trajectory in the t channel with intercept $1/2$,

$$\sigma^{Res}_{ab,ab} \tilde{\propto} (E \log E)^{-1}.$$

We now ask whether neglect of off-diagonal terms in (8) is an approximation corresponding to any reactions in the real world. We answer this question by looking at data. We first examine angle integrated cross sections for the reactions

(a) $\bar{p} + p \rightarrow \bar{n} + n$

(b) $\bar{p} + p \rightarrow \pi^+ + \pi^-$

(c) $\bar{p} + p \rightarrow \rho + \pi$

For the reaction (a) the amplitude is simply related to $\bar{p}p$ and $\bar{n}n$ elastic scattering

$$A = \frac{1}{\sqrt{2}} \left(A^{el}_{\bar{p}p} - A^{el}_{\bar{n}n} \right) \tag{10}$$

Similarly the amplitude $\pi^- p \rightarrow \pi^0 n$ (not listed above) has an amplitude

$$A = \frac{1}{\sqrt{2}} \left(A^{el}_{\pi^- p} - A^{el}_{\pi^+ p} \right) \tag{11}$$

These amplitudes should behave then much as the non-diffractive part of elastic scattering.

For reactions (b) and (c) however we expect a strong incoherent component since there is no simple relationship with forward elastic scattering. For some energy interval above threshold then we expect

$$\sigma(E) \sim 1/\rho(E) \sim e^{-E/T} \qquad (12)$$

Figure 1 shows data for reactions (a) and (b) above indicating the strong coherence of the reaction $\bar{p}p \rightarrow \bar{n}n$ and the strong incoherence of $\bar{p}p \rightarrow \pi\pi$. The dotted curves are calculated according to the formula described above.[4] Figure 2 shows data for the reaction $\bar{p}p \rightarrow \rho\pi$ along with a calculation of the incoherent component[4]. We see here a strong incoherent component falling into a weak coherent behaviour (baryon exchange).

We focus attention now on a different set of data. We look at the behaviour of two body reactions near 90° in the C.M. system. It is a remarkable fact that above $\sqrt{s} \sim 2$ GeV all s-channel non-exotic reactions show very similar behaviour; relatively flat angular distributions in the neighbourhood of 90°, rapid decrease with increasing energy and the same strength to within a factor of two or so, well down into the discrete resonance region. In contrast p-p elastic scattering is of the order of 50 times larger. K^+p scattering is probably also several times larger than K^-p scattering. (We need more data).

We explain the features of the non-exotic reactions as follows. Rapid oscillation in sign of the spherical harmonics removes interference terms in the Breit Wigner sum. As with forward or small angle non-elastic reactions not simply related to elastic scattering we are left with an incoherent sum of Breit-Wigners, even for elastic scattering. We have then[5] for θ near 90°

$$\sigma(\theta) \sim h(E)\, e^{-E/T} \sum (2\ell+1) P_\ell^2(\cos\theta) \qquad (13)$$

where $h(E)$ is a slowly varying function of E.

That non-exotic reactions (in the s-channel) should have absolute values of cross section which are all very similar can be seen as follows. Consider for example the reaction $\pi^-p \rightarrow \pi^0 n$. The amplitude for this reaction

$$A_{C.E} = \frac{1}{\sqrt{2}} \left(A^{el}_{\pi^- p} - A^{el}_{\pi^+ p} \right) \tag{14}$$

Because of our incoherence condition we have then a relationship between differential charge exchange and differential elastic scattering cross sections,

$$\sigma_{C.E.} \approx \frac{1}{2} \left(\sigma_{\pi^- p} + \sigma_{\pi^+ p} \right) \tag{15}$$

The flatness of the elastic differential cross sections near 90° is in agreement with our incoherent resonance sum description. Further as we expect $\sigma_{\pi^- p}$ and $\sigma_{\pi^+ p}$ are similar in value near 90° and so so too therefore is $\sigma_{C.E.}$.

Similarly using SU3 and other symmetries perhaps we can relate the large body of two body reaction data to elastic scattering. The statistical bootstrap result for two body widths obtained above leads us to expect that these are roughly equal in different channels. In fact all that we need know to relate $\bar{p}p$, $K^- p$ and πp scattering near 90° is the relative number of boson, strange baryon and nucleon resonances. These appear to equal to within a factor of two or so.[5]. Figures (3), (4), (5) and (6) show measured differential elastic scattering cross sections along with theoretical curves as described in reference (5). This rapid exponential fall is expected to persist until some coherent amplitude takes over. The photo-production of pions near 90° shows behaviour similar to that described above for elastic scattering and other two body hadronic reactions, being down by the fine structure constant.[6]

Reactions of Class II:

Let us consider now the class of peripheral reactions

$$a + b \rightarrow c + M \tag{16}$$

$$M \rightarrow d + e \tag{16a}$$

$$d + c + f \tag{16b}$$

Diagramatically we represent these as follows: The process described by (16a) is

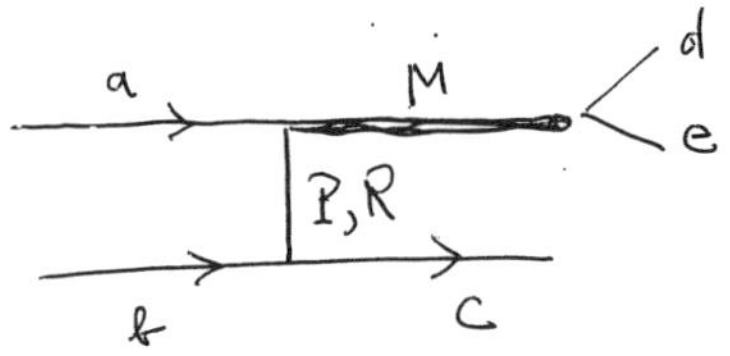

For process (16b) we take the fireballs to decay by a linear chain[3][7][8][9][10] as follows:

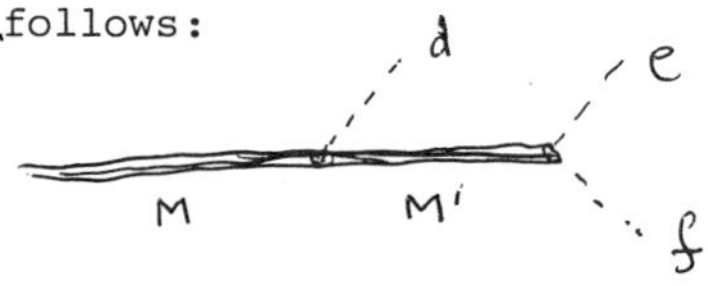

These reactions, in our picture are characterized by the production of a massive fireball (superposition of resonances) produced in a cone strongly peaked in the forward direction. This fireball then decays as described, outside the region of interaction mainly, at high enough energies. The production of the fireball we take to be described according to Regge considerations à la Mueller[11]. Our description can be considered then in the class of cluster models.

For Pomeron exchange, we can write, after integrating over momentum transfer, and to within logarithms of s,

$$\frac{d\sigma}{dM} \propto \beta_{PPR} \cdot \frac{1}{M^2} + \beta_{PPP} \cdot \frac{1}{M} \qquad (17a)$$

For exchanges of Regge trajectories with intercept 1/2

$$\frac{d\sigma}{dM} \approx \beta_{PRR} \cdot \frac{1}{s} \qquad (17b)$$

where we have neglected longitudinal momentum transfer effects. These results are valid for 1 $GeV^2 \ll M^2 \ll s$. These forms have a relatively weak dependence on mass when compared with $\rho(M)$.

To complete the description for the angle integrated rate for the reaction (16a) we must weight the expressions (17) with the appropriate branching ratio using considerations of the last section.[4] We note that in this scheme we are not considering angular momentum explicitly and so we do not have angular distribution results, although in principle we could include angular momentum. Our result then is the following

$$\frac{d\sigma_{de}(M)}{dM} \sim \left(1 \pm \frac{1}{s}\right)\left(1 \pm \frac{1}{M^2}\right) \frac{\rho_2(M, m_d, m_e)}{\rho(M)} \qquad (18)$$

where

$$\rho_2(M, m_d, m_e) = \frac{\pi}{2M^4}\sqrt{\lambda(M^2, m_d^2, m_e^2)}\left[M^4 - (m_d^2 - m_e^2)^2\right] \qquad (18a)$$

which is two body phase space for the particles M_d, M_c with invariant mass M.

Here

$$\lambda(x^2, y^2, z^2) = x^4 + y^4 + z^4 - 2x^2y^2 - 2x^2z^2 - 2y^2z^2 \qquad (18b)$$

Our answer, as Fermi conjectured, depends on two body phase space in an essential way when we make two particles. This phase space is very strongly cut off by competition with other channels. We recall that the number of channels $\propto \rho^{-1}(M)$.

Figures 7 and 8 show our calculation[10][12] of the A_1, A_3 enhancements for the reactions

$$\pi + p \to \rho\pi + p \qquad (19a)$$

$$\pi + p \to f\pi + p \qquad (19b)$$

Threshold Enhancements and Coherent Production

In order to test these ideas further it is of interest to go to higher mass enhancements if possible. As we go up in energy the density of fireball states increases and statistical methods which are of borderline validity in the cases of the last section should become more and more reliable. In this connection it is valuable to have data on complex nuclei. Complex nuclei select diffractive mechanisms if we confine ourselves to studying coherent production. Further we can use the whole periodic table if we wish to, therefore providing additional tests for our theory of hadronic production on nucleons. In this connection the A dependence of cross sections and momentum transfer dependence are particularly interesting. It is possible that nuclear structure effects could in principle complicate the picture but as we shall see this does not seem to be the case for coherent production.

We now focus attention on diffractive production of five charged pions on hydrogen and complex nuclei using incident pions. We are interested in data at high energies where the 5π mesons go strongly forward. We therefore rule out N^* production which can be treated as well of course. We describe the process according to the following diagram:

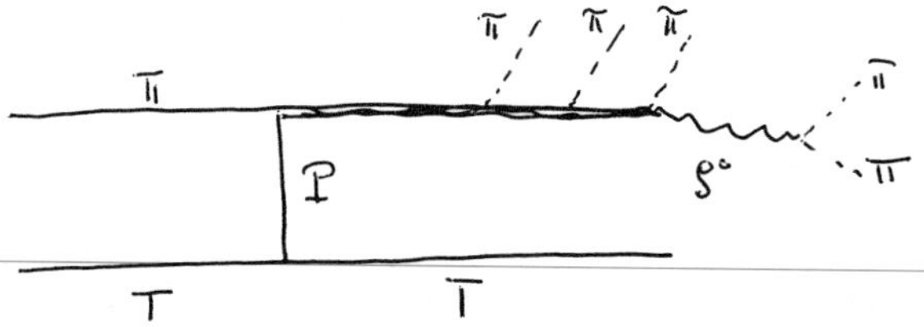

We first emit two pions successively and then a pion and a ρ^0 meson. The reason for this choice is the statistical bootstrap theory which favours decay of a heavy fireball into a pion of about 400 Mev and another heavy fireball. At the end of the chain instead of emitting two charged pions we emit a ρ^0 since the low mass two charged pion spectrum is dominated by the ρ^0 meson. Our meson spectrum in this problem consists of the charged pions, the ρ^0 meson and a continuum of non-exotic bosons. For neutral meson production the ω^0 would have to be included and maybe other mesons. Also higher multiplicity enhancements would easily get substantial contributions from more than one resonance in the decay chain. We have calculated that the emission of two ρ mesons in the 5π enhancement is a non-negligeable but not large contribution (of the order of 30%).

The calculation of the 5π production diagram above follows in a straightforward way using the methods outlined above. The result on hydrogen is a broad enhancement peaked at M ~ 2.1GeV. The results are shown in figure 9.

There is no data on hydrogen that we know of. There are however a series of measurements that have been made on complex nuclei.[13][14][15] Production on complex nuclei is normally analyzed assuming simply a "one step" optical model[16] as in the diagram below.

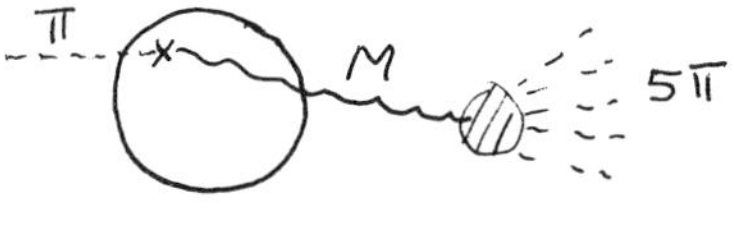

One has then for the coherent production amplitude on a nucleus[16]

$$F_A(M,\xi) = \mathcal{A}(s,0)\, N(M,\xi,\sigma'_{\pi},\sigma'_{M},A) \qquad (21)$$

$$\sigma' = \sigma^{tot}(1-i\alpha) \qquad (21a)$$

where $\mathcal{A}(s,0)$ is the production amplitude on a nucleon and N is an effective number depending on M and the mean free path of the fireball of Mass M in the nucleus with A nucleons. σ^{tot} is the total cross section of the incident pion or outgoing resonance. α is the ratio of real to imaginary part of the forward scattering amplitudes. There are corrections to this optical picture due to nuclear correlations. These are modest and I will not discuss them here.

Analyzing $\rho\pi$ data Bemporad et al find $\sigma_M \simeq 26$ mb independent of M. Analyzing 5π data, $\sigma_M \sim 17$ mb (but with substantial uncertainty). Since the enhancements are quite broad the independence of σ_M on M is rather remarkable. Figure 10 shows the measured 5π enhancement[13] with data for several complex nuclei superimposed. We also show the results of our calculation in figure 10. We have taken nuclear size into account using a simple gaussian form factor with mean radius 5 fm. The 5π peak which was at 2.1 GeV on hydrogen is now moved to 1.8 GeV on complex nuclei. Using our statistical theory we can also estimate the ratio of $\rho\pi\pi$ to $\rho\pi$ production. An experiment on Freon[14] yields the ratio $\sigma_{coh}(\rho\pi)/\sigma_{coh}(5\pi) = 40^{+26}_{-14}$. Our estimate using the Gaussian form factor and neglecting $\rho\rho\pi$ is 46. This number could change some if we include $\rho\rho\pi$ and if we do a more detailed correction due to nuclear size.

Experimental verification of our one step assumption for coherent production comes from the fact that in references (13) and (17) small t slopes for diffractive production on complex nuclei agree well with a one step assumption. Figure 11 shows the sensitivity of slope in coherent production to number of steps.

Electromagnetic Processes

In analogy with reactions of classes I and II we now consider electromagnetically induced reactions of the type

III $e^+ + e^- \to$ hadrons

IV $\gamma + p \to$ hadrons + p

For processes of class III we consider diagrams such as the following

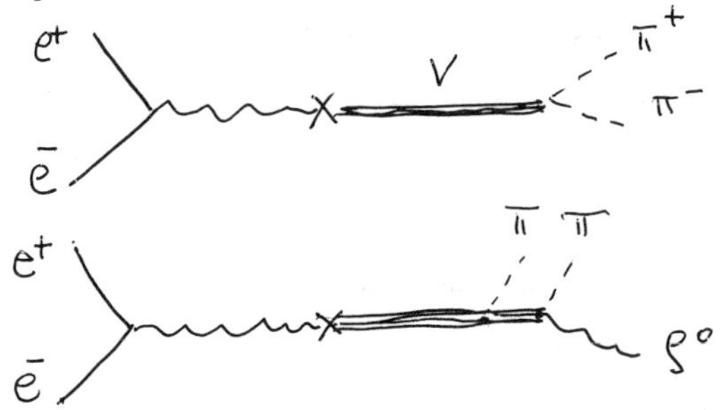

These processes have been treated as those of type I induced by hadrons. There are two differences between I and III type reactions. First one must use the appropriate formula for the density of vector meson states. Whereas the density of all hadronic states $\rho(M) \approx A/M^3\, e^{M/T}$, that for states of a given spin have an extra factor $M^{-3/2}$.

Further we must make an assumption about the vector dominance couplings. The statistical model does not tell us about vector dominance. However the fact that e^+e^- cross sections vary slowly with M dictates that the couplings are roughly inversely proportional to the square root of the density of vector meson states.

Recently Professor M. Pripstein and I have studied the question of possible similarities between the reactions.

$\bar{p}+p \rightarrow$ hadrons

$e^{+}+e^{-}$ hadrons

According to the statistical model the branching ratios for different hadronic final states should be very similar if we use only annihilation events for $\bar{p}$ p. We have made a compilation of $\bar{p}p$ data and compared the results with those for $e^{+}e^{-}$ recently presented at London by Professor Richter of SLAC. Comparing charged and neutral pion multiplicities as well as 2,4,6 and 8 prong cross sections the results for the two reactions are very similar. These comparisons are shown in the next two transparencies. We can understand the higher ratio of neutral to charged mesons as being the result of copious heavy boson emission by the s-channel fireballs. Detailed calculations are now in progress.

What is quite mysterious is shown in the next transparency. We see here that the $\bar{p}p$ total annihilation cross section has a similar energy dependence to that for the total $e^{+}e^{-}$ cross section (being roughly 10^{6} larger).

Since the data of Richter et al has not been published I will now show some calculations of electromagnetically induced hadron production using our statistical model compared with other data. Figure 12 shows a calculation of the pion form factor[18]. Figure 13 shows a calculation of the $\rho\pi\pi$ enhancement[19] (the ρ') along with some data.

Using the vector dominance assumption we have calculated the enhancement produced in the class IV reaction

$$\gamma + p \rightarrow \rho^{0}\pi^{+}\pi^{-} + p \qquad (24)$$
$$\quad\quad\quad\quad \hookrightarrow \pi^{+}\pi^{-}\pi^{+}\pi^{-}$$

This four pion enhancement is the statistical description of the bump similar to that observed in $e^+e^- \to \rho^0\pi^+\pi^- \to \pi^+\pi^-\pi^+\pi^-$. We assume that pomeron exchange is responsible for the reaction[18] and the results of the calculation are shown in figure 14 along with data. These same methods can be extended to calculate coherent photo-production on complex nuclei.

Final Remarks

The foregoing demonstrates a very general method for calculating the cross sections or branching ratios for a large number of exclusive channels in many reactions using a statistical approach. The main characteristic of the rates of production as a function of mass of the produced system is the rise from threshold and the rapid exponential fall off followed by some degree of coherence. It is essential that we treat resonances as resonances to get the proper behaviour. Four uncorrelated pions give a radically different threshold enhancement than $\rho\pi\pi$ for example.

The strong exponential fall with increasing mass of the produced system is a manifestation of competition from other channels. Although it is clearly possible to include such effects independent of a resonance superposition hypothesis we are inclined to propose that the picture presented above is a realistic one. We point out in this connection the recent evidence[20][21][22] for fluctuation phenomena in large angle π^- scattering at a few GeV. Such fluctuation phenomena and characteristic of a resonance superposition picture. Further experiments of this type would be very desireable. We note finally that the recent SLAC data on hadron production yields a thermodynamic spectrum of pions, kaons and anti-protons all corresponding to a hadronic temperature of 168 MeV which is very close to the temperature favoured by the statistical bootstrap theory.

FIG. 1

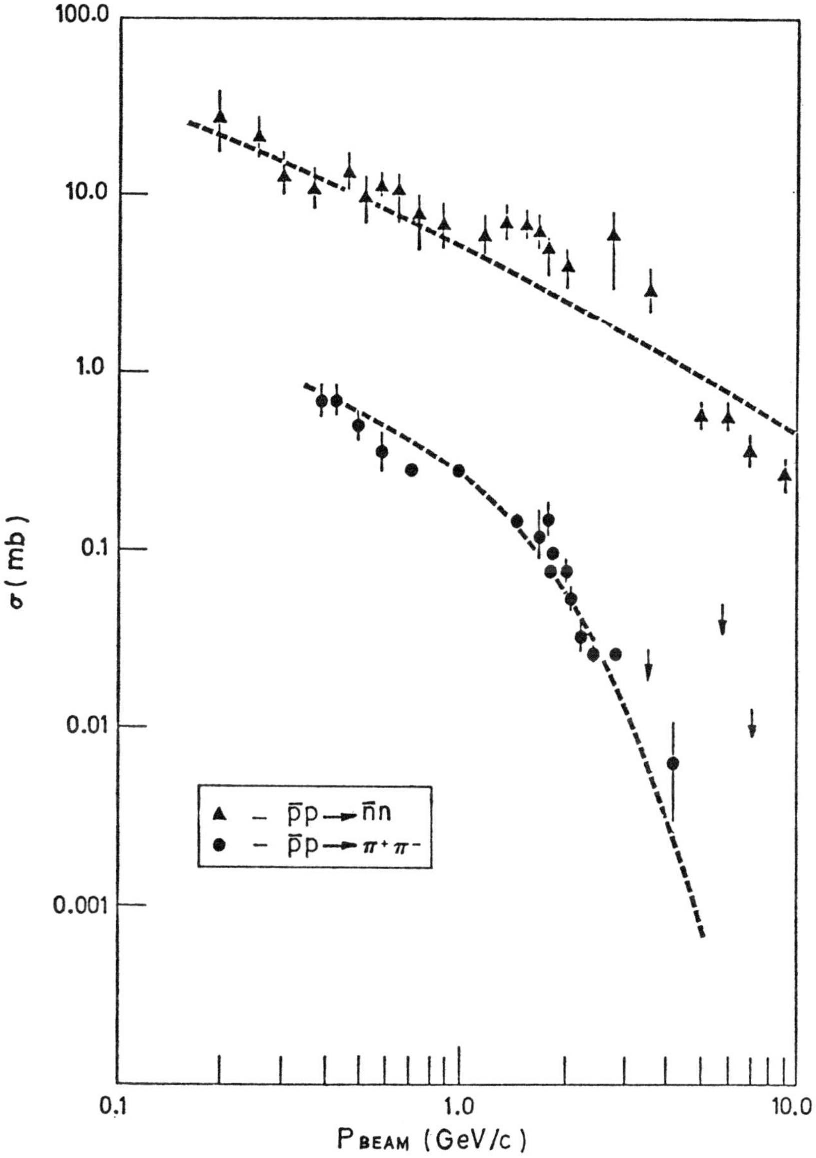

Figure 1 Data [23][24][25] for the reactions $\bar{p}p \rightarrow \bar{n}n$ and $\pi^+\pi^-$ along with our theoretical calculations.[4]

FIG. 2

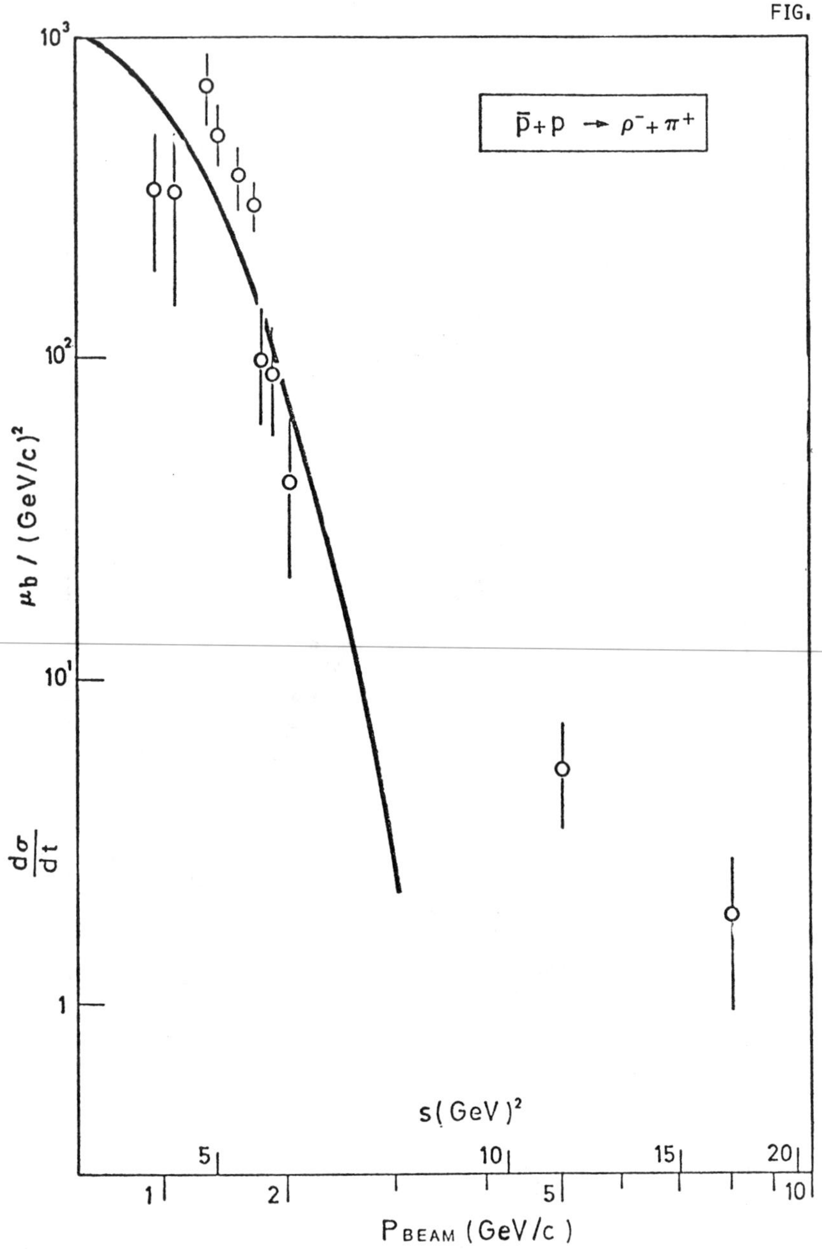

Figure 2 Data [25] for the reaction $\bar{p}p \rightarrow \rho\pi$ and our theoretical calculation [4] of the incoherent component.

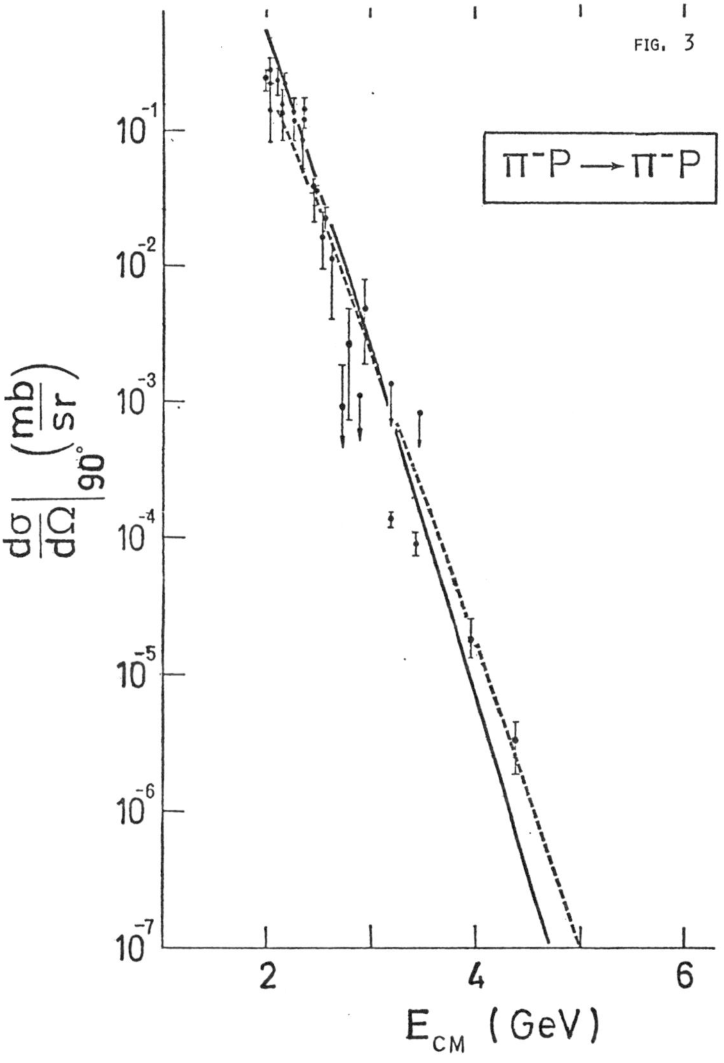

FIG 3 π^-p 90° elastic scattering calculation [5] and data.
The data compilation for figures 3-6 was made in reference 5.

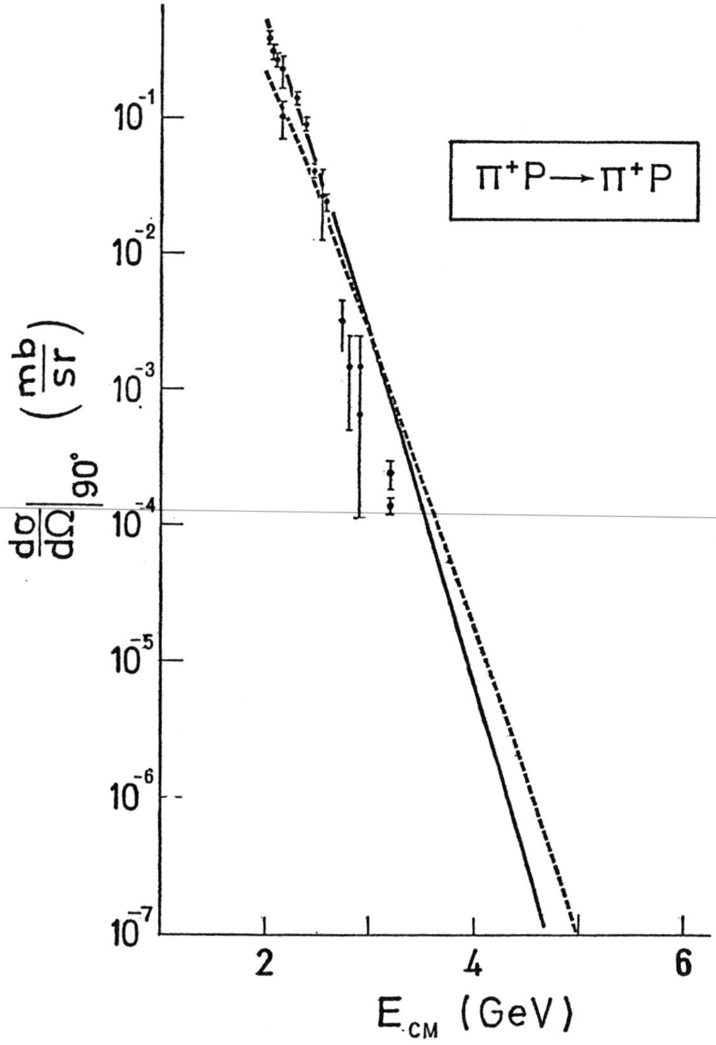

FIG. 4 π^- p 90° elastic scattering calculation (5) and data

FIG. 5

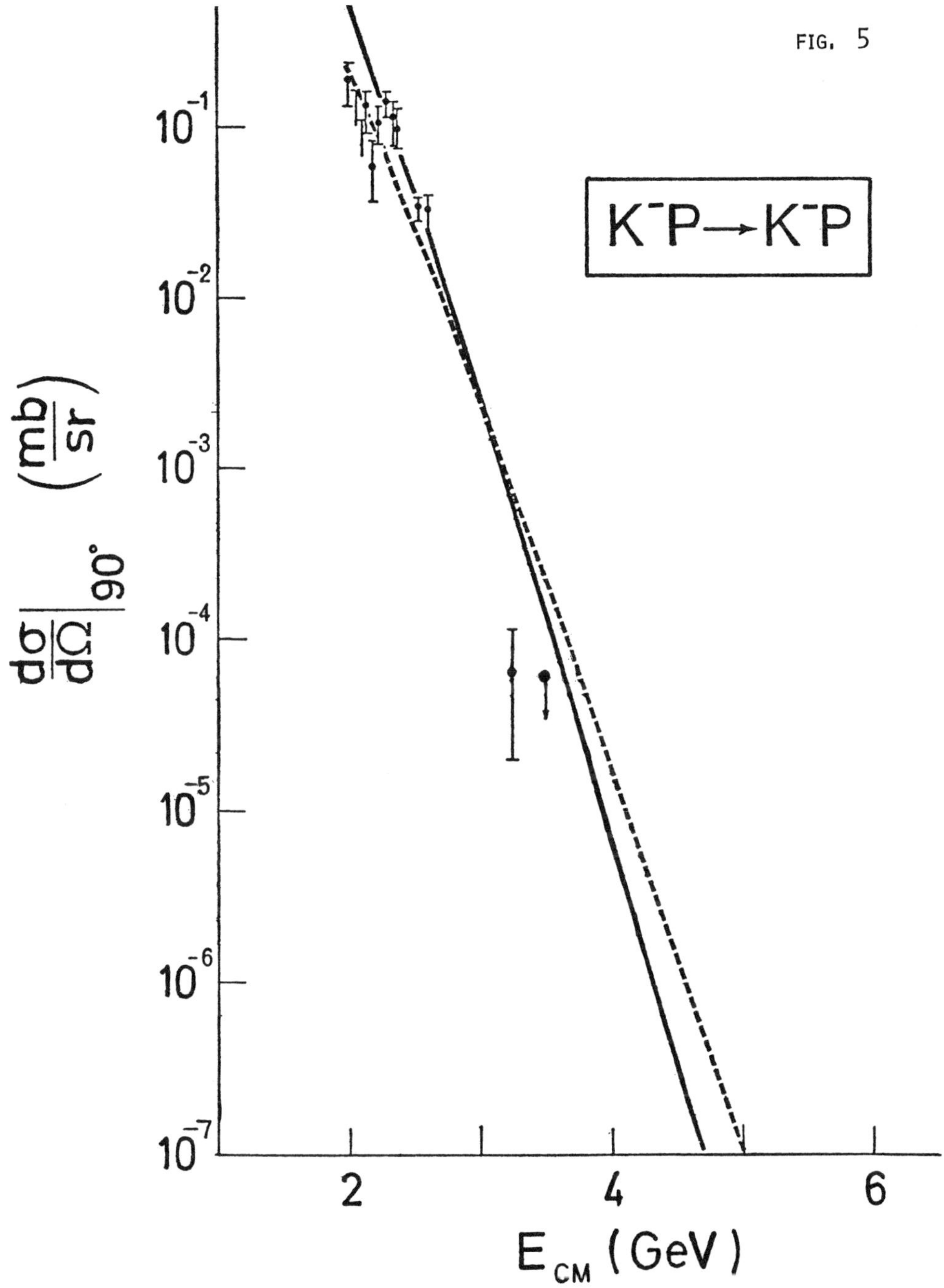

Figure 5 K^-p 90° elastic scattering calculation[5] and data.

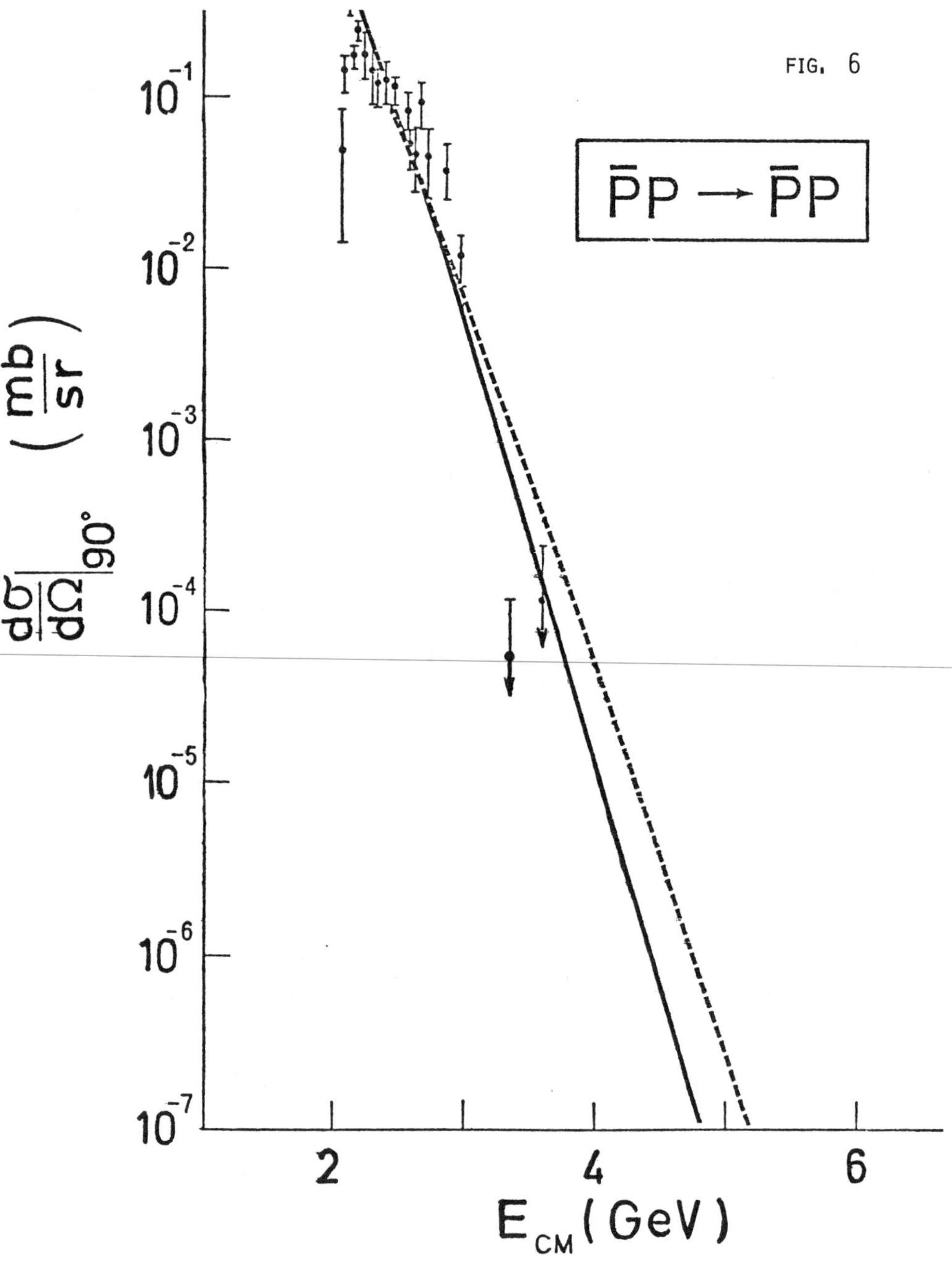

Figure 6 $\bar{p}p$ 90° elastic scattering calculation (5) and data.

FIG. 7

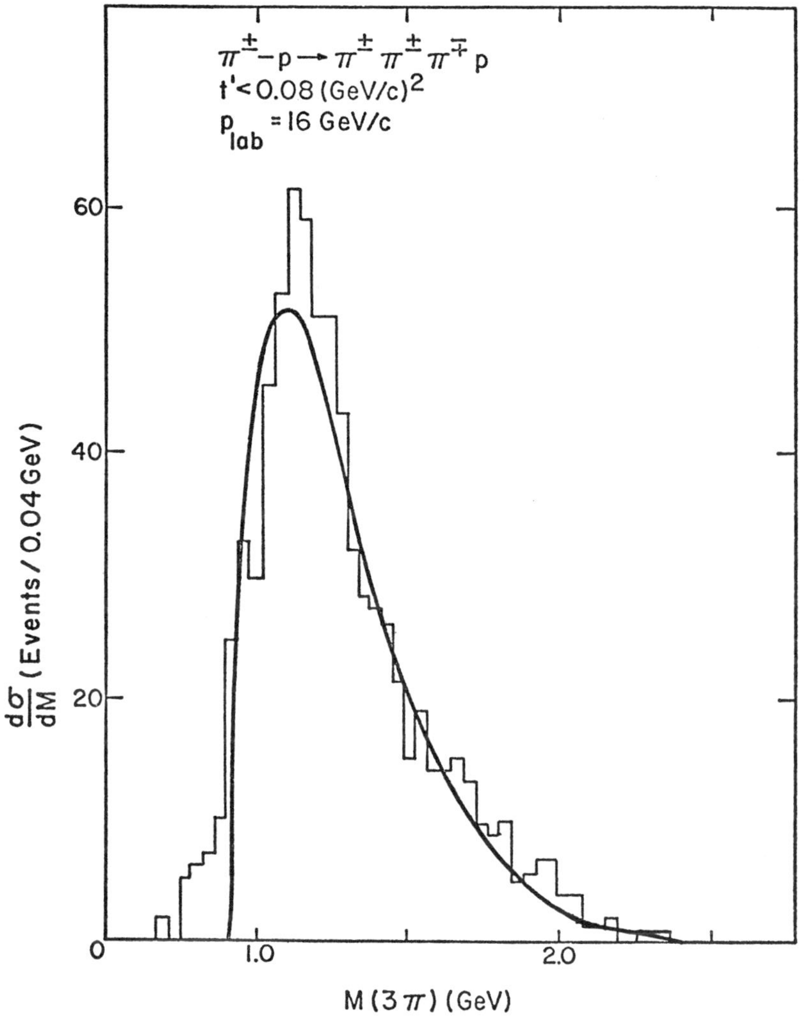

Figure 7 $\pi^{\pm}p \rightarrow \pi^{\pm}\pi^{\pm}\pi^{\mp}p$ mass spectrum[10] at 16 Gev/c. Data is from reference (26).

FIG. 8

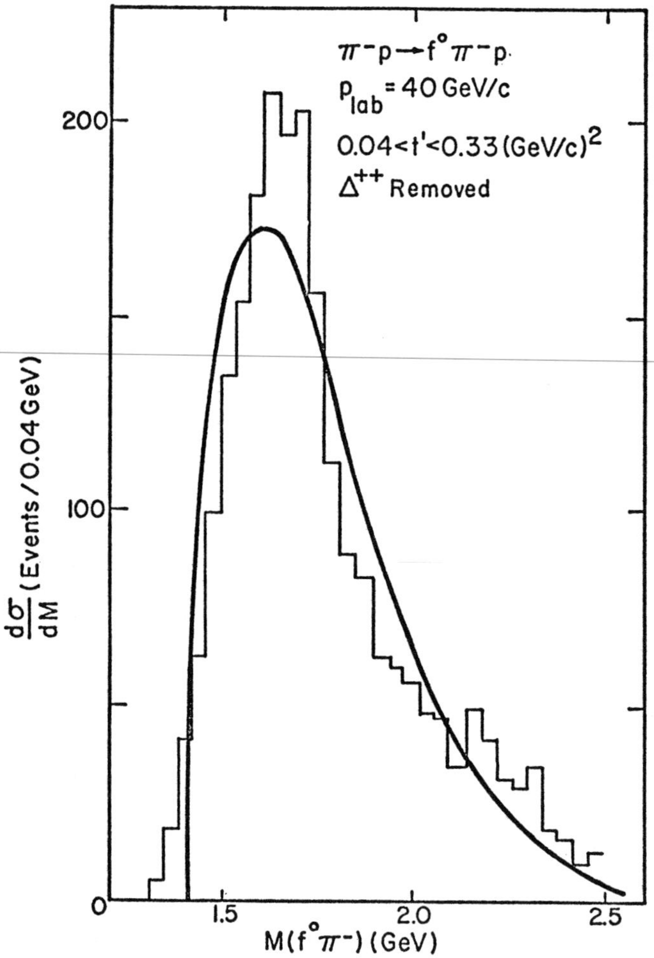

Figure 8 The A_3 mass distribution[10] ($f^o\pi$) in the reaction $\pi^- p \to \pi^+ \pi^- \pi^- p$ at 40 Gev/c. Data from reference (27).

FIG. 9

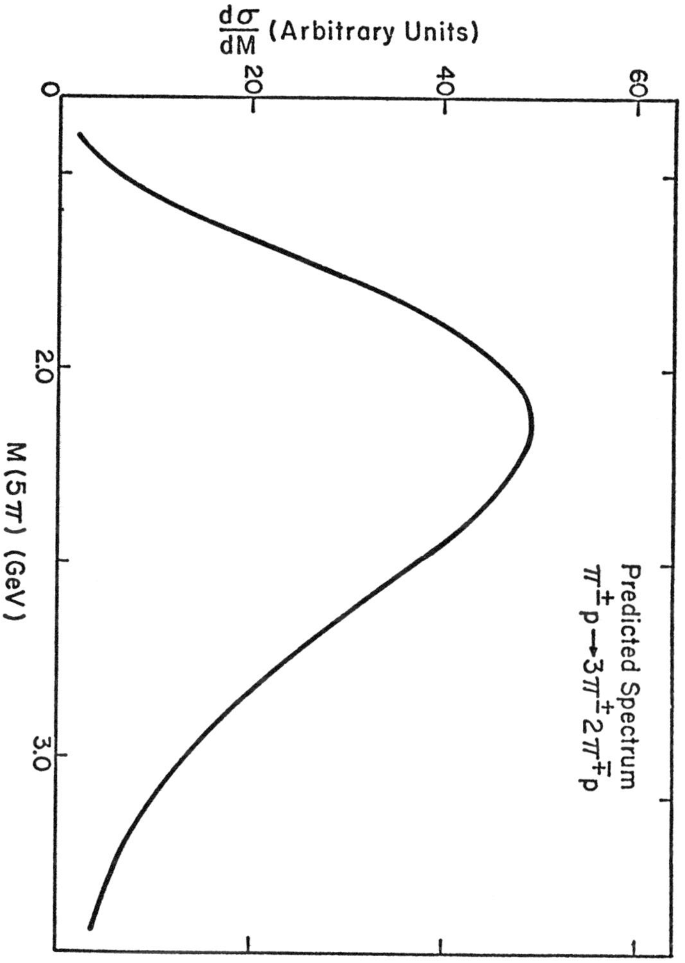

Figure 9 Calculated spectrum[10] for the reaction $\pi p \to \rho' \pi\pi\pi p$.

FIG. 10

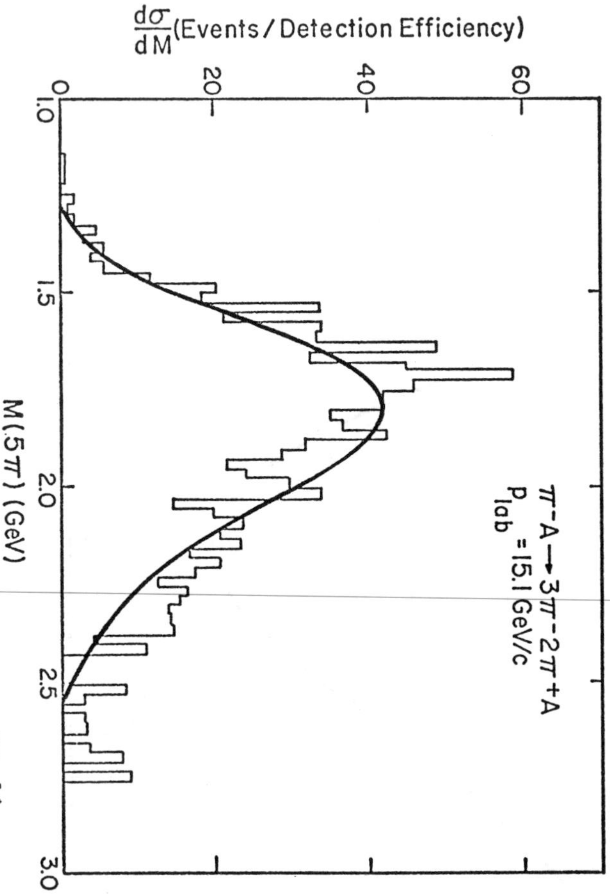

FIG. 10.

<u>Figure 10</u> Calculated spectrum[10] for the reaction $\pi A \rightarrow (\rho\pi\pi\pi) A$ on a series of complex nuclei. Data is from reference (13).

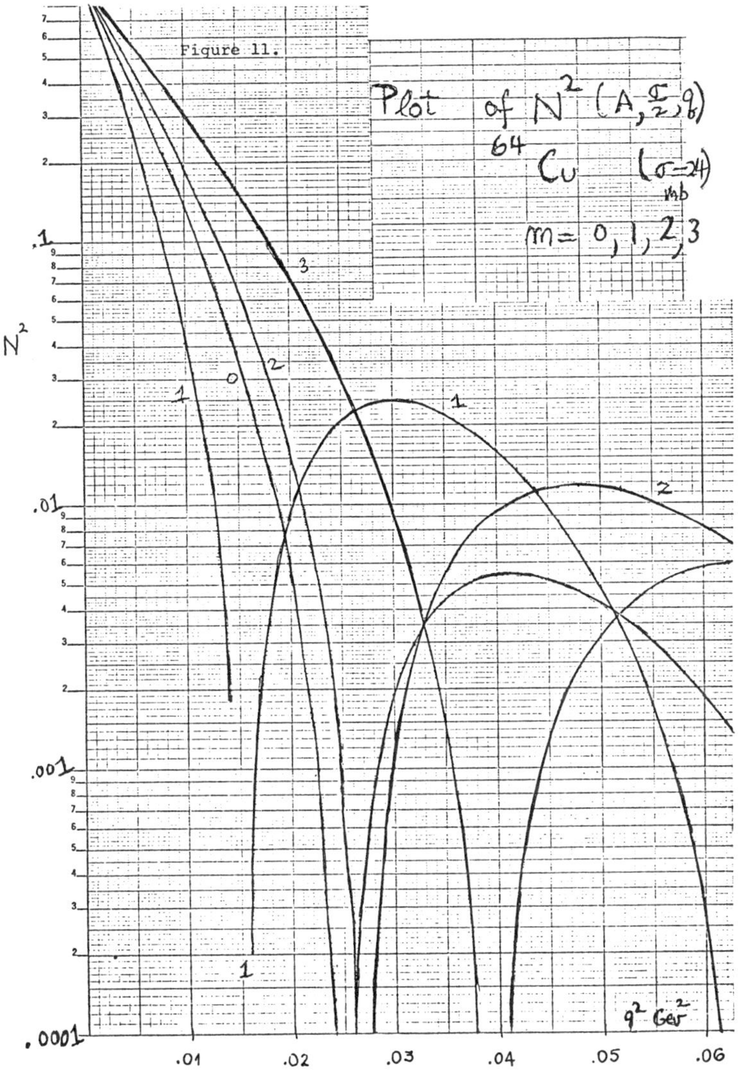

Figure 11 Sensitivity of diffraction pattern to number of steps. Calculation of B. Margolis and P. Valin (unpublished).

FIG. 12

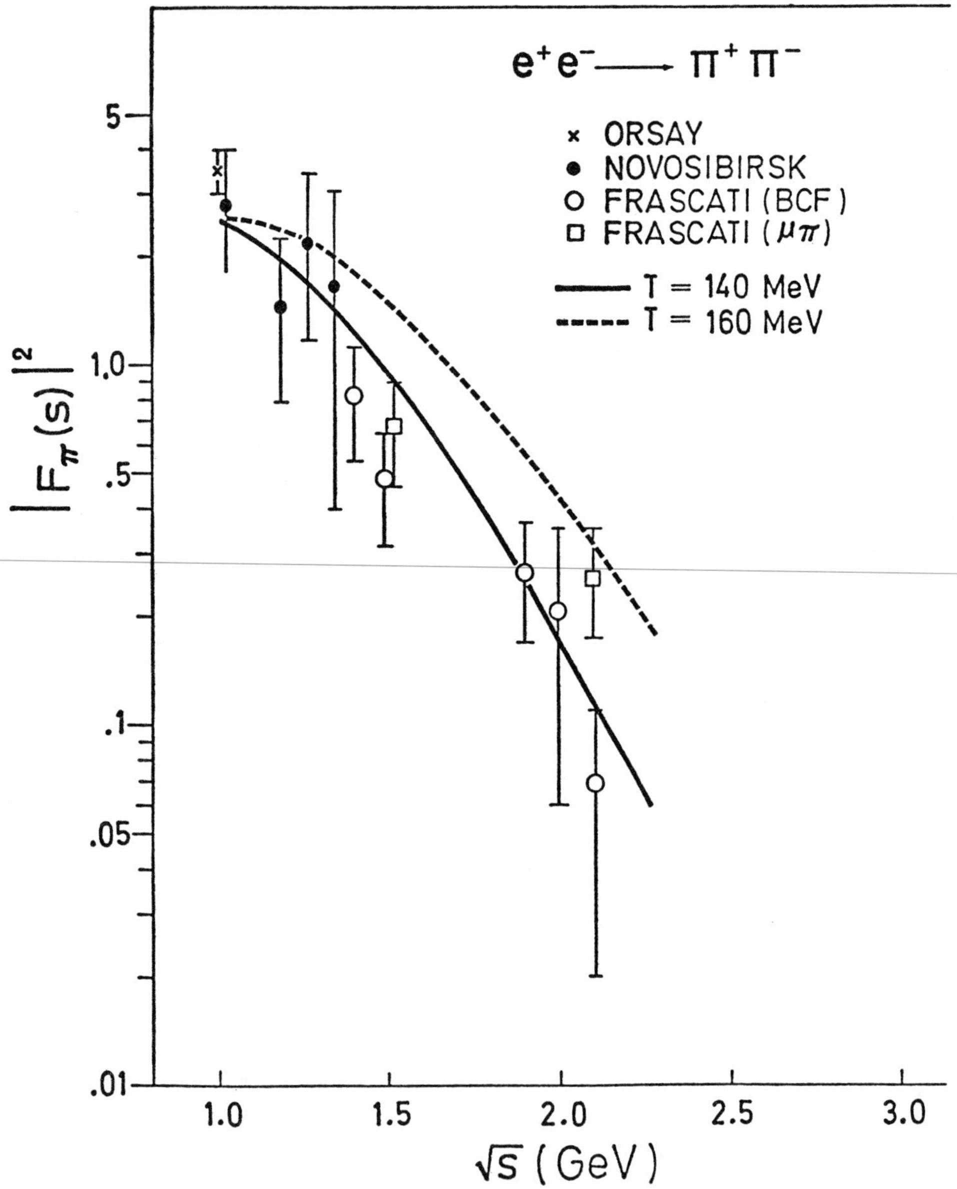

Figure 12 The pion form factor calculated[18] as outlined. The data is listed in reference (18).

FIG. 13

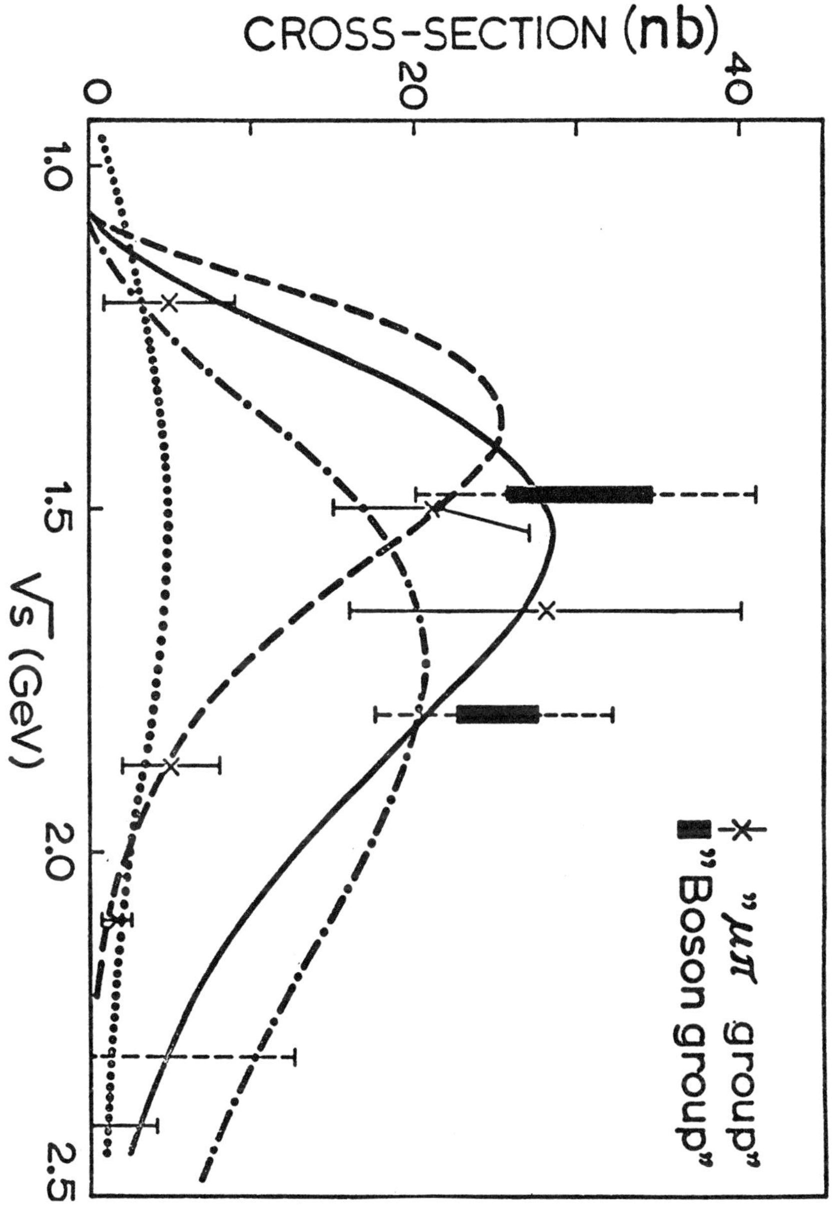

Figure 13 Calculation of $(e^+e^- \to \rho^0\pi^+\pi^-)$ from reference (19) for T=140 Mev---; T=100 Mev---, T-180Mev-.-.-. The shape of the cross section for four uncorrelated pions is given for T=140Mev...... Date is from references (28) and (29).

FIG. 14

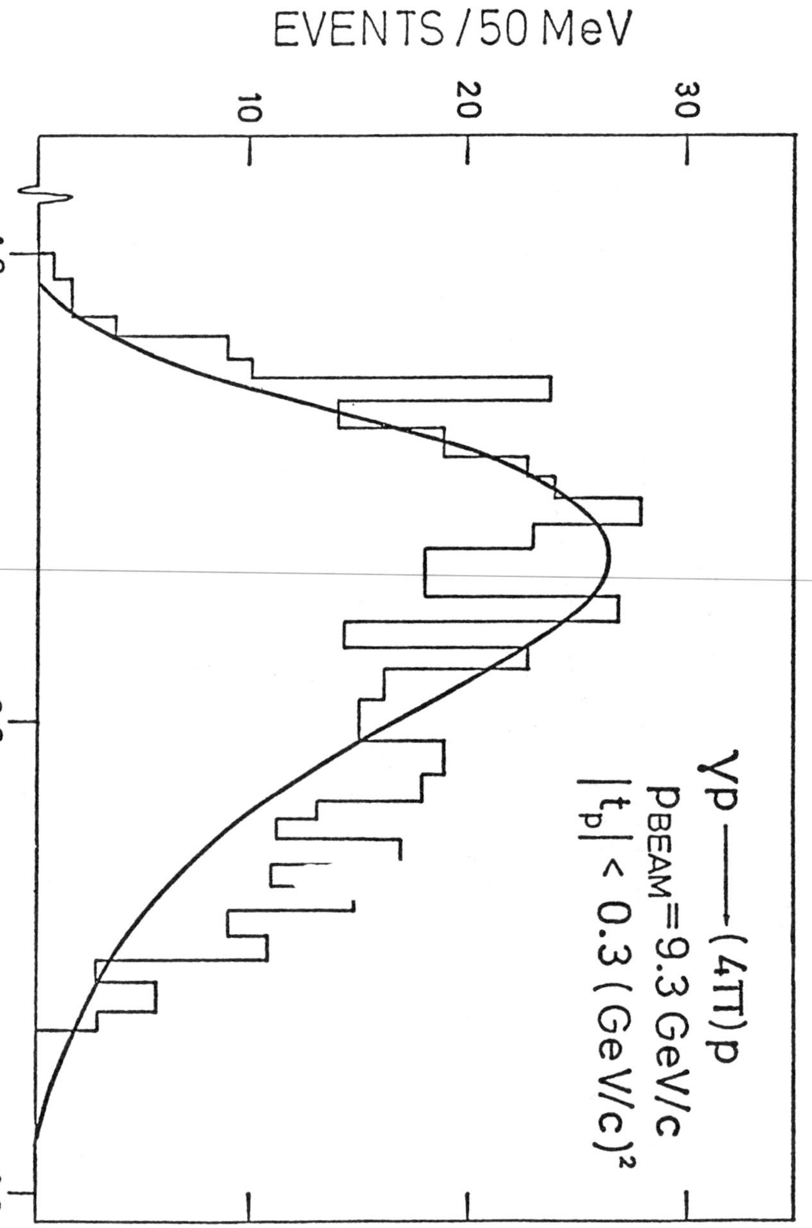

Figure 14 Calculation [18] of the reaction $\gamma p \to \rho \pi\pi p$ at 9.3 Gev/c

Data from reference (30).

REFERENCES

1) E. Fermi, Prog. Theor. Phys. 5, 570 (1950).

2) R. Hagedorn, Nuovo Cim. Suppl. 3, 147 (1965).

3) S. Frautschi, Phys. Rev. D3, 2821 (1971), See also W. Nahm, Nucl. Phys. B45, 525 (1972).

4) B. Margolis, W. J. Meggs and R. K. Logan, Phys. Rev. D8, 199 (1973)

5) G. Eilam, Y. Gell, B. Margolis and W. J. Meggs, Phys. Rev. D8, 2871 (1973).

6) R. L. Anderson et al., Phys. Rev. Lett. 30, 627 (1973).

7) C. J. Hamer and S. Frautschi Phys. Rev. D4, 2125 (1971).

8) R. Hagedorn and J. Ranft Nucl. Phys. B48, 157 (1972).

9) I. Montvay, Phys. Lett. 42B, 466 (1972).

10) B. Margolis and S. Rudaz, Phys. Rev. D9, 653 (1974).

11) A. Mueller, Phys. Rev. D2, 2963 (1970).

12) F. Johns, B. Margolis, W. J. Meggs and R. K. Logan, Phys. Rev. Lett. 29, 756 (1972).

13) C. Bemporad et al, Nucl. Phys. B33, 397 (1971); B42, 627 (1972)

14) R. Huson et al, Phys. Lett. 28B, 208 (1968).

15) K. Paler et al, Phys. Rev. Lett. 26, 1675 (1971).

16) K. S. Kolbig and B. Margolis, Nucl. Phys. B6, 85 (1968).

17) W. Carithers et al, Diffractive Production of $p\pi^-$ Systems on Nuclei.... (Submitted to London Conference July 1974, Rochester preprint.)

18) B. Margolis, W. J. Meggs and S. Rudaz, Phys. Rev. D8, 3945 (1973).

19) S. Rudaz, Nuovo Cim. Lett. 6, 292 (1973).

20) P. J. Carlson, Phys. Lett. B45, 161 (1973).

21) F. H. Schmidt et al, Phys. Lett. B45, 157 (1973).

22) R. Henzi, B. Margolis and W. A. Ross, Phys. Rev. D6, 1916 (1972).

23) Particle Data Group, Report No. L.B.L.-58 (unpublished.)

24) J. J. Yoh et al, Phys. Rev. Lett. 26, 922 (1571). D. Birnbaum et al,

REFERENCES - CONT'D

24) Phys. Rev. Lett. 23, 433 (1969).

25) A. Eide et al, Phys Lett. 23, 433 (1969)

26) J. Ballam et al, Phys. Rev. D4, 1946 (1971).

27) Study of A_3 in $\pi^- p \to \pi^+ \pi^- \pi^- p$, CIBS collaboration in Proc. of the Third Philadelphia Conference on Experimental Meson Spectoscopy-1972; P. 185

28) B. Barbarino et al, Nuovo Cim. Lett. 3, 689 (1972)

29) B. Bartoli et al, Frascati Report LNF-71/91 (1971).

30) Y. Eisenberg et al, Phys. Lett. 43B, 149 (1973)

DISCUSSION

BRODSKY (SLAC): What is your interpretation of P-P at 90°, considering that you can use crossing from non exotic channels to exotic ?

MARGOLIS: We have a long way to go to the cross sections, but I will try to answer your question. The difference here is related to whether we have two or more fire balls. When we calculate the density of two fire ball states we find the deflection of density in one fire ball state of the same mass. If you assume that the 90° cross section goes as 1/density of states, then the P-P will be higher because you have to make at least two fire balls.

Invited Talk

TWO-PARTICLE AMPLITUDE AND MANYPARTICLE NUCLEAR PROCESSES

V. V. KOMAROV

Institute of Nuclear Physics, Moscow State University

Moscow, USSR

A possibility to obtain two-particle potential parameters from the investigation of the differential cross sections of the manyparticle reactions is discussed in the talk.

The indefinities of the two-nucleon forces remain the base problem of two and many-particle scattering and the most important question in the nuclear physics. Therefore each attempt to approach the solution of the problem seems to be important.

Two-particle scattering experiments and phase analysis cannot help because of uncertainty of the extractions the two-nucleon potential parameters from that. Here is one from many other examples which shows that the two-nucleon experimental phase could be well fitted by different sets of variable parameters of the same potential I). The results of phase analysis of the nucleon-nucleon scattering within the energy limits from 0 to 100 MeV 2) were applied to the extraction of the Woods-Saxon potential parameters. It was done by exact solution of the Shrodinger equation for the free nucleon-nucleon scattering with the potential

$$V = - \frac{V_0}{1 - \exp(r - r_0/a)} ,$$

with the parameters Vo, r_0, a and r_c, where r_c is a hard core radius. The phase values in the energy region below 10 MeV were defined by scattering length and effective radius. It was shown thay by fixed parameters " r_c " and " a " there are domains on the (V_0, r_0) sheet where each point gives a different set of V_0, r_0, r_c and a values. The potential with these values of its parameters permits one to describe the behaviour of the nucleon-nucleon scattering phase within the same χ^2 deviation limits between theoretical and experimental phase curves. Fig. I. displays these domains of the Woods-Saxon potential parameters for the case of $\chi^2 \leq 2$ by a=0.I; 0,2; and 0.3 fm. There are fixed r_c solid curves on the figures. The domains of χ^2 equal meanings are shown by dashed lines. From this it follows that the experimental results of two-nucleon scattering could be well explained by the Woods-Saxon potential with the considerable variation of depth by proper values of r_0, a and r_c. It is easy to choose other examples of incompleteness the two-particle experiments for unique determination of two-particle potential parameters. By this situation it is obviously important to try to find some ways to overcome the uncertainty.

One of the possible ways which is now developing by some people is an analysis of the off-shell behaviour of the two-particle scattering amplitude $t(\bar{k}, \bar{k}', \varepsilon)$ as a function of relative momenta $\bar{k}$ and $\bar{k}'$ before and after scattering, and the relative energy ε. 3)

The sense of that analysis is obvious because if by phase analysis of the two-nucleon experiments there is one point of the amplitude meaning on the energy shell for each definite ε, the off-shell investigation gives a surface of $t(\bar{k}, \bar{k}', \varepsilon)$ variation in the k and k' space. This permits one to obtain some extra information about the two-particle potential parameters and we hope to avoid the above-mentioned uncertainty by performing some experiments on the off-shell $t(\bar{k}, \bar{k}', \varepsilon)$ behaviour.

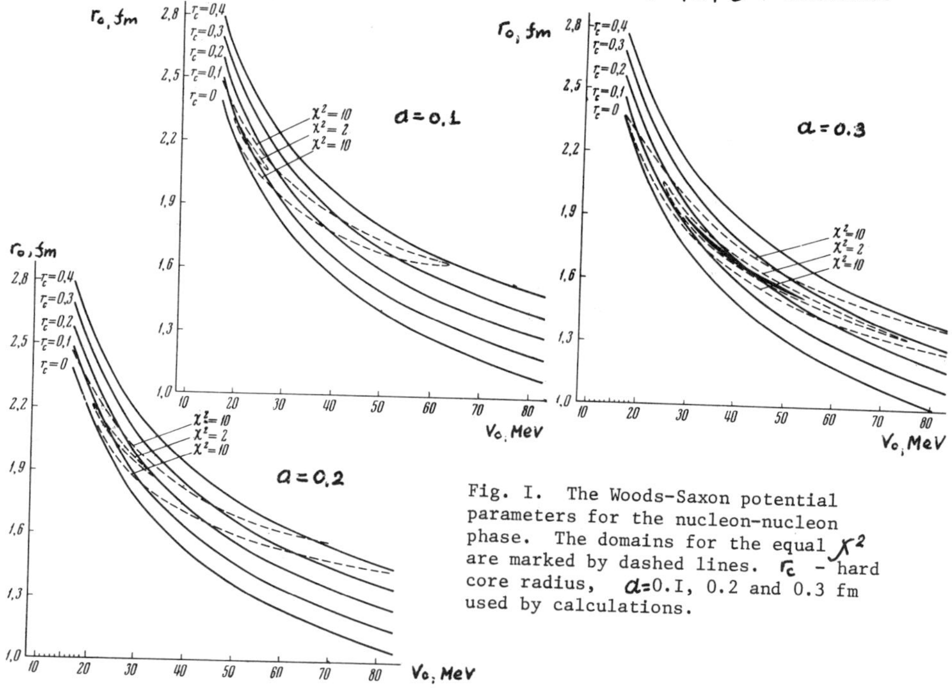

Fig. I. The Woods-Saxon potential parameters for the nucleon-nucleon phase. The domains for the equal χ^2 are marked by dashed lines. r_c - hard core radius, a=0.I, 0.2 and 0.3 fm used by calculations.

The off-shell behaviour of the two-particle amplitude may be obtained experimentally from many-body nuclear reactions. Indeed, if many-body reactions can be treated as the many-body scattering problem 4), then the integral equations of the scattering theory are applicable for the determination of the reaction amplitude. By assumption the two-body nuclear forces, the free term and the kernel of the many-body problem integral equation contains the function $t(\bar{K}, \bar{K}', \mathcal{E})$. This function is two-particle scattering amplitude analytical continue to the off-energy shell over variables $\bar{K}, \bar{K}'$ and $\mathcal{E}$.

This structure of the integral equations for many-particle reaction amplitudes opens the principal possibility of the determination of the two-particle amplitude behaviour and therefore the two-body force properties after the many-body nuclear reaction analysis. For the choice of the kinematical conditions of the proper experiments on the many-body nuclear reactione one should first investigate the main properties of the two-particle amplitude which follows from the mathematical determination of it.

Let us consider the analytical properties of the two-particle scattering amplitude $t(\bar{K}, \bar{K}', \mathcal{E})$ within the integral equation frames. It is well known that as a function of variable $\mathcal{E}$ the amplitude $t(\bar{K}, \bar{K}', \mathcal{E})$ is an analytical function on the whole of the complex $\mathcal{E}$ -sheet except for a few poles corresponding to bound states and a cut along the real axis. The properties of $t(\bar{K}, \bar{K}', \mathcal{E})$ as a function of $\mathcal{E}$ are defined by an integral equation which has a compact kernel containing a potential. The integral equation for the two-nucleon scattering amplitudes is of Fredholm type under the conditions that the potential is limiting and continuous 6). Thus the amplitude $t(\bar{K}, \bar{K}', \mathcal{E})$ analytical properties over $\mathcal{E}$ do not depend on the potential shape 5) but on the above conditions only.

For the nuclear force detailed investigation it is most important to know the analytical properties of $t(\bar{K}, \bar{K}', \mathcal{E})$ over the value of variables $\bar{K}, \bar{K}'$ and the angle between these vectors, $\vartheta(\bar{K}, \bar{K}')$. The behaviour of $t(\bar{K}, \bar{K}', \mathcal{E})$ as a function of $\bar{K}$ and $\bar{K}'$ is defined by introducing the shape of a two-particle potential. Of course, the amplitude $t(\bar{K}, \bar{K}', \mathcal{E})$. should be in the region of analyticity over $\mathcal{E}$. First we note that the analytical properties of $t(\bar{K}, \bar{K}'; \varepsilon)$ on the complex K-sheet are equal to those on the complex K'-sheet. It follows from the fact that the scattering operator $\hat{t}$ can be defined from the equation

$$\hat{t} = \hat{V} + \hat{V} G_0 \hat{t}$$

as well as from

$$\hat{t} = \hat{V} + \hat{t} G_0 \hat{V}$$

where $\hat{V}$ is an interaction operator and G_0 is a Green function operator for the free motion. For the function $t(\bar{K}, \bar{K}', \varepsilon)$ there also exist two equations, respectively:

$$t(\bar{K}, \bar{K}'; \varepsilon) = v(\bar{K}, \bar{K}') + \int v(\bar{K}, \bar{p})(\varepsilon - p^2/2m)^{-1} t(\bar{p}, \bar{K}'; \varepsilon)\, d\bar{p},$$

and

$$t(\bar{K}, \bar{K}'; \varepsilon) = v(\bar{K}, \bar{K}') + \int t(\bar{K}, \bar{p}, \varepsilon)(\varepsilon - p^2/2m)^{-1} v(\bar{p}, \bar{K}')\, d\bar{p},$$

where $v(\bar{K}, \bar{K}')$ is an interaction function in the momentum space.

Second, we note that if a potential is of limiting radius, $t(\bar{K}, \bar{K}'; \varepsilon)$ is an analytical function over $\bar{K}, \bar{K}'$ and $\vartheta(\bar{K}, \bar{K}')$ on complex sheets of each variables. For a potential of infinite radius one has to find the analytical regions for $t(\bar{K}, \bar{K}'; \varepsilon)$ for each particular case. For example, this region for the Yukawa potential is determined by the conditions:

a) over K and K': $|\mathrm{Im}\, K| < \mu/3$ and $|\mathrm{Im}\, K'| < \mu/3$ for arbitrary finite ReK and ReK', where μ is the Yukawa parameter.

b) over z: $|z| < z_0$, if $z_0 > 1$ and by existence of the above conditions for K and K'. $(z = \vartheta(\bar{K}, \bar{K}')$.

These properties are easy to be established from the Lippman-Shwinger equation and the spectral representation for the scattering two-particle amplitude 7).

All the above-mentioned analytical properties of the $t(\bar{K}, \bar{K}'; \varepsilon)$ help one to find the off-shell behaviour of this amplitude as a function of $\bar{K}$ and $\bar{K}'$ in the dependence of a two-particle potential shape. Bearing in mind that $t(\bar{K}, \bar{K}'; \varepsilon)$ is an analytical function over $\bar{K}$ and $\bar{K}'$ in the large regions, one may understand its behaviour in the dependence of a potential shape variety, using for it the Lippman-Shwinger equation for the partial amplitude

$$t_l(K, K'; \varepsilon) = v_l(K, K') + \frac{2}{\pi} \int_0^\infty \frac{p^2 dp}{m\varepsilon - p^2} t_l(K, p, \varepsilon)\, v_l(p, K'),$$

where

$$v_\ell(\kappa,\kappa') = \int_0^\infty r^2 \frac{J_{\ell+1/2}(\kappa r)}{\sqrt{\kappa r}} \cdot \frac{J_{\ell+1/2}(\kappa' r)}{\sqrt{\kappa' r}} v(r)\, dr$$

It is well known that the solution of this equation in analytical form is possible only for the square well potential, but an investigation permits one to draw definite conclusions about the solution for an arbitrary potential shape.

Since the equation for $t_\ell(\kappa,\kappa';\varepsilon)$ could be rewritten in the form:

$$t_\ell(\kappa,\kappa';\varepsilon) = v_\ell(\kappa,\kappa') + \frac{2}{\pi}\int_0^\infty \frac{p^2dp}{m\varepsilon-p^2} t_\ell(\kappa,p,\varepsilon)\, v_\ell(p,\kappa'),$$

the $t_\ell(\kappa,\kappa';\varepsilon)$ behaviour as a function of κ and κ' is defined mainly be the free term of the equations, i.e., by the function $v_\ell(\kappa,\kappa')$

It is possible to consider also some extremal regions of meanings, where approximate solutions could be found.

When $\kappa \to 0$ (or $\kappa r_0 \ll$ I, where r_0 is a nuclear force radius), one obtains the following approximate formulae

$$v_\ell(\kappa,\kappa') \simeq \kappa^\ell g_\ell(\kappa'),$$

and for

$$t_\ell(\kappa,\kappa';\varepsilon) \simeq \kappa^\ell \tau_\ell(\kappa';\varepsilon),$$

where

$$\tau_\ell(\kappa';\varepsilon) = g_\ell(\kappa') + \frac{2}{\pi}\int_0^\infty \frac{p^2dp}{m\varepsilon-p^2} g_\ell(p)\, t_\ell(p,\kappa';\varepsilon).$$

When $\kappa \to \infty$ (or $\kappa r_0 \gg 1$) the term $J_{\ell+1/2}(\kappa r)/\sqrt{\kappa r}$ spherical Bessel function is oscillating fast and decreasing.

One can see that in both limiting cases the $t_\ell(\kappa,\kappa';\varepsilon)$ behaviour over κ and κ' is independent of the potential shape and is determined only by the Bessel functions. This result is also valid for the small energy and for the half shell case, when $\kappa' = \sqrt{m\varepsilon}$. This situation will be used for the proposed experiments. Thus, for $t_\ell(\kappa,\kappa';\varepsilon)$ one may obtain by $\varepsilon \simeq 0$ and $\kappa' = \sqrt{m\varepsilon}$

a) when $\kappa r_0 \ll 1$, $t_\ell(\kappa,\sqrt{m\varepsilon},\varepsilon) \simeq \kappa^\ell \tau_\ell(\varepsilon)$.

For small energies $t_\ell(\kappa,\kappa';\varepsilon)$ depends on κ in a very simple way and the two-particle scattering s-wave is constant.

b) when $\kappa r_0 \gg 1$, $t_\ell(\kappa,\sqrt{m\varepsilon},\varepsilon)$ is oscillating and decreasing fast and it is also independent of the potential shape.

An important conclusion follows from the expression for $v_\ell(\kappa,\kappa')$ and both limiting cases, that the dependence $t_\ell(\kappa, \sqrt{m\varepsilon}, \varepsilon)$ on the potential shape (or parameters) should be by $\kappa r_0 \simeq 1$ condition.

These statements are proven also by calculations of ref. 9). The Pade approximation was used in these calculations for the Lippman-Shwinger equation, and $t_\ell(\kappa, \kappa', \varepsilon)$ behaviour was analysed for the Woods-Saxon and square well potentials. Figs. 2 and 3 display the results of these calculations of off-shell $t_\ell(\kappa, \kappa', \varepsilon)$, $(\kappa = \kappa', \kappa' \neq \sqrt{m\varepsilon})$ and half-shell $t_\ell(\kappa, \kappa', \varepsilon)$ $(\kappa \neq \kappa', \kappa' = \sqrt{m\varepsilon})$ for κ values from 0.I to 4.0 fm^{-I}, by fixed ε values 4.0, 8.0, 12.0, 20.0, 160 and 320 MeV. The Woods-Saxon parameters were used from the domains of uncertainty of fig. I. For the swuare well potential calculations the depth and width are also determined by the condition $\mathcal{K}^2 \leq 2$.

There are pairs of curves for $Re\, t_\ell(\kappa, \kappa', \varepsilon)$ with the same fixed ε and with the potential parameters from the same domain. The values of the parameters differ from each other as much as possible inside the same domain. The behaviour of $Im\, t_\ell(\kappa, \kappa', \varepsilon)$ is nearly the same, but its absolute value is ten times less.

In figs. 2 and 3, note that the curves go irrespective of ε and the potential shape for s-wave and $\kappa r_0 <$ I. For $\kappa r_0 >$ I there is also no difference between the curves calculated for the different potential shapes. This is in a good agreement with the above analysis of the $t_\ell(\kappa, \kappa', \varepsilon)$ behaviour. As follows from the analysis, the greatest difference in the curves calculated for the different potential parameters one may note for κ values of the $1/r_0$ order of magnitude, where r_0 is a potential radius. In other words, by $\kappa r_0 \simeq 1$ the curves are grouping in its form for the same potential parameters independently to the fixed energy value. The curvature increases for the potential depth decreasing and width increasing. While going from the $\kappa r_0 \ll 1$ to $\kappa r_0 >$ I, the $Re\, t_0(\kappa, \kappa', \varepsilon)$ curve began to bend for $\kappa r_0 \simeq 1$.

To all appearance, the behaviour of $t_0(\kappa, \kappa', \varepsilon)$ by $\kappa \simeq 0.5 \div 1.5\, fm^{-1}$ permits us to extract from the proper experiments the main tendency in the two-particle potential properties. The same behaviour is seen for $\ell =$ I and expected to be for $\ell > 1$ therefore it is possible to plan experiments for the wide energy ε range. We propose an experiment for a small ε when the reaction cross sections permit us to conclude about the $t_0(\kappa, \kappa', \varepsilon)$ behaviour.

Let us discuss now experimental possibilities of investigation of the off-shell

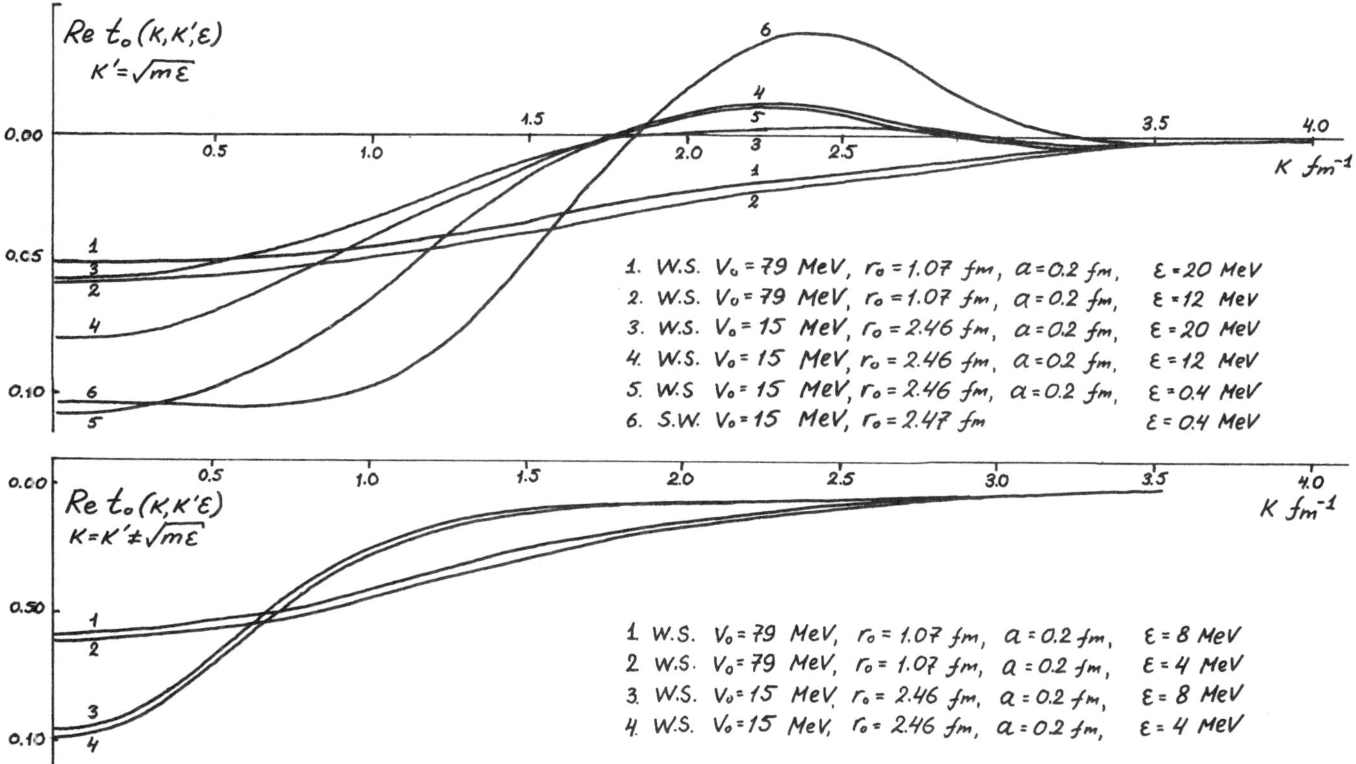

Fig. 2. The $t_e(\kappa,\kappa';\varepsilon)$ amplitude behaviour as a function of κ and κ', calculated from the Lippman-Shwinger equation by Pade approximation. Curves are for different values of the Woods-Saxon and square well potential parameters by fixed ε values.

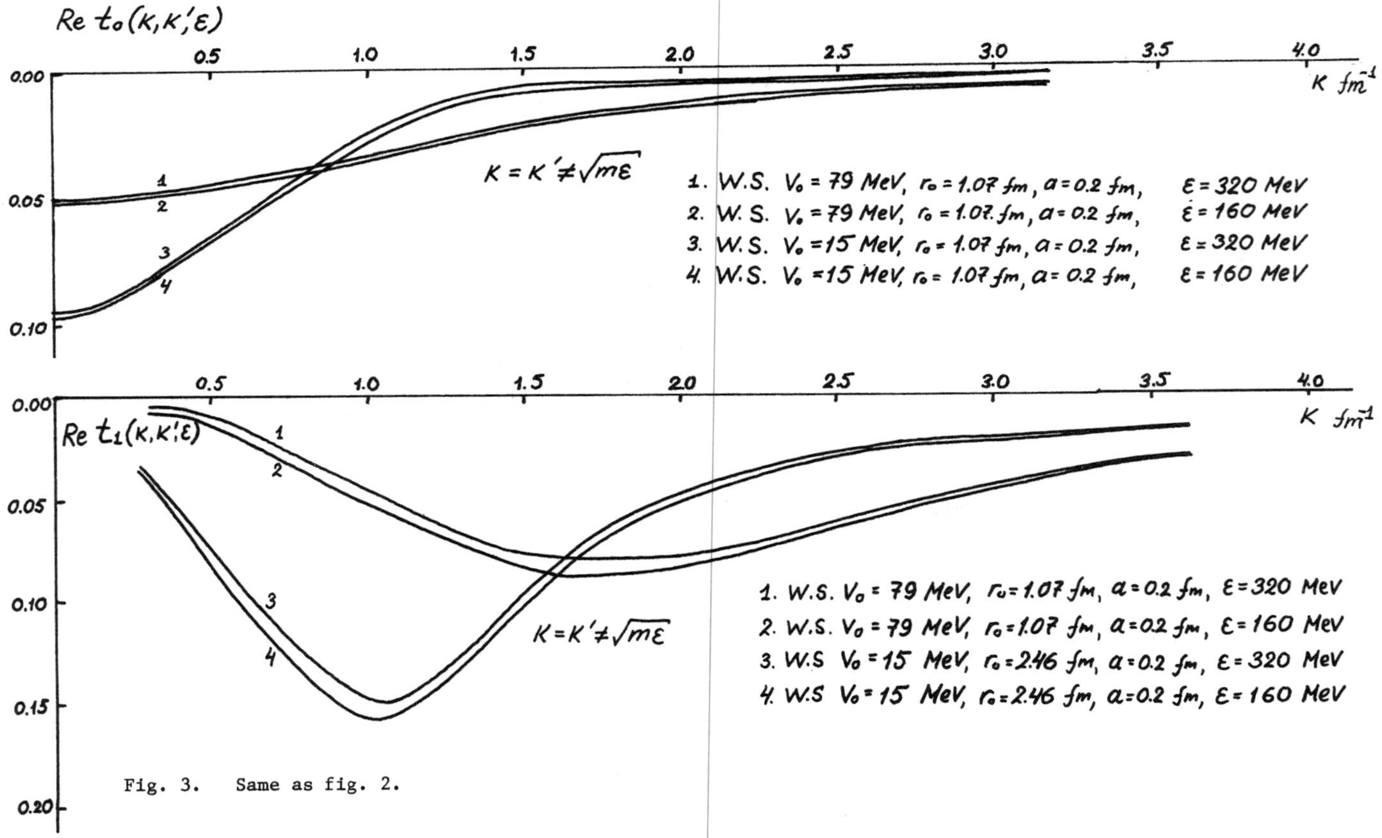

Fig. 3. Same as fig. 2.

or half-shell two-particle amplitude behaviour as a function of κ . Off and half shell information about the two-nucleon scattering amplitude may be obtained in different ways. We propose to study manyparticle nuclear reactions, in particular the N + d → 3N reaction, with regard to determining off and half shell information about the $t(\bar{\kappa}, \bar{\kappa}', \varepsilon)$ and therefore about two-nucleon potential parameters.

The N + d → 3N reaction amplitude by two-body force approximation could be represented as the sum of three subamplitudes:

$$T(N+d\to 3N) = T_{12,3}(\bar{K}_0,\alpha;\bar{K}_3,\bar{f}_{12},E) + T_{23,1}(\bar{K}_0,\alpha;\bar{K}_1,\bar{f}_{23},E) + T_{31,2}(\bar{K}_0,\alpha;\bar{K}_2,\bar{f}_{31},E),$$

where $T_{12,3}$, $T_{23,1}$ and $T_{31,2}$ differs by pair of interacting in the final state nucleons 12, 23, and 3I. In the formulae the following notation is used: $\bar{K}_0$ -incident nucleon momentum: α -deuteron binding constant; $\bar{f}_{12}$, $\bar{f}_{23}$, and $\bar{f}_{3I}$ - are relative motion momenta of the interacting in the final state nucleons; $\bar{K}_1$, $\bar{K}_2$ and $\bar{K}_3$ -are third nucleon momenta; and E-total energy of the three nucleon system. All momenta are in c.m.s. For the subamplitudes $T_{12,3}$, $T_{23,1}$, and $T_{31,2}$ there exist integralequations whose graphical form is presented in fig. 4.

Free terms and kernels of these equations contain the two-nucleon scattering amplitudes, $t_{ij}(\bar{f}_{ij}', \bar{f}_{ij}, \varepsilon_{ij})$.

For the purpose of numerical examples we now restrict our attention to the problem of finding a configuration of particle momenta which permits us to represent the reaction amplitude $T(N+d\to 3N)$ as the product of the two-nucleon amplitude and a constant:

$$T(N+d\to 3N) \to t_{ij}(\bar{f}_{ij}', \bar{f}_{ij}, \varepsilon_{ij})\cdot Const$$

Then we will find the region of momenta variations in the final state by which the differential cross section N + d → 3N reaction is determined by $\bar{f}_{ij}'$ variation in the factorised two- particle amplitude, by $f_{ij} = \sqrt{m\varepsilon_{ij}}$. The following configuration is worth to be considered for this aim. The relative momentum of the second and the third particles is close to zero $f_{23} \simeq 0$. The first particle moves in the final state in the backward direction about the incident nucleon, $\vartheta(\bar{K}_0, \bar{K}_1) \simeq 180^\circ$.

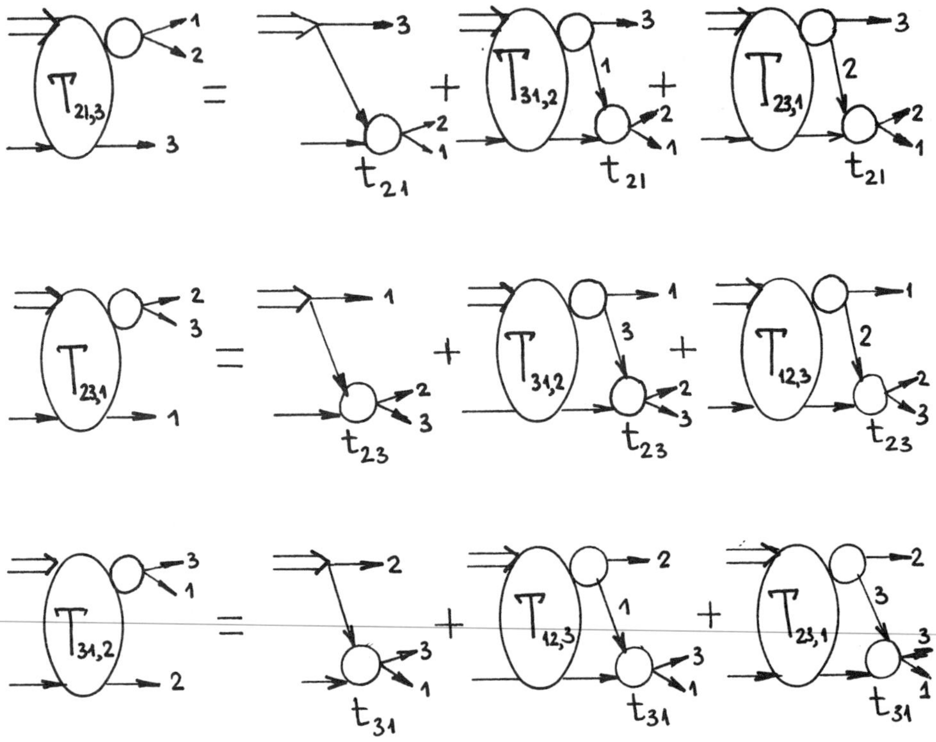

Fig. 4. The graphical representation of the integral equations for the three-particle scattering subamplitudes.

Fig. 5 displays the configuration, if the incident particle direction is along the axis from left to right.

It is convenient now to denote the relative momenta $\bar{K}_1$ and $\bar{f}_{23}$ by $-\bar{K}$ and $\bar{f}$ respectively, then the $\bar{K}_2$, $\bar{f}_{13}$, $\bar{K}_3$ and $\bar{f}_{12}$ momenta are in the form:

$$\bar{K}_2 = \frac{1}{2}\bar{K} - \bar{f}, \quad \bar{K}_3 = \frac{1}{2}\bar{K} + \bar{f}, \quad \bar{f}_{13} = \frac{3}{4}\bar{K} + \frac{\bar{f}}{2}, \quad \bar{f}_{12} = \frac{3}{4}\bar{K} - \frac{\bar{f}}{2},$$

or by $f \to 0$

$$\bar{K}_2 \simeq \frac{1}{2}\bar{K}, \quad \bar{K}_3 \simeq \frac{1}{2}\bar{K}, \quad \bar{f}_{31} \simeq \frac{3}{4}\bar{K}, \quad \bar{f}_{12} \simeq \frac{3}{4}\bar{K}.$$

After this we have to consider the behaviour of the subamplitudes $T_{23,I}$ $(\bar{K}_0, \alpha; -\bar{K}, \bar{f} \simeq 0, E)$, $T_{31,2}(K_0, \alpha; \frac{\bar{K}}{2}, \frac{3\bar{K}}{4}, E)$ and $T_{21,3}(K_0, \alpha; \frac{\bar{K}}{2}, \frac{3\bar{K}}{4}, E)$ in the dependence to a small variation of the first particle angle in the final state, say, $180^0 - \Delta^\circ \leqslant \vartheta(\bar{K}_0, \bar{K}) \leqslant 180^\circ + \Delta^\circ$. To find this dependence it is sufficient to study first iterations of the integral equations for the subamplitudes. An analysis of these iterations permit us to estimate the above mentioned dependence without a solution of the integral equations.

This analysis shows that for our particular case the T(N + d → 3N) amplitude is mainly determined by the subamplitude $T_{23,1}(K_0, \alpha; -\bar{K}, \bar{f} \simeq 0, E)$ because the $t_{23}(\bar{K}_0 - \frac{\bar{K}}{2}; f_{23} \simeq 0, \varepsilon_{23} \simeq 0)$ amplitude is in its pole meaning over ε_{23}, 4, 10).
And the subamplitude $T_{23,1}(\bar{K}_0, \alpha; -\bar{K}, \bar{f} \simeq 0, E)$ in its turn can be written down for the given configuration in the form:

$$T_{23,1}(\bar{K}_0 \alpha; -\bar{K}, \bar{f} \simeq 0, E) = C_1 \left[\alpha^2 + \left(\frac{\bar{K}_0}{2} - \bar{K}\right)^2\right]^{-1} t_{23}(\bar{K}_0 - \frac{\bar{K}}{2}, f_{23} \simeq 0, \varepsilon_{23} \simeq 0).$$

An s-wave is the most important for the $t_{23}(\bar{K}_0 - \frac{\bar{K}}{2}, f_{23} \simeq 0, \varepsilon \simeq 0)$ amplitude by its pole meaning over the variable ε_{23}

$$t_{23}(\bar{f}_{23}', \bar{f}_{23} \simeq 0, \varepsilon_{23} \simeq 0) \simeq t_0(f_{23}', f_{23} \simeq 0, \varepsilon_{23} \simeq 0).$$

Hence, the reaction amplitude T (N + d → 3N) is represented now in the form:

$$T(N + d \to 3N) \simeq C_1 \left[\alpha^2 + \left(\frac{\bar{K}_0}{2} - \bar{K}\right)^2\right]^{-1} t_{23}(\bar{K}_0 - \frac{\bar{K}}{2}, f_{23} \simeq 0, \varepsilon_{23} \simeq 0) + C_2,$$

where C_1 and C_2 are constants. This representation for T (N + d → 3N) differs from the Watson-Migdal approximation, because there is half-shell amplitude $t_{23}(\bar{K}_0 - \frac{\bar{K}}{2}; f_{23} \simeq 0, \varepsilon \simeq 0)$ depending on the energy and also on the off-shell momentum $K_0 - K/2$. By assuming that $t_{23}(\bar{K}_0 - \frac{\bar{K}}{2}; f_{23} \simeq 0, \varepsilon_{23} \simeq 0)$ coincide to the on shell amplitude, we come to the Watson-Migdal approximation. This might be under condition $|\bar{K}_0 - \bar{K}/2| r_0 \ll 1$. By that, as it was shown, the t_{23} doesn't depend on a potential shape.

The above investigated off and half-shell two-particle amplitude properties permit us to suppose the following behaviour of the N + d → 3N reaction differential cross-section.

For a small value of $|\bar{K}_0 - \bar{K}/2|$, i.e. $|K_0 - K/2| r_0 < 1$ the differential cross section is a smooth function by $\vartheta(\bar{K}_0 \bar{K})$ variation.

For $|\bar{K}_0 - \bar{K}/2| r_0 \simeq 1$ there should be a bend in the cross section curve. Thus, the curvature of the cross section apparently will show the radius and the depth of the nucleon-nucleon potential.

Bearing in mind the results of the off and half shell amplitude investigation for Woods-Saxon potential one has to measure the $N + d \rightarrow 3N$ differential cross section for the total energy $E = 70 \div 120$ MeV and for the $\vartheta(\bar{K}_0 \bar{K}) = 0^\circ \div 20^\circ$.

We would like to say in a conclusion that the analysis of the experimental angular distributions, similar to the proposed one, by exact solutions of the three-nucleon integral equations for a large variation of the two-nucleon potential, helps us to obtain detailed information about the two-nucleon interaction potential shape.

The proposed method could be generalised for the investigation of other two-particle nuclear systems.

REFERENCES

I. M.J. Zurina, J.T. Obukhovski, A.M. Popova, K.M. Ter-Martirosyan, Izvestia Acad. Sci. USSR, (Physics) 35, 1789, (1970)

2. Mac-Gregor et al. Phys. Rev. 169, II29 (1968), Ann. Rev. N.S. 10, 291 (1960)

3. M.K. Srivastava and D.W.L. Sprung, Nucl. Phys. A139 (1969) 605, A149 (1970) 113 I. Mc Carthy and P.C. Tandy, Nucl. Phys. A178 (1971)1. G.J. Stephenson Jr., E.F. Redish, G.M. Lerner and M.I. Haftel, Univ. Maryland, report No73-002. J.L. Ballot, M. L'Huillier and P. Benoist-Guental, Orsay. IPNO/TH, 74-23, July 1974.

4. V.V. Komarov, Int. Conf. Los-Angeles, 1972

5. S. Weinberg, Phys. Rev. 131, 440 (1963)

6. L.D. Faddeev, Proc. Math. Inst. Acad. Sci. USSR, LXIX, 1963

7. A.M. Popova, Yu.V. Popov, Theses of Symp. on Nucl. Spectr. Kharkov, 1974

8. V.S. Potapov, A.M. Brodski, V.V. Tolmachev, Vestnik MGU, Phys. 5, 17 (1969)

9. A.I. Vaskin, M.J. Zurina, V.N. Novozilova, A.M. Popova, Theses of Symp. on Nucl. Spectr. Kharkov, 1974

10. T.V. Gaivoronskaya, A.M. Popova, Izvestia Acad. Sci. USSR (Physics),37, II20 (1973).

DISCUSSION

LEVIN (Brown): In going from experimental data to half-shell t-matrices (and then to potentials), there is a unitary constraint which must be obeyed. If $t_\ell\,(q,k,\varepsilon=k^2)$ is the half shell, partial wave t- matrix, then the half shell unitary constraint is:

$$(1) \qquad \frac{\mathrm{Im}\left[t_\ell\,(q,k,\varepsilon=k^2)\right]}{\mathrm{Re}\left[t_\ell\,(q,k,\varepsilon=k^2)\right]} = \tan\ \delta_\ell(k^2)$$

where $\delta_\ell(k^2)$ is the on-shell phase shift. This holds for all q as long as one has a hermitian potential or a real symmetric k matrix. Any t_ℓ based directly on a potential or a wave function will obey (1). For $t_\ell\,(q,k,\ \varepsilon=k^2)$ derived phenomenologically, (1) provides a useful constraint. Eq. (1) can clearly be generalized to non-central interactions, and also to multi-channel processes.

SLOBODRIAN (Laval): About the unitary condition, how far off-shell is it valid? I can not believe it is valid everywhere.

LEVIN: Yes, valid eveywhere for all $q,k \to \infty$ as long as you believe in non-relativistic Schrödinger equation type of description.

SLOBODRIAN: Then I am sorry, but it is not valid everywhere.

LEVIN: As long as the simple amplitude you have obeys some kind of unitarity relation, then this is valid. If other processes come in and change the nature of the two-body amplitude, then it is something else. But as long as you can write down a standard type of unitarity relation based on a damping equation, this result holds. It is a very strange looking result. But it says this is independent of the off-shell portion.

SAUER (M.I.T.): The assumption of a hermitian energy independent potential, consistent with time-reversal invariance, yields the constraint on the half-shell matrix which Dr. F.S. Levin is talking about. However, under this assumption on the potential there are even more powerful constraints on the half-shell t- matrix. They are derived from the fact that scattering wave functions are orthogonal and, together with a possible bound state, complete. Since the half-shell t-matrix determines the scattering wave functions in full, orthogonality and completeness places constraints on the half-shell t-matrix.

KOMAROV: The use of the relationship mentioned by Prof. Levin for the analysis of experimental data could be somewhat difficult. In this case, I assure you that the imaginary part is 10 times less that the real part. About the type of potential, I tried to choose a very simple potential just to illustrate how it influences the t-matrix behaviour so that it could be done for every other potential.

NOYES (SLAC) Since Prof. Komarov is using a potential, his half-shell t-matrix is automatically unitary. The relationship mentioned by Prof. Levin will also hold for a relativistic Lippmann-Schwinger type two-body equation. It is an immediate consequence of what is sometimes called the Kowalski-Noyes representation.

KOMAROV: I considered the non-relativistic case only.

AMADO (Pennsylvania): There are important restrictions placed on the three and multi-body final state amplitudes, for example by unitarity, which Prof. Komarov's parametrization neglects. The correct implementation of these restrictions is very likely to mask the off-shell effects he wishes to investigate.

KOMAROV: That is right. On the possibility of this sort of investigation, we have to consider iterations of these equations. It is possible to do this in some particular cases, like cases when k_o, the incident particle is much larger than the binding energy constant or some other conditions. Nevertheless, it should help us to extract the main properties contained in these equations in the way I have at the Los Angeles Conference, for instance.

SLOBODRIAN: I would like to return to the question of hermiticity of the potential.

We have been discussing Bremsstrahlung phenomena. Hence there is absorptivity, even if small, and therefore the hamiltonian is non hermitian and we are dealing with an approximation in using the constraint mentioned by Prof. Levin.

Sensitivity of Deuteron Breakup Amplitudes for Several Potentials

M. I. Haftel and E. L. Petersen
Naval Research Laboratory, Washington, D. C. 20375

In this paper we discuss the sensitivity of breakup amplitudes for the various potentials discussed in the previous paper. Figures 1 a-e show plots of breakup amplitudes at 44.9-MeV incident proton energy for different potentials in various spin and angular momentum states. In these figures, p_f is the relative momentum of a two-body substate (referred to as d or d^*, depending on the spin coupling) in the final state. By conservation of energy, $p_f^2 + q_f^2 = E_{C.M.}$. The total spin is S, and L is the N-d (or N-d^*) relative orbital angular momentum, which, in the case of S-wave potentials, is the total orbital angular momentum. These plots exhibit two significant trends. First, only the $S = \frac{1}{2}$, L = 0 amplitudes vary appreciably if the potentials are phase-shift-equivalent (compare results for HA2-8.0, HB2-8.0, and HB2-11.0). The imaginary parts of the $S = \frac{1}{2}$, L = 0 amplitudes are appreciably more sensitive than the real parts except for $p \approx 0$, where the real part is sensitive as well. The $p \approx 0$ amplitudes are important in the final-state-interaction (FSI) region of phase-space. We note that for $S = \frac{1}{2}$, L = 0, fixing E_T as well as the phase-shifts eliminates only about 30% of the off-shell differences between the HA and HB form factors (compare amplitudes of HA2-8.0 with HB2-8.0 and with HB2-11.0). Further calculations indicate that fixing E_T has a greater effect in eliminating off-shell differences at lower energies (this eliminates about 75% of the off-shell effect at 14 MeV).

The quartet and L > 0 amplitudes are insensitive for the following reasons. In the quartet (S = 3/2), the spins of the protons must couple to one (triplet); thus, the wave-function is in a spin-isospin symmetric state under proton exchange. Hence, the three-nucleon spatial wave function must be antisymmetric under proton exchange. The protons then have a very low probability of approaching closely to each other. The Pauli principle here prevents a situation where all three nucleons can be close to each other, thus excluding configurations where off-shell effects should be strongest. This argument also applies to the M_{D1} (doublet) matrix element, which is a linear combination of doublet amplitudes describing the scattering when $S = \frac{1}{2}$, but the two protons are in a spin triplet state. Our further calculations indicate that it is the M_{D2} linear combination of doublet amplitudes, not the M_{D1}, that is sensitive to off-shell effects. For L > 0, the centrifugal barrier prevents a nucleon from approaching the center-of-mass of the final d (or d^*) very closely; hence, it is again improbable that all three particles can be near each other. This argument suggests that breakup amplitudes are sensitive to off-shell effects only in states where all three particles can approach each other, i.e., the $S = \frac{1}{2}$, L = 0 states. These are also the states where three-body forces should be the strongest, which may suggest an intrinsic difficulty in disentangling the effects of off-shell behavior and three-body forces in breakup.

The second trend in the amplitudes concerns potentials that differ on-shell (e.g., HA1-8.0, HA2-8.0). All amplitudes, including quartet and L > 0 are sensitive to on-shell changes. Since on-shell changes involve the long-range internucleon wave functions, the arguments limiting off-shell sensitivity for quartet and L > 0 are inapplicable. The results we show in this paper suggest that on-shell differences between potentials should affect breakup cross sections in all regions of phase-space while off-shell differences would only show up in restricted regions (i.e., where $S = \frac{1}{2}$, L = 0 is important).

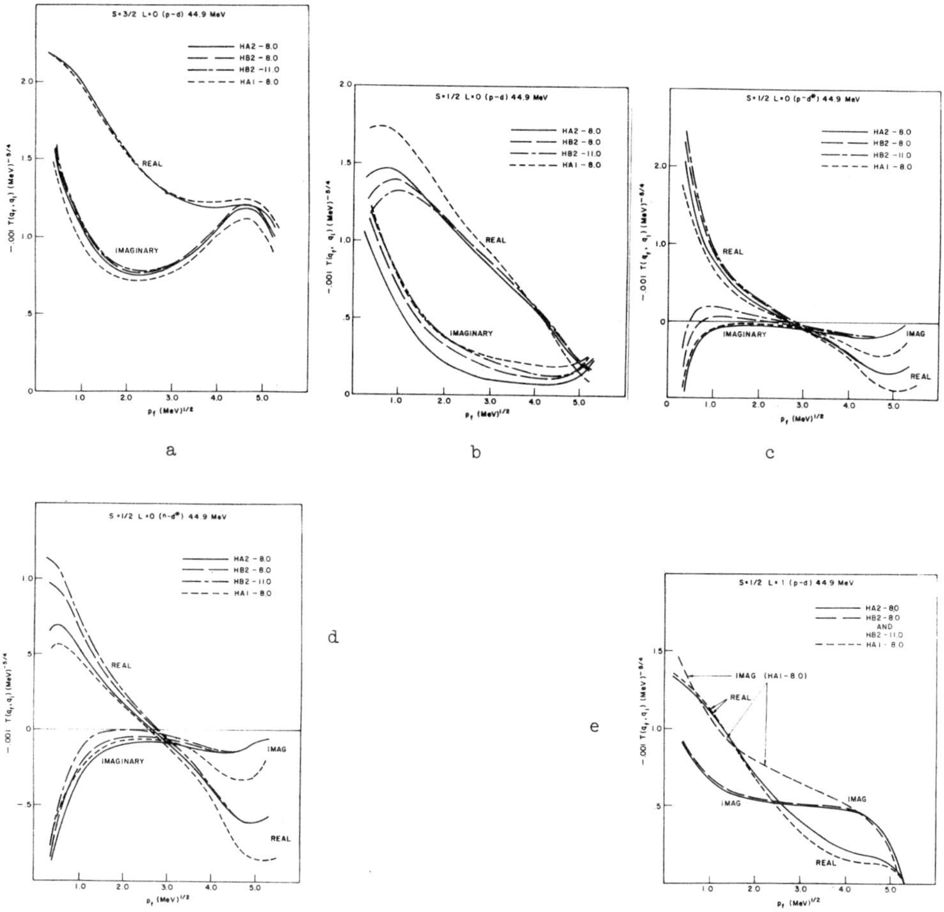

Figure 1. Breakup amplitudes $T(q_f; q_i)$ (these correspond to $3\sqrt{3}T_{iL}(q_i\sigma)/8$ of Eq. 17 of Ref. 1 with $q_i = \sqrt{E_{CM} + \kappa_o^2}$, $q_f = q_i$ of Ref. 1). The (N-d) or (N-d*) label identifies the third particle (N → p or n) and the reference pair spin state (d → spin 1, d* → spin 0).

References

1. W. Ebenhöh, Nucl. Phys. A191, 97 (1972).

Separable Potential Models for Use in Deuteron Breakup Calculations

M. I. Haftel and E. L. Petersen
Naval Research Laboratory, Washington, D. C. 20375

This paper discusses four S-wave potentials that are proposed for use in deuteron breakup calculations. We design our potentials for use in the separable potential code of Ebenhöh but consider alternatives to the Yamaguchi potentials employed heretofore. These potentials differ mainly in the off-shell momentum dependence, i.e., they have different form factors. Also, we employ an energy-dependent modification of the two-nucleon (N-N) T matrix[1] to assure phase-shift-equivalence of three of the potentials (HA2-8.0, HB2-8.0, HB2-11.0) and to adjust the triton binding energy (E_T = 8 or 11 MeV).

For separable potentials, the momentum space matrix elements have the form $\langle k|V|k'\rangle = -\lambda g(k)g(k')$ and the T matrix $\langle k|T(E)|k'\rangle = g(k)g(k')\tau(E)$ where

$$\tau(E) = -\left[\lambda^{-1} + 4\pi\int_0^\infty k^2dkg^2(k)/(E - k^2 + i\epsilon)\right]^{-1}$$

(We have suppressed angular momentum, spin and isospin labels.) Most previous calculations have used the Yamaguchi form factor $g(k) \sim (k^2 + \beta^2)^{-1}$, where λ, β are chosen to fit the N-N effective range parameters. Here we consider an alternate form factor $g(k) = (k_c^2 - k^2)/(k^2 + \beta^2)^2$. Different values of k_c^2 give different off-shell behavior, while λ, β (for a fixed k_c^2) are such as to give the experimental effective range parameters. Figure 1 depicts the Noyes-Kowalski $E = k^2 = 0$ half-shell function $f(p,k) = \langle p|T(k^2 + i\epsilon)|k\rangle/\langle k|T(k^2 + i\epsilon)|k\rangle$ for form factors HA ($k_c^2 = 4\ \text{fm}^{-2}$), HB ($k_c^2 = -2.89\ \text{fm}^{-2}$) as well as $f(p,k)$ for the Yamaguchi form factor (Y) and Reid-Soft-Core (RSC) potential. The form factor HA simulates a repulsive core while HB is similar to the "softer" Yamaguchi.

In general, separable potentials can differ in both their off-shell behavior (through $g(k)$) and on-shell values. We would like to fix the on-shell matrix elements since these are measured by N-N elastic scattering experiments. To accomplish phase-shift-equivalence, we employ the method of ref. 1 by replacing $\tau(E)$ by $\tau'(E) = \rho(E)\tau(E) \times [1 - 2i\pi^2k_0g^2(k_0)\tau(E)(1 - \rho(E))]^{-1}$ where $k_0 = \sqrt{E}$. The presence of the denominator means that the replacement $\tau(E) \rightarrow \tau'(E)$ corresponds to multiplying by $\rho(E)$ the on-shell K matrix element $K(E) = -\tan\delta(E)/(2\pi^2k_0)$, where $\delta(E)$ is the phase-shift predicted by $g(k)$. A real $\rho(E)$ guarantees a unitary two-body amplitude. In fact, the replacement of $\tau(E) \rightarrow \tau'(E)$ merely corresponds to an energy dependent λ, $(\lambda \rightarrow \lambda(E))$, in the potential. To fit the on-shell element to a given phase shift $\tilde{\delta}(E)$ (e.g. to gain phase-shift-equivalence between potentials), one simply chooses $\rho(E) = \tan\tilde{\delta}(E)/\tan\delta(E)$.

We modify the T matrix of the HB form factor to gain the phase-shifts of HA. Explicitly:

$$\rho(E) = 1 + \left[\frac{\tan\delta_{HA}(E)}{\tan\delta_{HB}(E)} - 1\right](1 + e^{-\gamma(E-E_0)})^{-1} \quad \text{for } E > E_c \qquad (1)$$
$$= 1 \text{ for } E \leq E_c$$

Potential HA1-8.0 (see Table 1) uses the HA form factor, but with $\rho(E) = 1 + (1 + e^{-\gamma E})^{-1}\Delta(E + \kappa^2)/(E_o + \kappa^2)$ where κ is defined by Ebenhöh, and $\Delta_t = -.0981$, $\Delta_s^{np} = .8274$, $\Delta_s^{pp} = .7073$. The form factors, triton binding energy, and $\rho(E)$ parameters for each potential proposed appear in Table 1. In this nomenclature, HA and HB refer to the form factors, 1 and 2 refer to a set of phase shifts, and 8 or 11 refer to the approximate triton binding energy. Note that by altering the E_o and E_c parameters, which essentially affect only the $E < 0$ T matrix, we can change the triton binding energy predictions. We illustrate the n-p and p-p cross sections predicted by all the potentials in Fig. 2. Note that HA1-8.0 gives results in good agreement with experiment up to 70 MeV.

Table 1

Potential	$k_c^2(\text{fm}^{-2})$	$\beta_t(\text{fm}^{-1})$	$\beta_s^{np}(\text{fm}^{-1})$	$\beta_s^{pp}(\text{fm}^{-1})$	$\gamma(\text{MeV}^{-1})$	E_o(MeV)	E_c(MeV)	E_T(MeV)
HA1-8.0	4.0	3.31806	2.24663	2.28694	.3333	25.0	$-\infty$	8.04
HA2-8.0	4.0	3.31806	2.24663	2.28694	$\rho(E) = 1$			8.04
HB2-8.0	-2.89	1.54130	1.37449	1.39032	.500	-45.5	-75.0	8.08
HB2-11.0	-2.89	1.54130	1.37449	1.39032	.500	-40.0	0.0	10.9

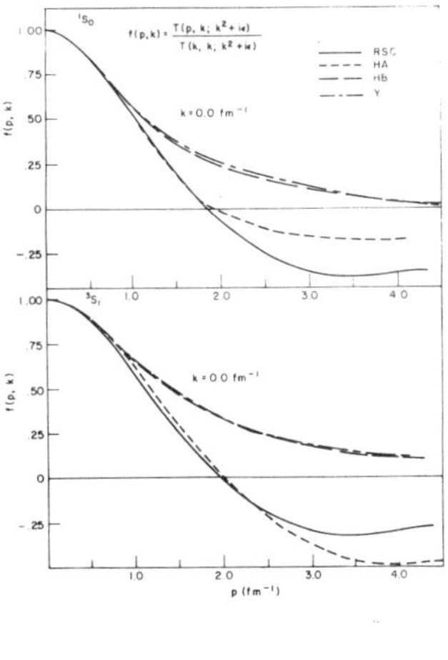

Figure 1. Noyes-Kowalski half-shell functions at zero energy for different potentials.

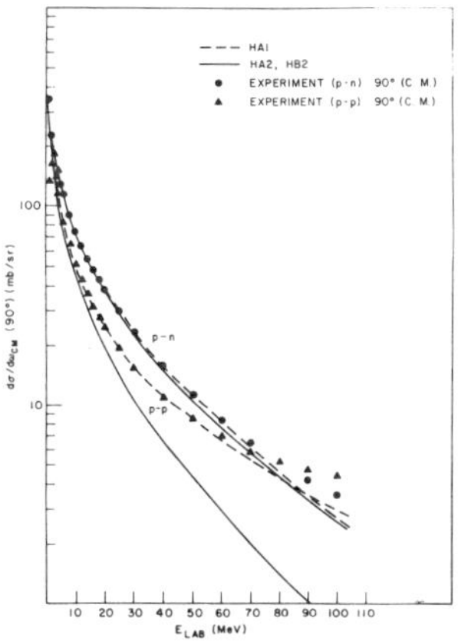

Figure 2. p-p and p-n 90° (C.M.) cross sections for various potentials.

References

1. E. L. Petersen et. al., Phys. Rev. C9, 508 (1974).

DEUTERON BREAKUP BY α-PARTICLES AT $E_{\alpha,lab}$ = 15.0 MeV

L Glantz, A Johansson, I Koersner and B Sundqvist
Tandem Accelerator Laboratory, University of Uppsala, Sweden

In recent years three-body theory, in particular the separable-potential model, has been quite successful in explaining a large amount of experimental data on various three-nucleon scattering reactions. It is of great interest to extend these comparisons to other few-body problems. The most likely system for which one could hope to obtain good experimental data as well as carry out realistic three-body calculations is probably the ^{6}Li compound system, when considered to consist of an "α" particle, a neutron and a proton. The breakup of the ^{6}Li compound system into ^{3}H + ^{3}He can occur only at an excitation energy 14.3 MeV higher than the ^{4}He + d breakup. A number of three-body calculations have been performed on the A = 6 system, but so far only for the bound states [1,2,3]. An experiment has been started to obtain information on the deuteron breakup with α particles at such a low energy that the three-body assumption could be expected to be valid. An α particle energy of 15 MeV was chosen, i e 5 MeV in the center-of-mass system, which is considered to be well below the 14.3 MeV mentioned above. The d(α,pα)n reaction has been studied experimentally at E_α = 18 MeV [4] and very recently at 21.9 and 23.7 MeV [5] and the α(d,pα)n reaction at E_d = 9 MeV [6].

The present measurements were kinematically complete and made with a $(CD_2)_n$ target and surface barrier detectors. A ΔE-E telescope was used when necessary to identify protons to eliminate coincidences from the reaction $^{12}C(\alpha,2\alpha)^8Be$. The absolute cross sections were determined by simultaneous counting of recoiling deuterons at θ_{lab} = 50.6°. The elastic α-d cross sections were taken from Senhouse and Tombrello [7].

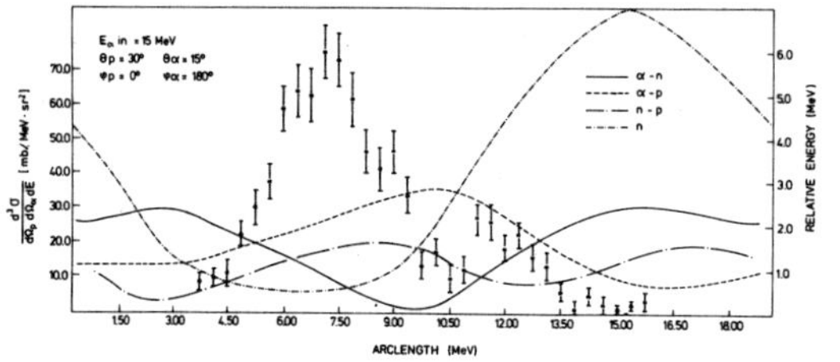

Fig 1 Correlation cross section for the d(α,pα)n reaction

An example of the results obtained is shown in Fig 1, where also the relative energies of the different two-body systems and the neutron laboratory energy are shown. The structure observed appears to be a superposition of effects of what has traditionally been assigned to quasifree α-p scattering (at low neutron laboratory energy) and strong final-state interactions in the α-n system (about the relative energy of 0.96 MeV) and in the α-p system (about the relative energy of 1.97 MeV). Clearly a Faddeev type of calculation is needed in this case. Further results will be presented and a comparison with the theory due to Nakamura attempted [8].

REFERENCES

1) V A Allessandrini, C A Garciá Canal, D Avalos, L Epele, H Fanchiotti and M A Gregorio, Proc Int Conf on The Three Body Problem in Nuclear and Particle Physics (ed I S C MacKee and P M Rolph), North Holland 1970, p 526

2) M S Shah and A N Mitra, Phys Rev C1 (1970) 35

3) A Ghovanlou and D R Lehman, Phys Rev C9 (1974) 1730

4) P A Assimakopoulos, E Beardsworth, D P Boyd and P F Donovan, Nucl Phys A144 (1970) 272

5) T Rausch, H Zell, D Wallenwein and W von Witsch, Nucl Phys A222 (1974) 429

6) H Nakamura, private communication

7) L S Senhouse Jr and T A Tombello, Nucl Phys 57 (1964) 624

8) H Nakamura, Nucl Phys A208 (1973) 207

Examination of Off Energy Shell Effects in Deuteron Breakup

E. L. Petersen and M. I. Haftel
Naval Research Laboratory, Washington, D. C. 20375

Since only the L = 0, S = ½, M_{D2} breakup amplitudes are sensitive to off-shell effects, one should investigate regions of phase-space where these amplitudes dominate the cross section. In Fig. 1, we plot predictions of $^2H(p,pp)n$ cross sections at 23 MeV for potentials HA2-8.0 and HB2-8.0 on the energy sharing locus of Jain, Rogers and Saylor.[1] This kinematic locus accentuates the amplitudes sensitive to off-shell effects. Figure 2 illustrates results for three on-shell equivalent potentials for θ_1 = 60°, θ_2 = -80° $^2H(p,pp)n$ at 45 MeV, a spectrum for which the L = 0 doublet M_{D2} amplitude is an important contribution. In both Figs. 1 and 2, there exist regions where some relatively large (up to about 40%) sensitivity occurs. However, where the sensitivity is relatively large, the cross sections are small and even a 40% difference would be difficult to distinguish experimentally. In fact, for the L = 0 amplitude to dominate, the cross section most probably must be small since one would be automatically ruling out the mechanism that accounts for the QFS enhancement (i. e., the coherent contributions of many partial waves). Furthermore, off-shell changes seem primarily to renormalize the cross sections (this has been observed in out-of-the-plane spectra also) making it more difficult to distinguish between potentials experimentally.

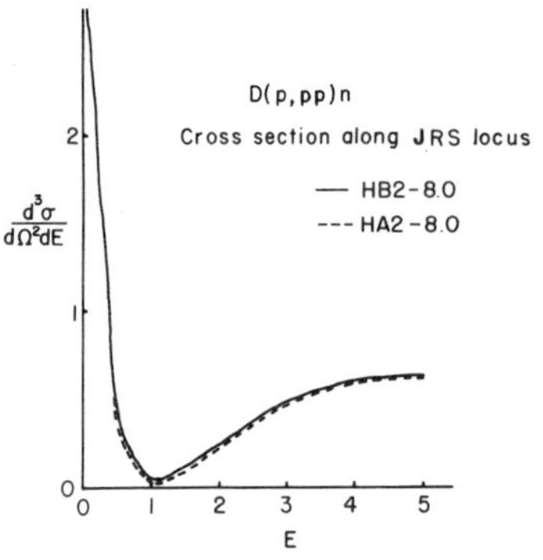

Figure 1. Changes along the Jain-Rogers-Saylor Energy Sharing Locus introduced by changing only the form factor

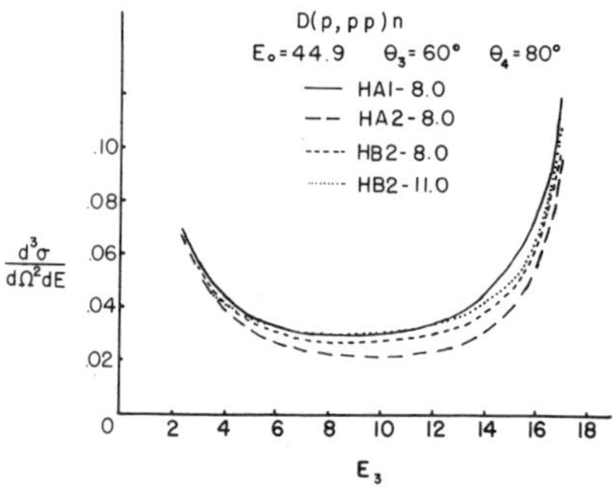

Figure 2. Changes of spectrum with
a) on-shell changes HA1-8.0 ↔ HA2-8.0
b) off-shell changes HA2-8.0 ↔ HB2-8.0
c) triton binding energy HB2-8.0 ↔ HB2-11.0

Is it possible to find regions of phase-space where the L = 0, S = ½ M_{D2} amplitude is important yet the cross section is not very small? In the final-state-interaction (FSI) region, the enhancement is due to the $^1S_0$ resonance in the N-N scattering amplitude at zero energy; therefore, unlike QFS, the L = 0 contribution could be important yet the cross section large. Also, a previous paper has shown that the real and imaginary parts of the S = ½, L = 0 breakup amplitudes are sensitive to off-shell effects at $p_f \approx 0$.

Since the relative importance of the L = 0 amplitude varies with the center-of-mass angle of the FSI pair, both the magnitude and shape of the FSI angular distribution should be sensitive to off-shell changes. Figure 3, which illustrates the FSI angular distribution at 26 MeV for various potentials, does indeed show changes in the magnitude and shape under off-shell changes in the force. The maximum sensitivity (about 13%) occurs in the vicinity of 130°, and this sensitivity is somewhat larger than the experimental statistical uncertainty. Also shown are data from Karlsruhe and the results of other calculations with different potentials.[2]

How does this FSI off-shell sensitivity vary with energy? Figure 4 illustrates the ratio of the FSI peaks as predicted by potentials HA2-8.0 and HB2-8.0 as a function of angle and energy. Even with the triton binding energy (E_T) and phase shifts fixed, sensitivity is as high as 20% at 45 MeV and seems to be increasing with energy. This result indicates that Brayshaw's remarks[3] on the connection between E_T and breakup results may apply only at very low projectile energies (compare the 45 MeV curve with the 14.4 MeV curve).

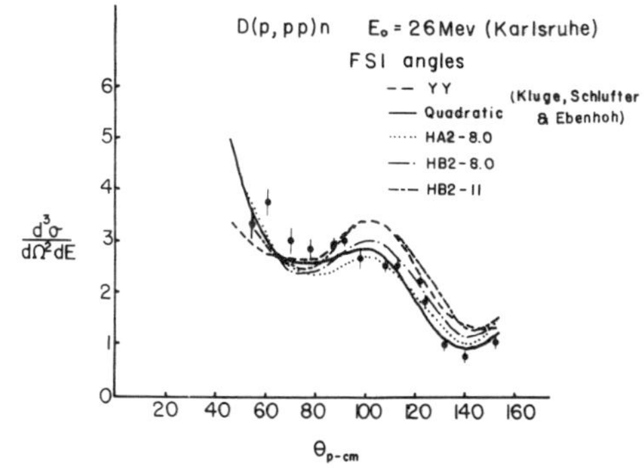

Figure 3. Peak cross section at final state angles

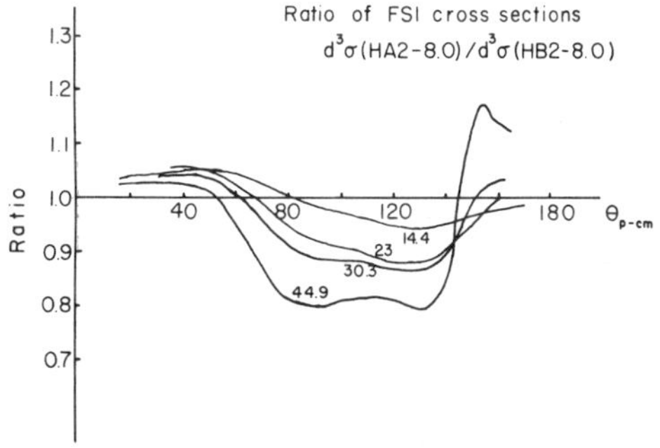

Figure 4. Effect of off-shell changes as a function of energy

<u>References</u>

1. M. Jain, J. G. Rogers and D. P. Saylor, Phys. Rev. Letters <u>31</u>, 838 (1973).
2. W. Kluge, R. Schlufter and W. Ebenhöh, preprint.
3. D. D. Brayshaw, Phys. Rev. Letters <u>32</u>, 382 (1974).

D(p,n) AND H(d,n) CROSS SECTIONS AT 21.9, 42.8 AND 45.8 MeV***

R. G. Graves, M. Jain*, L. C. Northcliffe and F. N. Rad**
Cyclotron Institute, Texas A&M University, College Station, Texas 77801

Absolute neutron production cross sections for the D(p,n)2p reaction at 21.9 MeV and the H(d,n)2p reaction at 42.8 MeV have been measured at 0 and 18 degrees in the laboratory to complement polarization data taken for these same reactions at about the same energies[1). In addition the 0 degree cross section for the D(p,n)2p reaction at 45.8 MeV was measured to compare with the available theoretical calculation by Oryu[2). The results are shown in Figs. 1 through 5. The D(p,n) cross sections exhibit the expected enhancement at high neutron energies due to the p-p final state interaction (FSI), and the 0 degree data exhibit a smaller enhancement at about one-half the incident proton energy due to quasi-free scattering (QFS). Because the D(p,n) reaction at 21.9 MeV and the H(d,n) reaction at 42.8 MeV have approximately the same center-of-mass energy, the neutron cross sections must be equal at $E_n \approx 6$ MeV, which for this experiment is too low to reliably determine if the the cross sections are indeed equal. (Also note that in the CM neutrons from the H(d,n) reaction are scattered in the forward direction whereas neutrons from the D(p,n) reaction are scattered in the backward direction). The H(d,n) spectra show a strong QFS peak and a lack of enhancement near the high energy end due to pp FSI. Such features are not unexpected as this reaction would be unlikely to proceed in a way in which the two final-state protons have low relative energy and hence would strongly interact.

Also shown in Figs. 1 through 5 (as solid lines) are the results of a theoretical calculation[3) based on the YY model of Cahill and Sloan[4). Basically, the results are an exact solution to the Faddeev equations using a separable S-wave spin-dependent nucleon-nucleon potential, with parameters determined from low-energy two-nucleon data. Coulomb effects have not been included. Also, the D(p,n) cross section at 45.8 MeV is compared with a theoretical calculation by Oryu (dashed line) who also solved the Faddeev equations but included a Coulomb effect in the p-p FSI. Generally, there is good agreement between theory and experiment.

It is hoped that these cross section measurements will help explain the origin of the nonzero polarization seen in these reactions[1). Presently there are no theories capable of calculating p-d neutron polarizations and one can only try to correlate peaks in the cross section data which are produced by known processes such as QFS and FSI with peaks, if any, in the polarization data. Although the large QFS peak in the H(d,n) data and the nonzero polarization are correlated, it is clear from the cross section data that there are other contributing reaction mechanisms which are undoubtly altering both the sign and the magnitude of the polarization one would expect if the dominant process were pure n-p scattering. Similarly, it appears that several different processes are contributing in the D(p,n) reaction and one cannot associate the nonzero polarization seen in this reaction with any particular process.

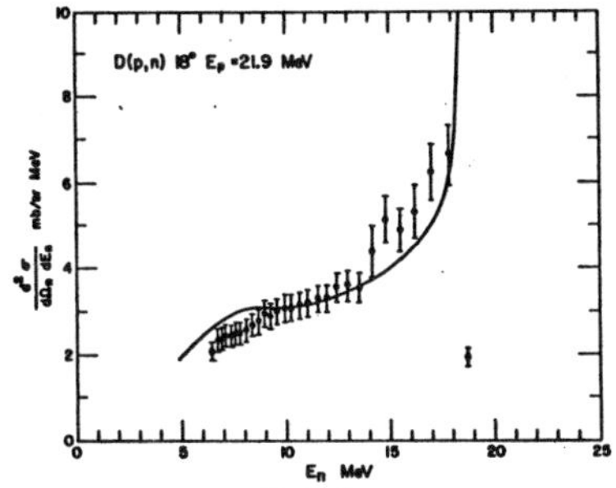

Figure 1.

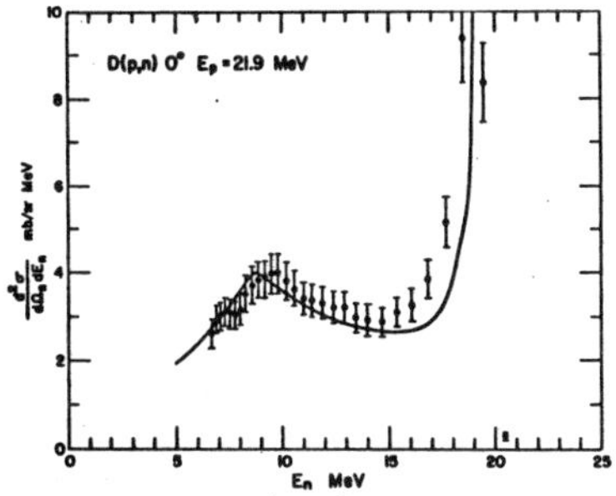

Figure 2.

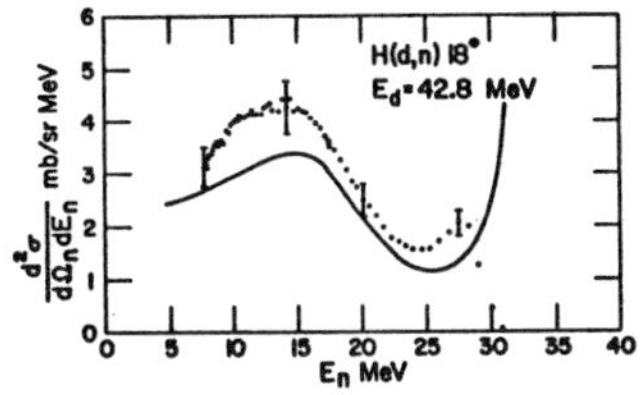

Figure 3.

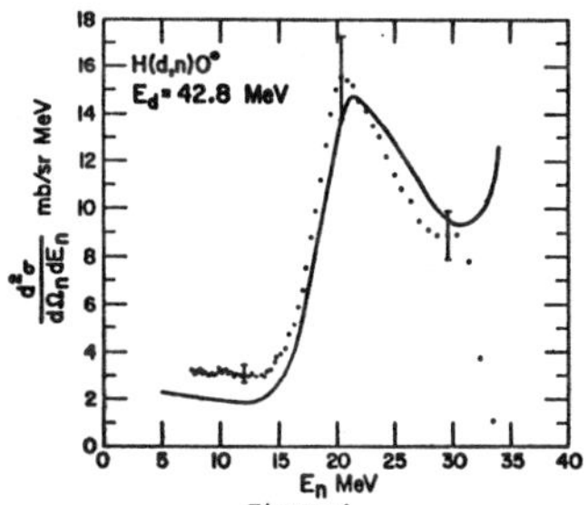

Figure 4.

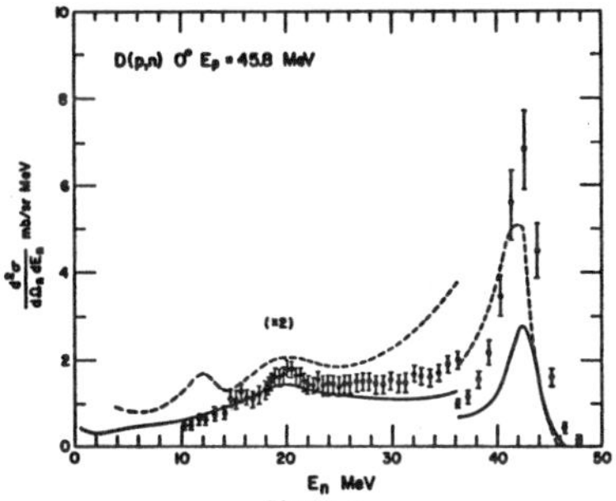

Figure 5.

---

* Present address: Los Alamos Scientific Laboratory, Los Alamos, New Mexico.

**Present address: Lawrence Berkeley Laboratory, University of California at Berkeley, Berkeley, California.

***Work supported by the National Science Foundation.

REFERENCES

1) F. N. Rad, L. C. Northcliffe, D. P. Saylor, J. G. Rogers and R. G. Graves, Phys. Rev. C8, 1248 (1973).

2) S. Oryu, Prog. Theor. Phys. 44, 1208 (1970).

3) M. Jain and G. D. Doolen, Phys. Rev. C8, 124 (1973).

4) R. T. Cahill and I. H. Sloan, Nucl. Phys. A165, 161 (1971).

$^3H(p,pn)d$ Knockout Reaction at 45.4 MeV*

D.J. Roberts, D.I. Bonbright, S.A. Elbakr, W.T.H. van Oers,
and J.W. Watson

Cyclotron Laboratory, University of Manitoba,
Winnipeg, Manitoba, R3T 2N2, Canada

A series of knockout experiments at 45 MeV incident proton energy has been undertaken at the University of Manitoba cyclotron on the nuclei ^3He and ^3H [1,2]. Information on the three-nucleon wavefunction has been sought, particularly the relative amounts of nucleon plus deuteron and nucleon plus singlet deuteron. The nucleon plus deuteron constituent is seen in three body breakup; the singlet deuteron contribution is inferred from four body breakup by considering the neutron and proton to be a "lump" particle, unbound by a small amount of energy (less than 2 MeV). In this connection, it is evident that the singlet deuteron can also be formed after breakup has occurred. It follows that a study of the reaction mechanism is essential.

An experimental approach to the above problem has been adopted. One wishes to compare as many knockout reactions on the three nucleon system as possible. One such comparison is presented here.

A study was made of the $^3H(p,pn)d$ reaction using a symmetric coplanar geometry. Protons and neutrons were detected in coincidence using an experimental technique which has been described previously[3]. The target consisted of tritium absorbed in scandium with an aluminum backing.

Figure 1 presents a comparison of the data on the $^3H(p,pn)d$ reaction with data on the reaction $^3He(p,pp)d$[2]. As the geometry chosen favored "quasi-free" scattering (the spectator deuteron has a minimal momentum of 20 MeV/c for the present experiment), one may use the impulse approximation to say that the cross section is a measure of the

*Work supported in part by the Atomic Energy Control Board of Canada.

relative momentum of the deuteron in ^3H and ^3He. (One can include a finite cutoff radius in the relative momentum function to approximate the effects of distortion.) One expects ^3He and ^3H to have quite similar wave functions [4], and it is seen that this is consistent with the data. To within statistics, both energy sharing spectra exhibit the same FWHM. It should be noted that the $^3He(p,2p)d$ cross section is larger in absolute magnitude.

Figure 2 is a comparison of results for the $^3H(p,pn)d^*$ and $^3He(p,pp)d^*$ reactions. Although the comparison is not as conclusive as before, it can be said that the shapes of both spectra are identical to within statistics. In this case, however, the cross section for the $^3H(p,pn)d^*$ reaction is comparable to the one for the $^3He(p,pp)d^*$ reaction. This is quite different from the comparison of absolute cross sections given in Fig. 1, and we plan to investigate this effect further.

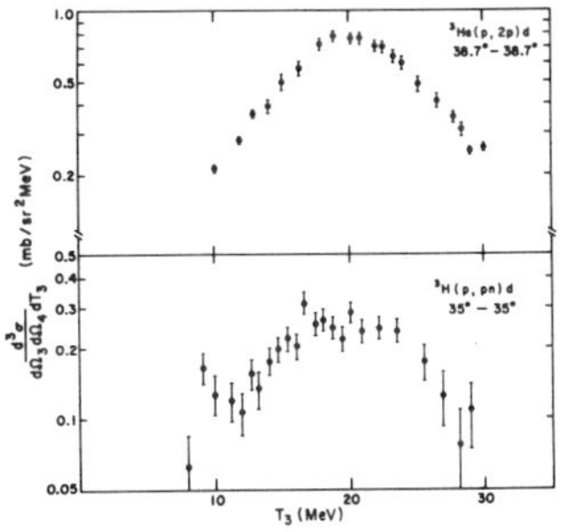

Figure 1

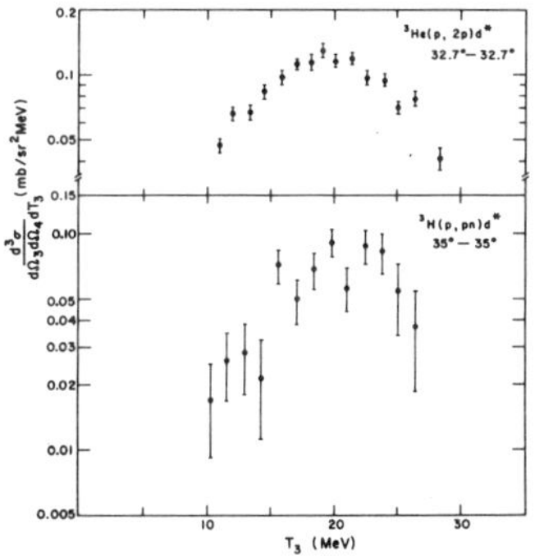

Figure 2

1) M. Jain, S.N. Bunker, C.A. Miller, J.M. Nelson, and W.T.H. van Oers, Lett. Nuovo Cimento, 8 (1973) 844

2) E.S.Y. Tin, Ph.D. Thesis, University of Manitoba, unpublished

3) C.A. Miller, J.W. Watson, D.I. Bonbright, F.J. Wilson, and D.O. Wells, Phys. Rev. Lett. 32 (1974) 684

4) T.A. Griffy and R.J. Oakes, Phys. Rev. 135 (1964) B1161

Potential-Dependence of Quasifree Scattering in Deuteron Breakup

E. L. Petersen and M. I. Haftel
Naval Research Laboratory, Washington, D. C. 20375
and
J. M. Lambert*
Georgetown University, Washington, D. C. 20007
and Naval Research Laboratory

In a previous publication,[1] the authors noted that S-wave Yamaguchi potentials give quasifree scattering (QFS) angular distributions whose shapes were potential independent but not in agreement with experiment. One suggestion was to consider off-shell variations by varying the form factor. In this paper we present theoretical predictions for quasifree scattering in deuteron breakup as obtained using the various potentials discussed in the previous paper. Fig. 1 illustrates the potential dependence of the $^2H(p,pp)n$ spectrum for $\theta_1 = -\theta_2 = 41.7^{\circ}$ at 23 MeV. The spectrum is independent (to within about 2%) of potential as long as the potentials differ only off-shell (e.g., compare HA2-8.0, HB2-8.0, HB2-11.0). On-shell senstivity is larger (compare HA1-8.0 and HA2-8.0). The sensitivity (or lack thereof) displayed in Fig. 1 also occurs for all QFS spectra we have investigated regardless of the angle or projectile energy (at least up to 45 MeV). Fig. 2, which shows the QFS angular distribution at 23 MeV, indicates that the shape of the QFS angular distributions are form factor independent. This is not surprising in view of the insensitivity of QFS spectra.

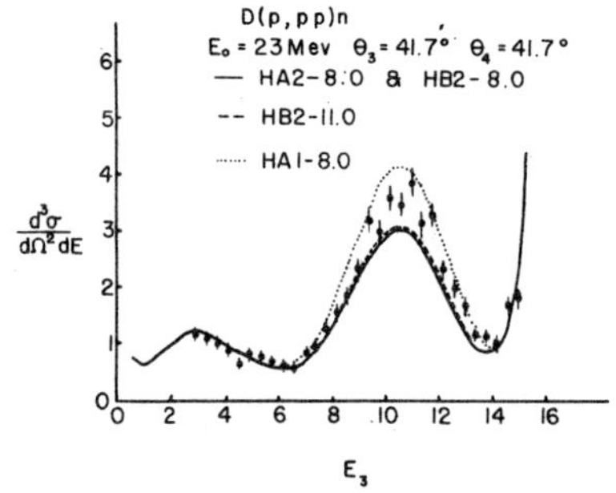

Figure 1. Potential variation in quasifree scattering spectrum

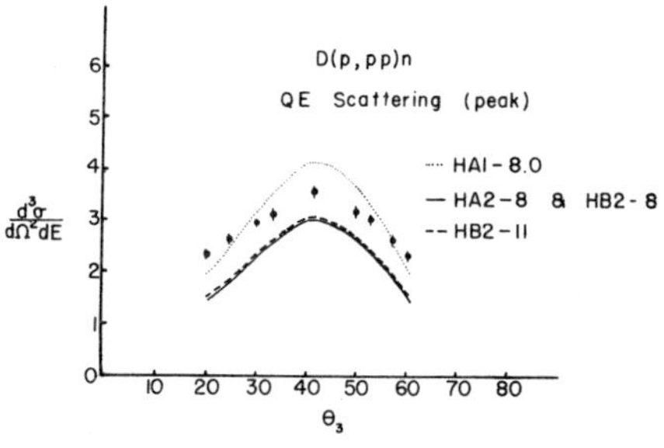

Figure 2. Potential variation in the quasifree scattering angular distribution at 23 MeV

We can understand the lack of off-shell sensitivity as follows. The previous paper demonstrated that the only breakup amplitudes sensitive to off-shell effects were the doublet ($S = \frac{1}{2}$), $L = 0$ amplitudes. Now, the quasifree enhancement, when viewed in terms of N-d (or N-d*) partial waves, occurs because of the coherent contributions of many partial waves. At 45 MeV in the symmetric QFS case, the $L = 0$ amplitude contributes less than 1% of the cross section. With the $L = 0$ contribution so small for QFS, not surprisingly, QFS cross sections are quite insensitive to off-shell changes. In the case of ^{2}H(p,pn)p QFS, the off-shell variations (less than 1%) are even much smaller than in the ^{2}H(p,pp)n case because the very insensitive quartet amplitude dominates the cross section at the QFS peak. In ^{2}H(p,pp)n, the quartet contribution at the peak is small (and even vanishes in the symmetric angle case). On-shell changes affect all partial waves and thus have large effects in QFS.

We can also examine the energy dependence due to on-shell changes by comparing potentials HA1-8.0 and HA2-8.0, Fig. 3. These have the same form factor and triton binding energy. Potential HA1-8.0 has an energy dependence that nearly predicts the free two-nucleon 90° scattering, although remaining only S-wave, while HA2-8.0 predicts a considerably smaller proton-proton cross section. Most of the data shown is for the nearly pure quasifree angles 34° - 43°. At 14 MeV, the predictions for this angle pair are 65% of those at the pure angles 39.4° - 39.4°.

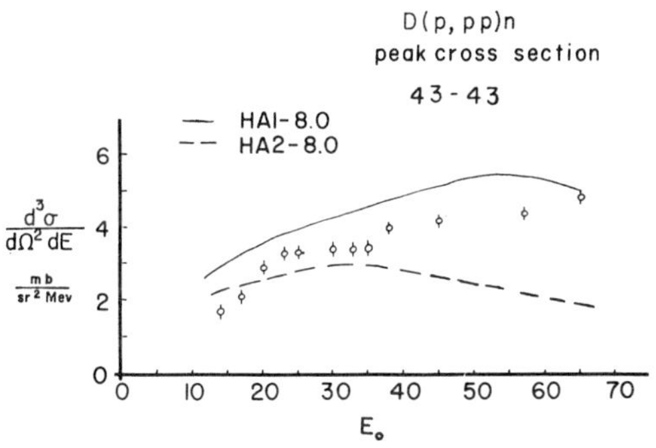

Figure 3. Energy dependence of quasifree scattering. The data is that of Margaziotis[2] and Chang.[3]

References

*Partially supported by NSF.

1. E. L. Petersen, et al., Phys. Rev. C, 9, 508 (1974).
2. D. J. Margaziotis, et al., Phys. Rev. C, 2, 2050 (1970).
3. V. K. C. Chang and P. G. Roos, University of Maryland Report 74-074 (1974).

THE TWO-BODY ELECTRO- AND PHOTO-DISINTEGRATION OF ^{3}HE

J.L. Matthews†, R.O. Owens and G. Ticcioni
Department of Natural Philosophy, Glasgow University, Scotland

During the past decade numerous experiments have been performed on the electro- and photo-disintegration of ^{3}He and the inverse (radiative capture) reaction.[1] Inconsistencies of as much as 30% in the energy-dependence and magnitude of the cross section are found in these data. There is however good evidence that the discrepancies in the photodisintegration measurements have now been resolved[2,3]: the total cross section rises to a peak value of 780 μb at E_γ = 13 MeV and then decreases smoothly with photon energy. There remains some disagreement between the photodisintegration[2] and the electrodisintegration[4,5] results (which themselves are not entirely consistent) when the latter are converted to equivalent photodisintegration cross sections using theoretical virtual photon spectra.[6,7] To aid in solving this problem, a measurement of the electro- and photo-disintegration processes in the same experiment would clearly be of value.

Such measurements have been performed at the Glasgow University linear accelerator laboratory. A ^{3}He gas target was irradiated with a 70 MeV electron beam and the energy spectrum of the emitted deuterons was measured with a magnetic spectrometer. Data were taken with and without a thin bremsstrahlung radiator in the beam to allow separate evaluation of the photo- and electro-disintegration yields. The cross sections at 90° (lab) were obtained over the photon (real or virtual) energy range 11-66 MeV with an average energy resolution of 1%. Angular distributions (30°-150°) were measured at E_γ = 13.5, 20, 35 and 50 MeV. Electrodisintegration data are most simply interpreted in terms of the electric dipole virtual γ spectrum.[7] Comparisons have been made between this and other methods of analysis, with consideration of the approximations involved.

In ref.[2], the two-body photodisintegration cross section was confronted with the best available theory[8], a Faddeev treatment of the three-body problem including Coulomb and tensor forces, short-range repulsion and multiple rescattering in the final state. Recently a new calculation has been reported[9] in which the three-body states are constructed using a charge-dependent interaction fitted to the low energy N-N scattering data. These two predictions are compared to the measured photodisintegration cross section in fig. 1, and neither is seen to be in satisfactory agreement. The calculation of Hendry and Phillips[8] greatly exceeds the peak cross section, and that of Gibson and Lehman[9], while lower at the peak, is too high throughout the energy range. Various theoretical explanations for these discrepancies have been proposed[8,9] and will be discussed.

---

†Present address: Department of Physics, Massachusetts Institute of Technology, Cambridge, Mass.

References

1. See ref.[3] for references to previous work
2. G. Ticcioni et al., Phys. Lett. 46B (1973) 369
3. J.L. Matthews et al., Nucl. Phys. A223 (1974) 221
4. S.K. Kundu, Y.M. Shin and G.D. Wait, Nucl. Phys. A171 (1971) 384
5. C.C. Chang, W.R. Dodge and J.J. Murphy, II, Phys. Rev. C 9 (1974) 1300
6. R.H. Dalitz and D.R. Yennie, Phys. Rev. 105 (1957) 1598; see also refs.[4,5]
7. W.C. Barber and T. Wiedling, Nucl. Phys. 18 (1960) 575
8. J.A. Hendry and A.C. Phillips, Nucl. Phys. A211 (1973) 533 and private communication
9. B.F. Gibson and D.R. Lehman, Bull. Am. Phys. Soc. II 19 (1974) 451 and private communication

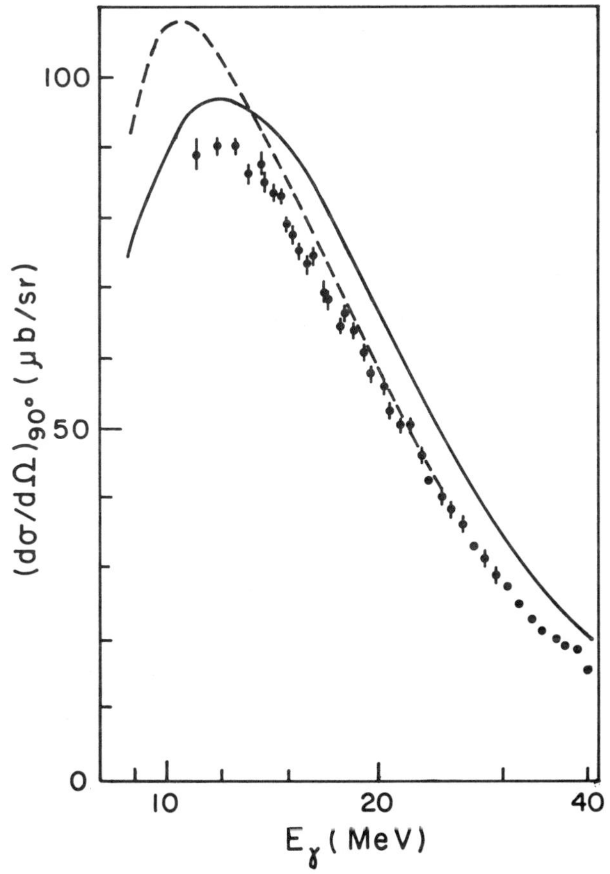

Fig. 1 Center-of-mass cross section at θ_{cm} = 90° for the $^3He(\gamma,d)p$ reaction. The errors shown on the experimental points are purely statistical; an additional systematic uncertainty of ± 5% is estimated. The solid curve is the calculation of Gibson and Lehman[9], the dashed curve that of Hendry and Phillips[2,8].

ELECTRODISINTEGRATION OF ^{3}He

P.T. Kan, G.A. Peterson, Z.M. Szalata, D.V. Webb and F.J. Kline
University of Massachusetts,* Amherst, Massachusetts 01002, U.S.A.

and

S.P. Fivozinsky, J.W. Lightbody, Jr. and S. Penner
Center for Radiation Research, National Bureau of Standards
Washington, D.C. 20234, U.S.A.

Energy spectra of electrons scattered through 92.5° from ^{3}He have been measured up to an excitation energy of 40 MeV by using electron beams of initial energies from 65 to 110 MeV from the National Bureau of Standards linear accelerator. The scattered electrons were momentum analyzed with a magnetic spectrometer having a triple coincidence detector hodoscope which included a plexiglass Čerenkov counter with low sensitivity to background neutrons.

A large sealed gas target[1] had a 14 cm x 8.3 cm x 4.1 cm aluminum frame with foil windows. The beam passed through 0.0015 cm stainless steel windows and the scattered electrons through 0.013 cm windows on the side. Pressure and temperature were continuously measured with a transducer and a thermistor. The number of counts within 1% statistical error was found to have no systematic variation with the incident beam current. The target background measured from an identical empty target was found to be low for this type of construction.

The uncorrected spectrum for an incident electron energy of 110 MeV is shown on the left in Fig. 1. The broken line indicates the empty target background multiplied by a factor of two, and the solid line is the calculated elastic radiation tail[2] which was subtracted from the spectrum before a bin-by-bin radiative unfolding[3] was applied to the continuum data.

The unfolded spectrum shown on the right in Fig. 1 is a sum of the two-body and three-body break-up cross sections, ^{3}He(e,e')dp and ^{3}He(e,e')ppn. The slope of the cross section undergoes a change at about 10 MeV, indicating perhaps the rise of the three-body continuum. The maximum of the cross section occurs at about 15 MeV, followed by a monotonically decreasing cross section. The errors shown in the spectrum are statistical only and there is no evidence of any narrow peaks. The form factors, which must be included in the radiative unfolding, will be extracted by an iterative and self-consistent method.

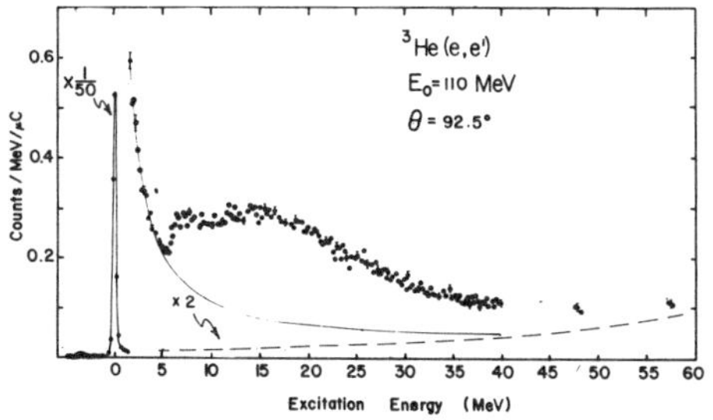

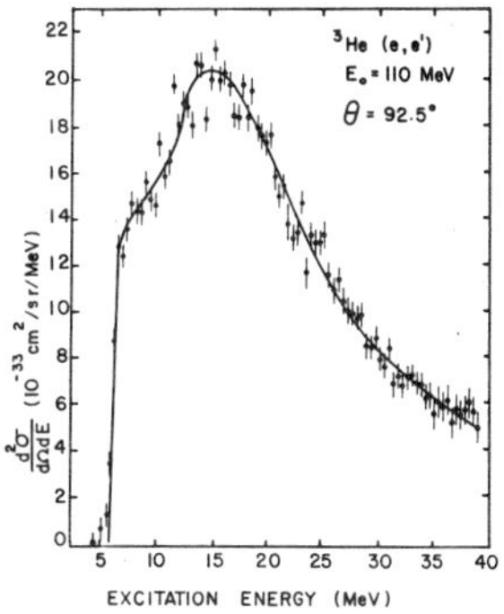

Figure 1.

^{3}He(e,e') spectra: uncorrected (left) and radiatively unfolded (right).

REFERENCES

*Work supported in part by the Office of Naval Research under Contract ONR N00014-67-A-0230-0009.

1. D.V. Webb, G.A. Peterson, P.T. Kan and Z.M. Szalata, to be published in Nucl. Instr. and Meth.

2. L.C. Maximon and D.B. Isabelle, Phys. Rev. 133B, 1344 (1964).

3. P.T. Kan, G.A. Peterson, D.V. Webb, S.P. Fivozinsky, J.W. Lightbody, Jr. and S. Penner, to be published in Phys. Rev.

NON-COPLANAR p-D BREAKUP CROSS SECTIONS*

J.M. Cameron, J.G. Rogers, and J. Soukup
University of Alberta, Edmonton, Alberta, Canada

D.I. Bonbright, A.M. McDonald, W.T.H. van Oers, and J.W. Watson
University of Manitoba, Winnipeg, Manitoba, Canada

W.M. Kloet
University of Maryland, College Park, Maryland, U.S.A.

J.A. Tjon
University of Utrecht, Utrecht, The Netherlands

Recently Kloet and Tjon [1] located regions of maximum sensitivity of three body calculations to the NN parameters by searching the entire four-dimensional phase space volume using various potential models. They found that the regions where the calculated cross section is most sensitive to the model parameters are where the models predict a local minimum in the cross section. These regions are also characterized by the dominance of the particular scattering state in which the two identical nucleons are in a state of zero spin [1]. A scheme in which the differential cross section is constrained by fixing the three NN relative energies and the momentum of one of the identical nucleons was recently proposed by Jain, Rogers and Saylor [2]. By making particular choices for the relative energies and fixed nucleon momentum, one defines a class of constant relative energy loci along which the model cross sections are completely dominated by the crucial $M^{\frac{1}{2}0}$ amplitude. (Here $M^{\frac{1}{2}0}$ denotes the amplitude describing the state in which the three nucleons have total spin 1/2 and the two identical nucleons have spin 0.) The loci are defined by equal relative energies for the two np pairs ($E_{35} = E_{45}$) and by the condition that symmetric coplanar scattering ($\theta_3 = \theta_4$, $\Delta\phi = 180°$) be possible on the locus.

Exact three body calculations were performed at 39.5 MeV along four of these symmetric constant relative energy loci using two different local potential models [1]

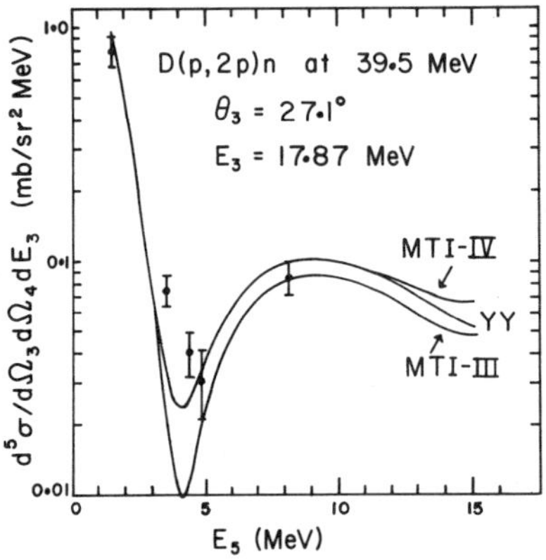

FIGURE 1

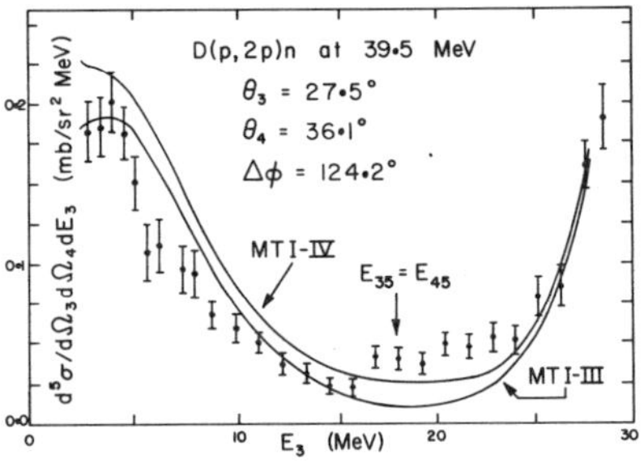

FIGURE 2

and one separable potential model [3]. The four loci were described by fixed proton emission angles of θ_3 = 27.1°, 39.0°, 46.5° and 53.5°. The general features of the calculated cross sections were the same at each angle and for all three models. The curves in figure 1 are typical of the calculated cross sections. Between the single scattering and multiple scattering region on each locus is a deep minimum caused by destructive interference between the single scattering part of $M^{\frac{1}{2}0}$ with the multiple scattering part. Such destructive interference effects are only possible in regions dominated by a single amplitude. Under any other conditions a destructive interference minimum in one amplitude would be filled in by incoherent contributions from the other two amplitudes. The curves labeled MTI-III and MTI-IV are local potential calculations with the Malfliet-Tjon potentials [1] with and without repulsive core in the $^3S_1$ state, respectively. The separable potential model (YY) gives almost identical results to the MTI-IV potential, however this was not the case for all the angles studied.

The data points shown in figure 1 were measured at the University of Manitoba sector-focused cyclotron using a special gas target arrangement. The gas scattering geometry adopted for the $\Delta E - E$ counter telescope arrangements resulted in an uncertainty of approximately $\pm$ 0.35 MeV in the neutron energy (E_5). Beam charge, target thickness and solid angle normalizations were obtained from elastic D(p,d)H scattering at 27.1°. The error bars shown in figure 1 include both the statistical and normalization uncertainties. The data indicate a minimum which is shifted towards larger neutron energies from that predicted by the three s-wave potential model calculations. More data are being collected along this constant relative energy locus.

Figure 2 shows one of the five energy sharing spectra which were obtained. The errors shown are from statistics only. The two curves are the results of the local potential calculations as indicated by the labels in the figure. The symmetric relative energy point, which is plotted also in figure 1, is indicated by the vertical arrow. The position of the minimum and the spectrum shape differ considerably from the model calculations.

BIBLIOGRAPHY

1) W.M. Kloet and J.A. Tjon, Nucl. Phys. A210 (1973) 380

2) M. Jain, J.G. Rogers and D.P. Saylor, Phys. Rev. Lett. 31 (1973) 838

3) M. Jain and G.D. Doolen, Phys. Rev. C8 (1973) 124

---

*Work supported in part by the Atomic Energy Control Board of Canada.

T=3/2 PHOTODISINTEGRATION OF $^3He^\dagger$

B. F. Gibson
Los Alamos Scientific Laboratory, Los Alamos, New Mexico 87544
and
D. R. Lehman
The George Washington University, Washington, D. C. 20006

The cross section for the photodisintegration of ^3He leading to the T=3/2 isopin component of the final state (the three-nucleon final state) has been recalculated. Our calculation was performed within the context of exact three-body theory. The N-N interactions were represented by s-wave, spin-dependent, separable potentials. The full T=3/2 photodisintegration amplitude was expressed as a sum of terms (see Fig. 1): the plane-wave Born amplitude for photodisintegration, terms involving only a single rescattering between pairs of nucleons, and terms combining the off-shell Born amplitude, an intermediate-state propagator, and the complete off-shell N-d and N-ϕ scattering amplitudes (obtained by solving an integral equation). This method of expressing the full disintegration amplitude in terms of the fully off-shell N-d and N-ϕ amplitudes and the off-shell Born amplitude can be applied to any weak process, as we are now doing for the radiative capture reaction: $\pi^- + {}^3H \to \gamma + 3n$. We have computed the off-shell scattering amplitudes by solving the corresponding integral equations using standard contour-rotation techniques[1,2] noting carefully the singularities in the propagators and the driving terms. Gaussian and Gegenbauer quadratures[3] were used in the angular and momentum variable integrations respectively.

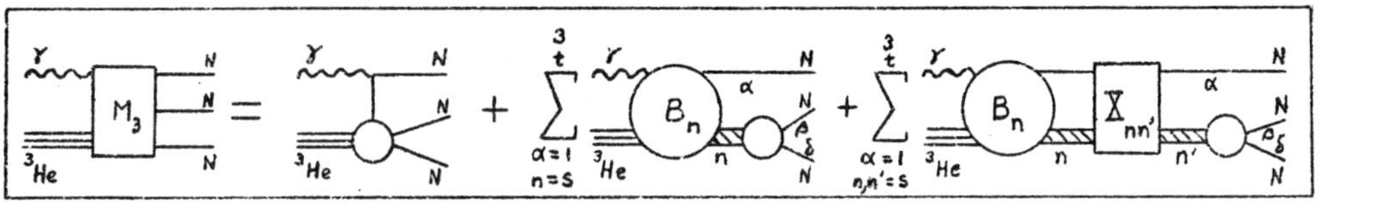

Fig. 1. Decomposition of the full photodisintegration amplitude in terms of the plane-wave Born amplitude, off-shell Born amplitude B_n, and off-shell N-d and N-ϕ amplitudes $X_{nn'}$.

We present here numerical results for 1) our calculation of the Barbour-Phillips model of Ref. 4 (Fig. 2a) in which an analytic form was assumed for the bound state having its one parameter adjusted to fit the ^3He rms radius and in which singlet and triplet separable potentials were used to numerically determine the final state and 2) our own model (Fig. 2b) in which a best fit to the low energy singlet and triplet scattering parameters was made to determine potentials which were then used to generate numerical solutions to the three-body equations for both the initial and final states (the triplet interaction strength was adjusted slightly for the bound state in order to obtain the correct ^3He binding energy[5]). These results are to be compared with the original work[4] of Barbour and Phillips and with that of Ripelle and Levinger[6] in a hyperspherical coordinate model.

It is evident that our plane wave Born result (- - - - -) and our full amplitude result (———)shown in Fig. 2a agree with the original calculation of Barbour and

Phillips shown in Fig. 2c. However, our result for the first rescattering approximation (—— ——) is somewhat smaller than theirs, implying that the inclusion of complete rescattering effects in the T=3/2 channel is not nearly so important as their curves might indicate. We note also that our next iteration after the first rescattering produces essentially the answer as the full calculation; i.e., differences were too small to be seen in Fig. 2a (and 2b). It is evident that neither of the calculations for the Barbour-Phillips model agrees with the Ripelle-Levinger curve (——-——), although it should be understood that different interactions are involved.

Comparison of our results for the Barbour-Phillips model (Fig. 2a) with the results of our consistent initial and final state model calculation (Fig. 2b) indicates that the peak cross section estimates for the three-body breakup reaction in Ref. 4 are possibly too large by some 15%. A similar result was found for the two-body breakup reaction where our model estimates[7] of the peak cross section were some 20% smaller than the estimates quoted in Ref. 4. This discrepancy is apparently due to the form of the bound state wave function assumed in the Barbour-Phillips model, one which overemphasizes the asymptotic shape of the wave function and correspondingly overestimates the magnitude of the cross section at low energy.

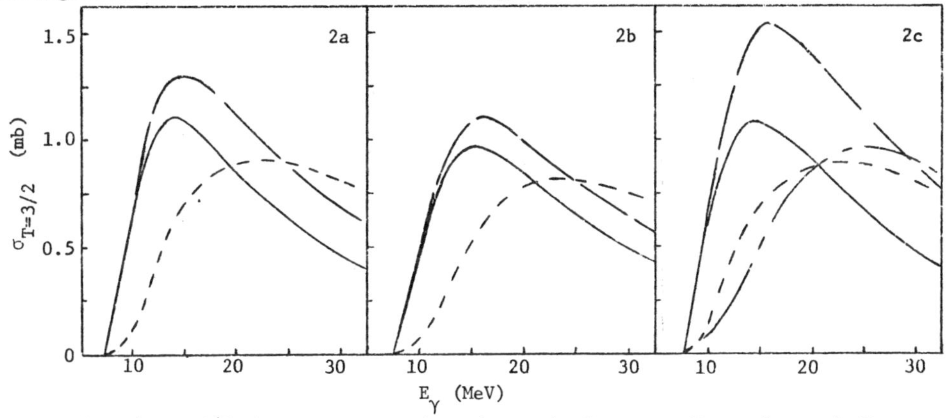

Fig. 2. T=3/2 photo cross section: (- - - - -) plane-wave Born; (—— ——) first rescattering approx.; (————) full amplitude solution. (2a) shows our calculation for the BP model; (2b) shows our model result; (2c) shows results of Ref. 4 and that of Ref. 6 (——-——).

REFERENCES

† Work supported in part by the U. S. Atomic Energy Commission.

1. J. H. Hetherington and L. M. Schick, Phys. Rev. 156 (1967) 1647.
2. R. Aaron and R. D. Amado, Phys. Rev. 150 (1966) 857.
3. T. Brady, M. Fuda, E. Harms, J. Levinger, and R. Stagat, Phys. Rev. 186 (1969) 1069.
4. I. M. Barbour and A. C. Phillips, Phys. Rev. C1 (1970) 165.
5. for example, see B. F. Gibson and G. J. Stephenson, Phys. Rev. C8 (1973) 1222.
6. M. Fabre de la Ripelle and J. S. Levinger, ORSAY preprint (1973) "Photodisintegration of ^{3}He into 3 Nucleons."
7. B. F. Gibson and D. R. Lehman, LASL preprint (1974) "Two-body Photodisintegration of ^{3}He and ^{3}H."

PHOTODISINTEGRATION CROSS SECTIONS OF ^{6}Li

M.A.K. Lodhi, Department of Physics, Texas Tech University, Lubbock, Texas, 79409, and M. L. Rustgi, Department of Physics, SUNY at Buffalo, Buffalo, New York, 14214.

The role of the valence nucleous in the independent particle shell model is assumed to be important. A long-range dynamical correlation has been invoked in the harmonic oscillator type single particle wave function of these nucleons. The calculated charge root mean square radius is found in good agreement with the experimental value as obtained from high energy electron scattering from ^{6}Li. The integrated cross section and integrated bremsstrahlung weighted cross section have also been calculated.

The main attempts in the analysis of the photodisintegration process below the pion threshold have been to understand the nature of the interaction between the electromagnetic field and the nucleus or inversely assuming this interaction known to obtain some knowledge about the nuclear wave function of the ground and allowed excited states. In the sum-rule approach, made popular by Levinger and Bethe[1], one sums over the excited states and expresses the various moments of the total photodisintegration cross section in terms of the ground state wave function of the nucleus. In this paper the Levinger-Bethe approach is applied to ^6Li to calculate the integrated cross section $\sigma_{int} = \int \sigma\, dw$ and integrated bremsstrahlung weighted $\sigma_b = \int \frac{\sigma}{w}\, dw$ cross sections and the results are compared with the experimental values of Costa et al.[2]

The structure of ^6Li has been well known to be peculiar for many years. However, within a reasonable approximation it can be described as consisting of four particles moving in a doubly closed 1s shell and two loosely bound nucleons in the 1p shell. The two valence nucleons may be considered to have a long range correlation producing a longer tail of the nucleus than what it would look in a simple shell-model picture. Under this assumption the wave function of ^6Li may be written as[3,4]

$$\Psi = \psi_o (1 + c\, r), \tag{1}$$

where c and r are some parameters and ψ_o is the simple L-S coupled shell-model wave function given by

$$\psi_o = (1s)^4\ (1p)^2, \tag{2}$$

with the quantum number L = 0, S = 1, J = 1, and T = 0 in the ground state. Assuming the two-body interaction to be of the form

$$V(r_{ij}) = V_o e^{-r_{ij}^2/r_o^2} P_{ij}\ . \tag{3}$$

The parameters c and r_o have been chosen from the earlier work[4-6] to reproduce certain specific measurements on ^6Li such as the charge form factor or the separation between the ground and excited states.

Following the work of Levinger and Bethe and employing the wave function given by (1), the expression for the integrated cross section is given by

$$\sigma_{int.} = \frac{2\pi^2 e^2 \hbar}{Mc} \left[\frac{NZ}{A} + \frac{M}{3\hbar} \frac{1}{N^2} \langle V(r_{ij}) r_{ij}^2 \rangle \right] \tag{4}$$

where

$$N^2 = 1 + \frac{5\sqrt{2}}{\sqrt{\pi}} \frac{c}{\alpha} + 5 \frac{c^2}{\alpha^2} , \tag{5}$$

and the matrix element $<V(r_{ij})r_{ij}^2>$ is calculated using the wave function (1) and the two-body interaction (3).

The numercial results for rms radius and σ_{int} are presented in Table 1 for five different wave functions labeled as 1-5. The first wave function reproduces the large diffraction minimum observed by Li et al.[5] The second wave function does what wave function 1 does but produces a smaller dip. The third wave function reproduces the form factor of ^{6}Li very well but ignores the diffraction minimum within the experimental errors. The fourth wave function gives the correct energy spacing between the ground state and the excited states. The fifth set of parameters is taken from the work of Elton and Lodhi.[6] The effects of correlation between the valence nucleons are studied simply by setting c = 0.

It is apparent from Table 1 that the observed σ_{int} are best reproduced by the first and second wave functions. The effect of correlations between the valence nucleons is found to be enormous on the bremsstrahlung weight cross section.

TABLE 1: Results for the bremsstrahlung weighted (σ_b) and integrated (σ_{int}) cross sections are given for various set of parameters. The experimental values for rms radius σ_b, and σ_{int} are 2.54 fm, 5.1 ± 0.8 mb and 130 ± 20 MeV-mb respectively.

	α (fm^{-1})	c (fm^{-1})	r_0 (fm)	V_0 (MeV)	$<r^2>^{1/2}_{ch}$ (fm)	σ_{int} (MeV-mb)
1.	0.56	0	1.36	60.0	2.40	140.22
	0.56	-0.257	1.36	60.0	2.49	123.60
2.	0.56	0	1.36	60.0	2.42	140.22
	0.56	-0.261	1.36	60.0	2.50	123.18
3.	0.603	0	1.36	60.0	2.25	145.74
	0.603	-0.320	1.36	60.0	2.37	125.86
4.	0.48	0	1.36	50.0	2.82	123.24
	0.48	-0.127	1.36	50.0	2.74	121.98
5.	0.54	0	1.36	47.5	2.50	127.74
	0.54	-0.190	1.36	47.5		121.98

REFERENCES

1. J. S. Levinger and H. A. Bethe, Phys. Rev. 78, 115 (1950).
2. S. Costa, R. Ferrero, C. Manfredotti, L. Pasqualini, and L. Roasio, Nuovo Cimneto 42B, 382 (1966).
3. G. C. Li, I. Sick, R. R. Whitney, and M. R. Yearian, Nucl. Phys. A162, 583 (1971).
4. M.A.K. Lodhi and R. W. Mires, Nucl. Phys. A149, 629 (1970).
5. L.R.B. Elton and M.A.K. Lodhi, Nucl Phys. 66, 209 (1965).
6. L. R. Elton, Introductory Nuclear Theory (Saunders, 1966).

Virtual Photon Analysis and a Direct Comparison with Electrodisintegration Theory of the $^3He(e,d)e'p$ Reaction†

D. M. Skopik, J. J. Murphy, II, Y. M. Shin, E. L. Tomusiak and K. F. Chong
Saskatchewan Accelerator Laboratory
Saskatoon, Saskatchewan S7N OWO

Although there have been a considerable number of experiments[1] performed in recent years directed towards determining the correct magnitude of the two-body photodisintegration cross section of ^3He, the experimental situation is far from satisfactory. In Fig. 1 we show the cross sections for the two-body disintegration process with typical error bars for each measurement. As can be seen there exists a wide range of magnitudes from which to choose and the data cannot be conveniently broken up into high and low results.

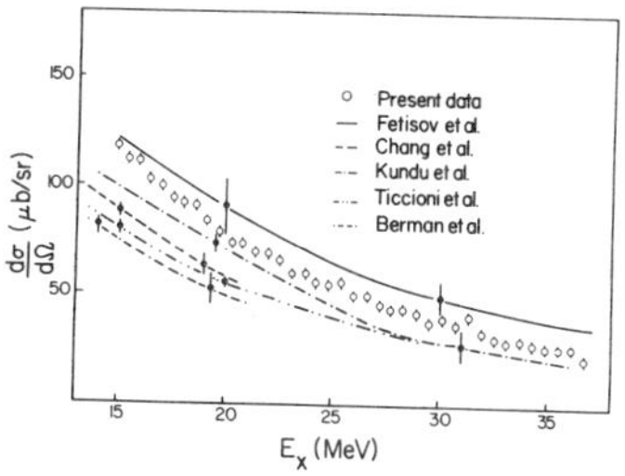

^3He two-body photodisintegration cross section at a laboratory angle of 90°. Typical error bars are shown.

Figure 1

In this paper we report results of measuring the $^3He(e,d)e'p$ reaction. The usual procedure in analyzing such electrodisintegration experiments is to extract the equivalent photon cross section by using the virtual photon theory.[2] This procedure has the advantage that it affords a direct comparison with experimental and theoretical results involving real photons. In contrast, the electrodisintegration cross sections can be calculated using a model and directly compared with experiment, if an integration over all possible states of the undetected outgoing electron is performed. The theoretical expression for the electrodisintegration cross section is given by

$$\frac{d^2\sigma}{d\Omega dE} = \frac{3\alpha^2 (2\pi)^6\hbar^2}{(\hbar ck)} k_D E_D \frac{1}{2} \sum_{\text{spin states}} \int \frac{d^3k'}{E_k q_\mu^4}$$

$$\times\left[2\,\mathrm{Re}\,(\mathbf{k}\cdot\mathbf{J}')(\mathbf{k}'\cdot\mathbf{J}')^* + \tfrac{1}{2} q_\mu^2\, \mathbf{J}\cdot\mathbf{J}'^*\right]$$

$$\times\left[\delta\left(E_k - E_{k'} - \frac{\hbar^2p^2}{2M} - \frac{\hbar^2}{2M_D}(\mathbf{k} - \mathbf{k}' - \mathbf{p})^2 - 5.495\right)\right]$$

where $\mathbf{k}$ and $\mathbf{k}'$ are the wave numbers of the incident and scattered electrons, $\mathbf{k}_D$ and and $\mathbf{p}$ are the momenta of the outgoing deuteron and proton. The four momentum transfer is

$$q_\mu{}^2 = \mathbf{q} \cdot \mathbf{q} - p_o{}^2 \text{ where } q_\mu = k_\mu - k'_\mu$$

and the current density $\mathbf{J}'$ is given by

$$\mathbf{J}' = \mathbf{J} - (\mathbf{q} \cdot \mathbf{J})\, \mathbf{q} \,/\, q_o{}^2$$

To compare our electrodisintegration results to theory we have computed the cross section with a simple model employing an Irving-Gunn ground state with a plane wave and a Hulthén deuteron wave function in the final state.

The results of this calculation for three incident energies are shown in Fig. 2 by the solid lines. At the lowest incident energy the agreement between experiment and theory is quite reasonable, but for the higher incident energies the agreement is satisfactory only for $E_d \gtrsim 5$ MeV. In Fig. 3 the angular distribution data are shown with the predictions of the model. Again we note good agreement for $E_o = 43$ MeV. For higher incident energies however, while the shape of the theoretical predictions appears to be adequate, the predicted magnitudes are not. The 90° differential cross sections $d^2\sigma/d\Omega dE$ measured in this experiment corresponds to the photodisintegration if $\mathbf{J}(\mathbf{q}) \simeq \mathbf{J}_T(\omega)$, i.e. the total nuclear current can be taken to be equal to the value of $\mathbf{J}_T(q)$ at the photon point. Since this condition is fulfilled only if the four-momentum transferred by the electrons to the nucleus is nearly the same as that of the real photon, the cross section at higher deuteron energies should approximate the photodisintegration data better than the cross section at lower deuteron energies. Kinematically at higher deuteron energies (or lower incident electron energies for a fixed deuteron energy in the angular distribution) the momentum transferred by the electron is restricted to lie in a narrower cone about the incident momentum vector. As a result the cross section near $\theta_e \simeq 0$ degrees is measured and the transverse contribution to the cross section dominates.

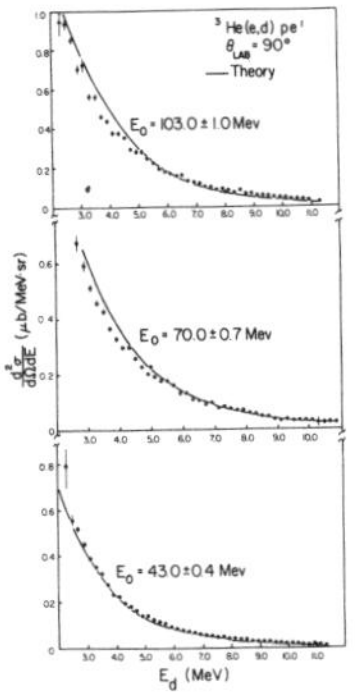

Deuteron energy distributions at 90° in the laboratory for three different incident energies. The error bars are statistical only. The solid curve is the calculation described in the text.

Figure 2

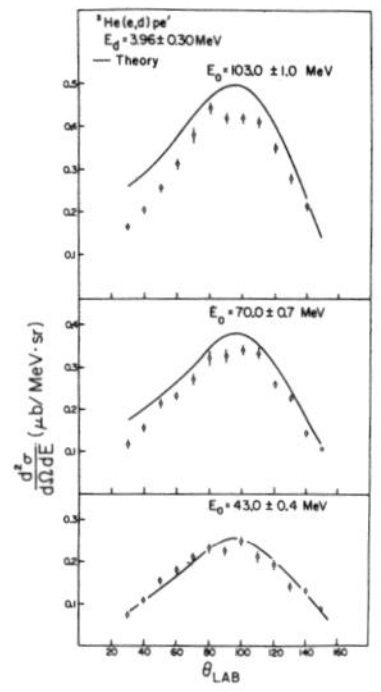

Angular distribution for 3.96 MeV deuterons at three different incident electron energies. The errors shown are statistical only. The solid curve is the calculation described in the text.

Figure 3

For each of the three 90° differential cross sections, we have also unfolded the equivalent photon cross sections using the virtual photon spectrum given in Ref. 2. No differences were found in the three measurements over the entire energy region investigated in the present work. This implies that the cross section due to the longitudinal nuclear current is very small since at low deuteron energies with high incident electron energy the kinematic conditions are such that the experiment would have been sensitive to contributions from the longitudinal nuclear current. (We point out here that the transverse part of the electrodisintegration model used above predicts our extracted photodisintegration cross section reasonably well. However this model fails in predicting the electrodisintegration cross section at low deuteron energies. Thus the discrepancies in the electrodisintegration theory probably arises from an inadequacy of the longitudinal current in the model. This is presently being investigated). Since there were no differences in the three sets of data analyzed using virtual photon theory, all three sets were combined and are shown in Fig. 1 by the open circles. These results are in good agreement with the earlier electrodisintegration measurements by Kundu et al. and lie within the error bars of Fetisov et al. Thus we conclude that while the question of the correct magnitude of the two-body disintegration of ^{3}He is in considerable disarray, our results tend to support the higher cross sections.

While only the direct reactions are shown in Fig. 1 we note that a recent capture measurement by Matthews et al.[3] agrees quite well with the cross section reported by Ticcioni et al., which is approximately 35% lower than our results. This discrepancy is not readily explainable.

†Work supported by the Atomic Energy Control Board of Canada

Bibliography

1) V. N. Fetisov, A. N. Gorbunov and A. T. Varfolomeev, Nuclear Phys. 71, 305 (1965); S. K. Kundu, Y. M. Shin and G. D. Wait, Nucl. Phys. A171, 384 (1971); C. C. Chang, W. R. Dodge and J. J. Murphy, II, Phys. Rev. C9, 1300 (1974); G. Ticcioni, S. N. Gardiner, J. L. Matthews and R. O. Owens, Phys. Lett. 46B, 369 (1973); B. L. Berman, L. J. Koester and J. H. Smith, Phys. Rev. 133, B117 (1964).

2) D. M. Skopik, Y. M. Shin, M. C. Phenneger and J. J. Murphy, II, Phys. Rev. C9, 531 (1974).

3) J. L. Matthews, T. Cruse, M. E. Williams, R. O. Owens and W. Savin, Nucl. Phys. A223, 221 (1974).

PROTON KNOCK-OUT REACTIONS AND CLUSTER STRUCTURE IN Li ISOTOPES

B.K.Jain[+] and A.K.Jain, Nuclear Physics Division, Bhabha Atomic Research Centre, Bombay 400085, India.

There have been many indications in support of cluster structure in light nuclei. Particular emphasis has however been laid on the intercluster motion when discussing clustering. It has been demonstrated earlier that the intercluster wave function can be studied best by cluster knock-out reactions[1]. It has also been demonstrated that the internal structure of the clusters is rather difficult to study using the cluster knock-out impulse approximation[2]. The wave functions derived from the variational calculations of the spectrum of Li isotopes contain parameters which show that the loosely bound clusters, d and t, should have different structure as compared to the corresponding free particles[3]. A more direct indication to the d- cluster contraction in ^{6}Li comes from the study of ^{6}Li(d,t p) ^{4}He reaction[4]. As proton knock-out reactions have been used successfully[5] for obtaining the proton momentum distribution, and hence the proton wave function, in light and medium mass nuclei we expect these reactions to be more sensitive to the internal structure of the clusters.

In this note we present the PWTA analysis of the ^{6}Li(e, e′p)^{5}He and ^{7}Li(p,2p) reactions using the cluster model formalism. The calculations are compared with the 700 MeV (e,e′p), ref.[6] and the 155 MeV (p,2p), ref.[7] data. Since our primary motivation is to investigate the sensitivity of these reactions to the cluster sizes, as a preliminary analysis, we have neglected the multiple scattering of the continuum protons.

The cross section for (p,2p) reaction has been written as [5]

$$\frac{d^3\sigma}{dE\,d\Omega_1\,d\Omega_2} = F_{Kin.} \cdot |t(q)|^2\ |g(Q)|^2$$

where F_{kin} is a kinematic factor depending purely on the kinematic variables and Q and q are the recoil momentum and momentum transfer respectively.

$$t(q) = \int e^{i\vec{q}\cdot\vec{r}}\ t(E,r)\ d\vec{r}$$

where $t(E,r) = t_L(r)\ \exp\left\{\frac{M\beta^2}{2\hbar^2}\left(E - t_L(r)\right)\right\}$, with parameters and other quantities, obtained from ref.8), fit the p - p scattering data at 90°. Nuclear structure information is contained in the form factor g(Q). For ^{6}Li and ^{7}Li this factor is calculated using the α-d and α-t cluster model wave functions [3] respectively. We have neglected the antisymmetrization of the cluster model wave functions; due to large cluster separation the exchange terms are expected to contribute negligibly. The bound proton wave function has, however, been matched to the exponential form corresponding to the separation energy. Internal wave functions have Gaussian forms with parameters α and $\bar{\alpha}$ respectively for the d and t clusters.

The results for ^{6}Li(e,e′p) are presented in fig.1 and for ^{7}Li(p,2p) in fig.2. It is to be remarked from the theoretical curves that the results are very sensitive to the size of the d and t clusters. Comparison of the theoretical curves with the experimental data in fig.1 suggest that the d cluster size in ^{6}Li is much smaller than the size of free deuteron. In fig.2, however, one finds that t cluster in ^{7}Li with $\bar{\alpha} = 0.346$ fm^{-2}, is not much different from free triton, with $\bar{\alpha} \approx 0.4$ fm^{-2}. The results are therefore in agreement with the other observations [4] and expectations[9].

From these results it is to be concluded that whereas cluster knock-out reactions may be considered most suitable for the study of intercluster behaviour [1,2] the proton knock-out reactions are most suited for investigating the intra cluster structures.

+ Present address: Physics Department, The University of Manitoba, Winnipeg, Canada.

1) A.K.Jain, N.Sarma and B.Banerjee, Nucl.Phys. A142(1970) 330.

2) A.K.Jain, J.Y.Grossiord.M.Chevallier, P.Gaillard, A.Guichard,M.Gusakow and J.R.Pizzi, Nucl.Phys. A216(1973) 519.

3) Y.C.Tang, K.Wildermuth and L.D.Pearlstein, Phys.Rev.123(1961) 548.

4) J.Y.Grossiord, C.Coste, A.Guichard, M.Gusakow, A.K.Jain, J.R.Pizzi, G.Bagieu and R.de Swiniarski, Phys. Rev.Lett.32(1974) 173.

5) B.K.Jain, Ph.D.Thesis, Unpublished, University of Surrey, 1968.

6) H.Hiramatsu, T.Kamae, H.Muramatsu, K.Nakamura, N.Izutsu and Y.Watase, Phys. Lett. 44B (1973) 50.

7) J.C.Roynette et al, Nucl.Phys. A95 (1967) 545.

8) O.Karmon and P.A.Deutchman, Phys.Lett.31B(1970) 113.

9) A.K.Jain and N.Sarma, Phys.Lett. 33B (1970) 271.

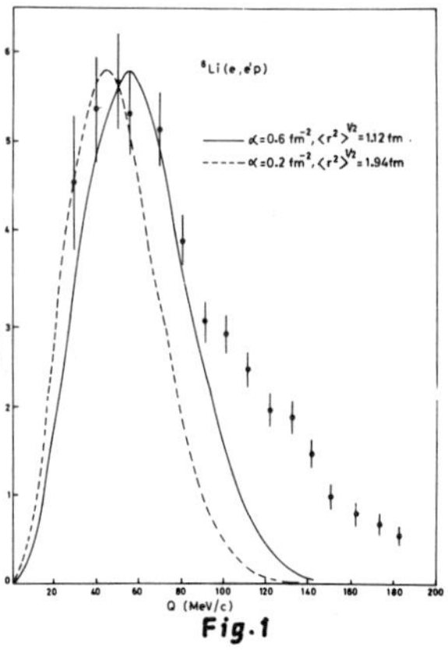

Fig.1

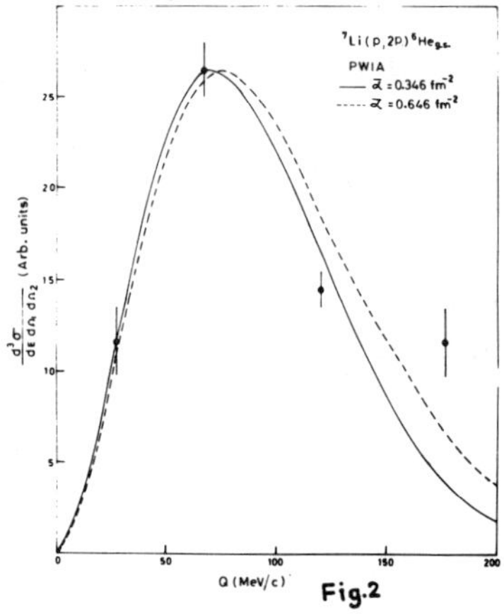

Fig.2

XI

EFFETS RELATIVISTES

RELATIVISTIC EFFECTS

LOW ENERGY RELATIVISTIC CORRECTIONS IN A QUASI-POTENTIAL MODEL

M. Bawin[+] and J.P. Lavine[++]

\+ Physique Nucléaire Théorique, Institut de Physique, Université de Liège, Sart Tilman, 4000 Liège 1, Belgium.

++ Département de Physique, Université Laval, Québec, G1K 7P4, Canada.

Recently there have been many treatments of two-body scattering with relativistic quasi-potential theories[1),2)]. We have used[3)] a three-dimensional, relativistic, two-body equation due to Todorov[4)] in an investigation of low energy relativistic corrections in nucleon-nucleon scattering. We examined this approach as an alternative to the Bethe-Salpeter equation in the ladder approximation (BSLA)[5)]. The BSLA equation (for two spinless particles) does not lead to a Klein-Gordon equation when one of the masses becomes infinite (static limit) and it does not have the eikonal form in the high energy limit[4),6)]. These defects are removed only by including all crossed graphs[7)].

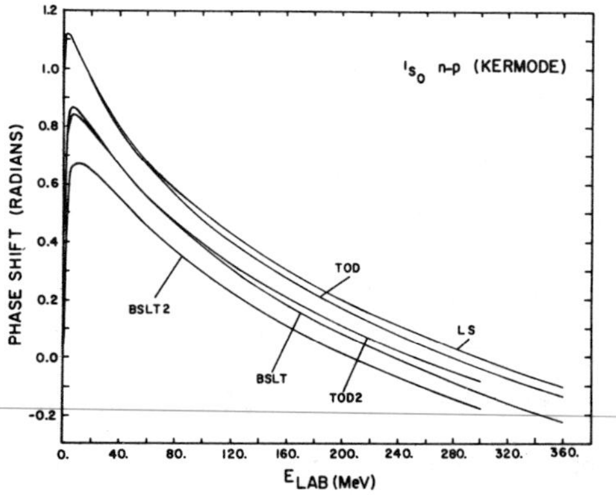

We study the importance of crossed graphs in the framework of the Todorov equation, since this equation possesses the correct static and eikonal limits. The BSLA equation includes retardation effects and a comparison with the instantaneous interaction approximation to the BSLA equation, the BSLT equation[8)], shows that the respective phase shifts differ by at least 10%. We phenomenologically include some effects of retardation in our potential by using a second-order term that contains the Bethe-Salpeter free Green function. Our equations and numerical techniques are described in reference 3, where results are given for the $^1S_0$ Reid soft-core potential[9)]. We found that the crossed graph effects decrease the magnitude of relativistic effects from the 30% of the BSLA[5)] to around 20% for our model.

Phase shift results are presented here for the $^1S_0$ neutron-proton potential of Kermode[10)]. This potential is a superposition of Yukawa terms and we use $-0.4285823771 \times 10^7$ and $+0.1764587996 \times 10^7$ for the 13th and 14th coupling constants (in MeV-fm) respectively. This potential has a strongly repulsive core region, but it is zero at both the origin and 0.57 fm. At larger radial distances the potential is attractive. It has a deeper minimum but it quickly becomes less attractive at intermediate distances than the $^1S_0$ Reid potential. We investigate how these shape differences affect conclusions on low energy relativistic corrections.

The figure shows the phase shifts for our model (TOD2) and for the Lippmann-Schwinger (LS) equation. We see that at low energies there are differences of 25% or less. The curve BSLT2 comes from using the BSLT equation with the second-order term added to the potential. BSLT2 is an excellent approximation to the BSLA equation for the $^1S_0$ Reid potential[11]. We remark that our TOD2 and BSLT2 results for the Kermode potential are only accurate to about 3% or so at low energies and become less accurate at higher energies. Hence, we see that the Kermode and the Reid potentials show similar relativistic corrections for the $^1S_0$ partial wave.

The approximate equality shown in the figure between the BSLT and the TOD2 results indicates an approximate cancellation between retardation and crossed graph effects. This was noted in reference 3 and also by Fronsdal and Huff[12]. This rough equality allows us to characterize the relativistic corrections by modifications to the inverse ranges of the Yukawa terms. For the $^1S_0$ Reid potential we benefit from the work of Brown, Jackson, and Kuo[13]. They leave the one-pion exchange (OPE) term alone, but change the other two inverse ranges: $2.8 \rightarrow 2.8(1.0233)\ fm^{-1}$ and $4.9 \rightarrow 4.9(1.0445) fm^{-1}$. The BSLT results for the modified potential agree with the LS phase shifts for the unmodified potential. The TOD2 results now fall into rough agreement with those of the LS equation. The Kermode potential is extremely sensitive to changes in the ranges, since it a superposition of 14 terms. We leave the OPE term unchanged, but we multiply the inverse ranges of all the other attractive terms by 0.999933 and the inverse ranges of the repulsive terms by 1.00015. These modifications bring the BSLT and the TOD2 phase shifts into rough agreement with the original potential's LS phase shifts. We note that the modified Kermode potential now has an attractive region between 0 and 0.03 fm.

Alternatively, we may adjust the coupling constants[5]. If all the coupling constants are multiplied by 1.09 then the low energy phase shift peak of BSLT or TOD2 is close to that for LS. The constant factor must be increased for the phase shifts at higher energies to be brought into agreement. For example, a factor of 1.25 is needed for energies around 180. MeV. This type of behaviour is expected from the form of the equations.(Work supported in part by the A.E.C.B. of Canada and the I.I.S.N. of Belgium).

REFERENCES

1) R.M. Woloshyn and A.D. Jackson, Nucl. Phys. B64, 269 (1973).
2) R.J. Yaes, Prog.Theoret. Phys. 50, 945 (1973). K. Holinde, K. Erkelenz, and R. Alzetta, Nucl. Phys. A194, 161 (1972). J. Hornstein and F. Gross, Phys. Letters 47B 205 (1973). C.J. Noble and K.C. Richards,Phys. Letters 47B, 319 (1973).
3) M. Bawin and J.P. Lavine, Nucl. Phys. B63, 513 (1973).
4) I.T. Todorov, Phys. Rev. D3, 2351 (1971).
5) M. Fortes and A.D. Jackson, Nucl. Phys. A175, 449 (1971).
6) C. Fronsdal and R.W. Huff, Phys. Rev. D3, 933 (1971).
7) W. Dittrich, Phys. Rev. D6, 2094 (1972).
H.D.I. Abarbanel and C. Itzykson, Phys. Rev. Letters, 23, 53 (1969).
R. Torgerson, Phys. Rev. 143, 1194 (1966).
8) R. Blankenbecler and R. Sugar, Phys. Rev. 142, 1051(1966).
A.A. Logunov and A.N. Tavkhelidze, Nuovo Cimento 29, 380 (1963).
9) R.V. Reid, Ann. of Phys. 50, 411 (1968).
10)M.W. Kermode, J. Phys. A5, L125 (1972).
11)M. Fortes and A.D. Jackson, Phys. Letters 38B, 277 (1972)
12)C. Fronsdal and R.W. Huff, Phys. Rev. D7, 3609 (1973).
13)G.E. Brown, A.D. Jackson, and T.T.S.Kuo, Nucl. Phys. A133, 481 (1969).

RADIATIVE NEUTRON CAPTURE NEAR THRESHOLD

Edward Dressler[†], Department of Physics, American University, Washington, D. C.

and

Franz Gross, Department of Physics, College of William and Mary, Williamsburg, Virginia

Progress has recently been made in understanding the discrepancy between theoretical and experimental values of the neutron radiative capture crossection $n + p \rightarrow \gamma + d$, near threshold. The amplitude in the non-relativistic impulse approximation is proportional to the overlap between the deuteron S state wave function, y, and the $^1S_0$ scattering state wave function, y and $G_M = 4.70$

$$M_{NR} = G_M \int dr\, u(r)\, y(r) \tag{1}$$

This gives a result for the cross section about 10% too small, which was not understood until recently when Riska and Brown,[1] and Gari and Huffman[2] accounted for the discrepancy as the sum of three effects: (1) the production of virtual nucleon resonances, (2) meson exchange currents, and (3) pair terms (virtual nucleon, anti-nucleon pair production by the photon). The last effect is the largest, accounting for about 7% of the correction to the cross section.

In this note we report preliminary results of an alternative way of calculating the large pair term contribution.[3] Our method, which has been used previously in the study of the deuteron form factor,[4] retains only that part of the relativistic impulse approximation which comes from the positive energy pole of the spectator nucleon, corresponding to all processes in which the spectator nucleon is on the mass shell. This approximation is quite accurate, preserves the original covariance, and leaves us with a generalization of Eq. (1), i.e. a 3-dimensional overlap integral between two relativistic wave functions in which only the interacting nucleon is virtual. Each of these wave functions (for the deuteron and the $^1S_0$ scattering state) can be written as a sum of a positive and negative energy part so that the relativistic amplitude is composed of four terms

$$M_R = M^{++} + M^{+-} + M^{-+} + M^{--}$$

where

$$M^{ab} = \int d^3p\, \psi_D^{(a)}(p)\, \psi_{^1S_0}^{(b)}(p) \tag{2}$$

$$a,b = + \text{ or } - .$$

The M^{++} term reduces to M_{NR}, while the M^{+-} and M^{-+} terms are a generalization of the pair terms mentioned above. The M^{--} terms are very small and will be neglected in the discussion that follows.

When the details are carried out, we obtain[3]

$$M^{+-} = \frac{F_1}{\sqrt{3}} \int_0^\infty dr\, Mr\, z(r)\, [u(r) + \frac{1}{\sqrt{2}} w(r)]$$

$$M^{-+} = \frac{F_1}{\sqrt{6}} \int_0^\infty dr\, Mr\, y(r)\, v_t(r) \tag{3}$$

where $F_1 = 1.0$, M is the nucleon mass, and w is the usual D state reduced wave function of the deuteron. The wave function v_t is one of two negative energy deuteron wave functions,[5] and z(r) is a negative energy $^1S_0$ wave function, which can be estimated by the formula[6]

$$\sqrt{3}\ 2MD\ z(r) = \left(\frac{D}{D_T}\right)^{\nu+\frac{1}{2}} \{(v_b + \frac{v_\ell}{r} + \frac{v_\ell^2 v_\ell'(1-\frac{v_\ell}{4})(\nu+\frac{1}{2})}{DD_T})\ y - v_\ell y'\} \quad , \tag{4}$$

Here $D = 1 - (1/2)\ v_\ell$, $D_T = D + (1/2)\ v_\ell^2$, and the nucleon-nucleon potentials v_b and v_ℓ are given elsewhere,[6,7] while $\nu = 1.56$ and the prime refers to differentiation with respect to r. If the nucleon-nucleon potentials are approximated by one pion exchange only, then $v_\ell = 0$, and Eq. (3) gives the same result obtained in Refs. 1 and 2.

A preliminary evaluation of the quantity $(M^{+-} + M^{-+})/M^{++}$ using an empirical wave function[9] for the $^1S_0$ state and Reid's soft core deuteron wave function gives about 1.5%, which is smaller than the 3.7% obtained in Ref. 2. Work is currently underway to obtain more reliable negative energy wave functions[10] and the results will probably be altered when this work is completed.

†Supported in part by AEC Laboratory Graduate Fellowship Grant AT-(40-1)-3807, Present address, Department of Physics, University of Saskatchewan, Saskatoon, Saskatchewan, Canada.

1) D. O. Riska and G. E. Brown, Phys. Letters 38B, 193 (1972).
2) M. Gari and A. H. Huffman, Phys. Rev. C7, 994 (1973).
3) Edward Dressler and Franz Gross, in preparation.
4) Franz Gross, Phys. Rev. 140, B410 (1965).
5) J. Hornstein and F. Gross, Physics Letters 47B, 205 (1973).
6) Etienne Delacroix and Franz, Gross, in preparation.
7) Franz Gross, Phys. Rev. D (scheduled for publication in the 1 July issue), SLAC PUB-1138.
8) In this limit our result is equal to that of Refs. 1 and 2 multiplied by λ where λ is the fraction of pseudoscalar and 1 - λ the fraction of pseudovector included in the π-NN coupling (see Ref. 7, where λ = 0.41). When λ= 1, we expect that the results should agree.
9) R. J. Adler, B. T. Chertok, and H. C. Miller, Phys. Rev. C2, 1 (1970).
10) Warren Buck and Franz Gross, in preparation.

RELATIVISTIC CORRECTIONS TO THE DEUTERON MAGNETIC MOMENT

Franz Gross, Department of Physics, College of William and Mary, Williamsburg, Virginia 23185.

It is well known that the magnetic moment of the deuteron can not be explained solely by non-relativistic theory unless one assumes a D state contribution of about 4%, which is uncomfortably smaller than the 6 - 7% predicted by most nuclear force models. If we assume that a D state probability of 6½% is correct (à la Reid)[1], the non-relativistic contribution to the magnetic moment is .843 nuclear magnetons, or about .014 less that the experimental value. In this note we show that this "discrepancy" may be removed by relativistic corrections arising from small relativistic components of the deuteron wave function.

Our method, which was developed some time ago[2], is to simplify the calculation of the deuteron form factor in the impulse approximation by retaining only those terms in which the spectator nucleon is on its mass shell. The covariant form factor we obtain includes three types of correction terms not included in the non-relativistic theory: (1) recoil corrections due to the center of mass motion of the deuteron, (2) corrections to the nucleon current, and (3) wave function corrections due to the presence of small negative energy components of the wave function arising from the fact that the interacting nucleon is off mass shell. In Ref. 2) the first two types of corrections were evaluated, but at that time we had no way of computing the small negative energy components.

Recently we have developed a method of calculating these negative energy components using a relativistic equation for the two nucleon system[4]. Our result for the magnetic moment, which is the static limit of the magnetic form factor, is:

$$\mu_d = (F_1 + F_2)\,(1 - 3/2\, P_d) + F_1\,(3/4\, P_d + \Delta) \qquad (1)$$

where $F_1 = 1$, $F_2 = -0.12$, P_d is the deuteron D state fraction and

$$\Delta = \int_0^\infty dr\, \frac{Mr}{\sqrt{3}}\, \{v_t\, (\frac{u}{\sqrt{2}} - w) - v_s (u + \frac{1}{\sqrt{2}} w)\} + v^2 \text{ terms} + (\frac{p}{M})^2 \text{ terms} \qquad (2)$$

Here u and w are the usual non-relativistic deuteron wave functions, and v_t and v_s are the negative energy P state wave functions which have recently been estimates[3]. The v^2 terms are those bilinear in v_t and v_s, which we neglect in this note. The $(\frac{p}{M})^2$ terms are the recoil terms, which are very small.

When these integrals are estimated using the negative energy wave functions of Ref. 3), we obtain

$$\Delta \cong .015 \qquad (3)$$

The recoil terms contribute only about -.003 to this result, so that the positive sign and great bulk of the contribution are due entirely to the first term of Eqn. (2) not previously estimated. Our result makes it possible to obtain agreement with the magnetic moment by assuming only a tiny amount of exchange current, which may be correct for the deuteron where selection rules eliminate most possible currents.

We call attention to the fact that this is one of several places where the negative energy components of the deuteron wave function seem to play an interesting role; others are radiative neutron capture[5)] and backward p - d scattering[6)]. Here, however, the "tail" of the negative energy wave function, which is due to one pion exchange, cancells, and the entire effect is due to the interior structure of these wave functions.

REFERENCES

1) R. V. Reid, Jr., Annals of Physics 50, 411 (1968).
2) Franz Gross, Phys. Rev. 140, B410 (1965); 142, 1025 (1966), E 152, 1517 (1966).
3) J. Hornstein and F. Gross, Physics Letters, 47B, 205 (1973).
4) Franz Gross, Phys. Rev. D (scheduled for publication in the 1 July issue), SLAC PUB-1138; see also Phys. Rev. 186, 1448 (1969).
5) Edward Dressler and Franz Gross, reported elsewhere at this conference.
6) J. Hornstein, F. Gross, R. A. Miller, and E. A. Remler, reported at the 1973 International Conference; see also E. A. Remler and R. A. Miller, Annals of Physics, 82, 189 (1974).

THRESHOLD ELECTRODISINTEGRATION OF THE DEUTERON CAUSED BY BACKWARD SCATTERING OF ELECTRONS

S.A.Smirnov and S.V.Trubnikov

Department of Quantum Field Theory, Steklov Mathematical Institute USSR Academy of Sciences, Vavilov str. 42, Moscow 117333, USSR

The deuteron electrodisintegration $e+d \to e+n+p$ yields important information about the neutron electromagnetic form factors and the structure of n-p interaction.In this paper we present the results of the numerical calculation of the scattering cross section in consistent relativistic formalism.The representation of Shirokov (RS) (ref. [I]) is the very convenient method for investigation of the elastic and inelastic ed-scattering, in particular the problems of exact account of the final state interaction (FSI) and relativistic effects.RS is a new relativistic integral representation for matrix elements of fields and currents, based on the general assumptions of the quantum field theory only.Using RS it is possible to express $\langle np | j^{\mu} | d \rangle$ (and $\langle d | j^{\mu} | d \rangle$) through the form factors of the physical nucleons and n-p phase shifts on the mass shell alone.The cross section is calculated for $q^2 \leqslant 5\,\mathrm{fm}^{-2}$, i.e. in kinematical situation of the experiments [2].For such values of q^2 the contributions of the deuteron D-state and meson exchange currents are small (see,e.g.,ref. [3]).Near the deuteron disintegration threshold only the S-state of the final n-p system was taken into account.In this case the cross section can be expressed in terms of the two inelastic invariant form factors $G_{s,t}$,which describe the magnetic dipole transitions to the final $^1S_0$ -, $^3S_1$-states of n-p system. RS leads to the following formula for G_u ("u" means "s" or "t"):

$$G_u(s,t)=\frac{\Gamma}{2\pi i}\operatorname*{Res}_{s'=M_D^2} B_t^{-1}(s')\left\{\int_{s_1(s,t)}^{s_2(s,t)} ds'' \frac{G_u^0(s,s'',t)\Delta B_t(s'')}{s''-M_D^2} - \frac{1}{2\pi i}\int_{4M^2}^{\infty} ds''' \frac{\Delta B_u(s''')}{s-s'''-i\varepsilon}\int_{s_1(s''',t)}^{s_2(s''',t)} ds'' \frac{G_u^0(s''',s'',t)\Delta B_t(s'')}{s''-M_D^2}\right\}, \quad (1)$$

$$s_{2,1}(s,t)=2M^2+(1/2M^2)(2M^2-t)(s-2M^2)\pm(1/2M^2)\sqrt{(-t)(4M^2-t)s(s-4M^2)}$$

here M and M_d are nucleon and deuteron masses, $t=-q^2<0$, s is the invariant mass of the final n-p system; Γ is the normalization constant, the first term in eq.(1) describes the Born approximation, the second one is caused by the FSI. $G_u^0(s,s',t)$ are the electromagnetic form factors of the noninteracting n-p system. The explicit formulae for G_u^0 in ref. [4] contain the new pure relativistic terms. $B_u(s)$ are the relativistic Jost functions:

$$B_s(s)=\tilde{B}_s(s),\ B_t(s)=\left(1-\frac{M_D^2-4M^2}{s-4M^2}\right)\tilde{B}_t(s),\ \tilde{B}_u(s)=exp\left\{\frac{1}{\pi}\int_{4M^2}^{\infty}d\hat{s}\,\frac{\delta_u(\hat{s})}{s-\hat{s}}\right\},$$

$$\Delta B_u(s)=B_u(s+i\varepsilon)-B_u(s-i\varepsilon),$$

$\delta_{s,t}(s)$ are the $^1S_0$- and $^3S_1$-phase shifts of n-p scattering. The experimental phase shifts were given in ref. [5]. Then the phase shifts were extrapolated to higher energies. The results are almost independent on high-energy tail of the phase shifts (see fig.1). The calculated cross section for $q^2\leqslant 5fm^{-2}$ coincides with the experimental one. Further the FSI term in eq.(1) dominates in the cross section near the deuteron disintegration threshold. For the S-wave deuteron the relativistic correction is positive and does not exceed 15% for $q^2\leqslant 10fm^{-2}$.

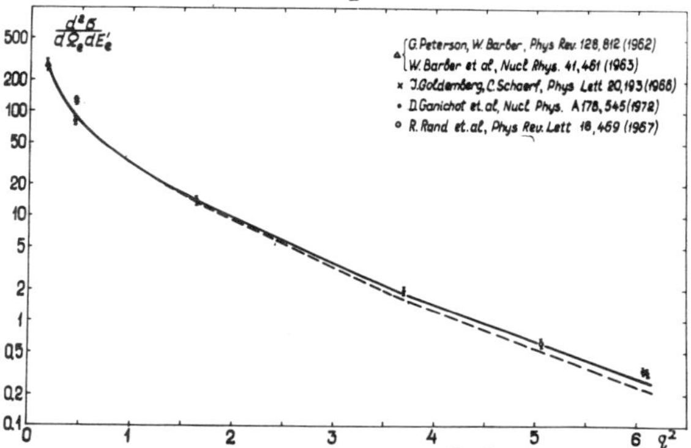

Fig.1
The calculated value of the cross section (in units 10^{-34} sm^2/ sr(Mev/c)) near threshold with the phase $\delta_u(\infty)=-\pi$ (solid line); the same value with the phase $\delta_u(\infty)=0$ (dashed line); q^2 is in fm^{-2}; $E'_e = 0.98E_{threshold}$. The experimental points are taken from ref. [2].

References

1) Yu.M.Shirokov, Nucl.Phys.B6(1968)159.
2) G.Peterson, W.Barber, Phys.Rev.128(1962)812; W.Barber, et al, Nucl.Phys. 41(1963)461; J.Goldemberg, C.Schaerf, Phys.Lett.20(1966)193; D.Ganichot et al, Nucl.Phys.A178(1972)545; R.E.Rand et al, Phys.Rev.Lett.18(1967) 469.
3) J.Hockert et al, Nucl.Phys.A217(1973)14; D.O.Riska, Asilomar Conf., 1973.
4) S.A.Smirnov and S.V.Trubnikov, Phys. Lett.B48(1974)105.
5) M.N.McGregor et al, Phys.Rev.169(1968)1149, 173(1968)1272, 182(1969) 1714.

XII

LES FAISCEAUX, L'EQUIPEMENT ET LES RESSOURCES NECESSAIRES A L'AVENIR

BEAMS, EQUIPMENT AND RESOURCES DEEMED NECESSARY IN THE FUTURE

LES INSTALLATIONS ACTUELLES ET FUTURES POUR L'ETUDE DES PROPRIETES NUCLEAIRES A ENERGIES INTERMEDIAIRES SUR LE SITE DE SACLAY

Claude G. Schuhl
Département de Physique Nucléaire
CEN Saclay, BP 2, 91190 Gif-sur-Yvette, France †

Le Département de Physique Nucléaire du CEA dispose, pour l'étude de la structure nucléaire, des accélérateurs suivants†† :

- Van de Graaff tandem de type "King" (depuis 1964, modifié en 1969). L'énergie maximale est de 16 MeV pour les protons, 24 MeV pour les particules alpha, 64 MeV pour les ions azote ou oxygène de charge 7 ;
- Cyclotron à énergie variable (depuis 1964). Il fournit des protons polarisés ou non jusqu'à 30 MeV, des faisceaux de particules alpha de 30 MeV et de ^{3}He de 40 MeV ;
- Accélérateur linéaire d'électrons de 600 MeV (depuis 1969). Cet accélérateur met à la disposition des scientifiques des faisceaux de négatons et de positons, de photons de freinage ou d'annihilation et des pions de basse énergie ;
- Le synchrotron Saturne avec des protons de 1 à 2 GeV,
- Un canal de réacteur nucléaire EL3.

Compte tenu du programme de la Conférence, nous étudierons les possibilités expérimentales de l'accélérateur linéaire de 600 MeV et l'évolution prévue du synchrotron Saturne. Ces deux accélérateurs rendent possible l'étude des sous-structures fines nucléaires.

I - L'ACCELERATEUR LINEAIRE D'ELECTRONS DE 600 MeV (ALS)

L'interaction électromagnétique possède l'avantage d'être bien connue théoriquement. Par contre le taux de réaction qu'elle produit est 100 à 1000 fois inférieur à celui obtenu avec des particules lourdes.

L'accélérateur linéaire a été conçu pour permettre l'étude des réactions transférant des énergies ou des impulsions élevées aux noyaux ; ceci dans l'idée de mieux comprendre les effets dus à la structure fine nucléaire. C'est pourquoi un effort a été réalisé pour accroître le cycle utile afin de réaliser des expériences de coïncidence. Enfin, pour permettre de couvrir le domaine de la résonance Δ(1236), l'énergie maximale à vide a été choisie de 600 MeV.

---

† Pour plus de détails concernant les activités du Département de Physique Nucléaire voir B.I.S.T. n° 179 mars 1973 (Commissariat à l'Energie Atomique) et le Compte Rendu d'activité 1972/1973, note CEA-N 1700.

†† Citons pour mémoire l'accélérateur linéaire d'électrons de 60 MeV (depuis 1964, modifié en 1969) auprès duquel les expériences photonucléaires avec photons monochromatiques d'annihilation en vol de positons se poursuivront jusqu'à la fin de 1974.

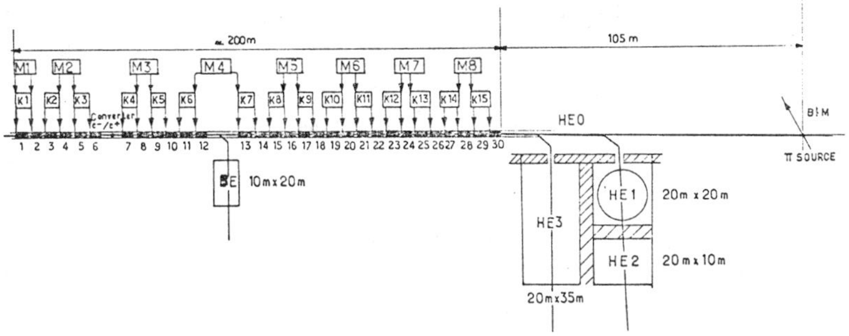

Fig. 1. - Disposition générale de l'accélérateur linéaire d'électrons de Saclay ($E_- = 600$ MeV).

La figure 1 représente une schématisation d'ensemble de l'accélérateur et des salles expérimentales.

1. Rappel des caractéristiques de l'accélérateur (fig. 1).

Le linéaire est formé de 30 sections. La puissance HF est fournie par 15 klystrons alimentés par 8 modulateurs. Le cycle utile est soit de 1 % (impulsions de 10 µs avec un taux de répétition de 1000 Hz) soit de 2 % (impulsion de 10 µs ou 20 µs avec un taux de répétition respectivement de 2000 Hz ou de 1000 Hz).

La longueur totale de la structure accélératrice est de 180 mètres. Entre la 6ème et la 7ème section est placé un convertisseur e^-/e^+. Les positons obtenus peuvent être accélérés dans les 24 sections suivantes :

TABLEAU I : Caractéristiques du Faisceau de Négatons

CYCLE UTILE	1 % (10 µs - 1000 Hz)		2 % (10 µs - 2000 Hz)	
VALEURS	CALCULEES	GARANTIES	CALCULEES	GARANTIES
Electrons				
Energie maximale (MeV)	640	554	453	380
Courant crête (mA)	20	20	15	15
obtenu à une énergie (MeV) avec une définition en énergie	585	500	412	340
- largeur à mi-hauteur (%)	0,4	1	0,35	1
- stabilité à long terme (%)	0,4	± 0,5	0,4	± 0,5
Emittance du faisceau à 500 MeV (rd.cm)	$4{,}5 \times 10^{-5}$	2×10^{-4}		
Courant crête maximal à 400 MeV (mA)	80	60	56	42
Courant moyen maximal (µA)	800	600	1120	840
Puissance moyenne maximale dans le faisceau (kW)	300	100	325	100

Les caractéristiques du faisceau de négatons sont données dans le tableau I. Notons qu'il a été possible de descendre en énergie aussi bas que 150 MeV avec une résolution de 0,1 % pour étudier la photoproduction des pions au seuil.

Parmi les possibilités du linéaire, nous voudrions insister sur les perspectives de développement des faisceaux de positons. A la sortie de la section 6, des négatons d'environ 85 MeV tombent sur une cible d'or d'environ une longueur de rayonnement. Des essais effectués récemment laissent à penser que le système cible de conversion - collimateur peut supporter un fort courant moyen d'électrons, supérieur à 300 µA. Par ailleurs, au niveau de la salle de basse énergie (fin de la 12ème section - 100 MeV de positons) un rendement de conversion de 3×10^{-3} a été mesuré. Nous espérons obtenir des faisceaux de positons d'énergie variant de 30 MeV à 450 MeV dont l'intensité sera de 0,3 à 1 µA moyen pour un $\Delta E/E \sim 1$ %.

L'accélérateur linéaire est sous la responsabilité de Monsieur F. Netter, Chef du Service de l'Accélérateur Linéaire (CEN de Saclay).

2. Les équipements expérimentaux lourds.

1. *La salle de basse énergie (BE) :* Energie maximale des négatons 200 MeV, des positons 100 MeV.

Elle est prévue pour étudier les réactions photonucléaires au moyen des photons monochromatiques d'annihilation des positons. Afin d'améliorer le rapport annihilation sur freinage le faisceau de rayons gamma utilisé fait un angle de 1°30' par rapport à la direction initiale des positons.

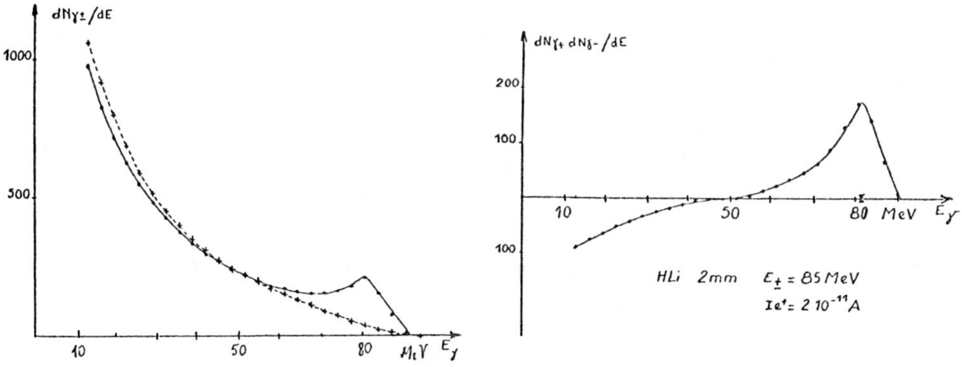

Fig. 2. - Spectre de photons monochromatiques de 81 MeV obtenus par annihilation en vol de positons de 85 MeV dans une cible de LiH. La figure de gauche montre les spectres en e^- et celle de droite la différence. Le détecteur est un cristal NaI(Tl).

La figure 2 représente l'un des premiers essais. La partie gauche montre les spectres mesurés dans un cristal NaI(Tl) obtenu avec des électrons de 85 MeV heurtant une cible de 2 mm de LiH. La courbe en pointillés est celle correspondant aux négatons, celle en trait plein aux positons. La partie droite est la différence (pour la même intensité d'électrons).

2. *La salle "photonucléaire" à haute énergie (HE3)* : Energie maximale des négatons 600 MeV, des positons 450 MeV (fig. 3).

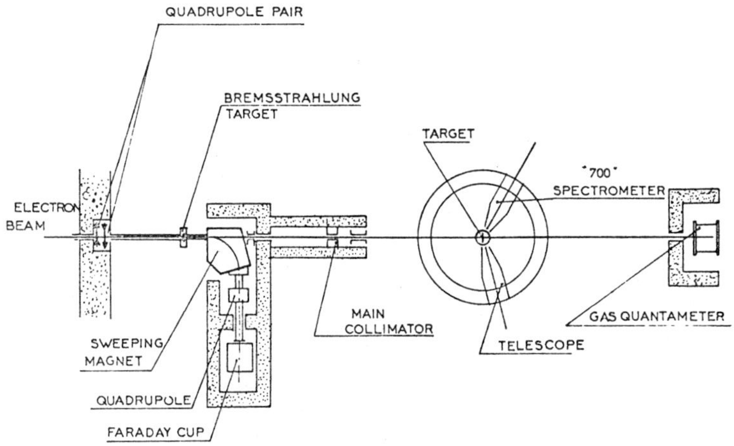

Fig. 3. - Schéma de principe de l'implantation de la source de photons de freinage et du carrousel (deux plate-formes sur les trois ont été représentées).

Deux sources de photons sont disponibles.

a) Une source de photons de freinage (Bremsstrahlung) produite dans un convertisseur de 10^{-2} longueur de rayonnement de cuivre, or ou platine. Les électrons qui traversent cette cible sont déviés par un aimant balai dans une cage de faraday redroidie par de l'eau.

Le faisceau de photons est collimaté pour obtenir une tache de 2 cm de diamètre sur la cible. A 300 MeV le flux est de l'ordre de 4×10^9 $h\nu$ s^{-1} pour un courant de 100 µA d'électrons de 400 MeV. La mesure du flux énergétique de photons s'effectue au moyen d'un quantamètre type Wilson.

b) Le faisceau d'annihilation. Si la source de photons de freinage peut être loin de la cible étudiée car le linéaire produit une intensité suffisante de négatons, il est nécessaire d'implanter la source d'annihilation le plus près possible de la cible utilisée. Un système plus compact comprenant une cible de lithium de 1 g cm^{-2} est situé à 3,5 m du centre du carrousel. Pour une intensité de 0,1 µA de positons de 300 MeV, le flux de photons monochromatiques sera de 10^8 s^{-1} dans une tache de 3 cm de diamètre vers l'avant.

c) Les systèmes de détection. La salle HE3 fut la seule en fonctionnement durant une année. Elle a permis de commencer l'étude de la diffusion élastique et inélastique d'électrons,des réactions (e,e'p) en coïncidence et l'étude de l'électroproduction des π^+ sur l'hydrogène s'y poursuit.

Autour de la chambre à cible, un carrousel formé de trois plate-formes tournantes.

Deux d'entre elles conservent leur aimant ("400" et "700"). Sur la troisième il est possible de plácer soit un aimant relativement léger (7 tonnes), soit un spectromètre à parcours.

TABLEAU II : Spectromètres Magnétiques de HE3

AIMANT	"400"	"700"
Type	double focalisation $n = 1/2 \quad \theta = 170°$	champ homogène $\theta = 90°$
Impulsion maximale	430 MeV/c	735 MeV/c
Angle solide	3,2 msr	2,3 msr
Grandissement	- 0,7	- 0,9
Angle du plan focal	60°	37°
Distance cible - plan focal	522 cm	482 cm
Plage d'analyse totale	$\Delta p/p = 0{,}09$	$\Delta p/p = 0{,}118$
Poids de l'aimant	320 000 N	415 000 N

Le spectromètre à parcours est constitué d'un ensemble de scintillateurs plastiques. L'angle solide atteint 30×10^{-3} sr. La résolution utilisée est de 5 MeV pour des protons de 150 MeV.

C'est avec cet ensemble que nous avons étudié les réactions $^3He(\gamma,p)d$, $^4He(\gamma,p)t$, $^4He(\gamma,n)$ ainsi que la photoproduction de pions sur les nucléons de l'hélion : $^4He(\gamma,p\pi^-)(ppn)$. Le groupe d'Orsay a étudié dans cette même salle la réaction $^3He(\gamma,\pi^+)t$.

L'aimant "0" permet de détecter des particules très vers l'avant ($\lesssim 10°$) jusqu'à 450 MeV/c.

3. *La salle "électron" à haute énergie (HE1)* (fig. 4).

Son but est double : permettre l'étude d'une part de la diffusion d'électrons à haute résolution dans le domaine d'énergie 150 à 600 MeV (dans un domaine de transfert variant de $q = 0{,}3$ à $6\ fm^{-1}$) et d'autre part des réactions (e,e'x) en coïncidence.

TABLEAU III : Spectromètres Magnétiques dans HE1

AIMANT	"600"	"900"
Type : focalisation	Simple $\theta = 153°$	Double $\theta = 169°7$
Impulsion maximale	630 MeV/c	900 MeV/c
Induction maximale	1,5 T	1,67 T
Angle solide	0,1 à 8 msr	0,1 à 5,6 msr
Admittance angulaire	$0{,}7\ p_o$ à $1{,}1\ p_o$	$0{,}95\ p_o$ à $1{,}05\ p_o$
Domaine angulaire	32° à 148°	25° à 155°
Précision angulaire	± 0,05°	± 0,05°
Résolution intrinsèque	3 à 7×10^{-4}	1 à 2×10^{-4}
Poids des aimants	1 700 000 N	1 800 000 N

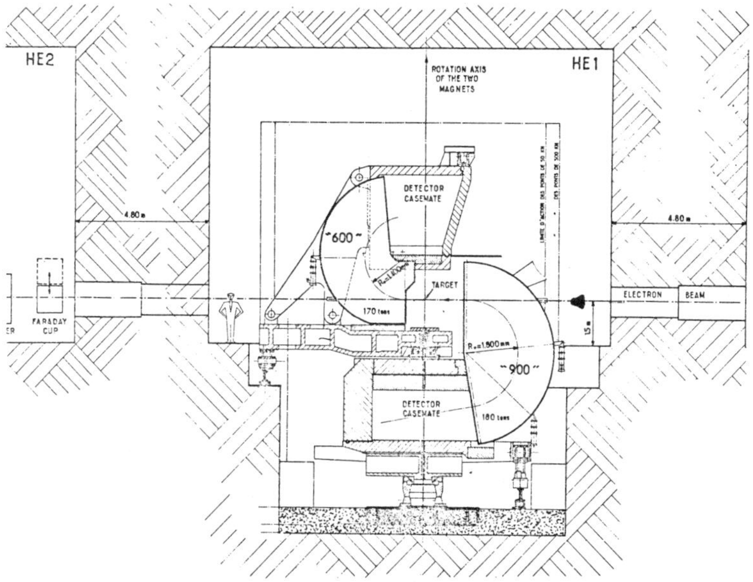

Fig. 4. - Coupe verticale de la salle de spectrométrie d'électrons (aimants "600" et "900").

Pour réaliser cet objectif, un ensemble de deux aimants tournant autour du même axe a été construit : l'aimant "900" particulièrement conçu pour optimaliser la résolution, l'aimant "600" pour avoir une acceptance élevée en impulsion. Le tableau III précise les caractéristiques de cet ensemble magnétique.

Avec l'ensemble de l'appareillage du "900" le bruit de fond correspond à une section efficace inférieure à 5×10^{-14} b.sr^{-1}. Jusqu'à maintenant une résolution globale de 3×10^{-4} a été obtenue à 400 MeV.

4. *Les faisceaux de pions et de muons* (fig. 5).

Un premier canal de pions ou/et de muons fonctionne depuis 1970. Les pions sont photoproduits par le spectre de freinage de négatons de 400 MeV. Le système de transport des pions est placé à 120° du faisceau de négatons incident. Il comprend deux aimants de 60° et 9 quadrupôles. La longueur totale du canal magnétique est de 8 mètres.

TABLEAU IV : Le Premier Canal de Pions

ENERGIE DES π (MeV)		20	40	60	80	100
Nombre de π^* par seconde pour $\Delta E/E = 8$ %	π^+	500	1900	2700	2700	2000
	π^-	700	2400	3300	3100	2000

*Ces données correspondent à 1 μA d'électrons incidents sur la cible de photoproduction.

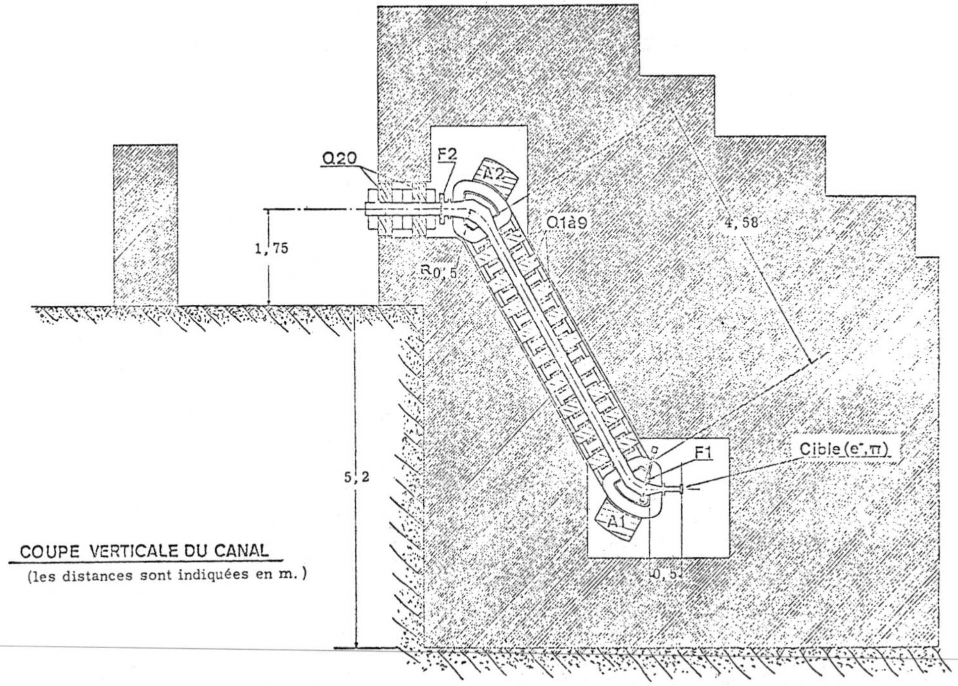

Fig. 5. - Schéma du premier canal de pions et muons.

Un second canal, très similaire, est en construction. Il devrait être terminé au printemps 1975.

Le tableau IV donne les caractéristiques du faisceau de pions.

Le nombre de muons stoppés dans une cible de 0,5 g.cm^{-2} et de 100 cm^2 de surface s'étire à 21 μ^+/s ou 31 μ^-/s par micro ampère d'électrons.

Compte tenu des intensités d'électrons que la machine peut fournir, soit de l'ordre de 300 µA de négatons de 420 MeV, le flux de π^- de 70 MeV atteint 10^6/s tandis que le nombre de μ^- stoppés dans 0,5 g.cm^2 est de l'ordre de 10^4/s.

Les principales expériences réalisées avec ce canal ont été : diffusion pion-nucléon avec ou sans mélange de charge, mesure du temps de vie du muon, réactions A(π,x)B, étude des X-muoniques des isotopes du samarium.

L'ensemble des équipements expérimentaux auprès de l'ALS, réalisé sous la direction de Christophe Tzara, dépend de Philippe Catillon, Chef de Service de Physique Nucléaire à Haute Energie (CEN de Saclay).

II - LE SYNCHROTRON "SATURNE"

L'étude de l'interaction proton-noyau dans la région du GeV est complémentaire de l'interaction électromagnétique. L'existence à Saclay du synchrotron Saturne nous a offert la possibilité de développer ce domaine de la physique.

1. Le spectromètre 1 GeV (SPES I).

Le Service de Physique Nucléaire à Moyenne Energie, dont le responsable est Monsieur Jacques Thirion, a pris en charge la réalisation d'un poste de spectroscopie nucléaire auprès de Saturne permettant des résolutions de 130 keV pour des particules de l'ordre du GeV.

Pour obtenir une telle résolution ($\sim 6 \times 10^{-5}$ en moment) SPES I est conçu sur le principe de la spectrométrie fine "compensée" (fig. 6a et 6b). Le poste d'étude comprend deux aimants :

. un aimant d'analyse du faisceau incident "A" avant la cible C ;

. un spectromètre d'analyse des particules diffusées après la cible.

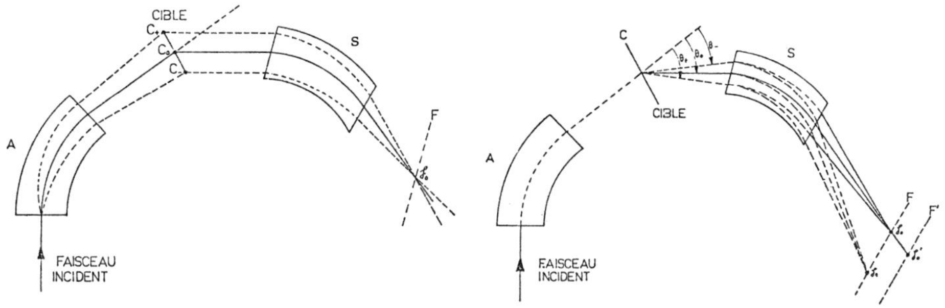

Fig. 6. - Principe de la spectrométrie fine compensée.
6a) Compensation de la dispersion de l'énergie incidente.
6b) Principe de focalisation des particules diffusées.
A : analyseur S : spectromètre

1. L'effet dû à la dispersion énergétique du faisceau incident est compensé (fig. 6a). L'aimant A disperse les particules incidentes selon leurs énergies. Si la dispersion sur la cible est adaptée aux propriétés dispersives du spectromètre S, les particules diffusées de façon identique en C^+, C^o et C^- seront regroupées en f_o.

2. La dispersion liée à la variation énergétique en fonction de l'angle de diffusion est corrigée sur toute l'ouverture du spectromètre et pour l'ouverture angulaire du faisceau (fig. 6b). Une particule diffusée élastiquement à θ_o sera courbée suivant l'axe moyen ($S\theta_o$). Une particule diffusée à angle plus petit θ^- (ou plus grand θ^+) devra avoir un trajet dans le champ magnétique plus long (ou plus court) pour recouper la trajectoire moyenne en f_o.

Par contre le point de regroupement des particules diffusées sur un état excité sera f_1.

Le tableau V résume les caractéristiques des aimants.

La figure 7 montre la résolution obtenue dans le cas de la diffusion de protons de 1,04 GeV à 6° par le carbone 12. Parmi les études réalisées, signalons la distribution angulaire de 3 à 47° dans le système du laboratoire de la diffusion élastique des

TABLEAU V : Poste de Spectrométrie Nucléaire à Saturne

	ANALYSEUR	SPECTROMETRE (SPES I)
Induction moyenne (proton de 1 GeV)	1,4 T	1,72 T
Rayon de courbure	4 m	3,3 m
Angle de duration	64°	97°
Angle inclinaison face d'entrée	23°5	0°
Angle inclinaison face sortie	0°	25°
Rayon de courbure des faces	∞	1,4 m
Entrefer utile	12 cm	14 cm
Largeur utile	46 cm	40 cm
Poids	600 000 N	900 000 N

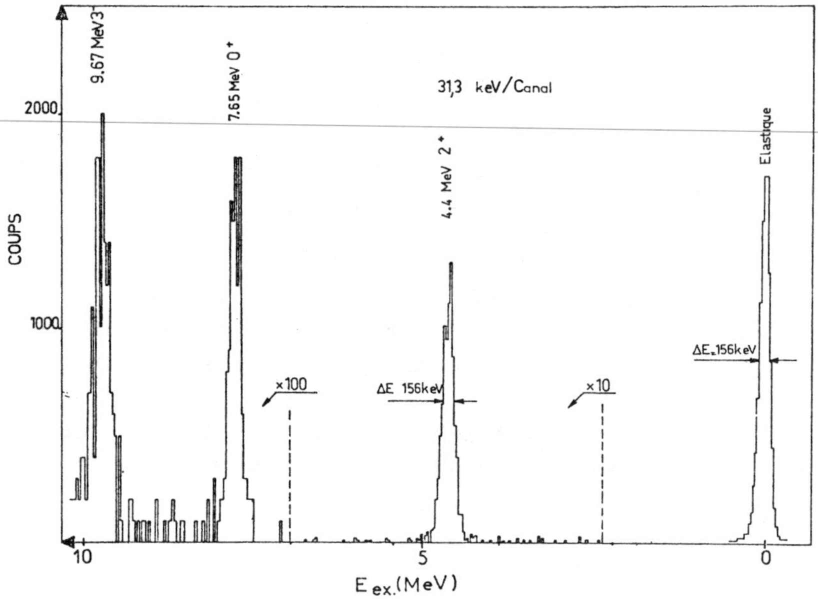

Fig. 7. - Spectre de protons de 1040 MeV diffusés à 6° par ^{12}C.

protons de 1,05 GeV par ^{4}He, en coopération de Saclay et Strasbourg.

3. Un nouveau spectromètre SPES II est en cours d'installation. Il possède un angle solide dix fois supérieur à SPES I ($\Omega = 2 \times 10^{-2}$ sr) et une bande d'analyse de moment cinq fois supérieure (0,82 p_o à 1,18 p_o). Il permettra d'inaugurer deux types d'expérience :

- Etude de processus tels que la production associée de mésons K et de baryons Λ (production d'hypernoyaux). Production également d'états nucléaires où un nucléon est remplacé par un isobare.

- Etude de processus à trois corps par le couplage des deux spectromètres SPES I et SPES II.

2. Le laboratoire national Saturne II.

Les résultats obtenus à Saturne ont souligné l'intérêt de poursuivre les études dans ce domaine. Mais Saturne est une vieille machine et il est indispensable de la rénover.

Dans ce but, l'Institut de Physique Nucléaire et de Physique des Particules (IN2P3) et le Commissariat à l'Energie Atomique (CEA) ont décidé de mettre en commun leurs efforts pour développer la spectroscopie nucléaire induite par hadron dans le domaine du GeV.

Il a été décidé de construire, à l'emplacement du synchrotron Saturne, un nouvel accélérateur appelé Saturne II sous la forme d'une "entreprise commune" des deux organismes. La direction scientifique est assurée par Monsieur J. Thirion (CEA) et Monsieur P. Radvanyi (IN2P3).

Saturne II sera un accélérateur à focalisation forte dont l'injection sera une injection électrostatique de 750 keV suivi d'un accélérateur linéaire (20 MeV de protons). Son énergie nominale variera de 500 MeV à 2,7 GeV. Il accélérera des protons et des deutérons polarisés ou non, des τ et des α et éventuellement des ^{14}N. La figure 8 représente un projet d'implantation de l'appareillage expérimental.

Le tableau suivant compare les propriétés de Saturne actuel et de la version Saturne II.

TABLEAU VI : Evolution du Synchrotron Saturne

		SATURNE ACTUEL	SATURNE II
Durée du cycle à 2 GeV	T	3 s	1 s
par cycle — Nombre de particules injectables	N_i	3×10^{12}	$7,5 \times 10^{12}$
par cycle — " " " accélérées	N_{acc}	$1,2 \times 10^{12}$	$2,5 \times 10^{12}$
par cycle — " " " extraites	N_{ext}	3×10^{10}	$1,1 \times 10^{12}$
(dans l'émittance $(\mathcal{E}_x)_{ext} = 3\pi$ mm.mrd à 1 GeV)			
Nombre de particules extraites par seconde (dans l'émittance $(\mathcal{E}_x)_{ext} = 3\pi$ mm.mrd)	dN_{ext}/dt	$10^{10}\ s^{-1}$	$1,1 \times 10^{12}\ s^{-1}$
Facteur d'utilisation à 2 GeV	$\Delta = \tau/T$	15 %	25 %
Dispersion des énergies dans le faisceau extrait — instantanée	$(\Delta E/E)_{inst}$	4×10^{-4}	qqs 10^{-4}
Dispersion des énergies dans le faisceau extrait — moyenne	$\overline{(\Delta E/E)}$	2×10^{-3}	

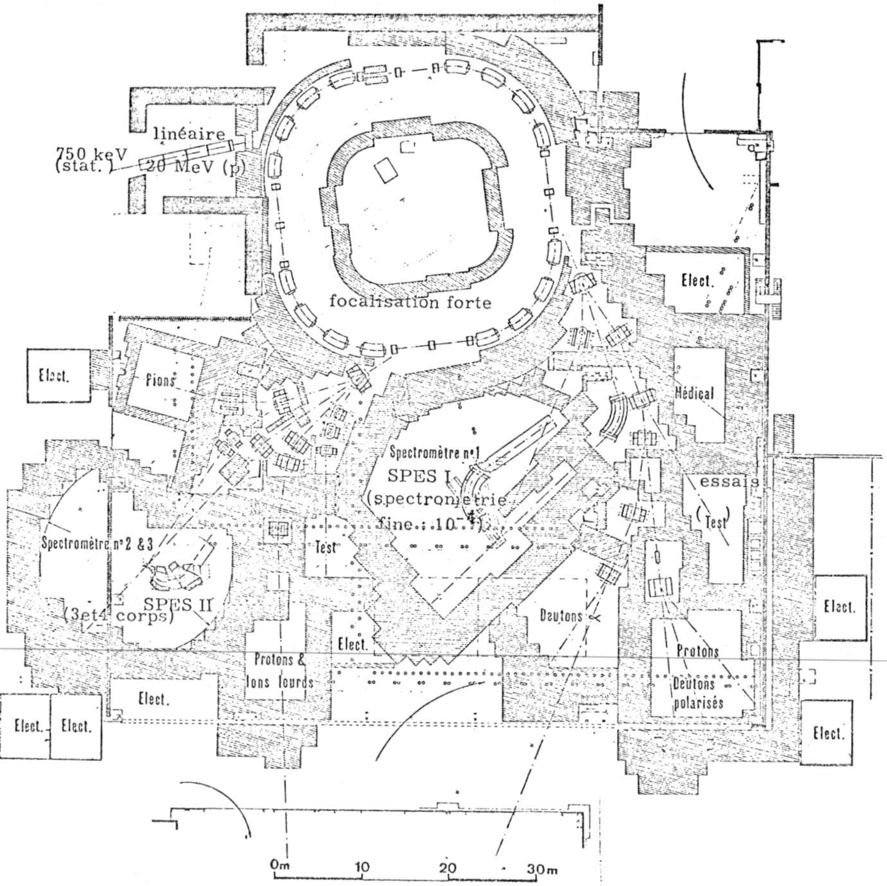

Fig. 8. - Projet d'implantation des aires expérimentales à Saturne II.

Après achèvement des travaux, l'exploitation de Saturne II sera assurée par un "Laboratoire National" géré en commun par l'IN2P3 et le CEA.

III - LES METHODES DE DETECTION

L'existence de plans focaux de grande longueur ($\sim$ 2 m) impose la mise au point de systèmes de détection appropriés :

- bonne précision de localisation ;
- capacité d'encaisser des flux de l'ordre de 10^6 particules/s.

La méthode employée consiste à construire des chambres d'ionisation multifils. Deux possibilités de lecture sont exploitées suivant les conditions expérimentales :

1) chaque fil est relié à une mémoire torique ;

2) les fils sont réunis l'un à l'autre par des résistances. La lecture s'effectue

par division de charge de place en place.

Dans le premier cas la précision est de l'ordre de grandeur de l'intervalle entre deux fils (2 mm en général) ; dans le second, en regroupant les fils par ensemble de 18 (soit 36 mm), la localisation atteint 0,5 mm.

LES INSTALLATIONS ACTUELLES ET FUTURES POUR L'ETUDE DES PROPRIETES NUCLEAIRES A ENERGIES INTERMEDIAIRES SUR LE SITE DE SACLAY

Claude G. Schuhl
Département de Physique Nucléaire
CEN Saclay, BP 2, 91190 Gif-sur-Yvette, France

L'étude des réactions nucléaires au moyen de rayonnement dont la longueur d'onde est inférieure à la distance entre nucléons dans le noyau s'effectue au CEN de Saclay avec deux accélérateurs : un accélérateur linéaire à électrons de 600 MeV et le synchrotron Saturne qui produit des protons de 3 GeV. Les implantations expérimentales en fonctionnement et en cours d'installation auprès de ces deux accélérateurs sont décrites. Un schéma de la version II de Saturne est également présenté.

ACTUAL AND FUTURE EXPERIMENTAL SET UP INSTALLED AT SACLAY FOR STUDYING NUCLEAR PROPERTIES AT INTERMEDIATE ENERGY

Claude G. Schuhl
Département de Physique Nucléaire
CEN Saclay, BP 2, 91190 Gif-sur-Yvette, France

Nuclear reactions with particles and radiation, the wavelength of which is more little than nucleon-nucleon distance in the nucleus are studied in "CEN de Saclay" around two accelerators : the 600 MeV lineac and the 3 GeV proton synchrotron (Saturne). The experimental facilities in place and in achievement near these two accelerators are described. A scheme of the new "Saturne" is also presented.

NEXT GENERATION EXPERIMENTAL FACILITIES FOR STUDIES OF ELECTROMAGNETIC INTERACTION OF FEW BODY PROBLEMS IN NUCLEAR PHYSICS*

Y. M. Shin

Saskatchewan Accelerator Laboratory, University of Saskatchewan,
Saskatoon, Canada, S7N OWO

The advantage of using electrons and photons as nuclear probes lies principally in the fact that their interaction with target nuclei is electromagnetic, which is the best understood process in physics. The relatively weak nature of the electromagnetic interaction guarantees the accuracy of perturbation calculations based on various models, and experimental results can be compared readily with theory.

The theoretical elastic electron-nucleus cross section in the first Born Approximation given by[1)]

$$\frac{d\sigma}{d\Omega} = \sigma_{NS}\left[\frac{q_\mu^{\,4}}{\underset{\sim}{q}^4} F_L^{\,2}(q) + \left[\left(\frac{q_\mu^{\,2}}{2q^2}\right) + \tan^2\frac{\theta}{2}\right] F_T^{\,2}(q)\right]$$

where

$$\sigma_{NS} = \left(\frac{e^2}{2E}\right)^2 \frac{\cos^2\theta/2}{\sin^4\theta/2} \, \frac{1}{(1 + 2E\sin^2 \tfrac{1}{2}\theta/M_T)},$$

$F_L(q)$ and $F_T(q)$ are the longitudinal and transverse form factors which contain nuclear information at a momentum transfer $\vec{q}$. The form factors $F_L(q)$ and $F_T(q)$ can be separated experimentally by measuring the cross section at different angles for a fixed momentum transfer q^2.

The measured form factors[2-5)] for the deuteron and ^{3}He are shown in Fig. 1.

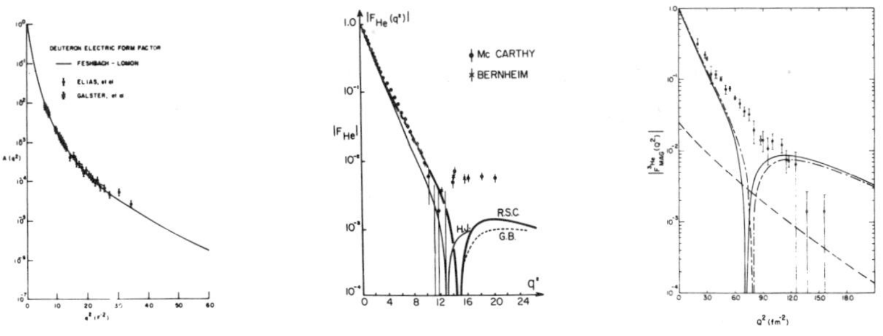

(a) (b) (c)

Fig. 1 Form factors of ^{2}H and ^{3}He. (a) Deuteron electric form factor.[2,3)] (b) ^{3}He charge form factor.[4,5)] (c) ^{3}He magnetic form factor.[4,5)] The theoretical curve is taken from Ref. 6 for the deuteron, Refs. 7, 8, 10 for the ^{3}He charge form factor,and Ref. 9 for the ^{3}He magnetic form factor.

Various curves drawn here are the theoretical form factors.[6-10] While Fig. 1 demonstrates the inadequacy of the present theoretical art for the trinucleon system, more experimental data and/or refinement of the existing data are required in the region $q^2 \gtrsim 9\ fm^{-2}$, i.e., corresponding to an incident electron energy of $\gtrsim$ 500 MeV.

Although the calculated form factor for the deuteron agrees quite well with experiment, questions remain concerning the amount of D-state in the ground state and exchange current contributions. In Fig. 2 theoretical predictions of the tensor polarization of the recoil deuterons are shown. Figure 2a shows the D-state probability dependence of the polarization,[11] while Fig. 2b indicates sensitivity of the polarization on the exchange current contribution.[12] Hence measurements on the tensor polarization for $q \geq 2\ fm^{-1}$ are quite important in obtaining information on the D-state probability and exchange current, for which an incident electron energy $\geq$ 500 MeV is required.

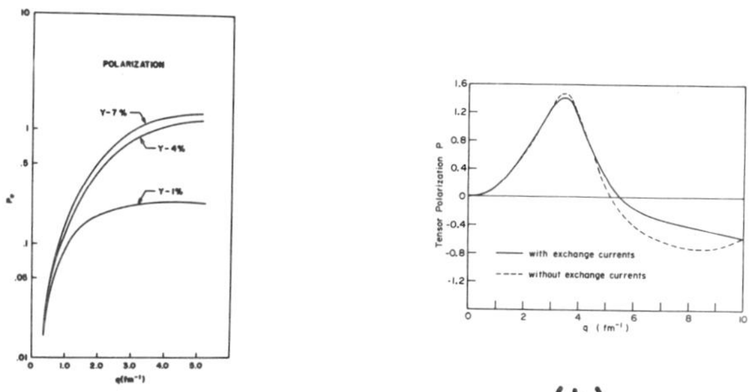

(a) (b)

Fig. 2 Tensor polarization of the recoil deuterons in the $^2H(e,d)e'$ reaction. (a) Dependence on the D-state probability.[11] (b) Exchange current dependence.[12]

Having defined the incident energy of electrons, the energy resolution of the beam and current must now be defined. The energy resolution should be determined by the kinematic conditions while the beam current may be determined by counting rate considerations.

The n-p break-up threshold energy is $\sim$ 2 MeV, while that for the trinucleon nuclei is $\sim$ 6 MeV; in order to distinguish the elastic process from the inelastic scattering an energy resolution of $\sim$ 0.1% is needed. With a 500 MeV incident electron beam and a 200 Atm deuterium gas target the counting rate at a momentum transfer $q^2 \gtrsim 10\ fm^{-2}$ is approximately $\sim$ 150 counts per hour at an average beam current of 1 μA. It takes approximately one hour per point with the best electron linear accelerator available in this energy range (Saclay) to have better than 3% statistical accuracy.

Therefore, for elastic electron scattering experiments in the momentum range $q^2 \gtrsim 10\ fm^2$ *an electron accelerator capable of delivering electron beams of energy 0.5 ~ 1 BeV with about 0.1% energy resolution and an average current in excess of 10 μA is desirable*, unless a new experimental technique, which will be discussed later, is used.

While elastic electron-nucleus scattering data provide an abundance of information about the ground state of nuclei, photo- and electro-disintegration experiments yield information concerning both the ground and continuum (or excited) states of the target nucleus under study.

An "ideal" experiment may be defined as one in which all observable quantities such as energy, momentum, angular distribution, polarization, etc., of all particles involved in the reaction are measured in order to understand the dynamics of the reaction. Some experiments with low energy heavy particle accelerators have come close to such an ideal experiment.

However, experimental studies of nuclear reactions induced by photons and electrons are far from satisfying such experimental conditions, primarily due to the following reasons.

(1) Lack of intense monochromatic photon beams has resulted in the use of bremsstrahlung beams in many cases. Because of the continuous nature of the bremsstrahlung, kinematic relations of the reaction under study cannot be established in most cases. Uncertainty in the shape of the bremsstrahlung intensity distribution and its monitoring have posed further experimental problems. Hence the best that experimentalists can do is to obtain a yield curve and unfold the cross section rather "artificially". Only in a few cases has complete kinematic information been obtained.

(2) For electron induced reactions the minimum number of particles involved in the final state is three, namely two nuclear fragments and the scattered electron. Hence only the excitation mechanism (if the scattered electrons are detected) can be studied.

A kinematically complete experiment for the electromagnetic break-up process implies coincidence measurements between pairs of particles involved in the reactions (either directly or indirectly), even for the two body decay mode. This combined with propensity of electrons to radiate in the presence of nuclei has been the source of difficulties in the study of electromagnetic interaction in the continuum.

Modern linear electron accelerators are capable of delivering a well defined electron beam with high intensity, but suffer from their low duty factor. The duty factor of a pulsed accelerator is the fraction of time during which the machine delivers a current. The accidental rate in a coincidence experiment is inversely proportional to the duty factor. Hence electron accelerators having a large duty factor ($D \simeq 1$) are most desired to perform coincidence measurements and much effort has been put into the development of such accelerators. In Table I some of the recent developments of the high duty factor accelerator are summarized.

Table I

High duty factor electron linear accelerators.

Institution	Energy (MeV)	Duty factor	ΔE/E (%)	I_{ave} (μA)
Saskatchewan	20 $\sim$ 250	> 0.8	0.08	$\sim$ 100 (proposed)
MIT	100 $\sim$ 400	0.02	0.025	10 (near completion)
Stanford	$\sim$ 688	1.00	0.01	100 (under construction)
Saclay	$\leq$ 600	$\sim$ 0.05	0.01	10 (operating)
Amsterdam	$\leq$ 500	$\sim$ 0.05	?	? (design funded)

At the University of Saskatchewan, a proposal has been made to build an electron pulse stretcher ring (Fig. 3). A 300 MeV, 1 μs electron beam pulse is injected into

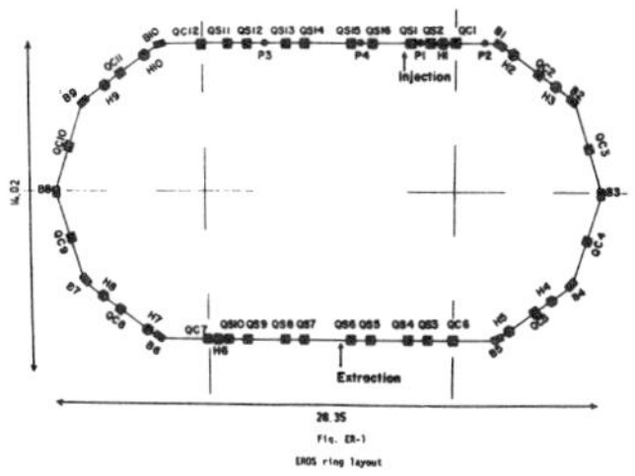

Fig. 3 Electron Ring of Saskatoon (EROS), B: bending magnet, Q: quadrupole, H: hexapole.

the ring (four-turn injection) and allowed to trickle out slowly while circulating $\sim$ 6600 times around the ring until the next pulse is injected. The building layout is shown in Fig. 4. Table II shows characteristics of the present and proposed machine.

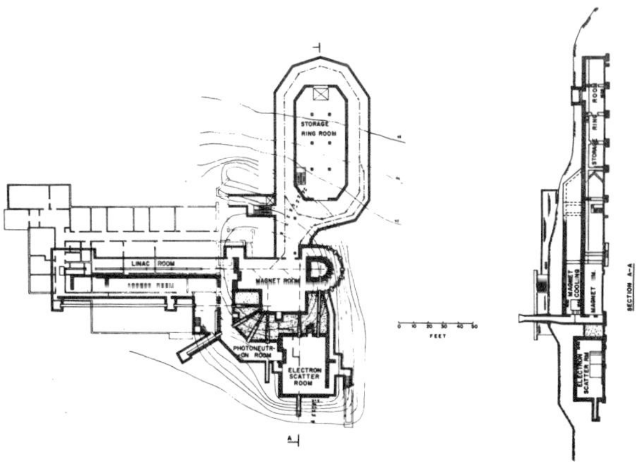

Fig. 4 Building layout of the EROS and 250 MeV accelerator.

Table II

Saskatchewan electron accelerator and proposed pulse stretcher.

	Present	Proposed (Ring)
Energy	∿ 250 MeV (by the end of 1974) (150 MeV)	$\leq$ 250 MeV
$\Delta E/E$	2%	0.8
Pulse length	1 μsec	-
Repetition rate	400 ∿ 1000 pps	-
Duty factor	∿ 0.0005	> 0.08
Emittance	Horizontal 3.2 mm mrad	3 mm mrad
	Vertical 3.2 mm mrad	30 mm mrad
I_{peak}	0.2 Amp	∿ 100 μA

What we can learn about the few nucleon system with such next generation accelerators depends on things which we have learned to date. Some of the outstanding problems are:

(1) The polarization of the emitted nucleon (for all energies) and angular distributions for $E_\gamma \gtrsim 50$ MeV in the photodisintegration of the deuteron.[13,14)]

(2) The photo- and electro-disintegration processes of ^{3}He and ^{3}H for all energies.[10)]

(3) Measurements of the n-n scattering length and effective range through final state interactions in the three body photodisintegration of the trinucleon system.

At Saskatoon an experimental program to map the electromagnetic properties of ^{3}He in the continuum, and to study effects of final state interaction in the three body decay process has been initiated. Figure 5 shows some of our coincidence measurements.[15)]

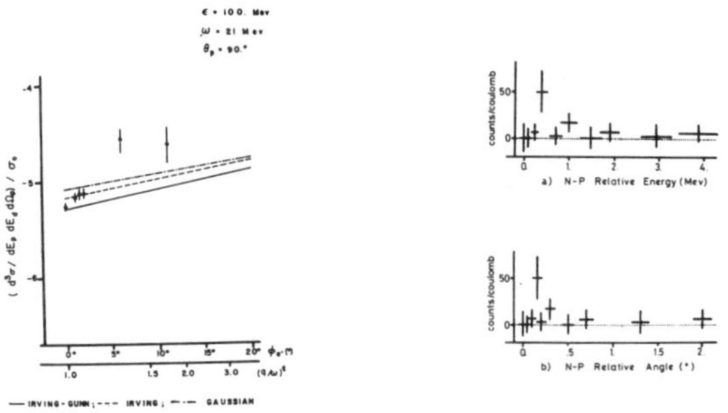

(a) (b)

Fig. 5 (a) Momentum transfer dependence of the cross section for the reaction ^{3}He(e,pd)e' at ω = 21 MeV.

(b) The (n-p) final state interaction in the reaction ^{3}He(γ,pp)n.

obtained by detecting nuclear fragments in coincidence using two magnetic spectrometers[16] (with our accelerator having a duty factor of 5×10^{-4} !!) The momentum dependence of the cross section at an energy transfer of 21 MeV is compared with a simple theoretical model involving the Irving-Gunn wave function and a plane wave final state (Fig. 5a). Figure 5b shows the importance of the n-p final state interaction in the three body decay channel of the $^3He(\gamma,2p)n$ interaction.

These data illustrate the kind of information one can obtain from coincidence measurements. Needless to say, many problems in the area of few nucleon nuclei exist and experimental results flowing from such coincidence measurements may provide solutions to the problems.

A high duty cycle intermediate energy electron accelerator must be matched with good experimental apparatus. For coincidence experiments, installation of two magnetic spectrometers is preferred; one which meets experimental criteria for conventional electron scattering and the other to analyze charged nuclear fragments. An energy loss spectrometer developed by the MIT group uses the full intensity of the raw beam ($\Delta E/E \simeq 1\%$) which is dispersed at the target position, and the scattered electrons are achromatically focussed at the focal plane of the spectrometer.[17] This system has the advantage of (i) utilizing a high intensity (raw) beam; and (ii) achieving the desired resolution ($\sim 10^{-4}$).

For the detection of heavy particles a spectrograph with a large effective solid angle is more essential than fine energy resolution (of course both are preferable) for studies of continuum regions of excitation in light nuclei. A spectrograph can be installed such as the Q3D being used at Chalk River and Heidelberg.[17] These spectrographs have $\Delta\Omega \sim 15$ ms and $\Delta E/E \sim 10^{-4}$.

In conclusion, however, what we would like to have in the near future and what we will obtain are two different things. We experimentalists must face this reality and be prepared to fully utilize whatever facility is available to us for the betterment of understanding nature.

BIBLIOGRAPHY

* Work supported by the Atomic Energy Control Board of Canada

1) T. de Forest and J. D. Walecka, Adv. in Phys. 57 (1966) 1
2) J. E. Elias et al., Phys. Rev. 177 (1969) 2075 and references cited therein
3) S. Falister et al., Nucl. Phys. 32B (1971) 221
4) J. S. McCarthy et al., Phys. Rev. Lett. 25 (1970) 884
5) M. Bernheim et al., Lett. Nuovo Cimento 5 (1972) 431
6) E. B. Dally, Proceedings of the International Conference on Photonuclear Reactions and Applications, Asilomar, ed. by B. L. Berman, Vol. 1 (1973) pp. 363-369
7) C. Gignoux and A. Laverne, Phys. Rev. Lett. 29 (1972) 436
8) A. N. Hennel and L. M. Delves, Phys. Lett. 40B (1972) 20
9) R. A. Brandenburg et al., Phys. Rev. Lett. 32 (1974) 1325

10) B. F. Gibson, Proceedings of the International Conference on Photonuclear Reactions and Applications, Asilomar, ed. by B. L. Berman, Vol. 1 (1973) pp. 373-383

11) T. J. Brady et al., Can. J. Phys. 52 (1974) 1322

12) M. Chemtob et al., Phys. Rev. C10 (1974) 344

13) Y. M. Shin, Proceedings of the International Conference on Photonuclear Reactions and Applications, Asilomar, ed. by B. L. Berman, Vol. 1 (1973) pp. 345-359

14) Y. M. Shin et al., Nucl. Phys. 154 (1970) 482

15) Y. M. Shin et al., Nucl. Inst. Meth. 94 (1971) 381

16) Y. M. Shin et al., Proceedings of the International Conference on Nuclear Structure Studies using Electrons and Photons, Sendai, ed. by K. Shoda (1972) pp. 481-488

17) S. Kowalski, Proceedings of the International Conference on Photonuclear Reactions and Applications, Asilomar, ed. by B. L. Berman, Vol. 1 (1973) pp. 227-242

EXPERIMENTAL FACILITIES FOR FEW NUCLEON STUDIES AT TRIUMF

J.M. Cameron

University of Alberta and TRIUMF
Edmonton, T6G 2J1

In this report an attempt is made to summarize the facilities planned for research in few body problems at TRIUMF. When asked to give a talk on equipment and resources I was reminded of a quotation I read recently which seemed somewhat apt.

> *It is not enough to set tasks, we must also solve the problem of the methods for carrying them out...Unless the problem of method is solved, talk about the task is useless...*
>
> Mao Tse-tung

The first thing we must have to solve our problem is a working accelerator and we believe that is now very close to being achieved. The TRIUMF machine characteristics are now widely known and have been summarized recently [1]. In Table 1 are noted some of the machine parameters which might be most interesting to potential users. The choice of acceleration of H^- ions results in a machine which is particularly suited to produce beams required for few body studies. The machine design allows extraction of up to four simultaneous beams, each independent in intensity and having an energy in the range 175 - 500 MeV. At an early stage in machine commissioning four phase selecting slits will be placed in the low energy beam and these should result in a beam of much better energy resolution (± 200 keV) than has ever before been available

Characteristics of TRIUMF H^- Accelerator		
Radius (inches)		310
Field (average) (Kg)		4.5
Energy (MeV)		175 - 500 continuous
Beams		2 (later 4) - independent in energy and ratio of up to 1 : 1000 in intensity
Resolution:	Mode (1) raw beam	± 600 keV
	Mode (2) with low energy phase selection slits	± 200 keV
Duty cycle:	macroscopic	100 %
	microscopic - mode (1)	12%; 20% (3rd)
	microscopic - mode (2)	1%; 8% (3rd)
Current		100 μa unpolarized 50 na polarized

(3rd) indicates that the correct mixture of third harmonic of the RF voltage is used.

TABLE 1

in this energy range. This same mode of operation can be used to obtain beams which have a time spread $\leq$ 1 nsec which combined with the programming available on the ion injection line will make high precision time-of-flight studies possible. In many few nucleon studies one is forced to a coincidence experiment if one wishes to define completely all kinematic variables. This in turn leads to requests for a high duty factor in the incident beam and here TRIUMF excels. As can be seen in Table 1, when appropriate amounts of the third harmonic of the RF voltage are added, an overall duty factor exceeding 25% can be achieved. Another special feature which it is hoped to have available at TRIUMF from the start is a polarized beam. A polarized source has been built and installed at TRIUMF and has produced up to 300 namps of polarized H^- ions. A prototype of this source built at the University of Alberta is shown in Fig.1.

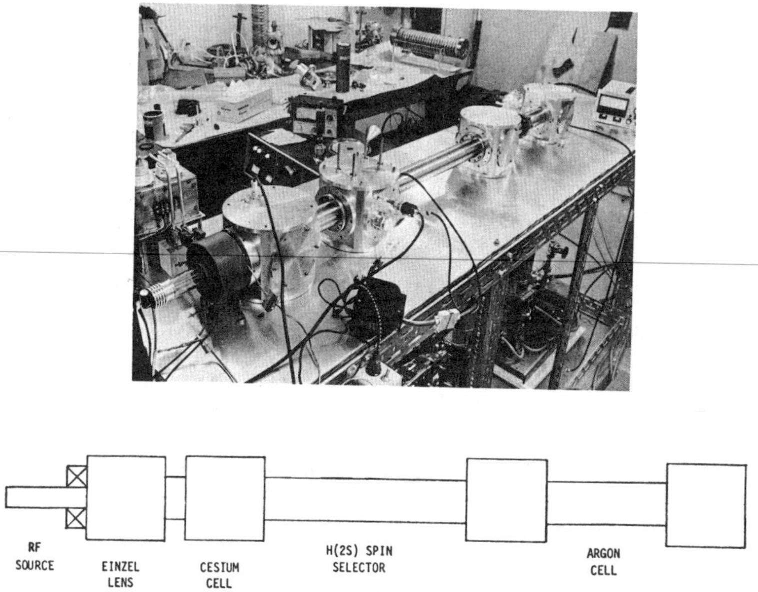

Prototype of the H^- polarized ion source

Figure 1

The source is a Lamb shift type using the 'Sona' zero crossing method. Polarization of 80% are typically achieved. A bunching system on the injection line is capable of increasing the beam accepted by the cyclotron by a factor of two which will be very useful for the polarized beam. Studies are under way to see if prebunching in the ion source region can be used to enhance further the fraction of beam accepted by the cyclotron, resulting hopefully in a few hundred nanoamperes of polarized beam being accelerated. Studies by the beam dynamics group have indicated very little depolarization need be expected during injection and acceleration of the polarized beam.

The next problem which must be solved, before our task is complete, is the installation of beam lines and experimental facilities. This stage is now in full swing at TRIUMF. A floor plan of the experimental areas is given in Fig. 2. A detailed account of the experimental facilities has been given by Pearce [2] and the initial experimental program has recently been reviewed by Measday [3]. All beam channels shown are presently under construction except for the high energy pion channel which will be

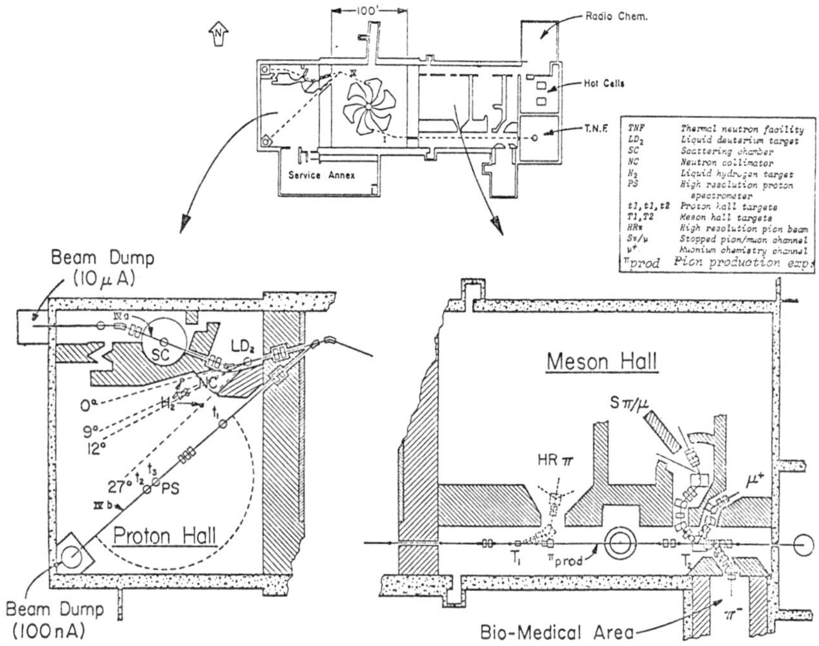

Experimental facility layout

Figure 2

installed later. The secondary beams which are planned for physics research are given in Table 2. During initial operation the beam current will be limited in intensity, and emphasis is being placed on using the proton and neutron beams. Because of the ease with which the energy can be varied, studies of the energy dependence of nucleon-nucleon and nucleon-nucleus interactions on light nuclei are being planned. I shall now discuss some of the equipment for few nucleon studies being installed at present and indicate a few things which we hope to have in the future.

Secondary Beams at TRIUMF		
Type		Intensity (sec^{-1})
NEUTRONS (unpolarized)	D(p,n)2p @ 0°	10^8 (into a 1° cone)
(polarized)	$D(\vec{p},\vec{n})2p$ @ 9°	10^6 (into a 1° cone); 75% polarization
π^+ (20 - 300 MeV)	Fast pion channel	$10^7 - 10^8$
π^- (20 - 300 MeV)	Fast pion channel	$10^6 - 10^7$
Stopped π^- ($5^- g^{-1} cm^2$)		5×10^8
Stopped μ^- ($5^{-1} g^{-1} cm^2$)		10^9

TABLE 2

Neutron Beam Line

Neutron beams will be produced using a liquid deuterium target as described by Robertson[4)] and shown schematically in Fig. 3. Polarized neutron beams can be obtained by polarization transfer at a scattering angle of 9° resulting in a mono-energetic neutron beam with polarization about 75%. It appears that the rotation

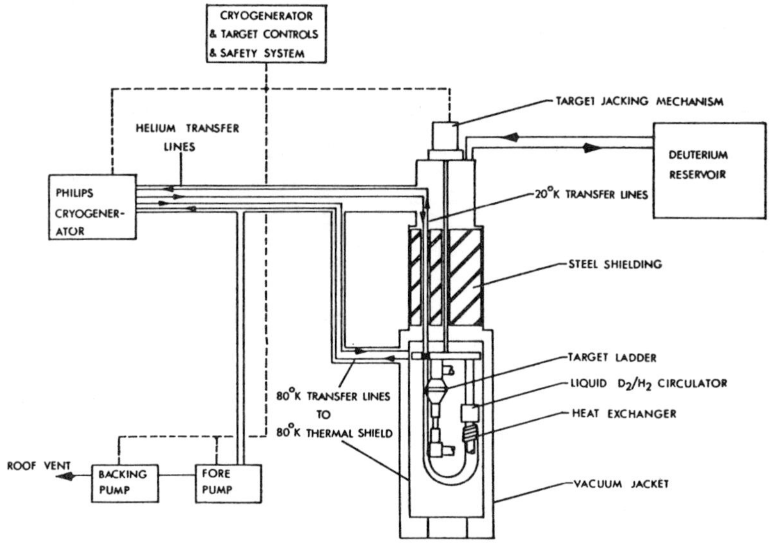

Schematic diagram of the liquid D_2/H_2 target

Figure 3

transfer parameter R may produce the highest polarization of the neutrons; thus a superconducting solenoid has been installed which is capable of rotating the proton spin axis onto the horizontal plane. A sophisticated proton polarimeter has been assembled by a group from England (BASQUE group). This will be used in a study of neutron-proton triple scattering parameters and is shown in Fig. 4.

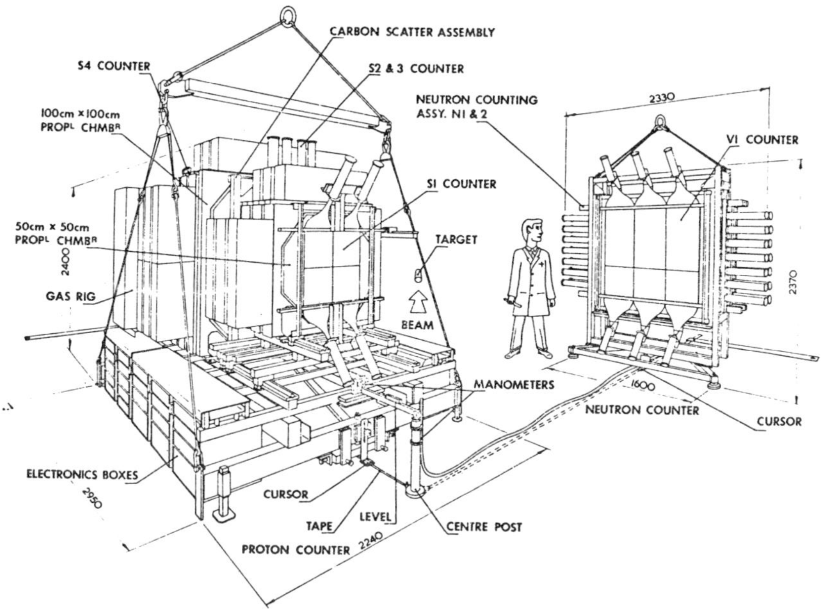

Equipment being installed to study the n-p interaction

Figure 4

Proton Interaction Studies

The study of proton interactions with light nuclei will be carried out on beam line IVb, also on beam line I during the initial period when the intensity is low. A major installation for proton studies is a large magnetic spectrometer. Due to budgetary limitations it has been decided first to install the quadrupole and one dipole of the QDD system and use these as a medium resolution spectrometer. This device would be capable of a resolution of $\pm$ 300 keV and have a solid angle of about 4 msr, with the quadrupole moved to 1 meter from the target. The later addition of the second dipole would result in a high resolution spectrometer with characteristics as given in Table 3. The QDD system of the high resolution spectrometer is shown in Fig. 5. While awaiting completion of the magnetic spectrometer, detection of protons

Proton-Magnetic Spectrometer

Type:	Quadrupole Dipole Dipole
Bend:	Vertical
Momentum:	1100 MeV/c (i.e. 500 MeV protons) at 14 kG
Resolution:	$\Delta p/p = \pm 0.01\%$ (i.e. $\pm$ 100 keV at 500 MeV)
Target to detectors:	15 m
Angular resolution:	2.2 mr
Solid angle:	1.1 π msr
Momentum acceptance:	$\pm$ 5.0% p_0

TABLE 3

will be done with NaI detectors, using multiwire proportional chambers to define accurately the scattering angle. Chamber development has been carried out at the University of Alberta and a number of small chambers have been completed. A 5" x 10"

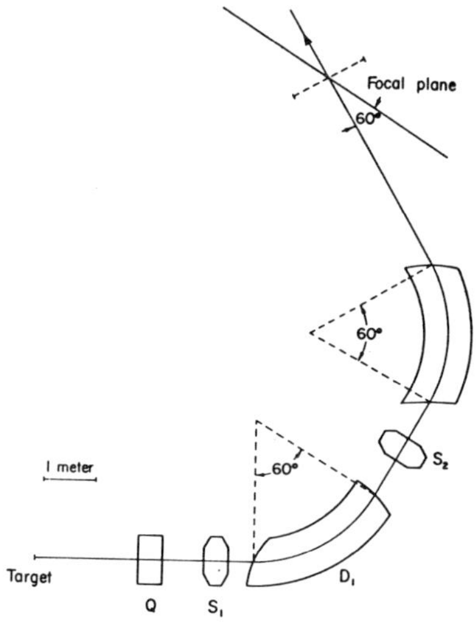

QDD system of the high resolution proton spectrometer.

Figure 5

chamber with all the associated electronics is shown in Fig. 6. Large chambers, 1 meter across, are presently being made for use with the magnetic spectrometer. A special wire chamber has been made for small angle studies of elastic scattering and for measurements of p-p bremsstrahlung far off the energy shell. The assembly, of

Multi Wire Chamber (5" x 10" - 64 wire x 128 wire)
Cover removed from Y position electronics

Figure 6

which this chamber is a part, is shown in Fig. 7. A number of experiments are being planned on light nuclei such as (p,pn) and ($\pi^{\pm}$,nn) in which outgoing neutrons will be

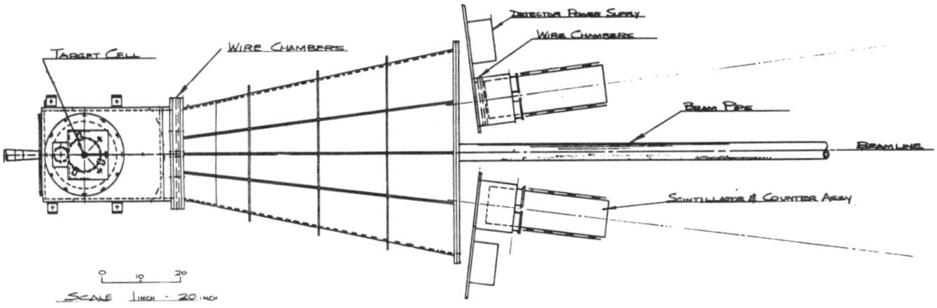

Figure 7

detected. Design of a large array of neutron detectors for use in such experiments is presently under way.

Scattering from light nuclei at intermediate energies often dictates the use of cryogenic targets and a number of such targets are planned. Initially two hydrogen targets and one ^{4}He target will be available. One or two ^{3}He targets should also be available for use within the next year.

While the intensity of the beam from TRIUMF is maintained at a low level, an experiment on pion production in the p+p interaction will be mounted on beam line I. The detector will be a modified alpha particle spectrograph built at U.B.C. some years ago. The specifications for this magnet are given in Table 4. These studies will be performed with both unpolarized and polarized proton beams.

Pion Spectrograph	
Radius	50 cm
Max pion energy	140 MeV
Detectors (initial)	plastic scintillator
(later)	drift chamber
Detector resolution (initial)	7 mm
(later)	1 mm
Solid angle	3 msr
Momentum acceptance	$\pm$ 20% p_0

TABLE 4

Pion Interaction Studies

There are proposals at TRIUMF to study many reactions induced by pions in which gamma-rays are produced. To do these experiments a large NaI crystal has been acquired. This detector is 18" diameter and 20" long. Checks of the crystal have verified the specified linearity of 4% across the face of the detector and a resolution of 7% has been obtained with 12 MeV gamma rays from the $^{11}B(p,\gamma)^{12}C$ reaction.

Future Developments

One area of development, which may prove very useful for studies involving detection of charged particles, is in High Purity Germanium detectors. Such detectors can now be fabricated up to 4.5 cm in diameter and with thicknesses of 1 to 2 cm. These detectors can be made with quite thin n-type windows and can be stacked to achieve the sort of depths, say 8 - 10 cm, required for medium energy applications. We are hoping to have some of this type of detector very soon and have been very encouraged by the results obtained with the detectors fabricated for the Carnegie-Mellon group at the Lawrence Berkeley Laboratory. These detectors have shown that, as expected, high resolution is obtainable and, perhaps more importantly, that the radiation damage problems can be overcome. We are planning further studies, in conjunction with personnel from L.B.L., on the problems to be encountered due to the high radiation environment in which these detectors must be used if they are to be useful for studies such as (p,2p).

The suitability of, and the interest within, TRIUMF to study the nucleon-nucleon interaction is furthering our interest in the development of polarized hydrogen targets. If funding is available, such targets may be expected to be available in two to three years.

The interest in pion production, especially further away from threshold, will require a larger and more sophisticated spectrometer and such a device is receiving consideration. As the present site of the pion production apparatus is only temporary, it is hoped that a suitable, more permanent, beam line will be developed. Such a low intensity beam line in the large experimental hall would also be useful for studies planned of total reaction cross sections on light nuclei. Such a line would also ensure more complete use of the facilities when polarized proton beams were being accelerated.

Obviously much development of the accelerator will be taking place over the next few years. One aim is to achieve separated turn acceleration which should result in an energy resolution of $\pm$ 50 keV with up to 30 μA of beam when the third harmonic of the RF voltage is added.

I should like in conclusion to thank all those who contributed facts and figures which I have used in this report and, even more so, all those who are doing the work.

References

1. Richardson, J.R., Proceedings of the VI International Cyclotron Conf., AIP Conf. Proc. No. 9, 126 (1972). See also other contributions to the same conference.

2. Pearce, R.M., TRIUMF Report, TRI-73-1.

3. Measday, D.F., Proceedings of 5th Int. Conf. High-Energy Phys. & Nucl. Struct., ed. by G. Tibell, North-Holland Publ. Co. (1974) p.426.

4. Robertson, L.P., *Extraction of Protons and Production of Secondary Beams from TRIUMF*, TRIUMF Report, VPP-70-2.

XIII

SOMMAIRE DE LA CONFERENCE

SUMMARY OF THE CONFERENCE

SOMMAIRE

Alberto Pignotti
Université de Buenos Aires

I have been asked to make some closing remarks at this Conference from the point of view of a particle physicist. As I look back over the previous twenty-one sessions, I recall some of the provocative things that have been mentioned. One of them is the realization pointed out by Margolis that the puzzling data on electron-positron annihilation into hadrons look, after all, surprisingly similar to the proton-antiproton annihilation data, and I am sure that this remark is going to keep busy a few particle physicists for a while. Another development that I find stimulating is the calculation reported by Brayshaw of the ω-meson as a resonant 3-pion state, giving the ρ parameters as an input and using a relativistic version of the boundary condition formalism. This is the materialization of an old idea that had up to now refused to be cast into a quantitative calculation and will have to stand the test of extension to other 3-pion systems to see whether it is just a happy coincidence or rather stands on solid ground. Incidentally, this example, as well as the discussion of final state interactions so clearly presented by Professor Amado, show the common ground that nuclear and particle physics share. Even though at times the language used may be different, the physical foundation is often the same, and one of the merits of this Conference is to have provided the occasion for such a merging of fields to happen.

But as we hear that at the Intersecting Storage Rings at CERN the equivalent proton lab. energy reaches nearly 2000 GeV and the average number of hadrons produced in such a collision is roughly 18, some of you will wonder what has high energy particle physics to do with nuclear physics and with the few body problem. It is to answer that question that I will address myself now.

In order to make a graphical representation of what goes on in a pp collision at those energies and show that not all of it is high energy after all, let me introduce the rapidity variable. What is the rapidity? In a collision the incident particle defines a direction in the lab. frame. There are infinitely many frames that differ from the lab. frame by a Lorentz boost in the direction of incidence. Each of these frames, as viewed from the laboratory, moves with a different rapidity, which is just the argument of the mentioned Lorentz transformation. Its definition is such that it is additive, just like the rotations around a given axis are additive. Thus, if frame 1 moves with rapidity y_{12} with respect to frame 2, and this in turn with y_{23} with respect to 3, 1 moves with $y_{13} = y_{12} + y_{23}$ with respect to 3. In other words, relativistically, velocities do not add but rapidities do (for collinear Lorentz transformations).

When a hadron emerges from a high energy collision we can thus describe its motion by the transverse components of its momentum $p_\perp$ and its rapidity y, i.e. the rapidity of the Lorentz frame in which it has only transverse motion. A set of particles will therefore be represented by a set of points in the $p_\perp$-y plane and , because of the additivity of y, by just a shift of the origin we find the description in the c.m. of projectile frames. In other words, distances in rapidity are invariant under the Lorentz transformations considered, and, as a consequence, the invariant mass of a 2-particle system depends on the difference of the rapidities of the two particles, once their transverse motion is given.

With the help of this variable let me now describe what happens in pp collisions at ISR energies, which were described by Professor Van Hove in his talk. The incident state is characterized by the diagram of fig. 1, drawn in a 'Lab' frame in which one of the incident particles is at rest and where $S \approx 2m^2 \cosh y^{inc} \approx m^2 \exp(y^{inc})$, i.e. $y^{inc} \approx \ln \frac{S}{m^2}$

fig. 1

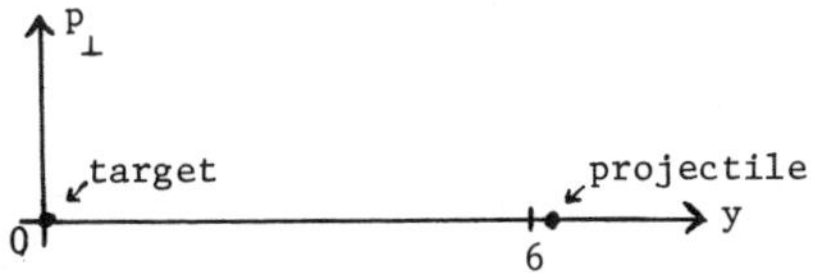

Roughly one out of five times this event is elastic, i.e. the final state looks like shown in fig. 2, where the probability of finding a large value of $p_\perp$ falls approximately like a gaussian with $p_\perp$.

fig. 2

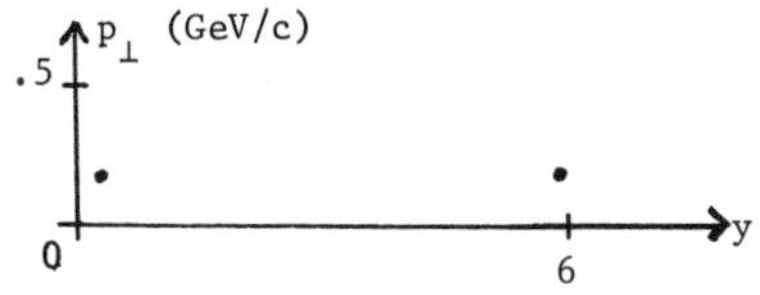

Roughly as often as there is an elastic event, there is a similar one in which however one of the protons dissociates into a few particle state, carrying the same proton numbers. Such events are called diffractive dissociation of one of the incident protons, and a typical one is described in fig. 3. Double diffraction dissociation is also expected to occur, but probably requires still higher energies to appear in its full strength.

fig. 3

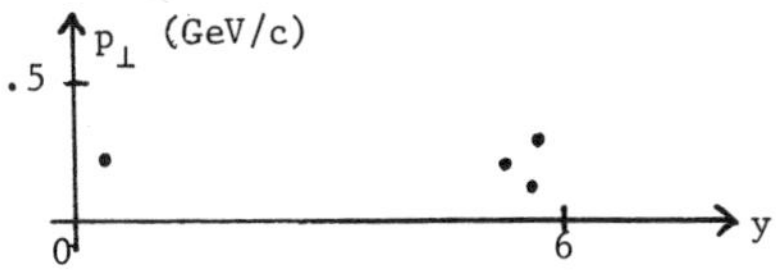

Finally, most of the time non diffractive events occur. They tend to populate evenly with mesons the available rapidity interval, and are rather large multiplicity events, but the transverse momenta of the final particles are still limited (see fig. 4)

fig. 4

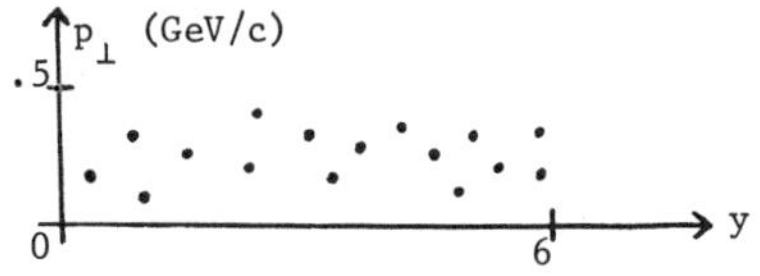

At first sight, we may think that here we have lost contact with the low energy and few-body physics, but the important realization is that in these processes the final particles tend to show short-range correlations in rapidity i.e. those near a given value of the rapidity behave independently of those at a distance of 2 or more units in rapidity space. Thus, one possible description, that perhaps may be somewhat naïve, but certainly has some amount of validity, is one in which particles are produced in clusters of limited size in rapidity $p_\perp$, typically 4 clusters of approximately 4 particles each at ISR (fig. 5).

fig. 5

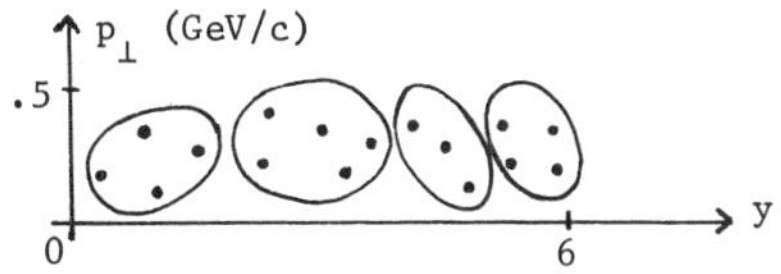

Particles in one cluster are correlated among themselves, but not with particles from other clusters. And as the energy varies, the basic cluster stays the same, and what changes is the average number of clusters that can be accomodated in the available rapidity. Thus, the basic dynamical ingredient, the driving force of the kernel of the integral equation is a <u>few body</u>, <u>medium energy</u> object. This is the basis of what is called Multipheral dynamics, initiated many years ago by Amati, Bertocchi, Fubini, Stanghellini and Tonin, and strikingly confirmed by recent ISR data. The essential role played by this medium-energy entity in high energy dynamics is, I believe, a crucial point in establishing the connection between medium and high energies, and the relevance of whatever we learn about the few-body problem for hadronic physics at high energy.

I would like to mention two more points in which particle and nuclear physics make contact. The first is the quark model, with the impressive successes in spectroscopy of hadronic states and deep inelastic scattering. Looked at from this point of view, basic particle physics processes become few body problems resembling nuclear physics, and Brodsky has discussed at this Conference how we can test the quark model and the underlying hadronic structure at very short distances by looking at those rare events which correspond to large values of $p_\perp$ -either in large- $p_\perp$ inclusive processes or fixed-angle 2-body scattering. In such events one expects inverse power behavior with $p_\perp$ and c.m. energy, respectively, as a consequence of scale invariance, and I refer you to Brodsky's talk for a discussion of the power involved and its implications. Lastly, I would like to mention the subject of very high collisions with nuclei, which provides a unique way of studying the interaction of short-lived hadronic systems and opens a lot of fascinating possibilities and questions, such as the rephrasing of more-or-less established properties of the hadron-hadron interactions for the hadron-nucleus or nucleus-nucleus case. The experimental study of the last one is going to be possible with the advent of heavy-ion high energy accelerators.

In clonclusion, I believe that this reunion of nuclear and particle physicists has been just a beginning of a fruitful cooperation, and that we shall have in future many opportunities to talk to each other. Actually, this Conference was advertised to be bilingual, and at some time I was afraid that 4 different languages would be spoken: nuclear and particle physicists both speaking French and English. Things actually worked out better than I expected, even though we did not have simultaneous translation between particle and nuclear jargons. Nevertheless, as a modest contribution towards an even better future communication between the fields, let me close by showing a little dictionary that I have compiled during the course of the Conference, and to which several colleagues have contributed, some of them without being aware of it:

Particle jargon:	Nuclear jargon:
particle	resonance
isobar model	sequential decay
hadron production	break up process
inclusive process	summed scattering
low energy	high energy
high energy	tremendous energy
qualitative agreemnet	great discrepancy
rapidity	that crazy variable

CONCLUDING REMARKS

H. Pierre NOYES
SLAC, Stanford University, Stanford California, USA

It is my honorable and dangerous task to try to identify among the many exciting ideas I have learned at this Conference a few which could point the way in which our field might go. What struck me forcefully was that once again nuclear and particle physics are finding a common ground. I believe that significant progress will come from strengthening this intimate relationship. In the area of nuclear and pionic phenomena I found again and again that both experiment and theory have been forced to the point where existing phenomenological methods of verification are inadequate. I will illustrate this shortly by several instances where only by including the pionic degrees of freedom in the systems, which many of us are accustomed to think only contain nucleons, can we hope to avoid triviality and frustration in our endeavours. At the same time we have learned many experimental and theoretical techniques which when joined might help us to meet this challenge. Thus we might be on the verge of an explosive development of new physics if we have the courage to seize the time. It has not been easy to reach this point. Much phenomenological work had to be done over the years when nuclear and particle physics appeared to be drifting apart. Experimental techniques such as polarization and complete kinematic experiments had to mature accompanied by sophisticated analyses and advanced computer programs. It is also clear that one catalyzing agent that has brought us together is the meson factories which will soon be flooding us with data.

This raises a question which, if I correctly interpret the duty held up to us by the Secretary General of the International Union of Pure and Applied Physics, I cannot ignore. As one organ of the conscience of the scientific community for our field, we must look beyond surface phenomena and our immediate interests to the various reasons why our work is supported. For instance, in describing some of the new laboratories, medical research facilities were mentioned. Significant by its omission is the fact that at least one of these accelerator installations envisages as well a weapons neutron test facility. I do not wish my words to be misinterpreted. This is just a dramatic instance of a fact known to practically all of us. Circumstances differ from individual to individual and country to country but all of us, myself included, draw support from our societies for reasons other than subjects which have been discussed at this Conference. In particular we all represent a privilieged class whose training separates us from the people and whose support disappears if we move too directly to aid the people against the interests of our ruling national elites. Some of us believe that we should welcome

* Work supported in part by the U.S. Atomic Energy Commission

this fact, others accept it as an unpleasant necessity, and still others abhor it. But too many simply ignore it. This can only make trouble for us in the future as it has in the past, unless we face this issue openly in meetings such as this to discuss how it affects our work. In raising this issue, I follow a precedent set by the banquet speaker at UCLA in 1972.

Having followed a much older tradition by drawing back the curtain so that the skeleton who shares our feast can join us, I turn to the exciting scientific challenges which lie before us. I trust I have not spoiled your appetite for them.

One unfortunate phrase which I have heard used at this Conference is "theorems of ignorance".In each instance where that phrase has been used, so far as I can see, all this means is that older phenomenological methods cannot resolve critical physical problems. But in each instance we also have in hand or could readily develop methods to turn this situation into a new and challenging opportunity. For example, the fact that many three body experiments are insensitive in some kinematic regions to pionic degrees of freedom, which some people call off-shell effects, has the immediate corollary that experiments in these regions can be used to measure two nucleon phase shifts which are still inaccessible to direct attack using conventional two nucleon experiments. Conversely, we have also seen that when the phase shifts are fixed there are other regions that are sensitive to the pionic degrees of freedom. This information will be difficult to extract but precious when obtained. The combined theoretical and experimental attack can clearly take us here into a new and exciting region. I wish to stress particularly that coulomb corrections can now be included so that p-d data can be analyzed unambiguously. It is time to start the code development -or in fact past time I would say- needed for this.

Of course some of us have believed for two years or more that the 1-2 MeV by which the so-called "realistic" potentials fail to bind the triton, and the structure observed at high momentum transfer in the electric form factor of He^3, can only be understood by including mesons explicitely. Work presented at this Conference should convince most of the rest that no physically acceptable modification of physically acceptable two nucleon interactions could resolve these discrepancies. We have long known that this must be true of magnetic effects in the two and three nucleon systems. If we can incorporate mesons directly into our picture of nuclei, we might be able to meet both problems simultaneously. This could also lead to a breakthrough in the old problem of how to connect the quadrupole moment, S-D asymptotic ratio, percentage D state, and magnetic moment of the deuteron so intimately that we can account for the many different phenomena which depend on these quantities.

This will not be easy. We have heard that the effective pion-nucleon scattering length changes in nuclear matter and the same will obviously be true of the charge and current distributions in the vicinity of the nucleon. This might explain the discrepancy between predicted coulomb energies of nuclei and experiment which increases systematically with A. We have also heard that pion capture leads to emerging clusters which seem to contain pions as well as nucleons. Two-nucleon interactions seem incapable of explaining simultaneously the density and binding energy of nuclear matter, though inclusion of the dominant 3-3 pion resonance seems to move the prediction in the right direction. The key to a successful attack on all these problems is obviously the same.

At a deeper level, we have learned that it is now possible to use the whole range of two nucleon phase shifts to extrapolate to the coupling constants of the pi-nucleon-nucleon system including coulombic effects. The same is true of the charged and neutral pion-nucleon coupling constants. It is to be emphasized that the constants so obtained have a unique theoretical and experimental significance, in contrast to the model-dependent scattering lenghts which have so long been a source of frustration. Splittings among these constants will describe the breaking of the isospin symmetry in an unambiguous way and present elementary particle theorists with a quantitative challenge they cannot ignore. Here again two-nucleon parameters from p-d and n-d experiments will play a critical role in achieving the required precision.

On the experimental side these precise extrapolation techniques provide a new and powerful way of parametrizing nucleon-nucleon phase shifts in the physical region. They can thus be used to point out those systematic discrepancies which undoubtebly still exist, but which still are hidden, in data from different laboratories. They can also help resolve questions at low energy about vacuum polarization and the ^3P waves in pp scattering. In particular they will be critical in isolating the $1/r^3$ interaction between the magnetic moments of the two protons form nuclear effects - a phenomenon which now appears to be capable of experimental test below 10 MeV.

At a still deeper level we are told that the parity violation, evidenced by the circular polarized gamma-rays from n-p capture, might serve as a probe of the nucleon distribution at short distances in the two nucleon system. The experiment needs repeating, and much work will have to be done before the theory is sufficiently well understood for unambiguous conclusions to be drawn, but the challenge has a rich reward if it can be met.

Until we can go below the hadrons to the quarks, or whatever underlies the hadronic physics we have been discussing, we are still dealing with phenomenology. We have heard at this Conference that very energetic electrons and neutrinos "see" these massless quarks in much the same way that alpha particles enabled Rutherford to "see" the massive nucleon. On the other hand, when two systems of quarks collide at ultra-high energies we are told that only some of them appear to be "active" in the collision. Here we might offer a suggestion coming from our own experience to our high energy brethren. Wick pointed out long ago that when two nuclear particles come close enough together for the active fluctuation to materialize what we now call the pion, they scatter independent of any specific interaction mechanism. If this is all that is available to quarks in the collisions between two quark systems, their "activity" might be explained by the creation of the clusters with which we are already familiar. This idea might unify our fields in still another way.

I conclude that, for those with courage to abandon old thought and a determination to carry through the difficult integrated theoretical and experimental programs successful attack on these problems will require, this Conference has helped to open up a host of exciting new directions for research in the few body problems in nuclear and particle physics.

INDEX DES AUTEURS — AUTHOR INDEX

- A -

- B -

- C -

- C -

- D -

- E -

- F -

- G -

- H -

- L -

- M -

- N -

- O -

- O -

- P -

- Q -

- R -

- S -

- S -

- T -

- V -

- W -

- W -

- Y -

- Z -

PARTICIPANTS

Aaron, Ronald
Physics Department
Northeastern University
Boston, Mass. 02115
U.S.A.

Afnan, Iraj R.
School of Physical Science
Flinders University
Bedford Park
South Australia 5042

Agnew, Lewis E.
Los Alamos Scientific Lab.
Los Alamos, N.M. 87544
U.S.A.

Allas, Richard
Naval Research Laboratory
Code 6611
Washington, D.C. 20375
U.S.A.

Alt, Erwin O.
Institut für Physik
Johannes Gutenberg Universität
D 6500 Mainz
W. Germany

Amado, R.D.
Department of Physics
University of Pennsylvania
Philadelphia, Pa 19104
U.S.A.

Amiot, Pierre
Département de physique
Université Laval
Québec, G1K 7P4

Amten, Lars
Tandem Accelerator Lab.
Box 533, S-751
21 Uppsala
Sweden

Asai, Juhachi
Department of Physics
University of Windsor
Windsor, Ontario

Aviles, Joseph B. (Jr.)
Naval Research Laboratory
Code 6611
Washington D.C. 20375
U.S.A.

Axen, D.
Department of Physics
University of British Columbia
Vancouver 8, B.C.

Bacher, Andrew
Indiana University
RR12 Box 241
Bloomington
U.S.A.

Baracca, Angelo
Istituto di Fisica Teorica
Universita di Firenze
50125 Firenze
Italie

Bawin, Michel
Inst. de physique nucléaire
Université de Liège, Sart Tilman
B 4000 Liège
Belgique

Beach, Louis A.
Naval Research Laboratory
Code 6611
Washington D.C. 20375
U.S.A.

Benayoun, Jean-Jacques
Institut des Sciences nucléaires
B.P. 257, Centre de Tri
38066 Grenoble
France

Benn, Johachim
Physik Institut
Der Universität Zürich
Zürich, Schönberggasse 9
Suisse

Bertozzi, William
Physics Department
Mass. Institute of Technology
Cambridge, Mass. 02139
U.S.A.

Bhakar, Balram Singh
Department of Physics
University of Manitoba
Winnipeg, Manitoba

Birchall, James
Lawrence Berkeley Laboratory
Berkeley, Ca 94720
U.S.A.

Bjork, C.W.
Los Alamos Scientific Laboratory
Los Alamos, N.M. 87544
U.S.A.

Bleszynski, M.
Instytut Fizyki Jadrowej
Radzikowskiego 152
31-342 Krakow
Poland

Bleuler, K.
Inst. f. Theor. Kernphysik
Universität Bonn
D-5300 Bonn, Nussallee 14-16
W. Germany

Boersma, H.J.
Natuurkundig Laboratorium
Der Vrije Universiteit
Amsterdam, Buitenveldert
The Netherlands

Bohannon, George
Department of Physics
Michigan State University
East Lansing, Michigan 48824
U.S.A.

Boivin, Michel
Département de physique
Université Laval
Québec, G1K 7P4

Bonbright, Daniel
Department of Physics
University of Manitoba
Winnipeg, Manitoba

Bondelid, Rollon O.
Naval Research Laboratory
Code 6610
Washington, D.C. 20375
U.S.A.

Bonner, B.E.
Los Alamos Scientific Lab
Los Alamos, N.M. 87544
U.S.A.

Bosco, Bernardino
Instituto di Fisica Teorica
Universita di Firenze
50125 Firenze
Italie

Bouchard, Charles-Albert
Département de physique
Université Laval
Québec, G1K 7P4

Boymond, Jean-Pierre
Physics Department
Princeton University
Princeton, N.J.
U.S.A.

Brady, F.P.
Department of Physics
University of California
Davis, California
U.S.A.

Brandenburg, Richard
Los Alamos Scientific Lab.
Los Alamos, N.M. 87544
U.S.A.

Brayshaw, David D.
Physics Department
University of Marvland
College Park, Maryland 20742
U.S.A.

Breunlich, Wolfgang
Inst. f. Radiumforschung
Osterr Akad. d. Wissenschaften
A-1090 Vienna, Boltzmanng. 3
Austria

Brodsky, Stanley J.
Stanford Linear Accelerator
Stanford University
Stanford, Ca 94305
U.S.A.

Brown, Ronald E.
School of Physics
University of Minnesota
Minneapolis, Minn. 55455
U.S.A.

Bugg, D.V.
Queen Mary College
Mile End Rd.
London, EI
England

Bussières de Nercy, André
Phys. nucl. à haute énergie
CEN Saclay, B.P. no 2
91190 Gif-sur-Yvette
France

Cahill, R.T.
Physical Sciences
Flinders University
Bedford Park 5042
Australia

Callerame, Joseph
Physics Department
Harvard University
Cambridge, Mass. 01238
U.S.A.

Cameron, John
Department of Physics
University of Alberta
Edmonton, Alberta

Carlson, Richard, F.
Department of Physics
University of Redlands
Redlands, Ca 92393
U.S.A.

Carruthers, Peter
Theoretical Division
Los Alamos Scientific Lab.
Los Alamos, N.M. 87544
U.S.A.

Celenza, Louis S.
Physics Department
Brooklyn College of Cuny
Brooklyn, N.Y. 11210
U.S.A.

Ceuleneer, René
Faculté des sciences
Univ. de l'Etat à Mons
7000 Mons, 19 av. Maistriau
Belgique

Chatterjee, A.
Calcutta Univ. and Saha Inst.of
Nuclear Physcis
92 Acharya Prafulla Chandra Rd.
Calcutta 700009, India

Cheng, Vianney K.
Department of Physics
Stanford University
Stanford, Ca 94305
U.S.A.

Chertok, Benson
Stanford Linear Accel. Center
Slac - P.O. Box 4349
Stanford, Ca 94305
U.S.A.

Clough, A.S.
Physics Department
University of Surrey
Guilford, Surrey
England

Coçu, Francis
Service de physique nucléaire
C.d'études Bruyères-le-Châtel
92120 Montrouge, B.P. no 61
France

Conzett, H.E.
Lawrence Berkeley Lab.
Univ. of California
Berkeley, Ca 94720
U.S.A.

Črepinšek, Ljubomir
Institute for Theoretical Physics
University Graz
A 8010 Graz, Universitätspentz 5
Austria

Cujec, B.
Département de physique.
Université Laval
Québec, G1K 7P4

Darling, Byron T.
Département de physique
Université Laval
Québec, G1K 7P4

Dautry, François
Department of Physics
State University of New York
Stony Brook, N.Y. 11794
U.S.A.

Davies, Walter G.
Atomic Energy of Canada Ltd.
Chalk River
Ontario

Day, B.D.
Argonne National Lab., Bld. 203
Argonne, Ill. 60439
U.S.A.

Delves, L.M.
Dept. of Computational Science
University of Liverpool
Liverpool
England

Deregel, Jean
D.Ph. PE, CEN-SACLAY
10 av. de Prédecelle
91470 Limours
France

Depommier, P.
Laboratoire de physique nucléaire
Université de Montréal
C.P. 6128
Montréal

Devins, D.W.
Department of Physics
Inidana University
Bloomington, Ind.
U.S.A.

Didelez, Jean-Pierre
Department of Physics
University of Maryland
College Park, Maryland 20740
U.S.A.

Eberhard, Philippe
Lawrence Berkeley Lab.
Berkeley, Ca 94270
U.S.A.

Edgington, J.A.
TRIUMF
University of British Columbia
Vancouver 8, B.C.

Elbaz, Edgard
Institut de physique nucléaire
43 boul. du 11 Novembre 1918
69621 Villeurbanne
France

Ellis, Dale
Physics Department,
McMaster University
Hamilton
Ontario L8S 4M1

Epstein, Geoffrey N.
Dept. of Theoretical Physics
University of Oxford
12 Parks Rd., Oxford
England

Faessler, Martin A.
Max-Planck Inst. f. Kernphysik
Heidelberg
W. Germany

Fayard, Claude
Institut de physique nucléaire
43 boul. du 11 Novembre 1918
69621 Villeurbanne
France

Ferguson, A.J.
Chalk River Nuclear Lab.
Nuclear Physics Branch
Chalk River, Ontario

Ferreira, Erasmo
Departemento de Fisica
Pontificia Univ. de Catolica
Rio de Janeiro Z.C. 19
Brésil

Ferreira, Lidia S.
Theoretical Phys. Dept.
12 Parks Rd.
Oxford,
England

Fiarman, Sidney
Department of Physics
Stanford University
Stanford, Ca 94305
U.S.A.

Fiedeldey, H.
Department of Physics
University of S. Africa
Pretoria 0001
South Africa

Fonseca, Antonio C.
Department of Physics
Univ. Notre-Dame
Notre-Dame, Ind. 46556
U.S.A.

Frascaria, Robert
Institut de physique nucléaire
B.P.no 1
91406 Orsay
France

Freed, Norman
Pennsylvania State University
104 Davey Lab.
University Park, Pa 16802
U.S.A.

Friar, J.L.
Department of Physics
Brown University
Providence
U.S.A.

Frois, Bernard
Physique nucl. à haute énergie
Cen Saclay, B.P. no 2
91190 Gif-sur-Yvette
France

Fujiwara, Noboru
Institut de physique nucléaire
Université de Paris-Sud, B.P.no 1
91 Orsay
France

Gibson, Benjamin F.
Los Alamos Scientific Lab.
Los Alamos, N.M. 87544
U.S.A.

Glavish, Hilton F.
Physics Department
Stanford University
Stanford, Ca 94305
U.S.A.

Green, A.M.
Res. Inst. of Theor. Physics
University of Helsinki
SF-00170 Helsinki
Finland

Gross, F.
Department of Physics
College of William and Mary
Williamburg, Va 23185
U.S.A.

Guaraldo, Carlo
Laborat. Nazionali di Frascati
Casella Postale 70
Frascati
Italie

Gurd, David
Triumf
University of British Columbia
Vancouver 8, B.C.

Gutay, László
Department of Physics
Purdue University
West Lafayette, Indiana 47907
U.S.A.

Glöckle, Walter
Inst.for Theoretical Physics
University Bochum
463 Bochum, Universitätsstr. 150
W.Germany

Gönczi, Laszlo
Tandem Accelerator Laboratory
Box 533
S-75121 Uppsala
Sweden

Gotow, Kazuo
Department of Physics
Virgina Polytechnic Inst.
Blacksburg, Va 24061
U.S.A.

Goulard, Gisèle
Département de physique
Collège militaire de St-Jean
St-Jean, JOJ 1RO

Greben, J.M.
Inst. for Theoretical Physics
University of Groningen
Groningen, P.O. Box 800
The Netherlands

Hackman, Roger
Virginia Polytechnical Institute
Blacksburg, Va 24061
U.S.A.

Haftel, Michael
Naval Research Laboratory
Code 6665
Washington D.C. 20375
U.S.A.

Hanna, S.S.
Physics Department
Stanford University
Stanford, CA 94305
U.S.A.

Harnois, Jean-Yves
Physics Department
University of Toronto
Toronto M5S 1A7

Hasinoff, Michael
Physics Department
University of British Columbia
Vancouver B.C. V67 1W5

Hayakawa, S.
Foster Radiation Lab.
McGill University
Montréal

Heiss, Peter
Institut für Theor. Physik
Universität Koln
Köln, Universitätstrasse 14
W. Germany

Heller, L.
Los Alamos Scientific Lab.
Los Alamos, N.M. 87544
U.S.A.

Hewitt, R.G.
Dept. of Theoretical Physics
The University of Sydney
Sydney, N.S.W. 2006
Australia

Hiebert, John C.
Cyclotron Institute
Texas A & M University
College Station TX 77843
U.S.A.

High, David Maynard
Département de physique
Université Laval
Québec, GlK 7P4

Hincks, E.P.
Division of Physics
National Research Council of Canada
Ottawa, KlA OR6

Ho Kim Q.
Département de physique
Université Laval
Québec, GlK 7P4

Houdayer, J. Alain
Foster Radiation Lab.
McGill University
Montréal

Hussein, Mahir S.
Instituto De Fisica
Universidade de Sao Paulo
C.P. 8-129, Sao Paulo
Brasil

Hwang, Chester F.
Los Alamos Scientific Lab.
Los Alamos, N.M. 87544
U.S.A.

Irshad, M.
Département de physique
Université Laval
Québec, GlK 7P4

Jaccard, Samuel
Triumf
Department of Physics
University of British Columbia
Vancouver 8, B.C.

Jain, B.K.
Bhabha Atomic Research Centre
Trombay, Bombay 400085
India

Jain, Mahavir
Texas A & M University
Los Alamos, N.M. 87544
U.S.A.

Jarmie, Nelson
Los Alamos Scientific Laboratory
Los Alamos, N.M. 87544
U.S.A.

Jaroszewicz, Thomasz
Institute of Nuclear Physics
Radzikowskiego 152
31-342 Krakow
Poland

Jeremie, Hannes
Département de Physique
Université de Montréal
C.P. 6128
Montréal

Johansson, Arne
Tamden Accelerator Laboratory
Box 533
S-75121 Uppsala
Sweden

Jones, Garth
Physics Department
University of British Columbia
Vancouver, B.C. V6T 1W5

Jovanovich, J.
Department of Physics
University of Manitoba
Winnipeg, Manitoba

Kan, Philip T.
Nuclear Physics, Grad.Res.Cent.
University of Massachusetts
Amherst, Mass. 01002
U.S.A.

Karlsson, Bengt
Institute of Theoretical Physics
Fack S 40220
S 40220 Göteborg 5
Sweden

Kermode, Mark W.
Department of Physics
Theory Group
McMaster University
Hamilton, Ontario L8S 4M1

Kim, E. Yeong
Los Alamos Scientific Lab.
Los Alamos, N.M. 87544
U.S.A.

King, S.P. Nicholas
Crocker Nuclear Laboratory
University of California
Davis, California
U.S.A.

Kloet, W.M.
Dept. of Physics and Astronomy
University of Maryland
College Park, Maryland 20742
U.S.A.

Kluge, Wolfgang
University of Karlsruhe
75 Karlsruhe, Postfach 3640
W. Germany

Kofoed-Hansen
CERN, Ch. 1211
Genève 23
Suisse

Kok, L.P.
Inst. f. Theoretical Physics
Groningen
The Netherlands

Komarov, V.V.
Institute of Nuclear Physics
Moscow State University
Moscow
USSR

Kowalski, K.L.
Department of Physics
Case Western Reserve University
Cleveland, Ohio 44106
U.S.A.

Krell, Max
Département de
Université de Sherbrooke
Sherbrooke, Qué.

Kuhn, Berthold
Zentralinstitut f. Kernforschung
Akademie Wissenschaften DDR
8051 Dresden
D.D.R. Germany

Lagu, A.V.
Department of Physics
Banaras Hindu University
Varanasi 221002
India

Lambert, James M.
Department of Physics
Georgetown University
Washington D.C. 20007
U.S.A.

Lamontagne, C.R.
Département de physique
Université Laval
Québec G1K 7P4

Lamothe, Micheline
Department of Physics
University of Toronto
Toronto M5S 1A7

Larue, Robert
Département de physique
Université Laval
Québec G1K 7P4

Lavine, J.P.
Département de physique
Université Laval
Québec, G1K 7P4

Lechanoine, C.
Université de Genève
32 boul. d'Yroy
Genève 4
Suisse

Lee, Jonathan
Department of Physics
McGill University
Montréal

Legrand, Dominique
Phys. nucl. à moyenne énergie
Cen-Saclay, B.P. no 2
Gif-sur-Yvette
France

Lehar, Frantz
CEN - Saclay, B.P. no 2
Gif-sur-Yvette
France

Levin, F.S.
Physics Department
Brown University
Providence
U.S.A.

Levinger, J.S.
Rensselaer Polytechnic Institute
TROY, N.Y. 12181
U.S.A.

L'Huillier, Marcelle
Institut de physique nucléaire
B.P. no 1, G-1406
Orsay
France

Lim, T.K.
Physics Department
Drexel University
Philadelphia, Pa 19104
U.S.A.

Liou, Ming-Kung
Department of Physics
Brooklyn College of the CUNY
Brooklyn
U.S.A.

Lisowski, Paul W.
Physics Department
Duke University
Durnam,
U.S.A.

Liu, Lon Chang
Physics Department
Brooklyn College,
Brooklyn, N.Y. 11210
U.S.A.

Locher, M.P.
Physics Department
University of Virginia & SIN
Charlottesville
U.S.A.

Lodhi, M.A.K.
Texas Tech. University
Lubbock
Texas 79409
U.S.A.

Loiseau, B.
Division de physique théorique
Institut de physique nucléaire
91406 Orsay, B.P. no 1
France

Lomon, Earle
Physics Department
Mass. Institute of Technology
Cambridge, Mass. 02139
U.S.A.

Lovas, Istvan
Institut f. Kernphysik
Kernforschungsanlage
D 5170 Jülich 1
W. Germany

Macek, Robert J.
Los Alamos Scientific Laboratory
Los Alamos , N.M. 87544
U.S.A.

Machleidt, R.
Inst. f. Theorische Kernphysik
Universität Bonn
D 53 Bonn, Nussalle 14-16
W. Germany

Mack, G.
Triangle Nuclear Lab.
Duke University
Durham, N.C. 27706
U.S.A.

Macq, P.
Inst. de physique corpusculaire
Université de Louvain
1348 Louvain
Belgique

Magnac-Valette, D.
Laboratoires des basses énergies
Centre de recherches nucléaires
67037 Strasbourg
France

Margaziotis, D.J.
Physics Department
California State University
Los Angeles, Ca
U.S.A.

Margolis, B.
Institute of Theoretical Physics
Department of Physics
McGill University
Montréal

Maschuw, Reinhard
Inst. f. Experimentalphysik
University of Hamburg
2000 Hamburg, Luruper Ch. 149
W. Germany

Masterson, Thomas G.
Department of Physics
University of British Columbia
Vancouver 8, B.C.

Matthews, J.L.
Laboratory for Nuclear Science
M.I.T.
Cambridge,Mass. 02139
U.S.A.

Mitra, Asoke
Physics Department
Delhi University
Delhi
India

Miljanić, Duro
T.W. Bonner Nucl. Lab.
Rice University
Houston, Texas
U.S.A.

Moeller, Karsten
Joint Inst. for Nucl. Research
Dubna, P.O. Box 79
101000 Moscow
USSR

Moravcsik, Michael J.
Inst. of Theoretical Science
University of Oregon
Eugene, Oregon 97403
U.S.A.

Moss, G.A.
Nuclear Research Centre
University of Alberta
Edmonton, Alberta

Murphy, James J.
Accelerator Laboratory
University of Saskatchewan
Saskatoon, Sask. S7N 0W0

Muench, Stephan
Physik Institut
Universität Zürich
Zürich, Schönberggasse 9
Suisse

McFarlane, W.K.
Physics Department
Temple University
Philadelphia, Pa 19122
U.S.A.

McKee, J.S.C.
Physics Department
University of Birmingham
Birmingham B1S 2TT
England

McManus, H.
Michigan State University
Physics Dept. Cyclotron Lab.
East Lansing,Michigan 148823
U.S.A.

McMillan, Malcolm,
Physics Department
University of British Columbia
Vancouver 8, B.C.

McNaughton, Michael W.
Crocker Nuclear Laboratory
University of California
Davis,Ca 95616
U.S.A.

Nichitiu, Florian
Joint Inst. for Nucl. Physics
Dubna, P.O. Box 79
101000 Moscow
USSR

Nisimura, Keigo
Department of Physics
Kyoto University
Sakyo-ku, Kyoto
Japan

Nogamy, Y.
Department of Physics
McMaster University
Hamilton, Ontario L8S 4M1

Northcliffe, L.C.
Cyclotron Institute
Texas A & M University
College Station, Texas 77843
U.S.A.

Noyes, Pierre H.
Stanford Linear Accel. Center
Stanford University
Stanford,
U.S.A.

O'Connel, J.S.
National Bureau of Standards
Washington, D.C. 20234
U.S.A.

Ohlsen, Gerald
Los Alamos Scientific Lab.
Los Alamos, N.M. 87544
U.S.A.

Olsen, Stephen L.
University of Rochester
Rochester, N.Y. 14627
U.S.A.

Oryu, Shinsho
Department of Physics
Science University of Tokyo
Noda-city Chiba 278
Japan

Owens, J.
RI Junior College
235A Promenade St.
Providence, R.I. 02908
U.S.A.

Perkins, Roger B.
Los Alamos Scientific Laboratory
Los Alamos, N.M. 87544
U.S.A.

Peterson, Edward L.
Naval Research Laboratory
Code 6611
Washington D.C. 20375
U.S.A.

Peterson, Gerald A.
Nucl. Physics,Grad.Res. Centre
University of Massachusetts
Amherst, Mass 01002
U.S.A.

Pianarosa, Piero
Département de physique
Université Laval
Québec, GlK 7P4

Picker, Harvey S.
300 Summit Street
Hartford Conn. 06106
U.S.A.

Pigeon, René
Département de physique
Université Laval
Québec GlK 7P4

Pignotti, Alberto
Facultad de Ciencias Exactas
Universidad de Buenos Aires
Nuñez, Buenos Aires
Argentine

Plattner, G.R.
Physikalisches Institut
Universität Basel
4056 Basel, Klingelbergstr 82
Suisse

Postma, Hans
Natuurkundig Laboratorium
State University of Groningen
Groningen,
The Netherlands

Preston, M.A.
Physics Department
McMaster University
Hamilton, Ontario

Pugh, H.G.
Physics Department
University of Maryland
College Park, Maryland 20742
U.S.A.

Quarati, Piero
Inst. di Fisica
Universita di Cagliari
09100 Cagliari
Italie

Rad, F.N.
Lawrence Berkeley Laboratory
Bldg. 88
Berkeley
U.S.A.

Radvanyi, Pierre
Institut de Physique nucléaire
B.P. no 1
91406 Orsay
France

Ramavataram, K.
Département de physique
Université Laval
Québec G1K 7P4

Ramavataram, S.
Département de physique
Université Laval
Québec G1K 7P4

Rao, L. Chitta
Département de physique
Université Laval
Québec G1K 7P4

Rao, K. Srinivasa
"Matscience" Institute
Madras 600020
India

Ratner, L.G.
Harrison M. Ràndall Lab. of Phys.
University of Michigan
Ann Arbor, Michigan 48104
U.S.A.

Rayet, Marc
Faculté des sciences
Université libre de Bruxelles
1050 Bruxelles
Belgique

Richardson, J. Reginald
Triumf
University of British Columbia
Vancouver 8, B.C.

Rinat, Abraham
Dept. of Nuclear Physics
Weizmann Institute of Science
Rehovot
Israel

Riska, D.O.
Department of Physics
State Univ. of New York
Stony Brook, N.Y. 11790
U.S.A.

Ritter, Hans Georg
Max Planck Inst. f. Kernphysik
Heidelberg
W. Germany

Robertson, Lyle P.
Triumf
University of
Vancouver , B.C.

Rogers, Joel G.
Nuclear Research Center
University of Alberta
Edmonton, Alta T6G 2J1

Rouben, Benjamin
Laborat. de Physique nucléaire
Université de Montréal
C.P. 6128
Montréal 101, P.Q.

Roy, René
Lawrence Berkeley Laboratory
Bldg. 88
Berkeley Ca 94720
U.S.A.

Saenz, A.W.
Naval Research Lab.
4555 Overlook Av.
Washington D.C. 20375
U.S.A.

Sandhas, Werner
Physikalisches Institut
Universitat Bonn
D 53 Bonn
W. Germany

Santhanam, T.S.
"Matscience Institute"
Madras 600020
India

Sasakawa, Tatuya
Département de physique
Université Tohoku
Katahiracho
980 Sendai, Japon

Sauer, Peter U.
M.I.T.
Cambridge, Mass. 02139
U.S.A.

Scharenberg, R.
Physics Department
Purdue University
W. Lafayette, Ind. 47907
U.S.A.

Schick, Lee H.
Dept. of Physics and Astronomy
University of Wyoming
Laramie, Wyo. 82070
U.S.A.

Schierholz, Gerrit
SLAC, Theory Division
Stanford, California 94305
U.S.A.

Schlüfter, Reinhard
Inst. f. Exper. Kernphysik
Der Universitat Karlsruhe
7500 Karlsruhe, P.B. 3640
W. Germany

Schmid, E.
Theoret. Physik
Universität TÜBINGEN
D-74 Tübingen, Hausserrt. 11ï
W. Germany

Schrieder, G.
Inst. f. Experimental Physik
Ruhr Universität
463 Bochum, Postfach 2148
W. Germany

Schuhl, Claude
Dépt de physique nucléaire
CEN-Saclay, B.P. no 2
91190 Gif-sur-Yvette
France

Schwager, Jörg
Inst. f. Theor. Physik der Universität
D-7400 Tübingen
W. Germany

Scrimaglio, Renato
C.N.E.N.
L.N.F. - Casella Postale 70
Frascati (Roma)
Italie

Seiler, F.
Physikalisches Institut
Universität Basel
4056 Basel, Klingelbergstr.82
Suisse

Seltz, Raymond
Lab. de basses énergies
Centre de rech. nucléaires
67037 Strasbourg
France

Serduke, Franklin
Physics Division
Argonne National Lab.
Argonne, Illinois 60439
U.S.A.

Shakin, Carl
Department of Physics
Brooklin College
Brooklin, N.Y. 11210
U.S.A.

Shannon, Darwin L.
U.C.L.A.
1117½ Westholme Ave.
Los Angeles, Calif. 90024
U.S.A.

Sharma, T.C.
Département de physique
Université Laval
Québec

Sherif, Helmy S.
Nuclear Research Centre
University of Alberta
Edmonton, Alberta

Schierholz, Gerrit
Stanford Linear Accelerator Center
Theory Division, Stanford
California 94305, U.S.A.

Shin, Y.M.
Accelerator Laboratory
University of Saskatchewan
Saskatoon, Saskatchewan

Shioyama, S.
Department of Physics
Science Univ. of Tokyo
Noda-City, Chiba
Japan

Shirato, Shoji
Bonner Nuclear Lab.
Rice University
Houston, Texas 77001
U.S.A.

Shively, Frank T.
Los Alamos Scientific Lab.
Los Alamos, N.M. 87544
U.S.A.

Shoucri, R
Département de physique
Université Laval
Québec, GlK 7P4

Silbar, R.R.
Schweizerisches Institut für
Nuklearforschung
Ch. 5234 Villigen
U.S.A.

Simmons, James E.
Physics Division, Mail Stop 456
Los Alamos Scientific Laboratory
Los Alamos, New Mexico 87544
U.S.A.

Simonius, Markus
Laboratorium fur Kernphysik
Eidg. Tech. Hochschule
Zürich
Suisse

Slaus, Ivo,
Institute "R. Bosković"
Bijenička 54, PP 1016
Yougoslavie

Slobodrian, Maria Luisa V.L.
Département de physique
Université Laval
Québec, GlK 7P4

Slobodrian, R.J.
Département de physique
Université Laval
Québec, GlK 7P4

Sourkes, Allan M.
Department of Physics
University of Manitoba
Winnipeg, Manitoba

Sprung, Donald
Physics Department
McMaster University
Hamilton, Ontario L8S 4M1

Srivastava, Mahendra
Department of Physics
McMaster University
Hamilton, Ontario L8S 4M1

Steinberg, Richard
Dept. of Physics and Astronomy
University of Maryland
College Park, Maryland 20742
U.S.A.

Stewart, Noël M.
Bedford College
University of London
London NW1
England

Stingl, Manfred
Inst. f. Theor. Physik
Universität Münster
44 Münster
W. Germany

Storm, D.
Nevis Laboratoires
Columbia University
P.O. Box 137, Irvington
N.Y. 10533, U.S.A.

St-Pierre, Claude
Département de physique
Université Laval
Québec, GlK 7P4

Strobel, George L.
Dept. of Phys. and Astronomy
The University of Georgia
Athens, Georgia 30602
U.S.A.

Svenne, Jaris D.
Department of Physics
University of Manitoba
Winnipeg, Manitoba

Sundqvist, B
Tandem Accelerator Lab.
Uppsala Universitet
S 75121 Uppsala
Sweden

Sunier, W. Jules
Los Alamos Scientific Lab
Los Alamos, N.M. 87544
U.S.A.

Swart, J.J. de
Dept. of Physics
University of Nijmegen
Nijmegen
The Netherlands

Szalata, Xenon M.
Nuclear Physics
University of Massachusetts
Amherst, Mass 01002
U.S.A.

Szöghy, Istvan
Département de physique
Université Laval
Québec, G1K 7P4

Teillac, Jean
Inst. national de phys. nucl.
11 rue Pierre et Marie Curie
75231 Paris
France

Thomann, Christoph
Physik Institut Universität
Schönberggasse 9
Zürich
Suisse

Thomas, A.W.
Department of Physics
Univ. of British Columbia
Vancouver B.C. V6T 1W5

Thompson, R.H.
Los Alamos Scientific Lab
Los Alamos, N.M. 87544
U.S.A.

Tjon, J.A.
Inst. for Theor. Physics
Utrecht, Sorbonnelaan 4
The Netherlands

Touma, Laila
Département de physique
Université Laval
Québec, G1K 7P4

Turchinetz, W.E.
MIT, Bates Linac Lab.
P.O. Bos 95
Middleton, Mass 01949
U.S.A.

Turcotte, Gilles
Département de physique
Université Laval
Québec, G1K 7P4

Van Hove, L.
CERN
1211 Genève 23
Suisse.

Van Oers, W.T.H.
Department of Physics
University of Manitoba
Winnipeg, Manitoba

Vavra, Jaroslav
Department of Physics
University of British Columbia
Vancouver, B.C.

Verwest, Bruce J.
State University of New York at
Stony Brook
Stony Brook, New York 11790
U.S.A.

Vinet, Luc
Centre de Rech. Mathématiques
Université de Montréal
C.P. 6128
Montréal

Vinh Mau, R.
Lab. de physique théorique
Université de Paris VI
75230 Paris
France

Viollier, Raoul D.
Physics Department
Stanford University
Stanford Ca 94305
U.S.A.

Vogt, Erich,
Department of Physics
University of British Columbia
Vancouver 8, B.C.

Walter, R.L.
Department of Physics
Duke University
Durham
U.S.A.

Warner, Robert E.
Nuclear Physics Branch
CRNL Chalk River
Ontario, K0J 1J0

Warren, J.B.
Physics Department
University of British Columbia
Vancouver, B.C.

Watson, John W.
Department of Physics
Cyclotron Lab.
University of Manitoba
Winnipeg

Weddigen, Christian
Kernforschungszentrum
D 75 Karlsruhe, Postfach 3640
W. Germany

Weiss, M.S.
Lawrence Livermore Laboratory
Livermore Cal. 99550
U.S.A.

Werren, David
Dépt de physique nucléaire
32 boul. d'Yvoy
Genève
Suisse

Wick, K.
Inst. f. Experimental Physik
Universitat Hamburg
Hamburg 50
W. Germany

Wigner, Eugene
Physics Department
Princeton University
Princeton, N.J. 08540
U.S.A.

Winternitz, Pavel
Centre de Rech. Mathématique
Université de Montréal
C.P. 6128
Montréal

Witsch, W.
Inst. f. Strahlen u. Kernphysik
Universitat Bonn
D 53 Bonn
W. Germany

Wolfe, David M.
Department of Physics
University of New Mexico
Alburquerque, N.M. 87106
U.S.A.

Wölfli, W.
Laboratorium für Kernphysik,ETH
8049 Zurich
Suisse

Woloshyn, Richard M.
Massachusetts Institute of Tech.
Cambridge, Mass. 02139
U.S.A.

Yuasa, T.
Institut de physique nucléaire
Université Paris-Sud
B.P. No 1, 91 Orsay
France

Zaikin, Dmitri
Institute of Nuclear Research
USSR Academy of Sciences
Moscow
USSR

Zamudio-Cristi, Jorge
Depto de Fisica
Universidad Simon Bolivar
Caracas 108
Venezuela

Zeiger, Enriques
SLAC
Stanford University
Stanford, California
U.S.A.

Zeitnitz, Bernhard
II Inst.f.Experimentalphysik
University of Hamburg
D 2 Hamburg
W. Germnay

Zielgelmann, Horst
Institut für Theoretische Physik
D 7400, Tübingen
W. Germany

Zingl, H.F.K.
Institut für Theoretische Physik
Univ. Graz
A 8010 Graz
Austria